FORMULAS AND EQUATIONS

Diode

$$I_D = I_S(e^{V_D/\phi_T} - 1) = Q_D/\tau_T$$

$$C_j = \frac{C_{j0}}{(1 - V_D/\phi_0)^m}$$

$$K_{eq} = \frac{-\phi_0^m}{(V_{high} - V_{low})(1 - m)} \times$$
$$[(\phi_0 - V_{high})^{1-m} - (\phi_0 - V_{low})^{1-m}]$$

MOS Transistor

$$V_T = V_{T0} + \gamma(\sqrt{|-2\phi_F + V_{SB}|} - \sqrt{|-2\phi_F|})$$

$$I_D = \frac{k'_n}{2}\frac{W}{L}(V_{GS} - V_T)^2(1 + \lambda V_{DS}) \text{ (sat)}$$

$$I_D = \upsilon_{sat}C_{ox}W\left(V_{GS} - V_T - \frac{V_{DSAT}}{2}\right)(1 + \lambda V_{DS})$$
$$\text{(velocity sat)}$$

$$I_D = k'_n\frac{W}{L}\left((V_{GS} - V_T)V_{DS} - \frac{V_{DS}^2}{2}\right) \text{ (triode)}$$

$$I_D = I_S e^{\frac{V_{GS}}{nkT/q}}\left(1 - e^{-\frac{V_{DS}}{kT/q}}\right) \text{ (subthreshold)}$$

Deep Submicron MOS Unified Model

$$I_D = 0 \text{ for } V_{GT} \leq 0$$

$$I_D = k'\frac{W}{L}\left(V_{GT}V_{min} - \frac{V_{min}^2}{2}\right)(1 + \lambda V_{DS}) \text{ for } V_{GT} \geq 0$$

with $V_{min} = \min(V_{GT}, V_{DS}, V_{DSAT})$

and $V_{GT} = V_{GS} - V_T$

MOS Switch Model

$$R_{eq} = \frac{1}{2}\left(\frac{V_{DD}}{I_{DSAT}(1 + \lambda V_{DD})} + \frac{V_{DD}/2}{I_{DSAT}(1 + \lambda V_{DD}/2)}\right)$$

$$\approx \frac{3}{4}\frac{V_{DD}}{I_{DSAT}}\left(1 - \frac{5}{6}\lambda V_{DD}\right)$$

Inverter

$$V_{OH} = f(V_{OL})$$

$$V_{OL} = f(V_{OH})$$

$$V_M = f(V_M)$$

$$t_p = 0.69R_{eq}C_L = \frac{C_L(V_{swing}/2)}{I_{avg}}$$

$$P_{dyn} = C_L V_{DD} V_{swing} f$$

$$P_{stat} = V_{DD} I_{DD}$$

Static CMOS Inverter

$$V_{OH} = V_{DD}$$

$$V_{OL} = GND$$

$$V_M \approx \frac{rV_{DD}}{1 + r} \text{ with } r = \frac{k_p V_{DSATp}}{k_n V_{DSATn}}$$

$$V_{IH} = V_M - \frac{V_M}{g} \qquad V_{IL} = V_M + \frac{V_{DD} - V_M}{g}$$

with $g \approx \dfrac{1 + r}{(V_M - V_{Tn} - V_{DSATn}/2)(\lambda_n - \lambda_p)}$

$$t_p = \frac{t_{pHL} + t_{pLH}}{2} = 0.69C_L\left(\frac{R_{eqn} + R_{eqp}}{2}\right)$$

$$P_{av} = C_L V_{DD}^2 f$$

Interconnect

Lumped RC: $t_p = 0.69 RC$

Distributed RC: $t_p = 0.38 RC$

RC-chain:

$$\tau_N = \sum_{i=1}^{N} R_i \sum_{j=i}^{N} C_j = \sum_{i=1}^{N} C_i \sum_{j=1}^{i} R_j$$

Transmission line reflection:

$$\rho = \frac{V_{refl}}{V_{inc}} = \frac{I_{refl}}{I_{inc}} = \frac{R - Z_0}{R + Z_o}$$

DIGITAL INTEGRATED CIRCUITS

A DESIGN PERSPECTIVE

Prentice Hall Electronics and VLSI Series

Charles S. Sodini, Series Editor

LEE, SHUR, FIELDLY, YTTERDAL   *Semiconductor Devices Modeling for VLSI*

LEUNG   *VLSI for Wireless Communications*

PLUMMER, DEAL, GRIFFIN   *Silicon VLSI Technology: Fundamentals, Practice, and Modeling*

RABAEY, CHANDRAKASAN, NIKOLIĆ   *Digital Integrated Circuits: A Design Perspective*, Second Edition

DIGITAL INTEGRATED CIRCUITS

A DESIGN PERSPECTIVE
SECOND EDITION

JAN M. RABAEY
ANANTHA CHANDRAKASAN
BORIVOJE NIKOLIĆ

PRENTICE HALL ELECTRONICS AND VLSI SERIES
CHARLES G. SODINI, SERIES EDITOR

Pearson Education, Inc.
Upper Saddle River, New Jersey 07458

Library of Congress Cataloging-in-Publication Data on file.

Vice President and Editorial Director, ECS: *Marcia J. Horton*
Publisher: *Tom Robbins*
Editorial Assistant: *Eric Van Ostenbridge*
Vice President and Director of Production and Manufacturing, ESM: *David W. Riccardi*
Executive Managing Editor: *Vince O'Brien*
Managing Editor: *David A. George*
Production Editor: *Daniel Sandin*
Director of Creative Services: *Paul Belfanti*
Creative Director: *Carole Anson*
Art and Cover Director: *Jayne Conte*
Art Editor: *Greg Dulles*
Manufacturing Manager: *Trudy Pisciotti*
Manufacturing Buyer: *Lisa McDowell*
Marketing Manager: *Holly Stark*

About the Cover: Detail of "Wet Orange," by Joan Mitchell (American, 1925–1992). Oil on canvas, 112×245 in. (284.5×622.3 cm). Carnegie Museum of Art, Pittsburgh, PA. Gift of Kaufmann's Department Store and the National Endowment for the Arts, 74.11. Photograph by Peter Harholdt, 1995.

© 2003, 1996 by Pearson Education, Inc.
Pearson Education, Inc.
Upper Saddle River, NJ 07458

Printed in the United States of America

10 9 8 7 6 5 4 3 2 1

ISBN 0-13-597444-5

Pearson Education Ltd., *London*
Pearson Education Australia Pty, Ltd., *Sydney*
Pearson Education Singapore, Pte. Ltd.
Pearson Education North Asia Ltd., *Hong Kong*
Pearson Education Canada Inc., *Toronto*
Pearson Educación de Mexico, S.A. de C.V.
Pearson Education—Japan, *Tokyo*
Pearson Education Malaysia, Pte. Ltd.
Pearson Education Inc., *Upper Saddle River, New Jersey*

To Kathelijn, Karthiyayani, Krithivasan,
and our Parents

"Qu'est-ce que l'homme dans la nature?
Un néant a l'égard de l'infini,
un tout al l'égard du néant,
un milieu entre rien et tout."

"What is man in nature?
Nothing in relation to the infinite,
everything in relation to nothing,
a mean between nothing and everything."

Blaise Pascal, *Pensées*, n. 4, 1670.

Preface

What is New?

Welcome to second edition of "*Digital Integrated Circuits: A Design Perspective.*" In the six years since the publication of the first, the field of digital integrated circuits has gone through some dramatic evolutions and changes. IC manufacturing technology has continued to scale to ever-smaller dimensions. Minimum feature sizes have scaled by a factor of almost ten since the writing of the first edition, and now are approaching the 100 nm realm. This scaling has a double impact on the design of digital integrated circuit. First of all, the complexity of the designs that can be put on a single die has increased dramatically. Dealing with the challenges this poses has led to new design methodologies and implementation strategies. At the same time, the plunge into the deep-submicron space causes devices to behave differently, and brings to the forefront a number of new issues that impact the reliability, cost, performance, and power dissipation of the digital IC. Addressing these issues in-depth is what differentiates this edition from the first.

A glance through the table of contents reveals extended coverage of issues such as deep-sub micron devices, circuit optimization, interconnect modeling and optimization, signal integrity, clocking and timing, and power dissipation. All these topics are illustrated with state-of-the-art design examples. Also, since MOS now represents more than 99% of the digital IC market, older technologies such as silicon bipolar and GaAs have been deleted (however, the interested reader can find the old chapters on these technologies on the web site of the book). Given the importance of methodology in today's design process, we have included *Design Methodology Inserts* throughout the text, each of which highlights one particular aspect of the design process. This new edition represents a major reworking of the book. The biggest change is the addition of two co-authors, Anantha and Bora, who have brought a broader insight into digital IC design and its latest trends and challenges.

Maintaining the Spirit of the First Edition

While introducing these changes, our intent has been to preserve the spirit and goals of the first edition—that is, to bridge the gap between the **circuit and system visions** on digital design. While starting from a solid understanding of the operation of electronic devices and an in-depth analysis of the nucleus of digital design—the inverter—we gradually channel this knowledge into the design of more complex modules such as gates, registers, controllers, adders, multipliers, and memories. We identify the compelling questions facing the designers of today's complex circuits: What are the dominant design parameters, what section of the design should he focus on and what details could she ignore? Simplification is clearly the only approach to address the increasing complexity of the digital systems. However, oversimplification can lead to circuit failure since global circuit effects such as timing, interconnect, and power consumption are ignored. To avoid this pitfall it is important to design digital circuits with both a circuits and a systems perspective in mind. This is the approach taken in this book, which brings the reader the knowledge and expertise needed to deal with complexity, using both analytical and experimental techniques.

How to Use This Book

The core of the text is intended for use in a *senior-level digital circuit design class*. Around this kernel, we have included chapters and sections covering the more advanced topics. In the course of developing this book, it quickly became obvious that it is difficult to define a subset of the digital circuit design domain that covers everyone's needs. On the one hand, a newcomer to the field needs detailed coverage of the basic concepts. On the other hand, feedback from early readers and reviewers indicated that an in-depth and extensive coverage of advanced topics and current issues is desirable and necessary. Providing this complete vision resulted in a text that exceeds the scope of a single-semester class. The more advanced material can be used as the basis for a *graduate class*. The wide coverage and the inclusion of state-of-the-art topics also makes the text useful as a reference work for professional engineers. It is assumed that students taking this course are familiar with the basics of logic design.

The organization of the material is such that the chapters can be taught or read in many ways, as long as a number of precedence relations are adhered to. The core of the text consists of Chapters 5, 6, 7, and 8. Chapters 1 to 4 can be considered as introductory. In response to popular demand, we have introduced a short treatise on semiconductor manufacturing in Chapter 2. Students with a prior introduction to semiconductor devices can traverse quickly through Chapter 3. We urge everyone to do at least that, as a number of important notations and foundations are introduced in that chapter. In addition, an original approach to the modeling of deep-submicron transistors enabling manual analysis, is introduced. To emphasize the importance of interconnect in today's digital design, we have moved the modeling of interconnect forward in the text to Chapter 4.

Chapters 9 to 12 are of a more advanced nature and can be used to provide a certain focus to the course. A course with a focus on the circuit aspects, for example, can supplement the core material with Chapters 9 and 12. A course focused on the digital system design should consider adding (parts of) Chapters 9, 10, and 11. All of these advanced chapters can be used to form the core of a graduate or a follow-on course. Sections considered *advanced* are marked with an *asterisk* in the text.

A number of possible paths through the material for a senior-level class are enumerated below. In the *instructor documentation,* provided on the book's web site, we have included a number of complete syllabi based on courses run at some academic institutions.

Basic circuit class (with minor prior device knowledge):
1, 2.1–3, 3, 4, 5, 6, 7, 8, (9.1–9.3, 12).

Somewhat more advanced circuit coverage:
1, (2, 3), 4, 5, 6, 7, 8, 9, 10.1–10.3, 10.5–10.6, 12.

Course with systems focus:
1, (2, 3), 4, 5, 6, 7, 8, 9, 10.1–10.4, 11, 12.1–12.2.

The *design methodology inserts* are, by preference, covered in concurrence with the chapter to which they are attached.

In order to maintain a consistent flow through each of the chapters, the topics are *introduced* first, followed by a detailed and in-depth discussion of the ideas. A *Perspective* section discusses how the introduced concepts relate to real world designs and how they might be impacted by future evolutions. Each chapter finishes with a *Summary,* which briefly enumerates the topics covered in the text, followed by *To Probe Further* and *Reference* sections. These provide ample references and pointers for a reader interested in further details on some of the material.

As the title of the book implies, one of the goals of this book is to stress the design aspect of digital circuits. To achieve this more practical viewpoint and to provide a real perspective, we have interspersed actual *design examples* and layouts throughout the text. These case studies help to answer questions, such as "How much area or speed or power is really saved by applying this technique?" To mimic the real design process, we are making extensive use of design tools such as circuit- and switch-level simulation as well as layout editing and extraction. Computer analysis is used throughout to verify manual results, to illustrate new concepts, or to examine complex behavior beyond the reach of manual analysis.

Finally, to facilitate the learning process, there are numerous examples included in the text. Each chapter contains a number of *problems or brain-teasers* (answers for which can be found in the back of the book), that provoke thinking and understanding while reading.

The Worldwide Web Companion

A worldwide web companion (*http://bwrc.eecs.berkeley.edu/IcBook/index.htm*) provides fully worked-out design problems and a complete set of overhead transparencies, extracting the most important figures and graphs from the text.

In contrast to the first edition, we have chosen **NOT to include problems sets** and **design problems** in the text. Instead we decided to make them available **on the book's web site**. This gives us the opportunity to dynamically upgrade and extend the problems, providing a more effective tool for the instructor. More than 300 challenging *exercises* are currently provided. The goal is to provide the individual reader an independent gauge for his understanding of the material and to provide practice in the use of some of the design tools. Each problem is keyed to the text sections it refers to (e.g., <1.3>), the design tools that must be used when solving the problem (e.g., SPICE) and a rating, ranking the problems on difficulty: (E) easy, (M) moderate, and

(C) challenging. Problems marked with a (D) include a design or research elements. Solutions to the problem sets are available only to instructors of academic institutions that have chosen to adopt our book for classroom use. They are available through the publisher on a password-protected web site.

Open-ended *design problems* help to gain the all-important insight into design optimization and trade-off. The use of design editing, verification and analysis tools is recommended when attempting these design problems. Fully worked out versions of these problems can be found on the web site.

In addition, the book's web site also offers samples of hardware and software laboratories, extra background information, and useful links.

Compelling Features of the Book

- Brings both circuit and systems views on design together. It offers a profound understanding of the design of complex digital circuits, while preparing the designer for new challenges that might be waiting around the corner.
- Design-oriented perspectives are advocated throughout. Design challenges and guidelines are highlighted. Techniques introduced in the text are illustrated with real designs and complete SPICE analysis.
- Is the first circuit design book that *focuses solely on deep-submicron devices*. To facilitate this, a simple transistor model for manual analysis, called the *unified MOS model*, has been developed.
- Unique in showing how to use the latest techniques to design complex high-performance or low-power circuits. Speed and power treated as equal citizens throughout the text.
- Covers crucial real-world system design issues such as signal integrity, power dissipation, interconnect, packaging, timing, and synchronization.
- Provides unique coverage of the latest design methodologies and tools, with a discussion of how to use them from a designers' perspective.
- Offers perspectives on how digital circuit technology might evolve in the future.
- Outstanding illustrations and a usable design-oriented four-color insert.
- To Probe Further and Reference sections provide ample references and pointers for a reader interested in further details on some of the material.
- Extensive instructional package is available over the internet from the author's web site. Includes design software, transparency masters, problem sets, design problems, actual layouts, and hardware and software laboratories.

The Contents at a Glance

A quick scan of the table of contents shows how the ordering of chapters and the material covered are consistent with the advocated design methodology. Starting from a model of the semiconductor devices, we will gradually progress upwards, covering the inverter, the complex logic gate (NAND, NOR, XOR), the functional (adder, multiplier, shifter, register) and the system module (datapath, controller, memory) levels of abstraction. For each of these layers, the dominant design parameters are identified and simplified models are constructed, abstracting away the nonessential details. While this layered modeling approach is the designer's best handle on complexity, it has some pitfalls. This is illustrated in Chapters 9 and 10, where topics with a global impact, such as interconnect parasitics and chip timing, are discussed. To further express the dichotomy between circuit and system design visions, we have divided the book contents into two major parts: Part II (Chapters 4–7) addresses mostly the circuit perspective of digital circuit design, while Part III (Chapters 8–12) presents a more system oriented vision. Part I (Chapters 1-4) provides the necessary foundation (design metrics, the manufacturing process, device and interconnect models).

Chapter 1 serves as a global *introduction*. After a historical overview of digital circuit design, the concepts of hierarchical design and the different abstraction layers are introduced. A number of fundamental metrics, which help to quantify cost, reliability, and performance of a design, are introduced.

Chapter 2 provides a short and compact introduction to the *MOS manufacturing process*. Understanding the basic steps in the process helps to create the three-dimensional understanding of the MOS transistor, which is crucial when identifying the sources of the device parasitics. Many of the variations in device parameters can also be attributed to the manufacturing process as well. The chapter further introduces the concept of design rules, which form the interface between the designer and the manufacturer. The chapter concludes with an overview of the chip packaging process, an often-overlooked but crucial element of the digital IC design cycle.

Chapter 3 contains a summary of the primary design building blocks, *the semiconductor devices*. The main goal of this chapter is to provide an intuitive understanding of the operation of the MOS as well as to introduce the device models, which are used extensively in the later chapters. Major attention is paid to the artifacts of modern submicron devices, and the modeling thereof. Readers with prior device knowledge can traverse this material rather quickly.

Chapter 4 contains a careful analysis of the *wire,* with interconnect and its accompanying parasitics playing a major role. We visit each of the parasitics that come with a wire (capacitance, resistance, and inductance) in turn. Models for both manual and computer analysis are introduced.

Chapter 5 deals with the nucleus of digital design, the *inverter*. First, a number of fundamental properties of digital gates are introduced. These parameters, which help to quantify the performance and reliability of a gate, are derived in detail for two representative inverter structures: the static complementary CMOS. The techniques and approaches introduced in this chapter are of crucial importance, as they are repeated over and over again in the analysis of other gate structures and more complex gate structures.

In **Chapter 6** this fundamental knowledge is extended to address the design of *simple and complex digital CMOS gates,* such as NOR and NAND structures. It is demonstrated that, depending upon the dominant design constraint (reliability, area, performance, or power), other

CMOS gate structures besides the complementary static gate can be attractive. The properties of a number of contemporary gate-logic families are analyzed and compared. Techniques to optimize the performance and power consumption of complex gates are introduced.

Chapter 7 discusses how memory function can be accomplished using either positive feedback or charge storage. Besides analyzing the traditional bistable flip-flops, other sequential circuits such as the mono- and astable multivibrators are also introduced. All chapters prior to Chapter 7 deal exclusively with combinational circuits, that is circuits without a sense of the past history of the system. *Sequential logic circuits*, in contrast, can remember and store the past state.

All chapters preceding **Chapter 8** present a circuit-oriented approach towards digital design. The analysis and optimization process has been constrained to the individual gate. In this chapter, we take our approach one step further and analyze how gates can be connected together to form the building blocks of a system. The system-level part of the book starts, appropriately, with a discussion of *design methodologies*. Design automation is the only way to cope with the ever-increasing complexity of digital designs. In Chapter 8, the prominent ways of producing large designs in a limited time are discussed. The chapter spends considerable time on the different implementation methodologies available to today's designer. Custom versus semi-custom, hardwired versus fixed, regular array versus ad-hoc are some of the issues put forward.

Chapter 9 revisits the impact of *interconnect wiring* on the functionality and performance of a digital gate. A wire introduces parasitic capacitive, resistive, and inductive effects, which are becoming ever more important with the scaling of the technology. Approaches to minimize the impact of these interconnect parasitics on performance, power dissipation and circuit reliability are introduced. The chapter also addresses some important issues such as supply-voltage distribution, and input/output circuitry.

In **Chapter 10** details how that in order to operate sequential circuits correctly, a strict ordering of the switching events has to be imposed. Without these *timing* constraints, wrong data might be written into the memory cells. Most digital circuits use a synchronous, clocked approach to impose this ordering. In Chapter 10, the different approaches to digital circuit timing and clocking are discussed. The impact of important effects such as clock skew on the behavior of digital synchronous circuits is analyzed. The synchronous approach is contrasted with alternative techniques, such as self-timed circuits. The chapter concludes with a short introduction to synchronization and clock-generation circuits.

In **Chapter 11**, the design of a variety of complex *arithmetic building blocks* such as adders, multipliers, and shifters, is discussed. This chapter is crucial because it demonstrates how the design techniques introduced in chapters 5 and 6 are extended to the next abstraction layer. The concept of the critical path is introduced and used extensively in the performance analysis and optimization. Higher-level performance models are derived. These help the designer to get a fundamental insight into the operation and quality of a design module, without having to resort to an in-depth and detailed analysis of the underlying circuitry.

Chapter 12 discusses in depth the different memory classes and their implementation. Whenever large amounts of data storage are needed, the digital designer resorts to special circuit modules, called *memories*. Semiconductor memories achieve very high storage density by compromising on some of the fundamental properties of digital gates. Instrumental in the design of reliable and fast memories is the implementation of the peripheral circuitry, such as the decoders, sense amplifiers, drivers, and control circuitry, which are extensively covered. Finally, as the primary issue in memory design is to ensure that the device works consistently under all operating circumstances, the chapter concludes with a detailed discussion of memory reliability. This chapter as well as the previous one are optional for undergraduate courses.

Acknowledgments

The authors would like to thank all those who contributed to the emergence, creation and correction of this manuscript. First of all, thanks to all the graduate students that helped over the years to bring the text to where it is today. Thanks also to the students of the eecs141 and eecs241 courses at Berkeley and the 6.374 course at MIT, who suffered through many of the experimental class offerings based on this book. The feedback from instructors, engineers, and students from all over the world has helped tremendously in focusing the directions of this new edition, and in fine-tuning the final text. The continuous stream of e-mails indicate to us that we are on the right track.

In particular, we would like to acknowledge the contributions of Mary-Jane Irwin, Vijay Narayanan, Eby Friedman, Fred Rosenberger, Wayne Burleson, Sekhar Borkar, Ivo Bolsens, Duane Boning, Olivier Franza, Lionel Kimerling, Josie Ammer, Mike Sheets, Tufan Karalar, Huifang Qin, Rhett Davis, Nathan Chan, Jeb Durant, Andrei Vladimirescu, Radu Zlatanovici, Yasuhisa Shimazaki, Fujio Ishihara, Dejan Markovic, Vladimir Stojanovic, SeongHwan Cho, James Kao, Travis Simpkins, Siva Narendra, James Goodman, Vadim Gutnik, Theodoros Konstantakopoulos, Rex Min, Vikas Mehrotra, and Paul-Peter Sotiriadis. Their help, input, and feedback are greatly appreciated. Obviously, we remain thankful to those who helped create and develop the first edition.

I am extremely grateful to the staff at Prentice Hall, who have been instrumental in turning a rough manuscript into an enjoyable book. First of all, I would like to acknowledge the help and constructive feedback of Tom Robbins, Publisher, Daniel Sandin, Production Editor, and David George, Managing Editor. A special word of thanks to Brenda Vanoni at Berkeley, for her invaluable help in the copy editing and the website creation process. The web expertise of Carol Sitea came in very handy as well.

I would like to highlight to role of computer aids in developing this manuscript. All drafts were completely developed on the FrameMaker publishing system (Adobe Systems). Graphs were mostly created using MATLAB. Microsoft Frontpage is the tool of choice for the web-page creation. For circuit simulations, we used HSPICE (Avant!). All layouts were generated using the Cadence physical design suite.

Finally, some words of gratitude to the people that had to endure the creation process of this book, Kathelijn, Karthiyayani, Krithivasan, and Rebecca. While the generation of a new edition brings substantially less pain than a first edition, we consistently underestimate what it takes, especially in light of the rest of our daily loads. They been a constant support, help and encouragement during the writing of this manuscript.

JAN M. RABAEY
ANANTHA CHANDRAKASAN
BORIVOJE NIKOLIĆ
Berkeley, Calistoga, Cambridge

Contents

Preface   **vii**

| *Part 1* | **The Fabrics** | **1** |

Chapter 1	**Introduction**	**3**
	1.1 A Historical Perspective	4
	1.2 Issues in Digital Integrated Circuit Design	6
	1.3 Quality Metrics of a Digital Design	15
	1.3.1 Cost of an Integrated Circuit	16
	1.3.2 Functionality and Robustness	18
	1.3.3 Performance	27
	1.3.4 Power and Energy Consumption	30
	1.4 Summary	31
	1.5 To Probe Further	31
	Reference Books	32
	References	33

Chapter 2	**The Manufacturing Process**	**35**
	2.1 Introduction	36
	2.2 Manufacturing CMOS Integrated Circuits	36
	2.2.1 The Silicon Wafer	37
	2.2.2 Photolithography	37
	2.2.3 Some Recurring Process Steps	41
	2.2.4 Simplified CMOS Process Flow	42
	2.3 Design Rules—The Contract between Designer and Process Engineer	47
	2.4 Packaging Integrated Circuits	51
	2.4.1 Package Materials	52
	2.4.2 Interconnect Levels	53
	2.4.3 Thermal Considerations in Packaging	59
	2.5 Perspective—Trends in Process Technology	61
	2.5.1 Short-Term Developments	61
	2.5.2 In the Longer Term	63
	2.6 Summary	64

| | 2.7 | To Probe Further | 64 |
| | | References | 64 |

***Design Methodology Insert A* IC LAYOUT**   **67**

| | A.1 | To Probe Further | 71 |
| | | References | 71 |

***Chapter 3* The Devices**   **73**

	3.1	Introduction	74
	3.2	The Diode	74
		3.2.1 A First Glance at the Diode—The Depletion Region	75
		3.2.2 Static Behavior	77
		3.2.3 Dynamic, or Transient, Behavior	80
		3.2.4 The Actual Diode—Secondary Effects	84
		3.2.5 The SPICE Diode Model	85
	3.3	The MOS(FET) Transistor	87
		3.3.1 A First Glance at the Device	87
		3.3.2 The MOS Transistor under Static Conditions	88
		3.3.3 The Actual MOS Transistor—Some Secondary Effects	114
		3.3.4 SPICE Models for the MOS Transistor	117
	3.4	A Word on Process Variations	120
	3.5	Perspective—Technology Scaling	122
	3.6	Summary	128
	3.7	To Probe Further	129
		References	130

***Design Methodology Insert B* Circuit Simulation**   **131**

| | | References | 134 |

***Chapter 4* The Wire**   **135**

	4.1	Introduction	136
	4.2	A First Glance	136
	4.3	Interconnect Parameters—Capacitance, Resistance, and Inductance	138
		4.3.1 Capacitance	138
		4.3.2 Resistance	144
		4.3.3 Inductance	148
	4.4	Electrical Wire Models	150
		4.4.1 The Ideal Wire	151
		4.4.2 The Lumped Model	151
		4.4.3 The Lumped *RC* Model	152
		4.4.4 The Distributed *rc* Line	156
		4.4.5 The Transmission Line	159

	4.5	SPICE Wire Models	170
		4.5.1 Distributed *rc* Lines in SPICE	170
		4.5.2 Transmission Line Models in SPICE	170
		4.5.3 Perspective: A Look into the Future	171
	4.6	Summary	174
	4.7	To Probe Further	174
		References	174

| **Part 2** | **A Circuit Perspective** | **177** |

Chapter 5	**The CMOS Inverter**	**179**
	5.1	Introduction	180
	5.2	The Static CMOS Inverter—An Intuitive Perspective	180
	5.3	Evaluating the Robustness of the CMOS Inverter:	
		The Static Behavior	184
		5.3.1 Switching Threshold	185
		5.3.2 Noise Margins	188
		5.3.3 Robustness Revisited	191
	5.4	Performance of CMOS Inverter: The Dynamic Behavior	193
		5.4.1 Computing the Capacitances	194
		5.4.2 Propagation Delay: First-Order Analysis	199
		5.4.3 Propagation Delay from a Design Perspective	203
	5.5	Power, Energy, and Energy Delay	213
		5.5.1 Dynamic Power Consumption	214
		5.5.2 Static Consumption	223
		5.5.3 Putting It All Together	225
		5.5.4 Analyzing Power Consumption Using SPICE	227
	5.6	Perspective: Technology Scaling and its Impact	
		on the Inverter Metrics	229
	5.7	Summary	232
	5.8	To Probe Further	233
		References	233

Chapter 6	**Designing Combinational Logic Gates in CMOS**	**235**
	6.1	Introduction	236
	6.2	Static CMOS Design	236
		6.2.1 Complementary CMOS	237
		6.2.2 Ratioed Logic	263
		6.2.3 Pass-Transistor Logic	269
	6.3	Dynamic CMOS Design	284
		6.3.1 Dynamic Logic: Basic Principles	284
		6.3.2 Speed and Power Dissipation of Dynamic Logic	287

	6.3.3	Signal Integrity Issues in Dynamic Design	290
	6.3.4	Cascading Dynamic Gates	295
6.4		Perspectives	303
	6.4.1	How to Choose a Logic Style?	303
	6.4.2	Designing Logic for Reduced Supply Voltages	303
6.5		Summary	306
6.6		To Probe Further	307
		References	308

Design Methodology Insert C **How to Simulate Complex Logic Circuits**   **309**

C.1	Representing Digital Data as a Continuous Entity	310
C.2	Representing Data as a Discrete Entity	310
C.3	Using Higher-Level Data Models	315
	References	317

Design Methodology Insert D **Layout Techniques for Complex Gates**   **319**

Chapter 7 **Designing Sequential Logic Circuits**   **325**

7.1		Introduction	326
	7.1.1	Timing Metrics for Sequential Circuits	327
	7.1.2	Classification of Memory Elements	328
7.2		Static Latches and Registers	330
	7.2.1	The Bistability Principle	330
	7.2.2	Multiplexer-Based Latches	332
	7.2.3	Master-Slave Edge-Triggered Register	333
	7.2.4	Low-Voltage Static Latches	339
	7.2.5	Static SR Flip-Flops—Writing Data by Pure Force	341
7.3		Dynamic Latches and Registers	344
	7.3.1	Dynamic Transmission-Gate Edge-triggered Registers	344
	7.3.2	C^2MOS—A Clock-Skew Insensitive Approach	346
	7.3.3	True Single-Phase Clocked Register (TSPCR)	350
7.4		Alternative Register Styles[*]	354
	7.4.1	Pulse Registers	354
	7.4.2	Sense-Amplifier-Based Registers	356
7.5		Pipelining: An Approach to Optimize Sequential Circuits	358
	7.5.1	Latch- versus Register-Based Pipelines	360
	7.5.2	NORA-CMOS—A Logic Style for Pipelined Structures	361
7.6		Nonbistable Sequential Circuits	364
	7.6.1	The Schmitt Trigger	364
	7.6.2	Monostable Sequential Circuits	367
	7.6.3	Astable Circuits	368
7.7		Perspective: Choosing a Clocking Strategy	370
7.8		Summary	371

| | 7.9 | To Probe Further | 372 |
| | | References | 372 |

| **Part 3** | **A System Perspective** | | **375** |

Chapter 8	**Implementation Strategies for Digital ICS**		**377**
	8.1	Introduction	378
	8.2	From Custom to Semicustom and Structured-Array Design Approaches	382
	8.3	Custom Circuit Design	383
	8.4	Cell-Based Design Methodology	384
		8.4.1 Standard Cell	385
		8.4.2 Compiled Cells	390
		8.4.3 Macrocells, Megacells and Intellectual Property	392
		8.4.4 Semicustom Design Flow	396
	8.5	Array-Based Implementation Approaches	399
		8.5.1 Prediffused (or Mask-Programmable) Arrays	399
		8.5.2 Prewired Arrays	404
	8.6	Perspective—The Implementation Platform of the Future	420
	8.7	Summary	423
	8.8	To Probe Further	423
		References	424

| *Design Methodology Insert E* | **Characterizing Logic and Sequential Cells** | | **427** |
| | References | | 434 |

| *Design Methodology Insert F* | **Design Synthesis** | | **435** |
| | References | | 443 |

Chapter 9	**Coping with Interconnect**		**445**
	9.1	Introduction	446
	9.2	Capacitive Parasitics	446
		9.2.1 Capacitance and Reliability—Cross Talk	446
		9.2.2 Capacitance and Performance in CMOS	449
	9.3	Resistive Parasitics	460
		9.3.1 Resistance and Reliability—Ohmic Voltage Drop	460
		9.3.2 Electromigration	462
		9.3.3 Resistance and Performance—RC Delay	464
	9.4	Inductive Parasitics[*]	469
		9.4.1 Inductance and Reliability— Voltage Drop	469
		9.4.2 Inductance and Performance—Transmission-line Effects	475
	9.5	Advanced Interconnect Techniques	480

	9.5.1 Reduced-Swing Circuits	480
	9.5.2 Current-Mode Transmission Techniques	486
9.6	Perspective: Networks-on-a-Chip	487
9.7	Summary	488
9.8	To Probe Further	489
	References	489

Chapter 10 **Timing Issues in Digital Circuits**   **491**

10.1	Introduction	492
10.2	Timing Classification of Digital Systems	492
	10.2.1 Synchronous Interconnect	492
	10.2.2 Mesochronous interconnect	493
	10.2.3 Plesiochronous Interconnect	493
	10.2.4 Asynchronous Interconnect	494
10.3	Synchronous Design—An In-depth Perspective	495
	10.3.1 Synchronous Timing Basics	495
	10.3.2 Sources of Skew and Jitter	502
	10.3.3 Clock-Distribution Techniques	508
	10.3.4 Latch-Based Clocking[*]	516
10.4	Self-Timed Circuit Design[*]	519
	10.4.1 Self-Timed Logic—An Asynchronous Technique	519
	10.4.2 Completion-Signal Generation	522
	10.4.3 Self-Timed Signaling	526
	10.4.4 Practical Examples of Self-Timed Logic	531
10.5	Synchronizers and Arbiters[*]	534
	10.5.1 Synchronizers—Concept and Implementation	534
	10.5.2 Arbiters	538
10.6	Clock Synthesis and Synchronization Using a Phase-Locked Loop[*]	539
	10.6.1 Basic Concept	540
	10.6.2 Building Blocks of a PLL	542
10.7	Future Directions and Perspectives	546
	10.7.1 Distributed Clocking Using DLLs	546
	10.7.2 Optical Clock Distribution	548
	10.7.3 Synchronous versus Asynchronous Design	549
10.8	Summary	550
10.9	To Probe Further	551
	References	551

Design Methodology Insert G **Design Verification**   **553**

	References	557

Chapter 11 **Designing Arithmetic Building Blocks**   **559**

    11.1   Introduction   560
    11.2   Datapaths in Digital Processor Architectures   560
    11.3   The Adder   561
           11.3.1 The Binary Adder: Definitions   561
           11.3.2 The Full Adder: Circuit Design Considerations   564
           11.3.3 The Binary Adder: Logic Design Considerations   571
    11.4   The Multiplier   586
           11.4.1 The Multiplier: Definitions   586
           11.4.2 Partial-Product Generation   587
           11.4.3 Partial-Product Accumulation   589
           11.4.4 Final Addition   593
           11.4.5 Multiplier Summary   594
    11.5   The Shifter   594
           11.5.1 Barrel Shifter   595
           11.5.2 Logarithmic Shifter   596
    11.6   Other Arithmetic Operators   596
    11.7   Power and Speed Trade-offs in Datapath Structures[*]   600
           11.7.1 Design Time Power-Reduction Techniques   601
           11.7.2 Run-Time Power Management   611
           11.7.3 Reducing the Power in Standby (or Sleep) Mode   617
    11.8   Perspective: Design as a Trade-off   618
    11.9   Summary   619
    11.10 To Probe Further   620
          References   621

Chapter 12 **Designing Memory and Array Structures**   **623**

    12.1   Introduction   624
           12.1.1 Memory Classification   625
           12.1.2 Memory Architectures and Building Blocks   627
    12.2   The Memory Core   634
           12.2.1 Read-Only Memories   634
           12.2.2 Nonvolatile Read-Write Memories   647
           12.2.3 Read-Write Memories (RAM)   657
            12.2.4 Contents-Addressable or Associative Memory (CAM)   670
    12.3   Memory Peripheral Circuitry[*]   672
           12.3.1 The Address Decoders   672
           12.3.2 Sense Amplifiers   679
           12.3.3 Voltage References   686
           12.3.4 Drivers/Buffers   689
           12.3.5 Timing and Control   689

12.4   Memory Reliability and Yield*   693
        12.4.1 Signal-to-Noise Ratio   693
        12.4.2 Memory Yield   698
12.5   Power Dissipation in Memories*   701
        12.5.1 Sources of Power Dissipation in Memories   701
        12.5.2 Partitioning of the Memory   702
        12.5.3 Addressing the Active Power Dissipation   702
        12.5.4 Data-Retention Dissipation   704
        12.5.5 Summary   707
12.6   Case Studies in Memory Design   707
        12.6.1 The Programmable Logic Array (PLA)   707
        12.6.2 A 4-Mbit SRAM   710
        12.6.3 A 1-Gbit NAND Flash Memory   712
12.7   Perspective: Semiconductor Memory Trends and Evolutions   714
12.8   Summary   716
12.9   To Probe Further   717
        References   718

Design Methodology Insert H   **Validation and Test
of Manufactured Circuits**   **721**
H.1   Introduction   721
H.2   Test Procedure   722
H.3   Design for Testability   723
        H.3.1 Issues in Design for Testability   723
        H.3.2 Ad Hoc Testing   725
        H.3.3 Scan-Based Test   726
        H.3.4 Boundary-Scan Design   729
        H.3.5 Built-in Self-Test (BIST)   730
H.4   Test-Pattern Generation   734
        H.4.1 Fault Models   734
        H.4.2 Automatic Test-Pattern Generation (ATPG)   736
        H.4.3 Fault Simulation   737
H.5   To Probe Further   737
        References   737

Problem Solutions   **739**

Index   **745**

PART

1

The Fabrics

"The complexity for minimum component costs has increased at a rate of roughly a factor of two per year. Certainly over the short term, this rate can be expected to continue, if not to increase. Over the longer term, the rate of increase is a bit more uncertain, although there is no reason to believe it will not remain nearly constant for at least 10 years. That means by 1975, the number of components per integrated circuit for minimum cost will be 65,000."

Gordon Moore,
Cramming more Components onto
Integrated Circuits, (1965).

1

Introduction

The evolution of digital circuit design

Compelling issues in digital circuit design

How to measure the quality of a design

Valuable references

1.1   A Historical Perspective

1.2   Issues in Digital Integrated Circuit Design

1.3   Quality Metrics of a Digital Design

    1.3.1   Cost of an Integrated Circuit

    1.3.2   Functionality and Robustness

    1.3.3   Performance

    1.3.4   Power and Energy Consumption

1.4   Summary

1.5   To Probe Further

1.1 A Historical Perspective

The concept of digital data manipulation has made a dramatic impact on our society. One has long grown accustomed to the idea of digital computers. Evolving steadily from mainframe and minicomputers, personal and laptop computers have proliferated into daily life. More significant, however, is a continuous trend towards digital solutions in all other areas of electronics. Instrumentation was one of the first noncomputing domains where the potential benefits of digital data manipulation over analog processing were recognized. Other areas such as control were soon to follow. Only recently have we witnessed the conversion of telecommunications and consumer electronics into the digital format. Increasingly, telephone data is transmitted and processed digitally over both wired and wireless networks. The compact disk has revolutionized the audio world, and digital video is following in its footsteps.

The idea of implementing computational engines using an encoded data format is by no means an idea of our times. In the early 19th century, Babbage envisioned large-scale mechanical computing devices, which he called *Difference Engines* [Swade93]. Although these engines use the decimal number system rather than the binary representation now common in modern electronics, the underlying concepts are very similar. The Analytical Engine, developed in 1834, was perceived as a general-purpose computing machine, with features strikingly close to modern computers. Besides executing the basic repertoire of operations (addition, subtraction, multiplication, and division) in arbitrary sequences, the machine operated in a two-cycle sequence, called "store" and "mill" (execute), not unlike today's computers. It even used pipelining to speed up the execution of the addition operation! Unfortunately, the complexity and the cost of the designs made the concept impractical. For instance, the design of Difference Engine I (part of which is shown in Figure 1-1) required 25,000 mechanical parts at a total cost of £17,470 (in 1834!).

The electrical solution turned out to be more cost effective. Early digital electronics systems were based on magnetically controlled switches (or relays). They were mainly used in the implementation of very simple logic networks. Such systems are still used in train safety systems. The age of digital electronic computing only started in full with the introduction of the vacuum tube. While originally used almost exclusively for analog processing, it soon was recognized that the vacuum tube was useful for digital computations as well. Not long thereafter, the first complete computers were realized. The era of the vacuum-tube-based computer culminated in the design of machines like the ENIAC (intended for computing artillery firing tables) and the UNIVAC I (the first successful commercial computer). To get an idea about *integration density*, the ENIAC was 80 feet long, 8.5 feet high, and several feet wide; it also incorporated 18,000 vacuum tubes. It became rapidly clear, however, that this design technology had reached its limits. Reliability problems and excessive power consumption made the implementation of larger engines economically and practically infeasible.

All changed with the invention of the *transistor* at Bell Telephone Laboratories in 1947 [Bardeen48], followed by the introduction of the bipolar junction transistor by Schockley in

Figure 1-1   Working part of Babbage's Difference Engine I (1832), the first known automatic calculator (from [Swade93], courtesy of the Science Museum of London).

1949 [Schockley49].[1] It took until 1956 before this led to the first bipolar digital logic gate, made of discrete components introduced by Harris [Harris56]. In 1958, Jack Kilby at Texas Instruments conceived the *integrated circuit (IC)*, in which all components, passive and active, are integrated on a single semiconductor substrate—he was awarded the Nobel prize for this breakthrough. It led to the introduction of the first set of integrated-circuit commercial logic gates, called the *Fairchild Micrologic family* [Norman60]. The first truly successful IC logic family, *TTL (Transistor-Transistor Logic)* was pioneered in 1962 [Beeson62]. Other logic families were devised with higher performance in mind. Examples of these are the current switching circuits that produced the first subnanosecond digital gates and culminated in the *ECL (Emitter-Coupled Logic)* family [Masaki74]. TTL had the advantage, however, of offering a higher integration density and was the basis of the first integrated circuit revolution. In fact, the manufacturing of TTL components is what spearheaded the first large semiconductor companies such as Fairchild, National, and Texas Instruments. The family was so successful that it composed the largest fraction of the digital semiconductor market until the 1980s.

Ultimately, bipolar digital logic lost the battle for hegemony in the digital design world for exactly the reasons that haunted the vacuum tube approach: the large power consumption per gate puts an upper limit on the number of gates that can be reliably integrated on a single die,

---

[1] An intriguing overview of the evolution of digital integrated circuits can be found in [Murphy93]. (Most of the data in this overview has been extracted from this reference). It is accompanied by some of the historically ground-breaking publications in the domain of digital ICs.

package, housing, or box. Although attempts were made to develop high integration density, low-power bipolar families (such as I^2L—*Integrated Injection Logic* [Hart72]), the MOS digital integrated circuit approach eventually held sway.

The basic principle behind the MOSFET transistor (originally called IGFET) was proposed in a patent by J. Lilienfeld (Canada) as early as 1925, and, independently, by O. Heil in England in 1935. Insufficient knowledge of the materials and gate stability problems, however, delayed the practical usability of the device for a long time. Once these were solved, MOS digital integrated circuits started to take off in full in the early 1970s. Remarkably, the first MOS logic gates introduced were of the CMOS variety [Wanlass63], and this trend continued until the late 1960s. The complexity of the manufacturing process delayed the full exploitation of these devices for two more decades. Instead, the first practical MOS integrated circuits were implemented in PMOS-only logic and were used in applications such as calculators. The second age of the digital integrated circuit revolution was inaugurated with the introduction of the first microprocessors by Intel in 1972 (the 4004) [Faggin72] and 1974 (the 8080) [Shima74]. These processors were implemented in NMOS-only logic, which has the advantage of higher speed over the PMOS logic. Simultaneously, MOS technology enabled the realization of the first high-density semiconductor memories. For instance, the first 4Kbit MOS memory was introduced in 1970 [Hoff70].

These events were at the start of a truly astounding evolution towards ever higher integration densities and speed performances, a revolution that currently is still in full swing. The road to the current levels of integration has not been without hindrances, however. In the late 1970s, NMOS-only logic started to suffer from the same plague that made high-density bipolar logic unattractive or infeasible: power consumption. This realization, combined with progress in manufacturing technology, finally tilted the balance towards the CMOS technology, and this is where it remains today. Interestingly enough, power consumption concerns are rapidly becoming dominant in CMOS design as well, and this time there does not seem to be a new technology around the corner to alleviate the problem.

Although the large majority of the current integrated circuits are implemented in the MOS technology, other technologies come into play when very high performance is at stake. An example of this is the BiCMOS technology that combines bipolar and MOS devices on the same die. When even higher performance is necessary, other technologies emerge such as silicon-germanium, and even superconducting technologies. These technologies only play a very small role in the overall digital integrated circuit design scene. With the ever increasing performance of CMOS, this role is bound to be reduced further with time—hence, the focus of this textbook is on CMOS only.

1.2 Issues in Digital Integrated Circuit Design

Integration density and performance of integrated circuits have gone through an astounding revolution in the last two decades. In the 1960s, Gordon Moore, then with Fairchild Corporation and later cofounder of Intel, predicted that the number of transistors that can be integrated on a

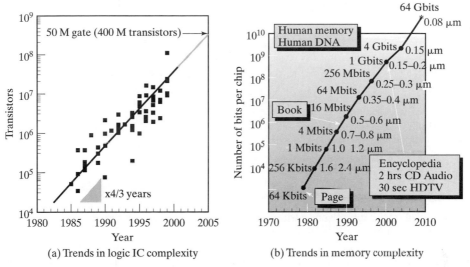

(a) Trends in logic IC complexity   (b) Trends in memory complexity

Figure 1-2   Evolution of integration complexity of logic ICs and memories as a function of time.

single die would grow exponentially with time. This prediction, later called *Moore's law*, has proven to be amazingly visionary [Moore65]. Its validity is best illustrated with the aid of a set of graphs. Figure 1-2 plots the integration density of both logic ICs and memory as a function of time. As can be observed, integration complexity doubles approximately every one to two years. As a result, memory density has increased by more than a thousandfold since 1970.

An intriguing case study is offered by the microprocessor. From its inception in the early 1970s, the microprocessor has grown in performance and complexity at a steady and predictable pace. The transistor counts for a number of landmark designs that are collected in Figure 1-3.

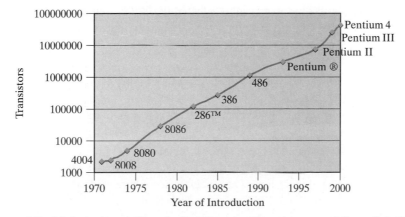

Figure 1-3   Historical evolution of microprocessor transistor count (from [Intel01]).

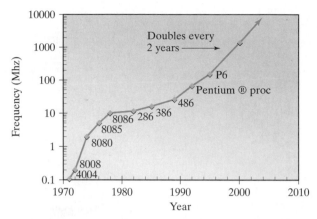

Figure 1-4   Microprocessor performance trends at the beginning of the 21st century. (Courtesy of Intel.)

The million-transistor/chip barrier was crossed in the late 1980s. Clock frequencies double every three years in the past decade and have reached into the GHz range. This is illustrated in Figure 1-4, which plots the microprocessor trends in terms of performance at the beginning of the 21st century. An important observation is that, as of now, these trends have not shown any signs of a slowdown.

It should not surprise the reader that this revolution has had a profound impact on how digital circuits are designed. Early designs were truly handcrafted. Every transistor was laid out and optimized individually and carefully fitted into its environment. This is adequately illustrated in Figure 1-5a, which shows the design of the Intel 4004 microprocessor. Obviously, this approach is not appropriate when more than a million devices have to be created and assembled. With the rapid evolution of the design technology, time to market is one of the crucial factors in the ultimate success of a component.

As a result, designers have increasingly adhered to rigid design methodologies and strategies that are more amenable to design automation. The impact of this approach is apparent from the layout of one of the later Intel microprocessors, the Pentium® 4, shown in Figure 1-5b. Instead of the individualized approach of the earlier designs, a circuit is constructed in a hierarchical way: a processor is a collection of modules, each of which consists of a number of cells on its own. Cells are reused as much as possible to reduce the design effort and to enhance the chances for a first-time-right implementation. The fact that this hierarchical approach is at all possible is the key ingredient for the success of digital circuit design and also explains why, for instance, very large-scale analog design has never caught on.

The obvious next question is why such an approach is feasible in the digital world and not (or to a lesser degree) in analog designs. The crucial concept here, and the most important one in dealing with the complexity issue, is *abstraction*. At each design level, the internal details of a complex module can be abstracted away and replaced by a *black-box view* or *model*. This model contains virtually all the information needed to deal with the block at the next level of hierarchy.

(a) The 4004 microprocessor

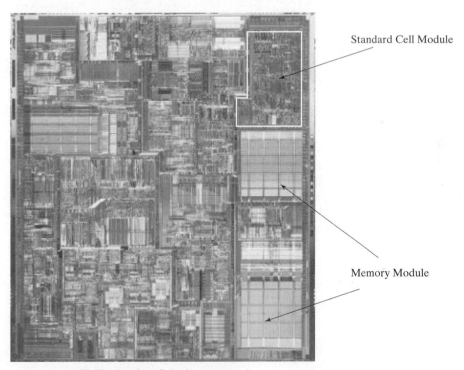

Standard Cell Module

Memory Module

(b) The Pentium ® 4 microprocessor

Figure 1-5   Comparing the design methodologies of the Intel 4004 (1971) and Pentium® 4 (2000) microprocessors (reprinted with permission from Intel).

For instance, once a designer has implemented a multiplier module, its performance can be defined very accurately and can be captured in a model. In general, the performance of this multiplier is only marginally influenced by the way it is utilized in a larger system. For all purposes, therefore, it can be considered a black box with known characteristics. As there exists no compelling need for the system designer to look inside this box, design complexity is substantially reduced. The impact of this *divide-and-conquer* approach is dramatic. Instead of having to deal with a myriad of elements, the designer has to consider only a handful of components, each of which are characterized in performance and cost by a small number of parameters.

This is analogous to a software designer using a library of software routines such as input/output drivers. Someone writing a large program does not bother to look inside those library routines. The only thing he cares about is the intended result of calling one of those modules. Imagine what writing software programs would be like if you had to fetch every bit individually from the disk and ensure its correctness instead of relying on handy "file open" and "get string" operators.

Abstraction levels typically used in digital circuit design are, in order of increasing abstraction, the device, circuit, gate, functional module (e.g., adder) and system levels (e.g., processor), as illustrated in Figure 1-6. A semiconductor device is an entity with a very complex behavior. No circuit designer will ever seriously consider the solid-state physics equations governing the behavior of the device when designing a digital gate. Instead, he will use a simplified model that adequately describes the input/output behavior of the transistor. For instance, an AND gate is adequately described by its Boolean expression ($Z = A.B$), its bounding box, the position of the input and output terminals, and the delay between the inputs and the output.

This design philosophy has been the enabler for the emergence of elaborate *computer-aided design* (CAD) frameworks for digital integrated circuits—without it the current design complexity would not have been achievable. Design tools include simulation at the various complexity levels, design verification, layout generation, and design synthesis. An overview of these tools and design methodologies is given in Chapter 8.

Furthermore, to avoid the redesign and reverification of frequently used cells, such as basic gates and arithmetic and memory modules, designers most often resort to *cell libraries*. These libraries not only contain the layouts, but also provide complete documentation and characterization of the behavior of the cells. The use of cell libraries is, for instance, apparent in the layout of the Pentium® 4 processor (Figure 1-5b). The integer and floating-point unit, just to name a few, contain large sections designed using one particular cell-based approach, called *standard cell*. Logic gates are placed in rows of cells of equal height and interconnected using routing channels. The layout of such a block can be generated automatically given that a library of cells is available.

The preceding analysis demonstrates that design automation and modular design practices have effectively addressed some of the complexity issues incurred in contemporary digital design. This leads to the following pertinent question: If design automation solves all our design problems, why should we be concerned with digital circuit design at all? Will the next-generation

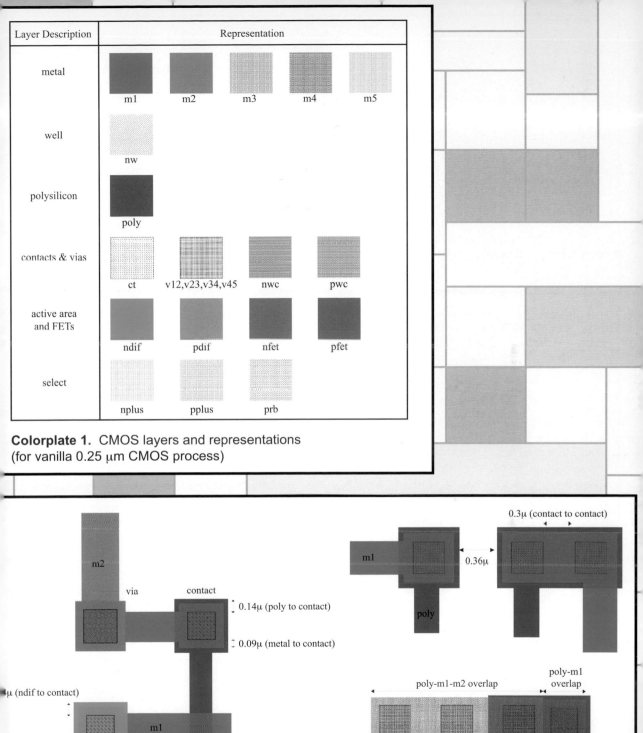

Layer Description	Representation				
metal	m1	m2	m3	m4	m5
well	nw				
polysilicon	poly				
contacts & vias	ct	v12,v23,v34,v45	nwc	pwc	
active area and FETs	ndif	pdif	nfet	pfet	
select	nplus	pplus	prb		

Colorplate 1. CMOS layers and representations (for vanilla 0.25 μm CMOS process)

Colorplate 4. Design rules regarding contacts and vias. Overlapping layers are marked by merged colors.

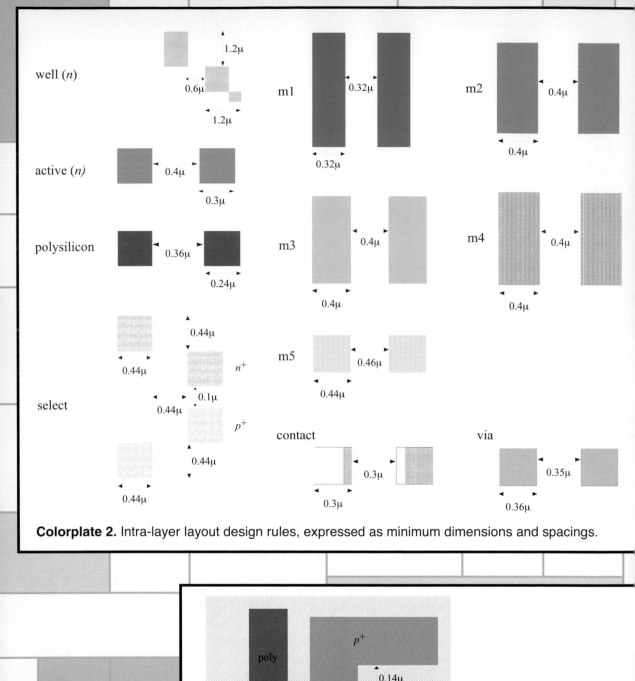

Colorplate 2. Intra-layer layout design rules, expressed as minimum dimensions and spacings.

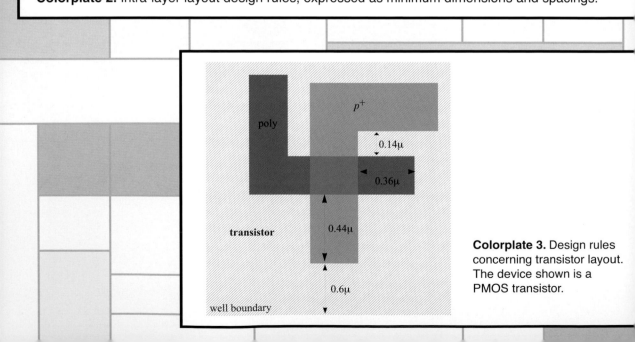

Colorplate 3. Design rules concerning transistor layout. The device shown is a PMOS transistor.

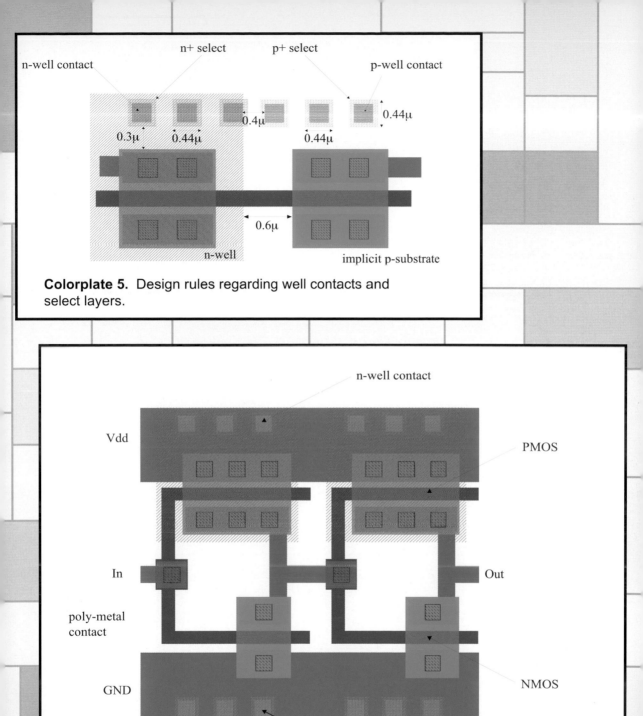

Colorplate 5. Design rules regarding well contacts and select layers.

Colorplate 6. Layout of inverter in 0.25 μm CMOS technology.

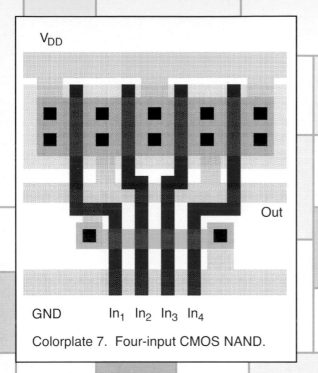

V_{DD}

Out

GND In_1 In_2 In_3 In_4

Colorplate 7. Four-input CMOS NAND.

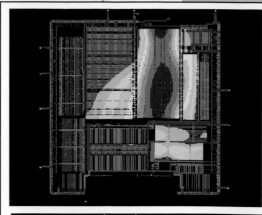

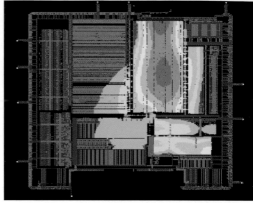

Colorplate 8. Simulated IR voltage drop in power-distribution network.

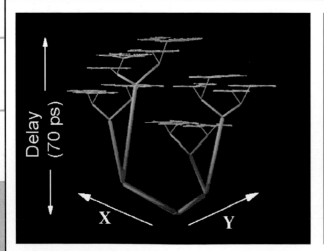

Colorplate 9. Visualizations of clock delay in a tree network driving different loads.

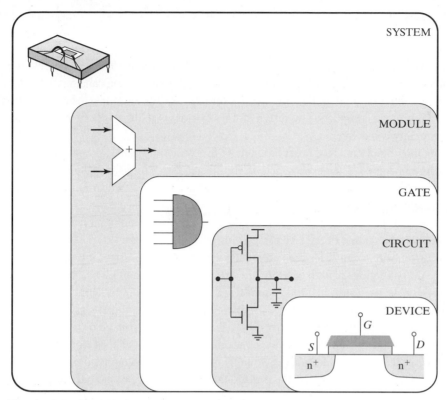

Figure 1-6   Design abstraction levels in digital circuits.

digital designer ever have to worry about transistors or parasitics, or is the smallest design entity he will ever consider be the gate and the module?

The truth is that the reality is more complex, and various reasons exist as to why an insight into digital circuits and their intricacies will still be an important asset for a long time to come:

- First, someone still has to *design and implement* the module libraries. Semiconductor technologies continue to advance from year to year. Until one has developed a foolproof approach towards "porting" a cell from one technology to another, each change in technology—which happens approximately every two years—requires a redesign of the library.
- Creating an adequate *model* of a cell or module requires an in-depth understanding of its internal operation. For instance, to identify the dominant performance parameters of a given design, one has to recognize the critical timing path first.
- The library-based approach works fine when the design constraints (speed, cost, or power) are not stringent. This is the case for a large number of *application-specific designs*, where

the main goal is to provide a more integrated system solution, and performance requirements are easily within the capabilities of the technology. Unfortunately, for a large number of other products such as microprocessors, success hinges on high performance, and designers therefore tend to push technology to its limits. At that point, the hierarchical approach tends to become somewhat less attractive. To resort to our previous analogy to software methodologies, a programmer tends to "customize" software routines when execution speed is crucial; compilers—or design tools—are not yet to the level of what human sweat or ingenuity can deliver.

• Even more important is the observation that the abstraction-based approach is only correct to a certain degree. The performance of, for instance, an adder can be substantially influenced by the way it is connected to its environment. The interconnection wires themselves contribute to delay because they introduce parasitic capacitances, resistances, and even inductances. The impact of the *interconnect parasitics* is bound to increase in the years to come with the scaling of the technology.

• Scaling tends to emphasize some other deficiencies of the abstraction-based model. Some design entities tend to be *global or external*. Examples of global factors are the clock signals, used for synchronization in a digital design, and the supply lines. Increasing the size of a digital design has a profound effect on these global signals. For instance, connecting more cells to a supply line can cause a voltage drop over the wire, which, in turn, can slow down all the connected cells. Issues such as clock distribution, circuit synchronization, and supply-voltage distribution are becoming more and more critical. Coping with them requires a profound understanding of the intricacies of digital circuit design.

• Another impact of technology evolution is that *new design issues* and constraints tend to emerge over time. A typical example of this is the periodical reemergence of power dissipation as a constraining factor, as was already illustrated in the historical overview. Another example is the changing ratio between device and interconnect parasitics. To cope with these unforeseen factors, one must at least be able to model and analyze their impact, requiring once again a profound insight into circuit topology and behavior.

• Finally, when things can go wrong, they often do. A fabricated circuit does not always exhibit the exact waveforms one might expect from advance simulations. Deviations can be caused by variations in the fabrication process parameters, or by the inductance of the package, or by a badly modeled clock signal. *Troubleshooting* a design requires circuit expertise.

For all these reasons, it is our profound belief that an in-depth knowledge of digital circuit design techniques and approaches is an essential asset for a digital-system designer. Even though she might not have to deal with the details of the circuit on a daily basis, the understanding will help her cope with unexpected circumstances and determine the dominant effects when analyzing a design.

Example 1.1   Clocks Defy Hierarchy

To illustrate some of the issues raised in the preceding discussion, let us examine the impact of deficiencies in one of the most important global signals in a design, the *clock*. The function of the clock signal in a digital design is to order the multitude of events happening in the circuit. This task can be compared to the function of a traffic light that determines which cars are allowed to move. It also makes sure that all operations are completed before the next one starts—a traffic light should be green long enough to allow a car or a pedestrian to cross the road. Under ideal circumstances, the clock signal is a periodic step waveform with transitions synchronized throughout the designed circuit (Figure 1-7a). In light of our analogy, changes in the traffic lights should be synchronized to maximize throughput while avoiding accidents. The importance of the *clock alignment* concept is illustrated with the example of two cascaded registers, both operating on the rising edge of the clock ϕ (Figure 1-7b). Under normal operating conditions, the input *In* gets sampled into the first register on the rising edge of ϕ and appears at the output exactly one clock period later. This is confirmed by the simulations shown in Figure 1-7c (signal *Out*).

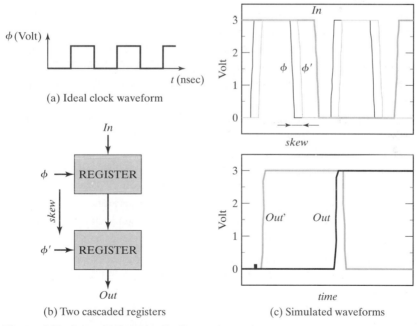

(a) Ideal clock waveform

(b) Two cascaded registers

(c) Simulated waveforms

Figure 1-7   Impact of clock misalignment.

Due to delays associated with routing the clock wires, it may happen that the clocks become misaligned with respect to each other. As a result, the registers are interpreting time indicated by the clock signal differently. Consider the case in which the clock signal for the second register is delayed—or skewed—by a value δ. The rising edge of the delayed clock ϕ' will postpone the sampling of the input of the second register. If the time it takes to propagate the output of the first register to the input of the second register is smaller than the clock delay, the latter will sample the wrong value. This causes the output to change prematurely, as clearly illustrated in the simulation, where the signal *Out'* goes high at the first rising edge of ϕ' instead of the second one. In terms of our traffic analogy, cars leaving the first traffic light hit the cars at the next light that have not yet left.

Clock misalignment, or *clock skew*, is an important example of how global signals may influence the functioning of a hierarchically designed system. Clock skew is actually one of the most critical design problems facing the designers of large high-performance systems.

Example 1.2   **Power Distribution Networks Defy Hierarchy**

While the clock signal is one example of a global signal that crosses the chip hierarchy boundaries, the power distribution network represents another. A digital system requires a stable DC voltage to be supplied to the individual gates. To ensure proper operation, this voltage should be stable within a few hundred millivolts. The power distribution system has to provide this stable voltage in the presence of very large current variations. The resistive nature of the on-chip wires and the inductance of the IC package pins make this a difficult proposition. For example, the average DC current to be supplied to a 100 W-1V microprocessor equals 100 A! The peak current can easily be twice as large, and current demand can readily change from almost zero to this peak value over a short time—in the range of 1 nsec or less. This leads to a current variation of 100 GA/s, which is a truly astounding number.

Consider the problem of the resistance of power distribution wires. For a current of 100 A, a wire resistance of merely 1.25 mΩ leads to a 5% drop in supply voltage (for a 2.5 V supply). Making the wires wider reduces the resistance and thus the voltage drop. While this sizing of the power network is relatively simple in a flat design approach, it is a lot more complex in a hierarchical design. For example, consider the two blocks shown in Figure 1-8a [Saleh01]. If power distribution for Block A is examined in isolation, the additional loading due to the presence of Block B is not taken into account. If power is routed through Block A to Block B, a larger IR drop will occur in Block B, since power is also being consumed by Block A before it reaches Block B.

Since the total IR drop is based on the resistance seen from the pin to the block, one could route around the block and feed power to each block separately, as shown in Figure 1-8b. Ideally, the main trunks should be large enough to handle all the current flowing through separate branches. Although routing power this way is easier to control and maintain, it also requires more area to implement. The large metal trunks of power have to

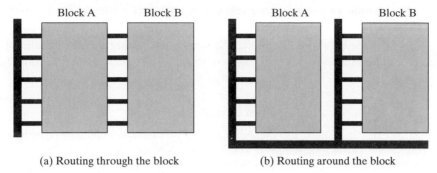

(a) Routing through the block   (b) Routing around the block

Figure 1-8 Power distribution network design.

be sized to handle all the current for each block. This requirement forces designers to set aside area for power busing that takes away from the available routing area.

As more and more blocks are added, the complex interactions between the blocks determine the actual voltage drops. For instance, it is not always easy to determine which way the current will flow when multiple parallel paths are available between the power source and the consuming gate. Also, currents into the different modules do rarely peak at the same time. All these considerations make the design of the power distribution a challenging job. It requires a design methodology approach that supersedes the artificial boundaries imposed by hierarchical design.

The purpose of this book is to provide *a bridge between the abstract vision of digital design and the underlying digital circuit and its peculiarities.* While starting from a solid understanding of the operation of electronic devices and an in-depth analysis of the nucleus of digital design—the inverter—we will gradually channel this knowledge into the design of more complex entities, such as complex gates, datapaths, registers, controllers, and memories. The persistent quest for a designer when designing each of the mentioned modules is to identify the dominant design parameters, to locate the section of the design he should focus his optimizations on, and to determine the specific properties that make the module under investigation (e.g., a memory) different from any others.

We also address other compelling (global) issues in modern digital circuit design such as *power dissipation, interconnect, timing,* and *synchronization.*

1.3 Quality Metrics of a Digital Design

This section defines a set of basic properties of a digital design. These properties help to quantify the quality of a design from different perspectives: cost, functionality, robustness, performance, and energy consumption. Which one of these metrics is most important depends upon the application. For instance, pure speed is a crucial property in a computer server. On the other hand, energy consumption is a dominant metric for hand heldmobile applications such as cell phones. The introduced properties are relevant at all levels of the design hierarchy—system, chip, module, and gate. To ensure consistency in the definitions throughout the design hierarchy stack, we propose a bottom-up approach: We start with defining the basic quality metrics of a simple inverter, and gradually expand these to the more complex functions such as gate, module, and chip.

1.3.1   Cost of an Integrated Circuit

The total cost of any product can be separated into two components: the recurring expenses or the *variable cost*, and the nonrecurring expenses or the *fixed cost*.

Fixed Cost

The fixed cost is independent of the sales volume or the number of products sold. An important component of the fixed cost of an integrated circuit is the effort in time and manpower it takes to produce the design. This design cost is strongly influenced by the complexity of the design, the aggressiveness of the specifications, and the productivity of the designer. Advanced design methodologies that automate major parts of the design process can help to boost the latter. Bringing down the design cost in the presence of an ever-increasing IC complexity is one of the major challenges always facing the semiconductor industry.

Additionally, one has to account for the *indirect costs*, the company overhead that cannot be billed directly to one product. It includes the company's research and development (R&D) costs, manufacturing equipment, marketing and sales costs, and building infrastructure, among others.

Variable Cost

The variable cost accounts for the cost that is directly attributable to a manufactured product, and is hence proportional to the product volume. Variable costs include the costs of the parts used in the product, assembly costs, and testing costs. The total cost of an integrated circuit is

$$\text{cost per IC} = \text{variable cost per IC} + \left(\frac{\text{fixed cost}}{\text{volume}}\right) \tag{1.1}$$

The impact of the fixed cost is more pronounced for small-volume products. This helps explain why it makes sense to have large design teams working for several years on a hugely successful product such as a microprocessor.

While the cost of producing a single transistor has dropped exponentially over the past few decades, the basic variable-cost equation has not changed:

$$\text{variable cost} = \frac{\text{cost of die} + \text{cost of die test} + \text{cost of packaging}}{\text{final test yield}} \tag{1.2}$$

As will be discussed further in Chapter 2, the IC manufacturing process groups a number of identical circuits onto a single *wafer* (see Figure 1-9). Upon completion of the fabrication, the wafer is chopped into *dies*, which are then individually packaged after being *tested*. We will

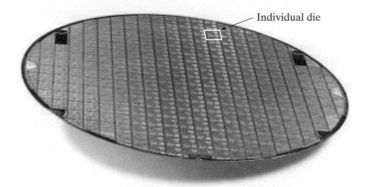

Figure 1-9 Finished wafer. Each square represents a die—in this case the AMD Duron™ microprocessor (reprinted with permission from AMD).

focus on the cost of the die in this discussion. The cost of packaging and testing is the topic of later chapters.

The die cost depends on the number of good dies on a wafer, and the percentage of those that are functional. The latter factor is called the *die yield*, given by

$$\text{cost of die} = \frac{\text{cost of wafer}}{\text{dies per wafer} \times \text{die yield}} \tag{1.3}$$

The number of dies per wafer is, in essence, the area of the wafer divided by the die area. The actual situation is somewhat more complicated as wafers are round, and chips are square. Dies around the perimeter of the wafer are therefore lost. The size of the wafer has been steadily increasing over the years, yielding more dies per fabrication run. Equation (1.3) also presents the first indication that the cost of a circuit is dependent upon the chip area—increasing the chip area simply means that fewer dies fit on a wafer.

The actual relation between cost and area is more complex, and depends upon the die yield. Both the substrate material and the manufacturing process introduce faults that can cause a chip to fail. Assuming that the defects are randomly distributed over the wafer and that the yield is inversely proportional to the complexity of the fabrication process, we obtain the equation

$$\text{die yield} = \left(1 + \frac{\text{defects per unit area} \times \text{die area}}{\alpha}\right)^{-\alpha} \tag{1.4}$$

where α is a parameter that depends upon the complexity of the manufacturing process, and it is roughly proportional to the number of masks. A good estimate for today's complex CMOS processes is $\alpha = 3$. The defects per unit area is a measure of the material and process-induced faults. A value between 0.5 and 1 defects/cm^2 is typical these days, but depends strongly upon the maturity of the process.

Example 1.3   Die Yield

Assume a wafer size of 12 inches, a die size of 2.5 cm², 1 defects/cm², and $\alpha = 3$. Determine the die yield of this CMOS process run.

The number of dies per wafer can be estimated with the following expression, which takes into account the lost dies around the perimeter of the wafer:

$$\text{dies per wafer} = \frac{\pi \times (\text{wafer diameter}/2)^2}{\text{die area}} - \frac{\pi \times \text{wafer diameter}}{\sqrt{2} \times \text{die area}}$$

This means there are 252 (= 296 – 44) potentially operational dies for this particular example. The die yield can be computed with the aid of Eq. (1.4), and equals 16%! This means that, on average only 40 of the dies will be fully functional.

The bottom line is that the number of functional dies per wafer, and hence the cost per die, is a strong function of the die area. While the yield tends to be excellent for the smaller designs, it drops rapidly once a certain threshold is exceeded. Bearing in mind the equations derived previously and the typical parameter values, we can conclude that die costs are proportional to the fourth power of the area:

$$\text{cost of die} = f(\text{die area})^4 \qquad\qquad (1.5)$$

The area is a function that is directly controllable by the designer(s), and it is the prime metric for cost. Small area is thus a desirable property for a digital gate. The smaller the gate, the higher the integration density and the smaller the die size. Smaller gates also tend to be faster and consume less energy—the total gate capacitance, which is one of the dominant performance parameters, often scales with the area.

The *number of transistors* in a gate is indicative of the expected implementation area but other parameters may have an impact. For instance, a complex interconnect pattern between the transistors can cause the wiring area to dominate. The *gate complexity*, as expressed by the number of transistors and the regularity of the interconnect structure, also has an impact on the design cost. Complex structures are more difficult to implement and tend to take more of the designer's valuable time. Simplicity and regularity is a precious property in cost-sensitive designs.

1.3.2   Functionality and Robustness

A prime requirement for a digital circuit is, obviously, that it performs the function it is designed for. The measured behavior of a manufactured circuit normally deviates from the expected response. One reason for this aberration is due to the variations in the manufacturing process. The dimensions and device parameters vary between runs or even on a single wafer or die. The electrical behavior of a circuit can be profoundly affected by those variations. The presence of disturbing noise sources on or off the chip is another source of deviations in circuit response.

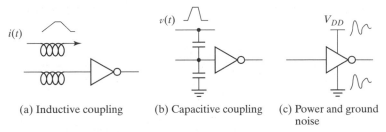

| (a) Inductive coupling | (b) Capacitive coupling | (c) Power and ground noise |

Figure 1-10   Noise sources in digital circuits.

The word *noise,* in the context of digital circuits, means *unwanted variations of voltages and currents at the logic nodes.* Noise signals can enter a circuit in many ways. Some examples of digital noise sources are depicted in Figure 1-10. For instance, two wires placed side by side in an integrated circuit form a coupling capacitor and a mutual inductance. Hence, a change in voltage or current on one of the wires can influence the signals on the neighboring wire. Noise on the power and ground rails of a gate also influences the signal levels in the gate.

Most noise in a digital system is internally generated, and the noise value is proportional to the signal swing. Capacitive and inductive cross talk, and the internally generated power supply noise are examples of such. Other noise sources such as input power supply noise are external to the system, and their value is not related to the signal levels. For these sources, the noise level is directly expressed in volts or amperes. Noise sources that are a function of the signal level are better expressed as a fraction or percentage of the signal level. Noise is a major concern in the engineering of digital circuits. How to cope with all these disturbances is one of the main challenges in the design of high-performance digital circuits and is a recurring topic in this book.

The steady-state parameters (also called the *static behavior*) of a gate measure how robust the circuit is with respect to both variations in the manufacturing process and noise disturbances. The definition and derivation of these parameters requires a prior understanding of how digital signals are represented in the world of electronic circuits.

Digital circuits (DC) perform operations on *logical* (or *Boolean*) variables. A logical variable x can only assume two discrete values:

$$x \in \{0,1\}$$

As an example, the inversion (i.e., the function that an inverter performs) implements the following compositional relationship between two Boolean variables x and y:

$$y = \bar{x} \colon \{x = 0 \Rightarrow y = 1; x = 1 \Rightarrow y = 0\} \tag{1.6}$$

A logical variable is, however, a mathematical abstraction. In a physical implementation, such a variable is represented by an electrical quantity. This is most often a node voltage that is not discrete, but can adopt a continuous range of values. This electrical voltage is turned into a discrete variable by associating *a nominal voltage level* with each logic state: $1 \Leftrightarrow V_{OH}$, $0 \Leftrightarrow V_{OL}$, where V_{OH} and V_{OL} represent the *high* and the *low* logic levels, respectively. Applying V_{OH}

to the input of an inverter yields V_{OL} at the output and vice versa. The difference between the two is called the *logic* or *signal swing* V_{sw}.

$$V_{OH} = \overline{(V_{OL})}$$
$$V_{OL} = \overline{(V_{OH})}$$

(1.7)

The Voltage-Transfer Characteristic

Assume now that a logical variable *in* serves as the input to an inverting gate that produces the variable *out*. The electrical function of a gate is best expressed by its *voltage-transfer characteristic* (VTC) (sometimes called the *DC transfer characteristic*), which plots the output voltage as a function of the input voltage $V_{out} = f(V_{in})$. An example of an inverter VTC is shown in Figure 1-11. The high and low nominal voltages, V_{OH} and V_{OL}, can readily be identified—$V_{OH} = f(V_{OL})$ and $V_{OL} = f(V_{OH})$. Another point of interest of the VTC is the *gate* or *switching threshold voltage* V_M (not to be confused with the threshold voltage of a transistor), that is defined as $V_M = f(V_M)$. V_M can also be found graphically at the intersection of the VTC curve and the line given by $V_{out} = V_{in}$. The gate threshold voltage presents the midpoint of the switching characteristics, which is obtained when the output of a gate is short-circuited to the input. This point will prove to be of particular interest when studying circuits with feedback (also called *sequential circuits*).

Even if an ideal nominal value is applied at the input of a gate, the output signal often deviates from the expected nominal value. These deviations can be caused by noise or by the loading on the output of the gate (i.e., by the number of gates connected to the output signal). Figure 1-12a illustrates how a logic level is represented in reality by a range of acceptable voltages, separated by a region of uncertainty, rather than by nominal levels alone. The regions of acceptable high and low voltages are delimited by the V_{IH} and V_{IL} voltage levels, respectively. By definition, these represent the points where the gain ($= dV_{out} / dV_{in}$) of the VTC equals −1 as shown in Figure 1-12b. The region between V_{IH} and V_{IL} is called the *undefined region*

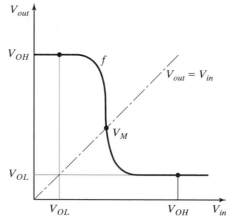

Figure 1-11   Inverter voltage-transfer characteristic.

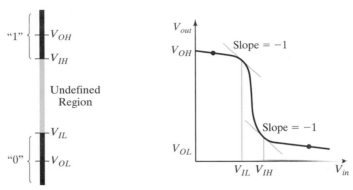

(a) Relationship between voltage and logic levels   (b) Definition of V_{IH} and V_{IL}

Figure 1-12   Mapping logic levels to the voltage domain.

(sometimes also called the *transition width,* or *TW*). Steady-state signals should avoid this region if proper circuit operation is to be ensured.

Noise Margins

For a gate to be robust and insensitive to noise disturbances, it is essential that the "0" and "1" intervals be as large as possible. A measure of the sensitivity of a gate to noise is given by the noise margins NM_L (*noise margin low*) and NM_H (*noise margin high*), which quantize the size of the legal "0" and "1," respectively, and set a fixed maximum threshold on the noise value:

$$NM_L = V_{IL} - V_{OL}$$
$$NM_H = V_{OH} - V_{IH}$$

(1.8)

The noise margins represent the levels of noise that can be sustained when gates are cascaded as illustrated in Figure 1-13. It is obvious that the margins should be larger than 0 for a digital circuit to be functional, and by preference, they should be as large as possible.

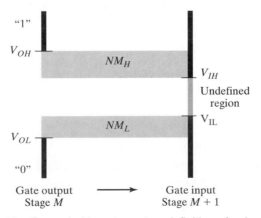

Figure 1-13   Cascaded inverter gates: definition of noise margins.

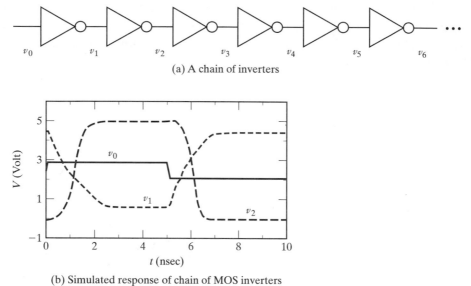

(a) A chain of inverters

(b) Simulated response of chain of MOS inverters

Figure 1-14   The regenerative property.

Regenerative Property

A large noise margin is desirable, but it is not the only requirement. Assume that a signal is disturbed by noise and differs from the nominal voltage levels. As long as the signal is within the noise margins, the gate that follows continues to function correctly, although its output voltage varies from the nominal one. This deviation is added to the noise injected at the output node and passed to the next gate. The effect of different noise sources may accumulate and eventually force a signal level into the undefined region. Fortunately, this does not happen if the gate possesses the *regenerative property*, which ensures that a disturbed signal gradually converges back to one of the nominal voltage levels after passing through a number of logical stages. This property can be understood as follows:

An input voltage v_{in} ($v_{in} \in$ "0") is applied to a chain of N inverters (Figure 1-14a). Assuming that the number of inverters in the chain is even, the output voltage v_{out} ($N \rightarrow \infty$) will equal V_{OL} if and only if the inverter possesses the regenerative property. Similarly, when an input voltage v_{in} ($v_{in} \in$ "1") is applied to the inverter chain, the output voltage will approach the nominal value V_{OH}.

Example 1.4   Regenerative Property

The concept of regeneration is illustrated in Figure 1-14b, which plots the simulated transient response of a chain of CMOS inverters. The input signal to the chain is a step waveform with a degraded amplitude, which could be caused by noise. Instead of swinging from rail to rail, v_0 only extends between 2.1 and 2.9 V. From the simulation, it can be observed that this deviation rapidly disappears while progressing through the chain; v_1, for instance, extends from 0.6 V to 4.45 V. Even further, v_2 already swings between the nominal V_{OL} and V_{OH}. The inverter used in this example clearly possesses the regenerative property.

The conditions under which a gate is regenerative can be intuitively derived by analyzing a simple case study. Figure 1-15a plots the VTC of an inverter $V_{out} = f(V_{in})$ as well as its inverse function $finv()$, which reverts the function of the x- and y-axis and is defined as

$$in = f(out) \Rightarrow in = finv(out) \tag{1.9}$$

Assume that a voltage v_0, deviating from the nominal voltages, is applied to the first inverter in the chain. The output voltage of this inverter equals $v_1 = f(v_0)$ and is applied to the next inverter. Graphically, this corresponds to $v_1 = finv(v_2)$. The signal voltage gradually converges to the nominal signal after a number of inverter stages, as indicated by the arrows. In Figure 1-15b the signal does not converge to any of the nominal voltage levels, but to an intermediate voltage level. Hence, the characteristic is nonregenerative. The difference between the two cases is due to the gain characteristics of the gates. To be regenerative, the VTC should have a transient region (or undefined region) with a gain *greater than* 1 in absolute value, bordered by the two legal zones, where the gain should be *less than* 1. Such a gate has two stable operating points. This clarifies the definition of the V_{IH} and the V_{IL} levels that form the boundaries between the legal and the transient zones.

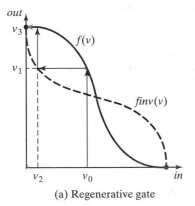

(a) Regenerative gate

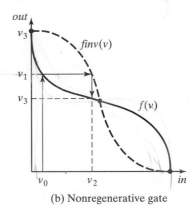

(b) Nonregenerative gate

Figure 1-15 Conditions for regeneration.

Noise Immunity

While the noise margin is a meaningful means for measuring the robustness of a circuit against noise, it is not sufficient. It expresses the capability of a circuit to "overpower" a noise source. *Noise immunity*, on the other hand, expresses the ability of the system to process and transmit information correctly in the presence of noise [Dally98]. Many digital circuits with low noise margins have very good noise immunity because they *reject a noise source* rather than over-power it. These circuits have the property that only a small fraction of a potentially damaging noise source is coupled to the important circuit nodes. More precisely, the transfer function between noise source and signal node is far less than 1. Circuits that do not posses this property are *susceptible* to noise.

To study the noise immunity of a gate, we have to construct a noise budget that allocates the power budget to the various noise sources. As discussed earlier, the noise sources can be divided into the following types of sources:

- those that are *proportional* to the signal swing V_{sw}. The impact on the signal node is expressed as $g\,V_{sw}$; and
- those that are *fixed*. The impact on the signal node equals $f\,V_{Nf}$, with V_{nf} the amplitude of the noise source, and f the transfer function *from noise* to signal node.

We assume, for the sake of simplicity, that the noise margin equals half the signal swing (for both H and L). To operate correctly, the noise margin has to be larger than the sum of the coupled noise values:

$$V_{NM} = \frac{V_{sw}}{2} \geq \sum_i f_i V_{Nfi} + \sum_j g_j V_{sw} \qquad (1.10)$$

Given a set of noise sources, we can derive the minimum signal swing necessary for the system to be operational:

$$V_{sw} \geq \frac{2 \sum_i f_i V_{Nfi}}{1 - 2 \sum_j g_j} \qquad (1.11)$$

This makes it clear that the signal swing (and the noise margin) has to be large enough to overpower the impact of the fixed sources ($f\,V_{Nf}$). On the other hand, the sensitivity to internal sources depends primarily upon the noise suppressing capabilities of the gate, this is the propor-tionality or gain factors g_j. In the presence of large gain factors, increasing the signal swing does not do any good to suppress noise, as the noise increases proportionally. In later chapters, we will discuss some differential logic families that suppress most of the internal noise, and thus can get away with very small noise margins and signal swings.

Directivity

The directivity property requires a gate to be *unidirectional*—that is, changes in an output level should not appear at any unchanging input of the same circuit. Otherwise, an output-signal transition reflects to the gate inputs as a noise signal, affecting the signal integrity.

In real gate implementations, full directivity can never be achieved. Some feedback of changes in output levels to the inputs cannot be avoided. Capacitive coupling between inputs and outputs is a typical example of such a feedback. It is important to minimize these changes so that they do not affect the logic levels of the input signals.

Fan-In and Fan-Out

The *fan-out* denotes *the number of load gates N that are connected to the output of the driving gate* (see Figure 1-16). Increasing the fan-out of a gate can affect its logic output levels. From the world of analog amplifiers, we know that this effect is minimized by making the input resistance of the load gates as large as possible (minimizing the input currents) and by keeping the output resistance of the driving gate small (reducing the effects of load currents on the output voltage). When the fan-out is large, the added load can deteriorate the dynamic performance of the driving gate. For these reasons, many generic and library components define a *maximum fan-out* to guarantee that the static and dynamic performance of the element meet specification.

The *fan-in* of a gate is defined as the *number of inputs* to the gate (see Figure 1-16b). Gates with large fan-in tend to be more complex, which often results in inferior static and dynamic properties.

The Ideal Digital Gate

Based on these observations, we can define the *ideal* digital gate from a static perspective. The ideal inverter model is important because it gives us a metric by which we can judge the quality of actual implementations.

Its VTC is shown in Figure 1-17 and has the following properties: infinite gain in the transition region, and gate threshold located in the middle of the logic swing, with high and low

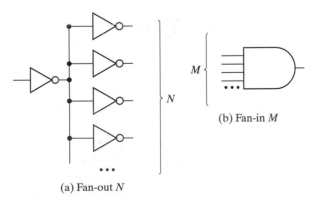

(b) Fan-in M

(a) Fan-out N

Figure 1-16 Definition of fan-out and fan-in of a digital gate.

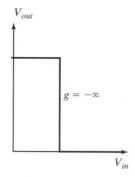

Figure 1-17   Ideal voltage-transfer characteristic.

noise margins equal to half the swing. The input and output impedances of the ideal gate are infinity and zero, respectively (i.e., the gate has unlimited fan-out). While this ideal VTC is unfortunately impossible in real designs, some implementations, such as the static CMOS inverter, come close.

Example 1.5   Voltage-Transfer Characteristic

Figure 1-18 shows an example of a voltage-transfer characteristic of an actual, but out-dated, gate structure. The values of the dc parameters are derived from inspection of the graph.

$$V_{OH} = 3.5 \text{ V}; \qquad\qquad V_{OL} = 0.45 \text{ V}$$

$$V_{IH} = 2.35 \text{ V}; \qquad\qquad V_{IL} = 0.66 \text{ V}$$

$$V_M = 1.64 \text{ V}$$

$$NM_H = 1.15 \text{ V}; \qquad\qquad NM_L = 0.21 \text{ V}$$

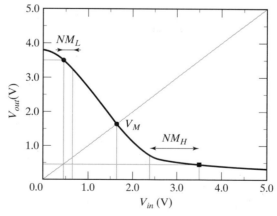

Figure 1-18   Voltage-transfer characteristic of an NMOS inverter of the 1970s.

The observed transfer characteristic, obviously, is far from ideal: it is asymmetrical, has a very low value for NM_L, and the voltage swing of 3.05 V is substantially below the maximum obtainable value of 5 V (which is the value of the supply voltage for this design).

1.3.3 Performance

From a system designer's perspective, the performance of a digital circuit expresses its computational ability. For instance, a microprocessor often is characterized by the number of instructions it can execute per second. This performance metric depends both on the architecture of the processor—for instance, the number of instructions it can execute in parallel—and the actual design of logic circuitry. While the former is crucially important, it is not the focus of this text. The reader may refer to the many excellent books on this topic [for instance, Hennessy02]. When focusing on the pure design, performance is most often expressed by the duration of the clock period (*clock cycle time*), or its rate (*clock frequency*). The minimum value of the clock period for a given technology and design is set by a number of factors such as the time it takes for the signals to propagate through the logic, the time it takes to get the data in and out of the registers, and the uncertainty of the clock arrival times. Each of these topics will be discussed in detail in this book. The performance of an individual gate lies at the core of the whole performance analysis, however.

The *propagation delay* t_p of a gate defines how quickly it responds to a change at its input(s). It expresses *the delay experienced by a signal when passing through a gate*. It is measured between the 50% transition points of the input and output waveforms, as shown in Figure 1-19 for

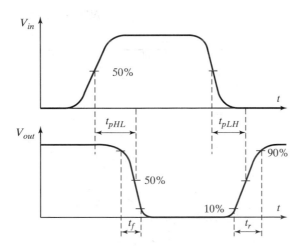

Figure 1-19 Definition of propagation delays and rise and fall times.

an inverting gate.[2] Because a gate displays different response times for rising or falling input waveforms, two definitions of the propagation delay are necessary. The t_{pLH} defines the response time of the gate for a *low-to-high* (or positive) output transition, while t_{pHL} refers to a *high-to-low* (or negative) transition. The propagation delay t_p is defined as the average of the two:

$$t_p = \frac{t_{pLH} + t_{pHL}}{2} \tag{1.12}$$

CAUTION: Observe that the propagation delay t_p, in contrast to t_{pLH} and t_{pHL}, is an artificial gate quality metric, and has no physical meaning per se. It is mostly used to compare different semiconductor technologies, circuit, or logic design styles.

The propagation delay is not only a function of the circuit technology and topology, but depends upon other factors as well. Most importantly, the delay is a function of the *slopes* of the input and output signals of the gate. To quantify these properties, we introduce the *rise and fall times*, t_r and t_f, which are metrics that apply to individual signal waveforms rather than to gates (see Figure 1-19), and they express how fast a signal transits between the different levels. The uncertainty over when a transition actually starts or ends is avoided by defining the rise and fall times between the 10% and 90% points of the waveforms, as shown in the figure. The rise/fall time of a signal is largely determined by the strength of the driving gate, and the load presented to it.

When comparing the performance of gates implemented in different technologies or circuit styles, it is important not to confuse the picture by including parameters such as load factors, fan-in, and fan-out. A uniform way of measuring the t_p of a gate so that technologies can be judged on an equal footing is desirable. The de facto standard circuit for delay measurement is the *ring oscillator*, which consists of an odd number of inverters connected in a circular chain (see Figure 1-20.) Due to the odd number of inversions, this circuit does not have a stable operating point and oscillates. The period T of the oscillation is determined by the propagation time of a signal transition through the complete chain, or $T = 2 \times t_p \times N$ with N the number of inverters in the chain. The factor 2 results from the observation that a full cycle requires both a low-to-high and a high-to-low transition. Note that this equation is only valid for $2Nt_p \gg t_f + t_r$. If this condition is not met, the circuit might not oscillate—one "wave" of signals propagating through the ring will overlap with a successor and eventually dampen the oscillation. Typically, a ring oscillator needs at least five stages to be operational.

---

[2]The 50% definition is inspired by the assumption that the switching threshold V_M is typically located in the middle of the logic swing.

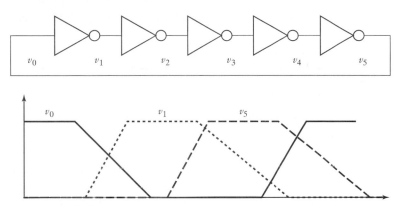

Figure 1-20 Ring oscillator circuit for propagation-delay measurement.

CAUTION: We must be extremely careful with results obtained from ring oscillator measurements. A t_p of 20 ps by no means implies that a circuit built with those gates will operate at 50 GHz. The oscillator results are primarily useful for quantifying the differences between various manufacturing technologies and gate topologies. The oscillator is an idealized circuit in which each gate has a fan-in and fan-out of exactly 1 and parasitic loads are minimal. In more realistic digital circuits, fan-ins and fan-outs are higher, and interconnect delays are non-negligible. The gate functionality is also substantially more complex than a simple invert operation. As a result, the achievable clock frequency, on average, is 50 to 100 times slower than the frequency obtained from ring oscillator measurements. This is an average observation; carefully optimized designs might approach use ideal frequency more closely.

Example 1.6   Propagation Delay of First-Order *RC* Network

Digital circuits are often modeled as first-order *RC* networks of the type shown in Figure 1-21. The propagation delay of such a network is thus of considerable interest.

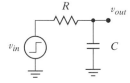

Figure 1-21   First-order *RC* network.

When applying a step input (with v_{in} going from 0 to *V*), the transient response of this circuit is known to be an exponential function, and is given by the following expression (where $\tau = RC$, the time constant of the network):

$$v_{out}(t) = (1 - e^{-t/\tau})\, V \tag{1.13}$$

The time to reach the 50% point is easily computed as $t = \ln(2)\tau = 0.69\tau$. Similarly, it takes $t = \ln(9)\tau = 2.2\tau$ to get to the 90% point. It is worth memorizing these numbers because they are used extensively throughout this text.

1.3.4   Power and Energy Consumption

The power consumption of a design determines how much energy is consumed per operation, and how much heat the circuit dissipates. These factors influence a great number of critical design decisions, such as the power supply capacity, the battery lifetime, supply line sizing, packaging and cooling requirements. Therefore, power dissipation is an important property of a design that affects feasibility, cost, and reliability. In the world of high-performance computing, power consumption limits, dictated by the chip package and the heat removal system, determine the number of circuits that can be integrated onto a single chip, and how fast they are allowed to switch.With the increasing popularity of mobile and distributed computation, energy limitations put a firm restriction on the number of computations that can be performed given a minimum time between battery recharges.

Depending on the design problem at hand, different dissipation measures must be considered. For instance, the peak power P_{peak} is important when studying supply line sizing. When addressing cooling or battery requirements, one is predominantly interested in the average power dissipation P_{av}. The measures are defined as

$$P_{peak} = i_{peak}V_{supply} = max[p(t)]$$

$$P_{av} = \frac{1}{T}\int_0^T p(t)dt = \frac{V_{supply}}{T}\int_0^T i_{supply}(t)dt \tag{1.14}$$

where $p(t)$ is the instantaneous power, i_{supply} is the current being drawn from the supply voltage V_{supply} over the interval $t \in [0,T]$, and i_{peak} is the maximum value of i_{supply} over that interval.

The dissipation can further be decomposed into *static* and *dynamic* components. The latter occurs only during transients, when the gate is switching. It is attributed to the charging of capacitors and temporary current paths between the supply rails; therefore it is proportional to the switching frequency: *the higher the number of switching events, the greater the dynamic power consumption.* On the other hand, the static component is present even when no switching occurs and is caused by static conductive paths between the supply rails or by leakage currents. It is always present, even when the circuit is in standby. Minimization of this consumption source is a worthwhile goal.

The propagation delay and the power consumption of a gate are related—the propagation delay is mostly determined by the speed at which a given amount of energy can be stored on the gate capacitors. The faster the energy transfer (or the higher the power consumption), the faster the gate. For a given technology and gate topology, the product of power consumption and propagation delay is generally a constant. This product is called the *power-delay product* (or PDP), and can be considered as a quality measure for a switching device. The PDP is simply the *energy consumed by the gate per switching event*. The ring oscillator is again the circuit of choice for measuring the PDP of a logic family.

An ideal gate is one that is fast and consumes little energy. The *energy-delay* product (E-D) is a combined metric that brings those two elements together, and is often used as the ultimate quality metric. Thus, it should be clear that the E-D is equivalent to *power-delay*^{2}.

Example 1.7   Energy Dissipation of First-Order *RC* Network

Let us again consider the first-order *RC* network (shown in Figure 1-21). When applying a step input (with V_{in} going from 0 to V), an amount of energy is provided by the signal source to the network. The total energy delivered by the source (from the start of the transition to the end) can be readily computed:

$$E_{in} = \int_0^\infty i_{in}(t)v_{in}(t)dt = V\int_0^\infty C\frac{dv_{out}}{dt}dt = (CV)\int_0^V dv_{out} = CV^2 \qquad (1.15)$$

It is interesting to observe that the energy needed to charge a capacitor from 0 to V volts with a step input is a function of the size of the voltage step and the capacitance, but is independent of the value of the resistor. We can also compute how much of the delivered energy gets stored on the capacitor at the end of the transition.

$$E_C = \int_0^\infty i_C(t)v_{out}(t)dt = \int_0^\infty C\frac{dv_{out}}{dt}v_{out}dt = C\int_0^V v_{out}dv_{out} = \frac{CV^2}{2} \qquad (1.16)$$

This is exactly half of the energy delivered by the source. For those who wonder what happened to the other half—a simple analysis shows that an equivalent amount gets dissipated as heat in the resistor during the transaction. It is left to the reader to demonstrate that during the discharge phase (for a step from V to 0), the energy originally stored on the capacitor gets dissipated in the resistor as well, and then turned into heat.

1.4 Summary

In this introductory chapter, we learned about the history and the trends in digital circuit design. We also introduced the important quality metrics used to evaluate the quality of a design: cost, functionality, robustness, performance, and energy/power dissipation.

1.5 To Probe Further

The design of digital integrated circuits has been the topic of many textbooks and monographs. To help the reader find more information on some selected topics, an extensive list of references follows. The state-of-the-art developments in the area of digital design are generally reported in technical journals or conference proceedings, the most important of which are listed.

Journals and Proceedings

IEEE Journal of Solid-State Circuits
IEICE Transactions on Electronics (Japan)
Proceedings of The International Solid-State and Circuits Conference (ISSCC)
Proceedings of the VLSI Circuits Symposium
Proceedings of the Custom Integrated Circuits Conference (CICC)
European Solid-State Circuits Conference (ESSCIRC)

Bibliography

MOS

M. Annaratone, *Digital CMOS Circuit Design*, Kluwer, 1986.

T. Dillinger, *VLSI Engineering*, Prentice Hall, 1988.

M. Elmasry, ed., *Digital MOS Integrated Circuits*, IEEE Press, 1981.

M. Elmasry, ed., *Digital MOS Integrated Circuits II*, IEEE Press, 1992.

L. Glasser and D. Dopperpuhl, *The Design and Analysis of VLSI Circuits*, Addison-Wesley, 1985.

A. Kang and Leblebici, *CMOS Digital Integrated Circuits*, 2nd Ed., McGraw-Hill, 1999.

C. Mead and L. Conway, *Introduction to VLSI Systems*, Addison-Wesley, 1980.

K. Martin, *Digital Integrated Circuit Design*, Oxford University Press, 2000.

D. Pucknell and K. Eshraghian, *Basic VLSI Design*, Prentice Hall, 1988.

M. Shoji, *CMOS Digital Circuit Technology*, Prentice Hall, 1988.

J. Uyemura, *Circuit Design for CMOS VLSI*, Kluwer, 1992.

H. Veendrick, Deep-Submicron C*MOS IC's: From Basics to ASICS*, Second Edition, Kluwer Academic Publishers, 2000.

N. Weste and K. Eshraghian, *Principles of CMOS VLSI Design*, Addison-Wesley, 1985, 1993.

High-Performance Design

K. Bernstein et al, *High Speed CMOS Design Styles*, Kluwer Academic, 1998.

A. Chandrakasan, F. Fox, and W. Bowhill, ed., *Design of High-Performance Microprocessor Circuits*, IEEE Press, 2000.

M. Shoji, *High-Speed Digital Circuits*, Addison-Wesley, 1996.

Low-Power Design

A. Chandrakasan and R. Brodersen, ed., *Low-Power Digital CMOS Design*, IEEE Press, 1998.

M. Pedram and J. Rabaey, ed., *Power-Aware Design Methodologies*, Kluwer Academic, 2002.

J. Rabaey and M. Pedram, ed., *Low-Power Design Methodologies*, Kluwer Academic, 1996.

G. Yeap, *Practical Low-Power CMOS Design*, Kluwer Academic, 1998.

Memory Design

K. Itoh, *VLSI Memory Chip Design*, Springer, 2001.

B. Keeth and R. Baker, *DRAM Circuit Design*, IEEE Press, 1999.

B. Prince, *Semiconductor Memories*, Wiley, 1991.

B. Prince, *High Performance Memories*, Wiley, 1996.

D. Hodges, *Semiconductor Memories*, IEEE Press, 1972.

Interconnections and Packaging

H. Bakoglu, *Circuits, Interconnections, and Packaging for VLSI*, Addison-Wesley, 1990.

W. Dally and J. Poulton, *Digital Systems Engineering*, Cambridge University Press, 1998.

E. Friedman, ed., *Clock Distribution Networks in VLSI Circuits and Systems,* IEEE Press, 1995.

J. Lau et al, ed., *Electronic Packaging: Design, Materials, Process, and Reliability*, McGraw-Hill, 1998.

Design Tools and Methodologies

V. Agrawal and S. Seth, *Test Generation for VLSI Chips*, IEEE Press, 1988.

D. Clein, *CMOS IC Layout*, Newnes, 2000.

G. De Micheli, *Synthesis and Optimization of Digital Circuits*, McGraw-Hill, 1994.

S. Rubin, *Computer Aids for VLSI Design*, Addison-Wesley, 1987.

J. Uyemura, *Physical Design of CMOS Integrated Circuits Using L-Edit*, PWS, 1995.

A. Vladimirescu, *The Spice Book*, John Wiley and Sons, 1993.

W. Wolf, *Modern VLSI Design*, Prentice Hall, 1998.

Bipolar and BiCMOS

A. Alvarez, *BiCMOS Technology and Its Applications*, Kluwer, 1989.

M. Elmasry, ed., *BiCMOS Integrated Circuit Design,* IEEE Press, 1994.

S. Embabi, A. Bellaouar, and M. Elmasry, *Digital BiCMOS Integrated Circuit Design*, Kluwer, 1993.

General

J. Buchanan, *CMOS/TTL Digital Systems Design*, McGraw-Hill, 1990.

H. Haznedar, *Digital Micro-Electronics*, Benjamin/Cummings, 1991.

D. Hodges and H. Jackson, *Analysis and Design of Digital Integrated Circuits*, 2nd ed., McGraw-Hill, 1988.

M. Smith, *Application-Specific Integrated Circuits*, Addison-Wesley, 1997.

R. K. Watts, *Submicron Integrated Circuits*, Wiley, 1989.

References

[Bardeen48] J. Bardeen and W. Brattain, "The Transistor, a Semiconductor Triode," *Phys. Rev.*, vol. 74, p. 230, July 15, 1948.

[Beeson62] R. Beeson and H. Ruegg, "New Forms of All Transistor Logic," *ISSCC Digest of Technical Papers,* pp. 10–11, Feb. 1962.

[Dally98] B. Dally, *Digital Systems Engineering,* Cambridge University Press, 1998.

[Faggin72] F. Faggin, M.E. Hoff, Jr, H. Feeney, S. Mazor, M. Shima, "The MCS-4 - An LSI MicroComputer System," 1972 IEEE Region Six Conference Record, San Diego, CA, pp. 1–6, April 1972.

[Harris56] J. Harris, "Direct-Coupled Transistor Logic Circuitry in Digital Computers," *ISSCC Digest of Technical Papers,* p. 9, Feb. 1956.

[Hart72] C. Hart and M. Slob, "Integrated Injection Logic—A New Approach to LSI," *ISSCC Digest of Technical Papers,* pp. 92–93, Feb. 1972.

[Hoff70] E. Hoff, "Silicon-Gate Dynamic MOS Crams 1,024 Bits on a Chip," *Electronics,* pp. 68–73, August 3, 1970.

[Intel01] "Moore's Law", *http://www.intel.com/research/silicon/mooreslaw.htm*

[Masaki74] A. Masaki, Y. Harada and T. Chiba, "200-Gate ECL Master-Slice LSI," *ISSCC Digest of Technical Papers,* pp. 62–63, Feb. 1974.

[Moore65] G. Moore, "Cramming more Components into Integrated Circuits," Electronics, Vol. 38, Nr 8, April 1965.

[Murphy93] B. Murphy, "Perspectives on Logic and Microprocessors," *Commemorative Supplement to the Digest of Technical Papers, ISSCC*, pp. 49–51, San Francisco, 1993.

[Norman60] R. Norman, J. Last and I. Haas, "Solid-State Micrologic Elements," *ISSCC Digest of Technical Papers,* pp. 82–83, Feb. 1960.

[Hennessy02] J. Hennessy, D. Patterson, and David Goldberg, *Computer Architecture A Quantitative Approach*, Third Edition, Morgan Kaufmann Publishers, 2002.

[Saleh01] R. Saleh, M. Benoit, and P, McCrorie, "Power Distribution Planning", Simplex Solutions, *http://www.simplex.com/wt/sec.php?page_name=wp_powerplan*

[Schockley49] W. Schockley, "The Theory of pn Junctions in Semiconductors and pn-Junction Transistors," *BSTJ,* vol. 28, p. 435, 1949.

[Shima74] M. Shima, F. Faggin and S. Mazor, "An N-Channel, 8-bit Single-Chip Microprocessor," *ISSCC Digest of Technical Papers,* pp. 56–57, Feb. 1974.

[Swade93] D. Swade, "Redeeming Charles Babbage's Mechanical Computer," *Scientific American,* pp. 86–91, February 1993.

[Wanlass63] F. Wanlass, and C. Sah, "Nanowatt logic Using Field-Effect Metal-Oxide Semiconductor Triodes," *ISSCC Digest of Technical Papers,* pp. 32–32, Feb. 1963.

Exercises

Please refer to **http://bwrc.eecs.berkeley.edu/IcBook** for up-to-date problem sets and exercises. By making the exercises electronically available rather than in print, we can provide a dynamic environment that tracks the rapid evolution of today's digital integrated circuit design technology.

2

The Manufacturing Process

Overview of manufacturing process

Design rules

IC packaging

Future Trends in Integrated Circuit Technology

2.1   Introduction

2.2   Manufacturing CMOS Integrated Circuits
    2.2.1   The Silicon Wafer
    2.2.2   Photolithography
    2.2.3   Some Recurring Process Steps
    2.2.4   Simplified CMOS Process Flow

2.3   Design Rules—Between the Designer and the Process Engineer

2.4   Packaging Integrated Circuits
    2.4.1   Package Materials
    2.4.2   Interconnect Levels
    2.4.3   Thermal Considerations in Packaging

2.5   Perspective—Trends in Process Technology
    2.5.1   Short-Term Developments
    2.5.2   In the Longer Term

2.6   Summary

2.7   To Probe Further

2.1 Introduction

Most digital designers will never be confronted with the details of the manufacturing process that lay at the core of the semiconductor revolution. Still, some insight into the steps that lead to an operational silicon chip comes in quite handy in understanding the physical constraints imposed on a designer of an integrated circuit, as well as the impact of the fabrication process on issues such as cost.

In this chapter, we briefly describe the steps and techniques used in a modern integrated circuit manufacturing process. It is not our aim to present a detailed description of the fabrication technology, which easily deserves a complete course [Plummer00]. Rather, we aim at presenting the general outline of the flow and the interaction between the various steps. We learn that a set of *optical masks* forms the central interface between the intrinsics of the manufacturing process and the design that the user wants to see transferred to the silicon fabric. The masks define the patterns that, when transcribed onto the different layers of the semiconductor material, form the elements of the electronic devices and the interconnecting wires. As such, these patterns have to adhere to some constraints, in terms of minimum width and separation, if the resulting circuit is to be fully functional. This collection of constraints is called the *design rule set*, and acts as the contract between the circuit designer and the process engineer. If the designer adheres to these rules, he gets a guarantee that his circuit will be manufacturable. An overview of the common design rules encountered in modern CMOS processes is given, as well as a perspective on the *IC packaging* options. The package forms the interface between the circuit implemented on the silicon die and the outside world, and as such has a major impact on the performance, reliability, longevity, and cost of the integrated circuit.

2.2 Manufacturing CMOS Integrated Circuits

A simplified cross section of a typical CMOS inverter is shown in Figure 2-1. The CMOS process requires that both *n*-channel (NMOS) and *p*-channel (PMOS) transistors be built in the same silicon material. To accommodate both types of devices, special regions called *wells* must be created in which the semiconductor material is opposite to the type of the channel. A PMOS transistor has to be created in either an *n*-type substrate or an *n*-well, while an NMOS device resides in either a *p*-type substrate or a *p*-well. The cross section shown in Figure 2-1 features an

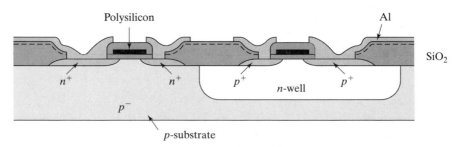

Figure 2-1   Cross section of an *n*-well CMOS process.

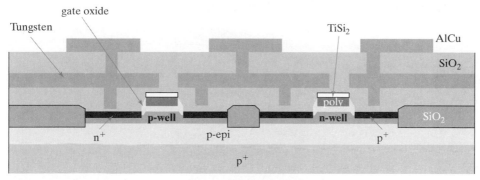

Figure 2-2   Cross section of modern dual-well CMOS process.

n-well CMOS process, where the NMOS transistors are implemented in the *p*-doped substrate, and the PMOS devices are located in the *n*-well. Modern processes are increasingly using a *dual-well* approach that uses both *n*- and *p*-wells, grown on top of an epitaxial layer, as shown in Figure 2-2.

The CMOS process requires a large number of steps, each of which consists of a sequence of basic operations. A number of these steps and/or operations are executed very repetitively in the course of the manufacturing process. Rather than immediately delving into a description of the overall process flow, we first discuss the starting material followed by a detailed perspective on some of the most frequently recurring operations.

2.2.1   The Silicon Wafer

The base material for the manufacturing process comes in the form of a single-crystalline, lightly doped *wafer*. These wafers have typical diameters between 4 and 12 inches (10 and 30 cm, respectively) and a thickness of, at most 1 mm. They are obtained by cutting a single-crystal ingot into thin slices (see Figure 2-3). A starting wafer of the p^--type might be doped around the levels of 2×10^{21} impurities/m^3. Often, the surface of the wafer is doped more heavily, and a single crystal *epitaxial layer* of the opposite type is grown over the surface before the wafers are handed to the processing company. One important metric is the defect density of the base material. High defect densities lead to a larger fraction of nonfunctional circuits, and consequently an increase in cost of the final product.

2.2.2   Photolithography

In each processing step, a certain area on the chip is masked out using the appropriate optical mask so that a desired processing step can be selectively applied to the remaining regions. The processing step can be any of a wide range of tasks, including oxidation, etching, metal and polysilicon deposition, and ion implantation. The technique to accomplish this selective masking, called *photolithography*, is applied throughout the manufacturing process. Figure 2-4 gives

Figure 2-3   Single-crystal ingot and sliced wafers (from [Fullman99]).

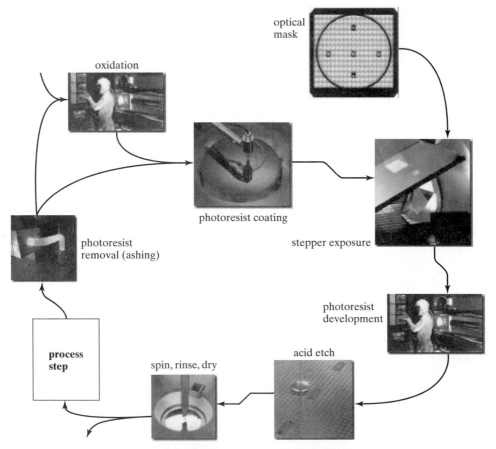

Figure 2-4   Typical operations in a single photolithographic cycle (from [Fullman99]).

a graphical overview of the different operations involved in a typical photolithographic process. The following steps can be identified:

1. *Oxidation layering*—this optional step deposits a thin layer of SiO_2 over the complete wafer by exposing it to a mixture of high-purity oxygen and hydrogen at approximately 1000°C. The oxide is used as an insulation layer and also forms transistor gates.

2. *Photoresist coating*—a light-sensitive polymer (similar to latex) is evenly applied to a thickness of approximately 1 μm by spinning the wafer. This material is originally soluble in an organic solvent, but has the property that the polymers cross-link when exposed to light, making the affected regions insoluble. A photoresist of this type is called *negative*. A positive photoresist has the opposite properties; originally insoluble, but soluble after exposure. By using both positive and negative resists, a single mask can sometimes be used for two steps, making complementary regions available for processing. Since the cost of a mask is increasing quite rapidly with the scaling of technology, reducing the number of masks surely is a high priority.

3. *Stepper exposure*—a glass mask (or reticle) containing the patterns that we want to transfer to the silicon is brought in close proximity to the wafer. The mask is opaque in the regions that we want to process, and transparent in the others (assuming a negative photoresist). The glass mask can be thought of as the negative of one layer of the microcircuit. The combination of mask and wafer is now exposed to ultraviolet light. Where the mask is transparent, the photoresist becomes insoluble.

4. *Photoresist development and bake*—the wafers are developed in either an acid or base solution to remove the nonexposed areas of photoresist. Once the exposed photoresist is removed, the wafer is "soft baked" at a low temperature to harden the remaining photoresist.

5. *Acid etching*—material is selectively removed from areas of the wafer that are not covered by photoresist. This is accomplished through the use of many different types of acid, base and caustic solutions as a function of the material that is to be removed. Much of the work with chemicals takes place at large wet benches where special solutions are prepared for specific tasks. Because of the dangerous nature of some of these solvents, safety and environmental impact is a primary concern.

6. *Spin, rinse, and dry*—a special tool (called SRD) cleans the wafer with deionized water and dries it with nitrogen. The microscopic scale of modern semiconductor devices means that even the smallest particle of dust or dirt can destroy the circuitry. To prevent this from happening, the processing steps are performed in ultraclean rooms where the number of dust particles per cubic foot of air ranges between 1 and 10. Automatic wafer handling and robotics are used whenever possible. This explains why the cost of a state-of-the-art fabrication facility easily reaches multiple billions of dollars. Even then, the wafers must be constantly cleaned to avoid contamination and to remove the leftover of the previous process steps.

7. *Various process steps*—the exposed area can now be subjected to a wide range of process steps, such as ion implantation, plasma etching, or metal deposition. These are the subjects of the subsequent section.

8. *Photoresist removal (or ashing)*—a high-temperature plasma is used to selectively remove the remaining photoresist without damaging device layers.

In Figure 2-5, we illustrate the use of the photolithographic process for one specific example, the patterning of a layer of SiO_2. The sequence of process steps shown in the figure patterns exactly one layer of the semiconductor material and may seem very complex. Yet, the reader has to bear in mind that the same sequence patterns the layer of **the complete surface of the wafer**. Hence, it is a very parallel process, transferring hundreds of millions of patterns to the semiconductor surface simultaneously. The concurrent and scalable nature of the optolithographical process is what makes the cheap manufacturing of complex semiconductor circuits possible, and lies at the core of the economic success of the semiconductor industry.

The continued scaling of the minimum feature sizes in integrated circuits puts an enormous burden on the developer of semiconductor manufacturing equipment. This is especially true for the optolithographical process. The dimensions of the features to be transcribed surpass the wavelengths of the optical light sources, so that achieving the necessary resolution and accuracy

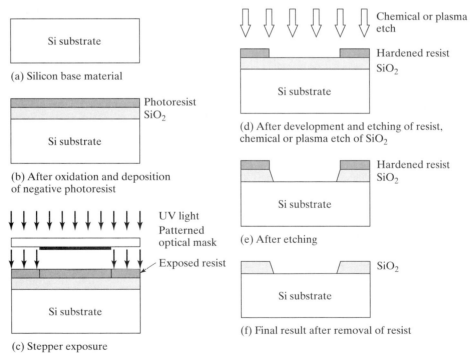

(a) Silicon base material

(b) After oxidation and deposition of negative photoresist

(c) Stepper exposure

(d) After development and etching of resist, chemical or plasma etch of SiO_2

(e) After etching

(f) Final result after removal of resist

Figure 2-5   Process steps for patterning of SiO_2.

becomes more and more difficult. So far, electrical engineering has extended the lifetime of this process at least until the 100 nm (or 0.1 μm) process generation. Techniques such as *optical mask correction* (OPC) prewarp the drawn patterns to account for the diffraction phenomena, encountered when printing close to the wavelength of the available optical source. This adds substantially to the cost of mask making. In the foreseeable future, other solutions that offer a finer resolution, such as extreme ultraviolet (EUV), X ray, or electron beam, may be needed. These techniques, while fully functional, are currently less attractive from an economic viewpoint.

2.2.3   Some Recurring Process Steps

Diffusion and Ion Implantation

Many steps of the integrated circuit manufacturing process require a change in the dopant concentration of some parts of the material. Examples include the creation of the source and drain regions, well and substrate contacts, the doping of the polysilicon, and the adjustments of the device threshold. Two approaches exist for introducing these dopants—diffusion and ion implantation. In both techniques, the area to be doped is exposed, while the rest of the wafer is coated with a layer of buffer material, typically SiO_2.

In *diffusion implantation*, the wafers are placed in a quartz tube embedded in a heated furnace. A gas containing the dopant is introduced in the tube. The high temperatures of the furnace, typically 900 to 1100 °C, cause the dopants to diffuse into the exposed surface both vertically and horizontally. The final dopant concentration is the greatest at the surface and decreases in a gaussian profile deeper in the material.

In *ion implantation*, dopants are introduced as ions into the material. The ion implantation system directs and sweeps a beam of purified ions over the semiconductor surface. The acceleration of the ions determines how deep they will penetrate the material, while the beam current and the exposure time determine the dosage. The ion implantation method allows for an independent control of depth and dosage. This is the reason that ion implantation has largely displaced diffusion in modern semiconductor manufacturing.

Ion implantation has some unfortunate side effects, however, the most important one being lattice damage. Nuclear collisions during the high energy implantation cause the displacement of substrate atoms, leading to material defects. This problem is largely resolved by applying a subsequent *annealing* step, in which the wafer is heated to around 1000°C for 15 to 30 minutes, and then allowed to cool slowly. The heating step thermally vibrates the atoms, which allows the bonds to reform.

Deposition

Any CMOS process requires the repetitive deposition of layers of a material over the complete wafer, to either act as buffers for a processing step, or as insulating or conducting layers. We have already discussed the oxidation process, which allows a layer of SiO_2 to be grown. Other materials require different techniques. For instance, silicon nitride (Si_3N_4) is used as a sacrificial buffer material during the formation of the field oxide and the introduction of the stopper implants. This silicon nitride is deposited everywhere using a process called *chemical vapor deposition* or CVD. This process is based on a gas-phase reaction, with energy supplied by heat at around 850°C.

Polysilicon, on the other hand, is deposited using a chemical deposition process, which flows silane gas over the heated wafer coated with SiO_2 at a temperature of approximately 650°C. The resulting reaction produces a noncrystalline or amorphous material called *polysilicon*. To increase the conductivity of the material, the deposition has to be followed by an implantation step.

The Aluminum interconnect layers typically are deployed using a process known as *sputtering*. The aluminum is evaporated in a vacuum, with the heat for the evaporation delivered by electron-beam or ion-beam bombarding. Other metallic interconnect materials such as Copper require different deposition techniques.

Etching

Once a material has been deposited, etching is used selectively to form patterns such as wires and contact holes. We already discussed the *wet etching* process, which makes use of acid or basic solutions. Hydrofluoric acid buffered with ammonium fluoride typically is used to etch SiO_2, for example.

In recent years, *dry* or *plasma etching* has advanced substantially. A wafer is placed into the etch tool's processing chamber and given a negative electrical charge. The chamber is heated to 100°C and brought to a vacuum level of 7.5 Pa, then filled with a positively charged plasma (usually a mix of nitrogen, chlorine, and boron trichloride). The opposing electrical charges cause the rapidly moving plasma molecules to align themselves in a vertical direction, forming a microscopic chemical and physical "sandblasting" action which removes the exposed material. Plasma etching has the advantage of offering a well-defined directionality to the etching action, creating patterns with sharp vertical contours.

Planarization

To reliably deposit a layer of material onto the semiconductor surface, it is essential that the surface be approximately flat. If special steps were not taken, this would definitely present problems in modern CMOS processes, where multiple patterned metal interconnect layers are superimposed onto each other. Therefore, a *chemical–mechanical planarization* (CMP) step is included before the deposition of an extra metal layer on top of the insulating SiO_2 layer. This process uses a slurry compound—a liquid carrier with a suspended abrasive component such as aluminum oxide or silica—to microscopically plane a device layer and to reduce the step heights.

2.2.4   Simplified CMOS Process Flow

The gross outline of a potential CMOS process flow is given in Figure 2-6. The process starts with the definition of the *active regions*—these are the regions where transistors will be constructed. All other areas of the die will be covered with a thick layer of silicon dioxide (SiO_2)

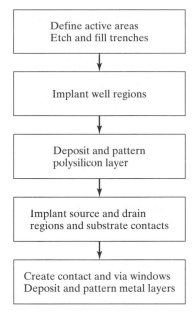

Figure 2-6   Simplified process sequence for the manufacturing of a n-dual-well CMOS circuit.

called the *field oxide*. This oxide acts as the insulator between neighboring devices, and it is either grown (as in the process of Figure 2-1) or deposited in etched trenches (Figure 2-2)—hence, the name *trench insulation*. Further insulation is provided by the addition of a reverse-biased *np*-diode, formed by adding an extra p^+ region called the *channel-stop implant* (or *field implant*) underneath the field oxide. Next, lightly doped *p*- and *n*-wells are formed through ion implantation. To construct an NMOS transistor in a *p*-well, heavily doped *n*-type *source* and *drain* regions are implanted (or diffused) into the lightly doped *p*-type substrate. A thin layer of SiO_2 called the *gate oxide* separates the region between the source and drain, and is itself covered by conductive polycrystalline silicon (or *polysilicon*, for short). The conductive material forms the *gate* of the transistor. PMOS transistors are constructed in an *n*-well in a similar fashion (just reverse *n*'s and *p*'s). Multiple insulated layers of metallic (most often Aluminum) wires are deposited on top of these devices to provide for the necessary interconnections between the transistors.

A more detailed breakdown of the flow into individual process steps and their impact on the semiconductor material is shown graphically in Figure 2-7. While most of the operations should be self-explanatory in light of the previous descriptions, some comments on individual operations are worthwhile. The process starts with a *p*-substrate surfaced with a lightly doped *p*-epitaxial layer (a). A thin layer of SiO_2 is then deposited, which will serve as the gate oxide for the transistors, followed by a deposition of a thicker sacrificial silicon nitride layer (b). A plasma etching step using the complementary of the active area mask creates the trenches used for insulating the devices (c). After providing the channel stop implant, the trenches are filled

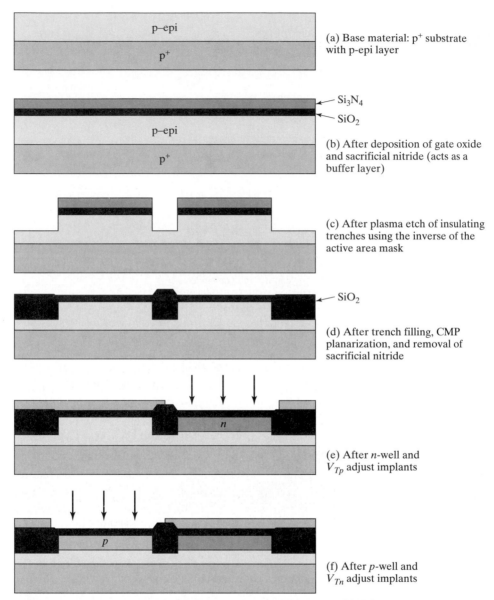

Figure 2-7 Process flow for the fabrication of an NMOS and a PMOS transistor in a dual-well CMOS process. Be aware that the drawings are stylized for understanding and that the aspects ratios are not proportioned to reality.

with SiO_2 followed by a number of steps to provide a flat surface (including inverse active pattern oxide etching, and chemical–mechanical planarization). At that point, the sacrificial nitride is removed (d). The n-well mask is used to expose only the n-well areas (the rest of the wafer is covered by a thick buffer material), after which an implant-annealing sequence is applied to

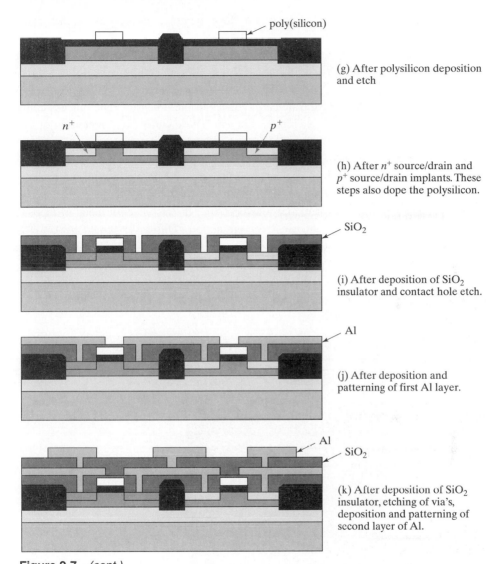

poly(silicon)

(g) After polysilicon deposition and etch

n^+ p^+

(h) After n^+ source/drain and p^+ source/drain implants. These steps also dope the polysilicon.

SiO_2

(i) After deposition of SiO_2 insulator and contact hole etch.

Al

(j) After deposition and patterning of first Al layer.

Al SiO_2

(k) After deposition of SiO_2 insulator, etching of via's, deposition and patterning of second layer of Al.

Figure 2-7 *(cont.)*

adjust the well-doping. This is followed by a second implant step to adjust the threshold voltages of the PMOS transistors. This implant only impacts the doping in the area just below the gate oxide (e). Similar operations (using other dopants) are performed to create the *p*-wells, and to adjust the thresholds of the NMOS transistors (f). A thin layer of polysilicon is chemically deposited and patterned with the aid of the polysilicon mask. Polysilicon is used both as gate electrode material for the transistors and as an interconnect medium (g). Consecutive ion implantations are used to dope the source and drain regions of the PMOS (p^+) and NMOS (n^+) transistors, respectively (h), after which the thin gate oxide not covered by the polysilicon is etched away.[1] The same implants also are used to dope the polysilicon on the surface, reducing its resistivity. Undoped polysilicon has a very high resistivity. Note that the polysilicon gate, which is patterned before the doping, actually defines the precise location of the channel region, and thus the location of the source and drain regions. This procedure, called the *self-aligned process*, allows for a very precise positioning of the two regions relative to the gate. Self-alignment is instrumental in reducing parasitic capacitances in the transistor. The process continues with the deposition of the metallic interconnect layers. These consist of a repetition of the following steps (i–k): deposition of the insulating material (most often SiO_2), etching of the contact or via holes, deposition of the metal (most often aluminum and copper, although tungsten often is used for the lower layers), and patterning of the metal. Intermediate planarization steps, using *chemical–mechanical polishing* or CMP, ensure that the surface remains reasonably flat, even in the presence of multiple interconnect layers. After the last level of metal is deposited, a final passivation or *overglass* is deposited for protection. This layer would be CVD SiO_2, although often an additional layer of nitride is deposited because it is more impervious to moisture. The final processing step etches openings to the pads used for bonding.

A cross section of the final artifact is shown in Figure 2-8. Observe how the transistors occupy only a small fraction of the total height of the structure. The interconnect layers take up the majority of the vertical dimension.

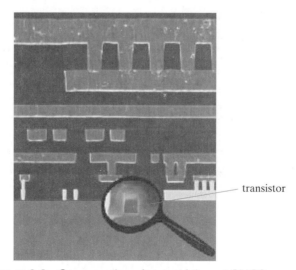

transistor

Figure 2-8   Cross section of state-of-the-art CMOS process.

---

[1] Most modern processes also include extra implants for the creation of the lightly doped drain regions (LDD), and the creation of gate spacers at this point. We have omitted these for the sake of simplicity.

2.3 Design Rules—Between the Designer and the Process Engineer

As processes become more complex, requiring the designer to understand the intricacies of the fabrication process and interpret the relations between the different masks is a sure road to trouble. The goal of defining a set of design rules is to allow for a ready translation of a circuit concept into an actual geometry in silicon. The design rules act as the interface or even the contract between the circuit designer and the process engineer.

Circuit designers generally want tighter, smaller designs, which lead to higher performance and higher circuit density. The process engineer, on the other hand, wants a reproducible and high-yield process. Consequently, design rules are a compromise that attempts to satisfy both sides.

The design rules provide a set of guidelines for constructing the various masks needed in the patterning process. They consist of minimum-width and minimum-spacing constraints and requirements between objects on the same or different layers.

The fundamental unity in the definition of a set of design rules is the *minimum line width*. It stands for the minimum mask dimension that can be safely transferred to the semiconductor material. In general, the minimum line width is set by the resolution of the patterning process, which is most commonly based on optical lithography. More advanced approaches use electron-beam EUV, or X-ray sources, all of which offer a finer resolution, but currently they are less attractive from an economical standpoint.

Even for the same minimum dimension, design rules tend to differ from company to company, and from process to process. This makes porting an existing design between different processes a time-consuming task. One approach to address this issue is to use advanced CAD techniques, which allow for migration between compatible processes. Another approach is to use *scalable design rules*. The latter approach, made popular by Mead and Conway [Mcad80], defines all rules as a function of a single parameter, most often called λ. The rules are chosen so that a design is easily ported over a cross section of industrial processes. Scaling of the minimum dimension is accomplished by simply changing the value of λ. This results in a *linear scaling* of all dimensions. For a given process, λ is set to a specific value, and all design dimensions are consequently translated into absolute numbers. Typically, the minimum line width of a process is set to 2λ. For instance, for a 0.25 µm process (i.e., a process with a minimum line width of 0.25 µm), λ equals 0.125 µm.

This approach, while attractive, suffers from two disadvantages:

1. Linear scaling is possible only over a limited range of dimensions (for instance, between 0.25 µm and 0.18 µm). When scaling over larger ranges, the relations between the different layers tend to vary in a nonlinear way that cannot be adequately covered by the linear scaling rules.
2. Scalable design rules are conservative: They represent a cross section over different technologies, and they must represent the worst case rules for the whole set. This results in overdimensioned and less dense designs.

For these and other reasons, scalable design rules normally are avoided by industry.[2] As circuit density is a prime goal in industrial designs, most semiconductor companies tend to use *micron rules*, which express the design rules in absolute dimensions and therefore can exploit the features of a given process to a maximum degree. Scaling and porting designs between technologies under these rules is more demanding and has to be performed either manually or using advanced CAD tools.

For this book, we have selected a "vanilla" 0.25 μm CMOS process as our preferred implementation medium. The rest of this section is devoted to a short introduction and overview of the design rules of this process, which fall in the micron-rules class. A complete design-rule set consists of the following entities: a set of layers, relations between objects on the same layer, and relations between objects on different layers. We discuss each of them in sequence.

Layer Representation

The layer concept translates the intractable set of masks currently used in CMOS into a simple set of conceptual layout levels that are easier to visualize by the circuit designer. From a designer's viewpoint, all CMOS designs are based on the following entities:

- *Substrates* and/or *wells*, which are *p*-type (for NMOS devices) and *n*-type (for PMOS)
- *Diffusion regions* (n^+ and p^+), which define the areas where transistors can be formed. These regions are often called the *active areas*. Diffusions of an inverse type are needed to implement contacts to the wells or to the substrate. These are called *select regions*.
- One or more *polysilicon* layers, which are used to form the gate electrodes of the transistors (but serve as interconnect layers as well).
- A number of *metal interconnect* layers.
- *Contact and via* layers, which provide interlayer connections.

A layout consists of a combination of polygons, each of which is attached to a certain layer. The functionality of the circuit is determined by the choice of the layers, as well as the interplay between objects on different layers. For example, an MOS transistor is formed by the cross section of the diffusion layer and the polysilicon layer. An interconnection between two metal layers is formed by a cross section between the two metal layers and an additional contact layer. To visualize these relations, each layer is assigned a standard color (or stipple pattern for a black-and-white representation). The different layers used in our CMOS process are represented in Colorplate 1 (color insert).

Intralayer Constraints

A first set of rules defines the minimum dimensions of objects on each layer, as well as the minimum spacings between objects on the same layer. All distances are expressed in μm. These constraints are presented in pictorial fashion in Colorplate 2.

---

[2]While not entirely accurate, lambda rules are still useful to estimate the impact of a technology scale on the area of a design.

Interlayer Constraints

Interlayer rules tend to be more complex. Because multiple layers are involved, it is harder to visualize their meaning or functionality. Understanding layout requires the capability of translating the two-dimensional picture of the layout drawing into the three-dimensional reality of the actual device. This takes some practice.

We present these rules in a set of separate groupings:

1. *Transistor Rules* (Colorplate 3). A transistor is formed by the overlap of the active and the polysilicon layers. From the intralayer design rules, it is already clear that the minimum length of a transistor equals 0.24 μm (the minimum width of polysilicon), while its width is at least 0.3 μm (the minimum width of diffusion). Extra rules include the spacing between the active area and the well boundary, the gate overlap of the active area, and the active overlap of the gate.

2. *Contact and Via Rules* (Colorplates 2 and 4). A contact (which forms an interconnection between metal and active or polysilicon) or a via (which connects two metal layers) is formed by overlapping the two interconnecting layers and providing a contact hole, filled with metal, between the two. In our process, the minimum size of the contact hole is 0.3 μm, while the polysilicon and diffusion layers have to extend at least 0.14 μm beyond the area of the contact hole. This sets the minimum area of a contact to 0.44 μm × 0.44 μm. This is larger than the dimensions of a minimum-size transistor! Excessive changes between interconnect layers in routing should therefore be avoided. The figure, furthermore, points out the minimum spacings between contact and via holes, as well as their relationship with the surrounding layers.

3. *Well and Substrate Contacts* (Colorplate 5). For robust digital circuit design, it is important for the well and substrate regions to be adequately connected to the supply voltages. Failing to do so results in a resistive path between the substrate contact of the transistors and the supply rails, and can lead to possibly devastating parasitic effects, such as latchup. It is therefore advisable to provide numerous substrate (well) contacts spread over the complete region. To establish an ohmic contact between a supply rail, implemented in metal1, and a p-type material, a p^+ diffusion region must be provided. This is enabled by the *select* layer, which reverses the type of diffusion. A number of rules regarding the use of the *select layer* are illustrated in Colorplate 5.

Consider an n-well process, which implements the PMOS transistors into an n-type well diffused in a p-type material. The nominal diffusion is p^+. To invert the polarity of the diffusion, an n-select layer is provided that helps to establish the n^+ diffusions for the well contacts in the n-region, as well as the n^+ source and drain regions for the NMOS transistors in the substrate.

Verifying the Layout

Ensuring that none of the design rules are violated is a fundamental requirement of the design process. Failing to do so will almost surely lead to a nonfunctional design. Doing so for a complex design that can contain millions of transistors is no simple task either, especially given the complexity of some design-rule sets. While design teams in the past used to spend numerous hours staring at room-size layout plots, most of this work is now done by computers. Computer-aided *Design-Rule Checking* (called *DRC*) is an integral part of the design cycle for virtually every chip produced today. A number of layout tools even perform *on-line DRC* and check the design in the background during the time of conception.

Example 2.1   Layout Example

An example of a complete layout containing an inverter is shown in Figure 2-9. To help the visualization process, a vertical cross section of the process along the design center is included, as well as a circuit schematic.

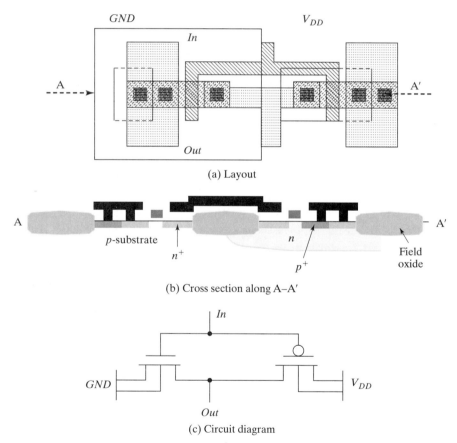

(a) Layout

(b) Cross section along A–A′

(c) Circuit diagram

Figure 2-9   A detailed layout example, including vertical process cross section and circuit diagram.

It is left as an exercise for the reader to determine the sizes of both the NMOS and the PMOS transistors.

2.4 Packaging Integrated Circuits

The IC package plays a fundamental role in the operation and performance of a component. Besides providing a means of bringing signal and supply wires in and out of the silicon die, it also removes the heat generated by the circuit and provides mechanical support. Finally, it also protects the die against environmental conditions such as humidity.

In addition, the packaging technology has a major impact on the performance and power dissipation of a microprocessor or signal processor. This influence is getting more pronounced as time progresses due to the reduction in internal signal delays and on-chip capacitance resulting from technology scaling. Currently, up to 50% of the delay of a high-performance computer is due to packaging delays, and this number is expected to rise. The search for higher performance packages with fewer inductive or capacitive parasitics has accelerated in recent years. The increasing complexity of what can be integrated on a single die also translates into a need for ever more input/output pins, as the number of connections going off-chip tends to be roughly proportional to the complexity of the circuitry on the chip. This relationship was first observed by E. Rent of IBM (published in [Landman71]), who translated it into an empirical formula that, appropriately, is called *Rent's rule*. This formula relates the number of input/output pins to the complexity of the circuit, as measured by the number of gates. It is written as

$$P = K \times G^\beta \qquad (2.1)$$

where K is the average number of I/Os per gate, G the number of gates, β the Rent exponent, and P the number of I/O pins to the chip. β varies between 0.1 and 0.7. Its value depends strongly upon the application area, architecture, and organization of the circuit, as demonstrated in Table 2-1. Clearly, microprocessors display a very different input/output behavior compared to memories.

The observed rate of pin-count increase for integrated circuits varies from 8% to 11% per year, and it has been projected that packages with more than 2000 pins will be required by 2010. For all these reasons, traditional dual-in-line, through-hole mounted packages have been replaced by other approaches, such as surface-mount, ball-grid array, and multichip module

Table 2-1 Rent's constant for various classes of systems ([Bakoglu90])

Application	β	K
Static memory	0.12	6
Microprocessor	0.45	0.82
Gate array	0.5	1.9
High-speed computer (chip)	0.63	1.4
High-speed computer (board)	0.25	82

techniques. It is useful for the circuit designer to be aware of the available options and their pros and cons.

Due to its multifunctionality, a good package must comply with a large variety of requirements:

- **Electrical requirements**—Pins should exhibit low capacitance (both interwire and to the substrate), resistance, and inductance. A large characteristic impedance should be tuned to optimize transmission line behavior. Observe that intrinsic integrated-circuit impedances are high.
- **Mechanical and thermal properties**—The heat removal rate should be as high as possible. Mechanical reliability requires a good matching between the thermal properties of the die and the chip carrier. Long-term reliability requires a strong connection from die to package, as well as from package to board.
- **Low Cost**—Cost is one of the more important properties to consider in any project. For example, while ceramics have a superior performance over plastic packages, they are also substantially more expensive. Increasing the heat removal capacity of a package also tends to raise the package cost. The least expensive plastic packaging can dissipate up to 1 W. Slightly more expensive, but still of somewhat low quality, plastic packages can dissipate up to 2 W. Higher dissipation requires more expensive ceramic packaging. Chips dissipating over 20 W require special heat sink attachments. Even more extreme techniques such as fans and blowers, liquid cooling hardware, or heat pipes are needed for higher dissipation levels.

  Packing density is a major factor in reducing board cost. The increasing pin count either requires an increase in the package size or a reduction in the pitch between the pins. Both have a profound effect on the packaging economics.

Packages can be classified in many different ways: by their main material, the number of interconnection levels, and the means used to remove heat. In this brief section, we provide only sketches of each of those issues.

2.4.1 Package Materials

The most common materials used for the package body are ceramic and polymers (plastics). The latter have the advantage of being substantially cheaper, but they suffer from inferior thermal properties. For example, the ceramic Al_2O_3 (alumina) conducts heat better than SiO_2 and the polyimide plastic by factors of 30 and 100, respectively. Furthermore, its thermal expansion coefficient is substantially closer to the typical interconnect metals. The disadvantage of alumina and other ceramics is their high dielectric constant, which results in large interconnect capacitances.

2.4.2 Interconnect Levels

The traditional packaging approach uses a two-level interconnection strategy. The die is first attached to an individual chip carrier or substrate. The package body contains an internal cavity where the chip is mounted. These cavities provide ample room for many connections to the chip leads (or pins). The leads compose the second interconnect level and connect the chip to the global interconnect medium, which normally is a PC board. Complex systems contain even more interconnect levels, since boards are connected together using backplanes or ribbon cables. The first two layers of the interconnect hierarchy are illustrated in the drawing of Figure 2-10. The sections that follow provide a brief overview of the interconnect techniques used at levels one and two of the interconnect hierarchy, and a brief discussion of some more advanced packaging approaches.

Interconnect Level 1—Die-to-Package Substrate

For a long time, *wire bonding* was the technique of choice to provide an electrical connection between die and package. In this approach, the backside of the die is attached to the substrate using glue with a good thermal conductance. Next, the chip pads are individually connected to the lead frame with aluminum or gold wires. The wire-bonding machine used for this purpose operates much like a sewing machine. An example of wire bonding is shown in Figure 2-11. Although the wire-bonding process is automated to a large degree, it has some major disadvantages:

1. Wires must be attached serially, one after the other. This leads to longer manufacturing times with increasing pin counts.
2. Larger pin counts make it substantially more challenging to find bonding patterns that avoid shorts between the wires.
3. The exact value of the parasitics is hard to predict because of the manufacturing approach and irregular outlay.

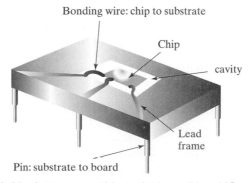

Figure 2-10 Interconnect hierarchy in traditional IC packaging.

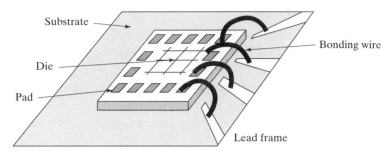

Figure 2-11   Wire bonding.

Bonding wires have inferior electrical properties, such as a high individual inductance (5 nH or more) and mutual inductance with neighboring signals. The inductance of a bonding wire is typically about 1 nH/mm, while the inductance per package pin ranges between 7 and 40 nH per pin, depending on the type of package as well as the positioning of the pin on the package boundary [Steidel83]. Typical values of the parasitic inductances and capacitances for a number of commonly used packages are summarized in Table 2-2.

New attachment techniques are being explored as a result of these deficiencies. In one approach, called *Tape Automated Bonding* (or TAB), the die is attached to a metal lead frame that is printed on a polymer film, typically polyimide (see Figure 2-12a). The connection between chip pads and polymer film wires is made using solder bumps (Figure 2-12b). The tape can then be connected to the package body using a number of techniques. One possible approach is to use pressure connectors.

The advantage of the TAB process is that it is highly automated. The sprockets in the film are used for automatic transport. All connections are made simultaneously. The printed approach helps to reduce the wiring pitch, which results in higher lead counts. Elimination of the long bonding wires improves the electrical performance. For instance, for a two-conductor layer, 48 mm

Table 2-2   Typical capacitance and inductance values of package and bonding styles (from [Steidel83], [Franzon93], and [Harper00]).

Package Type	Capacitance (pF)	Inductance (nH)
68-pin plastic DIP	4	35
68-pin ceramic DIP	7	20
300 pin Ball Grid Array	1–5	2–15
Wire bond	0.5–1	1–2
Solder bump	0.1–0.5	0.01–0.1

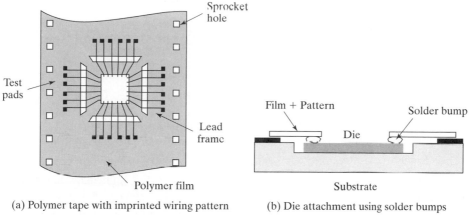

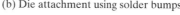

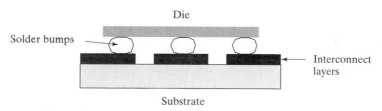

(a) Polymer tape with imprinted wiring pattern   (b) Die attachment using solder bumps

Figure 2-12   Tape-automated bonding (TAB).

Figure 2-13   Flip-chip bonding.

TAB Circuit, the following electrical parameters hold: $L \approx 0.3$–0.5 nH, $C \approx 0.2$–0.3 pF, and $R \approx 50$–200 Ω [Doane93, p. 420].

Another approach is to flip the die upside down and attach it directly to the substrate using solder bumps. This technique, called *flip-chip* mounting, has the advantage of a superior electrical performance (see Figure 2-13.) Instead of making all the I/O connections on the die boundary, pads can be placed at any position on the chip. This can help address the power- and clock-distribution problems, since the interconnect materials on the substrate (e.g., Cu or Au) typically are of better quality than the Al on the chip.

Interconnect Level 2—Package Substrate to Board

When connecting the package to the PC board, *through-hole mounting* has been the packaging style of choice. A PC board is manufactured by stacking layers of copper and insulating epoxy glass. In the through-hole mounting approach, holes are drilled through the board and plated with copper. The package pins are inserted and electrical connection is made with solder (see Figure 2-14a). The favored package in this class was the *dual-in-line* package or DIP, as in Figure 2-15-2. The packaging density of the DIP degrades rapidly when the number of pins exceeds 64. This problem can be alleviated by using the *pin-grid-array* (PGA) package that has

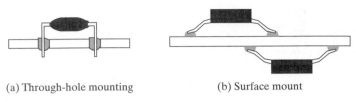

(a) Through-hole mounting   (b) Surface mount

Figure 2-14   Board-mounting approaches.

leads on the entire bottom surface instead of only on the periphery (Figure 2-15-3). PGAs can extend to large pin counts (over 400 pins are possible).

The through-hole mounting approach offers a mechanically reliable and sturdy connection. However, this comes at the expense of packaging density. For mechanical reasons, a minimum pitch of 2.54 mm between the through holes is required. Even under those circumstances, PGAs with large numbers of pins tend to substantially weaken the board. In addition, through holes limit the board packing density by blocking lines that might otherwise have been routed below them, which results in longer interconnections. PGAs with large pin counts therefore require extra routing layers to connect to the multitudes of pins. Finally, while the parasitic capacitance and inductance of the PGA are slightly lower than that of the DIP, their values are still substantial.

Many of the shortcomings of the through-hole mounting approach are solved by using the *surface-mount* technique. A chip is attached to the surface of the board with a solder connection without requiring any through holes (Figure 2-14b). Packing density is increased for the following reasons: (1) through holes are eliminated, which provides more wiring space; (2) the lead pitch is reduced; and (3) chips can be mounted on both sides of the board. In addition, the elimination of the through holes improves the mechanical strength of the board. On the negative side, the on-the-surface connection makes the chip-board connection weaker. Not only is it cumbersome to mount a component on a board, but also more expensive equipment is needed, since a simple soldering iron will no longer suffice. Finally, testing of the board is more complex, because the package pins are no longer accessible at the backside of the board. Signal probing becomes difficult or almost impossible.

A variety of surface-mount packages are currently in use with different pitch and pin-count parameters. Three of these packages are shown in Figure 2-15: the *small-outline package* with gull wings, the *plastic leaded package* (PLCC) with J-shaped leads, and the *leadless chip carrier*. An overview of the most important parameters for a number of packages is given in Table 2-3.

Even surface-mount packaging is unable to satisfy the quest for ever higher pin counts. This is worsened by the demand for power connections: today's high performance chips, operating at low supply voltages, require as many power and ground pins as signal I/Os! When more than 300 I/O connections are needed, solder balls replace pins as the preferred interconnect medium between package and board. An example of such a packaging approach, called ceramic

1 Bare die
2 DIP
3 PGA
4 Small-outline IC
5 Quad flat pack
6 PLCC
7 Leadless carrier

Figure 2-15 An overview of commonly used package types.

Table 2-3 Parameters of various types of chip carriers.

Package Type	Lead Spacing (Typical)	Lead Count (Maximum)
Dual in line	2.54 mm	64
Pin grid array	2.54 mm	> 300
Small-outline IC	1.27 mm	28
Leaded chip carrier (PLCC)	1.27 mm	124
Leadless chip carrier	0.75 mm	124

ball grid array (BGA), is shown in Figure 2-16. Solder bumps are used to connect both the die to the package substrate, and the package to the board. The area array interconnect of the BGA provides constant input/output density regardless of the number of total package I/O pins. A minimum pitch between solder balls of as low as 0.8 mm can be obtained, and packages with multiple thousands of I/O signals are feasible.

Multichip Modules—Die-to-Board

The deep hierarchy of interconnect levels in the package is becoming unacceptable in today's complex designs due to their higher levels of integration, large signal counts, and increased performance requirements. The trend, therefore, is toward reducing the number of levels. For the time being, attention is focused on the elimination of the first level in the packaging hierarchy. Removing one layer in the packaging hierarchy by mounting the die directly on the

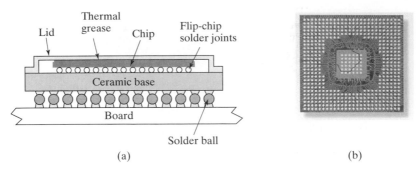

Figure 2-16   Ball grid array packaging; (a) cross section, (b) photo of package bottom.

wiring backplanes—board or substrate—offers a substantial benefit when performance or density is a major issue. This packaging approach is called the multichip module technique (or MCM), and results in a substantial increase in packing density, as well as improved performance overall.

A number of the previously mentioned die-mounting techniques can be adapted to mount dies directly on the substrate, including wire bonding, TAB, and flip-chip, although the latter two are preferable. The substrate itself can vary over a wide range of materials, depending upon the required mechanical, electrical, thermal, and economical requirements. Materials of choice are epoxy substrates (similar to PC boards), metal, ceramics, and silicon. Silicon has the advantage of presenting a perfect match in mechanical and thermal properties with respect to the die material.

The main advantages of the MCM approach are the increased packaging density and performance. An example of an MCM module implemented using a silicon substrate (commonly dubbed *silicon on silicon*) is shown in Figure 2-17. The module, which implements an avionics processor module and is fabricated by Rockwell International, contains 53 ICs and 40 discrete devices on a $2.2'' \times 2.2''$ substrate with aluminum polyimide interconnect. The interconnect wires are only an order of magnitude wider than what is typical for on-chip wires, since similar patterning approaches are used. The module itself has 180 I/O pins. Performance is improved by the elimination of the chip-carrier layer with its assorted parasitics, and through a reduction of the global wiring lengths on the die, a result of the increased packaging density. For instance, a solder bump has an assorted capacitance and inductance of only 0.1 pF and 0.01 nH, respectively. The MCM technology can also reduce power consumption significantly, since large output drivers—and associated dissipation—become superfluous due to the reduced load capacitance of the output pads. The dynamic power associated with the switching of the large load capacitances is simultaneously reduced.

While MCM technology offers some clear benefits, its main disadvantage is economic. This technology requires some advanced manufacturing steps that make the process expensive. The approach was until recently only justifiable when either dense housing or extreme performance is essential. In recent years, the economics have been shifting, and advanced multichip

Figure 2-17 Avionics processor module. *Courtesy of Rockwell Collins, Inc.*

packaging approaches have made inroad in several low-cost high-density applications as well. This trend is called the *system-in-a-package* (SIP) strategy.

2.4.3 Thermal Considerations in Packaging

As the power consumption of integrated circuits rises, it becomes increasingly important to efficiently remove the heat generated by the chips. A large number of failure mechanisms in ICs are accentuated by increased temperatures. Examples are leakage in reverse-biased diodes, electromigration, and hot-electron trapping. To prevent failure, the temperature of the die must be kept within certain ranges. The supported temperature range for commercial devices during operation equals 0° to 70°C. Military parts are more demanding and require a temperature range varying from –55° to 125°C.

The cooling effectiveness of a package depends on the thermal conduction (resistance) of the package material, which consists of the package substrate and body, the package composition, and the effectiveness of the heat transfer between package and cooling medium. Standard packaging approaches use still or circulating air as the cooling medium. The transfer efficiency can be improved by adding finned metal heat sinks to the package. More expensive packaging approaches, such as those used in mainframes or supercomputers, force air, liquids, or inert gases through tiny ducts in the package to achieve even greater cooling efficiencies.

Given the thermal resistance θ of the package, expressed in °C/W, we can derive the chip temperature by using the *heat flow equation*

$$\Delta T = T_{chip} - T_{env} = \theta Q, \tag{2.2}$$

with T_{chip} and T_{env} the chip and environment temperatures, respectively. Q represents the heat flow (in Watt). Observe how closely the heat flow equation resembles Ohm's Law. The heat flow and temperature differential are the equivalents of current and voltage difference, respectively. Thermal modeling of a chip, its package, and its environment is a complex task. We refer the reader to [Lau98 – Chapter 3] for a more detailed discussion on the topic.

Example 2.2   Thermal Conduction of Package

As an example, a 40-pin DIP has a thermal resistance of 38°C/W and 25°C/W for natural and forced convection of air. This means that a DIP can dissipate 2 watts (3 watts) of power with natural (forced) air convection, and still keep the temperature difference between the die and the environment below 75°C. For comparison, the thermal resistance of a ceramic PGA ranges from 15° to 30°C/W.

Since packaging approaches with decreased thermal resistance are prohibitively expensive, keeping the power dissipation of an integrated circuit within bounds is an economic necessity. The increasing integration levels and circuit performance make this task nontrivial. An interesting relationship in this context has been derived by Nagata [Nagata92]. It provides a bound on the integration complexity and performance as a function of the thermal parameters. We write

$$\frac{N_G}{t_p} \leq \frac{\Delta T}{\theta E} \tag{2.3}$$

where N_G is the number of gates on the chip, t_p the propagation delay, ΔT the maximum temperature difference between chip and environment, θ the thermal resistance between them, and E the switching energy of each gate.

Example 2.3   Thermal Bounds on Integration

For $\Delta T = 100°C$, $\theta = 2.5°C/W$ and $E = 0.1$ pJ, this results in $N_G/t_p \leq 4 \times 10^5$ (gates/nsec). In other words, the maximum number of gates on a chip, when all gates are operating simultaneously, must be less than 400,000 if the switching speed of each gate is 1 nsec. This is equivalent to a power dissipation of 40 W.

Fortunately, not all gates are operating simultaneously in real systems. The maximum number of gates can be substantially larger, based on the activity in the circuit. For example, it has been experimentally derived that the ratio between the average switching period and the propagation delay ranges from 20 to 200 in mini- and large-scale computers [Masaki92].

Nevertheless, Eq. (2.3) demonstrates that heat dissipation and thermal concerns present an important limitation on circuit integration. Design approaches for low power that reduce either E or the activity factor are rapidly gaining importance.

2.5 Perspective—Trends in Process Technology

Modern CMOS processes pretty much track the flow described in the previous sections, although a number of the steps might be reversed, a single well approach might be followed, a grown field oxide instead of the trench approach might be used, or extra steps such as LDD (*Lightly Doped Drain*) might be introduced. Also, it is quite common to cover the polysilicon interconnections as well as the drain and source regions with a *silicide* such as $TiSi_2$ to improve the conductivity (see Figure 2-2). This extra operation is inserted between steps i and j of our process. Some important modifications or improvements to the technology are currently under way or are on the horizon, and deserve some attention. Beyond these, we expect no dramatic changes from the described CMOS technology in the next decade.

2.5.1 Short-Term Developments

Copper and Low-k Dielectrics

A recurring theme throughout this book will be the increasing impact of interconnect on the overall design performance. Process engineers are continuously evaluating alternative options for the traditional Aluminum-conductor–SiO_2-insulator combination that has been the norm for the last several decades. In 1998, engineers at IBM introduced an approach that finally made the use of copper as an interconnect material in a CMOS process viable and economical [Geppert98]. Copper has a resistivity that is substantially lower than aluminum. It has the disadvantage of easy diffusion into silicon, which degrades the characteristics of the devices. Coating the copper with a buffer material such as titanium-nitride, preventing the diffusion, addresses this problem, but requires a special deposition process. The Dual Damascene process, introduced by IBM, (Figure 2-18) uses a metallization approach that fills trenches etched into the insulator, followed by a chemical–mechanical polishing step. This is in contrast with the traditional approach that first deposits a full metal layer, and removes the redundant material through etching.

In addition to the lower resistivity interconnections, insulator materials with a lower dielectric constant than SiO_2, and hence, lower capacitance have also found their way into the production process, starting with the 0.18-μm CMOS process generation.

Silicon on Insulator

Although it has been around a long time, there seems to be a good chance that Silicon-on-Insulator (SOI) CMOS might replace the traditional CMOS process, described in the previous sections (also known as the *bulk CMOS process*). The main difference lies in the start material: the SOI transistors are constructed in a very thin layer of silicon, deposited on top of a thick layer of insulating SiO_2 (see Figure 2-19). The primary advantages of the SOI process are reduced

Dual damascene IC process

* Oxide deposition

* Stud lithography and
  reactive ion etch

* Wire lithography and
  reactive ion etch

* Stud and wire
  metal deposition

* Metal chemical–
  mechanical polish

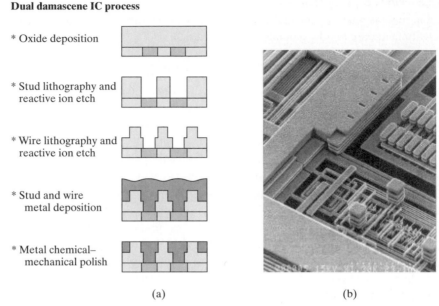

(a)   (b)

Figure 2-18   The damascene process (from [Geppert98]): process steps (a),
and microphotograph of interconnect after removal of insulator (b).

parasitics and better transistor on–off characteristics. It has, for example, been demonstrated by
researchers at IBM, that the porting of a design from a bulk CMOS to an SOI process—leaving
all other design and process parameters such as channel length and oxide thickness identical—
yields a performance improvement of 22% [Allen99]. Preparing a high quality SOI substrate at
an economical cost was long the main hindrance against a large-scale introduction of the process. This picture had changed by the end of the 1990s, and SOI is steadily moving into the
mainstream.

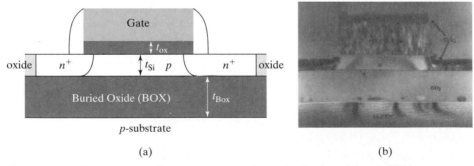

(a)   (b)

Figure 2-19   Silicon-on-insulator process—schematic diagram (a) and SEM
cross section (b). [Eaglesham99].

2.5.2 In the Longer Term

Extending the life of CMOS technology beyond the next decade, and going deeply below the 100 nm channel length region, however, will require redeveloping the process technology and the device structure. Already we are witnessing the emergence of a wide range of new devices (such as organic transistors, molecular switches, and quantum devices). While we cannot project what approaches will dominate in the next era, one interesting development is worth mentioning.

Truly Three-Dimensional Integrated Circuits

Getting signals in and out of the computation elements in a timely fashion is one of the main challenges presented by the continued increase in integration density. One way to address this problem is to introduce extra active layers, and to sandwich them between the metal interconnect layers, as shown in Figure 2-20. This enables us to position high density memory on top of the logic processors implemented in the bulk CMOS, reducing the distance between computation and storage, and thus also the delay [Souri00]. In addition, devices with different voltage, performance, or substrate material requirements can be placed in different layers. For instance, the top active layer can be reserved for the realization of optical transceivers, which may help to address the input/output requirements, or MEMS (Micro Electro-Mechanical Systems) devices providing sensing functions or radio frequency (RF) interfaces.

While this approach may seem to be promising, a number of major challenges and hindrances have to be resolved to make it truly viable. How to remove the dissipated heat is one of the more compelling questions, ensuring yield is another. Researchers are demonstrating major progress on these issues, and 3D integration might well be on the horizon. Before the true solution arrives, we might have to rely on some intermediate approaches. One alternative, called 2.5D *integration*, is to bond two fully processed wafers, on which circuits are fabricated on the

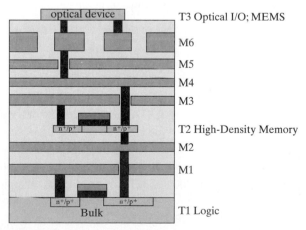

Figure 2-20 Example of true 3D integration. Extra active layers (T*), implementing high-density memory and I/O, are sandwiched between the metal interconnect layers (M*).

surface such that the chips completely overlap. Vias are etched to electrically connect both chips after metallization. The advantages of this technology lie in the similar electrical properties of devices on all active levels and the independence of processing temperature since all chips can be fabricated separately and later bonded. The major limitation of this technique is its lack of precision (best case alignment: +/– 2 μm), which restricts the interchip communication to global metal lines.

One picture that strongly emerges from these futuristic devices is that the line between chip, substrate, package, and board is blurring. Designers of these *systems on a die* or *systems in a package* will have to consider all these aspects simultaneously.

2.6 Summary

This chapter has presented a bird's-eye view of the manufacturing and packaging process of CMOS integrated circuits:

- The manufacturing process of integrated circuits requires many steps, each of which consists of a sequence of basic operations. A number of these steps and/or operations, such as photolithograpical exposure and development, material deposition, and etching, are executed very repetitively in the course of the manufacturing process.
- The *optical masks* forms the central interface between the intrinsics of the manufacturing process and the design that the user wants to see transferred to the silicon fabric.
- The *design-rules set* defines the constraints in terms of minimum width and separation that the IC design has to adhere to if the resulting circuit is to be fully functional. These design rules act as the contract between the circuit designer and the process engineer.
- The *package* forms the interface between the circuit implemented on the silicon die and the outside world, and as such has a major impact on the performance, reliability, longevity, and cost of the integrated circuit.

2.7 To Probe Further

Many books on semiconductor manufacturing have been published in the last few decades. An excellent overview of the state of the-art in CMOS manufacturing is *Silicon VLSI Technology* by J. Plummer, M. Deal, and P. Griffin [Plummer00]. A visual overview of the different steps in the manufacturing process can be found on the Web at [Fullman99]. Other sources for information are the IEEE Transactions on Electron Devices, and the Technical Digest of the IEDM conference. A number of great compendia are available for up-to-date and in-depth information about electronic packaging. [Doane93], [Harper00], and [Lau98] are good examples of such.

References

[Allen99] D. Allen, et al., "A 0.2 μm 1.8 V SOI 550 MHz PowerPC Microprocessor with Copper Interconnects," *Proceedings IEEE ISSCC Conference*, vol. XLII, pp. 438–439, February 1999.
[Bakoglu90] H. Bakoglu, *Circuits, Interconnections and Packaging for VLSI*, Addison-Wesley, 1990.

[Doane93] D. Doane, ed., *Multichip Module Technologies and Alternatives*, Van Nostrand-Reinhold, 1993.

[Eaglesham 99] D. Englesham, "0.18 µm CMOS and Beyond," *Proceedings 1999 Design Automation Conference*, pp. 703–708, June 1999.

[Franzon93] P. Franzon, "Electrical Design of Digital Multichip Modules," in [Doane93], pp 525–568, 1993.

[Fullman99] Fullman Kinetics, "The Semiconductor Manufacturing Process," *http://www.fullman-kinetics.com/semiconductors/semiconductors.html*, 1999.

[Geppert98] L. Geppert, "Technology—1998 Analysis and Forecast," *IEEE Spectrum,* vol. 35, no. 1, p. 23, January 1998.

[Harper00] C. Harper, Ed., *Electronic Packaging and Interconnection Handbook*, McGraw-Hill, 2000.

[Landman71] B. Landman and R. Russo, "On a Pin versus Block Relationship for Partitions of Logic Graphs," *IEEE Trans. on Computers*, vol. C-20, pp. 1469–1479, December 1971.

[Lau98] J. Lau et al., *Electronic Packaging—Design, Materials, Process, and Reliability*, McGraw-Hill, 1998.

[Masaki92] A. Masaki, "Deep-Submicron CMOS Warms Up to High-Speed Logic," *Circuits and Devices Magazine*, November 1992.

[Mead80] C. Mead and L. Conway, *Introduction to VLSI Systems*, Addison-Wesley, 1980.

[Nagata92] M. Nagata, "Limitations, Innovations, and Challenges of Circuits and Devices into a Half Micrometer and Beyond," *IEEE Journal of Solid State Circuits*, vol. 27, no. 4, pp. 465–472, April 1992.

[Plummer00] J. Plummer, M. Deal, and P. Griffin, *Silicon VLSI Technology*, Prentice Hall, 2000.

[Steidel83] C. Steidel, *Assembly Techniques and Packaging*, in [Sze83], pp. 551–598, 1983.

[Souri00] S. J. Souri, K. Banerjee, A. Mehrotra and K. C. Saraswat, "*Multiple Si Layer ICs: Motivation, Performance Analysis, and Design Implications,*" Proceedings 37th Design Automation Conference, pp. 213–220, June 2000.

IC LAYOUT

Creating a manufacturable layout

Verifying the layout

The increasing complexity of the integrated circuit has made the role of design-automation tools indispensable, and raises the abstractions the designer is working with to ever higher levels. Yet, when performance or design density is of primary importance, the designer has no other choice than to return to handcrafting the circuit topology and physical design. The labor-intensive nature of this approach, called *custom design*, translates into a high cost and a long time to market. Therefore, it can only be justified economically under the following conditions:

- The custom block can be reused many times, as a library cell, for instance.
- The cost can be amortized over a large volume. Microprocessors and semiconductor memories are examples of applications in this class.
- Cost is not among the prime design criteria.[1] Examples include space applications and scientific instrumentation.

With continuous progress in the design-automation arena, the share of custom design reduces from year to year. Even in high-performance microprocessors, large portions are

---

[1] This is becoming increasingly rare.

designed automatically using semicustom design approaches. Only the most performance-critical modules—such as the integer and floating-point execution units—are handcrafted.

Although the amount of design automation in the custom design process is minimal, some design tools have proven to be indispensable. Together with circuit simulators, these programs form the core of every design-automation environment, and they are the first tools an aspiring circuit designer will encounter.

Layout Editor

The layout editor is the premier working tool of the designer and exists primarily for the generation of a physical representation of a design, given a circuit topology. Virtually every design-automation vendor offers an entry in this field. The most well known is the MAGIC tool developed at the University of California at Berkeley [Ousterhout84], which has been widely distributed. Even though MAGIC did not withstand the evolution of software technology and user interface, some of its offspring did. Throughout this book, we will be using a layout tool called **max**, a MAGIC descendant developed by a company called MicroMagic [mmi00]. A typical **max** display is shown in Figure A-1 and illustrates the basic function of the layout editor—placing polygons on

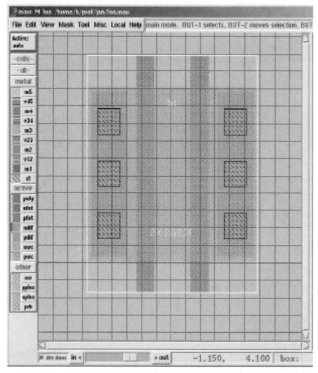

Figure A-1 View of a **max** display window. It plots the layout of two stacked NMOS transistor. The menu on the left side allows for the selection of the layer a particular poligon will be placed on.

different mask layers so that a functional physical design is obtained (scathingly called *polygon pushing*).

Symbolic Layout

Since physical design occupies a major fraction of the design time for a new cell or component, techniques to expedite this process have been in continual demand. The *symbolic-layout* approach has gained popularity over the years. In this design methodology, the designer only draws a shorthand notation for the layout structure. This notation indicates only the *relative* positioning of the various design components (transistors, contacts, wires). The *absolute* coordinates of these elements are determined automatically by the editor using a *compactor* [Hsueh79, Weste93]. The compactor translates the design rules into a set of constraints on the component positions, and solves a constrained optimization problem that attempts to minimize the area or another cost function.

An example of a symbolic notation for a circuit topology called a *sticks diagram* is shown in Figure A-2. The different layout entities are dimensionless, since only positioning is important. The advantage of this approach is that the designer does not have to worry about design rules, because the compactor ensures that the final layout is physically correct. Thus, she can avoid cumbersome polygon manipulations. Another plus of the symbolic approach is that cells can adjust themselves automatically to the environment. For example, automatic pitch matching of cells is an attractive feature in module generators. Consider the case of Figure A-3 (from [Croes88]), in which the original cells have different heights, and the terminal positions do not match. Connecting the cells would require extra wiring. The symbolic approach allows the cells to adjust themselves and connect without any overhead.

The disadvantage of the symbolic approach is that the outcome of the compaction phase often is unpredictable. The resulting layout can be less dense than what is obtained with the manual approach. This has prevented it from becoming a mainstream layout tool. Nonetheless, symbolic layout techniques have improved considerably over the years, and they have become very useful as a first-order drafting tool for new cells. More important, they form the solid underpining of the automatic cell-generation techniques, described later, in Chapter 8.

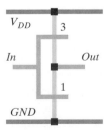

Figure A-2 Sticks representation of CMOS inverter. The numbers represent the (*Width/Length*)-ratios of the transistors.

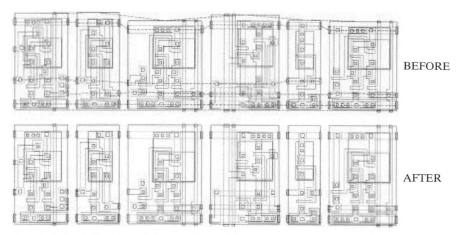

BEFORE

AFTER

Figure A-3   Automatic pitch matching of data path cells based on symbolic layout.

Design-Rule Checking

Design rules were introduced in Chapter 2 as a set of layout restrictions that ensure the manufactured design will operate as desired with no short or open circuits. A prime requirement of the
physical layout of a design is that it adhere to these rules. This can be verified with the aid of a
design-rule checker (DRC), which uses as inputs the physical layout of a design and a description of the design rules presented in the form of a *technology file*. Since a complex circuit can
contain millions of polygons that must be checked against each other, efficiency is the most
important property of a good DRC tool. The verification of a large chip can take hours or days of
computation time. One way of expediting the process is to preserve the design hierarchy at the
physical level. For example, if a cell is used multiple times in a design, it should be checked only
once. Besides speeding up the process, the use of hierarchy can make error messages more infor-
mative by retaining knowledge of the circuit structure.

DRC tools come in two formats: (1) The *on-line DRC* runs concurrent with the layout editor and flags design violations during the cell layout. For instance, **max** has a built-in design-rule
checking facility. An example of on-line DRC is shown in Figure A-4. (2) *Batch DRC* is used as
a postdesign verifier; it is run on a complete chip prior to shipping the mask descriptions to the
manufacturer.

Circuit Extraction

Another important tool in the custom-design methodology is the circuit extractor, which derives
a circuit schematic from a physical layout. By scanning the various layers and their interactions,
the extractor reconstructs the transistor network, including the sizes of the devices and the interconnections. The schematic produced can be used to verify that the artwork implements the
intended function. Furthermore, the resulting circuit diagram contains precise information on
the parasitics, such as the diffusion and wiring capacitances and resistances. This allows for a

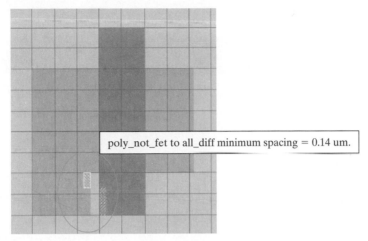

poly_not_fet to all_diff minimum spacing = 0.14 um.

Figure A-4 On-line design rule checking. The white dots indicate a design rule violation. The violated rule can be obtained with a simple mouse click.

more accurate simulation and analysis. The complexity of the extraction depends greatly upon the desired information. Most extractors extract the transistor network and the capacitances of the interconnect with respect to *GND* or other network nodes. Extraction of the wiring resistances already comes at a greater cost, yet it has become a necessity for virtually all high-performance circuits. Clever algorithms have helped to reduce the complexity of the resulting circuit diagrams. For very high-speed circuits, extraction of the inductance would be desirable as well. Unfortunately, this requires a three-dimensional analysis and is only feasible for small-sized circuits at present.

A.1 To Probe Further

More detailed information regarding the MAGIC and **max** layout editors can be found on the web site of this book. In-depth textbooks on layout generation and verification have been published, and can be of great help to the novice designer. To mention a few of them, [Clein00], [Uyemura95], and [Wolf94] offer some comprehensive and well-illustrated treatment and discussion.

References

[Clein00] D. Clein, *CMOS IC Layout—Concepts, Methodologies, and Tools*, Newnes, 2000.

[Croes88] K. Croes, H. De Man, and P. Six, "CAMELEON: A Process-Tolerant Symbolic Layout System," *Journal of Solid State Circuits*, vol. 23 no. 3, pp. 705–713, June 1988.

[Hsueh79] M. Hsueh and D. Pederson, "Computer-Aided Layout of LSI Building Blocks," *Proceedings ISCAS Conf.*, pp. 474–477, Tokyo, 1979.

[mmi00] MicroMagic, Inc, *http://www.micromagic.com*.

[Ousterhout84] J. Ousterhout, G. Hamachi, R. Mayo, W. Scott, and G. Taylor, "Magic: A VLSI Layout System,"
Proc. 21st Design Automation Conference, pp. 152–159, 1984.

[Uyemura95] J. Uyemura, *Physical Design of CMOS Integrated Circuits Using L-EDIT*, PWS Publishing Company,
1995.

[Weste93] N. Weste and K. Eshraghian, *Principles of CMOS VLSI Design—A Systems Perspective*, Addison-Wesley,
1993.

[Wolf94] W. Wolf, *Modern VLSI Design—A Systems Approach*, Prentice Hall, 1994.

3

The Devices

Qualitative understanding of MOS devices

Simple component models for manual analysis

Detailed component models for SPICE

Impact of process variations

3.1   Introduction

3.2   The Diode

    3.2.1   A First Glance at the Diode—The Depletion Region

    3.2.2   Static Behavior

    3.2.3   Dynamic, or Transient, Behavior

    3.2.4   The Actual Diode—Secondary Effects

    3.2.5   The SPICE Diode Model

3.3   The MOS(FET) Transistor

    3.3.1   A First Glance at the Device

    3.3.2   The MOS Transistor under Static Conditions

             Dynamic Behavior

    3.3.3   The Actual MOS Transistor—Some Secondary Effects

    3.3.4   SPICE Models for the MOS Transistor

3.4   A Word on Process Variations

3.5   Perspective—Technology Scaling

3.6   Summary

3.7   To Probe Further

3.1 Introduction

It is a well-known premise in engineering that the conception of a complex construction without a prior understanding of the underlying building blocks is a sure road to failure. This surely holds for digital circuit design as well. The basic building blocks in today's digital circuits are the silicon semiconductor devices—more specifically, the MOS transistors, and to a lesser degree, the parasitic diodes and the interconnect wires. The role of the semiconductor devices has been appreciated for a long time in the world of digital integrated circuits. On the other hand, interconnect wires have only recently started to play a dominant role as a result of the advanced scaling of the semiconductor technology.

Giving the reader the necessary *knowledge and understanding* of these components is the prime motivation for the next two chapters. It is not our intention to present an in-depth treatment of the physics of semiconductor devices and interconnect wires. For that purpose, we refer the reader to the many excellent textbooks on semiconductor devices, some of which are referenced at the end of the chapter. Our goal is to describe the functional operation of the devices, to highlight the properties and parameters that are particularly important in the design of digital gates, and to introduce notational conventions.

Another important function of this chapter is the introduction of *models*. Taking all the physical aspects of each component into account when designing complex digital circuits leads to an unnecessary complexity that quickly becomes intractable. Such an approach is similar to considering the molecular structure of concrete when constructing a bridge. To deal with this issue, an abstraction or model of the component behavior typically is employed. A range of models can be conceived for each component presenting a trade-off between accuracy and complexity. A simple first-order model is useful for manual analysis. It has limited accuracy, but helps us to understand the operation of the circuit and its dominant parameters. When more accurate results are needed, complex, second-order or higher models are employed in conjunction with computer-aided simulation. In this chapter, we present both first-order models for manual analysis, and higher order models for simulation for each component of interest.

Designers tend to take the component parameters offered in the models for granted. They should be aware, however, that these are only nominal values, and that the actual parameter values not only vary with operating temperature, but also over manufacturing runs, or even over a single wafer. To highlight this issue, we include a short discussion on *process variations* and their impact.

3.2 The Diode

Although diodes rarely occur directly in the schematic diagrams of present-day digital gates, they are still omnipresent. Each MOS transistor implicitly contains a number of reverse-biased diodes that directly influence the behavior of the device. In particular, the voltage-dependent capacitances contributed by these parasitic elements play an important role in the switching behavior of the MOS digital gate. Diodes are also used to protect the input devices of an IC against static charges. For these reasons, a brief review of the basic properties and device equations of the diode is appropriate. Rather than being comprehensive, we choose to focus on those aspects that prove to be influential in the design of digital MOS circuits, more precisely the operation in reverse-biased mode.[1]

3.2.1   A First Glance at the Diode—The Depletion Region

The *pn*-junction diode is the simplest of the semiconductor devices. Figure 3-1a shows a cross section of a typical *pn*-junction. It consists of two homogeneous regions of *p*- and *n*-type material, separated by a region of transition from one type of doping to another, which is assumed to be thin. Such a device is called a *step* or *abrupt junction*. The *p*-type material is doped with *acceptor* impurities (such as boron), which results in the presence of holes as the dominant or majority carriers. Similarly, the doping of silicon with *donor* impurities (such as phosphorus or arsenic) creates an *n*-type material, where electrons are the majority carriers. Aluminum contacts provide access to the *p*- and *n*-terminals of the device. The circuit symbol of the diode, as used in schematic diagrams, is introduced in Figure 3-1c.

To understand the behavior of the *pn*-junction diode, we often resort to a one-dimensional simplification of the device (Figure 3-1b). Bringing the *p*- and *n*-type materials together

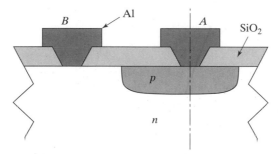

(a) Cross section of *pn*-junction in an IC process

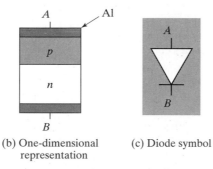

(b) One-dimensional   (c) Diode symbol
representation

Figure 3-1   Abrupt *pn*-junction diode and its schematic symbol.

---

[1]We refer the interested reader to the web site of the textbook for a comprehensive description of the diode operation.

causes a large concentration gradient at the boundary. The electron concentration changes from a high value in the *n*-type material to a very small value in the *p*-type material. The reverse is true for the hole concentration. This gradient causes electrons to *diffuse* from *n* to *p* and holes to diffuse from *p* to *n*. When the holes leave the *p*-type material, they leave behind immobile acceptor ions, which are negatively charged. Consequently, the *p*-type material is negatively charged in the vicinity of the *pn*-boundary. Similarly, a positive charge builds up on the *n*-side of the boundary as the diffusing electrons leave behind the positively charged donor ions. The region at the junction where the majority carriers have been removed, leaving the fixed acceptor and donor ions, is called the *depletion* or *space-charge region*. The charges create an electric field across the boundary, which is directed from the *n*- to the *p*-region. This field counteracts the diffusion of holes and electrons, because as it causes electrons to *drift* from *p* to *n* and holes to drift from *n* to *p*. Under equilibrium, the depletion charge sets up an electric field such that the drift currents are equal and opposite to the diffusion currents, resulting in a zero net flow.

The preceding analysis is summarized in Figure 3-2, which plots the current directions, the charge density, the electrical field, and the electrostatic field of the abrupt *pn*-junction under zero-bias conditions. In the device shown, the *p*-material is more heavily doped than the *n*, or $N_A > N_D$, with N_A and N_D the acceptor and donor concentrations, respectively. Hence, the charge concentration in the depletion region is higher on the *p*-side of the junction. Figure 3-2 also shows that under zero bias, there exists a voltage ϕ_0 across the junction called the *built-in potential*. This potential has the value

$$\phi_0 = \phi_T \ln\left[\frac{N_A N_D}{n_i^2}\right] \tag{3.1}$$

where ϕ_T is the *thermal voltage*

$$\phi_T = \frac{kT}{q} = 26\,\text{mV at 300 K} \tag{3.2}$$

The quantity n_i is the intrinsic carrier concentration in a pure sample of the semiconductor and equals approximately 1.5×10^{10} cm^{-3} at 300 K for silicon.

Example 3.1   Built-in Voltage of *pn*-junction

An abrupt junction has doping densities of $N_A = 10^{15}$ atoms/cm^3, and $N_D = 10^{16}$ atoms/cm^3. Calculate the built-in potential at 300 K.

From Eq. (3.1),

$$\phi_0 = 26\ln\left[\frac{10^{15} \times 10^{16}}{2.25 \times 10^{20}}\right] \text{mV} = 638 \text{ mV}$$

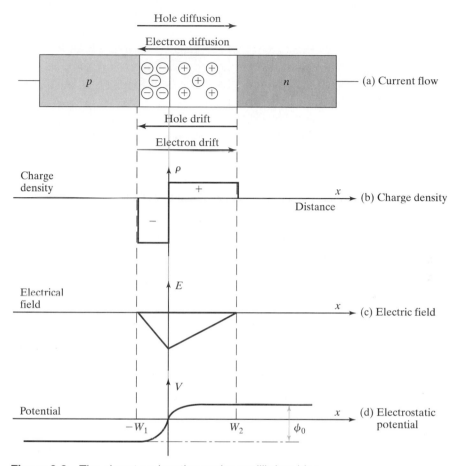

Figure 3-2 The abrupt *pn*-junction under equilibrium bias.

3.2.2 Static Behavior

The Ideal Diode Equation

Assume now that a forward voltage V_D is applied to the junction—the potential of the *p*-region is raised with respect to the *n*-zone. The applied potential lowers the potential barrier. Consequently, the flow of mobile carriers across the junction increases as the diffusion current dominates the drift component. These carriers traverse the depletion region and are injected into the neutral *n*- and *p*-regions, where they become minority carriers, as illustrated in Figure 3-3. Under the assumption that no voltage gradient exists over the neutral regions, which is approximately the case for most modern devices, these minority carriers will diffuse through the region as a result of the concentration gradient until they get recombined with a majority carrier. The net result is a current flowing through the diode from the *p*-region to the *n*-region, and the diode is said to be in the *forward-bias* mode.

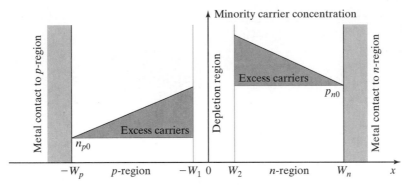

Figure 3-3 Minority carrier concentrations in the neutral region near an abrupt *pn*-junction under forward-bias conditions.

On the other hand, when a reverse voltage V_D is applied to the junction—this is, when the potential of the *p*-region is lowered with respect to the *n*-region—the potential barrier is raised. This results in a reduction in the diffusion current, and the drift current becomes dominant. A current flows from the *n*-region to the *p*-region. Since the number of minority carriers in the neutral regions (electrons in the *p*-zone, holes in the *n*-region) is very small, this drift current component can almost be ignored. (See Figure 3-4.) It is fair to state that in the *reverse-bias* mode, the diode operates as a nonconducting or blocking device. The device thus acts as a one-way conductor.

The most important property of the diode current is its *exponential dependence* upon the applied bias voltage. This is illustrated in Figure 3-5, which plots the diode current I_D as a function of the bias voltage V_D. The exponential behavior for positive-bias voltages is even more apparent in Figure 3-5b, where the current is plotted on a logarithmic scale. The current increases by a factor of 10 for every extra 60 mV (= 2.3 ϕ_T) of forward bias. At small voltage levels ($V_D < 0.15$ V), a deviation from the exponential dependence can be observed, which is due to the recombination of holes and electrons in the depletion region.

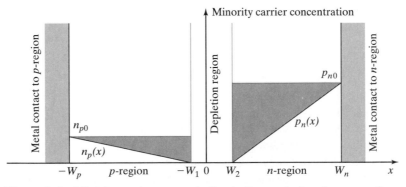

Figure 3-4 Minority carrier concentration in the neutral regions near the *pn*-junction under reverse-bias conditions.

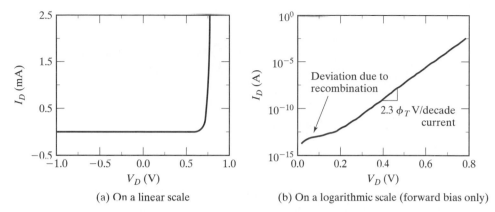

(a) On a linear scale   (b) On a logarithmic scale (forward bias only)

Figure 3-5   Diode current as a function of the bias voltage V_D.

The behavior of the diode for both forward and reverse-bias conditions is best described by the well-known *ideal diode equation*, which relates the current through the diode I_D to the diode bias voltage V_D:

$$I_D = I_S(e^{V_D/\phi_T} - 1) \qquad (3.3)$$

The exponential bahavior of the diode is illustrated in Figure 3-5, ϕ_T is the thermal voltage of Eq. (3.2) and is equal to 26 mV at room temperature. I_S represents a constant value called the *saturation current* of the diode. It is proportional to the area of the diode, and a function of the doping levels and widths of the neutral regions. Most often, I_S is determined empirically. It is worth mentioning that in actual devices, the reverse current is substantially larger than the saturation current I_S. This is due to the thermal generation of hole and electron pairs in the depletion region. The electric field that is present sweeps these carriers out of the region, causing an additional current component. For typical silicon junctions, the saturation current is nominally in the range of 10^{-17} A/μm^2, while the actual reverse currents are approximately three orders of magnitude higher. As a result, it is necessary to determine realistic values for the reverse diode leakage currents through actual device measurements.

Models for Manual Analysis

The derived current-voltage equations can be summarized in a set of simple models that are useful in the manual analysis of diode circuits. One model, shown in Figure 3-6a, is based on the ideal diode equation Eq. (3.3). While this model yields accurate results, its disadvantage is that it is strongly nonlinear, which prohibits a fast, first-order analysis of the dc-operation conditions of a network. An often-used, simplified model is derived by inspecting the diode current plot of Figure 3-5. For a "fully conducting" diode, the voltage drop over the diode V_D lies in a narrow range, approximately between 0.6 and 0.8 V. To a first degree, it is reasonable to assume that a conducting diode has a fixed voltage drop V_{Don} over it. Although the value of V_{Don} depends upon I_S, a value of 0.7 V typically is assumed, which gives rise to the model of Figure 3-6b, where a

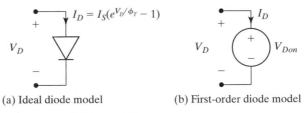

(a) Ideal diode model   (b) First-order diode model

Figure 3-6   Diode models.

conducting diode is replaced by a fixed voltage source, and a nonconducting diode is replaced by an open circuit.

Example 3.2   **Analysis of Diode Network**

Consider the simple network of Figure 3-7 and assume that $V_S = 3$ V, $R_S = 10$ kΩ, and $I_S = 0.5 \times 10^{-16}$ A. The diode current and voltage are related by the following network equation:

$$V_S - R_S I_D = V_D$$

Inserting the ideal diode equation and (painfully) solving the nonlinear equation using numerical or iterative techniques yields the following solution: $I_D = 0.224$ mA, and $V_D = 0.757$ V. The simplified model with $V_{Don} = 0.7$ V produces similar results ($V_D = 0.7$ V, $I_D = 0.23$ A) with far less effort. Hence, it makes considerable sense to use this model when determining a first-order solution of a diode network.

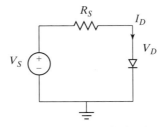

Figure 3-7   A simple diode circuit.

3.2.3   Dynamic, or Transient, Behavior

So far, we have mostly been concerned with the static, or steady-state, characteristics of the diode. Just as important in the design of digital circuits is the response of the device to changes in its bias conditions. The transient, or dynamic, response determines the maximum speed at which the device can be operated. Because the operation mode of the diode is a function of the amount of charge present in both the neutral and the space-charge regions, its dynamic behavior is strongly determined by how fast the charge can be moved around.

At this point, it is tempting to go into an in-depth analysis of the switching behavior of the diode in the forward-biasing mode, but we believe that would unnecessarily complicate the discussion. In fact, all diodes in an operational MOS digital integrated circuit are reverse biased, and they are supposed to remain in that condition under all circumstances. Only under exceptional conditions may forward biasing occur. A signal overshooting (or undershooting) the supply rail is an example of such. Due to its detrimental impact on the overall circuit operation, this should be avoided under all circumstances. Hence, we will devote our attention solely to what governs the dynamic response of the diode under reverse-biasing conditions—the depletion-region charge.

Depletion-Region Capacitance

In the ideal model, the depletion region is void of mobile carriers, and its charge is determined by the immobile donor and acceptor ions. The corresponding charge distribution under zero-bias conditions was plotted in Figure 3-2. This picture can easily be extended to incorporate the effects of biasing. At an intuitive level, the following observations can be easily verified: Under forward-bias conditions, the potential barrier is reduced, which means less space charge is needed to produce the potential difference. This corresponds to a reduced depletion-region width. On the other hand, under reverse conditions, the potential barrier is increased corresponding to an increased space charge and a wider depletion region. These observations are confirmed by the well-known depletion-region expressions shown following.[2] One observation is crucial—due to the global charge neutrality requirement of the diode, the total acceptor and donor charges must be numerically equal.

1. Depletion-region charge (V_D is positive for forward bias).

$$Q_j = A_D \sqrt{\left(2\varepsilon_{si}q\frac{N_A N_D}{N_A + N_D}\right)(\phi_0 - V_D)} \tag{3.4}$$

2. Depletion-region width.

$$W_j = W_2 - W_1 = \sqrt{\left(\frac{2\varepsilon_{si}}{q}\frac{N_A + N_D}{N_A N_D}\right)(\phi_0 - V_D)} \tag{3.5}$$

3. Maximum electric field.

$$E_j = \sqrt{\left(\frac{2q}{\varepsilon_{si}}\frac{N_A N_D}{N_A + N_D}\right)(\phi_0 - V_D)} \tag{3.6}$$

In the preceding equations, ε_{si} stands for the electrical permittivity of silicon and equals 11.7 times the permittivity of a vacuum, or 1.053×10^{-10} F/m. The ratio of the n- versus p-side of the depletion-region width is determined by the doping-level ratios: $W_2/(-W_1) = N_A/N_D$.

---

[2] A derivation of these expressions, which are valid for abrupt junctions, is either simple or can be found in any textbook on devices such as [Howe97].

From an abstract point of view, it is possible to visualize the depletion region as a capacitance, albeit one with very special characteristics. Because the space-charge region contains few mobile carriers, it acts as an insulator with a dielectric constant ε_{si} of the semiconductor material. The n- and p-regions act as the capacitor plates. A small change in the voltage applied to the junction dV_D causes a change in the space charge dQ_j. Hence, a depletion-layer capacitance can be defined as

$$C_j = \frac{dQ_j}{dV_D} = A_D \sqrt{\left(\frac{\varepsilon_{si}q}{2}\frac{N_A N_D}{N_A+N_D}\right)(\phi_0 - V_D)^{-1}}$$

$$= \frac{C_{j0}}{\sqrt{1-V_D/\phi_0}}$$

(3.7)

where C_{j0} is the capacitance under zero-bias conditions and is only a function of the physical parameters of the device. This gives us the relationship

$$C_{j0} = A_D \sqrt{\left(\frac{\varepsilon_{si}q}{2}\frac{N_A N_D}{N_A+N_D}\right)\phi_0^{-1}}$$

(3.8)

Notice that the same capacitance value is obtained when using the standard parallel-plate capacitor equation $C_j = \varepsilon_{si} A_D/W_j$ (with W_j given in Eq. (3.5)). Typically, the A_D factor is omitted, and C_j and C_{j0} are expressed as a capacitance/unit area.

The resulting junction capacitance is plotted in the function of the bias voltage in Figure 3-8 for a typical silicon diode found in MOS circuits. A strong *nonlinear dependence* can be observed. Note also that the capacitance decreases with an increasing reverse bias: a reverse bias of 5 V reduces the capacitance by a factor of more than two.

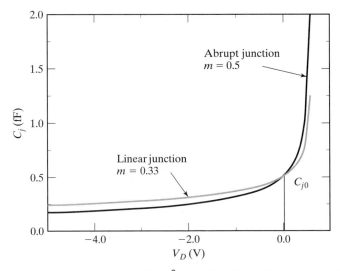

Figure 3-8   Junction capacitance (in fF/μm²) as a function of the applied bias voltage.

Example 3.3   Junction Capacitance

Consider the following silicon junction diode: $C_{j0} = 2 \times 10^{-3}$ F/m^2, $A_D = 0.5$ μm^2, and $\phi_0 = 0.64$ V. A reverse bias of -2.5 V results in a junction capacitance of 0.9×10^{-3} F/m^2 (0.9 fF/μm^2), or, for the total diode, a capacitance of 0.45 fF.

Equation (3.7) is only valid under the condition that the *pn*-junction is an *abrupt junction*, for which the transition from *n*- to *p*-material is instantaneous. This often is not the case in actual integrated-circuit *pn*-junctions, where the transition from *n*- to *p*-material can be gradual. In those cases, a linear distribution of the impurities across the junction is a better approximation than the step function of the abrupt junction. An analysis of the *linearly graded junction* shows that the junction capacitance equation of Eq. (3.7) still holds, but with a variation in order of the denominator. A more generic expression for the junction capacitance is

$$C_j = \frac{C_{j0}}{(1 - V_D/\phi_0)^m} \tag{3.9}$$

where *m* is called the *grading coefficient* and equals 1/2 for the abrupt junction and 1/3 for the linear or graded junction. Both cases are illustrated in Figure 3-8.

Large-Signal Depletion-Region Capacitance

Figure 3-8 raises awareness to the fact that the junction capacitance is a voltage-dependent parameter whose value varies widely between bias points. In digital circuits, operating voltages tend to move rapidly over a wide range. Under those circumstances, it is more attractive to replace the voltage-dependent, nonlinear capacitance C_j by an equivalent, linear capacitance C_{eq}. C_{eq} is defined such that, for a given voltage swing from voltages V_{high} to V_{low}, the same amount of charge is transferred as would be predicted by the nonlinear model

$$C_{eq} = \frac{\Delta Q_j}{\Delta V_D} = \frac{Q_j(V_{high}) - Q_j(V_{low})}{V_{high} - V_{low}} = K_{eq} C_{j0} \tag{3.10}$$

Combining Eq. (3.4) (extended to accommodate the grading coefficient *m*) and Eq. (3.10) yields the value of K_{eq}:

$$K_{eq} = \frac{-\phi_0^m}{(V_{high} - V_{low})(1 - m)} [(\phi_0 - V_{high})^{1-m} - (\phi_0 - V_{low})^{1-m}] \tag{3.11}$$

Example 3.4   Average Junction Capacitance

The diode of Example 3.3 is switched between 0 and -2.5 V. Compute the average junction capacitance ($m = 0.5$).

For the defined voltage range and for $\phi_0 = 0.64$ V, K_{eq} evaluates to 0.622. The average capacitance thus equals 1.24 fF/μm^2.

3.2.4   The Actual Diode—Secondary Effects

In practice, the diode current is less than what is predicted by the ideal diode equation. Not all applied bias voltage appears directly across the junction—there is always some voltage drop over the neutral regions. Fortunately, the resistivity of the neutral zones is generally small (between 1 and 100 Ω, depending upon the doping levels), and the voltage drop only becomes significant for large currents (>1 mA). This effect can be modeled by adding a resistor in series with the *n*- and *p*-region diode contacts.

In the preceding discussion, it was further assumed that under sufficient reverse bias, the reverse current reaches a constant value, which essentially is zero. When the reverse bias exceeds a certain level, called the *breakdown voltage*, the reverse current shows a dramatic increase, as shown in Figure 3-9. In the diodes found in typical CMOS processes, this increase is caused by the *avalanche breakdown*. The increasing reverse bias heightens the magnitude of the electrical field across the junction. Consequently, carriers crossing the depletion region are accelerated to high velocity. At a critical field E_{crit}, the carriers reach a high enough energy level that electron-hole pairs are created on collision with immobile silicon atoms. These carriers create, in turn, more carriers before leaving the depletion region. The value of E_{crit} is approximately 2×10^5 V/cm for impurity concentrations of the order of 10^{16} cm^{-3}. While avalanche breakdown in itself is not destructive and its effects disappear after the reverse bias is removed, maintaining a diode for a long time in avalanche conditions is not recommended because the high current levels and the associated heat dissipation might cause permanent damage to the structure. Observe that avalanche breakdown is not the only breakdown mechanism encountered in diodes. For highly doped diodes, another mechanism called *Zener breakdown* can occur. (Further discussion of this phenomenon is beyond the scope of this text.)

Finally, it is worth mentioning that the diode current is affected by the operating *temperature* in a dual way:

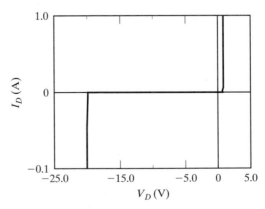

Figure 3-9   *I-V* characteristic of junction diode, showing breakdown under reverse-bias conditions (breakdown voltage = 20 V).

1. The thermal voltage ϕ_T, which appears in the exponent of the current equation, is linearly dependent on the temperature. An increase in ϕ_T causes the current to drop.
2. The saturation current I_S is also temperature dependent, because the thermal equilibrium carrier concentrations increase with increasing temperature. Theoretically, the saturation current approximately doubles every 5°C. Experimentally, the reverse current has been measured to double every 8°C.

This dual dependence has a significant impact on the operation of a digital circuit. First, current levels (and thus power consumption) can increase substantially. For instance, for a forward bias of 0.7 V at 300°K, the current increases approximately 6% per °C, and doubles every 12°C. For a fixed value of the current, the diode voltage V_D decreases with 2 mV for every °C. Second, integrated circuits rely heavily on reverse-biased diodes as isolators. Increasing the temperature causes the leakage current to increase and decreases the isolation quality.

3.2.5 The SPICE Diode Model

In the preceding sections, we have presented a model for manual analysis of a diode circuit. For more complex circuits, or when a more accurate modeling of the diode that takes into account second-order effects is required, manual circuit evaluation becomes intractable, and computer-aided simulation is necessary. While different circuit simulators have been developed over the last few decades, the SPICE program, developed at the University of California at Berkeley, is definitely the most successful [Nagel75]. Simulating an integrated circuit containing active devices requires a mathematical model for those devices. For the rest of the text it is called the *SPICE model*. The accuracy of the simulation depends directly upon the quality of this model. For instance, one cannot expect to see the result of a second-order effect in the simulation if this effect is not present in the device model. Creating accurate and computation-efficient SPICE models has been a long and tedious process, and it is by no means finished. Over time, every major semiconductor company has developed its own proprietary models, and they claim to have either better accuracy or greater computational efficiency and robustness. Fortunately, some consolidation in the modeling field has been observed as of late.

The standard SPICE model for a diode is simple, as shown in Figure 3-10. The steady-state characteristic of the diode is modeled by the nonlinear current source I_D, which is a modified version of the ideal diode equation:

$$I_D = I_S(e^{V_D/n\phi_T} - 1) \tag{3.12}$$

The extra parameter n is called the *emission coefficient*. It equals 1 for most common diodes, but it can be somewhat greater than 1 for others. The resistor R_s models the series resistance contributed by the neutral regions on both sides of the junction. For higher current levels, this resistance causes the internal diode voltage V_D to differ from the externally applied voltage, yielding a current that is lower than what would be expected from the ideal diode equation.

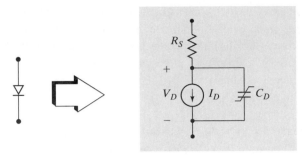

Figure 3-10   SPICE diode model.

The dynamic behavior of the diode is modeled by the nonlinear capacitance C_D, which combines the two different charge-storage effects in the diode: the space (or depletion-region) charge, and the excess minority carrier charge. We discussed only the former in this chapter, because the latter is an issue only under forward-biasing conditions. The nonlinear capacitance model is defined as follows:

$$C_D = \frac{C_{j0}}{(1 - V_D/\phi_0)^m} + \frac{\tau_T I_S}{\phi_T} e^{V_D/n\phi_T} \tag{3.13}$$

A listing of the parameters used in the diode model is given in Table 3-1. Besides the parameter name, symbol, and SPICE name, the table contains also the default value used by SPICE in case the parameter is left undefined. This table is by no means complete—other parameters are available to capture second-order effects such as breakdown, high-level injection,

Table 3-1   First-order SPICE diode model parameters.

Parameter Name	Symbol	SPICE Name	Units	Default Value
Saturation current	I_S	IS	A	1.0 E–14
Emission coefficient	n	N	–	1
Series resistance	R_S	RS	Ω	0
Transit time	τ_T	TT	s	0
Zero-bias junction capacitance	C_{j0}	CJ0	F	0
Grading coefficient	m	M	–	0.5
Junction potential	ϕ_0	VJ	V	1

and noise. For brevity, we limited the listing to the parameters of direct interest to this text. For a complete description of the device models and the usage of SPICE, we refer the reader to the numerous textbooks on the subject. (See, for example, [Banhzaf92], [Thorpe92], and [Vladimirescu93].)

3.3 The MOS(FET) Transistor

The metal–oxide–semiconductor field-effect transistor (MOSFET, or MOS, for short) is certainly the workhorse of contemporary digital design. Its major asset, from a digital perspective, is that it performs very well as a switch, and it introduces very few parasitic effects. Other important advantages are its integration density, combined with a relatively "simple" manufacturing process, which make it possible to produce large and complex circuits in an economical way.

Following the approach we took for the diode, we restrict ourselves in this section to a general overview of the transistor and its parameters. After a generic overview of the device, we present an analytical description of the transistor from both a static (steady-state) and a dynamic (transient) viewpoint. We conclude the discussion with an enumeration of some second-order effects and the introduction of the SPICE MOS transistor models.

3.3.1 A First Glance at the Device

The MOSFET is a four-terminal device. The voltage applied to the *gate* terminal determines if and how much current flows between the *source* and the *drain* ports. The *body* represents the fourth terminal of the transistor. Its function is secondary because it only serves to modulate the device characteristics and parameters.

At the most superficial level, the transistor can be thought of as a switch. When voltage is applied to the gate that is larger than a given value (the *threshold voltage V_T*), a conducting channel is formed between drain and source. In the presence of a voltage difference between the latter two, electrical current flows between them. The conductivity of the channel is modulated by the gate voltage—the larger the voltage difference between gate and source, the smaller the resistance of the conducting channel and the larger the current. When the gate voltage is lower than the threshold, no such channel exists, and the switch is considered open.

Two types of MOSFET devices can be identified. The NMOS transistor consists of n^+ drain and source regions embedded in a *p*-type substrate. The current is carried by electrons moving through an *n*-type channel between source and drain. This is in contrast to the *pn*-junction diode, where current is carried by both holes and electrons. MOS devices can also be made by using an *n*-type substrate and p^+ drain and source regions. In such a transistor, current is carried by holes moving through a *p*-type channel. The device is called a *p*-channel MOS, or PMOS transistor. In a complementary MOS technology (CMOS), both devices are present. The cross section of a contemporary dual-well CMOS process was presented in Chapter 2, and is reproduced here for convenience. (See Figure 3-11.)

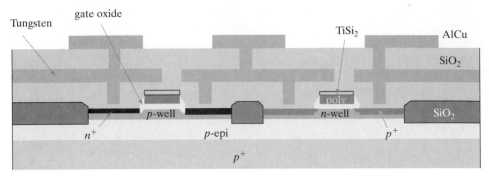

Figure 3-11   Cross section of contemporary dual-well CMOS process.

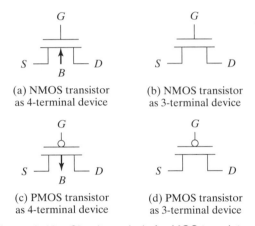

Figure 3-12   Circuit symbols for MOS transistors.

Circuit symbols for the various MOS transistors are shown in Figure 3-12. As mentioned earlier, the transistor is a four-port device, with gate, source, drain, and body terminals (Figure 3-12a and c). Since the body is generally connected to a dc supply that is identical for all devices of the same type (GND for NMOS, V_{dd} for PMOS), most often it is not shown on the schematics (Figure 3-12b and d). **If the fourth terminal is not shown, it is assumed that the body is connected to the appropriate supply.**

3.3.2   The MOS Transistor under Static Conditions

In the derivation of the static model of the MOS transistor, we concentrate on the NMOS device. All of these arguments are valid for PMOS devices as well, and they are discussed at the end of the section.

The Threshold Voltage

Consider first the case in which $V_{GS} = 0$, and drain, source, and bulk are connected to ground. The drain and source are connected by back-to-back pn-junctions (substrate–source and

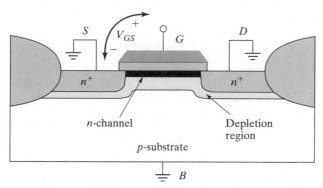

Figure 3-13   NMOS transistor for positive V_{GS}, showing depletion region and induced channel.

substrate–drain). Under the mentioned conditions, both junctions have a 0 V bias and can be considered off, which results in an extremely high resistance between drain and source.

Assume now that a positive voltage is applied to the gate (with respect to the source), as shown in Figure 3-13. The gate and substrate form the plates of a capacitor with the gate oxide as the dielectric. The positive gate voltage causes positive and negative charge to accumulate on the gate electrode and the substrate side, respectively. The latter manifests itself initially by repelling mobile holes. Hence, a depletion region is formed below the gate. This depletion region is similar to the one occurring in a *pn*-junction diode. Consequently, similar expressions hold for the width and the space charge per unit area. Compare the following expressions with Eq. (3.4) and Eq. (3.5):

$$W_d = \sqrt{\frac{2\varepsilon_{si}\phi}{qN_A}} \tag{3.14}$$

and

$$Q_d = \sqrt{2qN_A\varepsilon_{si}\phi} \tag{3.15}$$

In these equations, N_A is the substrate doping and ϕ is the voltage across the depletion layer (this is, the potential at the oxide–silicon boundary).

As the gate voltage increases, the potential at the silicon surface reaches a critical value at some point, and the semiconductor surface inverts to *n*-type material. This point marks the onset of a phenomenon known as *strong inversion,* which occurs at a voltage equal to twice the *Fermi Potential* (Eq. (3.16)) ($\phi_F \approx -0.3$ V for typical *p*-type silicon substrates):

$$\phi_F = \phi_T \ln\left(\frac{N_A}{n_i}\right) \tag{3.16}$$

Further increases in the gate voltage produce no further changes in the depletion layer width, but result in additional electrons in the thin inversion layer directly under the oxide. These are drawn into the inversion layer from the heavily doped *n*+ source region. Hence, a continuous *n*-type channel is formed between the source and drain regions, the conductivity of which is modulated by the gate-source voltage.

In the presence of an inversion layer, the charge stored in the depletion region is fixed and equals

$$Q_{B0} = \sqrt{2qN_A\varepsilon_{si}|2\phi_F|} \tag{3.17}$$

This picture changes somewhat when a substrate bias voltage V_{SB} is applied between source and body (V_{SB} is normally positive for *n*-channel devices). This causes the surface potential required for strong inversion to increase and to become $|-2\phi_F + V_{SB}|$. The charge stored in the depletion region can now be expressed as

$$Q_B = \sqrt{2qN_A\varepsilon_{si}(|(-2)\phi_F + V_{SB}|)}. \tag{3.18}$$

The value of V_{GS} where strong inversion occurs is called the *threshold voltage V_T*. V_T is a function of several components, most of which are material constants, such as the difference in work function between gate and substrate material, the oxide thickness, the Fermi voltage, the charge of impurities trapped at the surface between channel and gate oxide, and the dosage of ions implanted for threshold adjustment. From the preceding arguments, it should be clear that the source-bulk voltage V_{SB} has an impact on the threshold as well. Rather than relying on a complex (and hardly accurate) analytical expression for the threshold, we rely on an empirical parameter V_{T0}, which is the threshold voltage for $V_{SB} = 0$, and is mostly a function of the manufacturing process. The threshold voltage under different body-biasing conditions can then be determined by

$$V_T = V_{T0} + \gamma(\sqrt{|(-2)\phi_F + V_{SB}|} - \sqrt{|2\phi_F|}) \tag{3.19}$$

The parameter γ (gamma) is called the *body-effect coefficient*. It expresses the impact of changes in V_{SB}. Observe that the threshold voltage has a **positive** value for a typical **NMOS** device, while it has a **negative** value for a normal **PMOS** transistor.

In Figure 3.14, the effect of the well bias on the threshold voltage of an NMOS transistor is plotted for typical values of $|-2\phi_F| = 0.6$ V and $\gamma = 0.4$ $V^{0.5}$. A negative bias on the well or substrate causes the threshold to increase from 0.45 V to 0.85 V. Note also that V_{SB} always has to be larger than -0.6 V in an NMOS. If not, the source-body diode becomes forward biased, which deteriorates the transistor operation.

Example 3.5   Threshold Voltage of a PMOS Transistor

A PMOS transistor has a threshold voltage of -0.4 V, while the body-effect coefficient equals -0.4. Compute the threshold voltage for $V_{SB} = -2.5$ V. $2\phi_F = 0.6$ V.

Using Eq. (3.19), we obtain $V_T(-2.5$ V$) = -0.4 - 0.4 \times ((2.5 + 0.6)^{0.5} - 0.6^{0.5})$ V $= -0.79$ V, which is twice the threshold under zero-bias conditions!

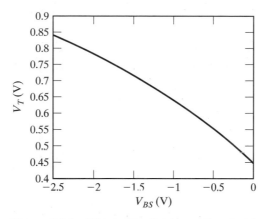

Figure 3-14   Effect of body-bias on threshold.

Resistive Operation

Assume now that $V_{GS} > V_T$, and that a small voltage V_{DS} is applied between drain and source. The voltage difference causes a current I_D to flow from drain to source. (See Figure 3-15.) Using a simple analysis, a first-order expression of the current as a function of V_{GS} and V_{DS} can be obtained.

At a point x along the channel, the voltage is $V(x)$, and the gate-to-channel voltage at that point equals $V_{GS} - V(x)$. Under the assumption that this voltage exceeds the threshold voltage all along the channel, the induced channel charge per unit area at point x can be computed by using the following equation:

$$Q_i(x) = -C_{ox}[V_{GS} - V(x) - V_T]\qquad(3.20)$$

Here, C_{ox} stands for the capacitance per unit area presented by the gate oxide, and it equals

$$C_{ox} = \frac{\varepsilon_{ox}}{t_{ox}}\qquad(3.21)$$

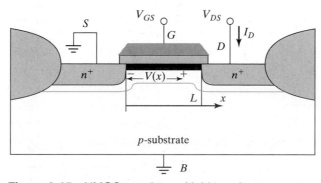

Figure 3-15   NMOS transistor with bias voltages.

with $\varepsilon_{ox} = 3.97 \times \varepsilon_o = 3.5 \times 10^{-11}$ F/m the oxide permittivity and t_{ox} the thickness of the oxide. The latter is 10 nm (= 100 Å) or smaller for contemporary processes. For an oxide thickness of 5 nm, this translates into an oxide capacitance of 7 fF/μm^2.

The current is given as the product of the drift velocity of the carriers υ_n and the available charge. Due to charge conservation, it is a constant over the length of the channel. W is the width of the channel in a direction perpendicular to the current flow. The equation is

$$I_D = -\upsilon_n(x)Q_i(x)W \tag{3.22}$$

The electron velocity is related to the electric field through a parameter called the *mobility* μ_n (expressed in m^2/V $\cdot$ s). The mobility is a complex function of crystal structure and local electrical field. In general, an empirical value is used:

$$\upsilon_n = -\mu_n \xi(x) = \mu_n \frac{dV}{dx} \tag{3.23}$$

Combining Eq. (3.20) through Eq. (3.23) yields

$$I_D dx = \mu_n C_{ox} W(V_{GS} - V - V_T)dV . \tag{3.24}$$

Integrating the equation over the length of the channel L yields the voltage-current relation of the transistor:

$$I_D = k'_n \frac{W}{L}\left[(V_{GS} - V_T)V_{DS} - \frac{V_{DS}^2}{2}\right] = k_n\left[(V_{GS} - V_T)V_{DS} - \frac{V_{DS}^2}{2}\right] \tag{3.25}$$

Here, k'_n is called the *process transconductance parameter* and is given by

$$k'_n = \mu_n C_{ox} = \frac{\mu_n \varepsilon_{ox}}{t_{ox}} . \tag{3.26}$$

The product of the process transconductance k'_n and the (W/L) ratio of an (NMOS) transistor is called the *gain factor* k_n of the device. For smaller values of V_{DS}, the quadratic factor in Eq. (3.25) can be ignored, and we observe a linear dependence between V_{DS} and I_D. The operation region where Eq. (3.25) holds is thus called the *resistive* or *linear region*. One of its main properties is that it displays a continuous conductive channel between source and drain regions.

NOTICE: The W and L parameters in Eq. (3.25) represent the *effective channel width and length* of the transistor. These values differ from the dimensions drawn on the layout due to effects such as lateral diffusion of the source and drain regions (L), and the encroachment of the isolating field oxide (W). For the remainder of this book, W and L will always stand for the effective dimensions, while a d subscript will be used to indicate the drawn size. The following expressions relate the two parameters, with ΔW and ΔL being parameters of the manufacturing process:

$$W = W_d - \Delta W$$
$$L = L_d - \Delta L \tag{3.27}$$

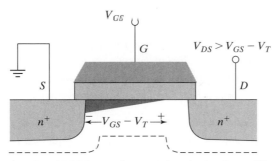

Figure 3-16   NMOS transistor under pinch-off conditions.

The Saturation Region

As the value of the drain-source voltage is further increased, the assumption that the channel voltage is larger than the threshold all along the channel ceases to hold. This happens when $V_{GS} - V(x) < V_T$. At that point, the induced charge is zero, and the conducting channel disappears or is *pinched off*. This is illustrated in Figure 3-16, which shows (in an exaggerated fashion) how the channel thickness gradually is reduced from source to drain until pinch-off occurs. No channel exists in the vicinity of the drain region. Obviously, for this phenomenon to occur, it is essential that the pinch-off condition be met at the drain region, or

$$V_{GS} - V_{DS} \leq V_T \tag{3.28}$$

Under those circumstances, the transistor is in the *saturation* region, and Eq. (3.25) no longer holds. The voltage difference over the induced channel (from the pinch-off point to the source) remains fixed at $V_{GS} - V_T$, and consequently, the current remains constant (or saturates). Replacing V_{DS} by $V_{GS} - V_T$ in Eq. (3.25) yields the drain current for the saturation mode. It is worth observing that, to the first degree, the current is no longer a function of V_{DS}. Notice also the *squared dependency* of the drain current with respect to the control voltage V_{GS}:

$$I_D = \frac{k_n' W}{2 L}(V_{GS} - V_T)^2 \tag{3.29}$$

Channel-Length Modulation

Equation 3.29 suggests that the transistor in the saturation mode acts as a perfect current source—the current between drain and source terminals is constant and independent of the applied voltage over the terminals. This is not entirely correct. The effective length of the conductive channel is actually modulated by the applied V_{DS}: increasing V_{DS} causes the depletion region at the drain junction to grow, reducing the length of the effective channel. As can be observed from Eq. (3.29), the current increases when the length L is decreased. A more accurate description of the current of the MOS transistor is therefore given by

$$I_D = I_D'(1 + \lambda V_{DS}) \tag{3.30}$$

where I_D' is the expression for the current derived earlier (this is, Eq. (3.29)), and λ is an empirical parameter called the *channel-length modulation*. Analytical expressions for λ have proven to be complex and inaccurate. In general, λ is proportional to the inverse of the channel length. In shorter transistors, the drain-junction depletion region presents a larger fraction of the channel, and the channel-modulation effect is more pronounced. It is therefore advisable to resort to long-channel transistors if a high-impedance current source is needed.

Velocity Saturation

The behavior of transistors with very short channel lengths (called *short-channel devices*) deviates considerably from the resistive and saturated models just presented. The main culprit for this deficiency is the *velocity saturation* effect. Eq. (3.23) states that the velocity of the carriers is proportional to the electrical field, independent of the value of that field. In other words, the carrier mobility is a constant. However, at high (horizontal) field strengths, the carriers fail to follow this linear model. In fact, when the electrical field along the channel reaches a critical value ξ_c, the velocity of the carriers tends to saturate due to scattering effects (collisions suffered by the carriers). This is illustrated in Figure 3-17.

The saturation velocity for electrons and holes is approximately the same: 10^5 m/s. The critical field at which saturation occurs depends upon the doping levels and the vertical electrical field applied. For electrons, it varies between 1 and 5 V/μm. This means that in an NMOS device with a channel length of 0.25 μm, only about two volts between drain and source are needed to reach the saturation point. This condition is easily met in current short-channel devices. A somewhat higher electrical field is needed for holes in an *n*-type silicon to achieve saturation. Velocity-saturation effects are therefore less pronounced in PMOS transistors.

This effect has a profound impact on the operation of the transistor. We illustrate this with a first-order derivation of the device characteristics under velocity-saturating conditions [Ko89]. The velocity as a function of the electrical field, plotted in Figure 3-17, can be roughly approximated by the following conditional equation:

$$\upsilon = \frac{\mu_n \xi}{1 + \xi/\xi_c} \quad \text{for} \quad \xi \leq \xi_c$$
$$= \upsilon_{sat} \quad \text{for} \quad \xi \geq \xi_c \tag{3.31}$$

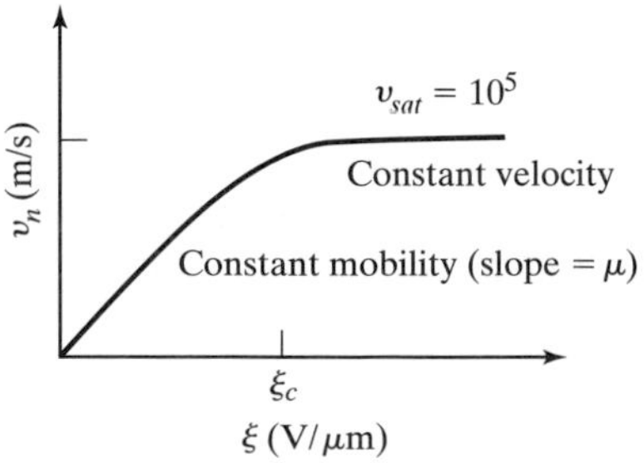

Figure 3-17 Velocity-saturation effect.

The continuity requirement between the two regions dictates that $\xi_c = 2\upsilon_{sat}/\mu_n$. If we reevaluate Eq. (3.20) and Eq. (3.22) in light of the revised velocity formula, we are led to a modified expression for the drain current in the resistive region:

$$I_D = \frac{\mu_n C_{ox}}{1 + (V_{DS}/\xi_c L)} \left(\frac{W}{L}\right) \left[(V_{GS} - V_T)V_{DS} - \frac{V_{DS}^2}{2}\right]$$

$$= \mu_n C_{ox}\left(\frac{W}{L}\right)\left[(V_{GS} - V_T)V_{DS} - \frac{V_{DS}^2}{2}\right]\kappa(V_{DS}) \tag{3.32}$$

The $\kappa(V)$ factor measures the degree of velocity saturation and is defined as follows:

$$\kappa(V) = \frac{1}{1 + (V/\xi_c L)} \tag{3.33}$$

V_{DS}/L can be interpreted as the average field in the channel. In case of long-channel devices (large values of L) or small values of V_{DS}, κ approaches 1, and Eq. (3.32) simplifies to the traditional current equation for the resistive operation mode. For short-channel devices, κ is less than 1, which means that the delivered current is less than what would normally be expected.

When increasing the drain-source voltage, the electrical field in the channel ultimately reaches the critical value, and the carriers at the drain become velocity saturated. The saturation drain voltage V_{DSAT} can be calculated by equating the current at the drain under saturation conditions to the current given by Eq. (3.32) for $V_{DS} = V_{DSAT}$ [Toh88]. The former is derived from Eq. (3.22), assuming that the drift velocity is saturated and equals υ_{sat}. We have

$$I_{DSAT} = \upsilon_{sat}C_{ox}W(V_{GT} - V_{DSAT})$$

$$= \kappa(V_{DSAT})\mu_n C_{ox}\frac{W}{L}\left[V_{GT}V_{DSAT} - \frac{V_{DSAT}^2}{2}\right] \tag{3.34}$$

where V_{GT} is a shorthand notation for $V_{GS} - V_T$. After cancelling, we obtain

$$V_{DSAT} = \kappa(V_{GT})V_{GT} \tag{3.35}$$

Further increasing the drain-source voltage does not yield more current (to a first degree), and the transistor current saturates at I_{DSAT}. This leads to the following two observations:

- For a short-channel device and for large enough values of V_{GT}, $\kappa(V_{GT})$ is substantially less than 1, and thus $V_{DSAT} < V_{GT}$. The device enters saturation before V_{DS} reaches $V_{GS} - V_T$. Short-channel devices therefore experience an extended saturation region, and tend to operate more often in saturation conditions than their long-channel counterparts. (See Figure 3-18.)

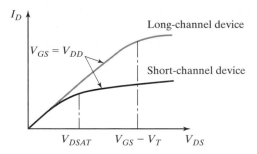

Figure 3-18 Short-channel devices display an extended saturation region due to velocity saturation.

- The saturation current I_{DSAT} displays a *linear dependence* with respect to the gate-source voltage V_{GS}, which is in contrast with the squared dependence in the long- channel device. This reduces the amount of current a transistor can deliver for a given control voltage. On the other hand, reducing the operating voltage does not have such a significant effect in submicron devices as it would have in a long-channel transistor.

The preceding equations ignore the fact that a larger portion of the channel becomes velocity-saturated with a further increase of V_{DS}. From a modeling perspective, it appears as though the effective channel is shortening with increasing V_{DS}, similar in effect to the channel-length modulation. The resulting increase in current is easily accommodated by introducing an extra $(1 + \lambda \times V_{DS})$ multiplier. For those interested in an in-depth perspective on the short-channel effects in MOS transistors, see [Ko89],[3] among others.

Velocity Saturation—Revisited

Unfortunately, the drain-current equations Eq. (3.32) and Eq. (3.34) are complex expressions of V_{GS} and V_{DS}, which makes them rather unwieldy for a first-order manual analysis. A substantially simpler model can be obtained by making two assumptions:

1. The velocity saturates abruptly at ξ_c, and it is approximated by the following expression:

$$\begin{aligned}
\upsilon &= \mu_n \xi &&\text{for}\quad \xi \le \xi_c \\
&= \upsilon_{sat} = \mu_n \xi_c &&\text{for}\quad \xi \ge \xi_c
\end{aligned} \qquad (3.36)$$

2. The drain-source voltage V_{DSAT} at which the critical electrical field is reached and velocity saturation comes into play is *constant* and is approximated as

$$V_{DSAT} \approx L\xi_c = \frac{L\upsilon_{sat}}{\mu_n} \qquad (3.37)$$

From Eq. (3.35), we derive that this assumption is reasonable for larger values of V_{GT}.

---

[3]For the sake of simplicity, we have deliberately chosen to ignore effects such as the mobility degradation due to the vertical electrical field. We feel this could be done safely without impairing the subsequent modeling effort.

Under these circumstances, the current equations for the resistive region remain unchanged from the long-channel model. Once V_{DSAT} is reached, the current abruptly saturates. The value for I_{DSAT} at that point can be derived by plugging the saturation voltage into the current equation for the resistive region (Eq. (3.25)):

$$
\begin{aligned}
I_{DSAT} &= I_D(V_{DS} = V_{DSAT}) \\
&= \mu_n C_{ox}\frac{W}{L}\left((V_{GS} - V_T)V_{DSAT} - \frac{V_{DSAT}^2}{2}\right) \\
&= \upsilon_{sat}C_{ox}W\left(V_{GS} - V_T - \frac{V_{DSAT}}{2}\right)
\end{aligned}
\tag{3.38}
$$

This model is truly first-order and empirical. The simplified velocity model causes substantial deviations in the transition zone between linear and velocity-saturated regions. Yet, by carefully choosing the model parameters, decent matching can be obtained with empirical data in the other operation regions, as will be shown later in this text. Most importantly, the equations are coherent with the familiar long-channel equations, and they provide the digital designer with a much needed tool for intuitive understanding and interpretation.

Drain Current versus Voltage Charts

The behavior for the MOS transistor in the different operation regions is best understood by analyzing its I_D-V_{DS} curves, which plot I_D versus V_{DS} with V_{GS} as a parameter. Figure 3-19 shows these charts for two NMOS transistors, implemented in the same technology and with the same W/L ratio. One would therefore expect both devices to display identical *I-V* characteristics, The main difference, however, is that the first device has a long channel length ($L_d = 10 \ \mu$m), while

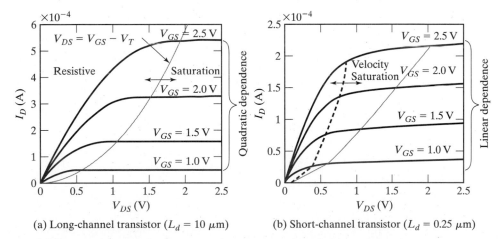

(a) Long-channel transistor ($L_d = 10 \ \mu$m)   (b) Short-channel transistor ($L_d = 0.25 \ \mu$m)

Figure 3-19 *I-V* characteristics of long- and a short-channel NMOS transistors in a 0.25 μm CMOS technology. The (*W/L*) ratio of both transistors is identical and equals 1.5. Observe the difference in the *y*-axis scale.

the second transistor is a short-channel device ($L_d = 0.25$ μm), and experiences velocity saturation.

Consider first the long-channel device. In the resistive region, the transistor behaves like a voltage-controlled resistor, while in the saturation region, it acts as a voltage-controlled current source (when the channel-length modulation effect is ignored). The transition between both regions is delineated by the $V_{DS} = V_{GS} - V_T$ curve. The squared dependence of I_D as a function of V_{GS} in the saturation region—typical for a long-channel device—is clearly observable from the spacing between the different curves. The linear dependence of the saturation current with respect to V_{GS} is apparent in the short-channel device of Figure 3-19b. Notice also how velocity saturation causes the device to saturate for substantially smaller values of V_{DS}. The striped line shows where velocity saturation kicks in as predicted by Eq. (3.35), while the dashed line indicates when traditional saturation would have occurred ($V_{DS} = V_{GT}$). This results in a substantial drop in current drive for high voltage levels. For instance, at ($V_{GS} = 2.5$ V, $V_{DS} = 2.5$ V), the drain current of the short transistor is only 40% of the corresponding value of the longer device (220 μA versus 540 μA).

The difference in dependence upon V_{GS} between long- and short-channel devices is even more pronounced in another set of simulated charts that plot I_D as a function of V_{GS} for a fixed value greater than V_{DS}, which ensures saturation. (See Figure 3-20.) A quadratic versus linear dependence is apparent for larger values of V_{GS}.

All the derived equations hold for the PMOS transistor as well. The only difference is that **for PMOS devices, the polarities of all voltages and currents are reversed**. This is illustrated in Figure 3-21, which plots the I_D-V_{DS} characteristics of a minimum-size PMOS transistor in our

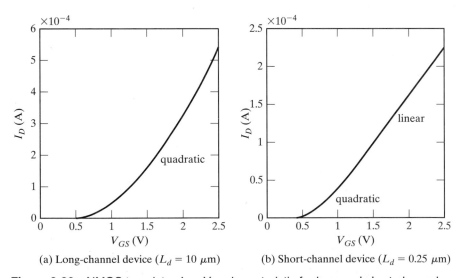

(a) Long-channel device ($L_d = 10$ μm)   (b) Short-channel device ($L_d = 0.25$ μm)

Figure 3-20   NMOS transistor $I_D - V_{GS}$ characteristic for long and short-channel devices (0.25 μm CMOS technology). $W/L = 1.5$ for both transistors and $V_{DS} = 2.5$ V.

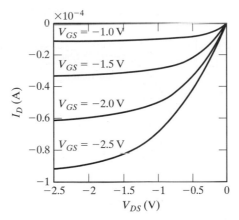

Figure 3-21 *I-V* characteristics of ($W_d = 0.375$ μm, $L_d = 0.25$ μm) PMOS transistor in 0.25 μm CMOS process. Due to the smaller mobility, the maximum current is only 42% of what is achieved by a similar NMOS transistor.

generic 0.25 μm CMOS process. The curves are in the third quadrant as I_D, V_{DS}, and V_{GS} are all negative. Notice that the effects of velocity saturation are less pronounced than in the NMOS devices. This can be attributed to the higher value of the critical electrical field, resulting from the smaller mobility of holes versus electrons.

Subthreshold Conduction

A closer inspection of the I_D-V_{GS} curves of Figure 3-20 reveals that the current does not drop abruptly to 0 at $V_{GS} = V_T$. It becomes apparent that the MOS transistor is already partially conducting for voltages below the threshold voltage. This effect is called *subthreshold* or *weak-inversion* conduction. The onset of strong inversion means that ample carriers are available for conduction, but it does not imply that no current at all can flow for gate-source voltages below V_T, although the current levels are small under those conditions. In summary, the transition from the "on" to the "off" condition is not abrupt, but gradual.

To study this effect in somewhat more detail, we redraw the curve of Figure 3-20b on a logarithmic scale, as shown in Figure 3-22. This confirms that the current does not drop to zero immediately for $V_{GS} < V_T$, but actually decays in an exponential fashion, similar to the operation of a bipolar transistor.[4] In the absence of a conducting channel, the n^+ (source)–p (bulk)–n^+ (drain) terminals actually form a parasitic bipolar transistor. The current in this region can be approximated by the following expression:

$$I_D = I_S e^{\frac{V_{GS}}{nkT/q}} \left(1 - e^{-\frac{V_{DS}}{kT/q}} \right) (1 + \lambda V_{DS}) \tag{3.39}$$

where I_S and n are empirical parameters, with $n \geq 1$, and typically ranging around 1.5.

[4]See the web site for this book for further information on bipolar transistors.

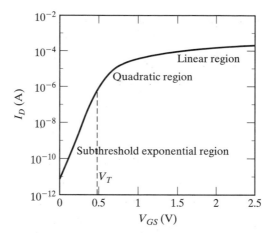

Figure 3-22 I_D current versus V_{GS} (on logarithmic scale), showing the exponential characteristic of the subthreshold region.

In most digital applications, the presence of subthreshold current is undesirable because it deviates from the ideal switchlike behavior that we prefer to assume for the MOS transistor. We would rather have the current drop as fast as possible once the gate-source voltage falls below V_T. The (inverse) rate of decline of the current with respect to V_{GS} below V_T is thus a quality measure of a device. It is often quantified by the *slope factor S,* which measures by how much V_{GS} has to be reduced for the drain current to drop by a factor of 10. From Eq. (3.39), we find that

$$S = n\left(\frac{kT}{q}\right)\ln(10) \tag{3.40}$$

where S is expressed in mV/decade. For an ideal transistor with the sharpest possible roll off, $n = 1$, and $(kT/q)\ln(10)$ evaluates to 60 mV/decade at room temperature, which means that the subthreshold current drops by a factor of 10 for a reduction in V_{GS} of 60 mV. Unfortunately, n is greater than 1 for actual devices and the current falls at a reduced rate (90 mV/decade for $n = 1.5$). The current roll-off is further decreased by a rise in the operating temperature—most integrated circuits operate at temperatures considerably above room temperature. The value of n is determined by the intrinsic device topology and structure. Reducing its value therefore requires a different process technology, silicon on insulator, for example.

Subthreshold current has some important repercussions. In general, we want the current through the transistor to be as close as possible to zero at $V_{GS} = 0$. This is especially important in *dynamic circuits*, which rely on the storage of charge on a capacitor and whose operation can be severely degraded by subthreshold leakage. Achieving this in the presence of subthreshold current requires a firm lower bound on the value of the threshold voltage of the devices.

Example 3.6   Subthreshold Slope

For the example of Figure 3-22, a slope of 89.5 mV/decade is observed (between 0.2 and 0.4 V). This is equivalent to an n-factor of 1.49.

In Summary—Models for Manual Analysis

The preceding discussions made it clear that the deep-submicron transistor is a complex device. Its behavior is heavily nonlinear and is influenced by a large number of second-order effects. Fortunately, accurate circuit-simulation models have been developed that make it possible to predict the behavior of a device with amazing precision over a large range of device sizes, shapes, and operation modes, as we discuss later in this chapter. While excellent from an accuracy perspective, these models fail to provide a designer with an intuitive understanding of the behavior of a circuit and its dominant design parameters. Such an understanding is necessary in the design analysis and optimization process. By necessity, a designer who misses a clear vision of what drives and governs the circuit operation must rely on a lengthy trial-and-error optimization process, which most often leads to an inferior solution.

The obvious question now is how to abstract the behavior of our MOS transistor into a simple and tangible analytical model that does not lead to hopelessly complex equations, yet captures the essentials of the device. It turns out that the first-order expressions derived earlier in the chapter can be combined into a single expression that meets these goals. The model presents the transistor as a single current source (as shown in Figure 3-23), the value of which is given by the figure. The reader can verify that, depending upon the operating condition, the model simplifies into either Eq. (3.25), Eq. (3.29), or Eq. (3.38) (corrected for channel-length modulation).[5] We have

$$I_D = 0 \ \ \text{for} \ \ V_{GT} \leq 0$$

$$I_D = k'\frac{W}{L}\left(V_{GT}V_{min} - \frac{V_{min}^2}{2}\right)(1 + \lambda V_{DS}) \ \text{for} \ V_{GT} \geq 0$$

$$\text{with} \ \ V_{min} = \min(V_{GT}, V_{DS}, V_{DSAT}),$$

$$V_{GT} = V_{GS} - V_T,$$

$$\text{and} \ \ V_T = V_{T0} + \gamma(\sqrt{|-2\phi_F + V_{SB}|} - \sqrt{|-2\phi_F|})$$

Figure 3-24 shows how the **unified model** divides the overall operation space of the transistor into three regions—linear, velocity saturated, and saturated—and identifies the boundaries between those regions.

Besides being a function of the voltages at the four terminals of the transistor, the model employs a set of five parameters: V_{TO}, γ, V_{DSAT}, k', and λ. In principle, it is possible to determine these parameters from the process technology and from the device physics equations. The

---

[5]The unified model also introduces a channel-length modulation correction in the linear region. This factor, which is small, serves solely for modeling purposes and is not grounded in physical considerations Be further aware that the "min" function turns into a "max" operation for PMOS transistors.

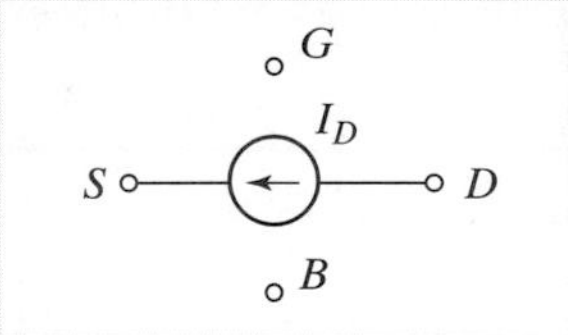

$$I_D = 0 \text{ for } V_{GT} \le 0$$

$$I_D = k'\frac{W}{L}\left(V_{GT}V_{min} - \frac{V_{min}^2}{2}\right)(1 + \lambda V_{DS}) \text{ for } V_{GT} \ge 0$$

with $V_{min} = \min(V_{GT}, V_{DS}, V_{DSAT})$,

$$V_{GT} = V_{GS} - V_T,$$

and $V_T = V_{T0} + \gamma(\sqrt{|-2\phi_F + V_{SB}|} - \sqrt{|-2\phi_F|})$

Figure 3-23   A unified MOS model for manual analysis.

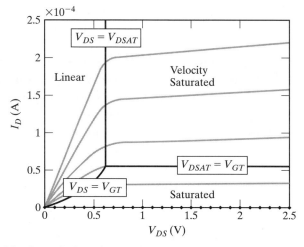

Figure 3-24   Boundaries of operation regions, as defined in the unified model for manual analysis.

complexity of the device makes this a precarious task, however. A more rewarding approach is to choose the values such that a good matching with the actual device characteristics is obtained. More significantly, the model should match the best in the regions that matter the most. In digital circuits, this in the region of high V_{GS} and V_{DS}. The performance of an MOS digital circuit is primarily determined by the maximum available current (this is, the current obtained for $V_{GS} = V_{DS} =$ supply voltage). A good matching in this region is therefore essential. **Finally, observe that for a typical NMOS device all five parameters are positive, while they have negative values for a typical PMOS transistor.**

Example 3.7   Manual Analysis Model for 0.25 μm CMOS Process[6]

Based on the simulated I_D–V_{DS} and I_D–V_{GS} plots of a ($W = 0.375$ μm, $L = 0.25$ μm) transistor implemented in our generic 0.25 micron CMOS process (Figure 3-19, Figure 3-20), we have derived a set of device parameters to match well in the ($V_{DS} = 2.5$ V, $V_{GS} = 2.5$ V) region—2.5 V being the typical supply voltage for this process. The resulting characteris-

---

[6]A MATLAB implementation of the model is available on the web site of the textbook.

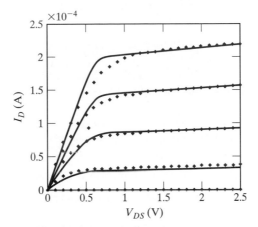

Figure 3-25 Correspondence between simple model (solid line) and SPICE simulation (dotted) for minimum-size NMOS transistor ($W = 0.375$ μm, $L = 0.25$ μm). Observe the discrepancy in the transition zone between resistive and velocity saturation.

tics are plotted in Figure 3-25 for the NMOS transistor, and compared with the simulated values. Overall, a good correspondence can be observed with the exception of the transition region between resistive and velocity-saturation. This discrepancy, which occurs as a result of the simple velocity model of Eq. (3.36), is acceptable because it occurs in the lower value range of V_{DS}. It demonstrates that our model, while simple, manages to give a fair indication of the overall behavior of the device.

Design Data—Transistor Model for Manual Analysis

Table 3-2 tabulates the obtained parameter values for the minimum-sized NMOS and a similarly sized PMOS device in our generic 0.25 μm CMOS process. These values will be used as generic model parameters in later chapters.

Table 3-2 Parameters for manual model of generic 0.25 μm CMOS process (minimum length device).

	V_{T0} (V)	γ ($V^{0.5}$)	V_{DSAT} (V)	k' (A/V^2)	λ (V^{-1})
NMOS	0.43	0.4	0.63	115×10^{-6}	0.06
PMOS	−0.4	−0.4	−1	$−30 \times 10^{-6}$	−0.1

A word of caution—The model presented here is derived from the characteristics of a single device with a minimum channel length and width. Trying to extrapolate this behavior to devices with substantially different values of W and L probably will lead to sizable errors. Fortunately, digital circuits typically use only minimum-length devices because they lead to the smallest implementation area. Matching for these transistors usually turns out to be acceptable. It is, however, advisable to use a different set of model parameters for devices with dramatically different widths. ■

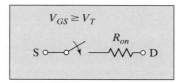

Figure 3-26   NMOS transistor modeled as a switch.

The presented *current-source model* will prove to be very useful in the analysis of the basic properties and metrics of a simple digital gate, yet its nonlinearity makes it intractable for anything more complex. Hence, we introduce an even simpler model that is both linear and straightforward. It is based on the underlying assumption in most digital designs that the transistor is nothing more than a *switch* with an infinite "off" resistance, and a finite "on" resistance R_{on}. (See Figure 3.26.)

The main problem with this model is that R_{on} is time varying, nonlinear and dependent on the operation point of the transistor. When studying digital circuits in the transient mode—which means while switching between different logic states—it is attractive to assume that R_{on} is a constant and linear resistance R_{eq}, chosen so that the final result is similar to what would be obtained with the original transistor. A reasonable approach in that respect is to use the average value of the resistance over the operation region of interest, or even simpler, the average value of the resistances at the end points of the transition. The latter assumption works well if the resistance does not experience any strong nonlinearities over the range of the averaging interval. Accordingly, we have

$$R_{eq} = \text{average}_{t = t_1 \ldots t_2}(R_{on}(t)) = \frac{1}{t_2 - t_1}\int_{t_1}^{t_2} R_{on}(t)dt = \frac{1}{t_2 - t_1}\int_{t_1}^{t_2} \frac{V_{DS}(t)}{I_D(t)}dt$$

$$\approx \frac{1}{2}(R_{on}(t_1) + R_{on}(t_2)) \tag{3.41}$$

Example 3.8   Equivalent Resistance when (Dis)Charging a Capacitor

One of the most common scenarios in contemporary digital circuits is the discharging of a capacitor from V_{DD} to GND through an NMOS transistor with its gate voltage set to V_{DD}, or vice versa—the charging of the capacitor to V_{DD} through a PMOS with its gate at GND. Of special interest is the point at which the voltage on the capacitor reaches the midpoint ($V_{DD}/2$)—this is by virtue of the definition of the propagation delay, as introduced in Chapter 2. Assuming that the supply voltage is substantially greater than the velocity-saturation voltage V_{DSAT} of the transistor, it is fair to state that the transistor stays in velocity saturation for the entire duration of the transition. This scenario is plotted in Figure 3-27 for the case of an NMOS discharging a capacitor from V_{DD} to $V_{DD}/2$.

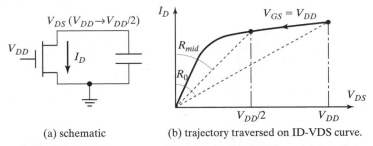

| (a) schematic | (b) trajectory traversed on ID-VDS curve. |

Figure 3-27 Discharging a capacitor through an NMOS transistor: Schematic (a) and I-V trajectory (b). The instantaneous resistance of the transistor equals (V_{DS}/I_D) and is visualized by the angle with respect to the y-axis.

With the aid of Eq. (3.41), we can derive the value of the equivalent resistance, which averages the resistance of the device over the interval:

$$R_{eq} = \frac{1}{-V_{DD}/2} \int_{V_{DD}}^{V_{DD}/2} \frac{V}{I_{DSAT}(1+\lambda V)} dV \approx \frac{3}{4} \frac{V_{DD}}{I_{DSAT}} \left(1 - \frac{7}{9}\lambda V_{DD}\right)$$

$$\text{with } I_{DSAT} = k'\frac{W}{L}\left((V_{DD} - V_T)V_{DSAT} - \frac{V_{DSAT}^2}{2}\right)$$

(3.42)

A similar result can be obtained by just averaging the values of the resistance at the endpoints of the transition region (and simplifying the result using a Taylor expansion):

$$R_{eq} = \frac{1}{2}\left(\frac{V_{DD}}{I_{DSAT}(1+\lambda V_{DD})} + \frac{V_{DD}/2}{I_{DSAT}(1+\lambda V_{DD}/2)}\right) \approx \frac{3}{4}\frac{V_{DD}}{I_{DSAT}}\left(1 - \frac{5}{6}\lambda V_{DD}\right)$$

(3.43)

We draw three worthwhile conclusions from these expressions:

- The resistance is inversely proportional to the (W/L) ratio of the device. Doubling the transistor width cuts the resistance in half.
- For $V_{DD} \gg V_T + V_{DSAT}/2$, the resistance becomes virtually independent of the supply voltage. This is confirmed in Figure 3-28, which plots the (simulated) equivalent resistance as a function of the supply voltage V_{DD}. Only a minor improvement in resistance, attributable to the channel-length modulation, can be observed when raising the supply voltage.
- Once the supply voltage approaches V_T, the resistance dramatically increases.

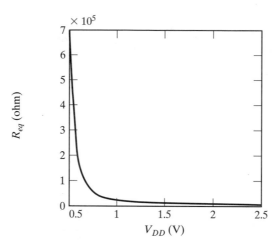

Figure 3-28 Simulated equivalent resistance of a minimum size NMOS transistor in 0.25 μm CMOS process as a function of V_{DD} ($V_{GS} = V_{DD}$, $V_{DS} = V_{DD} \rightarrow V_{DD}/2$).

Design Data — Equivalent Resistance Model

Table 3-3 enumerates the equivalent resistances obtained by simulation of our generic 0.25 μm CMOS process. These values will come in handy when analyzing the performance of CMOS gates in later chapters.

Table 3-3 Equivalent resistance R_{eq} ($W/L = 1$) of NMOS and PMOS transistors in 0.25 μm CMOS process (with the drawn $L = L_{min}$). For larger devices, divide R_{eq} by W/L.

V_{DD} (V)	1	1.5	2	2.5
NMOS (kΩ)	35	19	15	13
PMOS (kΩ)	115	55	38	31

Dynamic Behavior

The dynamic response of a MOSFET transistor is solely a function of the time it takes to (dis)charge the parasitic capacitances that are intrinsic to the device and the extra capacitance introduced by the interconnecting lines and load (the subject of Chapter 4). A profound understanding of the nature and the behavior of these intrinsic capacitances is essential for the designer of high-quality digital integrated circuits. They originate from three sources: the basic MOS structure, the channel charge, and the depletion regions of the reverse-biased *pn*-junctions of drain and source. Aside from the MOS structure capacitances, all capacitors are nonlinear and vary with the applied voltage, which makes analyzing them difficult. We discuss each of the components in the paragraphs to follow.

MOS Structure Capacitances

The gate of the MOS transistor is isolated from the conducting channel by the gate oxide, which has a capacitance per unit area equal to $C_{ox} = \varepsilon_{ox} / t_{ox}$. We learned earlier that from an I-V perspective, it is useful to have C_{ox} as large as possible, or to keep the oxide thickness very thin. The total value of this capacitance is called the *gate capacitance* C_g, and it can be decomposed into two elements, each with a different behavior. Obviously, one part of C_g contributes to the channel charge. Another part is solely due to the topological structure of the transistor, and is discussed in the remainder of this section.

Consider the transistor structure of Figure 3-29. Ideally, the source and drain diffusion should end right at the edge of the gate oxide. In reality, both source and drain tend to extend somewhat below the oxide by an amount x_d, called the *lateral diffusion*. Hence, the effective channel of the transistor L becomes shorter than the drawn length L_d (or the length the transistor was originally designed for) by a factor of $\Delta L = 2x_d$. This also gives rise to a parasitic capacitance between gate and source (drain) that is called the *overlap capacitance*. This capacitance is linear and has a fixed value:

$$C_{GSO} = C_{GDO} = C_{ox}x_dW = C_oW \qquad (3.44)$$

Since x_d is technology determined, it is customary to combine it with the oxide capacitance to yield the overlap capacitance per unit transistor width C_o (more specifically, C_{gso} and C_{gdo}).

Channel Capacitance

Perhaps the most significant MOS parasitic circuit element, the gate-to-channel capacitance C_{GC}, varies in both magnitude and in its division into three components C_{GCS}, C_{GCD}, and C_{GCB} (being the gate-to-source, gate-to-drain, and gate-to-body capacitances, respectively), depending

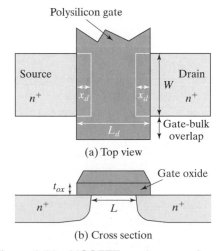

(a) Top view

(b) Cross section

Figure 3-29 MOSFET overlap capacitance.

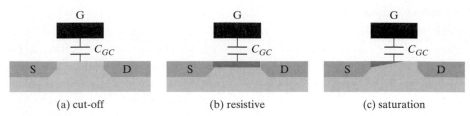

(a) cut-off   (b) resistive   (c) saturation

Figure 3-30 The gate-to-channel capacitance and how the operation region influences its distribution over the three other device terminals.

upon the operation region and terminal voltages. This varying distribution is explained with the simple diagrams of Figure 3-30. When the transistor is in the cut-off region (a), no channel exists, and the total capacitance C_{GC} appears between gate and body. In the resistive region (b), an inversion layer is formed, which acts as a conductor between source and drain. Consequently, $C_{GCB} = 0$ as the body electrode is shielded from the gate by the channel. Symmetry dictates that the capacitance distributes evenly between source and drain. Finally, in the saturation mode (c), the channel is pinched off. The capacitance between gate and drain is approximately zero, and so is the gate-body capacitance. All the capacitance is therefore between gate and source.

The actual value of the total gate-channel capacitance and its distribution over the three components is best understood with the aid of several charts. The first plot (Figure 3-31a) captures the evolution of the capacitance as a function of V_{GS} for $V_{DS} = 0$. For $V_{GS} = 0$, the transistor is off, no channel is present and the total capacitance, equal to WLC_{ox}, appears between gate and body. When increasing V_{GS}, a depletion region forms under the gate. This appears as an increase in the thickness of the gate dielectric, which results in a reduction in capacitance. Once the transistor turns on ($V_{GS} = V_T$), a channel is formed and C_{GCB} drops to 0. With $V_{DS} = 0$, the device operates in the resistive mode, and the capacitance divides equally between source and drain ($C_{GCS} = C_{GCD} = WLC_{ox}/2$). The large fluctuation of the channel capacitance around $V_{GS} = V_T$ is worth remembering. A designer looking for a well-behaved linear capacitance should avoid operation in this region.

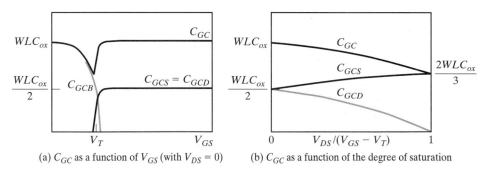

(a) C_{GC} as a function of V_{GS} (with $V_{DS} = 0$)   (b) C_{GC} as a function of the degree of saturation

Figure 3-31   Distribution of the gate-channel capacitance as a function of V_{GS} and V_{DS} (from [Dally98]).

Table 3-4 Average distribution of channel capacitance of MOS transistor for different operation regions.

Operation Region	C_{GCB}	C_{GCS}	C_{GCD}	C_{GC}	C_G
Cutoff	$C_{ox}WL$	0	0	$C_{ox}WL$	$C_{ox}WL + 2C_oW$
Resistive	0	$C_{ox}WL/2$	$C_{ox}WL/2$	$C_{ox}WL$	$C_{ox}WL + 2C_oW$
Saturation	0	$(2/3)C_{ox}WL$	0	$(2/3)C_{ox}WL$	$(2/3)C_{ox}WL + 2C_oW$

Once the transistor is on, the distribution of its gate capacitance depends upon the degree of saturation, measured by the $V_{DS}/(V_{GS} - V_T)$ ratio. As illustrated in Figure 3-31b, C_{GCD} gradually drops to 0 for increasing levels of saturation, while C_{GCS} increases to 2/3 $C_{ox}WL$. This also means that the total gate capacitance is getting smaller with an increased level of saturation.

The preceding discussion established that the gate-capacitance components are nonlinear and varying with the operating voltages. To make a first-order analysis possible, we adopt a simplified piecewise-linear model with a constant capacitance value in each region of operation. The assumed values are summarized in Table 3-4.

Example 3.9   Using a Circuit Simulator to Extract Capacitance

Determining the value of the parasitic capacitances of an MOS transistor for a given operation mode is a labor-intensive task, and requires the knowledge of a number of technology parameters that are often not explicitly available. Fortunately, once a SPICE model of the transistor is attained, a simple simulation can provide the data you need. Assume we want to know the value of the total gate capacitance of a transistor in a given technology as a function of V_{GS} (for $V_{DS} = 0$). A simulation of the circuit of Figure 3-32a will give us exactly this information. In fact, the relation

$$I = C_G(V_{GS})\frac{dV_{GS}}{dt}$$

is valid, and can be rewritten as

$$C_G(V_{GS}) = I / \left(\frac{dV_{GS}}{dt}\right)$$

A transient simulation gives us V_{GS} as a function of time, which can be translated into a capacitance with some simple mathematical manipulations. Figure 3-32b plots the simulated gate capacitance of a minimum-size 0.25 μm NMOS transistor as a function of

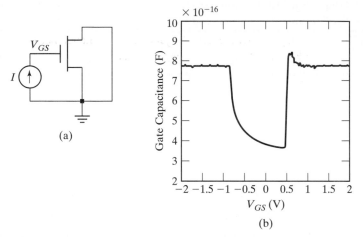

Figure 3-32   Simulating the gate capacitance of an MOS transistor; (a) circuit configuration used for the analysis, (b) resulting capacitance plot for minimum-size NMOS transistor in 0.25 μm technology.

V_{GS}. The graphs clearly show the drop of the capacitance when V_{GS} approaches V_T and the discontinuity at V_T, which was predicted in Figure 3-31.

Junction Capacitances

A final capacitive component is contributed by the reverse-biased source-body and drain-body *pn*-junctions. The depletion-region capacitance is nonlinear and decreases when the reverse bias is raised, as discussed earlier. To understand the components of the junction capacitance (often called the *diffusion capacitance*), we must look at the source (drain) region and its surroundings. Figure 3-33 shows that the junction consists of two components:

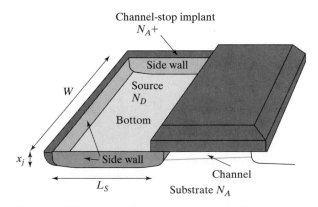

Figure 3-33   Detailed view of source junction.

- The *bottom-plate* junction, which is formed by the source region (with doping N_D) and the substrate with doping N_A. The total depletion region capacitance for this component is given by $C_{bottom} = C_j W L_S$, where C_j is the junction capacitance per unit area, as given by Eq. (3.9). As the bottom-plate junction typically is of the abrupt type, the grading coefficient m approaches 0.5.
- The *side-wall* junction, formed by the source region with doping N_D and the p^+ channel-stop implant with doping level N_A^+. The doping level of the stopper is larger than that of the substrate, resulting in a larger capacitance per unit area. The side-wall junction typically is graded, and its grading coefficient varies from 0.33 to 0.5. Its capacitance value is given by $C_{sw} = C_{sw}' x_j (W + 2 \times L_s)$. Notice that no side-wall capacitance is counted for the fourth side of the source region, as this represents the conductive channel.[7] Since x_j, the junction depth, is a technology parameter, it is normally combined with C_{jsw}' into a capacitance per unit perimeter $C_{jsw} = C_{jsw}' x_j$. An expression for the total junction capacitance can now be derived as follows:

$$C_{diff} = C_{bottom} + C_{sw} = C_j \times AREA + C_{jsw} \times PERIMETER$$
$$= C_j L_S W + C_{jsw}(2L_S + W) \qquad (3.45)$$

Since these are all small-signal capacitances, we normally linearize them and use average capacitances similar to those of Eq. (3.10).

Problem 3.1   Using a Circuit Simulator to Determine the Drain Capacitance

Using circuit simulation (in the style of Figure 3.32), derive a simple circuit that helps you determine the drain capacitance of an NMOS transistor in the different operation modes.

Capacitive Device Model

All the preceding contributions can be combined in a single capacitive model for the MOS transistor, which is shown Figure 3-34. Its components are readily identified on the basis of earlier discussions. The equation is

$$C_{GS} = C_{GCS} + C_{GSO}; \; C_{GD} = C_{GCD} + C_{GDO}; \; C_{GB} = C_{GCB}$$
$$C_{SB} = C_{Sdiff}; \; C_{DB} = C_{Ddiff} \qquad (3.46)$$

It is essential for the designers of high-performance and low-energy circuits to be very familiar with this model, as well as to have an intuitive feeling for the relative values of its components.

---

[7]To be entirely correct, we should take the diffusion capacitance of the source(drain)-to-channel junction into account. Due to the doping conditions and the small area, this component can nearly always be ignored in a first-order analysis. Detailed SPICE models include a factor C_{JSWG} to account for this junction.

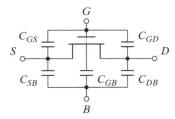

Figure 3-34   MOSFET capacitance model.

Example 3.10   MOS Transistor Capacitances

Consider an NMOS transistor with the following parameters: $t_{ox} = 6$ nm, $L = 0.24$ μm, $W = 0.36$ μm, $L_D = L_S = 0.625$ μm, $C_O = 3 \times 10^{-10}$ F/m, $C_{j0} = 2 \times 10^{-3}$ F/m^2, and $C_{jsw0} = 2.75 \times 10^{-10}$ F/m. Determine the zero-bias value of all relevant capacitances.

The gate capacitance per unit area is easily derived as $(\varepsilon_{ox} / t_{ox})$ and equals 5.7 fF/μm^2. The gate-to-channel C_{GC} is then $WLC_{ox} = 0.49$ fF. To find the total gate capacitance, we have to add the source and drain overlap capacitors, each of which is $WC_O = 0.105$ fF. This leads to a total gate capacitance of 0.7 fF.

The diffusion capacitance consists of the bottom and the side-wall capacitances. The former is equal to $C_{j0} L_D W = 0.45$ fF, while the side-wall capacitance under zero-bias conditions evaluates to $C_{jsw0} (2L_D + W) = 0.44$ fF. This results in a total drain- (source)-to-bulk capacitance of 0.89 fF.

The diffusion capacitance seems to dominate the gate capacitance. This is a worst case condition, however. When increasing the value of the reverse bias over the junction—as is the normal operation mode in MOS circuits—the diffusion capacitance is substantially reduced. Also, clever design can help reduce the value of L_D (L_S). In general, the contribution of diffusion capacitances is at most equal to, and very often smaller than the gate capacitance.

Design Data—MOS Transistor Capacitances

Table 3-5 summarizes the parameters needed to estimate the parasitic capacitances of the MOS transistors in our generic 0.25 μm CMOS process.

Table 3-5 Capacitance parameters of NMOS and PMOS transistors in 0.25 μm CMOS process.

	C_{ox} (fF/μm^2)	C_O (fF/μm)	C_j (fF/μm^2)	m_j	ϕ_b (V)	C_{jsw} (fF/μm)	m_{jsw}	ϕ_{bsw} (V)
NMOS	6	0.31	2	0.5	0.9	0.28	0.44	0.9
PMOS	6	0.27	1.9	0.48	0.9	0.22	0.32	0.9

Source–Drain Resistance

The performance of a CMOS circuit may further be affected by another set of parasitic elements, being the resistances in series with the drain and source regions, as shown in Figure 3-35a. This effect becomes more pronounced when transistors are scaled down, because this leads to shallower junctions and smaller contact openings. The resistance of the drain (source) region can be expressed as

$$R_{S,D} = \frac{L_{S,D}}{W}R_{\square} + R_C, \tag{3.47}$$

where R_C is the contact resistance, W is the width of the transistor, and $L_{S,D}$ is the length of the source or drain region (Figure 3-35b). $R_{\square}$ is the *sheet resistance* per square of the drain–source diffusion, and ranges from 20 to 100 $\Omega/\square$. Observe that the resistance of a square of material is constant, independent of its size.

   The series resistance causes a deterioration in the device performance, because it reduces the drain current for a given control voltage. Keeping its value as small as possible is thus an important design goal for both the device and the circuit engineer. One option, popular in many contemporary processes, is to cover the drain and source regions with a low-resistivity material such as titanium or tungsten. This process is called *silicidation* and effectively reduces the sheet resistance to values in the range of 1 to 4 $\Omega/\square$.[8] Making the transistor wider than needed is another possibility, which should be apparent from Eq. (3.47). With a process that includes silicidation and proper attention to layout, parasitic resistance is not important. The reader should be aware, however, that careless layout may lead to resistances that severely degrade device performance.

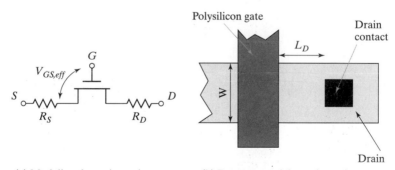

   (a) Modeling the series resistance   (b) Parameters of the series resistance

Figure 3-35   Series drain and source resistance.

---

[8] Silicidation is also used to reduce the resistance of the polysilicon gate. This will be further discussed in Chapter 4.

3.3.3   The Actual MOS Transistor—Some Secondary Effects

The operation of a contemporary transistor may show some important deviations from the model we have presented thus far. These divergences become especially pronounced once the dimensions of the transistor reach the deep submicron realm. At that point, the assumption that the operation of a transistor is adequately described by a one-dimensional model—where it is assumed that all current flows on the surface of the silicon and the electrical fields are oriented along that plane—is no longer valid. Two- or even three-dimensional models are more appropriate. We already touched on an example of such in Section 3.2.2 when we discussed the mobility degradation.

The understanding of some of these second-order effects and their impact on the device behavior is essential in the design of today's digital circuits—therefore, it merits some discussion. First, a word of warning: trying to take all those effects into account in a manual, first-order analysis results in intractable and opaque circuit models. It is therefore advisable to analyze and design MOS circuits first using the ideal model. The impact of the nonidealities can be studied in a second round using computer-aided simulation tools with more precise transistor models.

Threshold Variations

Equation (3.19) states that the threshold voltage is only a function of the manufacturing technology and the applied body bias V_{SB}. The threshold, therefore, can be considered a constant over all NMOS (PMOS) transistors in a design. As the device dimensions are reduced, this model gets inaccurate, and the threshold potential becomes a function of L, W, and V_{DS}. Two-dimensional second-order effects that were ignorable for long-channel devices suddenly gain significance.

In the traditional derivation of the V_{TO}, for example, it is assumed that the channel depletion region is solely due to the applied gate voltage, and that all depletion charge beneath the gate originates from the MOS field effects. This ignores the depletion regions of the source and reverse-biased drain junction, which become relatively more important with shrinking channel lengths. Since a part of the region below the gate is already depleted (by the source and drain fields), a smaller threshold voltage suffices to cause strong inversion. In other words, V_{TO} decreases with L for short-channel devices (Figure 3-36a). A similar effect can be obtained by raising the drain–source (bulk) voltage, as this increases the width of the drain-junction depletion region. Consequently, the threshold decreases with increasing V_{DS}. This effect, known as

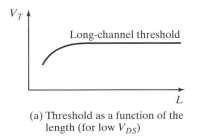
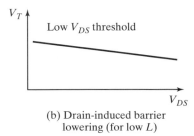

(a) Threshold as a function of the   (b) Drain-induced barrier
    length (for low V_{DS})   lowering (for low L)

Figure 3-36   Threshold variations.

drain-induced barrier lowering, or *DIBL,* causes the threshold potential to be a function of the operating voltages (Figure 3-36b). For high enough values of the drain voltage, the source and drain regions can even be shorted together, and normal transistor operation ceases to exist. The sharp increase in current that results from this effect, which is called *punch-through,* may cause permanent damage to the device and should be avoided. Punch-through defines an upper bound on the drain-source voltage that can be applied over the transistor.

Since the majority of the transistors in a digital circuit are designed at the minimum channel length, the variation of the threshold voltage as a function of the length is almost uniform over the complete design, and is therefore not much of an issue except for the increased subthreshold leakage currents. More troublesome is the DIBL, because this effect varies with the operating voltage. For example, this is a problem in dynamic memories. The leakage current of a cell—this is, the subthreshold current of the access transistor—becomes a function of the voltage on the bit line, and depends upon the applied data patterns. From the cell perspective, DIBL manifests itself as a data-dependent noise source.

We should mention that the threshold of the MOS transistor is also subject to *narrow-channel* effects. The depletion region of the channel does not stop abruptly at the edges of the transistor, but extends somewhat under the isolating field oxide. The gate voltage must support this extra depletion charge to establish a conducting channel. This effect is ignorable for wide transistors, but becomes significant for small values of W, where it results in an increase of the threshold voltage. For small geometry transistors, with small values of L and W, the effects of short and narrow channels may tend to cancel each other out.

Hot-Carrier Effects

Besides varying over a design, threshold voltages in short-channel devices also have the tendency to *drift over time*. This is the result of the *hot-carrier* effect [Hu92]. Over the last several decades, device dimensions have been scaled down continuously, but power supply and the operating voltages have not scaled accordingly. The resulting increase in the electrical field strength causes an increasing velocity of the electrons, which can leave the silicon and tunnel into the gate oxide upon reaching a sufficiently high level of energy. Electrons trapped in the oxide change the threshold voltage, typically increasing the thresholds of NMOS devices, while decreasing the V_T of PMOS transistors. For an electron to become hot, an electrical field of at least 10^4 V/cm is necessary. This condition is easily met in devices with channel lengths at or below 1 μm. The hot-electron phenomenon can lead to a long-term reliability problem, where a circuit might degrade or fail after being in use for some time. This is illustrated in Figure 3-37, which shows the degradation in the *I-V* characteristics of an NMOS transistor after it has been subjected to extensive operation. State-of-the-art MOSFET technologies therefore use specially engineered drain and source regions to ensure that the peaks in the electrical fields are bounded, thus preventing carriers from reaching the critical values necessary to become hot. The reduced supply voltage that is typical for deep submicron technologies can be attributed, in part, to the necessity of keeping hot-carrier effects under control.

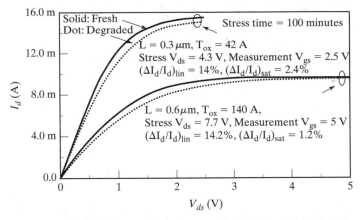

Figure 3-37   Hot-carrier effects cause the *I-V* characteristics of an NMOS transistor to degrade from extensive usage (from [McGaughy98]).

CMOS Latchup

The MOS technology contains a number of intrinsic bipolar transistors. These are especially troublesome in CMOS processes, where the combination of wells and substrates results in the formation of parasitic *n-p-n-p* structures. Triggering these thyristorlike devices leads to a shorting of the V_{DD} and V_{SS} lines, usually resulting in a destruction of the chip, or at best a system failure that can only be resolved by a power-down.

Consider the *n*-well structure of Figure 3-38a.[9] The *n-p-n-p* structure is formed by the source of the NMOS, the *p*-substrate, the *n*-well, and the source of the PMOS. A circuit equivalent is shown in Figure 3-38b. When one of the two bipolar transistors gets forward biased

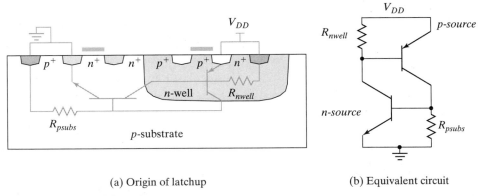

(a) Origin of latchup   (b) Equivalent circuit

Figure 3-38   CMOS latchup.

---

[9]This discussion on the mechanisms causing latchup is high level by necessity. Interested readers can find in-depth discussions in device textbooks such as [Streetman95].

(e.g., due to current flowing through the well, or substrate), it feeds current into the base of the other transistor. This positive feedback increases the current until the circuit either fails or burns out.

From the preceding analysis, the message to the designer is clear—to avoid latchup, the resistances R_{nwell} and R_{psubs} should be minimized. This can be achieved by providing numerous well and substrate contacts placed close to the source connections of the NMOS/PMOS devices. Devices carrying a lot of current (such as transistors in the I/O drivers) should be surrounded by *guard rings*. These circular well/substrate contacts, positioned around the transistor, reduce the resistance even further and reduce the gain of the parasitic bipolars. The latchup effect was especially critical in early CMOS processes. In recent years, process innovations and improved design techniques have all but eliminated the risks for latchup.

3.3.4   SPICE Models for the MOS Transistor

The complexity of the behavior of the short-channel MOS transistor and its many parasitic effects has led to the development of a wealth of models for varying degrees of accuracy and computing efficiency. In general, more accuracy also means more complexity and, as a result, increased run time. In this section, we briefly discuss the characteristics of the more popular MOSFET models, and we describe how to instantiate a MOS transistor in a circuit description.

SPICE Models

SPICE has three built-in MOSFET models selected by the LEVEL parameter in the model card. Unfortunately, all of these models have been rendered obsolete by the progression to short-channel devices. They should only be used for first-order analysis, and we therefore limit ourselves to a short discussion of their main properties:

- The LEVEL 1 SPICE model implements the *Shichman–Hodges model*, which is based on the square law, long-channel expressions derived earlier in this chapter. It does not handle short-channel effects.
- The LEVEL 2 model is a geometry-based model, which uses detailed-device physics to define its equations. It handles effects such as velocity saturation, mobility degradation, and drain-induced barrier lowering. Unfortunately, including all 3D effects of an advanced submicron process in a pure physics-based model becomes complex and inaccurate.
- LEVEL 3 is a semiempirical model. It relies on a mixture of analytical and empirical expressions, and it uses measured device data to determine its main parameters. It works quite well for channel lengths down to 1 µm.

In response to the inadequacy of the built-in models, SPICE vendors and semiconductor manufacturers have introduced a wide range of accurate, albeit proprietary, models. A complete description of each would take more space than we can allow, so we refer the interested reader to the extensive literature on this topic [e.g., Vladimirescu93].

The BSIM3V3 SPICE Model

The confusing situation of having to use a different model for each manufacturer has fortunately been partially resolved by the adoption of the BSIM3v3 model as an industrywide standard for the modeling of deep-submicron MOSFET transistors. The **B**erkeley **S**hort-Channel **IGFET M**odel (or BSIM in short) provides a model that is analytically simple and is based on a 'small' number of parameters, which normally are extracted from empirical data. Its popularity and accuracy make it the natural choice for all the simulations presented in this book.

A full-fledged BSIM3v3 model (denoted as LEVEL 49) contains over 200 parameters, the majority of which are related to the modeling of second-order effects. Fortunately, understanding the intricacies of all these parameters is not a requirement for the digital designer. We present only an overview of the parameter categories. (See Table 3-6.) The *Bin* category deserves extra attention. Providing a single set of parameters that is acceptable over all possible device dimensions is nearly impossible. Instead, the manufacturer provides a set of models, each of which is valid for a limited region delineated by LMIN, LMAX, WMIN, and WMAX (called a bin). It typically is left to the user to select the correct bin for a particular transistor.

We refer the interested reader to the BSIM3v3 documentation provided on the web site of this textbook (http://bwrc.eecs.berkeley.edu/IcBook) for a complete description of the model parameters and equations. The LEVEL-49 models for our generic 0.25 μm CMOS process can be found there as well.

Table 3-6   BSIM3-V3 model parameter categories, and some important parameters.

Parameter Category	Description
Control	Selection of level and models for mobility, capacitance, and noise LEVEL, MOBMOD, CAPMOD
DC	Parameters for threshold and current calculations VTH0, K1, U0, VSAT, RSH,
AC & Capacitance	Parameters for capacitance computations CGS(D)O, CJ, MJ, CJSW, MJSW
dW and dL	Derivation of effective channel length and width
Process	Process parameters such as oxide thickness and doping concentrations TOX, XJ, GAMMA1, NCH, NSUB
Temperature	Nominal temperature and temperature coefficients for various device parameters TNOM
Bin	Bounds on device dimensions for which model is valid LMIN, LMAX, WMIN, WMAX
Flicker Noise	Noise model parameters

Table 3-7   SPICE transistor parameters.

Parameter Name	Symbol	SPICE Name	Units	Default Value
Drawn Length	L	L	m	–
Effective Width	W	W	m	–
Source Area	$AREA$	AS	m^2	0
Drain Area	$AREA$	AD	m^2	0
Source Perimeter	$PERIM$	PS	m	0
Drain Perimeter	$PERIM$	PD	m	0
Squares of Source Diffusion		NRS	–	1
Squares of Drain Diffusion		NRD	–	1

Transistor Instantiation

The parameters that can be specified for an individual transistor are enumerated in Table 3-7. Not all these parameters have to be defined for each transistor. SPICE assumes default values (which often are zero!) for the missing factors.

WARNING: It is hard to expect accuracy from a simulator when the circuit description provided by the designer does not contain the necessary details. For example, you must accurately specify the area and the perimeter of the source and drain regions of the devices when performing a performance analysis. Lacking this information, which is used for the computation of the parasitic capacitances, your transient simulation will not be very useful. Similarly, often it is necessary to painstakingly define the value of the drain and source resistance. The NRS and NRD values multiply the sheet resistance specified in the transistor model for an accurate representation of the parasitic series source and drain resistance of each transistor.

Example 3.11   SPICE Description of a CMOS Inverter

In this example, a SPICE description of a CMOS inverter consisting of an NMOS and a PMOS transistor is given. Transistor M1 is an NMOS device of model-type (and bin) *nmos.*1 with its drain, gate, source, and body terminals connected to nodes *nvout*, *nvin*, 0, and 0, respectively. Its gate length is the minimum allowed in this technology (0.25 μm). The '+' character at the start of line 2 indicates that this line is a continuation of the previous one.

The PMOS device of type *pmos.1*, connected between nodes *nvout, nvin, nvdd,* and *nvdd* (D, G, S, and B, respectively), is three times wider, which reduces the series resistance, but increases the parasitic diffusion capacitances as the area and perimeter of the drain and source regions go up.

Finally, the *.lib* line refers to the file that contains the transistor models:

> M1 nvout nvin 0 0 nmos.1 W=0.375U L=0.25U
>
> +AD=0.24P PD=1.625U AS=0.24P PS=1.625U NRS=1 NRD=1
>
> M2 nvout nvin nvdd nvdd pmos.1 W=1.125U L=0.25U
>
> +AD=0.7P PD=2.375U AS=0.7P PS=2.375U NRS=0.33 NRD=0.33
>
> .lib 'c:\Design\Models\cmos025.l'

3.4 A Word on Process Variations

The preceding discussions assumed that a device is adequately modeled by a single set of parameters. In reality, the parameters of a transistor vary from wafer to wafer, or even between transistors on the same die, depending upon the position. This observed random distribution between supposedly identical devices is primarily the result of two factors:

1. Variations in the process parameters, such as impurity concentration densities, oxide thicknesses, and diffusion depths, caused by nonuniform conditions during the deposition and/or the diffusion of the impurities. These result in diverging values for sheet resistances, and transistor parameters such as the threshold voltage.
2. Variations in the dimensions of the devices, mainly resulting from the limited resolution of the photolithographic process. This causes deviations in the (W/L) ratios of MOS transistors and the widths of interconnect wires.

Observe that quite a number of these deviations are totally uncorrelated. For instance, variations in the length of an MOS transistor are unrelated to variations in the threshold voltage because both are set by different process steps.

- These process variations impact the parameters that determine the circuit performance. Consider, for instance, the transistor current, and its parameters.
- The *threshold voltage* V_T can vary for numerous reasons: changes in oxide thickness, substrate, polysilicon and implant impurity levels, and the surface charge. Accurate control of the threshold voltage is an important goal for many reasons. While in the past, thresholds could vary by as much as 50%, state-of-the-art digital processes now manage to control the thresholds to within 25–50 mV.
- The main cause for variations in the process transconductance k_n' is changes in oxide thickness. Variations can also occur in the mobility, but to a lesser degree.
- Variations in W and L. These are mainly caused by the lithographic process. Observe that variations in W and L are totally uncorrelated because the first is determined in the field-

oxide step, while the second is defined by the polysilicon definition and the source and drain diffusion processes.

The measurable effect of the process variations may be a substantial deviation of the circuit behavior from the nominal or expected response, and this could be in either positive or negative directions. This poses an important economic dilemma for the designer. Assume, for instance, that you are asked to design a microprocessor running at a clock frequency of 3 GHz. It is important, economically, that the majority of the manufactured dies meet that performance requirement. One way to achieve that goal is to design the circuit assuming worst case values for all possible device parameters. Although safe, this approach is prohibitively conservative and results in severely overdesigned and therefore uneconomical circuits.

To help the designer make a decision on how much margin to provide, the device manufacturer commonly provides fast and slow device models in addition to the nominal ones. These result in larger (or smaller) currents than expected, corresponding to the 3σ variation.

Example 3.12   MOS Transistor Process Variations

To illustrate the possible impact of process variations on the performance of an MOS device, consider a minimum-size NMOS device in our generic 0.25 μm CMOS process. (In Chapter 5, we will establish that the speed of the device is proportional to the drain current that can be delivered.)

Assume initially that $V_{GS} = V_{DS} = 2.5$ V. From earlier simulations, we know that this produces a drain current of 220 μA. The nominal model is now replaced by the fast and slow models, which modify the length and width ($\pm10\%$), threshold (±60 mV), and oxide thickness ($\pm5\%$) parameters of the device. Simulations produce the following data:

> Fast: $I_d = 265$ μA: +20%
> Slow: $I_d = 182$ μA: −17%

Let us now proceed one step further. The supply voltage delivered to a circuit is by no means a constant, either. For instance, the voltage delivered by a battery can drop off substantially towards the end of its lifetime. In practice, a variation of 10% in the supply voltage may well be expected:

> Fast + $V_{dd} = 2.75$ V: $I_d = 302$ μA: +37%
> Slow + $V_{dd} = 2.25$ V: $I_d = 155$ μA: −30%

The current levels and the associated circuit performance can thus vary by nearly 100% between the extreme cases. To guarantee that the fabricated circuits meet the performance requirements under all circumstances, we must make the transistor 42% (= 220 μA/ 155 μA) wider then would be required in the nominal case. This translates into a severe area penalty.

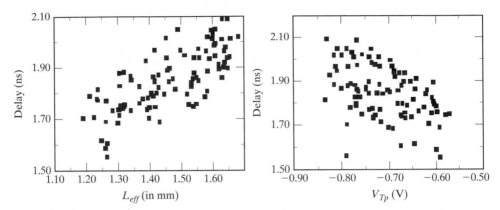

Figure 3-39 Distribution plots of speed of adder circuit as a function of varying device parameters, as obtained by a Monte Carlo analysis. The circuit is implemented in a 2 μm (nominal) CMOS technology (Courtesy of Eric Boskin, UCB, and ATMEL corp.).

Fortunately, in reality these worst (or best) case conditions occur only very rarely. The probability that all parameters would assume their worst case values simultaneously is very low, and most designs will display a performance centered around the nominal design. The art of *designing for manufacturability* is to center the nominal design so that the vast majority of the fabricated circuits (e.g., 98%) will meet the performance specifications, while keeping the area overhead minimal.

Specialized design tools to help meet this goal are available. For example, the Monte Carlo analysis approach [Jensen91] simulates a circuit over a wide range of randomly chosen values for the device parameters. The result is a distribution plot of design parameters (such as speed, or sensitivity to noise) that can help to determine if the nominal design is economically viable. Examples of such distribution plots are shown in Figure 3-39, which graphically demonstrates the impact of variations in the effective transistor channel length and the PMOS transistor thresholds on the speed of an adder cell. As can be seen in the figure, technology variations can have a substantial impact on the performance parameters of a design.

One important conclusion from the preceding discussion is that SPICE simulations should be treated with care. The device parameters presented in a model represent average values, measured over a batch of manufactured wafers. Actual implementations are bound to differ from the simulation results, and for reasons other than imperfections in the modeling approach. Furthermore, be aware that temperature variations on the die can present yet another source of parameter deviations. Optimizing an MOS circuit with SPICE to a resolution level of a picosecond or a microvolt is not worth the effort.

3.5 Perspective—Technology Scaling

Over the last few decades, we have observed a spectacular increase in integration density and computational complexity in digital integrated circuits. As discussed earlier, applications that were considered implausible yesterday are already forgotten today. Underlying this revolution

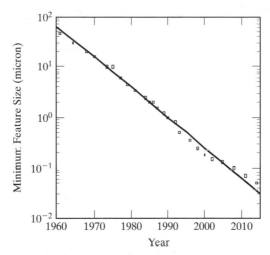

Figure 3-40   Evolution of (average) minimum channel length of MOS
transistors over time. Dots represent observed or projected (2000 and beyond)
values. The continuous line represents a scaling scenario that reduces
the minimum feature with a factor 2 every 5 years.

are the advances in device manufacturing technology, which allow for a steady reduction in the minimum feature size, such as the minimum transistor channel length realizable on a chip. To illustrate this point, Figure 3-40 plots the evolution of the (average) minimum device dimensions, starting from the 1960s and projecting into the 21st century. We observe a reduction rate of approximately 13% per year, halving every 5 years. Moreover, there is no real sign of a slow-down in sight, and the breathtaking pace may continue well into the foreseeable future.

A pertinent question is how this continued reduction in feature size influences the operating characteristics and properties of the MOS transistor, and indirectly, the critical digital design metrics, such as switching frequency and power dissipation. A first-order projection of this behavior is called a *scaling analysis,* the topic of this section. In addition to the minimum device dimension, we have to consider the supply voltage as a second independent variable in such a study. Different scaling scenarios result due to these two independent variables changing with respect to each other. [See Dennard74, Baccarani84.]

Three different models are analyzed in Table 3-8. To make the results tractable, we assume that all device dimensions scale by the same factor S (with $S > 1$ for a reduction in size). This includes the width and length of the transistor, the oxide thickness, and the junction depths. Similarly, we assume that all voltages, including the supply voltage and the threshold voltages, scale by a same ratio U. The relations governing the scaling behavior of the dependent variables are tabulated in column 2. Observe that this analysis only considers short-channel devices with a linear dependence between control voltage and saturation current (as expressed by Eq. (3.38)). We discuss each scenario, in turn.

Table 3-8 Scaling scenarios for short-channel devices.

Parameter	Relation	Full Scaling	General Scaling	Fixed-Voltage Scaling
W, L, t_{ox}		$1/S$	$1/S$	$1/S$
V_{DD}, V_T		$1/S$	$1/U$	1
N_{SUB}	V/W_{depl}^2	S	S^2/U	S^2
$Area$/Device	WL	$1/S^2$	$1/S^2$	$1/S^2$
C_{ox}	$1/t_{ox}$	S	S	S
C_{gate}	$C_{ox}WL$	$1/S$	$1/S$	$1/S$
k_n, k_p	$C_{ox}W/L$	S	S	S
I_{sat}	$C_{ox}WV$	$1/S$	$1/U$	1
$Current\ Density$	$I_{sat}/Area$	S	S^2/U	S^2
Ron	V/I_{sat}	1	1	1
$Intrinsic\ Delay$	$R_{on}C_{gate}$	$1/S$	$1/S$	$1/S$
P	$I_{sat}V$	$1/S^2$	$1/U^2$	1
$Power\ Density$	$P/Area$	1	S^2/U^2	S^2

Full Scaling (Constant Electrical Field Scaling)

In this ideal model, voltages and dimensions are scaled by the same factor S. The goal is to keep the electrical field patterns in the scaled device identical to those in the original device. Keeping the electrical fields constant ensures the physical integrity of the device and avoids breakdown or other secondary effects. This scaling leads to greater device density (*Area*), higher performance (*Intrinsic Delay*), and reduced power consumption (*P*). The effects of full scaling on the device and circuit parameters are summarized in the third column of Table 3-8. We use the intrinsic time constant, which is the product of the gate capacitance and the "on" resistance, as a measure for the performance. The analysis shows that the "on" resistance remains constant due to the simultaneous scaling of voltage swing and current level. The improved performance is solely due to the reduced capacitance. The results demonstrate clearly the beneficial effects of scaling—the speed of the circuit increases in a linear fashion, while the power of the gate scales down quadratically![10]

Fixed-Voltage Scaling

In reality, full scaling is not a feasible option. In order to keep new devices compatible with existing components, voltages cannot be scaled arbitrarily—providing for multiple supply voltages adds considerably to the cost of a system. As a result, voltages have not been scaled down along with feature sizes, and designers adhere to well-defined standards for supply voltages and signal levels. As illustrated in Figure 3-41, 5 V was the de facto standard for all digital components up to the early 1990s, and a *fixed-voltage scaling model* was followed.

New standards such as 3.3 V and 2.5 V only began to take a hold with the introduction of the 0.5 μm CMOS technology. Today, we see a much closer tracking between voltage and device dimension. The reason for this change in operation model can be explained, in part, with the aid of the fixed-voltage scaling model, summarized in the fifth column of Table 3-8. In a velocity-saturated device, keeping the voltage constant while scaling the device dimensions

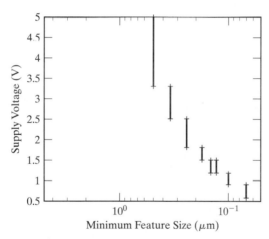

Figure 3-41 Evolution of min and max supply-voltage in digital integrated circuits as a function of feature size.

---

[10]Some assumptions were made when deriving Table 3.8:
  1. It was assumed that the carrier mobilities are not affected by the scaling.
  2. The substrate doping N_{sub} was scaled so that the maximum depletion-layer width was reduced by a factor S.
  3. It also was assumed that the delay of the device is mainly determined by the intrinsic capacitance (the gate capacitance), and that other device capacitances, (such as the diffusion capacitances), scale appropriately. This assumption is approximately true for the full-scaling case, but not for fixed-voltage scaling, where C_{diff} scales as $1/\sqrt{S}$.

does not give a performance advantage over the full-scaling model, but instead comes with a major power penalty. The gain of an increased current is simply off-set by the higher voltage level, which only hurts the power dissipation. This scenario is very different from the situation that existed when transistors were operating in the long-channel mode, and the current was a quadratic function of the voltage (recall Eq. (3.29)). Keeping the voltage constant under these circumstances gives a distinct performance advantage, because it causes a net reduction in "on" resistance. Other reasons for scaling the supply voltage with the technology include physical phenomena such as the hot-carrier effect and oxide breakdown. These effects contributed substantially in making the fixed-voltage scaling model unsustainable.

Problem 3.2   Scaling of Long-channel Devices

Demonstrate that for a long-channel transistor, full-voltage scaling results in a reduction of the intrinsic delay with a factor S^2, while increasing the power dissipation per device by S. Reconstruct Table 3-8, assuming that the current is a quadratic function of the voltage (Eq. (3.29)).

WARNING: The image described in the previous section represents a first-order model. Increasing the supply voltage still offers some performance benefit for short-channel transistors. This is apparent in Figure 3-28 and Table 3-3, which show a reduction of the equivalent "on" resistance with increasing supply voltage—even for the high voltage range. Yet, this effect, which is mostly due to the channel-length modulation, is secondary and far smaller than what would be obtained in long-channel devices.

     The reader should remember this warning throughout the study of scaling. The goal is to discover first-order trends, which implies ignoring second-order effects such as mobility degradation, series resistance, etc.

General Scaling

We observe in Figure 3-41 that the supply voltages, while moving downwards, are not scaling as fast as the technology. For instance, for the technology scaling from 0.5 μm to 0.1 μm, the maximum supply voltage only reduces from 5 V to 1.5 V. The obvious question, then, is why not stick to the full-scaling model, when keeping the voltage higher does not yield any convincing benefits? This departure is motivated by the following points:

- Some of the intrinsic device voltages—such as the silicon bandgap and the built-in junction potential—are material parameters and thus cannot be scaled.
- The scaling potential of the transistor threshold voltage is limited. Making the threshold too low makes it difficult to turn off the device completely. This is aggravated by the large process variation in the value in the threshold, even on the same wafer.

     For these and other reasons, a more general scaling model is needed, with dimensions and voltages scaled independently. This general scaling model is shown in the fourth column of Table 3-8, where device dimensions are scaled by a factor S, while voltages are reduced by a factor U. When the voltage is held constant, $U = 1$, the scaling model reduces to the fixed-voltage model. Note that the general-scaling model offers a performance scenario identical to the full- and the fixed-scaling models, while its power dissipation lies between the two models (for $S > U > 1$).

Verifying the Model

To summarize, in Table 3.9 we have combined the characteristics of some of the most recent CMOS processes and projections of future ones. Observe how the operating voltages are being continuously reduced with diminishing device dimensions. As predicted by the scaling model, the maximum drive current per unit width remains approximately constant. Maintaining this level of drive in the presence of a reduced supply voltage requires an aggressive lowering of the threshold voltage, which translates into a rapid increase of the subthreshold leakage current.

On the basis of the data in the table, we reasonably conclude that both integration density and performance will continue to increase. How long they will continue to increase is unknown, and is the cause of major speculation. Experimental sub-20-nm CMOS devices have proven to be operational in the laboratories and to display current characteristics that are surprisingly close to present-day transistors. These transistors, while working along similar concepts as the current MOS devices, look very different from the structures we are familiar with, and require some substantial *device engineering*. Figure 3-42, for example, shows a potential transistor structure, the Berkeley FinFET dual-gated transistor, which has proven to be operational up to very small channel lengths.

Another option is the *vertical transistor*. Even while the addition of many metal layers has turned the integrated circuit into a truly three-dimensional artifact, the transistor itself is still laid out mostly in a horizontal plane. This forces the device designer to jointly optimize packing density and performance parameters. By rotating the device so that the drain is on the top and the source is on the bottom, these concerns are separated: packing density still is dominated by horizontal dimensions, while performance issues are mostly determined by vertical spacings. (See Figure 3-43.) Operational devices of this type have been fabricated with channel lengths substantially below 0.1 μm [Eaglesham00].

Table 3-9 MOSFET technology projection for high performance logic (from [SIA01]).

Year of Introduction	2001	2003	2005	2007	2010	2013	2016
Drawn channel length (nm)	90	65	45	35	25	18	13
Physical channel length (nm)	65	45	32	25	18	13	9
Gate oxide (nm)	2.3	2.0	1.9	1.4	1.2	1.0	0.9
V_{DD} (V)	1.2	1.0	0.9	0.7	0.6	0.5	0.4
NMOS I_{Dsat} (μA/μm)	900	900	900	900	1200	1500	1500
NMOS I_{leak} (μA/μm)	0.01	0.07	0.3	1	3	7	10

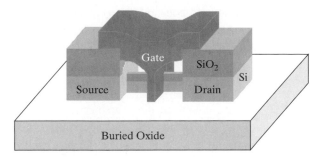

Figure 3-42   FinFET dual-gated transistor with 25 nm channel length [Huang99].

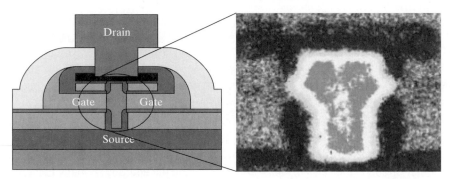

Figure 3-43   Vertical transistor with dual gates. The photo on the right shows an enlarged view of the channel area.

Integrated circuits integrating more then one billion transistors clocked at speeds of tens of GHz's seem to be well under way. Whether this will continue remains an open question. Even though it might be technologically feasible, other factors have to be considered in such an undertaking. We may wonder whether such a part can be manufactured in an economical way. Current semiconductor plants cost over $5 billion, and this price is expected to rise substantially with smaller feature sizes. Design considerations also play a role. The level of power consumption of such a component might be prohibitive, and the growing role of interconnect parasitics might put an upper bound on performance. Finally, system considerations may determine what level of integration is really desirable. All in all, it is obvious that the field of semiconductor circuit design will continue to face exciting challenges in the future.

3.6 Summary

In this chapter, we have presented a comprehensive overview of the operation of the MOSFET transistor, the semiconductor device at the core of virtually all contemporary digital integrated circuits. Besides attempting to provide an intuitive understanding of its behavior, we have presented a variety of modeling approaches, ranging from simple models—useful for a first-order manual analysis of the circuit operation—to complex SPICE models. These models will be used extensively in later chapters, where we look at the fundamental building blocks of digital circuits. We started off with a short discussion of the semiconductor diode, one of the most dominant parasitic circuit elements in CMOS designs.

- The static behavior of the junction diode is well described by the ideal diode equation, which states that the current is an *exponential function of the applied voltage bias*.
- In reverse-biased mode, the depletion-region space charge of the diode can be modeled as a *nonlinear voltage-dependent capacitance*. This is particularly important as the omnipresent source-body and drain-body junctions of the MOS transistors all operate in this mode. A linearized large-scale model for the depletion capacitance was introduced for manual analysis.
- The MOS(FET) transistor is a voltage-controlled device, in which the controlling gate terminal is insulated from the conducting channel by an SiO_2 capacitor. Based on the value of the gate-source voltage with respect to a threshold voltage VT, three operation regions have been identified: *cut-off, linear,* and *saturation*. One of the most enticing properties of the MOS transistor, which makes it particularly amenable to digital design, is that it approximates a voltage-controlled switch—when the control voltage is low, the switch is nonconducting (open); for a high control voltage, a conducting channel is formed, and the switch can be considered closed. This two-state operation matches the concepts of binary digital logic.
- The continuing reduction of the device dimensions to the submicron range has introduced some substantial deviations from the traditional long-channel MOS transistor model. The most important one is the *velocity saturation* effect, which changes the dependence of the transistor current with respect to the controlling voltage from *quadratic to linear*. Models for this effect, as well as other second-order parasitics, were introduced. One particularly important effect is the *subthreshold conduction*, which causes devices to conduct current even when the control voltage drops below the threshold.
- The dynamic operation of the MOS transistor is dominated by the *device capacitors*. The main contributors are the gate capacitance and the capacitance formed by the depletion regions of the source and drain junctions. The minimization of these capacitances is the prime requirement in high-performance MOS design.
- SPICE models and their parameters have been introduced for all devices. It was observed that these models represent an average behavior and can vary over a single wafer or die.
- The MOS transistor is expected to dominate the digital integrated circuit scene for at least the next decade. Continued scaling will lead to device sizes of approximately 0.03 micron by the year 2010, and logic circuits integrating more than 1 billion transistors on a die.

3.7 To Probe Further

Semiconductor devices have been discussed in numerous books, reprint volumes, tutorials, and journal articles. The *IEEE Transactions on Electron Devices* is the premier journal, and most of the state-of-the-art devices and their modeling are discussed there. The proceedings of the

International Electron Devices Meeting (IEDM) provide another valuable resource. The books (such as [Streetman95] and [Pierret96]) and journal articles that follow contain excellent discussions of the semiconductor devices of interest or refer to specific topics brought up in the course of this chapter.

References

[Baccarani84] G. Baccarani, M. Wordeman, and R. Dennard, "Generalized Scaling Theory and Its Application to 1/4 Micrometer MOSFET Design," *IEEE Trans. Electron Devices*, ED-31(4): p. 452, 1984.

[Banzhaf92] W. Bhanzhaf, *Computer Aided Analysis Using PSPICE,* 2nd ed., Prentice Hall, 1992.

[Dennard74] R. Dennard et al., "Design of Ion-Implanted MOSFETS with Very Small Physical Dimensions," *IEEE Journal of Solid-State Circuits,* SC-9, pp. 256–258, 1974.

[Eaglesham99] D. Eaglesham, "*0.18 μm CMOS and Beyond,*" Proceedings 1999 Design Automation Conference, pp. 703–708, June 1999.

[Howe97] R. Howe and S. Sodini, *Microelectronics: An Integrated Approach,* Prentice Hall, 1997.

[Hu92] C. Hu, "IC Reliability Simulation," *IEEE Journal of Solid State Circuits*, vol. 27, no. 3, pp. 241–246, March 1992.

[Huang99] X. Huang, W. C. Lee, C. Kuo, D. Hisamoto, L. Chang, J. Kedzierski, E. Anderson, H. Takeuchi, Y. K. Choi, K. Asano, V. Subramanian, T. J. King, J. Bokor, and C. Hu, "Sub 50-nm FinFET: PMOS," *International Electron Devices Meeting*, pp. 67–70, 1999.

[Jensen91] G. Jensen et al., "Monte Carlo Simulation of Semiconductor Devices," *Computer Physics Communications*, 67, pp. 1–61, August 1991.

[Ko89] P. Ko, "Approaches to Scaling," in *VLSI Electronics: Microstructure Science,* vol. 18, chapter 1, pp. 1–37, Academic Press, 1989.

[McGaughy98] J. F. Chen, B.W. McGaughy, and C. Hu, "Statistical Variation of NMOSFET Hot-carrier Lifetime and Its Impact on Digital Circuit Reliability," *International Electron Device Meeting Technical Digest (IEDM)*, pp. 29–32, 1995.

[Nagel75] L. Nagel, "SPICE2: a Computer Program to Simulate Semiconductor Circuits," Memo ERL-M520, Dept. Elect. and Computer Science, University of California at Berkeley, 1975.

[Pierret96] R. Pierret, *Semiconductor Device Fundamentals*, Addison-Wesley, 1996.

[SIA01] *International Technology Roadmap for Semiconductors*, http://www.sematech.org, 2001.

[Streetman95] B. Streetman, *Solid State Electronic Devices*, Prentice Hall, 1995.

[Thorpe92] T. Thorpe, *Computerized Circuit Analysis with SPICE*, John Wiley and Sons, 1992.

[Toh88] K. Toh, P. Ko, and R. Meyer, "An Engineering Model for Short-Channel MOS Devices," *IEEE Journal of Solid State Circuits*, vol. 23, no. 4, pp. 950–958, August 1988.

[Vladimirescu93] A. Vladimirescu, *The SPICE Book*, John Wiley and Sons, 1993.

Exercises

http://bwrc.eecs.berkeley.edu/IcBook provides up-to-date and challenging problem sets and exercises on diodes and MOS devices.

Circuit Simulation

> *Properties of Circuit Simulation*
>
> *Circuit Simulation Models*

The primary expectation of a designer with respect to design automation is the availability of accurate and fast analysis tools. The first computer-aided design (CAD) tool to gain wide acceptance was the SPICE circuit simulator, which is undoubtedly the most widely used computer aid for the design of (digital) circuits ever. SPICE was originally developed at the University of California at Berkeley [Nagel75], which released three major versions over the years (Spice-1, 2, and 3). Subsequently, numerous SPICE derivatives (such as PSPICE or HSPICE) have been introduced, bringing commercial support to a university-conceived tool. In recent years, we have witnessed the emergence of a number of competing circuit simulators (such as SPECTRE [Spectre], and ELDO [Eldo]), which bring with them better performance or accuracy through the adoption of new algorithms and modern programming techniques.

In the course of this text, we use circuit simulation extensively to illustrate the basic concepts of digital circuits and to validate our manual models. The assignments associated with the book also rely heavily on circuit simulation. Hence, you should be very familiar with the capabilities and peculiarities of simulation and analysis at this level of abstraction.

Properties of Circuit Simulation

We begin with a summary of some important properties of circuit simulation:

- When analyzing a digital network using a circuit simulator, the resulting voltage and current *signals* are represented as *continuous waveforms*.
- In a transient analysis, *time* seems to be a *continuous* variable, and for all practical purposes, it can be thought of as such. In reality, the simulator executed on a digital computer evaluates only a limited number of time points and obtains the intermediate data points by interpolation.

Executing a transient simulation means solving a set of differential equations at each time point. To make matters worse, the accurate modeling of semiconductor devices requires the introduction of nonlinearities, such as the current equations for MOS devices or the diode-capacitance model. For instance, Figure B-1 shows the network that must be solved when performing a transient analysis on a CMOS inverter. All capacitors and current sources in this model display a strong nonlinear behavior. The analytical solution of the set of nonlinear differential equations describing such a network is generally based on iteration, which is computationally expensive. At each time step, an initial guess is made of the node values based on the values of the previous time step. This estimate is iteratively refined until some predefined error criterion is met. The tighter the error bound, the better the accuracy, but the more iterations needed.

Simulation Modes

SPICE offers a range of analysis modes, each of them targeting a particular aspect of design. Two simulation modes are of particular interest to the digital designer and thus deserve some explanation.

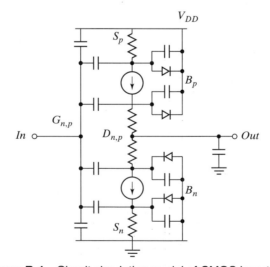

Figure B-1   Circuit simulation model of CMOS inverter.

DC Sweep   The DC analysis of a circuit simulator calculates the DC (or static) bias of a circuit. When computing the voltage-transfer characteristic of a gate, we are interested in a set of bias points over a range of input values. The DC-sweep function provides a convenient way of producing such a curve. When you "sweep" a source, the simulator starts with one value for it, calculates the DC bias point, then increments the value and does another bias-point calculation. This proceeds until a final value is reached. The net effect is that a single command yields the same result as many individual bias operations. A typical dc-sweep command line looks like the following:

.DC VIN 0 0.5 2.5

It sweeps the voltage source VIN from 0 to 2.5 V with increments of 0.5 V.

Transient Analysis   This is the most frequently used analysis mode in digital circuit simulation. Transient analysis simulates the operation of your circuit as time progresses. It is the ideal tool for determining transient parameters, such as propagation delay, rise and fall times, and power dissipation. The analysis is controlled along the following line:

.TRAN 1ps 250ps

It specifies the time interval of the simulation (250 ps), as well as when data was recorded (every ps). The latter is not directly related to the exact time points the simulator actually computes— the simulator itself decides this based on the amount of activity in the circuits.

SPICE (or equivalent simulators) provide a whole range of extra tools to help the designer get worthwhile answers efficiently. The web site for this book offers a wide range of useful tips and information for the interested reader.

Device Models

While most circuit simulators use similar solution techniques, the added value lies in the robustness, efficiency, and accuracy of the device models. For instance, it is possible to come up with very elaborate and accurate models that are completely useless because of poor convergence behavior. The complexity of the behavior of the short-channel MOS transistor and its many parasitic effects has led to the development of a wealth of models for varying degrees of accuracy and computing efficiency, some of which were introduced in the previous chapter.

This realization enticed semiconductor manufacturing companies to come up with their own proprietary models for their transistors. This led to a smorgasboard of models, whose quality was hard to judge—every company naturally claimed that their model was the best. This situation also made it difficult for a customer to move between processes and manufacturers. Fortunately, the Berkeley Short-Channel IGFET Model (or BSIM), developed at the University of California at Berkeley in the 1990s, changed the landscape drastically. BSIM3v3 was adopted as the industry standard, and now it is used extensively by all semiconductor manufacturers. A more complete model, BSIM4, was released in 2000. BSIM models that address alternative processes such as SOI (BSIMSOI) also have been developed. More information about BSIM can be found at the Berkeley web site ([Bsim34]).

CAUTION: Novice designers often tend to have an almost religious faith in simulation results. Thus, it is not rare to see designers optimizing propagation delays of CMOS circuits into the subpicosecond range. One should be aware that the simulation can diverge from reality due to a number of error sources, such as inaccuracies in the device models, deviations in the device parameters, and parasitic resistances and capacitances. The actual and predicted circuit behaviors might further diverge because of process variations over the die or temperature variations. Designers should, therefore, allow for a substantial margin between the design constraints and simulation results. The size of that margin obviously depends on the modeling effort that was performed—how precise and in-depth the device models are and how accurate the extracted parasitics are. Margins of up to 10% are quite normal.

   While circuit simulation has been the workhorse of the digital designer for a long time, the computational complexity makes it impractical for larger circuits. Therefore, it is mostly used for the analysis of the critical parts of a design. Higher level simulators are employed for the overall analysis. These simulators trade off simulation performance against accuracy or abstraction. They are discussed in a later Design Methodology Insert.

To Probe Further

SPICE has been the topic of many textbooks [e.g., Thorpe92, Vladimirescu93]. In addition, a lot of information such as user manuals and user guides can be found on the Web [e.g., Rabaey96].

References

[BSIM34] "BSIM, The Berkeley Short-Channel IGFET Model," *http://www-device.eecs.berkeley.edu/~bsim3*.

[Eldo] "Eldo AMS Simulation," *http://www.mentor.com/eldo/overview.html,* Mentor Graphics.

[Huang99] X. Huang, W. C. Lee, C. Kuo, D. Hisamoto, L. Chang, J. Kedzierski, E. Anderson, H. Takeuchi, Y. K. Choi, K. Asano, V. Subramanian, T.J. King, J bokor, and C. Hu, "Sub 50-nm FinFET: PMOS," International Electron Devices Meeting, pp. 67–70, 1999.

[Nagel75] L. Nagel, "SPICE2: a Computer Program to Simulate Semiconductor Circuits," Memo ERL-M520, Dept. Elect. and Computer Science, University of California at Berkeley, 1975.

[Rabaey96] J. Rabaey (et al), "The Spice Page," *http://bwrc.eecs.berkeley.edu/ICbook/spice*, 1996.

[Spectre] "Spectre Circuit Simulator," *http://www.cadence.com/datasheets/spectre_cir_sim.html,* Cadence.

[Thorpe92] T. Thorpe, *Computerized Circuit Analysis with SPICE*, John Wiley and Sons, 1992.

[Vladimirescu93] A. Vladimirescu, *The SPICE Book*, John Wiley and Sons, 1993.

C H A P T E R

4

The Wire

Determining and quantifying interconnect parameters

Introducing circuit models for interconnect wires

Detailed wire models for SPICE

Technology scaling and its impact on interconnect

4.1   Introduction

4.2   A First Glance

4.3   Interconnect Parameters—Capacitance, Resistance, and Inductance
    4.3.1   Capacitance
    4.3.2   Resistance
    4.3.3   Inductance

4.4   Electrical Wire Models
    4.4.1   The Ideal Wire
    4.4.2   The Lumped Model
    4.4.3   The Lumped *RC* Model
    4.4.4   The Distributed *rc* Line
    4.4.5   The Transmission Line

4.5   SPICE Wire Models
    4.5.1   Distributed rc Lines in SPICE
    4.5.2   Transmission Line Models in SPICE
    4.5.3   Perspective: A Look into the Future

4.6   Summary

4.7   To Probe Further

4.1 Introduction

Throughout most of the history of integrated circuits, on-chip interconnect wires were almost like second class citizens, only considered in special cases or when performing high-precision analysis. With the introduction of deep submicron semiconductor technologies, this picture has undergone rapid changes. The parasitic effects introduced by the wires display a scaling behavior that differs from the active devices such as transistors, and they tend to gain in importance as device dimensions are reduced and circuit speed is increased. In fact, they start to dominate some of the relevant metrics of digital integrated circuits such as speed, energy consumption, and reliability. This situation is aggravated by the fact that improvements in technology make the production of ever larger die sizes economically feasible, which results in an increase in the average length of an interconnect wire and in the associated parasitic effects. A careful and in-depth analysis of the role and behavior of the interconnect wire in a semiconductor technology is, therefore, not only desirable, but essential.

4.2 A First Glance

The designer of an electronic circuit has multiple choices in realizing the interconnections between the various devices that make up the circuit. State-of-the-art processes offer multiple layers of aluminum or copper, and at least one layer of polysilicon. Even the heavily doped n^+ or p^+ diffusion layers typically used for the realization of source and drain regions can be employed for wiring purposes. These wires appear in the schematic diagrams of electronic circuits as simple lines with no apparent impact on the circuit performance. From our discussion of the integrated circuit manufacturing process, it should be clear that this picture is too simplistic, and that the wiring of today's integrated circuits forms a complex geometry that introduces capacitive, resistive, and inductive parasitics. All three have multiple effects on the circuit's behavior:

1. They all cause an increase in propagation delay, or, equivalently, a drop in performance.
2. They all have an impact on the energy dissipation and the power distribution.
3. They all cause the introduction of extra noise sources, which affect the reliability of the circuit.

A designer can decide to play it safe and include all these parasitic effects in her analysis and design optimization process. This conservative approach is not very constructive, however, and most often it is not even feasible. First of all, a "complete" model is dauntingly complex and is only applicable to very small topologies. Hence, it is totally useless for today's integrated circuits, with their millions of circuit nodes. Furthermore, this approach has the disadvantage of "not seeing the forest for the trees," so to speak. The circuit behavior at a given circuit node is only determined by a few dominant parameters. Bringing all possible effects to bear may obscure the picture and turn the optimization and design process into a "trial-and-error" operation, rather than an enlightened and focused search.

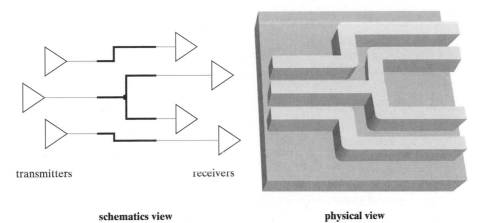

transmitters   receivers

schematics view   **physical view**

Figure 4-1 Schematic and physical views of wiring of bus network. The latter shows only a limited area (as indicated by the emphasis in the schematics).

Thus, it is important for the designer to have a clear insight into the parasitic wiring effects, their relative importance, and their models. This is best illustrated with a simple example, as shown in Figure 4-1. Each wire in a bus network connects a transmitter (or transmitters) to a set of receivers and is implemented as a chain of wire segments of various lengths and geometries. Assume that all segments are implemented on a single interconnect layer and isolated from the silicon substrate and from each other by a layer of dielectric material. (Be aware that the reality may be far more complex.)

A full-fledged circuit model, that takes into account the parasitic capacitance, resistance, and the inductance of the interconnections is shown in Figure 4-2a. Observe that these extra circuit elements are not located in a single physical point, but are distributed over the length of the wire. This is necessary when the length of the wire becomes significantly greater than its width. In addition, interwire parasitics are present, creating coupling effects between the different bus signals that were not present in the original schematics.

Analyzing the behavior of this schematic, which only models a small part of the circuit, is slow and cumbersome. Fortunately, substantial simplifications often can be made, including the following:

- Inductive effects can be ignored if the resistance of the wire is substantial enough—this is the case for long aluminum wires with a small cross section, for example, or if the rise and fall times of the applied signals are slow.
- When the wires are short, the cross section of the wire is large, or the interconnect material used has a low resistivity, a capacitance-only model can be used (see Figure 4-2b).
- Finally, when the separation between neighboring wires is large, or when the wires only run together for a short distance, interwire capacitance can be ignored, and all the parasitic capacitance can be modeled as capacitance to ground.

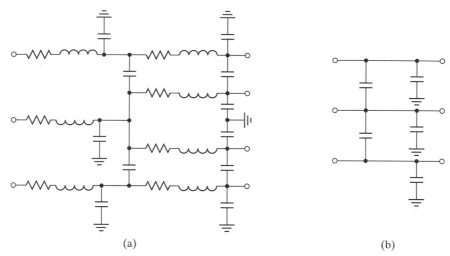

(a)   (b)

Figure 4-2   Wire models for the circuit of Figure 4-1. Model (a) considers most of the wire parasitics (with the exception of interwire resistance and mutual inductance), while model (b) only considers capacitance.

Obviously, the latter cases are the easiest to model, analyze, and optimize. The experienced designer knows to differentiate between dominant and secondary effects. The goal of this chapter is to present the basic techniques of estimating the values of the various interconnect parameters, simple models to evaluate their impact, and a set of rules of thumb for deciding when and where a particular model or effect should be considered.

4.3 Interconnect Parameters—Capacitance, Resistance, and Inductance

4.3.1   Capacitance

An accurate modeling of the wire capacitance(s) in a state-of-the-art integrated circuit is a non-trivial task, and even today it is the subject of advanced research. The task is further complicated by the fact that the interconnect structure of contemporary integrated circuits is three-dimensional, as was clearly demonstrated in the integrated-circuit cross section of Figure 2-8. The capacitance of such a wire is a function of its shape, its environment, its distance to the substrate, and the distance to surrounding wires. Rather than getting lost in complex equations and models, a designer typically will use an advanced extraction tool to get precise values of the interconnect capacitances of a completed layout. Most semiconductor manufacturers also provide empirical data for the various capacitance contributions, as measured from a number of test dies. Yet, some simple, first-order models come in handy to provide a basic understanding of the nature of interconnect capacitance and its parameters, and of how wire capacitance will evolve with future technologies.

Consider first a simple rectangular wire placed above the semiconductor substrate, as shown in Figure 4-3. If the width of the wire is substantially larger than the thickness of the

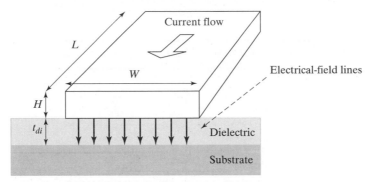

Figure 4-3   Parallel-plate capacitance model of interconnect wire.

insulating material, it may be assumed that the electrical-field lines are orthogonal to the capacitor plates, and that its capacitance can be modeled by the *parallel-plate capacitor model* (also called *area capacitance*). Under those circumstances, the total capacitance of the wire can be approximated as[1]

$$C_{int} = \frac{\varepsilon_{di}}{t_{di}}WL \tag{4.1}$$

where W and L are, respectively, the width and length of the wire, and t_{di} and ε_{di} represent the thickness of the dielectric layer and its permittivity. SiO_2 is the dielectric material of choice in integrated circuits, although some materials with lower permittivity, and thus lower capacitance, are coming into use. Examples of the latter are organic polyimides and aerogels. ε is typically expressed as the product of two terms, this is $\varepsilon = \varepsilon_r \varepsilon_0$. $\varepsilon_0 = 8.854 \times 10^{-12}$ F/m is the permittivity of free space, and ε_r the relative permittivity of the insulating material. Table 4-1 presents the relative permittivity of several dielectrics used in integrated circuits. In summary, the important message from Eq. (4.1) is that the capacitance is proportional to the overlap between the conductors and inversely proportional to their separation.

As with earlier examples, this model is too simplistic in actuality. To minimize the resistance of the wires while scaling technology, it is desirable to keep the cross section of the wire ($W \times H$) as large as possible. (This will become apparent in a later section.) On the other hand, small values of W lead to denser wiring and less area overhead. As a result, over the years, we have witnessed a steady reduction in the W/H ratio such that it has even dropped below unity in advanced processes. This is clearly visible on the process cross section of Figure 2-8. Under those circumstances, the parallel-plate model assumed earlier becomes inaccurate. The capacitance between the side walls of the wires and the substrate, called the *fringing capacitance*, can no longer be ignored and contributes to the overall capacitance. This effect is illustrated in Figure 4-4a. Presenting an exact model for this geometry is difficult. Therefore, as good

---

[1]To differentiate between distributed (per unit length) wire parameters versus total lumped values, we will use lower-case to denote the former and uppercase for the latter.

Table 4-1 Relative permittivity of some typical dielectric materials.

Material	ε_r
Free space	1
Aerogels	~1.5
Polyimides (organic)	3-4
Silicon dioxide	3.9
Glass-epoxy (PC board)	5
Silicon Nitride (Si_3N_4)	7.5
Alumina (package)	9.5
Silicon	11.7

engineering practice dictates, we use a simplified model that approximates the capacitance as the sum of two components (Figure 4-4b): a parallel-plate capacitance determined by the orthogonal field between a wire of width w and the ground plane, in parallel with the fringing capacitance

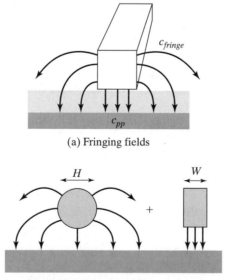

(a) Fringing fields

(b) Model of fringing-field capacitance

Figure 4-4 The fringing-field capacitance. The model decomposes the capacitance into two contributions: a parallel-plate capacitance, and a fringing capacitance, modeled by a cylindrical wire with a diameter equal to the thickness of the wire.

modeled by a cylindrical wire with a dimension equal to the interconnect thickness H. The resulting approximation, which is simple and works fairly well in practice, is

$$c_{wire} = c_{pp} + c_{fringe} = \frac{w\varepsilon_{di}}{t_{di}} + \frac{2\pi\varepsilon_{di}}{\log(t_{di}/H)} \tag{4.2}$$

where $w = W - H/2$ is a good approximation for the width of the parallel-plate capacitor. Numerous more accurate models (e.g., [Vdmeijs84]) have been developed over time, but these tend to be substantially more complex, and therefore beyond the scope of this discussion.

To illustrate the importance of the fringing-field component, Figure 4-5 plots the value of the wiring capacitance as a function of (W/t_{di}) (or indirectly of (W/H)). For larger values of (W/H), the total capacitance approaches the parallel-plate model. For (W/H) smaller than 1.5, the fringing component actually becomes the dominant component. The fringing capacitance can increase the overall capacitance by a factor of more than 10 for small line widths. It is interesting to observe that the total capacitance levels off to a constant value of approximately 1 pF/cm for line widths smaller than the insulator thickness. In other words, the capacitance is no longer a function of the width.

We have restricted our analysis thus far to the case of a single rectangular conductor placed over a ground plane. This structure, called a *microstripline*, used to be a good model for semiconductor interconnections when the number of interconnect layers was restricted to 1 or 2. Today's processes offer many more layers of interconnect, which are packed quite densely in addition. In this scenario, the assumption that a wire is completely isolated from its surrounding structures and only capacitively coupled to ground becomes untenable. This is illustrated in Figure 4-6, where the capacitance components of a wire embedded in an interconnect hierarchy

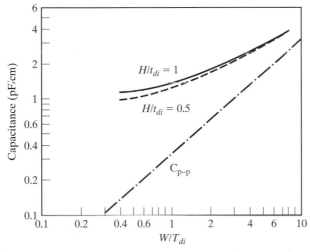

Figure 4-5   Capacitance of interconnect wire as a function of (W/t_{di}), including fringing-field effects (from [Schaper83]). Two values of H/t_{di} are considered. Silicon dioxide with $\varepsilon_r = 3.9$ is used as dielectric.

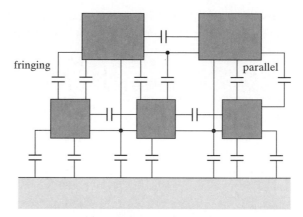

Figure 4-6   Capacitive coupling between wires in interconnect hierarchy.

are identified. Each wire is not only coupled to the grounded substrate, but also to the neighboring wires on the same layer and on adjacent layers. To a first order, this does not change the total capacitance connected to a given wire. The main difference is that not all of its capacitive components terminate at the grounded substrate—a large number of them connect to other wires, which have dynamically varying voltage levels. Later, we will see that these *floating capacitors* form not only a source of noise (cross talk), but also can have a negative impact on the performance of the circuit.

In sum, interwire capacitances become a dominant factor in multilayer interconnect structures. This effect is more pronounced for wires in the higher interconnect layers, as these wires are farther away from the substrate. The increasing contribution of the interwire capacitance to the total capacitance with decreasing feature sizes is best illustrated by Figure 4-7. In this graph, which

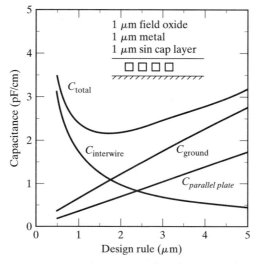

Figure 4-7   Interconnect capacitance as a function of design rules. It consists of a capacitance to ground and an interwire capacitance (from [Schaper83]).

plots the capacitive components of a set of parallel wires routed above a ground plane, it is assumed that dielectric and wire thickness are held constant, while scaling all other dimensions. When W becomes smaller than $1.75\ H$, the interwire capacitance starts to dominate.

Interconnect Capacitance Design Data

A set of typical interconnect capacitances for a standard 0.25 μm CMOS process are given in Table 4-2. The process supports one layer of polysilicon and five layers of aluminum. The first four metal layers have the same thickness and use a similar dielectric, while the wires at the fifth metal layer are almost twice as thick and are embedded in a dielectric with a higher permittivity. When placing the wires over the thick field oxide that is used to isolate different transistors, use the "Field" column in the table. Wires routed over the active area have a higher capacitance, as seen in the "Active" column. Be aware that the values presented are only averages. To obtain more accurate results for actual structures, complex three-dimensional models that take the environment of the wire into account should be used.

Table 4-3 tabulates representative values for the capacitances between parallel wires placed on the same layer with a minimum spacing (as dictated by the design rules). Observe that these numbers include both the parallel plate and fringing components. Once again, the capacitances are a strong function of the topology. For example, a ground plane placed on a neighboring layer terminates a large fraction of the fringing field and effectively reduces the interwire capacitance (although the total capacitance seen by the wire might go up slightly). The polysilicon wires experience a reduced interwire capacitance due to the smaller thickness of the wires. On the other hand, the thick Al5 wires display the highest inter-wire

Table 4-2 Wire area and fringe capacitance values for typical 0.25 μm CMOS process. The table rows represent the top plate of the capacitor, and the columns represent the bottom plate. The area capacitances are expressed in aF/μm^2, while the fringe capacitances (given in the shaded rows) are in aF/μm.

	Field	Active	Poly	Al1	Al2	Al3	Al4
Poly	88						
	54						
Al1	30	41	57				
	40	47	54				
Al2	13	15	17	36			
	25	27	29	45			
Al3	8.9	9.4	10	15	41		
	18	19	20	27	49		
Al4	6.5	6.8	7	8.9	15	35	
	14	15	15	18	27	45	
Al5	5.2	5.4	5.4	6.6	9.1	14	38
	12	12	12	14	19	27	52

Table 4-3 Inter-wire capacitance per unit wire length for different interconnect layers of typical 0.25 µm CMOS process. The capacitances are expressed in aF/µm, and are for minimally spaced wires.

Layer	Poly	Al1	Al2	Al3	Al4	Al5
Capacitance	40	95	85	85	85	115

capacitance. Therefore, it is advisable to either separate wires at this level by an amount that is larger than the minimum allowed, or to use it for global signals that are not that sensitive to interference. The supply rails are an example of the latter. ∎

Example 4.1   Capacitance of Metal Wire

Some global signals, such as clocks, are distributed all over the chip. The length of those wires can be substantial. For die sizes between 1 and 2 cm, wires can reach a length of 10 cm and have associated wire capacitances of substantial value. Consider an aluminum wire of 10 cm long and 1 µm wide, routed on the first aluminum layer. We can compute the value of the total capacitance by using the data presented in Table 4-2:

Area (parallel-plate) capacitance: $(0.1 \times 10^6 \ \mu m^2) \times 30 \ aF/\mu m^2 = 3 \ pF$

Fringing capacitance: $2 \times (0.1 \times 10^6 \ \mu m) \times 40 \ aF/\mu m = 8 \ pF$

Total capacitance: $11 \ pF$

Notice the factor 2 in the computation of the fringing capacitance, which takes the two sides of the wire into account.

Suppose now that a second wire is routed alongside the first one, separated by only the minimum allowed distance. From Table 4-3, we can determine that this wire will couple to the first with a capacitance equal to

$$C_{inter} = (0.1 \times 10^6 \ \mu m) \times 95 \ aF/\mu m = 9.5 \ pF$$

which is almost as large as the total capacitance to ground!

A similar exercise shows that moving the wire to Al4 would reduce the capacitance to ground to 3.45 pF (0.65 pF area and 2.8 pF fringe), while the interwire capacitance would remain approximately the same, at 8.5 pF.

4.3.2   Resistance

The resistance of a wire is proportional to its length L, and inversely proportional to its cross section A. The resistance of a rectangular conductor in the style of Figure 4-3 can be expressed as

$$R = \frac{\rho L}{A} = \frac{\rho L}{HW} \tag{4.3}$$

Table 4-4 Resistivity of commonly used conductors (at 20°C).

Material	$\rho\,(\Omega\text{-m})$
Silver (Ag)	1.6×10^{-8}
Copper (Cu)	1.7×10^{-8}
Gold (Au)	2.2×10^{-8}
Aluminum (Al)	2.7×10^{-8}
Tungsten (W)	5.5×10^{-8}

where the constant ρ is the resistivity of the material (in Ω-m). The resistivities of some commonly used conductive materials are tabulated in Table 4-4. Aluminum is the interconnect material most often used in integrated circuits because of its low cost and its compatibility with the standard integrated-circuit fabrication process. Unfortunately, it has a large resistivity compared with materials such as copper. With ever-increasing performance targets, top-of-the-line processes are increasingly using copper as the conductor of choice.

Since H is a constant for a given technology, Eq. (4.3) can be rewritten as

$$R = R_\square \frac{L}{W} \tag{4.4}$$

where

$$R_\square = \frac{\rho}{H} \tag{4.5}$$

is the *sheet resistance* of the material, having units of $\Omega/\square$ (pronounced as "ohms per square"). This expresses that the resistance of a square conductor is independent of its absolute size, as is apparent from Eq. (4.4). To obtain the resistance of a wire, simply multiply the sheet resistance by its ratio (L/W).

Interconnect Resistance Design Data

Typical values of the sheet resistance of various interconnect materials are given in Table 4-5.

Table 4-5 Sheet resistance values for a typical 0.25 μm CMOS process.

Material	Sheet Resistance ($\Omega/\square$)
n- or p-well diffusion	1000–1500
n^+, p^+ diffusion	50–150
n^+, p^+ diffusion with silicide	3–5
n^+, p^+ polysilicon	150–200
n^+, p^+ polysilicon with silicide	4–5
Aluminum	0.05–0.1

From this table, we conclude that aluminum is the preferred material for the wiring of long interconnections. Polysilicon should only be used for local interconnect. Although the sheet resistance of the diffusion layer (n^+, p^+) is comparable to that of polysilicon, the use of diffusion wires should be avoided due to its large capacitance and the associated RC delay. ∎

Advanced processes also offer silicided polysilicon and diffusion layers. A silicide is a compound material formed using silicon and a refractory metal. This creates a highly conductive material that can withstand high-temperature process steps without melting. Examples of silicides include WSi_2, $TiSi_2$, $PtSi_2$, and $TaSi$. WSi_2, for instance, has a resistivity ρ of 130 $\mu\Omega$-cm, which is approximately eight times lower than polysilicon. The silicides are most often used in a configuration called a *polycide*, which is a simple layered combination of polysilicon and a silicide. A typical polycide consists of a lower level of polysilicon with an upper coating of silicide and combines the best properties of both materials—good adherence and coverage (from the poly) and high conductance (from the silicide). A MOSFET fabricated with a polycide gate is shown in Figure 4-8. The advantage of the silicided gate is a reduced gate resistance. Similarly, silicided source and drain regions reduce the source and drain resistance of the device.

Transitions between routing layers add extra resistance to a wire, called the *contact resistance*. The preferred routing strategy is thus to keep signal wires on a single layer whenever possible and to avoid excess contacts or via's. It is possible to reduce the contact resistance by making the contact holes larger. Unfortunately, current tends to concentrate around the perimeter in a larger contact hole. This effect, called *current crowding*, puts a practical upper limit on the size of the contact. The following *contact resistances* (for minimum-size contacts) are typical for a 0.25 μm process: 5–20 Ω for metal or polysilicon to n^+, p^+, and metal to polysilicon; 1–5 Ω for via's (metal-to-metal contacts).

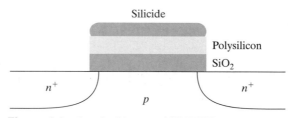

Figure 4-8   A polycide-gate MOSFET.

Example 4.2   Resistance of a Metal Wire

Consider again the aluminum wire of Example 4.1, which is 10 cm long and 1 μm wide, and is routed on the first aluminum layer. Assuming a sheet resistance for aluminum of 0.075 $\Omega/\square$, we can compute the total resistance of the wire as follows:

$$R_{wire} = 0.075 \ \Omega/\square \times (0.1 \times 10^6 \ \mu m) / (1 \ \mu m) = 7.5 \ k\Omega$$

Implementing the wire in polysilicon with a sheet resistance of 175 $\Omega/\square$ raises the overall resistance to 17.5 MΩ, which is clearly unacceptable. Silicided polysilicon with a sheet resistance of 4 Ω/□ offers a better alternative, but still translates into a wire with a 400-kΩ resistance.

So far, we have considered the resistance of a semiconductor wire to be linear and constant. This is definitely the case for most semiconductor circuits. At very high frequencies however, an additional phenomenon—called the *skin effect*—comes into play such that the resistance becomes frequency dependent. High-frequency currents tend to flow primarily on the surface of a conductor, with the current density falling off exponentially with depth into the conductor. The *skin depth* δ is defined as the depth at which the current falls off to a value of e^{-1} of its nominal value, and is given by

$$\delta = \sqrt{\frac{\rho}{\pi f \mu}} \qquad (4.6)$$

where f is the frequency of the signal, and μ is the permeability of the surrounding dielectric (typically equal to the permeability of free space, or $\mu = 4\pi \times 10^{-7}$ H/m). For aluminum at 1 GHz, the skin depth equals 2.6 μm. The obvious question now is if this is something we should be concerned about when designing state-of-the-art digital circuits.

The effect can be approximated by assuming that the current flows uniformly in an outer shell of the conductor with thickness δ, as illustrated in Figure 4-9 for a rectangular wire. Assuming that the overall cross section of the wire is now limited to approximately $2(W + H)\delta$, we obtain the following expression for the resistance (per unit length) at high frequencies $(f > f_s)$:

$$r(f) = \frac{\sqrt{\pi f \mu \rho}}{2(H + W)} \qquad (4.7)$$

The increased resistance at higher frequencies may cause an extra attenuation—and thus distortion—of the signal being transmitted over the wire. To determine the onset of the skin effect, we can find the frequency f_s, where the skin depth is equal to half the largest dimension (W or H) of the conductor. Below f_s, the whole wire is conducting current, and the resistance is equal to (constant) low-frequency resistance of the wire. From Eq. (4.6), we find the value of f_s:

$$f_s = \frac{4\rho}{\pi\mu(\max(W, H))^2} \qquad (4.8)$$

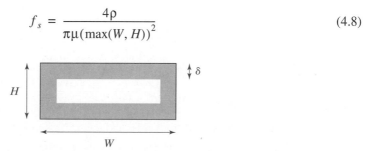

Figure 4-9   The skin effect reduces the flow of the current to the surface of the wire.

Example 4.3   Skin Effect and Aluminum Wires

We determine the impact of the skin effect on contemporary integrated circuits by analyzing an aluminum wire with a resistivity of 2.7×10^{-8} Ω-m, embedded in a SiO_2 dielectric with a permeability of $4\pi \times 10^{-7}$ H/m. From Eq. (4.8), we find that the largest dimension of wire should be at least 5.2 μm for the effect to be noticeable at 1 GHz. This is confirmed by the more accurate simulation results of Figure 4-10, which plots the increase in resistance due to skin effects for aluminum conductors of varying widths. A 30% increase in resistance can be observed at 1 GHz for a 20-μm wire, while the increase for a 1-μm wire is less than 2%.

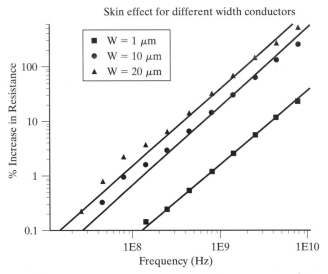

Figure 4-10   Skin-effect-induced increase in resistance as a function of frequency and wire width. All simulations were performed for a wire thickness of 0.7 μm [Sylvester97].

To summarize, the skin effect is only an issue for wider wires. Since clocks tend to carry the highest frequency signals on a chip and also are fairly wide to limit resistance, the skin effect is likely to have its first impact on these lines. This is a real concern for GHz-range design, as clocks determine the overall performance of the chip (cycle time, instructions per second, etc.). Another design concern is that the adoption of better conductors such as copper may move the onset of skin effects to lower frequencies.

4.3.3   Inductance

Integrated-circuit designers tend to dismiss inductance as something they heard about in their physics classes, but that has no impact on their field. This was definitely the case in the first decades of integrated digital circuit design. Yet, with the adoption of low-resistive interconnect materials and the increase of switching frequencies to the super-GHz range, inductance starts to play an important role, even on a chip. Consequences of on-chip inductance include ringing and overshoot effects, reflections of signals due to impedance mismatch, inductive coupling between lines, and switching noise due to (Ldi/dt) voltage drops.

The inductance of a section of a circuit can always be evaluated with the aid of its definition, which states that a changing current passing through an inductor generates a voltage drop

$$\Delta V = L \frac{di}{dt} \tag{4.9}$$

It is possible to compute the inductance of a wire directly from its geometry and its environment. A simpler approach relies on the fact that the capacitance c and the inductance l (per unit length) of a wire are related by the expression

$$cl = \varepsilon\mu \tag{4.10}$$

where ε and μ, respectively, are the permittivity and permeability of the surrounding dielectric. The caveat is that for this expression to be valid, the conductor must be completely surrounded by a uniform dielectric medium. This is not often the case. Yet, even when the wire is embedded in different dielectric materials, it is possible to adopt "average" dielectric constants so that Eq. (4.10) still can be used to get an approximate value of the inductance.

Some other interesting relations, obtained from Maxwell's laws, can be pointed out. The constant product of permeability and permittivity also defines the speed v at which an electromagnetic wave can propagate through the medium, which is given by

$$v = \frac{1}{\sqrt{lc}} = \frac{1}{\sqrt{\varepsilon\mu}} = \frac{c_0}{\sqrt{\varepsilon_r\mu_r}} \tag{4.11}$$

where c_0 equals the speed of light (30 cm/ns) in a vacuum. The propagation speeds for a number of materials used in the fabrication of electronic circuits are tabulated in Table 4-6. The propagation speed in SiO_2 is two times slower than in a vacuum.

Table 4-6 Dielectric constants and wave-propagation speeds for various materials used in electronic circuits. The relative permeability μ_r of most dielectrics is approximately equal to 1.

Dielectric	ε_r	Propagation Speed (cm/ns)
Vacuum	1	30
SiO_2	3.9	15
PC board (epoxy glass)	5.0	13
Alumina (ceramic package)	9.5	10

Example 4.4   Inductance of a Semiconductor Wire

Consider an Al1 (i.e., first layer of aluminum) wire implemented in the 0.25 micron CMOS technology and routed on top of the field oxide. From Table 4-2, we can derive the capacitance of the wire per unit length:

$$c = (W \times 30 + 2 \times 40) \text{ aF/}\mu\text{m}$$

From Eq. (4.10), we can derive the inductance per unit length of the wire, assuming that a uniform dielectric consisting of SiO_2 (make sure to use the correct units!):

$$l = (3.9 \times 8.854 \times 10^{-12}) \times (4 \pi \, 10^{-7})/c$$

For wire widths of 0.4 μm, 1 μm and 10 μm, this leads to the following:

$$W = 0.4 \, \mu\text{m:} \quad c = 92 \text{ aF/}\mu\text{m;} \quad l = 0.47 \text{ pH/}\mu\text{m}$$
$$W = 1 \, \mu\text{m:} \quad c = 110 \text{ aF/}\mu\text{m;} \quad l = 0.39 \text{ pH/}\mu\text{m}$$
$$W = 10 \, \mu\text{m:} \quad c = 380 \text{ aF/}\mu\text{m;} \quad l = 0.11 \text{ pH/}\mu\text{m}$$

Assuming a sheet resistance of 0.075 Ω/q, we can also determine the resistance of the wire:

$$r = 0.075/W \, \Omega/\mu\text{m}$$

It is interesting to observe that the inductive part of the wire impedance becomes equal in value to the resistive component at a frequency of 30.6 GHz (for a 1-μm-wide wire), as can be obtained from solving the following expression:

$$\omega l = 2\pi f l = r$$

For extrawide wires, this frequency reduces to approximately 11 GHz. For wires with a smaller capacitance and resistance (such as the thicker wires located at the upper interconnect layers), this frequency can become as low as 500 MHz, especially when better interconnect materials such as copper are being used. Yet, these numbers indicate that inductance only is an issue in integrated circuits at the high end of the performance curve.

4.4 Electrical Wire Models

In previous sections, we have introduced the electrical properties of the interconnect wire— capacitance, resistance, and inductance—and presented some simple relations and techniques to derive their values from the interconnect geometries and topologies. These parasitic elements have an impact on the electrical behavior of the circuit and influence its delay, power dissipation, and reliability. To study these effects requires the introduction of electrical models that estimate and approximate the real behavior of the wire as a function of its parameters. These models vary from very simple to very complex depending upon the effects that are being studied and the required accuracy. In this section, we first derive models for manual analysis. How to cope with interconnect wires in the SPICE circuit simulator is the topic that follows next.

4.4.1 The Ideal Wire

In schematics, wires occur as simple lines with no attached parameters or parasitics. These wires have no impact on the electrical behavior of the circuit. A voltage change at one end of the wire propagates immediately to its other ends, even if they are some distance away. Hence, it may be assumed that the same voltage is present at every segment of the wire at every point in time, and that the whole wire is an *equipotential region*. While this *ideal-wire model* is simplistic, it has its value, especially in the early phases of the design process when the designer wants to concentrate on the properties and the behavior of the transistors that are being connected. Also, when studying small circuit components such as gates, the wires tend to be very short and their parasitics ignorable. Taking these into account would make the analysis unnecessarily complex. More often, though, wire parasitics play an important role, and more complex models should be considered.

4.4.2 The Lumped Model

The circuit parasitics of a wire are distributed along its length and are not lumped into a single position. Yet, when only a single parasitic component is dominant, when the interaction between the components is small, or when looking at only one aspect of the circuit behavior, it is often useful to lump the different fractions into a single circuit element. The advantage of this approach is that the effects of the parasitics can then be described by an ordinary differential equation. As we will see later, the description of a distributed element requires partial differential equations.

As long as the resistive component of the wire is small, and the switching frequencies are in the low to medium range, it is meaningful to consider only the capacitive component of the wire, and to lump the distributed capacitance into a single capacitor, as shown in Figure 4-11. Observe that in this model the wire still represents an equipotential region, and that the wire itself does not introduce any delay. The only impact on performance is introduced by the loading effect of the capacitor on the driving gate. This capacitive-lumped model is simple, yet effective, and is the model of choice for the analysis of most interconnect wires in digital integrated circuits.

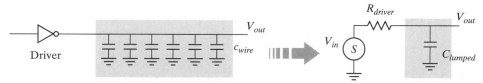

Figure 4-11 Distributed versus lumped capacitance model of wire. $C_{lumped} = L \times c_{wire}$, with L the length of the wire and c_{wire} the capacitance per unit length. The driver is modeled as a voltage source and a source resistance R_{driver}.

Example 4.5   Lumped Capacitance Model of Wire

For the circuit of Figure 4-11, assume that a driver with a source resistance of 10 kΩ is used to drive a 10-cm-long, 1-μm-wide Al1 wire. In Example 4.1, we found that the total lumped capacitance for this wire equals 11 pF.

The operation of this simple RC network is described by the following ordinary differential equation (similar to the expression derived in Example 1):

$$C_{lumped} \frac{dV_{out}}{dt} + \frac{V_{out} - V_{in}}{R_{driver}} = 0$$

When applying a step input (with V_{in} going from 0 to V), the transient response of this circuit is known to be an exponential function, and is given by the expression

$$V_{out}(t) = (1 - e^{-t/\tau}) V$$

where $\tau = R_{driver} \times C_{lumped}$ is the time constant of the network.

The time to reach the 50% point is easily computed as $t = \ln(2)\tau = 0.69\tau$. Similarly, it takes $t = \ln(9)\tau = 2.2\tau$ to get from the 10% to the 90% point. Plugging in the numbers for this specific example yields

$$t_{50\%} = 0.69 \times 10 \text{ K}\Omega \times 11 \text{ pF} = 76 \text{ ns}$$
$$t_{90\%} = 2.2 \times 10 \text{ K}\Omega \times 11 \text{ pF} = 242 \text{ ns}$$

These numbers are not even acceptable for the lowest performance digital circuits. Techniques to deal with this bottleneck, such as reducing the source resistance of the driver, will be introduced in Chapters 5 and 9.

While the lumped capacitor model is the most popular, sometimes it is useful to present lumped models of a wire with respect to either resistance and inductance. This is often the case when studying the supply distribution network. Both the resistance and inductance of the supply wires can be interpreted as parasitic noise sources that introduce voltage drops and bounces on the supply rails.

4.4.3   The Lumped *RC* Model

On-chip metal wires of more than a few millimeters of length have a significant resistance. The equipotential assumption, presented in the lumped-capacitor model, is no longer adequate, and a resistive–capacitive model has to be adopted.

A first approach lumps the total wire resistance of each wire segment into one single *R,* and similarly combines the global capacitance into a single capacitor *C*. This simple model, called the *lumped RC model,* is pessimistic and inaccurate for long interconnect wires, which are more adequately represented by a *distributed rc model*. Yet, before analyzing the distributed model, it is worthwhile to spend some time on the analysis and the modeling of lumped *RC* networks for the following reasons:

- The distributed rc model is complex, and no closed-form solutions exist. The behavior of the distributed rc line can be adequately modeled by a simple RC network.
- A common practice in the study of the transient behavior of complex transistor-wire networks is to reduce the circuit to an RC network. Having a means to analyze such a network effectively and to predict its first-order response would add a great asset to the designers tool box.

In Example 4.5, we analyzed a single resistor-single capacitor network. The behavior of such a network is fully described by a single differential equation, and its transient waveform is a modeled by an exponential with a single time constant (or network pole). Unfortunately, deriving the correct waveforms for a network with a larger number of capacitors and resistors rapidly becomes hopelessly complex: describing its behavior requires a set of ordinary differential equations, and the network now contains many time constants (or poles and zeros). Short of running a full-fledged SPICE simulation, delay calculation methods such as the *Elmore delay formula* come to the rescue [Elmore48].

Consider the resistor-capacitor network of Figure 4-12. This circuit is called an *RC tree,* and it has the following properties:

- The network has a single input node (labeled s in Figure 4-12).
- All the capacitors are between a node and the ground.
- The network does not contain any resistive loops (which makes it a tree).

An interesting result of this particular circuit topology is that a unique resistive path exists between the source node s and any node i of the network. The total resistance along this path is called the *path resistance* R_{ii}. For example, the path resistance between the source node s and node 4 in the example of Figure 4-12 is

$$R_{44} = R_1 + R_3 + R_4$$

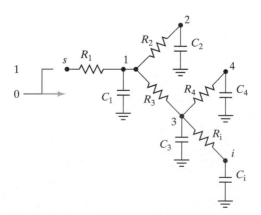

Figure 4-12   Tree-structured *RC* network.

The definition of the path resistance can be extended to address the *shared path resistance* R_{ik}, which represents the resistance shared among the paths from the root node s to nodes k and i:

$$R_{ik} = \sum R_j \Rightarrow (R_j \in [path(s \to i) \cap path(s \to k)]) \tag{4.12}$$

For the circuit of Figure 4-12, $R_{i4} = R_1 + R_3$, while $R_{i2} = R_1$.

Assume now that each of the N nodes of the network is initially discharged to GND, and that a step input is applied at node s at time $t = 0$. The Elmore delay at node i is then given by the following expression:

$$\tau_{Di} = \sum_{k=1}^{N} C_k R_{ik} \tag{4.13}$$

The Elmore delay is equivalent to the first-order time constant of the network (or the first moment of the impulse response). The designer should be aware that this time constant represents a simple approximation of the actual delay between source node and node i. Still, in most cases, this approximation has proven to be quite reasonable and acceptable. It offers the designer a powerful mechanism for providing a quick estimate of the delay of a complex network.

Example 4.6   *RC* Delay of a Tree-Structured Network

Using Eq. (4.13), we can compute the Elmore delay for node i in the network of Figure 4-12: $\tau_{Di} = R_1 C_1 + R_1 C_2 + (R_1 + R_3)C_3 + (R_1 + R_3)C_4 + (R_1 + R_3 + R_i)C_i$

As a special case of the *RC* tree network, let us consider the simple, nonbranched *RC* chain (or ladder) shown in Figure 4-13. This network is worth analyzing because it is a structure that is often encountered in digital circuits, and also because it represents an approximate model of a resistive–capacitive wire. The Elmore delay of this chain network can be derived with the aid of Eq. (4.13):

$$\tau_{DN} = \sum_{i=1}^{N} C_i \sum_{j=1}^{i} R_j = \sum_{i=1}^{N} C_i R_{ii} \tag{4.14}$$

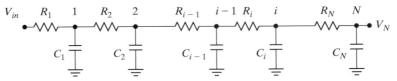

Figure 4-13   *RC* chain.

In other words, the shared-path resistance is replaced by the path resistance. As an example, consider node 2 in the RC chain of Figure 4-13. Its time constant consists of two components contributed by nodes 1 and 2. The component of node 1 consists of $C_1 R_1$, where R_1 is the total resistance between the node and the source, and the contribution of node 2 equals $C_2(R_1 + R_2)$. The equivalent time constant at node 2 equals $C_1 R_1 + C_2(R_1 + R_2)$. τ_i of node i can be derived in a similar way:

$$\tau_{Di} = C_1 R_1 + C_2(R_1 + R_2) + \dots + C_i(R_1 + R_2 + \dots + R_i)$$

Example 4.7   Time Constant of Resistive–Capacitive Wire

The model presented in Figure 4-13 can be used as an approximation of a resistive–capacitive wire. The wire with a total length of L is partitioned into N identical segments, each with a length of L/N. Therefore, the resistance and capacitance of each segment are given by rL/N and cL/N, respectively. Using the Elmore formula, we can compute the dominant time constant of the wire as

$$\tau_{DN} = \left(\frac{L}{N}\right)^2 (rc + 2rc + \dots + Nrc) = (rcL^2)\frac{N(N+1)}{2N^2} = RC\frac{N+1}{2N} \qquad (4.15)$$

where $R \, (= rL)$ and $C \, (= cL)$ are the total lumped resistance and capacitance of the wire. For very large values of N, this model asymptotically approaches the distributed rc line. Equation (4.15) then simplifies to the following expression:

$$\tau_{DN} = \frac{RC}{2} = \frac{rcL^2}{2} \qquad (4.16)$$

Equation (4.16) leads to two important conclusions:

- The delay of a wire is **a quadratic function of its length**! This means that doubling the length of the wire quadruples its delay.
- The delay of the distributed rc line is **one half of the delay** that would have been predicted by the lumped RC model. The latter combines the total resistance and capacitance into single elements, and has a time constant equal to RC (as is also obtained by setting $N = 1$ in Eq. (4.15)). This confirms the observation made earlier that the lumped model presents a pessimistic view on the delay of resistive wire.

WARNING: Be aware that an RC chain is characterized by a number of time constants. The Elmore expression determines the value of only the dominant one and thus is only a first-order approximation.

The Elmore delay formula has proven to be extremely useful. Besides making it possible to analyze wires, the formula can also be used to approximate the propagation delay of complex transistor networks. In the switch model, transistors are replaced by their equivalent linearized on-resistance. The evaluation of the propagation delay is then reduced to the analysis of the resulting RC network. More precise minimum and maximum bounds on the voltage waveforms in an RC tree have further been established [Rubinstein83]. These bounds have formed the basis for most computer-aided timing analyzers at the switch and functional level [Horowitz83]. An interesting result [Lin84] is that the exponential voltage waveform with the Elmore delay as time constant is always situated between these minimum and maximum bounds, which demonstrates the validity of the Elmore approximation.

4.4.4 The Distributed *rc* Line

In the previous discussion, we showed that the lumped RC model is a pessimistic model for a resistive–capacitive wire, and that a distributed rc model (Figure 4-14a) is more appropriate. As before, L represents the total length of the wire, while r and c stand for the resistance and capacitance per unit length. A schematic representation of the distributed rc line is given in Figure 4-14b.

The voltage at node i of this network can be determined by solving the following set of partial differential equations:

$$c\Delta L \frac{\partial V_i}{\partial t} = \frac{(V_{i+1} - V_i) + (V_{i-1} - V_i)}{r\Delta L} \qquad (4.17)$$

The correct behavior of the distributed rc line is then obtained by reducing ΔL asymptotically to 0. For $\Delta L \rightarrow 0$, Eq. (4.17) becomes the well-known *diffusion equation*:

$$rc\frac{\partial V}{\partial t} = \frac{\partial^2 V}{\partial x^2} \qquad (4.18)$$

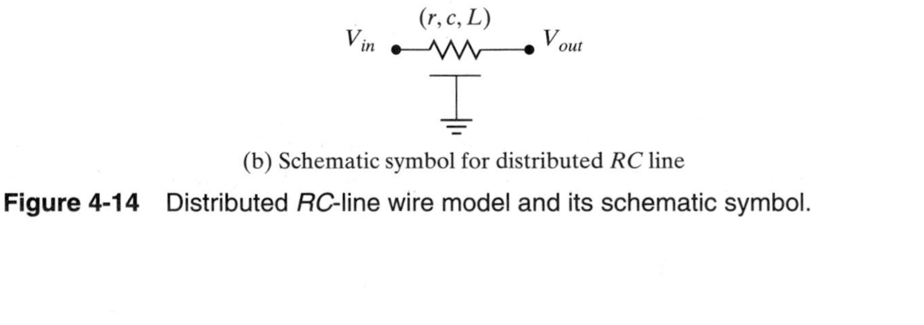

(a) Distributed model

(b) Schematic symbol for distributed RC line

Figure 4-14   Distributed RC-line wire model and its schematic symbol.

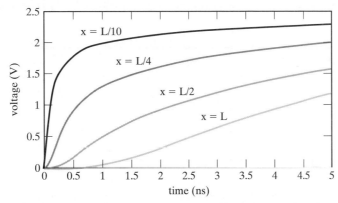

Figure 4-15 Simulated step response of resistive–capacitive wire as a function of time and place.

In this equation, V is the voltage at a particular point in the wire, and x is the distance between this point and the signal source. No closed-form solution exists for this equation, but approximative expressions such as the formula presented in Eq. (4.19) can be derived [Bakoglu90].

$$V_{out}t = 2\,erfc\left(\sqrt{\frac{RC}{4t}}\right) \qquad\qquad t \ll RC$$

$$= 1.0 - 1.366e^{-2.5359\frac{t}{RC}} + 0.366e^{-9.4641\frac{t}{RC}} \qquad t \gg RC$$

(4.19)

These equations are difficult to use for ordinary circuit analysis. It is known, however, that the distributed rc line can be approximated by a lumped RC-ladder network, which can be easily used in computer-aided analysis.

Figure 4-15 shows the response of a wire to a step input, plotting the waveforms at different points in the wire as a function of time. Observe how the step waveform "diffuses" from the start to the end of the wire, and the waveform rapidly degrades, resulting in a considerable delay for long wires. Driving these rc lines and minimizing the delay and signal degradation is one of the trickiest problems in modern digital integrated circuit design. Accordingly, it receives considerable attention throughout this book.

Some of the important reference points in the step response of the lumped and the distributed RC model of the wire are tabulated in Table 4-7. For example, the propagation delay (defined at 50% of the final value) of the lumped network not surprisingly equals $0.69\,RC$. The distributed network, on the other hand, has a delay of only $0.38\,RC$, with R and C the total resistance and capacitance of the wire. This result, obtained from simulation, is in line with Eq. (4.16).

Example 4.8   *RC* Delay of Aluminum Wire

Let us consider again the 10 cm long, 1 µm wide, All wire of Example 4.1. In Example 4.4, we derived the following values for r and c:

$$c = 110 \text{ aF/µm};\; r = 0.075 \text{ Ω/µm}$$

Table 4-7 Step response of lumped and distributed *RC* networks—points of interest.

Voltage Range	Lumped *RC* Network	Distributed *RC* Network
$0 \to 50\%$ (t_p)	0.69 *RC*	0.38 *RC*
$0 \to 63\%$ (τ)	*RC*	0.5 *RC*
$10\% \to 90\%$ (t_r)	2.2 *RC*	0.9 *RC*
$0\% \to 90\%$	2.3 *RC*	1.0 *RC*

Using the entry of Table 4-7, we derive the propagation delay of the wire:

$$t_p = 0.38 \, RC = 0.38 \times (0.075 \, \Omega/\mu m) \times (110 \, aF/\mu m) \times (10^5 \, \mu m)^2 = 31.4 \, ns$$

We can also deduce the propagation delays of an identical wire implemented in polysilicon and Al5 (the 5th metal layer). The values of the capacitances are obtained from Table 4-2, while the resistances are assumed to be, respectively, 150 Ω/μm and 0.0375 Ω/μm for Poly and Al5:

Poly: $t_p = 0.38 \times (150 \, \Omega/\mu m) \times (88 + 2 \times 54 \, aF/\mu m) \times (10^5 \, \mu m)^2 = 112 \, \mu s!$

Al5: $t_p = 0.38 \times (0.0375 \, \Omega/\mu m) \times (5.2 + 2 \times 12 \, aF/\mu m) \times (10^5 \, \mu m)^2 = 4.2 \, ns$

Obviously, the choice of the interconnect material and layer has a dramatic effect on the delay of the wire.

An important question for a designer to answer when analyzing an interconnect network is whether the effects of *RC* delays should be considered, or whether she can get away with a simpler lumped capacitive model. A simple rule of thumb proves to be very useful here.

Design Rules of Thumb

• *rc* delays should be considered only when t_{pRC} is comparable or larger than the t_{pgate} of the driving gate.

This defines a critical length

$$L_{crit} = \sqrt{\frac{t_{pgate}}{0.38rc}} \tag{4.20}$$

RC delays become dominant for interconnect wires longer than this critical length L_{crit}. The actual value of L_{crit} depends on the sizing of the driving gate and the chosen interconnect material.

• *rc* delays should only be considered when the rise (fall) time at the line input is smaller than *RC*, the rise (fall) time of the line.

In other words, they should be considered only when

$$t_{rise} < RC \tag{4.21}$$

where R and C are the total resistance and capacitance of the wire, respectively. When this condition is not met, the change in signal is slower than the propagation delay of the wire, and a lumped capacitive model suffices.

∎

Example 4.9   *RC* versus Lumped *C*

The presented rule can be illustrated with the aid of the simple circuit shown in Figure 4-16. It is assumed here that the driving gate can be modeled as a voltage source with a finite source resistance R_s. Applying the Elmore formula,[2] the total propagation delay of the network can be approximated by the expression

$$\tau_D = R_s C_w + \frac{R_w C_w}{2} = R_s C_w + 0.5 r_w c_w L^2$$

and

$$t_p = 0.69 R_s C_w + 0.38 R_w C_w$$

with $R_w = rL$ and $C_w = cL$. The delay introduced by the wire resistance becomes dominant when $(R_w C_w)/2 \ge R_s C_w$, or when $L \ge 2R_s/r$. Assume now a driver with a source resistance of 1 kΩ driving an Al1 wire of 1 μm wide ($r = 0.075$ Ω/μm). This leads to a critical length of 2.67 cm.

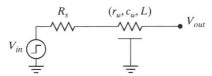

Figure 4-16   *rc*-line of length L driven by source with resistance equal to R_s.

4.4.5   The Transmission Line

When the switching speeds of the circuits become sufficiently fast, and the quality of the interconnect material becomes high enough that the resistance of the wire is kept within bounds, the inductance of the wire starts to dominate the delay behavior, and transmission line effects must be considered. More precisely, this is the case when the rise and fall times of the signal become comparable to the time of flight of the signal waveform across the line, as determined by the speed of light. With the advent of copper interconnect and the high switching speeds enabled by the deep-submicron technologies, transmission line effects must already be considered in the fastest CMOS designs.

---

[2]Hint: Replace the wire by the lumped RC network of Figure 4-13, and apply the Elmore equation on the resulting network.

In this section, we first analyze the transmission line model. Next, we apply the model to the current semiconductor technology and determine when those effects should be actively considered in the design process.

Transmission Line Model

Similar to the resistance and capacitance of an interconnect line, the inductance is distributed over the wire. A distributed *rlc* model of a wire, known as the *transmission line model*, becomes the most accurate approximation of the actual behavior. The transmission line has the prime property that a signal propagates over the interconnection medium as a *wave*. This is in contrast to the distributed *rc* model, in which the signal *diffuses* from the source to the destination governed by Eq. (4.18), the diffusion equation. In the wave mode, a signal propagates by alternatively transferring energy from the electric to the magnetic fields, or, equivalently, from the capacitive to the inductive modes.

Consider the point x along the transmission line of Figure 4-17 at time t. The following set of equations holds:

$$\frac{\partial v}{\partial x} = -ri - l\frac{\partial i}{\partial t}$$

$$\frac{\partial i}{\partial x} = -gv - c\frac{\partial v}{\partial t} \qquad (4.22)$$

When we assume that the leakage conductance g equals 0, which is true for most insulating materials, and eliminate the current i, we get the *wave propagation equation*, written as

$$\frac{\partial^2 v}{\partial x^2} = rc\frac{\partial v}{\partial t} + lc\frac{\partial^2 v}{\partial t^2} \qquad (4.23)$$

where r, c, and l are the resistance, capacitance, and inductance per unit length, respectively.

To better understand the behavior of the transmission line, we first assume that the resistance of the line is small. In this case, a simplified capacitive/inductive model, called the *lossless transmission line*, is appropriate. This model is applicable for wires at the printed circuit board level. Due to the high conductivity of the copper interconnect material used there, the resistance of the transmission line can be ignored. On the other hand, resistance plays an important role in integrated circuits, and a more complex model, called the *lossy transmission line* should be considered. (The lossy model is only discussed briefly, at the end of the chapter.)

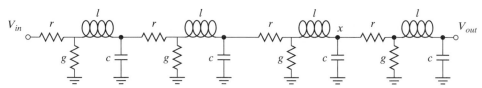

Figure 4-17   Lossy transmission line.

The Lossless Transmission Line

For the lossless line, Eq. (4.23) simplifies to the *ideal wave* equation:

$$\frac{\partial^2 v}{\partial x^2} = lc\frac{\partial^2 v}{\partial t^2} = \frac{1}{v^2}\frac{\partial^2 v}{\partial t^2} \tag{4.24}$$

A step input applied to a lossless transmission line propagates along the line with a speed υ, given by Eq. (4.11) and repeated as follows:

$$\upsilon = \frac{1}{\sqrt{lc}} = \frac{1}{\sqrt{\varepsilon\mu}} = \frac{c_0}{\sqrt{\varepsilon_r\mu_r}} \tag{4.25}$$

Even though the values of both l and c depend on the geometric shape of the wire, their product is a constant and is only a function of the surrounding media. The propagation delay per unit wire length (t_p) of a transmission line is the inverse of the speed:

$$t_p = \sqrt{lc} \tag{4.26}$$

Let us now analyze how a wave propagates along a lossless transmission line. Suppose that a voltage step V has been applied at the input and has propagated to point x of the line, as in Figure 4-18. All currents are equal to 0 at the right side of x, while the voltage over the line equals V at the left side. An additional capacitance cdx must be charged for the wave to propagate over an additional distance dx. This requires the current to be

$$i = \frac{dq}{dt} = c\frac{dx}{dt}v = c\upsilon v = \sqrt{\frac{c}{l}}v \tag{4.27}$$

since the propagation speed of the signal dx/dt equals υ. This means that the signal sees the remainder of the line as a real impedance Z_0, with

$$Z_0 = \frac{V}{I} = \sqrt{\frac{l}{c}} = \frac{\sqrt{\varepsilon\mu}}{c} = \frac{1}{c\upsilon} \tag{4.28}$$

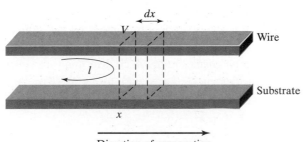

Figure 4-18 Propagation of voltage step along a lossless transmission line.

This impedance, called the *characteristic impedance* of the line, is a function of the dielectric medium and the geometry of the conducting wire and isolator (Eq. (4.28)), and it is independent of the length of the wire and the frequency. That a line of arbitrary length has a constant, real impedance is a great feature because it simplifies the design of the driver circuitry. Typical values of the characteristic impedance of wires in semiconductor circuits range from 10 to 200 Ω.

Example 4.10   Propagation Speeds of Signal Waveforms

The information in Table 4-6 shows that it takes 1.5 ns for a signal wave to propagate from source to destination on a 20-cm wire deposited on an epoxy printed circuit board. If transmission line effects were an issue on silicon integrated circuits, it would take 0.67 ns for the signal to reach the end of a 10-cm wire.

WARNING: The characteristic impedance of a wire is a function of the overall interconnect topology. The electromagnetic fields in complex interconnect structures tend to be irregular, and they are strongly influenced by issues such as the current return path. Providing a general answer to the latter problem has thus far proven illusive, and no closed-formed analytical solutions are available. Hence, accurate inductance and characteristic impedance extraction is still an active research topic. For some simplified structures, approximations have been derived. For instance, the characteristic impedances of a triplate strip line (a wire embedded in between two ground planes) and a semiconductor microstripline (wire above a semiconductor substrate) are approximated by the following two equations:

$$Z_0(\text{triplate}) \approx 94\Omega \sqrt{\frac{\mu_r}{\varepsilon_r}} \ln\left(\frac{2t + W}{H + W}\right) \tag{4.29}$$

and

$$Z_0(\text{microstrip}) \approx 60\Omega \sqrt{\frac{\mu_r}{0.475\varepsilon_r + 0.67}} \ln\left(\frac{4t}{0.536W + 0.67H}\right) \tag{4.30}$$

Termination

The behavior of the transmission line is strongly influenced by the termination of the line. The termination determines how much of the wave is reflected upon arrival at the end of the wire. This is expressed by the *reflection coefficient* ρ, which determines the relationship between the voltages and currents of the incident and reflected waveforms. We have

$$\rho = \frac{V_{refl}}{V_{inc}} = \frac{I_{refl}}{I_{inc}} = \frac{R - Z_0}{R + Z_o} \tag{4.31}$$

where R is the value of the termination resistance. The total voltages and currents at the termination end are the sum of incident and reflected waveforms:

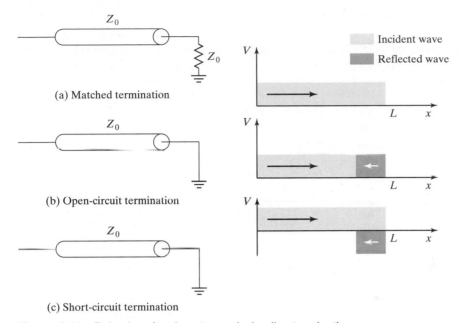

(a) Matched termination

(b) Open-circuit termination

(c) Short-circuit termination

Figure 4-19   Behavior of various transmission line terminations.

$$V = V_{inc}(1 + \rho)$$
$$I = I_{inc}(1 - \rho) \tag{4.32}$$

Three interesting cases can be distinguished, as illustrated in Figure 4-19. In case (a), the terminating resistance is equal to the characteristic impedance of the line. The termination appears as an infinite extension of the line, and no waveform is reflected. This is also demonstrated by the value of ρ, which equals 0. In case (b), the line termination is an open circuit ($R = \infty$), and $\rho = 1$. The total voltage waveform after reflection is twice the incident one, as predicted by Eq. (4.32). Finally, in case (c), where the line termination is a short circuit, $R = 0$, and $\rho = -1$. The total voltage at the end of the wire after reflection equals zero.

The transient behavior of a complete transmission line can now be examined. It is influenced by the characteristic impedance of the line, the series impedance of the source Z_S, and the loading impedance Z_L at the destination end, as shown in Figure 4-20.

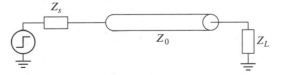

Figure 4-20   Transmission line with terminating impedances.

First, consider the case in which the wire is open at the destination end, or $Z_L = \infty$, and $\rho_L = 1$. An incoming wave is completely reflected without phase reversal. Under the assumption that the source impedance is resistive, three possible scenarios are sketched in Figure 4-21: $R_S = 5Z_0$, $R_S = Z_0$, and $R_S = 1/5Z_0$.

1. **Large source resistance—$R_S = 5Z_0$ (Figure 4-21a)**

Only a small fraction of the incoming signal V_{in} of 5 V is injected into the transmission line. The amount injected is determined by the resistive divider formed by the source resistance and the characteristic impedance Z_0:

$$V_{source} = (Z_0 / (Z_0 + R_S)) V_{in} = 1/6 \times 5 \text{ V} = 0.83 \text{ V} \tag{4.33}$$

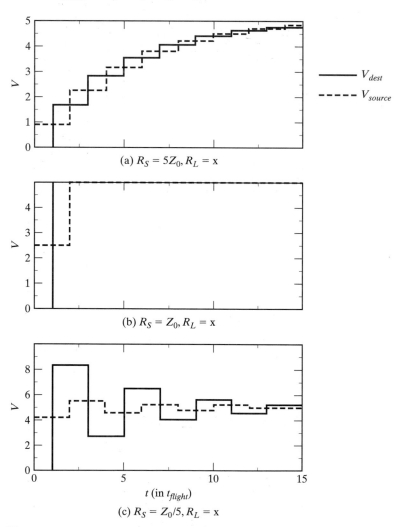

(a) $R_S = 5Z_0, R_L = \text{x}$

(b) $R_S = Z_0, R_L = \text{x}$

(c) $R_S = Z_0/5, R_L = \text{x}$

Figure 4-21   Transient response of transmission line.

This signal reaches the end of the line after L/υ seconds, where L stands for the length of the wire and is fully reflected, which effectively doubles the amplitude of the wave ($V_{dest} = 1.67$ V). The time it takes for the wave to propagate from one end of the wire to the other is called the *time of flight*, $t_{flight} = L/\upsilon$. Nearly the same thing happens when the wave reaches the source node again. The incident waveform is reflected with an amplitude determined by the source reflection coefficient, which equals two-thirds for this particular case:

$$\rho_S = \frac{5Z_0 - Z_0}{5Z_0 + Z_0} = \frac{2}{3} \tag{4.34}$$

The voltage amplitude at source and destination nodes gradually reaches its final value of V_{in}. The overall rise time, however, is many times greater than L/υ.

When multiple reflections are present, as in this case, keeping track of waves on the line and total voltage levels rapidly becomes cumbersome. A graphical construction called a *lattice diagram* often is used to keep track of the data (see Figure 4-22). The diagram contains the values of the voltages at the source and destination ends, as well as the values of the incident and reflected wave forms. The line voltage at a termination point equals the sum of the previous voltage, the incident, and the reflected waves.

2. **Small source resistance**—$R_S = Z_0/5$ (Figure 4-21c)

A large portion of the input is injected in the line. Its value is doubled at the destination end, which causes a severe overshoot. At the source end, the phase of the signal is reversed ($\rho_S = -2/3$). The signal bounces back and forth and exhibits severe ringing. It takes multiple L/υ's before it settles.

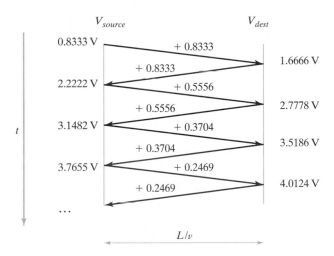

Figure 4-22 Lattice diagram for $R_S = 5Z_0$ and $RL = \infty$. $V_{step} = 5$ V (as in Figure 4-21a).

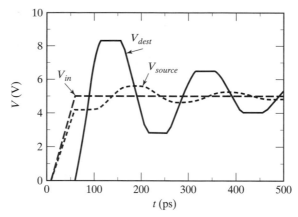

Figure 4-23 Simulated transient response of lossless transmission line for finite input rise times ($R_S = Z_0/5$, $t_r = t_{flight}$).

3. **Matched source resistance**—$R_S = Z_0$ (Figure 4-21b)

Half of the input signal is injected at the source. The reflection at the destination end doubles the signal, and the final value is reached immediately. It is obvious that this is the most effective case.

Note that the preceding analysis is an ideal one because we assume that the input signal has a zero rise time. In real conditions, the signals are substantially smoother, as demonstrated in the simulated response of Figure 4-23 (for $R_S = Z_0/5$ and $t_r = t_{flight}$).

Problem 4.1   Transmission Line Response

Derive the lattice diagram of the preceding transmission line for $R_S = Z_0/5$, $R_L = \infty$, and $V_{step} = 5$ V. Also try the reverse picture—assume that the series resistance of the source equals zero, and consider different load impedances.

Example 4.11   Capacitive Termination

Loads in MOS digital circuits tend to be of a capacitive nature. One might wonder how this influences the transmission line behavior and when the load capacitance should be taken into account.

The characteristic impedance of the transmission line determines the current that can be supplied to charge capacitive load C_L. From the load's point of the view, the line behaves as a resistance with value Z_0. The transient response at the capacitor node, therefore, displays a time constant $Z_0 C_L$. This is illustrated in Figure 4-24, which shows the simulated transient response of a series-terminated transmission line, with a characteristic impedance of 50 Ω loaded by a capacitance of 2 pF. The response shows how the output rises to its final value with a time constant of 100 ps (= 50 $\Omega \times 2$ pF) after a delay equal to the line's time of flight.

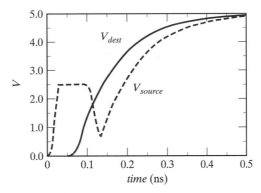

Figure 4-24 Capacitively terminated transmission line: $R_S = 50\ \Omega$, $R_L = \infty$, $C_L = 2$ pF, $Z_0 = 50\ \Omega$, $t_{flight} = 50$ ps.

This asymptotic response causes some interesting artifacts. After 2 t_{flight}, a seemingly unexpected voltage dip occurs at the source node that can be explained as follows. Upon reaching the destination node, the incident wave is reflected. This reflected wave also approaches its final value asymptotically. Since V_{dest} equals 0 initially (instead of the expected jump to 5 V), the reflection equals –2.5 V, rather than the expected 2.5 V. This forces the transmission line temporarily to 0 V, as shown in the simulation. This effect gradually disappears as the output node converges to its final value.

The propagation delay of the line equals the sum of the time of flight of the line (50 ps) and the time it takes to charge the capacitance (0.69 $Z_0 C_L = 69$ ps). This is exactly what the simulation yields. In general, we can say that the capacitive load should only be considered in the analysis when its value is comparable to, or larger than, the total capacitance of the transmission line [Bakoglu90].

Lossy Transmission Line

While board and module wires are thick enough and wide enough to be treated as lossless transmission lines, the same is not entirely true for on-chip interconnect, where the resistance of the wire is an important factor. The lossy transmission line model should be applied instead. Going into great detail about the behavior of a lossy line is not a subject of this text, so we only discuss the effects of resistive loss on the transmission line behavior in a qualitative fashion.

The response of a lossy *RLC line* to a unit step combines wave propagation with a diffusive component. This is demonstrated in Figure 4-25, which plots the response of the *RLC* transmission line as a function of distance from the source. The step input still propagates as a wave through the line. However, the amplitude of this traveling wave is attenuated along the line:

$$\frac{V_{step}(x)}{V_{step}(0)} = e^{-\frac{r}{2Z_0}x}$$

(4.35)

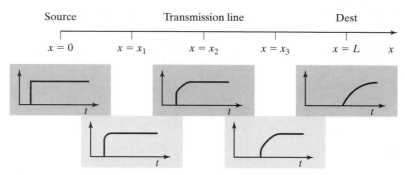

Figure 4-25   Step response of lossy transmission line.

The arrival of the wave is followed by a diffusive relaxation to the steady-state value at point x. The farther it is from the source, the more the response resembles the behavior of a distributed RC line. In fact, the resistive effect becomes dominant, and the line behaves as a distributed RC line when R (= rL, the total resistance of the line) $\gg 2 Z_0$. When $R = 5 Z_0$, only 8% of the original step reaches the end of the line. At that point, the line is more appropriately modeled as a distributed rc line.

Be aware that the actual wires on chips, boards, or substrates behave in a far more complex way than predicted by this simple analysis. For example, branches on wires, often called *transmission line taps,* cause extra reflections and can affect both signal shape and delay. Since the analysis of these effects is very involved, the only meaningful approach is to use computer analysis and simulation techniques. For a more extensive discussion of these effects, we refer the reader to [Bakoglu90] and [Dally98].

Design Rule of Thumb

Once again, we have to ask ourselves the question, When is it appropriate to consider transmission line effects? From the preceding discussion, we can derive two important constraints:

• **Transmission line effects should be considered when the rise or fall time of the input signal (t_r, t_f) is smaller than the time of flight of the transmission line (t_{flight}).**

This leads to the following rule of thumb, which determines when transmission line effects should be considered:

$$t_r(t_f) < 2.5 t_{flight} = 2.5 \frac{L}{\upsilon} \tag{4.36}$$

For on-chip wires with a maximum length of 1 cm, we only worry about transmission line effects when $t_r < 150$ ps. At the board level, where wires can reach a length of up to 50 cm, we should account for the delay of the transmission line when $t_r < 8$ ns. This condition is easily achieved with state-of-the-art processes and packaging technologies. Ignoring the inductive component of the propagation delay can easily result in overly optimistic delay predictions.

• **Transmission line effects should only be considered when the total resistance of the wire is limited:**

$$R < 5Z_0 \qquad (4.37)$$

If this is not the case, the distributed RC model is sufficient.

Both constraints can be summarized in the following set of bounds on the wire length:

$$\frac{t_r}{2.5}\frac{1}{\sqrt{lc}} < L < \frac{5}{r}\sqrt{\frac{l}{c}} \qquad (4.38)$$

• **The transmission line is considered lossless when the total resistance is substantially smaller than the characteristic impedance, or when**

$$R < \frac{Z_0}{2} \qquad (4.39)$$

∎

Example 4.12   When to Consider Transmission Line Effects

Consider again our Al1 wire. Using the data from Example 4.4 and Eq. (4.28), we can approximate the value of Z_0 for various wire widths:

$$W = 0.1 \ \mu m: \quad c = 92 \ aF/\mu m; \quad Z_0 = 74 \ \Omega$$
$$W = 1.0 \ \mu m: \quad c = 110 \ aF/\mu m; \quad Z_0 = 60 \ \Omega$$
$$W = 10 \ \mu m: \quad c = 380 \ aF/\mu m; \quad Z_0 = 17 \ \Omega$$

For a wire with a width of 1 μm, we can derive the maximum length of the wire for which we should consider transmission line effects by using Eq. (4.37):

$$L_{max} = \frac{5Z_0}{r} = \frac{5 \times 60 \ \Omega}{0.075 \ \Omega/\mu m} = 4000 \ \mu m$$

From Eq. (4.36), we find a corresponding maximum rise (or fall) time of the input signal equal to

$$t_{rmax} = 2.5 \times (4000 \ \mu m)/(15 \ cm/ns) = 67 \ ps$$

This is hard to accomplish using current technologies. For these wires, a lumped capacitance model is more appropriate. Transmission line effects are more plausible in wider wires. For a 10-μm wide wire, we find a maximum length of 11.3 mm, which corresponds to a maximum rise time of 188 ps.

Assume now that a copper wire is implemented on level 5 with a characteristic impedance of 200 Ω, and a resistance of 0.025 Ω/μm. The resulting maximum wire length equals 40 mm. If rise times are smaller than 670 ps, transmission line effects will occur.

Be aware however that the values for Z_0 derived in this example are only approximations. In actual designs, more complex expressions or empirical data should be used.

4.5 SPICE Wire Models

In previous sections, we have discussed the various interconnect parasitics, introducing simple models for each of them. Yet, only through detailed simulation will the designer understand the full and precise impact of these effects. In this section, we introduce the models that SPICE provides for the capacitive, resistive, and inductive parasitics.

4.5.1   Distributed *rc* Lines in SPICE

Because of the importance of the distributed *rc* line in today's design, most circuit simulators have built-in distributed *rc* models of high accuracy. For instance, the Berkeley SPICE3 simulator supports an uniform-distributed *rc*-line model (URC). This model approximates the *rc* line as a network of lumped *RC* segments with internally generated nodes. Parameters include the length of the wire *L* and (optionally) the number of segments used in the model.

Example 4.13   SPICE3 URC Model

A typical example of a SPICE3 instantiation of a distributed rc line is the following:

U1 N1=1 N2=2 N3=0 URCMOD L=50m N=6

.MODEL URCMOD URC(RPERL=75K CPERL=100pF)

N1 and N2 represent the terminal nodes of the line, while N3 is the node the capacitances are connected to. RPERL and CPERL stand for the resistance and capacitance per meter.

If your simulator does not support a distributed *rc* model, or if the computational complexity of these models slows down your simulation too much, you can construct a simple, yet accurate, model yourself by approximating the distributed *rc* by a lumped *RC* network with a limited number of elements. Figure 4-26 shows some of these approximations ordered along increasing precision and complexity. The accuracy of the model is determined by the number of stages. For instance, the error of the π3 model is less than 3%, which generally is sufficient for most applications.

4.5.2   Transmission Line Models in SPICE

SPICE supports a lossless transmission line model. The line characteristics are defined by the characteristic impedance Z_0, while the length of the line can be defined in either of two forms. A first approach is to directly define the *transmission delay TD*, which is equivalent to the time of flight. Alternatively, a frequency *F* may be given together with *NL*, the dimensionless, normalized electrical length of the transmission line, which is measured with respect to the wavelength in the line at the frequency *F*. The following relation is valid:

$$NL = F \cdot TD \tag{4.40}$$

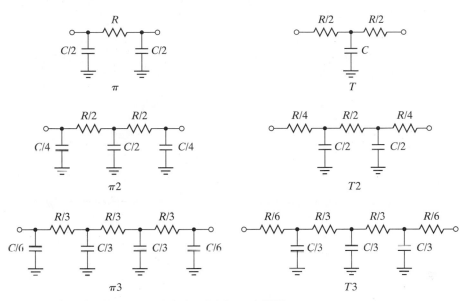

Figure 4-26 Simulation models for distributed *RC* line.

No lossy transmission line model is currently provided. When necessary, loss can be added by breaking up a long transmission line into shorter sections and adding a small series resistance in each section to model the transmission line loss. Be careful when using this approximation. First, the accuracy is still limited, and secondly the simulation speed might be severely affected, since SPICE chooses a time step that is less than or equal to half of the value of *TD*. For small transmission lines, this time step might be much smaller than what is needed for transistor analysis.

4.5.3 Perspective: A Look into the Future

As with the approach we followed for the MOS transistor, it is worthwhile to explore how the wire parameters will evolve with further scaling of the technology. As transistor dimensions are reduced, the interconnect dimensions must also be reduced to take full advantage of the scaling process.

A straightforward approach is to scale all dimensions of the wire by the same factor *S* as the transistors (*ideal scaling*). This might not be possible for at least one dimension of the wire: its length. It can be surmised that the length of *local interconnections*—wires that connect closely grouped transistors—scales in the same way as these transistors. On the other hand, *global interconnections*, which provide the connectivity between large modules and the input/output circuitry, display a different scaling behavior. Examples of such wires are clock signals and data and instruction buses. Figure 4-27 contains a histogram showing the distribution of the wire lengths in an actual microprocessor design, which contains approximately 90,000 gates

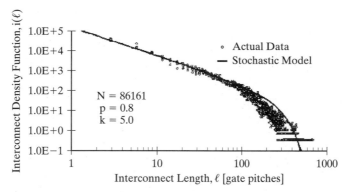

Figure 4-27 Distribution of wire lengths in an advanced microprocessor as a function of the gate pitch.

[Davis98]. While most of the wires tend to be only a couple of gate pitches long, a substantial number of them are much longer, reaching lengths of up to 500 gate pitches.

The average length of these long wires is proportional to the die size (or complexity) of the circuit. An interesting trend is that while transistor dimensions have continued to shrink over the last several decades, the chip sizes have actually increased gradually over the same period. In fact, the size of the typical die (which is the square root of the die area) was increasing by 6% per year, doubling about every decade. Chips have scaled from 2 mm × 2 mm in the early 1960s to approximately 2 cm × 2 cm in 2000. While this increase in die size is bound to taper off, some growth is still to be expected in the coming years.

When studying the scaling behavior of the wire length, therefore, we have to differentiate between local and global wires. In the subsequent analysis, we will consider three models: local wires ($S_L = S > 1$), constant length wires ($S_L = 1$), and global wires ($S_L = S_C < 1$).

Let us now assume now that all other wire dimensions of the interconnect structure (W, H, t) scale with a technology factor S. This leads to the scaling behavior illustrated in Table 4-8. Note that this is only a first-order analysis intended to look at overall trends. Effects

Table 4-8 Ideal scaling of wire properties.

Parameter	Relation	Local Wire	Constant Length	Global Wire
W, H, t		$1/S$	$1/S$	$1/S$
L		$1/S$	1	$1/S_C$
C	LW/t	$1/S$	1	$1/S_C$
R	L/WH	S	S^2	S^2/S_C
CR	L^2/Ht	1	S^2	S^2/S_C^2

such a fringing capacitance are ignored, and breakthroughs in semiconductor technology, such as new interconnect and dielectric materials, also are not considered.

The surprising conclusion from this exercise is that scaling of the technology does not reduce wire delay (as personified by the RC time constant). A constant delay is predicted for local wires, while the delay of the global wires increases 50% per year (for $S = 1.15$ and $S_C = 0.94$). This is in dramatic contrast with the gate delay, which reduces from year to year. This explains why wire delays are starting to play a predominant role in today's digital integrated circuit design.

The ideal scaling approach clearly has problems, however, as it causes a rapid increase in wire resistance. This explains why other interconnect scaling techniques are attractive. One option is to scale the wire thickness at a different rate. The "constant-resistance" model of Table 4-9 explores the impact of not scaling the wire thickness at all. While this approach seemingly has a positive impact on the performance, it causes the fringing and interwire capacitance components to come to the foreground. As a result, we introduce an extra capacitance scaling factor ε_c (> 1), which captures the increasingly horizontal nature of the capacitance when wire widths and pitches are shrunk while the height is kept constant.

This scaling scenario offers a slightly more optimistic perspective, as long as we assume that $\varepsilon_c < S$. Still, delay is bound to increase substantially for intermediate and long wires independent of the scaling scenario. To keep these delays from becoming excessive, interconnect technology has to be drastically improved. One option is to use better interconnect (Cu) and insulation materials (polymers and air). The other option is to differentiate between local and global wires. In the former, integration density and low capacitance are crucial; in the latter, keeping the resistance under control is essential. To address these conflicting demands, modern interconnect topologies combine a dense and thin wiring grid at the lower metal layers with fat and widely spaced wires at the higher levels, as shown in Figure 4-28. Even with these advances, it is obvious that interconnect will play a dominant role in both high performance and low energy circuits for years to come. The continuation of the digital semiconductor roadmap may

Table 4-9 "Constant resistance" scaling of wire properties.

Parameter	Relation	Local Wire	Constant Length	Global Wire
W, t		$1/S$	$1/S$	$1/S$
H		1	1	1
L		$1/S$	1	$1/S_C$
C	$\varepsilon_c LW/t$	ε_c/S	ε_c	ε_c/S_C
R	L/WH	1	S	S/S_C
CR	L^2/Ht	ε_c/S	$\varepsilon_c S$	$\varepsilon_c S/S_C^2$

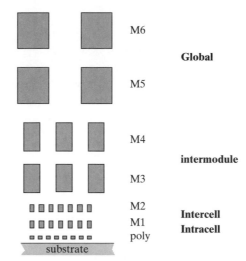

M6

Global

M5

M4

intermodule

M3

M2
M1
poly

Intercell
Intracell

substrate

Figure 4-28   Interconnect hierarchy of 0.25 µm CMOS process, drawn to scale.

well rely on the introduction of alternative and innovative approaches, such as the 3D integration introduced in Chapter 2.

4.6 Summary

This chapter has presented a careful and in-depth analysis of the role and the behavior of the interconnect wire in modern semiconductor technology. The main goal is to identify the dominant parameters that set the values of the wire parasitics—capacitance, resistance, and inductance—and to present adequate wire models that will aid us in the further analysis and optimization of complex digital circuits.

4.7 To Probe Further

The subject of interconnect and its modeling is a hotly debated topic, receiving major attention in journals and conferences. A number of textbooks and reprint volumes also have been published. [Bakoglu90], [Tewksbury94], and [Dally98] present in-depth coverage of interconnect issues, and they are valuable resources for further browsing.

References

[Bakoglu90] H. Bakoglu, *Circuits, Interconnections and Packaging for VLSI*, Addison-Wesley, 1990.

[Dally98] B. Dally, *Digital Systems Engineering,* Cambridge University Press, 1998.

[Davis98] J. Davis and J. Meindl, "Is Interconnect the Weak Link?" *IEEE Circuits and Systems Magazine*, pp. 30–36, March 1998.

[Elmore48] E. Elmore, "The Transient Response of Damped Linear Networks with Particular Regard to Wideband Amplifiers," *Journal of Applied Physics*, pp. 55–63, January 1948.

[Horowitz83] M. Horowitz, "Timing Models for MOS Circuits," Ph.D. diss., Stanford University, 1983.

[Lin84] T. Lin and C. Mead, "Signal delay in general RC networks," *IEEE Transactions on Computer-Aided Design*, vol. 3, no. 4, pp. 321–349, 1984.

[Rubinstein83] J. Rubinstein, P. Penfield, and M. Horowitz, "Signal Delay in *RC* Networks," *IEEE Transactions on Computer-Aided Design*, vol. CAD-2, pp. 202–211, July 1983.

[Schaper83] L. Schaper and D. Amey, "Improved Electrical Performance Required for Future MOS Packaging," *IEEE Trans. on Components, Hybrids and Manufacturing Technology*, vol. CHMT-6, pp. 282–289, September 1983.

[Sylvester97] D. Sylvester, "High-Frequency VLSI Interconnect Modeling," *Project Report EE241*, UCB, May 1997.

[Tewksbury94] S. Tewksbury, ed., *Microelectronics System Interconnections—Performance and Modeling*, IEEE Press, 1994.

[Vdmeijs84] N. Van De Meijs and J. Fokkema, "VLSI Circuit Reconstruction from Mask Topology," *Integration*, vol. 2, no. 2, pp. 85–119, 1984.

2

A Circuit Perspective

"A good scientist is a person with original ideas. A good engineer is a person who makes a design that works with as few original ideas as possible,"

Freeman Dyson
in "Disturbing the Universe", (1979).

"A design is what the designer has when time and money run out,"

James Poole
in "The Fifth 637 Best Things Anybody Ever Said", (1993).

The CMOS Inverter

Quantification of integrity, performance, and energy metrics of an inverter
Optimization of an inverter design

5.1   Introduction

5.2   The Static CMOS Inverter—An Intuitive Perspective

5.3   Evaluating the Robustness of the CMOS Inverter—The Static Behavior

    5.3.1   Switching Threshold

    5.3.2   Noise Margins

    5.3.3   Robustness Revisited

5.4   Performance of CMOS Inverter: The Dynamic Behavior

    5.4.1   Computing the Capacitances

    5.4.2   Propagation Delay: First-Order Analysis

    5.4.3   Propagation Delay from a Design Perspective

5.5   Power, Energy, and Energy Delay

    5.5.1   Dynamic Power Consumption

    5.5.2   Static Consumption

    5.5.3   Putting It All Together

    5.5.4   Analyzing Power Consumption by Using SPICE

5.6   Perspective: Technology Scaling and its Impact on the Inverter Metrics

5.7   Summary

5.8   To Probe Further

5.1 Introduction

The inverter is truly the nucleus of all digital designs. Once its operation and properties are clearly understood, designing more intricate structures such as logic gates, adders, multipliers, and microprocessors is greatly simplified. The electrical behavior of these complex circuits can be almost completely derived by extrapolating the results obtained for inverters. The analysis of inverters can be extended to explain the behavior of more complex gates such as NAND, NOR, or XOR, which in turn form the building blocks for modules such as multipliers and processors.

In this chapter, we focus on a single incarnation of the inverter gate—the static CMOS inverter. This is certainly the most popular inverter at present, and therefore deserves special attention. We analyze the gate with respect to the different design metrics that were outlined in Chapter 1:

- *cost*, expressed by the complexity and area
- *integrity and robustness*, expressed by the static (or steady-state) behavior
- *performance*, determined by the dynamic (or transient) response
- *energy efficiency,* set by the energy and power consumption

Using this analysis, we develop a model of the gate and identify its design parameters. We develop methods to choose the parameter values so that the resulting design meets the desired specifications. While each of these parameters can easily be quantified for a given technology, we also discuss how they are affected by *scaling of the technology.*

While the chapter focuses uniquely on the CMOS inverter, in the next chapter, we see that the same methodology also applies to other gate topologies.

5.2 The Static CMOS Inverter—An Intuitive Perspective

Figure 5-1 shows the circuit diagram of a static CMOS inverter. Its operation is readily understood with the aid of the simple switch model of the MOS transistor that we introduced in Chapter 3 (see Figure 3-26). The transistor is nothing more than a switch with an infinite off-

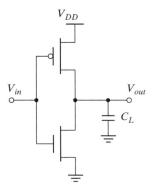

Figure 5-1 Static CMOS inverter. V_{DD} stands for the supply voltage.

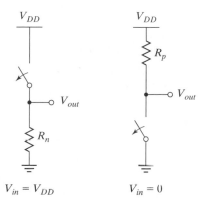

Figure 5-2   Switch models of CMOS inverter.

resistance (for $|V_{GS}| < |V_T|$) and a finite on-resistance (for $|V_{GS}| > |V_T|$). This leads to the following interpretation of the inverter. When V_{in} is high and equal to V_{DD}, the NMOS transistor is on and the PMOS is off. This yields the equivalent circuit of Figure 5-2a. A direct path exists between V_{out} and the ground node, resulting in a steady-state value of 0 V. On the other hand, when the input voltage is low (0 V), NMOS and PMOS transistors are off and on, respectively. The equivalent circuit of Figure 5-2b shows that a path exists between V_{DD} and V_{out}, yielding a high output voltage. The gate clearly functions as an inverter.

A number of other important properties of static CMOS can be derived from this switch level view:

- The high and low output levels equal V_{DD} and *GND*, respectively; in other words, the voltage swing is equal to the supply voltage. This results in high noise margins.
- The logic levels are not dependent upon the relative device sizes, so that the transistors can be minimum size. Gates with this property are called *ratioless*. This is in contrast with *ratioed logic*, where logic levels are determined by the relative dimensions of the composing transistors.
- In steady state, there always exists a path with finite resistance between the output and either V_{DD} or *GND*. A well-designed CMOS inverter, therefore, has a *low output impedance*, which makes it less sensitive to noise and disturbances. Typical values of the output resistance are in kΩ range.
- The *input resistance* of the CMOS inverter is extremely high, as the gate of an MOS transistor is a virtually perfect insulator and draws no dc input current. Since the input node of the inverter only connects to transistor gates, the steady-state input current is nearly zero. A single inverter can theoretically drive an infinite number of gates (or have an infinite fan-out) and still be functionally operational; however, increasing the fan-out also increases the propagation delay, as will become clear shortly. Although fan-out does not have any effect on the steady-state behavior, it degrades the transient response.

- No direct path exists between the supply and ground rails under steady-state operating conditions (i.e., when the input and outputs remain constant). The absence of current flow (ignoring leakage currents) means that the gate does not consume any static power.

SIDELINE: The preceding observation, while seemingly obvious, is of crucial importance, and is one of the primary reasons CMOS is the digital technology of choice at present. The situation was very different in the 1970s and early 1980s. All early microprocessors—such as the Intel 4004—were implemented in a pure NMOS technology. The lack of complementary devices (such as the NMOS and PMOS transistor) in such a technology makes the realization of inverters with zero static power nontrivial. The resulting static power consumption puts a firm upper bound on the number of gates that can be integrated on a single die; hence, the forced move to CMOS in the 1980s, when scaling of the technology allowed for higher integration densities.

The nature and the form of the voltage-transfer characteristic (VTC) can be graphically deduced by superimposing the current characteristics of the NMOS and the PMOS devices. Such a graphical construction is traditionally called *a load-line plot*. It requires that the *I-V* curves of the NMOS and PMOS devices are transformed onto a common coordinate set. We have selected the input voltage V_{in}, the output voltage V_{out} and the NMOS drain current I_{DN} as the variables of choice. The PMOS *I-V* relations can be translated into this variable space by the following relations (the subscripts n and p denote the NMOS and PMOS devices, respectively):

$$I_{DSp} = -I_{DSn}$$
$$V_{GSn} = V_{in}; \quad V_{GSp} = V_{in} - V_{DD} \qquad (5.1)$$
$$V_{DSn} = V_{out}; \quad V_{DSp} = V_{out} - V_{DD}$$

The load-line curves of the PMOS device are obtained by a mirroring around the *x*-axis and a horizontal shift over V_{DD}. This procedure is outlined in Figure 5-3, where the subsequent steps to adjust the original PMOS *I-V* curves to the common coordinate set V_{in}, V_{out}, and I_{Dn} are illustrated.

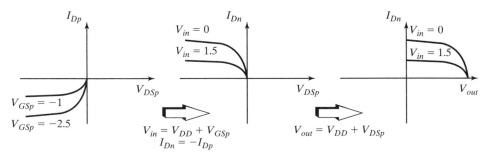

Figure 5-3 Transforming PMOS *I-V* characteristic to a common coordinate set (assuming $V_{DD} = 2.5$ V).

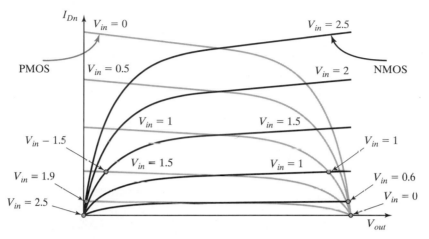

Figure 5-4   Load curves for NMOS and PMOS transistors of the static CMOS inverter (V_{DD} = 2.5 V). The dots represent the dc operation points for various input voltages.

The resulting load lines are plotted in Figure 5-4. For a dc operating point to be valid, the currents through the NMOS and PMOS devices must be equal. Graphically, this means that the dc points must be located at the intersection of corresponding load lines. A number of those points (for V_{in} = 0, 0.5, 1, 1.5, 2, and 2.5 V) are marked on the graph. As can be seen, all operating points are located either at the high or low output levels. The VTC of the inverter thus exhibits a very narrow transition zone. This results from the high gain during the switching transient, when both NMOS and PMOS are simultaneously on and in saturation. In that operation region, a small change in the input voltage results in a large output variation. All these observations translate into the VTC shown in Figure 5-5.

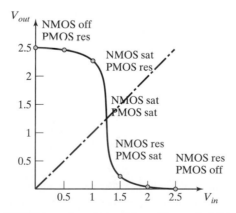

Figure 5-5   VTC of static CMOS inverter, derived from Figure 5-4 (V_{DD} = 2.5 V). For each operation region, the modes of the transistors are annotated—off, res(istive), or sat(urated).

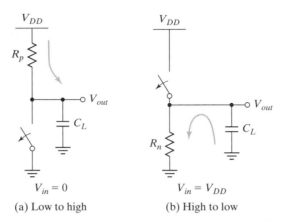

(a) Low to high   (b) High to low

Figure 5-6   Switch model of dynamic behavior of static CMOS inverter.

Before going into the analytical details of the operation of the CMOS inverter, a qualitative analysis of the transient behavior of the gate is appropriate. This response is dominated mainly by the output capacitance of the gate, C_L, which is composed of the drain diffusion capacitances of the NMOS and PMOS transistors, the capacitance of the connecting wires, and the input capacitance of the fan-out gates. Assuming temporarily that the transistors switch instantaneously, we can get an approximate idea of the transient response by using the simplified switch model again. Let us first consider the low-to-high transition (see Figure 5-6a). The gate response time is simply determined by the time it takes to charge the capacitor C_L through the resistor R_p. In Example 4.5, we learned that the propagation delay of such a network is proportional to the time constant $R_p C_L$. **Hence, a fast gate is built either by keeping the output capacitance small or by decreasing the on-resistance of the transistor.** The latter is achieved by increasing the W/L ratio of the device. Similar considerations are valid for the high-to-low transition (Figure 5-6b), which is dominated by the $R_n C_L$ time constant. The reader should be aware that the on-resistance of the NMOS and PMOS transistor is not constant; rather, it is a nonlinear function of the voltage across the transistor. This complicates the exact determination of the propagation delay. (An in-depth analysis of how to analyze and optimize the performance of the static CMOS inverter is offered in Section 5.4.)

5.3 Evaluating the Robustness of the CMOS Inverter—The Static Behavior

In the preceding qualitative discussion, the overall shape of the voltage-transfer characteristic of the static CMOS inverter was sketched, and the values of V_{OH} and V_{OL}—which are evaluated to V_{DD} and GND, respectively—were derived. It remains to determine the precise values of V_M, V_{IH}, and V_{IL}, as well as the noise margins.

5.3.1 Switching Threshold

The switching threshold V_M is defined as the point where $V_{in} = V_{out}$. Its value can be obtained graphically from the intersection of the VTC with the line given by $V_{in} = V_{out}$ (see Figure 5-5). In this region, both PMOS and NMOS are always saturated, since $V_{DS} = V_{GS}$. An analytical expression for V_M is obtained by equating the currents through the transistors. We solve for the case in which the supply voltage is high enough so that the devices can be assumed to be velocity-saturated (or $V_{DSAT} < V_M - V_T$). Furthermore, we ignore the channel length modulation effects. We have

$$k_n V_{DSATn}\left(V_M - V_{Tn} - \frac{V_{DSATn}}{2}\right) + k_p V_{DSATp}\left(V_M - V_{DD} - V_{Tp}\frac{V_{DSATp}}{2}\right) = 0 \tag{5.2}$$

Solving for V_M yields

$$V_M = \frac{\left(V_{Tn} + \frac{V_{DSATn}}{2}\right) + r\left(V_{DD} + V_{Tp} + \frac{V_{DSATp}}{2}\right)}{1 + r} \quad \text{with} \quad r = \frac{k_p V_{DSATp}}{k_n V_{DSATn}} = \frac{\upsilon_{satp} W_p}{\upsilon_{satn} W_n} \tag{5.3}$$

assuming identical oxide thicknesses for PMOS and NMOS transistors. For large values of V_{DD} (compared with threshold and saturation voltages), Eq. (5.3) can be simplified:

$$V_M \approx \frac{r V_{DD}}{1 + r} \tag{5.4}$$

Equation (5.4) states that the switching threshold is set by the ratio r, which compares the relative driving strengths of the PMOS and NMOS transistors. It is generally desirable for V_M to be located around the middle of the available voltage swing (or at $V_{DD}/2$), since this results in comparable values for the low and high noise margins. This requires r to be approximately 1, which is equivalent to sizing the PMOS device so that $(W/L)_p = (W/L)_n \times (V_{DSATn}k'_n)/(V_{DSATp}k'_p)$. To move V_M upwards, a larger value of r is required, which means making the PMOS wider. Increasing the strength of the NMOS, on the other hand, moves the switching threshold closer to GND.

From Eq. (5.2), we derive the required ratio of PMOS to NMOS transistor sizes such that the switching threshold is set to a desired value V_M:

$$\frac{(W/L)_p}{(W/L)_n} = \frac{k'_n V_{DSATn}(V_M - V_{Tn} - V_{DSATn}/2)}{k'_p V_{DSATp}(V_{DD} - V_M + V_{Tp} + V_{DSATp}/2)} \tag{5.5}$$

When using this expression, make sure that the assumption that both devices are velocity saturated still holds for the chosen operation point.

Problem 5.1   Inverter Switching Threshold for Long-Channel Devices, or Low-Supply Voltages

The preceding expressions were derived under the assumption that the transistors are velocity saturated. When the PMOS and NMOS are long-channel devices, or when the supply voltage is low, velocity saturation does not occur $(V_M - V_T < V_{DSAT})$. Under these circumstances, the following equation holds for V_M:

$$V_M = \frac{V_{Tn} + r(V_{DD} + V_{Tp})}{1 + r} \quad \text{with } r = \sqrt{\frac{-k_p}{k_n}} \qquad (5.6)$$

Derive this equation.

Design Technique—Maximizing the Noise Margins

When designing static CMOS circuits, it is advisable to balance the driving strengths of the transistors by making the PMOS section wider than the NMOS section if maximizing the noise margins and obtaining symmetrical characteristics are desired. The required ratio is given by Eq. (5.5).   ■

Example 5.1   Switching Threshold of CMOS Inverter

We derive the sizes of PMOS and NMOS transistors such that the switching threshold of a CMOS inverter, implemented in our generic 0.25 μm CMOS process, is located in the middle between the supply rails. We use the process parameters presented in Example 3.7, and assume a supply voltage of 2.5 V. The minimum size device has a width-to-length ratio of 1.5. With the aid of Eq. (5.5), we find that

$$\frac{(W/L)_p}{(W/L)_n} = \frac{115 \times 10^{-6}}{30 \times 10^{-6}} \times \frac{0.63}{1.0} \times \frac{(1.25 - 0.43 - 0.63/2)}{(1.25 - 0.4 - 1.0/2)} = 3.5$$

Figure 5-7 plots the values of switching threshold as a function of the PMOS-to-NMOS ratio, as obtained by circuit simulation. The simulated PMOS-to-NMOS ratio of 3.4 for a 1.25-V switching threshold confirms the value predicted by Eq. (5.5).

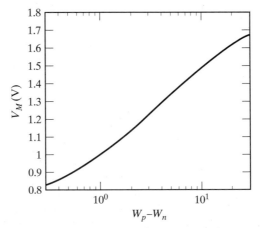

Figure 5-7   Simulated inverter switching threshold versus PMOS-to-NMOS ratio (0.25-μm CMOS, $V_{DD} = 2.5$ V).

An analysis of the curve of Figure 5-7 leads to some interesting observations:

1. V_M is relatively insensitive to variations in the device ratio. This means that small variations of the ratio (e.g., making it 3 or 2.5) do not disturb the transfer characteristic that much. It is therefore an accepted practice in industrial designs to set the width of the PMOS transistor to values smaller than those required for exact symmetry. For the preceding example, setting the ratio to 3, 2.5, and 2 yields switching thresholds of 1.22 V, 1.18 V, and 1.13 V, respectively.

2. The effect of changing the W_p-to-W_n ratio is to shift the transient region of the VTC. Increasing the width of the PMOS or the NMOS moves V_M toward V_{DD} or GND, respectively. This property can be very useful, as asymmetrical transfer characteristics are actually desirable in some designs. This is demonstrated by the example of Figure 5-8. The incoming signal V_{in} has a very noisy zero value. Passing this signal through a symmetrical inverter would lead to erroneous values (Figure 5-8a). This can be addressed by raising the threshold of the inverter, which results in a correct response (Figure 5-8b). Later in the text we will see other circuit instances in which inverters with asymmetrical switching thresholds are desirable. Changing the switching threshold by a considerable amount, however, is not easy, especially when the ratio of supply voltage to transistor threshold is relatively small (2.5/0.4 = 6, for our particular example). To move the threshold to 1.5 V requires a transistor ratio of 11, and further increases are prohibitively expensive. Observe that Figure 5-7 is plotted in a semilog format.

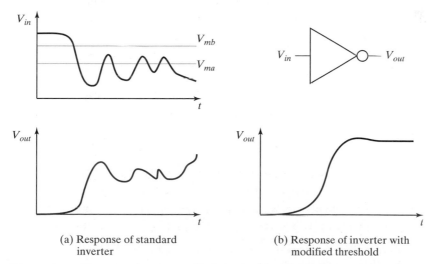

(a) Response of standard inverter

(b) Response of inverter with modified threshold

Figure 5-8   Changing the inverter threshold can improve the circuit reliability.

5.3.2   Noise Margins

By definition, V_{IH} and V_{IL} are the operational points of the inverter where $\dfrac{dV_{out}}{dV_{in}} = -1$. In the terminology of the analog circuit designer, these are the points where the gain g of the amplifier, formed by the inverter, is equal to -1. While it is indeed possible to derive analytical expressions for V_{IH} and V_{IL}, these tend to be unwieldy and provide little insight in what parameters are instrumental in setting the noise margins.

A simpler approach is to use a piece-wise linear approximation for the VTC, as shown in Figure 5-9. The transition region is approximated by a straight line, the gain of which equals the gain g at the switching threshold V_M. The crossover with the V_{OH} and the V_{OL} lines is used to define V_{IH} and V_{IL} points. The error introduced is small and well within the range of what is required for an initial design. This approach yields the following expressions for the width of the transition region $V_{IH} - V_{IL}$, V_{IH}, V_{IL}, and the noise margins NM_H and NM_L:

$$V_{IH} - V_{IL} = -\frac{(V_{OH} - V_{OL})}{g} = \frac{-V_{DD}}{g}$$

$$V_{IH} = V_M - \frac{V_M}{g} \qquad V_{IL} = V_M + \frac{V_{DD} - V_M}{g} \tag{5.7}$$

$$NM_H = V_{DD} - V_{IH} \qquad NM_L = V_{IL}$$

These expressions make it increasingly clear that a high gain in the transition region is very desirable. In the extreme case of an infinite gain, the noise margins simplify to $V_{OH} - V_M$ and $V_M - V_{OL}$ for NM_H and NM_L, respectively, and span the complete voltage swing.

It remains for us to determine the midpoint gain of the static CMOS inverter. We assume once again that both PMOS and NMOS are velocity saturated. It is apparent from Figure 5-4 that the gain is a strong function of the slopes of the currents in the saturation region. The channel-

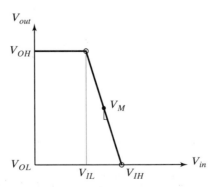

Figure 5-9   A piecewise linear approximation of the VTC simplifies the derivation of V_{IL} and V_{IH}.

length modulation factor therefore cannot be ignored in this analysis—doing so would lead to an infinite gain. The gain can now be derived by differentiating the current Eq. (5.8), which is valid around the switching threshold, with respect to V_{in}:

$$k_n V_{DSATn}\left(V_{in} - V_{Tn} - \frac{V_{DSATn}}{2}\right)(1 + \lambda_n V_{out}) +$$

$$k_p V_{DSATp}\left(V_{in} - V_{DD} - V_{Tp} - \frac{V_{DSATp}}{2}\right)(1 + \lambda_p V_{out} - \lambda_p V_{DD}) = 0 \qquad (5.8)$$

Differentiating and solving for dV_{out}/dV_{in}, yields

$$\frac{dV_{out}}{dV_{in}} = -\frac{k_n V_{DSATn}(1 + \lambda_n V_{out}) + k_p V_{DSATp}(1 + \lambda_p V_{out} - \lambda_p V_{DD})}{\lambda_n k_n V_{DSATn}(V_{in} - V_{Tn} - V_{DSATn}/2) + \lambda_p k_p V_{DSATp}(V_{in} - V_{DD} - V_{Tp} - V_{DSATp}/2)} \qquad (5.9)$$

Ignoring some second-order terms and setting $V_{in} = V_M$ produces the gain expression,

$$g = -\frac{1}{I_D(V_M)}\frac{k_n V_{DSATn} + k_p V_{DSATp}}{\lambda_n - \lambda_p}$$

$$\approx \frac{1 + r}{(V_M - V_{Tn} - V_{DSATn}/2)(\lambda_n - \lambda_p)} \qquad (5.10)$$

with $I_D(V_M)$ the current flowing through the inverter for $V_{in} = V_M$. The gain is almost purely determined by technology parameters, especially the channel-length modulation. It can only be influenced in a minor way by the designer through the choice of the supply voltage and the transistor sizes.

Example 5.2   Voltage Transfer Characteristic and Noise Margins of CMOS Inverter

Assume an inverter in the generic 0.25-μm CMOS technology designed with a PMOS-to-NMOS ratio of 3.4 and with the NMOS transistor minimum size ($W = 0.375$ μm, $L = 0.25$ μm, $W/L = 1.5$). We first compute the gain at V_M (= 1.25 V):

$$I_D(V_M) = 1.5 \times 115 \times 10^{-6} \times 0.63 \times (1.25 - 0.43 - 0.63/2) \times (1 + 0.06 \times 1.25) = 59 \times 10^{-6}\ \text{A}$$

$$g = -\frac{1}{59 \times 10^{-6}}\frac{1.5 \times 115 \times 10^{-6} \times 0.63 + 1.5 \times 3.4 \times 30 \times 10^{-6} \times 1.0}{0.06 + 0.1} = -27.5 \quad \text{(Eq. 5.10a)}$$

This yields the following values for V_{IL}, V_{IH}, NM_L, NM_H:

$$V_{IL} = 1.2\ \text{V},\ V_{IH} = 1.3\ \text{V},\ NM_L = NM_H = 1.2$$

Figure 5-10 plots the simulated VTC of the inverter, as well as its derivative, the gain. A close to ideal characteristic is obtained. The actual values of V_{IL} and V_{IH} are 1.03 V and 1.45 V, respectively, which leads to noise margins of 1.03 V and 1.05 V. These values are lower than those predicted, for two reasons:

- Eq. (5.10) overestimates the gain. As observed in Figure 5-10b, the maximum gain (at V_M) equals only 17. This reduced gain would yield values for V_{IL} and V_{IH} of 1.17 V, and 1.33 V, respectively[1].
- The most important deviation is due to the piecewise linear approximation of the VTC, which is optimistic with respect to the actual noise margins.
  The expressions obtained are, however, perfectly useful as first-order estimations, as well as means of identifying the relevant parameters and their impact.

To conclude this example, we also extracted from simulations the output resistance of the inverter in the low- and high-output states. Low values of 2.4 kΩ and 3.3 kΩ, respectively, were observed. The output resistance is a good measure of the sensitivity of the gate with respect to noise induced at the output, and is preferably as low as possible.

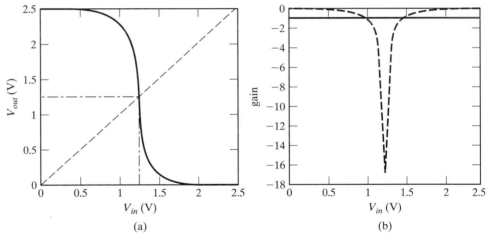

Figure 5-10   Simulated Voltage Transfer Characteristic (a) and voltage gain (b) of CMOS inverter (0.25-μm CMOS, V_{DD} = 2.5 V).

SIDELINE: Surprisingly (or perhaps not so surprisingly), the static CMOS inverter can also be used as an analog amplifier, as it has a fairly high gain in its transition region. This region is very narrow, however, as is apparent in the graph of Figure 5-10b. It also receives poor marks on other amplifier properties such as supply noise rejection. Still, this observation can be used to demonstrate one of the major differences between analog and digital design. Where the analog designer would bias the amplifier in the middle of the transient region so that a maximum linearity is obtained, the digital designer will operate the device in the regions of extreme nonlinearity, resulting in well-defined and well-separated high and low signals.

---

[1] In addition, Eq. (5.10) is not entirely valid for this particular example. The attentive reader will observe that for the operating conditions at hand, the PMOS operates in saturation mode, not velocity saturation. The impact on the result is minor, however.

Problem 5.2   Inverter Noise Margins for Long-Channel Devices

Derive expressions for the gain and noise margins assuming that PMOS and NMOS are long-channel devices (or that the supply voltage is low), so that velocity saturation does not occur.

5.3.3   Robustness Revisited

Device Variations

While we design a gate for nominal operation conditions and typical device parameters, we should always be aware that the actual operating temperature might vary over a large range, and that the device parameters after fabrication probably will deviate from the nominal values we used in our design optimization process. Fortunately, the dc characteristics of the static CMOS inverter turn out to be rather insensitive to these variations, and the gate remains functional over a wide range of operating conditions. This already became apparent in Figure 5-7, which shows that variations in the device sizes have only a minor impact on the switching threshold of the inverter. To further confirm the assumed robustness of the gate, we have resimulated the voltage transfer characteristic by replacing the nominal devices by their worst or best case incarnations. Two corner cases are plotted in Figure 5-11: a better-than-expected NMOS, combined with an inferior PMOS, and the opposite scenario. Comparing the resulting curves with the nominal response shows that the operation of the gate is by no means affected, and that the variations mainly cause a shift in the switching threshold. This robust behavior, which ensures functionality of the gate over a wide range of conditions, has contributed in a big way to the popularity of the static CMOS gate.

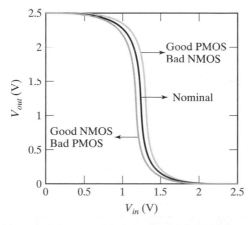

Figure 5-11   Impact of device variations on static CMOS inverter VTC. The "good" device has a smaller oxide thickness (– 3 nm), a smaller length (– 25 nm), a higher width (+ 30 nm), and a smaller threshold (– 60 mV). The opposite is true for the "bad" transistor.

Scaling the Supply Voltage

In Chapter 3, we observed that continuing technology scaling forces the supply voltages to reduce at rates similar to the device dimensions. At the same time, device threshold voltages are virtually kept constant. You may wonder about the impact of this trend on the integrity parameters of the CMOS inverter. Do inverters keep on working when the voltages are scaled, and are there potential limits to the supply scaling?

A first hint on what might happen was offered in Eq. (5.10), which indicates that the gain of the inverter in the transition region actually increases with a reduction of the supply voltage! Note that for a fixed transistor ratio r, V_M is approximately proportional to V_{DD}. Plotting the (normalized) VTC for different supply voltages not only confirms this conjecture, but even shows that the inverter is well and alive for supply voltages close to the threshold voltage of the composing transistors (see Figure 5-12a). At a voltage of 0.5 V—which is just 100 mV above the threshold of the transistors—the width of the transition region measures only 10% of the supply voltage (for a maximum gain of 35), while it widens to 17% for 2.5 V. So, given this improvement in dc characteristics, why do we not choose to operate all of our digital circuits at these low supply voltages? Three important reasons come to mind:

- Reducing the supply voltage indiscriminately has a positive impact on the energy dissipation, but is absolutely detrimental to the delay of the gate, as we will learn in the next sections.
- The dc characteristic becomes increasingly sensitive to variations in the device parameters, such as the transistor threshold, once supply voltages and intrinsic voltages become comparable.
- Scaling the supply voltage means reducing the signal swing. While this typically helps to reduce the internal noise in the system (such as caused by crosstalk), it makes the design more sensitive to external noise sources that do not scale.

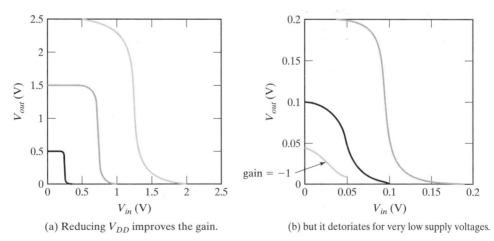

(a) Reducing V_{DD} improves the gain.   (b) but it detoriates for very low supply voltages.

Figure 5-12   VTC of CMOS inverter as a function of supply voltage (0.25-μm CMOS technology).

To provide an insight into the question on potential limits to the voltage scaling, we have plotted the voltage transfer characteristic of the same inverter for the even lower supply voltages of 200 mV, 100 mV, and 50 mV in Figure 5-12b. The transistor thresholds are kept at the same level. Amazingly enough, we still obtain an inverter characteristic, even though the supply voltage is not large enough to turn the transistors on! The explanation can be found in the subthreshold operation of the transistors. The subthreshold currents are sufficient to switch the gate between low and high levels, as well as to provide enough gain to produce acceptable VTCs. The low value of the switching currents ensures a very slow operation, but this might be acceptable for some applications (such as watches, for example).

At around 100 mV, we start observing a major deterioration of the gate characteristic. V_{OL} and V_{OH} are no longer at the supply rails and the transition region gain approaches 1. The latter turns out to be a fundamental showstopper. To achieve sufficient gain for use in a digital circuit, it is necessary that the supply be at least two times $\phi_T = kT/q$ (= 25 mV at room temperature), the thermal voltage introduced in Chapter 3 [Swanson72]. It turns out that below this same voltage, thermal noise becomes an issue as well, potentially resulting in unreliable operation. We express this relation as

$$V_{DDmin} > 2...4\frac{kT}{q} \tag{5.11}$$

Equation (5.11) presents a true lower bound on supply scaling. It suggests that the only way to get CMOS inverters to operate below 100 mV is to reduce the ambient temperature—or in other words, to cool the circuit.

Problem 5.3   Minimum Supply Voltage of CMOS Inverter

Once the supply voltage drops below the threshold voltage, the transistors operate in the subthreshold region, and display an exponential current–voltage relationship (as expressed in Eq. (3.39)). Derive an expression for the gain of the inverter under these circumstances (assume symmetrical NMOS and PMOS transistors, and a maximum gain at $V_M = V_{DD}/2$). The resulting expression demonstrates that the minimum voltage is a function of the slope factor n of the transistor:

$$g = -\left(\frac{1}{n}\right)\left(e^{V_{DD}/2\phi_T} - 1\right) \tag{5.12}$$

According to this expression, the gain drops to -1 at $V_{DD} = 48$ mV (for $n = 1.5$ and $\phi T = 25$ mV).

5.4 Performance of CMOS Inverter: The Dynamic Behavior

The qualitative analysis presented earlier concluded that the propagation delay of the CMOS inverter is determined by the time it takes to charge and discharge the load capacitor C_L through the PMOS and NMOS transistors, respectively. This observation suggests that **getting C_L as small as possible is crucial to the realization of high-performance CMOS circuits**. It is thus worthwhile to first study the major components of the load capacitance before embarking on an in-depth analysis of the propagation delay of the gate. In addition to this detailed analysis, this section also presents a summary of techniques that a designer might use to optimize the performance of the inverter.

5.4.1 Computing the Capacitances

Manual analysis of MOS circuits where each capacitor is considered individually is virtually impossible. The problem is exacerbated by the many nonlinear capacitances in the MOS transistor model. To make the analysis tractable, we assume that all capacitances are lumped together into one single capacitor C_L, located between V_{out} and *GND*. Be aware that this is a considerable simplification of the actual situation, even in the case of a simple inverter.

Figure 5-13 shows the schematic of a cascaded inverter pair. It includes all the capacitances influencing the transient response of node V_{out}. It is initially assumed that the input V_{in} is driven by an *ideal voltage source with zero rise and fall times*. Accounting only for capacitances connected to the output node, C_L breaks down into the following components.

Gate-Drain Capacitance C_{gd12}

M_1 and M_2 are either in cut-off or in the saturation mode during the first half (up to 50% point) of the output transient. Under these circumstances, the only contributions to C_{gd12} are the overlap capacitances of both M_1 and M_2. The channel capacitance of the MOS transistors does not play a role here, as it is located either completely between gate and bulk (cut-off) or gate and source (saturation) (see Chapter 3).

The lumped capacitor model now requires that this floating gate-drain capacitor be replaced by a capacitance to ground. This is accomplished by taking the so-called Miller effect into account. During a low-high or high-low transition, the terminals of the gate-drain capacitor are moving in opposite directions (see Figure 5-14). The voltage change over the floating capacitor is thus twice the actual output voltage swing. To present an identical load to the out-

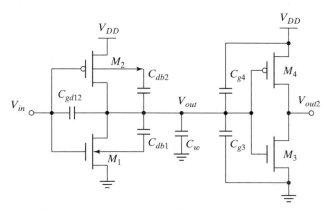

Figure 5-13   Parasitic capacitances, influencing the transient behavior of the cascaded inverter pair.

Figure 5-14 The Miller effect—A capacitor experiencing identical but opposite voltage swings at both its terminals can be replaced by a capacitor to ground, whose value is two times the original value.

put node, the capacitance to ground must have a value that is twice as large as the floating capacitance.

We use the following equation for the gate-drain capacitors: $C_{gd} = 2\,C_{GD0}W$ (with C_{GD0} the overlap capacitance per unit width as used in the SPICE model). For an in-depth discussion of the Miller effect, please refer to textbooks such as [Sedra87, p. 57].[2]

Diffusion Capacitances C_{db1} and C_{db2}

The capacitance between drain and bulk is due to the reverse-biased *pn*-junction. Such a capacitor is, unfortunately, quite nonlinear and depends heavily on the applied voltage. We argued in Chapter 3 that the best approach to simplifying the analysis is to replace the nonlinear capacitor by a linear one with the same change in charge for the voltage range of interest. A multiplication factor K_{eq} is introduced to relate the linearized capacitor to the value of the junction capacitance under zero-bias conditions.

$$C_{eq} = K_{eq}C_{j0} \tag{5.13}$$

with C_{j0} the junction capacitance per unit area under zero-bias conditions. For convenience, we repeat Eq. (3.11) here, written as

$$K_{eq} = \frac{-\phi_0^m}{(V_{high} - V_{low})(1-m)}[(\phi_0 - V_{high})^{1-m} - (\phi_0 - V_{low})^{1-m}] \tag{5.14}$$

with ϕ_0 the built-in junction potential and m the grading coefficient of the junction. Observe that the junction voltage is defined to be negative for reverse-biased junctions.

Example 5.3 K_{eq} for a 2.5-V CMOS Inverter

Consider the inverter of Figure 5-13 designed in the generic 0.25-μm CMOS technology. The relevant capacitance parameters for this process were summarized in Table 3-5.

Let us first analyze the NMOS transistor (C_{db1} in Figure 5-13). The propagation delay is defined by the time between the 50% transitions of the input and the output. For the CMOS inverter, this is the time instance where V_{out} reaches 1.25-V, as the output

---

[2]The Miller effect discussed in this context is a simplified version of the general analog case. In a digital inverter, the large-scale gain between input and output always equals −1.

voltage swing goes from rail to rail or equals 2.5 V. We therefore linearize the junction capacitance over the interval $\{2.5\ V,\ 1.25\ V\}$ for the high-to-low transition, and $\{0,\ 1.25\ V\}$ for the low-to-high transition.

During the high-to-low transition at the output, V_{out} initially equals 2.5 V. Because the bulk of the NMOS device is connected to GND, this translates into a reverse voltage of 2.5 V over the drain junction or $V_{high} = -2.5$ V. At the 50% point, $V_{out} = 1.25$ V or $V_{low} = -1.25$ V. Evaluating Eq. (5.14) for the bottom plate and sidewall components of the diffusion capacitance yields the following data:

Bottom plate: $K_{eq}\ (m = 0.5,\ \phi_0 = 0.9) = 0.57$

Sidewall: $K_{eqsw}\ (m = 0.44,\ \phi_0 = 0.9) = 0.61$

During the low-to-high transition, V_{low} and V_{high} equal 0 V and -1.25 V, respectively, resulting in higher values for K_{eq}:

Bottom plate: $K_{eq}\ (m = 0.5,\ \phi_0 = 0.9) = 0.79$

Sidewall: $K_{eqsw}\ (m = 0.44,\ \phi_0 = 0.9) = 0.81$

The PMOS transistor displays a reverse behavior, as its substrate is connected to 2.5 V. Hence, for the high-to-low transition ($V_{low} = 0$, $V_{high} = -1.25$ V), we have

Bottom plate: $K_{eq}\ (m = 0.48,\ \phi_0 = 0.9) = 0.79$

Sidewall: $K_{eqsw}\ (m = 0.32,\ \phi_0 = 0.9) = 0.86$

Finally, for the low-to-high transition ($V_{low} = -1.25$ V, $V_{high} = -2.5$ V), we have

Bottom plate: $K_{eq}\ (m = 0.48,\ \phi_0 = 0.9) = 0.59$

Sidewall: $K_{eqsw}\ (m = 0.32,\ \phi_0 = 0.9) = 0.7$

By using this approach, the junction capacitance can be replaced by a linear component and treated as any other device capacitance. The result of the linearization is a minor error in the voltage and current waveforms. The logic delays are not significantly influenced by this simplification.

Wiring Capacitance C_w

The capacitance due to the wiring depends on the length and width of the connecting wires, and is a function of the distance of the fan-out from the driving gate and the number of fan-out gates. As argued in Chapter 4, this component is growing in importance with the scaling of the technology.

Gate Capacitance of Fan-Out C_{g3} and C_{g4}

We assume that the fan-out capacitance equals the total gate capacitance of the loading gates M_3 and M_4. Hence,

$$
\begin{aligned}
C_{fan\text{-}out} &= C_{gate}(\text{NMOS}) + C_{gate}(\text{PMOS}) \\
&= (C_{GSOn} + C_{GDOn} + W_n L_n C_{ox}) + (C_{GSOp} + C_{GDOp} + W_p L_p C_{ox})
\end{aligned}
\tag{5.15}
$$

This expression simplifies the actual situation in two ways:

- It assumes that all components of the gate capacitance are connected between V_{out} and GND (or V_{DD}), and it ignores the Miller effect on the gate–drain capacitances. This has a relatively minor effect on the accuracy, since we can safely assume that the connecting gate does not switch before the 50% point is reached, and V_{out2} thus remains constant in the interval of interest.

- A second approximation is that the channel capacitance of the connecting gate is constant over the interval of interest. This is not exactly the case as we discovered in Chapter 3. The total channel capacitance is a function of the operation mode of the device, and varies from approximately (2/3) WLC_{ox} (saturation) to the full WLC_{ox} (linear and cutoff). A drop in overall gate capacitance also occurs just before the transistor turns on, as in Figure 3-31. During the first half of the transient, it may be assumed that one of the load devices is always in linear mode, while the other transistor evolves from the off mode to saturation. Ignoring the capacitance variation results in a pessimistic estimation with an error of approximately 10%, which is acceptable for a first-order analysis.

Example 5.4   Capacitances of a 0.25-μm CMOS Inverter

A minimum-size, symmetrical CMOS inverter has been designed in the 0.25-μm CMOS technology. The layout is shown in Figure 5-15. The supply voltage V_{DD} is set to 2.5 V. From the layout, we derive the transistor sizes, diffusion areas, and perimeters. This data is summarized in Table 5-1. As an example, we will derive the drain area and perimeter for the NMOS transistor. The drain area is formed by the metal-diffusion contact, which has an area of $4 \times 4 \; \lambda^2$, and the rectangle between contact and gate, which has an area of $3 \times 1 \; \lambda^2$. This results in a total area of 19 λ^2, or 0.30 μm^2 (as $\lambda = 0.125$ μm). The perimeter of the drain area is rather involved and consists of the following components (going counterclockwise): $5 + 4 + 4 + 1 + 1 = 15 \; \lambda$ or PD = $15 \times 0.125 = 1.875$ μm. Notice that the gate side of the drain perimeter is not included, as this is not considered a part of the sidewall. The drain area and perimeter of the PMOS transistor are derived similarly (the rectangular shape makes the exercise considerably simpler): AD = $5 \times 9 \; \lambda^2 = 45 \; \lambda^2$, or 0.7 μm^2; PD = $5 + 9 + 5 = 19 \; \lambda$, or 2.375 μm.

Table 5-1   Inverter transistor data.

	W/L	AD (μm^2)	PD (μm)	AS (μm^2)	PS (μm)
NMOS	0.375/0.25	0.3 (19 λ^2)	1.875 (15λ)	0.3 (19 λ^2)	1.875 (15λ)
PMOS	1.125/0.25	0.7 (45 λ^2)	2.375 (19λ)	0.7 (45 λ^2)	2.375 (19λ)

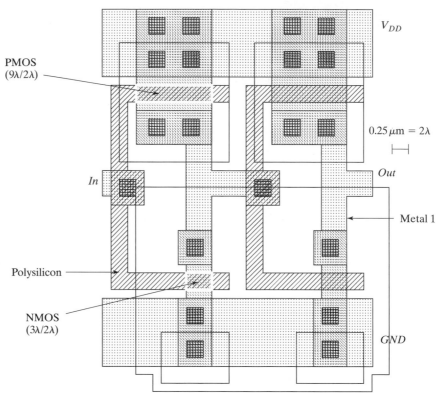

Figure 5-15 Layout of two chained, minimum-size inverters using SCMOS Design Rules (see also Color-plate 6).

This physical information can be combined with the approximations derived earlier to come up with an estimation of C_L. The capacitor parameters for our generic process were summarized in Table 3-5 and are repeated here for convenience:

Overlap capacitance: CGD0(NMOS) = 0.31 fF/μm; CGDO(PMOS) = 0.27 fF/μm
Bottom junction capacitance: CJ(NMOS) = 2 fF/μm^2; CJ(PMOS) = 1.9 fF/μm^2
Sidewall junction capacitance: CJSW(NMOS) = 0.28 fF/μm; CJSW(PMOS) = 0.22 fF/μm
Gate capacitance: C_{ox}(NMOS) = C_{ox}(PMOS) = 6 fF/μm^2

Finally, we should also consider the capacitance contributed by the wire connecting the gates and implemented in metal 1 and polysilicon. A layout extraction program typically delivers precise values for this parasitic capacitance. Inspection of the layout helps us form a first-order estimate. It yields that the metal-1 and polysilicon areas of the wire, which are not over active diffusion, equal 42 λ^2 and 72 λ^2, respectively. With the aid of the interconnect parameters of Table 4-2, we find the wire capacitance (observe that we ignore the fringing

Table 5-2 Components of C_L (for high-to-low and low-to-high transitions).

Capacitor	Expression	Value (fF) (H → L)	Value (fF) (L → H)
C_{gd1}	$2\ \mathrm{CGD0_n}\ W_n$	0.23	0.23
C_{gd2}	$2\ \mathrm{CGD0_p}\ W_p$	0.61	0.61
C_{db1}	$K_{eqn}\ AD_n\ CJ + K_{eqswn}\ PD_n\ CJSW$	0.66	0.90
C_{db2}	$K_{eqp}\ AD_p\ CJ + K_{eqswp}\ PD_p\ CJSW)$	1.5	1.15
C_{g3}	$(\mathrm{CGD0_n + CGSO_n})\ W_n + C_{ox}\ W_n\ L_n$	0.76	0.76
C_{g4}	$(\mathrm{CGD0_p + CGSO_p})\ W_p + C_{ox}\ W_p\ L_p$	2.28	2.28
C_w	From Extraction	0.12	0.12
C_L	Σ	6.1	6.0

capacitance in this simple exercise; due to the short length of the wire, this contribution can be ignored compared with the other entries):

$$C_{wire} = 42/8^2\ \mu m^2 \times 30\ aF/\mu m^2 + 72/8^2\ \mu m^2 \times 88\ aF/\mu m^2 = 0.12\ fF$$

The results of bringing all the components together are summarized in Table 5-2. We use the values of K_{eq} derived in Example 5.3 for the computation of the diffusion capacitances. Notice that the load capacitance is almost evenly split between its two major components: the *intrinsic capacitance*, composed of diffusion and overlap capacitances, and the *extrinsic load capacitance*, contributed by wire and connecting gate.

5.4.2 Propagation Delay: First-Order Analysis

One way to compute the propagation delay of the inverter is to integrate the capacitor (dis)charge current. This results in the expression

$$t_p = \int_{v_1}^{v_2} \frac{C_L(v)}{i(v)}\,dv \tag{5.16}$$

with i the (dis)charging current, v the voltage over the capacitor, and v_1 and v_2 the initial and final voltage, respectively. An exact computation of this equation is intractable, as both $C_L(v)$ and $i(v)$ are nonlinear functions of v. Instead, we fall back to the simplified switch model of the inverter introduced in Figure 5-6 to derive a reasonable approximation of the propagation delay adequate for manual analysis. The voltage dependencies of the "on" resistance and the load capacitor are addressed by replacing both by a constant linear element with a value averaged over the interval of interest. The preceding section derived precisely this value for the load capacitance. An expression for the average "on" resistance of the MOS transistor was already derived in Example 3.8 and is repeated here for convenience:

$$R_{eq} = \frac{1}{V_{DD}/2} \int_{V_{DD}/2}^{V_{DD}} \frac{V}{I_{DSAT}(1 + \lambda V)} dV \approx \frac{3}{4} \frac{V_{DD}}{I_{DSAT}} \left(1 - \frac{7}{9}\lambda V_{DD}\right) \tag{5.17}$$

with $I_{DSAT} = k'\frac{W}{L}\left((V_{DD} - V_T)V_{DSAT} - \frac{V_{DSAT}^2}{2}\right)$

Deriving the propagation delay of the resulting circuit is now straightforward—it is nothing more than the analysis of a first-order linear RC network, identical to the exercise of Example 4.5. We learned there that the propagation delay of such a network, excited by a voltage step, is proportional to the time constant of the network, formed by pull-down resistor and load capacitance. Hence,

$$t_{pHL} = \ln(2)R_{eqn}C_L = 0.69R_{eqn}C_L \tag{5.18}$$

Similarly, we can obtain the propagation delay for the low-to-high transition. We write

$$t_{pLH} = 0.69R_{eqp}C_L \tag{5.19}$$

with R_{eqp} the equivalent on resistance of the PMOS transistor over the interval of interest. This analysis assumes that the equivalent load-capacitance is identical for both the high-to-low and low-to-high transitions. This was shown to be approximately the case in the example of the previous section. The overall propagation delay of the inverter is defined as the average of the two values:

$$t_p = \frac{t_{pHL} + t_{pLH}}{2} = 0.69C_L\left(\frac{R_{eqn} + R_{eqp}}{2}\right) \tag{5.20}$$

Very often, it is desirable for a gate to have identical propagation delays for both rising and falling inputs. This condition can be achieved by making the "on" resistance of the NMOS and PMOS approximately equal. Remember that this condition is identical to the requirement for a symmetrical VTC.

Example 5.5   Propagation Delay of a 0.25 μm CMOS Inverter

To derive the propagation delays of the CMOS inverter of Figure 5-15, we make use of Eq. (5.18) and Eq. (5.19). The load capacitance C_L was already computed in Example 5.4, while the equivalent "on" resistances of the transistors for the generic 0.25-μm CMOS process were derived in Table 3-3. For a supply voltage of 2.5 V, the normalized "on" resistances of NMOS and PMOS transistors equal 13 kΩ and 31 kΩ, respectively. From the layout, we determine the (W-to-L) ratios of the transistors to be 1.5 for the NMOS, and

4.5 for the PMOS. We assume that the difference between drawn and effective dimensions is small enough to be ignorable. This leads to the following values for the delays:

$$t_{pHL} = 0.69 \times \left(\frac{13 \text{ k}\Omega}{1.5}\right) \times 6.1 \text{ fF} = 36 \text{ ps}$$

$$t_{pLH} = 0.69 \times \left(\frac{31 \text{ k}\Omega}{4.5}\right) \times 6.0 \text{ fF} = 29 \text{ ps}$$

and

$$t_p = \left(\frac{36 + 29}{2}\right) = 32.5 \text{ ps}$$

The accuracy of this analysis is checked by performing a SPICE transient simulation on the circuit schematic, extracted from the layout of Figure 5-15. The computed transient response of the circuit is plotted in Figure 5-16, and determines the propagation delays to be 39.9 ps and 31.7 ps for the HL and LH transitions, respectively. The manual results are good, considering the many simplifications made during their derivation. Notice in particular the overshoots on the simulated output signals. These are caused by the gate-drain capacitances of the inverter transistors, which couple the steep voltage step at the input node directly to the output before the transistors can even start to react to the changes at the input. These overshoots clearly have a negative impact on the performance of the gate and explain why the simulated delays are larger than the estimations.

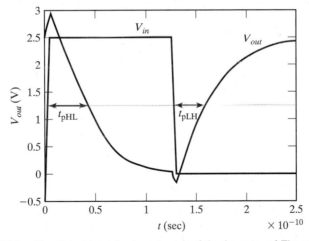

Figure 5-16 Simulated transient response of the inverter of Figure 5-15.

WARNING: This example might give the impression that manual analysis always leads to close approximations of the actual response, which is not necessarily the case. Large deviations often can be observed between first- and higher order models. The purpose of the manual analysis is to get a basic insight into the behavior of the circuit and to determine the dominant parameters. A detailed simulation is indispensable when quantitative data is required. Consider the preceding example a stroke of good luck.

The obvious question a designer asks at this point is how to manipulate or optimize the delay of a gate. To provide an answer to this question, it is necessary to make the parameters governing the delay explicit by expanding R_{eq} in the delay equation. Combining Eq. (5.18) and Eq. (5.17), and assuming for the time being that the channel-length modulation factor λ is ignorable, yields the following expression for t_{pHL} (a similar analysis holds for t_{pLH}):

$$t_{pHL} = 0.69 \frac{3}{4} \frac{C_L V_{DD}}{I_{DSATn}} = 0.52 \frac{C_L V_{DD}}{(W/L)_n k'_n V_{DSATn}(V_{DD} - V_{Tn} - V_{DSATn}/2)} \qquad (5.21)$$

In the majority of designs, the supply voltage is chosen high enough so that $V_{DD} \gg V_{Tn} + V_{DSATn}/2$. Under these conditions, the delay becomes virtually independent of the supply voltage:

$$t_{pHL} \approx 0.52 \frac{C_L}{(W/L)_n k'_n V_{DSATn}} \qquad (5.22)$$

Observe that this is a first-order approximation, and that increasing the supply voltage yields an observable, albeit small, improvement in performance due to a nonzero channel-length modulation factor. This analysis is confirmed in Figure 5-17, which plots the propagation delay of the inverter as a function of the supply voltage. It comes as no surprise that this curve is virtu-

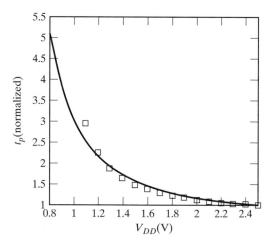

Figure 5-17 Propagation delay of CMOS inverter as a function of supply voltage (normalized with respect to the delay at 2.5 V). The dots indicate the delay values predicted by Eq. (5.21). Observe that this equation is only valid when the devices are velocity saturated. Hence, the deviation at low supply voltages.

ally identical in shape to the one of Figure 3-28, which charts the equivalent "on" resistance of the MOS transistor as a function of V_{DD}. While the delay is relatively insensitive to supply variations for higher values of V_{DD}, a sharp increase can be observed starting around $\approx 2V_T$. This operation region clearly should be avoided if achieving high performance is a primary design goal.

Design Techniques

From the preceding discussion, we deduce that the propagation delay of a gate can be minimized in the following ways:

- *Reduce C_L.* Remember that three major factors contribute to the load capacitance: the internal diffusion capacitance of the gate itself, the interconnect capacitance, and the fan-out. Careful layout helps to reduce the diffusion and interconnect capacitances. **Good design practice requires keeping the drain diffusion areas as small as possible.**
- *Increase the W/L ratio of the transistors.* This is the most powerful and effective performance optimization tool in the hands of the designer. Proceed with caution, however, when applying this approach. Increasing the transistor size also raises the diffusion capacitance and hence C_L. In fact, once the intrinsic capacitance (i.e. the diffusion capacitance) starts to dominate the extrinsic load formed by wiring and fan-out, increasing the gate size no longer helps in reducing the delay. It only makes the gate larger in area. This effect is called *self-loading*. In addition, wide transistors have a larger gate capacitance, which increases the fan-out factor of the driving gate and adversely affects its speed as well.
- *Increase V_{DD}.* As illustrated in Figure 5-17, the delay of a gate can be modulated by modifying the supply voltage. This flexibility allows the designer to trade off energy dissipation for performance, as we will see in a later section. However, increasing the supply voltage above a certain level yields only very minimal improvement and thus should be avoided. Also, reliability concerns (oxide breakdown, hot-electron effects) enforce firm upper bounds on the supply voltage in deep submicron processes. ∎

Problem 5.4   Propagation Delay as a Function of (dis)charge Current

So far, we have expressed the propagation delay as a function of the equivalent resistance of the transistors. Another approach would be to replace the transistor by a current source with a value equal to the average (dis)charge current over the interval of interest. Derive an expression of the propagation delay using this alternative approach.

5.4.3   Propagation Delay from a Design Perspective

Some interesting design considerations and trade-offs can be derived from the delay expressions we have derived so far. Most importantly, they lead to a general approach toward transistor sizing that will prove to be extremely useful.

NMOS-to-PMOS Ratio

So far, we have consistently widened the PMOS transistor so that its resistance matches that of the pull-down NMOS device. This typically requires a ratio of 3 to 3.5 between PMOS and

NMOS width. The motivation behind this approach is to create an inverter with a symmetrical VTC and to equate the high-to-low and low-to-high propagation delays. However, this does not imply that this ratio also yields the minimum overall propagation delay. If symmetry and reduced noise margins are not of prime concern, it is actually possible to speed up the inverter by reducing the width of the PMOS device!

The reasoning behind this statement is that, while widening the PMOS improves the t_{pLH} of the inverter by increasing the charging current, it also degrades the t_{pHL} by causing a larger parasitic capacitance. When two contradictory effects are present, a transistor ratio must exist that optimizes the propagation delay of the inverter.

This optimum ratio can be derived using a simple analysis technique. Consider two identical cascaded CMOS inverters. The approximate load capacitance of the first gate is given by

$$C_L = (C_{dp1} + C_{dn1}) + (C_{gp2} + C_{gn2}) + C_W \tag{5.23}$$

where C_{dp1} and C_{dn1} are the equivalent drain diffusion capacitances of PMOS and NMOS transistors of the first inverter and C_{gp2} and C_{gn2} are the gate capacitances of the second gate. C_W represents the wiring capacitance.

When the PMOS devices are made β times larger than the NMOS ones ($\beta = (W/L)_p/(W/L)_n$), all transistor capacitances scale in approximately the same way, or $C_{dp1} \approx \beta\, C_{dn1}$, and $C_{gp2} \approx \beta\, C_{gn2}$. Equation (5.23) can then be rewritten as

$$C_L = (1 + \beta)(C_{dn1} + C_{gn2}) + C_W \tag{5.24}$$

Based on Eq. (5.20), the following expression for the propagation delay can be derived,

$$
\begin{aligned}
t_p &= \frac{0.69}{2}((1 + \beta)(C_{dn1} + C_{gn2}) + C_W)\left(R_{eqn} + \frac{R_{eqp}}{\beta}\right) \\
&= 0.345((1 + \beta)(C_{dn1} + C_{gn2}) + C_W)R_{eqn}\left(1 + \frac{r}{\beta}\right)
\end{aligned}
\tag{5.25}
$$

Here, r ($= R_{eqp}/R_{eqn}$) represents the resistance ratio of identically sized PMOS and NMOS transistors. The optimal value of β can be found by setting $\frac{\partial t_p}{\partial \beta}$ to 0, which yields

$$\beta_{opt} = \sqrt{r\left(1 + \frac{C_w}{C_{dn1} + C_{gn2}}\right)} \tag{5.26}$$

This means that when the wiring capacitance is negligible ($C_{dn1} + C_{gn2} \gg C_W$), β_{opt} equals $\sqrt{r}$, in contrast to the factor r normally used in the noncascaded case. If the wiring capacitance dominates, larger values of β should be used. The surprising result of this analysis is that smaller device sizes (and thus a smaller design area) yield a faster design at the expense of symmetry and noise margin.

Example 5.6   Sizing of CMOS Inverter Loaded by an Identical Gate

Consider again our standard design example. From the values of the equivalent resistances (Table 3-3), we find that a ratio β of 2.4 (= 31 kΩ / 13 kΩ) would yield a symmetrical transient response. Eq. (5.26) now predicts that the device ratio for an optimal performance should equal 1.6. These results are verified in Figure 5-18, which plots the simulated propagation delay as a function of the transistor ratio β. The graph clearly illustrates how a changing β trades off between t_{pLH} and t_{pHL}. The optimum point occurs around $\beta = 1.9$, which is somewhat higher than predicted. Observe also that the rising and falling delays are identical at the predicted point of β equal to 2.4. This is the preferred operation point when the worst case delay is the prime concern.[3]

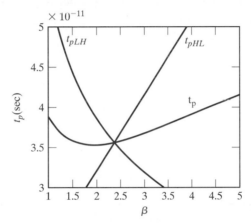

Figure 5-18 Propagation delay of CMOS inverter as a function of the PMOS-to-NMOS transistor ratio β.

Sizing Inverters for Performance

In this analysis, we assume a symmetrical inverter, which is an inverter where PMOS and NMOS are sized such that the rise and fall delays are identical. The load capacitance of the inverter can be divided into an intrinsic and an extrinsic component, or $C_L = C_{int} + C_{ext}$. C_{int} represents the self-loading or intrinsic output capacitance of the inverter, and is associated with the diffusion capacitances of the NMOS and PMOS transistors as well as the gate–drain overlap (Miller) capacitances. C_{ext} is the extrinsic load capacitance, attributable to fan-out and wiring

---

[3]You probably wonder why we do not always consider the worst of the rising and falling delays as the prime performance measure of a gate. When cascading inverting gates to form a more complex logic network, you quickly realize that the average of the two is a more meaningful measure. A rising transition on one gate is followed by a falling transition on the next.

capacitance. Assuming that R_{eq} stands for the equivalent resistance of the gate, we can express the propagation delay as

$$
\begin{aligned}
t_p &= 0.69 R_{eq}(C_{int} + C_{ext}) \\
&= 0.69 R_{eq} C_{int}(1 + C_{ext}/C_{int}) = t_{p0}(1 + C_{ext}/C_{int})
\end{aligned}
\tag{5.27}
$$

where $t_{p0} = 0.69\ R_{eq} C_{int}$ represents the delay of the inverter only loaded by its own intrinsic capacitance ($C_{ext} = 0$), and is called the *intrinsic or unloaded delay*.

The next question is how transistor sizing impacts the performance of the gate. To answer this question, we must establish the relationship between the various parameters in Eq. (5.27) and a sizing factor S, which relates the transistor sizes of our inverter to a reference gate—typically a minimum-sized inverter. The intrinsic capacitance C_{int} consists of the diffusion and Miller capacitances, both of which are proportional to the width of the transistors. Hence, $C_{int} = SC_{iref}$. The resistance of the gate relates to the reference gate as $R_{eq} = R_{ref}/S$. We can now rewrite Eq. (5.27) as

$$
\begin{aligned}
t_p &= 0.69(R_{ref}/S)(SC_{iref})(1 + C_{ext}/(SC_{iref})) \\
&= 0.69 R_{ref} C_{iref}\left(1 + \frac{C_{ext}}{SC_{iref}}\right) = t_{p0}\left(1 + \frac{C_{ext}}{SC_{iref}}\right)
\end{aligned}
\tag{5.28}
$$

From this analysis, we draw two important conclusions:

- The intrinsic delay of the inverter t_{p0} is independent of the sizing of the gate, and is determined purely by technology and inverter layout. When no load is present, an increase in the drive of the gate is totally offset by the increased capacitance.
- Making S infinitely large yields the maximum obtainable performance gain, eliminating the impact of any external load, and reducing the delay to the intrinsic one. Yet, any sizing factor S that is sufficiently larger than (C_{ext}/C_{int}) produces similar results at a substantial gain in silicon area.

Example 5.7   Device Sizing for Performance

Let us explore the performance improvement that can be obtained by device sizing in the design of Example 5.5. We find from Table 5-2 that $C_{int}/C_{ext} \approx 1.05$ ($C_{int} = 3.0$ fF, $C_{ext} = 3.15$ fF). This would predict a maximum performance gain of 2.05. A scaling factor of 10 allows us to get within 10% of this optimal performance, while larger device sizes only yield ignorable performance gains.

This is confirmed by simulation results, which predict a maximum obtainable performance improvement of 1.9 ($t_{p0} = 19.3$ ps). In Figure 5-19, we observe that the majority of the improvement is already obtained for $S = 5$, and that sizing factors larger than 10 barely yield any extra gain.

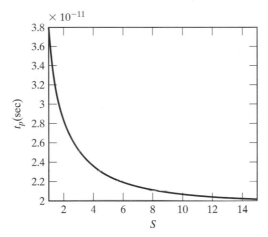

Figure 5-19 Increasing inverter performance by sizing the NMOS and PMOS transistor with an identical factor S for a fixed fan-out (inverter of Figure 5-15).

Sizing a Chain of Inverters

While sizing up an inverter reduces its delay, it also increases its input capacitance. Gate sizing in an isolated fashion without taking into account its impact on the delay of the preceding gates is a purely academic enterprise. Therefore, a more relevant problem is determining the optimum sizing of a gate when **embedded in a real environment**. A simple chain of inverters is a good first case to study. To determine the input loading effect, the relationship between the input gate capacitance C_g and the intrinsic output capacitance of the inverter has to be established. Both are proportional to the gate sizing. Hence, the following relationship holds, independently of gate sizing:

$$C_{int} = \gamma C_g \tag{5.29}$$

In Eq. (5.29), γ is a proportionality factor that is only a function of technology and is close to 1 for most submicron processes, as we observed in the preceding examples. Rewriting Eq. (5.28), we obtain

$$t_p = t_{p0}\left(1 + \frac{C_{ext}}{\gamma C_g}\right) = t_{p0}(1 + f/\gamma) \tag{5.30}$$

which establishes that the delay of an inverter is **only a function of the ratio between its external load capacitance and its input capacitance**. This ratio is called the *effective fan-out f*.

Let us consider the circuit of Figure 5.20. The goal is to minimize the delay through the inverter chain, with the input capacitance C_{g1} of the first inverter—typically a minimally-sized gate—and the load capacitance C_L at the end of the chain fixed.

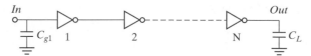

Figure 5-20 Chain of N inverters with fixed input and output capacitance.

Given the delay expression for the j-th inverter stage,[4]

$$t_{p,j} = t_{p0}\left(1 + \frac{C_{g,j+1}}{\gamma C_{g,j}}\right) = t_{p0}(1 + f_j/\gamma) \tag{5.31}$$

we can derive the total delay of the chain:

$$t_p = \sum_{j=1}^{N} t_{p,j} = t_{p0}\sum_{j=1}^{N}\left(1 + \frac{C_{g,j+1}}{\gamma C_{g,j}}\right), \text{ with } C_{g,N+1} = C_L \tag{5.32}$$

This equation has $N-1$ unknowns, being $C_{g,2}, C_{g,3}, \ldots, C_{g,N}$. The minimum delay can be found by taking $N-1$ partial derivatives, and equating them to 0, or $\partial t_p/\partial C_{g,j} = 0$. The result is a set of constraints,

$$C_{g,j+1}/C_{g,j} = C_{g,j}/C_{g,j-1} \qquad \text{with } (j = 2 \ldots N) \tag{5.33}$$

In other words, the optimum size of each inverter is the geometric mean of its neighbors sizes:

$$C_{g,j} = \sqrt{C_{g,j-1}C_{g,j+1}} \tag{5.34}$$

This means that each inverter is sized up by the same factor f with respect to the preceding gate, has the same effective fan-out ($f_j = f$), and thus the same delay. With $C_{g,1}$ and C_L given, we can derive the sizing factor as

$$f = \sqrt[N]{C_L/C_{g,1}} = \sqrt[N]{F} \tag{5.35}$$

and the minimum delay through the chain as

$$t_p = Nt_{p0}(1 + \sqrt[N]{F}/\gamma) \tag{5.36}$$

F represents the *overall effective fan-out* of the circuit and equals $C_L/C_{g,1}$. Observe how the relationship between t_p and F is a strong function of the number of stages. As expected, the relationship is linear when only 1 stage is present. Introducing a second stage turns it into a square root function, and so on. The obvious question now is how to choose the number of stages so that the delay is minimized for a given value of F.

---

[4]This expression ignores the wiring capacitance, which is a fair assumption for the time being.

Choosing the Right Number of Stages in an Inverter Chain

Evaluation of Eq. (5.36) reveals the trade-offs in choosing the number of stages for a given F $(= f^N)$. When the number of stages is too large, the first component of the equation, which represents the intrinsic delay of the stages, becomes dominant. If the number of stages is too small, the effective fan-out of each stage becomes large, and the second component is dominant. The optimum value can be found by differentiating the minimum delay expression by the number of stages and setting the result to 0. We obtain

$$\gamma + \sqrt[N]{F} - \frac{\sqrt[N]{F}\ln F}{N} = 0$$

or equivalently (5.37)

$$f = e^{(1+\gamma/f)}$$

Equation (5.35) has only a closed-form solution for $\gamma = 0$—that is when the self-loading is ignored and the load capacitance only consists of the fan-out. Under these simplified conditions, it is found that the optimal number of stages equals $N = \ln(F)$, and the effective fan-out of each stage is set to $f = e = 2.71828$. This optimal buffer design scales consecutive stages in an exponential fashion, and is thus called an exponential horn [Mead80]. When self-loading is included, Eq. can only be solved numerically. The results are plotted in Figure 5-21a. For the typical case of $\gamma \approx 1$, the optimum tapering factor turns out to be close to 3.6. Figure 5-21b plots the (normalized) propagation delay of the inverter chain as a function of the effective fan-out for $\gamma = 1$. Choosing values of the fan-out that are higher than the optimum does not effect the delay very much and reduces the required number of buffer stages and the implementation area. A common practice is to **select an optimum fan-out of 4**. The use of too many stages ($f < f_{opt}$), on the other hand, has a substantial negative impact on the delay, and should be avoided.

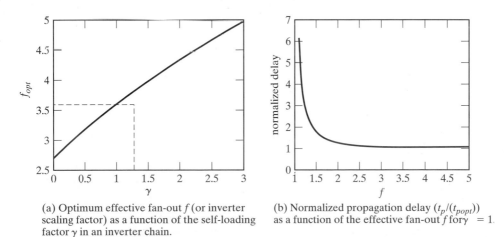

(a) Optimum effective fan-out f (or inverter scaling factor) as a function of the self-loading factor γ in an inverter chain.

(b) Normalized propagation delay ($t_p/(t_{popt})$) as a function of the effective fan-out f for $\gamma = 1$.

Figure 5-21 Optimizing the number of stages in an inverter chain.

Example 5.8   The Impact of Introducing Buffer Stages

Table 5-3 enumerates the values of $t_{p,opt}/t_{p0}$ for the unbuffered design, the dual stage, and optimized inverter chain for a variety of values of F (for $\gamma = 1$). Observe the impressive speedup obtained with cascaded inverters when driving very large capacitive loads.

Table 5-3 t_{opt}/t_{p0} versus x for various driver configurations.

F	Unbuffered	Two Stage	Inverter Chain
10	11	8.3	8.3
100	101	22	16.5
1000	1001	65	24.8
10,000	10,001	202	33.1

The preceding analysis can be extended to not only cover chains of inverters, but also networks of inverters that contain actual fan-out, an example of which is shown in Figure 5-22. We merely have to adjust the expression for C_{ext} to incorporate the additional fan-out factors.

Problem 5.5   Sizing an Inverter Network

Determine the sizes of the inverters in the circuit of Figure 5-22, such that the delay between nodes *Out* and *In* is minimized. You may assume that $C_L = 64\ C_{g,1}$.

Hints: Determine first the ratios between the devices that minimize the delay. You should find that the following relationship must hold:

$$\frac{4C_{g,2}}{C_{g,1}} = \frac{4C_{g,3}}{C_{g,2}} = \frac{C_L}{C_{g,3}}$$

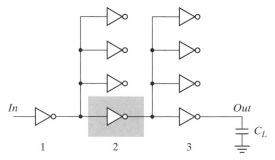

Figure 5-22   Inverter network, in which each gate has a fan-out of 4 gates, distributing a single input to 16 output signals in a treelike fashion.

Finding the actual gate sizes ($C_{g,3} = 2.52C_{g,2} = 6.35C_{g,1}$) is a relatively straightforward task (with $2.52 = 16^{1/3}$). Straightforward sizing of the inverter chain, without taking the extra fan-out into account, would have led to a sizing factor of 4 instead of 2.52.

The Rise–Fall Time of the Input Signal

All of the preceding expressions were derived under the assumption that the input signal to the inverter abruptly changed from 0 to V_{DD} or vice versa. Only one of the devices is assumed to be on during the (dis)charging process. In reality, the input signal changes gradually and, temporarily, PMOS and NMOS transistors conduct simultaneously. This affects the total current available for (dis)charging and impacts the propagation delay. Figure 5-23 plots the propagation delay of a minimum-size inverter as a function of the input signal slope—as obtained from SPICE. It can be observed that t_p increases (approximately) linearly with increasing input slope, once $t_s > t_p$ ($t_s = 0$).

While it is possible to derive an analytical expression describing the relationship between input signal slope and propagation delay, the result tends to be complex and of limited value. From a design perspective, it is more valuable to relate the impact of the finite slope on the performance directly to its cause, which is the limited driving capability of the preceding gate. If the latter would be infinitely strong, its output slope would be zero, and the performance of the gate under examination would be unaffected. The strength of this approach is that it realizes that a gate is never designed in isolation, and that its performance is affected by both the fan-out and the driving strength of the gate(s) feeding into its inputs. This leads to a revised expression for the propagation delay of an inverter i in a chain of inverters [Hedenstierna87]:

$$t_p^i = t_{step}^i + \eta t_{step}^{i-1} \tag{5.38}$$

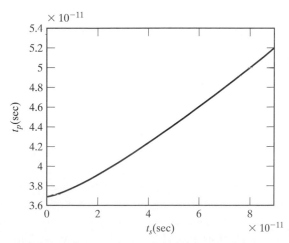

Figure 5-23 t_p as a function of the input signal slope (10–90% rise or fall time) for minimum-size inverter with fan-out of a single gate.

Eq. (5.38) states that the propagation delay of inverter i equals the sum of the delay of the same gate for a step input (t_{step}^i) (i.e. zero input slope) augmented with a fraction of the step-input delay of the preceding gate ($i - 1$). The fraction η is an empirical constant, which typically has values around 0.25. This expression has the advantage of being very simple, while exposing all relationships necessary for the delay computations of complex circuits.

Example 5.9   Delay of Inverter Embedded in Network

Consider, for example, the circuit of Figure 5-22. With the aid of Eq. (5.31) and Eq. (5.38), we can derive an expression for the delay of the stage-2 inverter, marked by the gray box:

$$t_{p,2} = t_{p0}\left(1 + \frac{4C_{g,3}}{\gamma C_{g,2}}\right) + \eta t_{p0}\left(1 + \frac{4C_{g,2}}{\gamma C_{g,1}}\right)$$

An analysis of the overall propagation delay in the style of Problem 5.5, leads to the following revised sizing requirements for minimum delay:

$$\frac{4(1 + \eta)C_{g,2}}{C_{g,1}} = \frac{4(1 + \eta)C_{g,3}}{C_{g,2}} = \frac{C_L}{C_{g,3}}$$

If we assume $\eta = 0.25$, f_2 and f_1 evaluate to 2.47.

Design Challenge

It is advantageous to keep the signal rise times smaller than or equal to the gate propagation delays. This proves to be true not only for performance, but also for power consumption considerations, as will be discussed later. Keeping the rise and fall times of the signals small and of approximately equal values is one of the major challenges in high-performance design; it is often called *slope engineering*. ■

Problem 5.6   Impact of Input Slope

Determine if reducing the supply voltage increases or decreases the influence of the input signal slope on the propagation delay. Explain your answer.

Delay in the Presence of (Long) Interconnect Wires

The interconnect wire has played a minimal role in our analysis thus far. When gates get farther apart, the wire capacitance and resistance can no longer be ignored, and may even dominate the transient response. Earlier delay expressions can be adjusted to accommodate these extra contributions by employing the wire modeling techniques introduced in the previous chapter. The analysis detailed in Example 4.9 is directly applicable to the problem at hand. Consider the circuit of Figure 5-24, where an inverter drives a single fan-out through a wire of length L. The driver is represented by a single resistance R_{dr}, which is the average between R_{eqn} and R_{eqp}. C_{int} and C_{fan} account for the intrinsic capacitance of the driver, and the input capacitance of the fan-out gate, respectively.

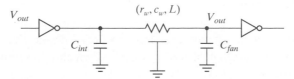

Figure 5-24 Inverter driving single gate through wire of length L.

The propagation delay of the circuit can be obtained by applying the Elmore delay expression:

$$t_p = 0.69 R_{dr} C_{int} + (0.69 R_{dr} + 0.38 R_w) C_w + 0.69 (R_{dr} + R_w) C_{fan}$$
$$= 0.69 R_{dr} (C_{int} + C_{fan}) + 0.69 (R_{dr} c_w + r_w C_{fan}) L + 0.38 r_w c_w L^2 \tag{5.39}$$

The 0.38 factor accounts for the fact that the wire represents a distributed delay. C_w and R_w stand for the total capacitance and resistance of the wire, respectively. The delay expression contain a component that is linear with the wire length, as well a quadratic one. It is the latter that causes the wire delay to rapidly become the dominant factor in the delay budget for longer wires.

Example 5.10   Inverter Delay in Presence of Interconnect

Consider the circuit of Figure 5-24, and assume the device parameters of Example 5.5: $C_{int} = 3$ fF, $C_{fan} = 3$ fF, and $R_{dr} = 0.5(13/1.5 + 31/4.5) = 7.8$ kΩ. The wire is implemented in metal1 and has a width of 0.4 μm—the minimum allowed. This yields the following parameters: $c_w = 92$ aF/μm, and $r_w = 0.19$ Ω/μm (Example 4.4). With the aid of Eq. (5.39), we can compute at what wire length the delay of the interconnect becomes equal to the intrinsic delay caused purely by device parasitics. Solving the following quadratic equation yields a single (meaningful) solution:

$$6.6 \times 10^{-18} L^2 + 0.5 \times 10^{-12} L = 32.29 \times 10^{-12}$$
$$\text{or, } L = 65 \ \mu\text{m}$$

Observe that the extra delay is due solely to the linear factor in the equation—more specifically, to the extra capacitance introduced by the wire. The quadratic factor (the distributed wire delay) becomes dominant only at much larger wire lengths (> 7 cm). This can be attributed to the high resistance of the (minimum-size) driver transistors. A different balance emerges when wider transistors are used. Analyze, for instance, the same problem with the driver transistors 100 times wider.

5.5 Power, Energy, and Energy Delay

So far, we have seen that the static CMOS inverter with its almost ideal VTC—symmetrical shape, full logic swing, and high noise margins—offers a superior robustness, which simplifies the design process considerably and opens the door for design automation. Another major attractor for static CMOS is the almost complete absence of power consumption in steady-state operation mode. It is this combination of robustness and low static power that has made static CMOS the technology of choice of most contemporary digital designs. The power dissipation of a CMOS circuit is instead dominated by the dynamic dissipation resulting from charging and discharging capacitances.

5.5.1   Dynamic Power Consumption

Dynamic Dissipation due to Charging and Discharging Capacitances

Each time the capacitor C_L gets charged through the PMOS transistor, its voltage rises from 0 to V_{DD}, and a certain amount of energy is drawn from the power supply. Part of this energy is dissipated in the PMOS device, while the remainder is stored on the load capacitor. During the high-to-low transition, this capacitor is discharged, and the stored energy is dissipated in the NMOS transistor.[5]

A precise measure for this energy consumption can be derived. Let us first consider the low-to-high transition. We assume, initially, that the input waveform has zero rise and fall times—in other words, the NMOS and PMOS devices are never on simultaneously. Therefore, the equivalent circuit of Figure 5-25 is valid. The values of the energy E_{VDD}, taken from the supply during the transition, as well as the energy E_C, stored on the capacitor at the end of the transition, can be derived by integrating the instantaneous power over the period of interest:

$$E_{VDD} = \int_0^\infty i_{VDD}(t)V_{DD}dt = V_{DD}\int_0^\infty C_L\frac{dv_{out}}{dt}dt = C_LV_{DD}\int_0^{V_{DD}} dv_{out} = C_LV_{DD}^2 \tag{5.40}$$

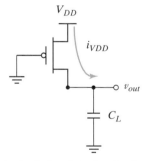

Figure 5-25   Equivalent circuit during the low-to-high transition.

---

[5]Observe that this model is a simplification of the actual circuit. In reality, the load capacitance consists of multiple components, some of which are located between the output node and GND, others between output node and V_{DD}. The latter experience a charge–discharge cycle that is out of phase with the capacitances to GND (i.e., they get charged when V_{out} goes low and discharged when V_{out} rises.) While this distributes the energy delivery by the supply over the two phases, it does not affect the overall dissipation, and the results presented in this section are still valid.

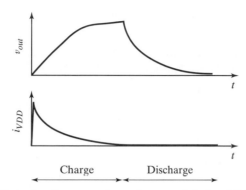

Figure 5-26   Output voltages and supply current during (dis)charge of C_L.

and

$$E_C = \int_0^\infty i_{VDD}(t)v_{out}dt = \int_0^\infty C_L\frac{dv_{out}}{dt}v_{out}dt = C_L\int_0^{V_{DD}} v_{out}dv_{out} = \frac{C_LV_{DD}^2}{2} \tag{5.41}$$

The corresponding waveforms of $v_{out}(t)$ and $i_{VDD}(t)$ are pictured in Figure 5-26.

These results can also be derived by observing that during the low-to-high transition, C_L is loaded with a charge C_LV_{DD}. Providing this charge requires an energy from the supply equal to $C_LV_{DD}^2$ ($= Q \times V_{DD}$). The energy stored on the capacitor equals $C_LV_{DD}^2/2$. This means that only half of the energy supplied by the power source is stored on C_L. The other half has been dissipated by the PMOS transistor. Notice that this energy dissipation is independent of the size (and hence the resistance) of the PMOS device! During the discharge phase, the charge is removed from the capacitor, and its energy is dissipated in the NMOS device. Once again, there is no dependence on the size of the device. In summary, each switching cycle (consisting of an L→H and an H→L transition) takes a fixed amount of energy, equal to $C_LV_{DD}^2$. In order to compute the power consumption, we have to take into account how often the device is switched. If the gate is switched **on and off** $f_{0\to1}$ times per second, the power consumption is given by

$$P_{dyn} = C_LV_{DD}^2 f_{0\to1} \tag{5.42}$$

where $f_{0\to1}$ represents the frequency of energy-consuming transitions (these are $0 \to 1$ transitions for static CMOS).

Advances in technology result in ever-higher values of $f_{0\to1}$ (as t_p decreases). At the same time, the total capacitance on the chip (C_L) increases as more and more gates are placed on a single die. Consider, for instance, a 0.25-μm CMOS chip with a clock rate of 500 MHz and an average load capacitance of 15 fF/gate, assuming a fan-out of 4. The power consumption per gate for a 2.5-V supply then equals approximately 50 μW. For a design with 1 million gates, and assuming that a transition occurs at every clock edge, this would result in a power consumption of 50 W! This evaluation, fortunately, presents a pessimistic perspective. In reality, not all gates in the complete IC switch at the full rate of 500 Mhz. The actual activity in the circuit is substantially lower.

Example 5.11   Capacitive Power Dissipation of Inverter

The capacitive dissipation of the CMOS inverter of Example 5.4 is now easily computed. In Table 5-2, the value of the load capacitance was determined to equal 6 fF. For a supply voltage of 2.5 V, the amount of energy needed to charge and discharge that capacitance equals

$$E_{dyn} = C_L V_{DD}^2 = 37.5 \text{ fJ}$$

Assume that the inverter is switched at the (hypothetically) maximum possible rate $(T = 1/f = t_{pLH} + t_{pHL} = 2t_p)$. For a t_p of 32.5 ps (Example 5.5), we find that the dynamic power dissipation of the circuit is

$$P_{dyn} = E_{dyn}/(2t_p) = 580 \text{ μW}$$

Of course, an inverter in an actual circuit is rarely switched at this maximum rate, and even if it would be, the output does not swing from rail to rail. The power dissipation will thus be substantially lower. For a rate of 4 GHz ($T = 250$ ps), the dissipation reduces to 150 μW. This is confirmed by simulations, which yield a power consumption of 155 μW.

Computing the dissipation of a complex circuit is complicated by the $f_{0 \to 1}$ factor, also called the *switching activity*. While the switching activity is easily computed for an inverter, it turns out to be far more complex in the case of more complex gates and circuits. One concern is that the switching activity of a network is a function of the nature and the statistics of the input signals: If the input signals remain unchanged, no switching happens, and the dynamic power consumption is zero! On the other hand, rapidly changing signals provoke plenty of switching and therefore dissipation. Other factors influencing the activity are the overall network topology and the function to be implemented. We can accommodate this by writing

$$P_{dyn} = C_L V_{DD}^2 f_{0 \to 1} = C_L V_{DD}^2 P_{0 \to 1} f = C_{EFF} V_{DD}^2 f \qquad (5.43)$$

where f now presents the maximum possible event rate of the inputs (which is often the clock rate) and $P_{0 \to 1}$ the probability that a clock event results in a $0 \to 1$ (or power-consuming) event at the output of the gate. $C_{EFF} = P_{0 \to 1} C_L$ is called the *effective capacitance* and represents the average capacitance switched every clock cycle. For our example, an activity factor of 10% $(P_{0 \to 1} = 0.1)$ reduces the average consumption to 5 W.

Example 5.12   Switching Activity

Consider the waveforms in Figure 5.27, where the upper waveform represents the idealized clock signal, and the bottom one shows the signal at the output of the gate. Power consuming transitions occur 2 out of 8 times, which is equivalent to a transition probability of 0.25 (or 25%).

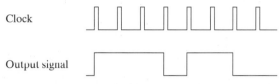

Clock

Output signal

Figure 5-27   Clock and signal waveforms.

Low Energy–Power Design Techniques

With the increasing complexity of digital integrated circuits, it is anticipated that the power problem will only worsen in future technologies. This is one of the reasons that lower supply voltages are becoming more and more attractive. **Reducing V_{DD} has a quadratic effect on P_{dyn}.** For instance, reducing V_{DD} from 2.5 V to 1.25 V for our example drops the power dissipation from 5 W to 1.25 W. This assumes that the same clock rate can be sustained. Figure 5-17 demonstrates that this assumption is not that unrealistic as long as the supply voltage is substantially higher than the threshold voltage. A large performance penalty occurs once V_{DD} approaches $2\,V_T$.

When a lower limit on the supply voltage is set by external constraints (as often happens in real-world designs), or when the performance degradation due to lowering the supply voltage is intolerable, the only means of reducing the dissipation is by lowering the effective capacitance. This can be achieved by addressing both of its components: the physical capacitance and the switching activity.

A *reduction in the switching activity* can only be accomplished at the logic and architectural abstraction levels, and will be discussed in more detail in Chapter 11. *Lowering the physical capacitance* is a worthwhile goal overall, and it also may help to improve the performance of the circuit. As most of the capacitance in a combinational logic circuit is due to transistor capacitances (gate and diffusion), it makes sense to keep those contributions to a minimum when designing for low power. This means that transistors should be kept to *minimal size* whenever possible or reasonable. This definitely affects the performance of the circuit, but the effect can be offset by using logic or architectural speedup techniques. The only instances where transistors should be sized up is when the load capacitance is dominated by extrinsic capacitances (such as fan-out or wiring capacitance). This is contrary to common design practices used in cell libraries, where transistors are generally made large to accommodate a range of loading and performance requirements.

These observations lead to an interesting design challenge. Assume we have to minimize the energy dissipation of a circuit with a specified lower bound on the performance. An attractive approach is to lower the supply voltage as much as possible, and to compensate the loss in performance by increasing the transistor sizes. Yet, the latter causes the capacitance to increase. It may be foreseen that at a low enough supply voltage, the latter factor may start to dominate and cause energy to increase with a further drop in the supply voltage.

■

Example 5.13 Transistor Sizing for Energy Minimization

To analyze the transistor sizing for a minimum energy problem, we examine the simple case of a static CMOS inverter driving an external load capacitance C_{ext}, as in Figure 5.28. To take the input loading effects into account, we assume that the inverter itself is driven by a minimum-sized device. The goal is to minimize the energy dissipation of the complete circuit, while maintaining a lower bound on performance. The degrees of freedom are the size factor f of the inverter and the supply voltage V_{dd} of the circuit. The propagation delay of the optimized circuit should not be larger than that of a reference circuit, chosen to have as parameters $f = 1$ and $V_{dd} = V_{ref}$.

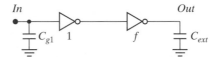

Figure 5-28 CMOS inverter driving an external load capacitance C_{ext}, while being driven by a minimum sized gate.

Using the approach introduced in Section 5.4.3 (*Sizing a Chain of Inverters*), we can derive the following expression for the propagation delay of the circuit:

$$t_p = t_{p0}\left(\left(1 + \frac{f}{\gamma}\right) + \left(1 + \frac{F}{f\gamma}\right)\right) \tag{5.44}$$

here $F = (C_{ext}/C_{g1})$ is the overall effective fan-out of the circuit, and t_{p0} is the intrinsic delay of the inverter. Its dependence upon V_{DD} is approximated by the following expression, derived from Eq. (5.21):

$$t_{p0} \sim \frac{V_{DD}}{V_{DD} - V_{TE}} \tag{5.45}$$

The energy dissipation for a single transition at the input is easily found once the total capacitance of the circuit is known:

$$E = V_{dd}^2 C_{g1}((1 + \gamma)(1 + f) + F) \tag{5.46}$$

The performance constraint now states that the propagation delay of the scaled circuit should be equal (or smaller) to the delay of the reference circuit ($f = 1$, $V_{dd} = V_{ref}$). To simplify the subsequent analysis, we make the assumption that the intrinsic output capacitance of the gate equals its gate capacitance, or $\gamma = 1$. Hence,

$$\frac{t_p}{t_{pref}} = \frac{t_{p0}\left(2 + f + \frac{F}{f}\right)}{t_{p0ref}(3 + F)} = \left(\frac{V_{DD}}{V_{ref}}\right)\left(\frac{V_{ref} - V_{TE}}{V_{DD} - V_{TE}}\right)\left(\frac{2 + f + \frac{F}{f}}{3 + F}\right) = 1 \tag{5.47}$$

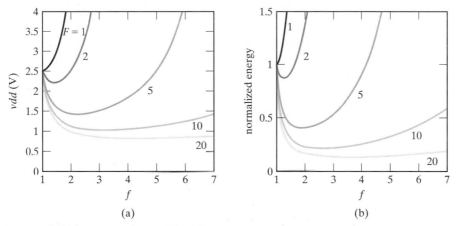

Figure 5-29 Sizing of an inverter for energy minimization. (a) Required supply voltage as a function of the sizing factor f *for different values of the overall effective fan-out F*; (b) Energy of scaled circuit (normalized with respect to the reference case) as a function of f. $V_{ref} = 2.5$ V, $V_{TE} = 0.5$ V.

Equation (5.47) establishes a relationship between the sizing factor f and the supply voltage, plotted in Figure 5-29a for different values of F. Those curves show a clear minimum. Increasing the size of the inverter from the minimum initially increases the performance, and hence allows for a lowering of the supply voltage. This is fruitful until the optimum sizing factor of $f = \sqrt{F}$ is reached, which should not surprise those who read the previous sections carefully. Further increases in the device sizes only increase the self-loading factor, deteriorate the performance, and require an increase in supply voltage. Also, observe that for the case of $F = 1$, the reference case is the best solution; any resizing just increases the self-loading.

With the $V_{DD}(f)$ relationship in hand, we can derive the energy of the scaled circuit (normalized with respect to the reference circuit) as a function of the sizing factor f:

$$\frac{E}{E_{ref}} = \left(\frac{V_{DD}}{V_{ref}}\right)^2 \left(\frac{2 + 2f + F}{4 + F}\right) \tag{5.48}$$

Finding an analytical expression for the optimal sizing factor is possible, but yields a complex and messy equation. A graphical approach is just as effective. The resulting charts are plotted in Figure 5-29b, from which a number of conclusions can be drawn:[6]

- **Device sizing, combined with supply voltage reduction, is a very effective approach in reducing the energy consumption of a logic network.** This is especially true for networks with large effective fan-outs, where energy reductions with almost a factor of 10 can be obtained. The gain is also sizable for smaller values of F. The only exception is the $F = 1$ case, where the minimum size device is also the most effective one.

---

[6]We will revisit some of these conclusions in Chapter 11 in a broader context.

- Oversizing the transistors beyond the optimal value comes at a hefty price in energy. This is, unfortunately, a common approach in many of today's designs.
- The optimal sizing factor for energy is smaller than the one for performance, especially for large values of F. For example, for a fan-out of 20, f_{opt}(energy) = 3.53, while f_{opt}(performance) = 4.47. Increasing the device sizes only leads to a minimal supply reduction once V_{DD} starts approaching V_{TE}, thus leading to very minimal energy gains.

Dissipation Due to Direct-Path Currents

In actual designs, the assumption of the zero rise and fall times of the input wave forms is not correct. The finite slope of the input signal causes a direct current path between V_{DD} and GND for a short period of time during switching, while the NMOS and the PMOS transistors are conducting simultaneously. This is illustrated in Figure 5-30. Under the (reasonable) assumption that the resulting current spikes can be approximated as triangles and that the inverter is symmetrical in its rising and falling responses, we can compute the energy consumed per switching period as follows:

$$E_{dp} = V_{DD}\frac{I_{peak}t_{sc}}{2} + V_{DD}\frac{I_{peak}t_{sc}}{2} = t_{sc}V_{DD}I_{peak} \qquad (5.49)$$

We compute the average power consumption as

$$P_{dp} = t_{sc}V_{DD}I_{peak} f = C_{sc}V_{DD}^2 f \qquad (5.50)$$

The direct-path power dissipation is proportional to the switching activity, similar to the capacitive power dissipation. t_{sc} represents the time both devices are conducting. For a linear input slope, this time is reasonably well approximated by Eq. (5.51) where t_s represents the 0–100% transition time,

$$t_{sc} = \frac{V_{DD} - 2V_T}{V_{DD}}t_s \approx \frac{V_{DD} - 2V_T}{V_{DD}} \times \frac{t_{r(f)}}{0.8} \qquad (5.51)$$

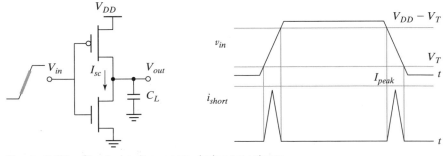

Figure 5-30   Short-circuit currents during transients.

I_{peak} is determined by the saturation current of the devices and is hence directly proportional to the sizes of the transistors. The peak current is also **a strong function of the ratio between input and output slopes**. This relationship is best illustrated by the following simple analysis: Consider a static CMOS inverter with a $0 \rightarrow 1$ transition at the input. Assume first that the load capacitance is very large, so that the output fall time is significantly larger than the input rise time (Figure 5-31a). Under those circumstances, the input moves through the transient region before the output starts to change. As the source-drain voltage of the PMOS device is approximately 0 during that period, the device shuts off without ever delivering any current. The short-circuit current is close to zero in this case. Consider now the reverse case, where the output capacitance is very small, and the output fall time is substantially smaller than the input rise time (Figure 5-31b). The drain-source voltage of the PMOS device equals V_{DD} for most of the transition period, guaranteeing the maximal short-circuit current (equal to the saturation current of the PMOS). This clearly represents the worst case condition. The conclusions of the preceding analysis are confirmed in Figure 5-32, which plots the short-circuit current through the NMOS transistor during a low-to-high transition as a function of the load capacitance.

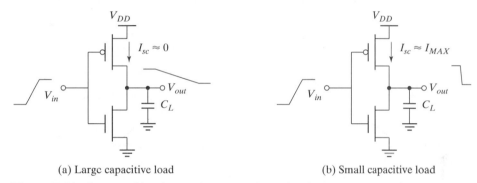

(a) Large capacitive load   (b) Small capacitive load

Figure 5-31   Impact of load capacitance on short-circuit current.

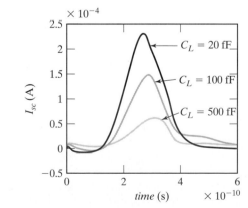

Figure 5-32   CMOS inverter short-circuit current through NMOS transistor as a function of the load capacitance (for a fixed input slope of 500 ps).

This analysis leads to the conclusion that the short-circuit dissipation is minimized by making the output rise/fall time larger than the input ris/fall time. On the other hand, making the output rise/fall time too large slows down the circuit and can cause short-circuit currents in the fan-out gates. This presents a perfect example of how local optimization and forgetting the global picture can lead to an inferior solution.

Design Techniques

A more practical rule, which optimizes the power consumption in a global way, can be formulated ([Veendrick84]):

> The power dissipation due to short-circuit currents is minimized by matching the rise/fall times of the input and output signals. At the overall circuit level, this means that rise/fall times of all signals should be kept constant within a range.

Making the input and output rise times of a gate identical is not the optimum solution for that particular gate on its own, but keeps the overall short-circuit current within bounds. This is shown in Figure 5-33, which plots the short-circuit energy dissipation of an inverter (normalized with respect to the zero-input rise time dissipation) as a function of the ratio r between input and output rise/fall times. When the load capacitance is too small for a given inverter size ($r > 2...3$ for $V_{DD} = 5$ V), the power is dominated by the short-circuit current. For very large capacitance values, all power dissipation is devoted to charging and discharging the load capacitance. When the rise/fall times of inputs and outputs are equalized, most power dissipation is associated with the dynamic power, and only a minor fraction ($< 10\%$) is devoted to short-circuit currents.

Observe also that the impact of **short-circuit current is reduced when we lower the supply voltage**, as is apparent from Eq. (5.51). In the extreme case, when $V_{DD} < V_{Tn} + |V_{Tp}|$, short-circuit dissipation is completely eliminated, because both devices are never on simultaneously. With threshold voltages scaling at a slower rate than the supply voltage, short-circuit power dissipation is becoming less important

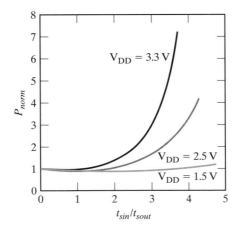

Figure 5-33 Power dissipation of a static CMOS inverter as a function of the ratio between input and output rise/fall times. The power is normalized with respect to zero input rise-time dissipation. At low values of the slope ratio, input/output coupling leads to some extra dissipation.

in deep submicron technologies. At a supply voltage of 2.5 V and thresholds around 0.5 V, an input/output slope ratio of 2 is needed to cause a 10% degradation in dissipation. ∎

Finally, it is worth observing that the short-circuit power dissipation can be modeled by adding a load capacitance $C_{sc} = t_{sc}I_{peak}/V_{DD}$ in parallel with C_L, as is apparent in Eq. (5.50). The value of this short-circuit capacitance is a function of V_{DD}, the transistor sizes, and the input/output slope ratio.

5.5.2 Static Consumption

The static (or steady-state) power dissipation of a circuit is expressed by the relation

$$P_{stat} = I_{stat}V_{DD} \tag{5.52}$$

where I_{stat} is the current that flows between the supply rails in the absence of switching activity.

Ideally, the static current of the CMOS inverter is equal to zero, as the PMOS and NMOS devices are never on simultaneously in steady-state operation. There is, unfortunately, a leakage current flowing through the reverse-biased diode junctions of the transistors, located between the source or drain and the substrate, as shown in Figure 5-34. This contribution is, in general, very small and can be ignored. For the device sizes under consideration, the leakage current per unit drain area typically ranges between 10–100 pA/μm^2 at room temperature. For a die with 1 million gates, each with a drain area of 0.5 μm^2 and operated at a supply voltage of 2.5 V, the worst case power consumption due to diode leakage equals 0.125 mW, which clearly is not much of an issue.

However, be aware that the junction leakage currents are caused by thermally generated carriers. Their value increases with increasing junction temperature, and this occurs in an exponential fashion. At 85°C (a commonly imposed upper bound for junction temperatures in commercial hardware), the leakage currents increase by a factor of 60 over their room-temperature values. Keeping the overall operation temperature of a circuit low is consequently a desirable goal. As the temperature is a strong function of the dissipated heat and its removal mechanisms, this can only be accomplished by limiting the power dissipation of the circuit or by using chip packages that support efficient heat removal.

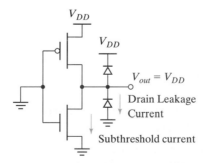

Figure 5-34 Sources of leakage currents in CMOS inverter (for $V_{in} = 0$ V).

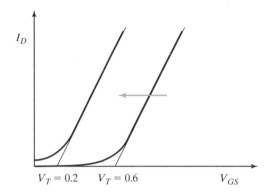

Figure 5-35   Decreasing the threshold increases the subthreshold current at $V_{GS} = 0$.

An emerging source of leakage current is the subthreshold current of the transistors. As discussed in Chapter 3, an MOS transistor can experience a drain-source current, even when V_{GS} is smaller than the threshold voltage (see Figure 5-35). The closer the threshold voltage is to zero volts, the larger the leakage current at $V_{GS} = 0$ V and the larger the static power consumption. To offset this effect, the threshold voltage of the device has generally been kept high enough. Standard processes feature V_T values that are never smaller than 0.5–0.6 V and that in some cases are even substantially higher (~ 0.75 V).

This approach is being challenged by the reduction in supply voltages that typically goes with deep submicron technology scaling, as became apparent in Figure 3-41. We concluded earlier (Figure 5-17) that scaling the supply voltages while keeping the threshold voltage constant results in an important loss in performance, especially when V_{DD} approaches 2 V_T. One approach to address this performance issue is to scale the device thresholds down as well. This moves the curve of Figure 5-17 to the left, which means that the performance penalty for lowering the supply voltage is reduced. Unfortunately, the threshold voltages are lower bounded by the amount of allowable subthreshold leakage current, as demonstrated in Figure 5-35. The choice of the threshold voltage thus represents a trade-off between performance and static power dissipation. The continued scaling of the supply voltage predicted for the next generations of CMOS technologies, however, forces the threshold voltages ever downwards, and makes subthreshold conduction a major source of power dissipation. Process technologies that contain devices with sharper turn-off characteristics will therefore become more attractive. An example of the latter is the SOI (Silicon-on-Insulator) technology whose MOS transistors have slope factors that are close to the ideal 60 mV/decade.

**Example 5.14   Impact of Threshold Reduction on Performance
                         and Static Power Dissipation**

Consider a minimum size NMOS transistor in the 0.25-μm CMOS technology. In Chapter 3, we derived that the slope factor S for this device equals 90 mV/decade. The off-current (at $V_{GS} = 0$) of the transistor for a V_T of approximately 0.5 V equals 10^{-11} A

(Figure 3-22). Reducing the threshold by 200 mV to 0.3 V multiplies the off-current of the transistors with a factor of 170! Assuming a million gate design with a supply voltage of 1.5 V, this translates into a static power dissipation of $10^6 \times 170 \times 10^{-11} \times 1.5 =$ 2.6 mW. A further reduction of the threshold to 100 mV results in an unacceptable dissipation of almost 0.5 W! At that supply voltage, the threshold reductions correspond to a performance improvement of 25% and 40%, respectively.

This lower bound on the thresholds is in some sense artificial. The idea that the leakage current in a static CMOS circuit has to be zero is a misconception. Certainly, the presence of leakage currents degrades the noise margins, because the logic levels are no longer equal to the supply rails, but as long as the noise margins are within range, this is not a compelling issue. The leakage currents, of course, cause an increase in static power dissipation. This is offset by the drop in supply voltage, which is enabled by the reduced thresholds at no cost in performance, and results in a quadratic reduction in dynamic power. For a 0.25-µm CMOS process, the following circuit configurations obtain the same performance: 3-V supply–0.7-V V_T; and 0.45-V supply–0.1-V V_T. The dynamic power consumption of the latter is, however, 45 times smaller [Liu93]! Choosing the correct values of supply and threshold voltages once again requires a trade-off. The optimal operation point depends upon the activity of the circuit. In the presence of a sizable static power dissipation, it is essential that nonactive modules are *powered down*, lest static power dissipation would become dominant. Power-down (also called *standby*) can be accomplished by disconnecting the unit from the supply rails, or by lowering the supply voltage.

5.5.3 Putting It All Together

The total power consumption of the CMOS inverter is now expressed as the sum of its three components:

$$P_{tot} = P_{dyn} + P_{dp} + P_{stat} = (C_L V_{DD}^2 + V_{DD} I_{peak} t_s) f_{0 \to 1} + V_{DD} I_{leak} \qquad (5.53)$$

In typical CMOS circuits, the capacitive dissipation is by far the dominant factor. The direct-path consumption can be kept within bounds by careful design, and thus should not be an issue. Leakage is ignorable at present, but this might change in the not-too-distant future.

The Power-Delay Product, or Energy per Operation

In Chapter 1, we introduced the *power-delay product* (PDP) as a quality measure for a logic gate:

$$PDP = P_{av} t_p \qquad (5.54)$$

The *PDP* presents a measure of energy, as is apparent from the units (W × s = Joule). Assuming that the gate is switched at its maximum possible rate of $f_{max} = 1/(2t_p)$, and ignoring the contributions of the static- and direct-path currents to the power consumption, we find that

$$PDP = C_L V_{DD}^2 f_{max} t_p = \frac{C_L V_{DD}^2}{2} \qquad (5.55)$$

Here, *PDP* stands for the **average energy consumed per switching event** (i.e., for a $0 \rightarrow 1$, or a $1 \rightarrow 0$ transition). Remember that earlier we had defined E_{av} as the average energy per switching cycle (or per energy-consuming event). As each inverter cycle contains a $0 \rightarrow 1$, and a $1 \rightarrow 0$ transition E_{av} thus is twice the *PDP*.

Energy-Delay Product

The validity of the *PDP* as a quality metric for a process technology or gate topology is questionable. It measures the energy needed to switch the gate, which is an important property. For a given structure, however, this number can be made arbitrarily low by reducing the supply voltage. From this perspective, the optimum voltage to run the circuit would be the lowest possible value that still ensures functionality. This comes at the expense of performance, as discussed earlier. A more relevant metric should combine a measure of performance and energy. The energy-delay product (or *EDP*) does exactly that:

$$EDP = PDP \times t_p = P_{av} t_p^2 = \frac{C_L V_{DD}^2}{2} t_p \qquad (5.56)$$

It is worth analyzing the voltage dependence of the EDP. Higher supply voltages reduce delay, but harm the energy, and the opposite is true for low voltages. An optimum operation point should therefore exist. Assuming that NMOS and PMOS transistors have comparable threshold and saturation voltages, we can simplify the propagation delay expression Eq. (5.21) as

$$t_p \approx \frac{\alpha C_L V_{DD}}{V_{DD} - V_{Te}} \qquad (5.57)$$

where $V_{Te} = V_T + V_{DSAT}/2$, and α is a technology parameter. Combining Eq. (5.56) and Eq. (5.57)[7] yields

$$EDP = \frac{\alpha C_L^2 V_{DD}^3}{2(V_{DD} - V_{TE})} \qquad (5.58)$$

The optimum supply voltage can be obtained by taking the derivative of Eq. (5.58) with respect to V_{DD}, and equating the result to 0. The result is

$$V_{DDopt} = \frac{3}{2} V_{TE} \qquad (5.59)$$

The remarkable outcome from this analysis is the low value of the supply voltage that simultaneously optimizes performance and energy. For submicron technologies with thresholds in the range of 0.5 V, the optimum supply is situated around 1 V.

---

[7]This equation is only accurate as long as the devices remain in velocity saturation, which is probably not the case for the lower supply voltages. This introduces some inaccuracy in the analysis, but will not distort the overall result.

Example 5.15   Optimum Supply Voltage for 0.25-µm CMOS Inverter

From the technology parameters for our generic CMOS process presented in Chapter 3, the value of V_{TE} can be derived as follows:

$$V_{Tn} = 0.43 \text{ V}, V_{Dsatn} = 0.63 \text{ V}, V_{TEn} = 0.74 \text{ V}$$
$$V_{Tp} = -0.4 \text{ V}, V_{Dsatp} = -1 \text{ V}, V_{TEp} = -0.9 \text{ V}$$
$$V_{TE} \approx (V_{TEn} + |V_{TEp}|)/2 = 0.8 \text{ V}$$

Hence, $V_{DDopt} = (3/2) \times 0.8 \text{ V} = 1.2 \text{ V}$. The simulated graphs of Figure 5-36, which plot normalized delay, energy, and energy-delay product, confirm this result. The optimum supply voltage is predicted to equal 1.1 V. The charts clearly illustrate the trade-off between delay and energy.

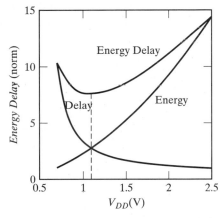

Figure 5-36   Normalized delay, energy, and energy-delay plots for CMOS inverter in 0.25-µm CMOS technology.

WARNING: While the preceding example demonstrates that a supply voltage exists that minimizes the energy-delay product of a gate, this voltage does not necessarily represent the optimum voltage for a given design problem. For instance, some designs require a minimum performance, which requires a higher voltage at the expense of energy. Similarly, a lower energy design is possible by operating at a lower voltage and by obtaining the overall system performance through the use of architectural techniques such as pipelining or concurrency.

5.5.4   Analyzing Power Consumption by Using SPICE

A definition of the average power consumption of a circuit was provided in Chapter 1, and is repeated here for the sake of convenience. We write

$$P_{av} = \frac{1}{T}\int_0^T p(t)dt = \frac{V_{DD}}{T}\int_0^T i_{DD}(t)dt \tag{5.60}$$

Figure 5-37 Equivalent circuit to measure average power in SPICE.

with T the period of interest, and V_{DD} and i_{DD} the supply voltage and current, respectively. Some implementations of SPICE provide built-in functions to measure the average value of a circuit signal. For instance, the HSPICE .MEASURE TRAN I(VDD) AVG command computes the area under a computed transient response (I(VDD)) and divides it by the period of interest. This is identical to the definition given in Eq. (5.60). Other implementations of SPICE are, unfortunately, not as powerful. This is not as bad as it seems, as long as one realizes that SPICE is actually a differential equation solver. A small circuit can easily be conceived that acts as an integrator and whose output signal is nothing but the average power.

Consider, for instance, the circuit of Figure 5-37. The current delivered by the power supply is measured by the current-controlled current source and integrated on the capacitor C. The resistance R is only provided for DC-convergence reasons and should be chosen as high as possible to minimize leakage. A clever choice of the element parameter ensures that the output voltage P_{av} equals the average power consumption. The operation of the circuit is summarized in Eq. under the assumption that the initial voltage on the capacitor C is zero:

$$C\frac{dP_{av}}{dt} = ki_{DD}$$

or   (5.61)

$$P_{av} = \frac{k}{C}\int_0^T i_{DD}dt$$

Equating Eq. (5.60) and Eq. yields the necessary conditions for the equivalent circuit parameters: $k/C = V_{DD}/T$. Under these circumstances, the equivalent circuit shown presents a convenient means of tracking the average power in a digital circuit.

Example 5.16   Average Power of Inverter

The average power consumption of the inverter of Example 5.4 is analyzed using the above technique for a toggle period of 250 ps (T = 250 ps, k = 1, V_{DD} = 2.5 V, hence C = 100 pF). The resulting power consumption is plotted in Figure 5-38, showing an average power consumption of approximately 157.3 μW. The .MEAS AVG command yields a value of 160.3 μW, which demonstrates the approximate equivalence of both methods. These numbers are equivalent to an energy of 39 fJ (which is close to the 37.5 fJ derived in

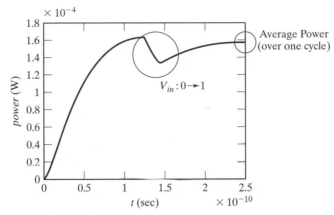

Figure 5-38 Deriving the power consumption by using SPICE.

Example 5.11). Observe the slightly negative dip during the high-to-low transition. This is due to the injection of current into the supply, when the output briefly overshoots V_{DD} as a result of the capacitive coupling between input and output (as is apparent from in the transient response of Figure 5-16).

5.6 Perspective: Technology Scaling and its Impact on the Inverter Metrics

In Section 3.5, we have explored the impact of the scaling of technology on some of the important design parameters, such as area, delay, and power. For the sake of clarity, we repeat here some of the most important entries of the scaling table (Table 3-8).

The validity of these theoretical projections can be verified by looking back and observing the trends during the past few decades. From Figure 5-39, we can see that the gate delay indeed decreases exponentially at a rate of 13% per year, or halving every five years. This rate is on course with the prediction of Table 5-4, since S averages approximately 1.15 as we had already

Table 5-4 Scaling scenarios for short-channel devices (S and U represent the technology and voltage scaling parameters, respectively).

Parameter	Relation	Full Scaling	General Scaling	Fixed-Voltage Scaling
Area–Device	$W–L$	$1/S^2$	$1/S^2$	$1/S^2$
Intrinsic Delay	$R_{on}C_{gate}$	$1/S$	$1/S$	$1/S$
Intrinsic Energy	$C_{gate}V^2$	$1/S^3$	$1/SU^2$	$1/S$
Intrinsic Power	*Energy–Delay*	$1/S^2$	$1/U^2$	1
Power Density	$P–Area$	1	S^2/U^2	S^2

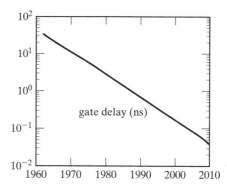

Figure 5-39   Scaling of the gate delay (from [Dally98]).

observed in Figure 3-40. The delay of a two-input NAND gate with a fan-out of four has gone from tens of nanoseconds in the 1960s to a tenth of a nanosecond in the year 2000, and is projected to be a few tens of picoseconds by 2010.

Reducing power dissipation has only been a second-order priority until recently. Hence, statistics on dissipation per gate or design are only marginally available. An interesting chart is shown in Figure 5-40, which plots the power density measured over a large number of designs produced between 1980 and 1995. Although the variation is large—even for a fixed technology—it shows the power density increasing approximately with S^2. This is in correspondence with the fixed-voltage scaling scenario presented in Table 5-4. For more recent years, we expect a scenario more in line with the full-scaling model—which predicts a constant power density—due to the accelerated supply-voltage scaling and the increased attention to power-reducing design techniques. Even under these circumstances, power dissipation per chip will continue to increase due to the ever-larger die sizes.

The scaling model presented has one major flaw, however. The performance and power predictions produce purely "intrinsic" numbers that take only device parameters into account. In Chapter 4, we concluded that the interconnect wires exhibit a different scaling behavior, and that wire parasitics may come to dominate the overall performance. Similarly, charging and discharging the wire capacitances may dominate the energy budget. To get a crisper perspective, one has to construct a combined model that considers device and wire scaling models simultaneously. The impact of the wire capacitance and its scaling behavior is summarized in Table 5-5. We adopt the fixed-resistance model introduced in Chapter 4. We furthermore assume that the resistance of the driver dominates the wire resistance, which is definitely the case for short to medium-long wires.

The model predicts that the interconnect-caused delay (and energy) gain in importance with the scaling of technology. This impact is limited to an increase with εc for short wires ($S = S_L$), but it becomes increasingly more significant for medium-range and long wires ($S_L < S$). These conclusions

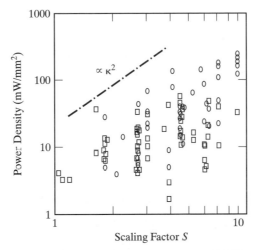

Figure 5-40 Evolution of power density in micro- and DSP processors, as a function of the scaling factor S ([Kuroda95]). S is normalized to 1 for a 4-µm process.

Table 5-5 Scaling scenarios for wire capacitance. S and U represent the technology and voltage scaling parameters, respectively, while S_L stands for the wire-length scaling factor. ε_c represents the impact of fringing and interwire capacitances.

Parameter	Relation	General Scaling
Wire Capacitance	WL/t	ε_c/S_L
Wire Delay	$R_{on}C_{int}$	ε_c/S_L
Wire Energy	$C_{int}V^2$	$\varepsilon_c/S_L U^2$
Wire Delay / Intrinsic Delay		$\varepsilon_c S/S_L$
Wire Energy / Intrinsic Energy		$\varepsilon_c S/S_L$

have been confirmed by a number of studies, an example of which is shown in Figure 5-41. How the ratio of wire over intrinsic contributions will actually evolve is debatable, as it depends upon a wide range of independent parameters, such as system architecture, design methodology, transistor sizing, and interconnect materials. The doomsday scenario that interconnect may cause CMOS performance to saturate in the very near future may very well be exaggerated. Yet, it is clear that increased attention to interconnect is an absolute necessity, and may change the way the next-generation circuits are designed and optimized (e.g., [Sylvester98]).

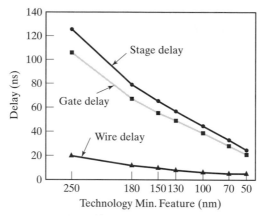

Figure 5-41   Evolution of wire delay-to-gate delay ratio with respect to technology (from [Fisher98]).

5.7 Summary

This chapter presented a rigorous and in-depth analysis of the static CMOS inverter. The key characteristics of the gate are summarized as follows:

- The static CMOS inverter combines a pull-up PMOS section with a pull-down NMOS device. The PMOS is normally made wider than the NMOS due to its lower current-driving capabilities.
- The gate has an almost ideal voltage-transfer characteristic. The logic swing is equal to the supply voltage and is not a function of the transistor sizes. The noise margins of a symmetrical inverter (where PMOS and NMOS transistor have equal current-driving strength) approach $V_{DD}/2$. The steady-state response is not affected by fan-out.
- Its propagation delay is dominated by the time it takes to charge or discharge the load capacitor C_L. To a first order, it can be approximated as follows:

$$t_p = 0.69 C_L \left(\frac{R_{eqn} + R_{eqp}}{2} \right)$$

Keeping the load capacitance small is the most effective means of implementing high-performance circuits. Transistor sizing may help to improve performance as long as the delay is dominated by the extrinsic (or load) capacitance of fan-out and wiring.

- The power dissipation is dominated by the dynamic power consumed in charging and discharging the load capacitor. It is given by $P_{0 \to 1} C_L V_{DD}^2 f$. The dissipation is proportional to the activity in the network. The dissipation due to the direct-path currents occurring during switching can be limited by careful tailoring of the signal slopes. The static dissipation usually can be ignored, but might become a major factor in the future as a result of sub-threshold currents.

- Scaling the technology is an effective means of reducing the area, propagation delay and power consumption of a gate. The impact is even more striking if the supply voltage is scaled simultaneously.
- The interconnect component is gradually taking a larger fraction of the delay and performance budget.

5.8 To Probe Further

The operation of the CMOS inverter has been the topic of numerous publications and textbooks. Virtually every book on digital design devotes a substantial number of pages to the analysis of the basic inverter gate. An extensive list of references was presented in Chapter 1. Some references of particular interest that we quoted in this chapter follow.

References

[Dally98] W. Dally and J. Poulton, *Digital Systems Engineering*, Cambridge University Press, 1998.

[Fisher98] P. D. Fisher and R. Nesbitt, "The Test of Time: Clock-Cycle Estimation and Test Challenges for Future Microprocessors," *IEEE Circuits and Devices Magazine,* 14(2), pp. 37–44, 1998.

[Hedenstierna87] N. Hedenstierna and K. Jeppson, "CMOS Circuit Speed and Buffer Optimization," *IEEE Transactions on CAD*, vol. CAD-6, no. 2, pp. 270–281, March 1987.

[Kuroda95] T. Kuroda and T. Sakurai, "Overview of low-power ULSI circuit techniques," *IEICE Trans. on Electronics*, vol. E78-C, no. 4, pp. 334-344, April 1995.

[Liu93] D. Liu and C. Svensson, "Trading speed for low power by choice of supply and threshold voltages," *IEEE Journal of Solid-State Circuits*, vol. 28, no.1, pp. 10–17, Jan. 1993, p.10–17.

[Mead80] C. Mead and L. Conway, *Introduction to VLSI Systems*, Addison-Wesley, 1980.

[Sedra87] A. Sedra and K. Smith, *MicroElectronic Circuits*, Holt, Rinehart and Winston, 1987.

[Swanson72] R. Swanson and J. Meindl, "Ion-Implanted Complementary CMOS transistors in Low-Voltage Circuits," *IEEE Journal of Solid-State Circuits*, vol. SC-7, no. 2, pp.146–152, April 1972.

[Sylvester98] D. Sylvester and K. Keutzer, "Getting to the Bottom of Deep Submicron," *Proceedings ICCAD Conference*, pp. 203, San Jose, November 1998.

[Veendrick84] H. Veendrick, "Short-Circuit Dissipation of Static CMOS Circuitry and its Impact on the Design of Buffer Circuits," *IEEE Journal of Solid-State Circuits*, vol. SC-19, no. 4, pp. 468–473, 1984.

Exercises and Design Problems

REMINDER: Please refer to **http://bwrc.eecs.berkeley.edu/IcBook** for up-to-date problem sets, design problems, and exercises. By making the exercises electronically available instead of in print, we can provide a dynamic environment that tracks the rapid evolution of today's digital integrated circuit design technology.

6

Designing Combinational Logic Gates in CMOS

In-depth discussion of logic families in CMOS—
static and dynamic, pass-transistor, nonratioed and ratioed logic

Optimizing a logic gate for area, speed, energy, or robustness

Low-power and high-performance circuit-design techniques

6.1   Introduction

6.2   Static CMOS Design
    6.2.1   Complementary CMOS
    6.2.2   Ratioed Logic
    6.2.3   Pass-Transistor Logic

6.3   Dynamic CMOS Design
    6.3.1   Dynamic Logic: Basic Principles
    6.3.2   Speed and Power Dissipation of Dynamic Logic
    6.3.3   Signal Integrity Issues in Dynamic Design
    6.3.4   Cascading Dynamic Gates

6.4   Perspectives
    6.4.1   How to Choose a Logic Style?
    6.4.2   Designing Logic for Reduced Supply Voltages

6.5   Summary

6.6   To Probe Further

6.1 Introduction

The design considerations for a simple inverter circuit were presented in the previous chapter. We now extend this discussion to address the synthesis of arbitrary digital gates, such as NOR, NAND, and XOR. The focus is on *combinational logic* or *nonregenerative* circuits—that is, circuits having the property that at any point in time, the output of the circuit is related to its current input signals by some Boolean expression (assuming that the transients through the logic gates have settled). No intentional connection from outputs back to inputs is present.

This is in contrast to another class of circuits, known as *sequential* or *regenerative*, for which the output is not only a function of the current input data, but also of previous values of the input signals (see Figure 6-1). This can be accomplished by connecting one or more outputs intentionally back to some inputs. Consequently, the circuit "remembers" past events and has a sense of *history*. A sequential circuit includes a combinational logic portion and a module that holds the state. Example circuits are registers, counters, oscillators, and memory. Sequential circuits are the topic of the next chapter.

There are numerous circuit styles to implement a given logic function. As with the inverter, the common design metrics by which a gate is evaluated are area, speed, energy, and power. Depending on the application, the emphasis will be on different metrics. For example, the switching speed of digital circuits is the primary metric in a high-performance processor, while in a battery operated circuit, it is energy dissipation. Recently, power dissipation also has become an important concern and considerable emphasis is placed on understanding the sources of power and approaches to dealing with power. In addition to these metrics, robustness to noise and reliability are also very important considerations. We will see that certain logic styles can significantly improve performance, but they usually are more sensitive to noise.

6.2 Static CMOS Design

The most widely used logic style is static complementary CMOS. The static CMOS style is really an extension of the static CMOS inverter to multiple inputs. To review, the primary advantage of the CMOS structure is robustness (i.e., low sensitivity to noise), good performance, and low power consumption with no static power dissipation. Most of those properties are carried over to large fan-in logic gates implemented using a similar circuit topology.

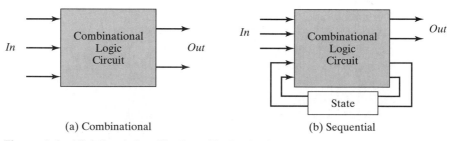

(a) Combinational   (b) Sequential

Figure 6-1   High-level classification of logic circuits.

The complementary CMOS circuit style falls under a broad class of logic circuits called *static* circuits in which **at every point in time, each gate output is connected to either V_{DD} or V_{SS} via a low-resistance path**. Also, the outputs of the gates assume at all times the value of the Boolean function implemented by the circuit (ignoring, the transient effects during switching periods). This is in contrast to the *dynamic* circuit class, which relies on temporary storage of signal values on the capacitance of high-impedance circuit nodes. The latter approach has the advantage that the resulting gate is simpler and faster. Its design and operation are, however, more involved and prone to failure because of increased sensitivity to noise.

In this section, we sequentially address the design of various static circuit flavors, including complementary CMOS, ratioed logic (pseudo-NMOS and DCVSL), and pass-transistor logic. We also deal with issues of scaling to lower power supply voltages and threshold voltages.

6.2.1 Complementary CMOS

Concept

A static CMOS gate is a combination of two networks—the *pull-up network* (PUN) and the *pull-down network* (PDN), as shown in Figure 6-2. The figure shows a generic *N*-input logic gate where all inputs are distributed to both the pull-up and pull-down networks. The function of the PUN is to provide a connection between the output and V_{DD} anytime the output of the logic gate is meant to be 1 (based on the inputs). Similarly, the function of the PDN is to connect the output to V_{SS} when the output of the logic gate is meant to be 0. The PUN and PDN networks are constructed in a mutually exclusive fashion such that *one and only one* of the networks is conducting in steady state. In this way, once the transients have settled, a path always exists between V_{DD} and the output F for a high output ("one"), or between V_{SS} and F for a low output ("zero"). This is equivalent to stating that the output node is always a *low-impedance* node in steady state.

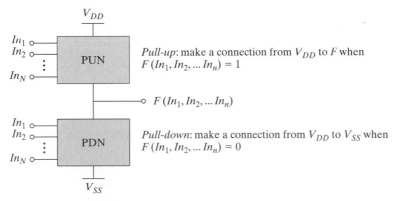

Figure 6-2 Complementary logic gate as a combination of a PUN (pull-up network) and a PDN (pull-down network).

In constructing the PDN and PUN networks, the designer should keep the following observations in mind:

- A transistor can be thought of as a switch controlled by its gate signal. An NMOS switch is *on* when the controlling signal is high and is *off* when the controlling signal is low. A PMOS transistor acts as an inverse switch that is *on* when the controlling signal is low and *off* when the controlling signal is high.
- The PDN is constructed using NMOS devices, while PMOS transistors are used in the PUN. The primary reason for this choice is that NMOS transistors produce "strong zeros," and PMOS devices generate "strong ones." To illustrate this, consider the examples shown in Figure 6-3. In Figure 6-3a, the output capacitance is initially charged to V_{DD}. Two possible discharge scenarios are shown. An NMOS device pulls the output all the way down to GND, while a PMOS lowers the output no further than $|V_{Tp}|$—the PMOS turns *off* at that point and stops contributing discharge current. NMOS transistors are thus the preferred devices in the PDN. Similarly, two alternative approaches to charging up a capacitor are shown in Figure 6-3b, with the output initially at GND. A PMOS switch succeeds in charging the output all the way to V_{DD}, while the NMOS device fails to raise the output above $V_{DD} - V_{Tn}$. This explains why PMOS transistors are preferentially used in a PUN.
- A set of rules can be derived to construct logic functions (see Figure 6-4). NMOS devices connected in series correspond to an AND function. With all the inputs high, the series combination conducts and the value at one end of the chain is transferred to the other end. Similarly, NMOS transistors connected in parallel represent an OR function. A conducting path exists between the output and input terminal if at least one of the inputs is high. Using similar arguments, construction rules for PMOS networks can be formulated. A series con-

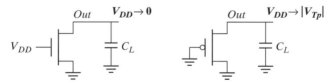

(a) Pulling down a node by using NMOS and PMOS switches

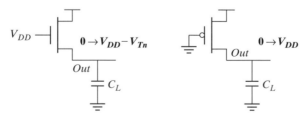

(b) Pulling down a node by using NMOS and PMOS switches

Figure 6-3   Simple examples illustrate why an NMOS should be used as a pull-down, and a PMOS should be used as a pull-up device.

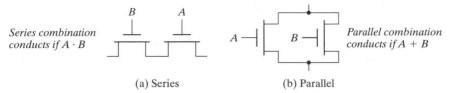

(a) Series   (b) Parallel

Figure 6-4 NMOS logic rules—series devices implement an AND, and parallel devices implement an OR.

nection of PMOS conducts if both inputs are low, representing a NOR function ($\overline{A} \cdot \overline{B} = \overline{A + B}$), while PMOS transistors in parallel implement a NAND ($\overline{A + B} = \overline{A} \cdot \overline{B}$).

• Using De Morgan's theorems ($\overline{A + B} = \overline{A} \cdot \overline{B}$ and $\overline{A \cdot B} = \overline{A} + \overline{B}$), it can be shown that the pull-up and pull-down networks of a complementary CMOS structure are *dual* networks. This means that a parallel connection of transistors in the pull-up network corresponds to a series connection of the corresponding devices in the pull-down network, and vice versa. Therefore, to construct a CMOS gate, one of the networks (e.g., PDN) is implemented using combinations of series and parallel devices. The other network (i.e., PUN) is obtained using the duality principle by walking the hierarchy, replacing series subnets with parallel subnets, and parallel subnets with series subnets. The complete CMOS gate is constructed by combining the PDN with the PUN.

• The complementary gate is naturally *inverting*, implementing only functions such as NAND, NOR, and XNOR. The realization of a noninverting Boolean function (such as AND OR, or XOR) in a single stage is not possible, and requires the addition of an extra inverter stage.

• The number of transistors required to implement an *N*-input logic gate is 2*N*.

Example 6.1   Two-Input NAND Gate

Figure 6-5 shows a two-input NAND gate ($F = \overline{A \cdot B}$). The PDN network consists of two NMOS devices in series that conduct when both *A* and *B* are high. The PUN is the dual

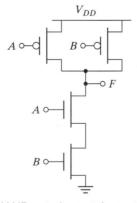

Figure 6-5 Two-input NAND gate in complementary static CMOS style.

network, and it consists of two parallel PMOS transistors. This means that F is 1 if $A = 0$ or $B = 0$, which is equivalent to $F = \overline{A \cdot B}$. The truth table for the simple two input NAND gate is given in Table 6-1. It can be verified that the output F is always connected to either V_{DD} or GND, but never to both at the same time.

Table 6-1 Truth Table for two-Input NAND.

A	B	F
0	0	1
0	1	1
1	0	1
1	1	0

Example 6.2   Synthesis of Complex CMOS Gate

Using complementary CMOS logic, consider the synthesis of a complex CMOS gate whose function is $F = \overline{D + A \cdot (B + C)}$. The first step in the synthesis of the logic gate is to derive the pull-down network as shown in Figure 6-6a by using the fact that NMOS devices in series implements the AND function and parallel device implements the OR function. The next step is to use duality to derive the PUN in a hierarchical fashion. The PDN network is broken into smaller networks (i.e., subset of the PDN) called subnets that simplify the derivation of the PUN. In Figure 6-6b, the subnets (SN) for the pull-down net-

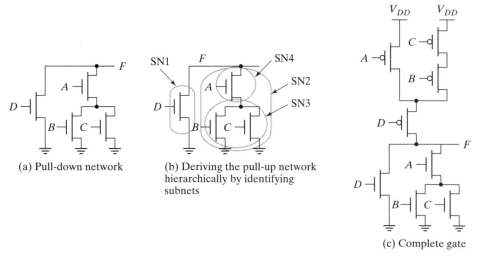

(a) Pull-down network

(b) Deriving the pull-up network hierarchically by identifying subnets

(c) Complete gate

Figure 6-6   Complex complementary CMOS gate.

work are identified. At the top level, SN1 and SN2 are in parallel, so that in the dual network they will be in series. Since SN1 consists of a single transistor, it maps directly to the pull-up network. On the other hand, we need to sequentially apply the duality rules to SN2. Inside SN2, we have SN3 and SN4 in series, so in the PUN they will appear in parallel. Finally, inside SN3, the devices are in parallel, so they appear in series in the PUN. The complete gate is shown in Figure 6-6c. The reader can verify that for every possible input combination, there always exists a path to either V_{DD} or GND.

Static Properties of Complementary CMOS Gates

Complementary CMOS gates inherit all the nice properties of the basic CMOS inverter. They exhibit rail-to-rail swing with $V_{OH} = V_{DD}$ and $V_{OL} =$ GND. The circuits also have no static power dissipation, since the circuits are designed such that the pull-down and pull-up networks are mutually exclusive. The analysis of the DC voltage transfer characteristics and the noise margins is more complicated than for the inverter, as these parameters **depend upon the data input patterns** applied to gate.

Consider the static two-input NAND gate shown in Figure 6-7. Three possible input combinations switch the output of the gate from high to low: (a) $A = B = 0 \rightarrow 1$, (b) $A = 1, B = 0 \rightarrow 1$, and (c) $B = 1, A = 0 \rightarrow 1$. The resulting voltage transfer curves display significant differences. The large variation between case (a) and the others (b and c) is explained by the fact that in the former case, both transistors in the pull-up network are on simultaneously for $A = B = 0$, representing a strong pull-up. In the latter cases, only one of the pull-up devices is on. The VTC is shifted to the left as a result of the weaker PUN.

The difference between (b) and (c) results mainly from the state of the internal node *int* between the two NMOS devices. For the NMOS devices to turn on, both gate-to-source voltages

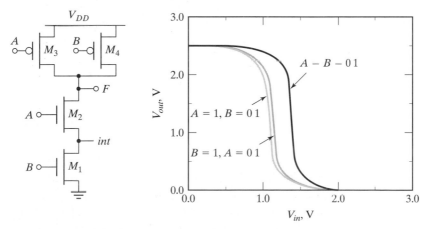

Figure 6-7 The VTC of a two-input NAND is data dependent. NMOS devices are 0.5 μm/0.25 μm while the PMOS devices are sized at 0.75 μm/0.25 μm.

must be above V_{Tn}, with $V_{GS2} = V_A - V_{DS1}$ and $V_{GS1} = V_B$. The threshold voltage of transistor M_2 will be higher than transistor M_1 due to the body effect. The threshold voltages of the two devices are given by the following equations:

$$V_{Tn2} = V_{Tn0} + \gamma((\sqrt{|2\phi_f| + V_{int}}) - \sqrt{|2\phi_f|}) \tag{6.1}$$

$$V_{Tn1} = V_{Tn0} \tag{6.2}$$

For case (b), M_3 is turned *off*, and the gate voltage of M_2 is set to V_{DD}. To a first order, M_2 may be considered as a resistor in series with M_1. Since the drive on M_2 is large, this resistance is small and has only a small effect on the voltage transfer characteristics. In case (c), transistor M_1 acts as a resistor, causing a V_T increase in M_2 due to body effect. The overall impact is quite small, as seen from the plot.

Design Consideration

The important point to take away from the preceding discussion is that the **noise margins are input/pattern dependent**. In Example 6.2, a glitch on only one of the two inputs has a larger chance of creating a false transition at the output than if the glitch were to occur on both inputs simultaneously. Therefore, the former condition has a lower low-noise margin. A common practice when characterizing gates such as NAND and NOR is to connect all the inputs together. Unfortunately, this does not represent the worst case static behavior; the data dependencies should be carefully modeled. ∎

Propagation Delay of Complementary CMOS Gates

The computation of propagation delay proceeds in a fashion similar to the static inverter. For the purpose of delay analysis, each transistor is modeled as a resistor in series with an ideal switch. The value of the resistance is dependent on the power supply voltage and an equivalent large signal resistance, scaled by the ratio of device width over length, must be used. The logic is transformed into an equivalent RC network that includes the effect of internal node capacitances. Figure 6-8 shows the two-input NAND gate and its equivalent RC switch level model. Note that the internal node capacitance C_{int}—attributable to the source/drain regions and the gate overlap capacitance of M_2 and M_1—is included here. While complicating the analysis, the capacitance of the internal nodes can have quite an impact in some networks such as large fan-in gates. In a first pass, we ignore the effect of the internal capacitance.

A simple analysis of the model shows that, similarly to the noise margins, **the propagation delay depends on the input patterns**. Consider, for instance, the low-to-high transition. Three possible input scenarios can be identified for charging the output to V_{DD}. If both inputs are driven low, the two PMOS devices are on. The delay in this case is $0.69 \times (R_p/2) \times C_L$, since the two resistors are in parallel. This is not the worst case low-to-high transition, which occurs when only one device turns on, and is given by $0.69 \times R_p \times C_L$. For the pull-down path, the output is discharged only if both A and B are switched high, and the delay is given by $0.69 \times (2R_N) \times C_L$ to a first order. In other words, adding devices in series slows down the circuit, and devices must be made wider to avoid a performance penalty. When sizing the transistors in a gate with multiple inputs, we should pick the combination of inputs that triggers the worst case conditions.

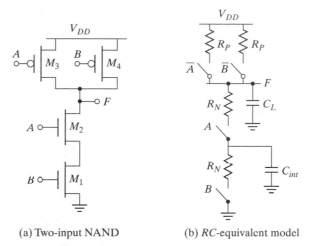

(a) Two-input NAND   (b) *RC*-equivalent model

Figure 6-8 Equivalent *RC* model for a two-input NAND gate.

For the NAND gate to have the same pull-down delay (t_{phl}) as a minimum-sized inverter, the NMOS devices in the PDN stack must be made twice as wide so that the equivalent resistance of the NAND pull-down network is the same as the inverter. The PMOS devices can remain unchanged.[1]

This first-order analysis assumes that the extra capacitance introduced by widening the transistors can be ignored. This is not a good assumption, in general, but it allows for a reasonable first cut at device sizing.

Example 6.3   Delay Dependence on Input Patterns

Consider the NAND gate of Figure 6-8a. Assume NMOS and PMOS devices of 0.5 μm/ 0.25 μm and 0.75 μm/0.25 μm, respectively. This sizing should result in approximately equal worst case rise and fall times (since the effective resistance of the pull-down is designed to be equal to the pull-up resistance).

Figure 6-9 shows the simulated low-to-high delay for different input patterns. As expected, the case in which both inputs transition go low ($A = B = 1 \rightarrow 0$) results in a smaller delay, compared with the case in which only one input is driven low. Notice that the worst case low-to-high delay depends upon which input (*A* or *B*) goes low. The reason for this involves the internal node capacitance of the pull-down stack (i.e., the source of M_2). For the case in which $B = 1$ and *A* transitions from $1 \rightarrow 0$, the pull-up PMOS device only has to charge up the output node capacitance (M_2 is turned off). On the other hand, for the case in which $A = 1$ and *B* transitions from $1 \rightarrow 0$, the pull-up PMOS device

---

[1] In deep-submicron processes, even larger increases in the width are needed due to the on-set of velocity saturation. For a two-input NAND, the NMOS transistors should be made 2.5 times as wide instead of 2 times.

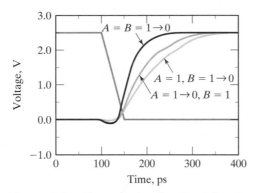

Input Data Pattern	Delay (ps)
$A = B = 0 \rightarrow 1$	69
$A = 1, B = 0 \rightarrow 1$	62
$A = 0 \rightarrow 1, B = 1$	50
$A = B = 1 \rightarrow 0$	35
$A = 1, B = 1 \rightarrow 0$	76
$A = 1 \rightarrow 0, B = 1$	57

Figure 6-9   Example showing the delay dependence on input patterns.

has to charge up the sum of the output and the internal node capacitances, which slows down the transition.

The table in Figure 6-9 shows a compilation of various delays for this circuit. The first-order transistor sizing indeed provides approximately equal rise and fall delays. An important point to note is that the high-to-low propagation delay depends on the initial state of the internal nodes. For example, when both inputs transition from $0 \rightarrow 1$, it is important to establish the state of the internal node. The worst case happens when the internal node is initially charged up to $V_{DD} - V_{Tn}$, which can be ensured by pulsing the A input from $1 \rightarrow 0 \rightarrow 1$, while input B only makes the $0 \rightarrow 1$ transition. In this way, the internal node is initialized properly.

The important point to take away from this example is that estimation of delay can be fairly complex, and requires a careful consideration of internal node capacitances and data patterns. Care must be taken to model the worst case scenario in the simulations. A brute force approach that applies all possible input patterns may not always work, because it is important to consider the state of internal nodes.

The CMOS implementation of a NOR gate ($F = \overline{A + B}$) is shown in Figure 6-10. The output of this network is high, if and only if both inputs A and B are low. The worst case pull-down transition happens when only one of the NMOS devices turns on (i.e., if either A or B is high). Assume that the goal is to size the NOR gate such that it has approximately the same delay as an inverter with the following device sizes: NMOS of 0.5 μm/0.25 μm and PMOS of 1.5 μm/ 0.25 μm. Since the pull-down path in the worst case is a single device, the NMOS devices (M_1 and M_2) can have the same device widths as the NMOS device in the inverter. For the output to be pulled high, both devices must be turned on. Since the resistances add, the devices must be made two times larger compared with the PMOS in the inverter (i.e., M_3 and M_4 must have a size of 3 μm/0.25 μm). Since PMOS devices have a lower mobility relative to NMOS devices, stacking devices in series must be avoided as much as possible. A NAND implementation is clearly preferred over a NOR implementation for implementing generic logic.

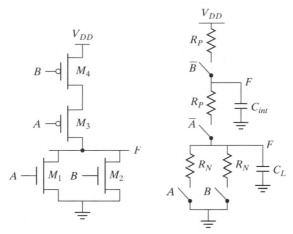

Figure 6-10 Sizing of a NOR gate.

Problem 6.1   Transistor Sizing in Complementary CMOS Gates

Determine the sizes of the transistors in Figure 6-6c such that it has approximately the same t_{plh} and t_{phl} as an inverter with the following sizes: NMOS: 0.5 µm/0.25 µm and a PMOS: 1.5 µm/0.25 µm.

So far in the analysis of propagation delay, we have ignored the effect of internal node capacitances. This is often a reasonable assumption for a first-order analysis. However, in more complex logic gates with large *fan-ins*, the internal node capacitances can become significant. Consider a four-input NAND gate, as drawn in Figure 6-11, which shows the equivalent *RC* model of the gate, including the internal node capacitances. The internal capacitances consist of the junction capacitances of the transistors, as well as the gate-to-source and gate-to-drain capacitances. The latter are turned into capacitances to ground using the Miller equivalence. The delay analysis for such a circuit involves solving distributed *RC* networks, a problem we already encountered when analyzing the delay of interconnect networks. Consider the pull-down delay of the circuit. The output is discharged when all inputs are driven high. The proper initial conditions must be placed on the internal nodes (i.e., the internal nodes must be charged to $V_{DD} - V_{TN}$) before the inputs are driven high.

The propagation delay can be computed by using the Elmore delay model:

$$t_{pHL} = 0.69(R_1 \cdot C_1 + (R_1 + R_2) \cdot C_2 + (R_1 + R_2 + R_3) \cdot C_2 + (R_1 + R_2 + R_3 + R_4) \cdot C_L) \quad (6.3)$$

Notice that the resistance of M_1 appears in all the terms, which makes this device especially important when attempting to minimize delay. Assuming that all NMOS devices have an equal size, Eq. (6.3) simplifies to

$$t_{pHL} = 0.69R_N(C_1 + 2 \cdot C_2 + 3 \cdot C_3 + 4 \cdot C_L) \quad (6.4)$$

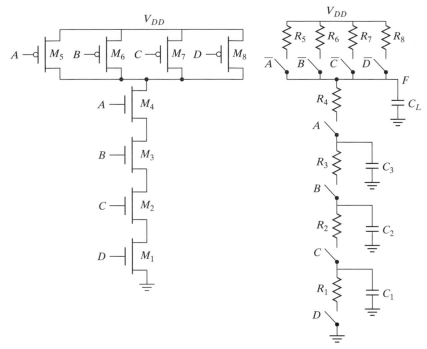

Figure 6-11　Four-input NAND gate and its *RC* model.

Example 6.4　A Four-Input Complementary CMOS NAND Gate

In this example, we evaluate the *intrinsic (or unloaded) propagation delay* of a four-input NAND gate (without any loading) is evaluatedusing hand analysis and simulation. The layout of the gate is shown in Figure 6-12. Assume that all NMOS devices have a *W/L* of 0.5 μm/0.25 μm, and all PMOS devices have a device size of 0.375 μm/0.25 μm. The devices are sized such that the worst case rise and fall times are approximately equal to a first order (ignoring the internal node capacitances).

By using techniques similar to those employed for the CMOS inverter in Chapter 5, the capacitance values can be computed from the layout. Notice that in the pull-up path, the PMOS devices share the drain terminal, in order to reduce the overall parasitic contribution. Using our standard design rules, we find that the area and perimeter for various devices can be easily computed, as shown in Table 6-2.

In this example, we focus on the pull-down delay, and the capacitances will be computed for the high-to-low transition at the output. While the output makes a transition from V_{DD} to 0, the internal nodes only transition from $V_{DD} - V_{Tn}$ to GND. We need to linearize the internal junction capacitances for this voltage transition, but, to simplify the analysis, we use the same K_{eff} for the internal nodes as for the output node.

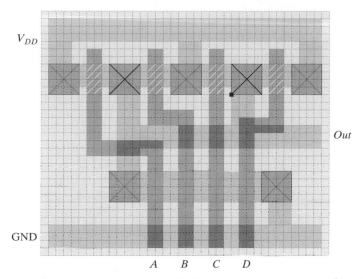

Figure 6-12 Layout a four-input NAND gate in complementary CMOS. See also Colorplate 7.

Table 6-2 Area and perimeter of transistors in four-input NAND gate.

Transistor	W (μm)	AS (μm^2)	AD (μm^2)	PS (μm)	PD (μm)
1	0.5	0.3125	0.0625	1.75	0.25
2	0.5	0.0625	0.0625	0.25	0.25
3	0.5	0.0625	0.0625	0.25	0.25
4	0.5	0.0625	0.3125	0.25	1.75
5	0.375	0.297	0.172	1.875	0.875
6	0.375	0.172	0.172	0.875	0.875
7	0.375	0.172	0.172	0.875	0.875
8	0.375	0.297	0.172	1.875	0.875

It is assumed that the output connects to a single, minimum-size inverter. The effect of intracell routing, which is small, is ignored. The various contributions are summarized in Table 6-3. For the NMOS and PMOS junctions, we use $K_{eq} = 0.57$, $K_{eqsw} = 0.61$, and $K_{eq} = 0.79$, $K_{eqsw} = 0.86$, respectively. Notice that the gate-to-drain capacitance is multiplied by a factor of two for all internal nodes as well as the output node, to account for the Miller effect. (This ignores the fact that the internal nodes have a slightly smaller swing due to the threshold drop.)

Table 6-3   Computation of capacitances for high-to-low transition at the output. The table shows the intrinsic delay of the gate without extra loading. Any *fan-out* capacitance would simply be added to the C_L term.

Capacitor	Contributions (H → L)	Value (fF) (H → L)
C1	$C_{d1} + C_{s2} + 2 * C_{gd1} + 2 * C_{gs2}$	$(0.57 * 0.0625 * 2 + 0.61 * 0.25 * 0.28) +$ $(0.57 * 0.0625 * 2 + 0.61 * 0.25* 0.28) +$ $2 * (0.31 * 0.5) + 2 * (0.31 * 0.5) = 0.85$ fF
C2	$C_{d2} + C_{s3} + 2 * C_{gd2} + 2 * C_{gs3}$	$(0.57 * 0.0625 * 2 + 0.61 * 0.25 * 0.28) +$ $(0.57 * 0.0625 * 2 + 0.61 * 0.25* 0.28) +$ $2 * (0.31 * 0.5) + 2 * (0.31 * 0.5) = 0.85$ fF
C3	$C_{d3} + C_{s4} + 2 * C_{gd3} + 2 * C_{gs4}$	$(0.57 * 0.0625 * 2+ 0.61 * 0.25 * 0.28) +$ $(0.57 * 0.0625 * 2+ 0.61 * 0.25* 0.28) +$ $2 * (0.31 * 0.5) + 2 * (0.31 * 0.5) = 0.85$ fF
CL	$C_{d4} + 2 * C_{gd4} + C_{d5} + C_{d6} + C_{d7}$ $+ C_{d8} + 2 * C_{gd5} + 2 * C_{gd6}$ $+ 2 * C_{gd7} + 2 * C_{gd8}$ $= C_{d4} + 4 * C_{d5} + 4 * 2 * C_{gd6}$	$(0.57 * 0.3125 * 2 + 0.61 * 1.75 *0.28) +$ $2 * (0.31 * 0.5)+ 4 * (0.79 * 0.171875* 1.9+ 0.86$ $* 0.875 * 0.22)+ 4 * 2 * (0.27 * 0.375) = 3.47$ fF

Using Eq. (6.4), we compute the propagation delay, as follows:

$$t_{pHL} = 0.69\left(\frac{13\text{K}\Omega}{2}\right)(0.85 \text{ fF} + 2 \cdot 0.85 \text{ fF} + 3 \cdot 0.85 \text{ fF} + 4 \cdot 3.47 \text{ fF}) = 85 \text{ ps}$$

The simulated delay for this particular transition was found to be 86 ps! The hand analysis gives a fairly accurate estimate, given all of the assumptions and linearizations that were made. For example, we assume that the gate–source (or gate–drain) capacitance only consists of the overlap component. This is not entirely the case, because, during the transition, some other contributions come in place depending upon the operating region. Once again, the goal of hand analysis is not to provide a totally accurate delay prediction, but rather to give intuition into what factors influence the delay and to aid in initial transistor sizing. Accurate timing analysis and transistor optimization is usually done using SPICE. The simulated worst case low-to-high delay time for this gate was 106 ps.

While complementary CMOS is a very robust and simple approach for implementing logic gates, there are two major problems associated with using this style as the complexity of the gate (i.e., *fan-in*) increases. First, the number of transistors required to implement an N fan-in gate is $2N$. This can result in a significantly large implementation area.

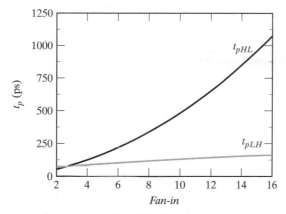

Figure 6-13 Propagation delay of CMOS NAND gate as a function of fan-in. A fan-out of one inverter is assumed, and all pull-down transistors are minimal size.

The second problem is that propagation delay of a complementary CMOS gate deteriorates rapidly as a function of the fan-in. In fact, the *unloaded intrinsic delay* of the gate is, at worst, a *quadratic function of the fan-in*.

- The large number of transistors ($2N$) increases the overall capacitance of the gate. For an N-input gate, the *intrinsic capacitance* increases linearly with the fan-in. Consider, for instance, the NAND gate of Figure 6-11. Given the linear increase in the number of PMOS devices connected to the output node, we expect the low-to-high delay of the gate to increase linearly with fan-in—while the capacitance goes up linearly, the pull-up resistance remains unchanged.
- The series connection of transistors in either the PUN or PDN of the gate causes an additional slowdown. We know that the *distributed RC network* in the PDN of Figure 6-11 comes with a delay that is quadratic in the number of elements in the chain. The high-to-low delay of the gate should hence be a quadratic function of the fan-in.

Figure 6-13 plots the (intrinsic) propagation delay of a NAND gate as a function of fan-in assuming a fixed fan-out of one inverter (NMOS: 0.5 μm and PMOS: 1.5 μm). As predicted, t_{pLH} is a linear function of fan-in, while the simultaneous increase in the pull-down resistance and the load capacitance cause an approximately quadratic relationship for t_{pHL}. Gates with a *fan-in* greater than or equal to 4 become excessively slow and must be avoided.

Design Techniques for Large Fan-in

The designer has a number of techniques at his disposition to reduce the delay of large fan-in circuits:

- **Transistor Sizing** The most obvious solution is to increase the transistor sizes. This lowers the resistance of devices in series and lowers the time constants. However, increasing the transistor sizes results in larger parasitic capacitors, which not only affect the *propagation delay* of the gate in question, but

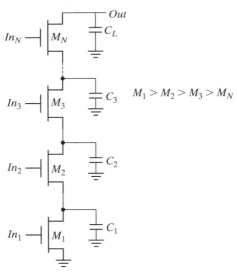

Figure 6-14   Progressive sizing of transistors in large transistor chains copes with the extra load of internal capacitances.

also present a larger load to the preceding gate. This technique should therefore be used with caution. If the load capacitance is dominated by the intrinsic capacitance of the gate, widening the device only creates a "self-loading" effect, and the *propagation delay* is unaffected. Sizing is only effective when the load is dominated by the fan-out. A more comprehensive approach toward sizing transistors in complex CMOS combinational networks is discussed in the next section.

• **Progressive Transistor Sizing**   An alternate approach to uniform sizing (in which each transistor is scaled up uniformly), is to use progressive transistor sizing (Figure 6-14). Referring back to Eq. (6.3), we see that the resistance of M_1 (R_1) appears N times in the delay equation, the resistance of M_2 (R_2) appears $N-1$ times, etc. From the equation, it is clear that R_1 should be made the smallest, R_2 the next smallest, etc. Consequently, a progressive scaling of the transistors is beneficial: $M_1 > M_2 > M_3 > M_N$. This approach reduces the dominant resistance, while keeping the increase in capacitance within bounds. For an excellent treatment on the optimal sizing of transistors in a complex network, we refer the interested reader to [Shoji88, pp. 131–143]. You should be aware, however, of one important pitfall of this approach. While progressive resizing of transistors is relatively easy in a schematic diagram, it is not as simple in a real layout. Very often, design-rule considerations force the designer to push the transistors apart, which causes the internal capacitance to grow. This may offset all the gains of the resizing!

• **Input Reordering**   Some signals in complex combinational logic blocks might be more critical than others. Not all inputs of a gate arrive at the same time (due, for instance, to the propagation delays of the preceding logical gates). An input signal to a gate is called *critical* if it is the last signal of all inputs to assume a stable value. The path through the logic which determines the ultimate speed of the structure is called the *critical path*.

     Putting the critical-path transistors closer to the output of the gate can result in a speed up, as demonstrated in Figure 6-15. Signal In_1 is assumed to be a critical signal. Suppose further that In_2 and In_3 are high, and that In_1 undergoes a $0 \to 1$ transition. Assume also that C_L is initially charged high. In case (a), no path to GND exists until M_1 is turned on, which, unfortunately, is the last event to happen. The delay between the arrival of In_1 and the output is therefore determined by the time it takes to discharge C_L, C_1, and C_2. In the second case, C_1 and C_2 are already

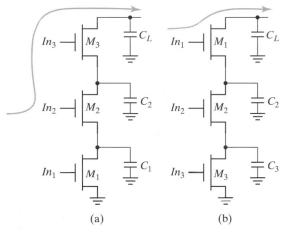

(a)   (b)

Figure 6-15 Influence of transistor ordering on delay.
Signal In_1 is the critical signal.

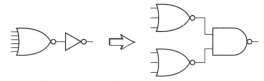

Figure 6-16 Logic restructuring can reduce the gate fan-in.

discharged when In_1 changes. Only C_L still has to be discharged, resulting in a smaller delay.

• **Logic Restructuring** Manipulating the logic equations can reduce the fan-in requirements and thus reduce the gate delay, as illustrated in Figure 6-16. The quadratic dependency of the gate delay on *fan-in* makes the six-input NOR gate extremely slow. Partitioning the NOR gate into two three-input gates results in a significant speedup, which by far offsets the extra delay incurred by turning the inverter into a two-input NAND gate.   ∎

Optimizing Performance in Combinational Networks

Earlier, we established that minimization of the propagation delay of a gate in isolation is a purely academic effort. The sizing of devices should happen in its proper context. In Chapter 5, we developed a methodology to do so for inverters. We also found that an optimal fan-out for a chain of inverters driving a load C_L is $(C_L/C_{in})^{1/N}$, where N is the number of stages in the chain, and C_{in} the input capacitance of the first gate in the chain. If we have an opportunity to select the number of stages, we found out that we would like to keep the fan-out per stage around 4. Can this result be extended to determine the size of any combinational path for minimal delay? By extending our previous approach to address complex logic networks, we find out that this is indeed possible [Sutherland99].[2]

---

[2] The approach introduced in this section is commonly called logical effort, and was formally introduced in [Sutherland99], which presents an extensive treatment of the topic. The treatment offered here represents only a glance over of the overall approach.

Table 6-4 Estimates of intrinsic delay factors of various logic types, assuming simple layout styles, and a fixed PMOS–NMOS ratio.

Gate type	p
Inverter	1
n-input NAND	n
n-input NOR	n
n-way multiplexer	$2n$
XOR, NXOR	$n2^{n-1}$

To do so, we modify the basic delay equation of the inverter that we introduced in Chapter 5, namely,

$$t_p = t_{p0}\left(1 + \frac{C_{ext}}{\gamma C_g}\right) = t_{p0}(1 + f/\gamma) \tag{6.5}$$

to

$$t_p = t_{p0}(p + gf/\gamma) \tag{6.6}$$

with t_{p0} still representing the intrinsic delay of an inverter and f the *effective fan-out*, defined as the ratio between the external load and the input capacitance of the gate. In this context, f is also called the *electrical effort,* and p represents the ratio of the intrinsic (or unloaded) delays of the complex gate and the simple inverter, and is a function of gate topology, as well as layout style. The more involved structure of the multiple-input gate causes its intrinsic delay to be higher than that of an inverter. Table 6-4 enumerates the values of p for some standard gates, assuming simple layout styles, and ignoring second-order effects such as internal node capacitances.

The factor g is called the *logical effort*, and represents the fact that, for a given load, complex gates have to work harder than an inverter to produce a similar response. In other words, the logical effort of a logic gate tells how much worse it is at producing output current than an inverter, given that each of its inputs may present only the same input capacitance as the inverter. Equivalently, logical effort is how much more input capacitance a gate presents to deliver the same output current as an inverter. Logical effort is a useful parameter, because it depends only on circuit topology. The logical efforts of some common logic gates are given in Table 6-5.

Table 6-5   Logic efforts of common logic gates, assuming a PMOS–NMOS ratio of 2.

Gate Type	Number of Inputs			
	1	2	3	n
Inverter	1			
NAND		4/3	5/3	(n + 2)/3
NOR		5/3	7/3	(2n + 1)/3
Multiplexer		2	2	2
XOR		4	12	

Example 6.5   Logical Effort of Complex Gates

Consider the gates shown in Figure 6-17. Assuming PMOS–NMOS ratio of 2, the input capacitance of a minimum-sized symmetrical inverter equals three times the gate capacitance of a minimum-sized NMOS (called C_{unit}). We size the two-input NAND and NOR such that their equivalent resistances equal the resistance of the inverter (using the techniques described earlier). This increases the input capacitance of the two-input NAND to 4 C_{unit}, or 4/3 the capacitance of the inverter. The input capacitance of the two-input NOR is 5/3 that of the inverter. Equivalently, for the same input capacitance, the NAND and NOR gate have 4/3 and 5/3 less driving strength than the inverter. This affects the delay component that corresponds to the load, increasing it by this same factor, called the *logical effort*. Hence, $g_{NAND} = 4/3$, and $g_{NOR} = 5/3$.

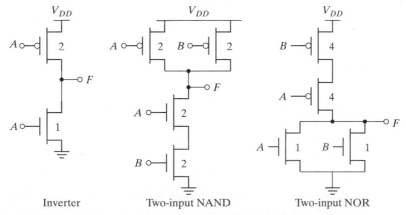

Figure 6-17   Logical effort of two-input NAND and NOR gates.

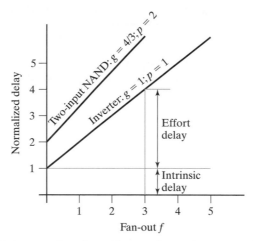

Figure 6-18   Delay as a function of fan-out for an inverter and a two-input NAND.

The delay model of a logic gate, as represented in Eq. (6.6), is a simple linear relationship. Figure 6-18 shows this relationship graphically: the delay is plotted as a function of the fan-out for an inverter and for a two-input NAND gate. The slope of the line is the logical effort of the gate; its intercept is the intrinsic delay. The graph shows that we can adjust the delay by adjusting the effective fan-out (by transistor sizing) or by choosing a logic gate with a different logical effort. Observe also that fan-out and logical effort contribute to the delay in a similar way. We call the product of the two $h = fg$, the *gate effort*.

The total delay of a path through a combinational logic block can now be expressed as

$$t_p = \sum_{j=1}^{N} t_{p,j} = t_{p0} \sum_{j=1}^{N} \left(p_j + \frac{f_j g_j}{\gamma} \right) \tag{6.7}$$

We use a similar procedure as we did for the inverter chain in Chapter 5 to determine the minimum delay of the path. By finding $N-1$ partial derivatives and setting them to zero, we find that **each stage should bear the same gate effort**:

$$f_1 g_1 = f_2 g_2 = \ldots = f_N g_N \tag{6.8}$$

The logical effort along a path in the network compounds by multiplying the logical efforts of all the gates along the path, yielding the *path logical effort G*:

$$G = \prod_{1}^{N} g_i \tag{6.9}$$

We also can define a *path effective fan-out* (or *electrical effort*) *F*, which relates the load capacitance of the last gate in the path to the input capacitance of the first gate:

$$F = \frac{C_L}{C_{g1}} \tag{6.10}$$

To relate F to the effective fan-outs of the individual gates, we must introduce another factor to account for the logical fan-out within the network. When fan-out occurs at the output of a node, some of the available drive current is directed along the path we are analyzing, and some is directed off the path. We define the *branching effort* b of a logical gate on a given path to be

$$b = \frac{C_{\text{on-path}} + C_{\text{off-path}}}{C_{\text{on-path}}} \tag{6.11}$$

where $C_{\text{on-path}}$ is the load capacitance of the gate along the path we are analyzing and $C_{\text{off-path}}$ is the capacitance of the connections that lead off the path. Note that the branching effort is, if the path does not branch (as in a chain of gates). The *path branching effort* is defined as the product of the branching efforts at each of the stages along the path, or

$$B = \prod_1^N b_i \tag{6.12}$$

The path electrical effort can now be related to the electrical and branching efforts of the individual stages:

$$F = \prod_1^N \frac{f_i}{b_i} = \frac{\prod f_i}{B} \tag{6.13}$$

Finally, the total path effort H can be defined. Using Eq. (6.13), we write

$$H = \prod_1^N h_i = \prod_1^N g_i f_i = GFB \tag{6.14}$$

From here on, the analysis proceeds along the same lines as the inverter chain. The gate effort that minimizes the path delay is

$$h = \sqrt[N]{H} \tag{6.15}$$

and the minimum delay through the path is

$$D = t_{p0} \left(\sum_{j=1}^N p_j + \frac{N(\sqrt[N]{H})}{\gamma} \right) \tag{6.16}$$

Note that the path intrinsic delay is a function of the types of logic gates in the path and is not affected by the sizing. The size factors of the individual gates in the chain s_i can then be derived by working from front to end (or vice versa). We assume that a unit-size gate has a driving capability equal to a minimum-size inverter. Based on the definition of the logical effort, this means that its input capacitance is g times larger than that of the reference inverter, which equals C_{ref}. With s_1 the sizing factor of the first gate in the chain, the input capacitance of the chain C_{g1}

equals $g_1 s_1 C_{ref}$. Including the branching effort, we know that the input capacitance of gate 2 is (f_1/b_1) larger, or

$$g_2 s_2 C_{ref} = \left(\frac{f_1}{b_1}\right) g_1 s_1 C_{ref} \tag{6.17}$$

For gate i in the chain, this yields

$$s_i = \left(\frac{g_1 s_1}{g_i}\right) \prod_{j=1}^{i-1} \left(\frac{f_j}{b_j}\right) \tag{6.18}$$

Example 6.6   Sizing Combinational Logic for Minimum Delay

Consider the logic network of Figure 6-19, which may represent the critical path of a more complex logic block. The output of the network is loaded with a capacitance which is five times larger than the input capacitance of the first gate, which is a minimum-sized inverter. The effective fan-out of the path thus equals $F = C_L/C_{g1} = 5$. Using the entries in Table 6-5, we find the path logical effort as follows:

$$G = 1 \times \frac{5}{3} \times \frac{5}{3} \times 1 = \frac{25}{9}$$

Since there is no branching, $B = 1$. Hence, $H = GFB = 125/9$, and the optimal stage effort h is $\sqrt[4]{H} = 1.93$. Taking into account the gate types, we derive the following fan-out factors: $f_1 = 1.93$; $f_2 = 1.93 \times (3/5) = 1.16$; $f_3 = 1.16$; $f_4 = 1.93$. Notice that the inverters are assigned larger than the more complex gates because they are better at driving loads.

Finally, we derive the gate sizes (with respect to the minimum-sized versions) using Eq. (6.18). This leads to the following values: $a = f_1 g_1/g_2 = 1.16$; $b = f_1 f_2 g_1/g_3 = 1.34$; and $c = f_1 f_2 f_3 g_1/g_4 = 2.60$.

These calculations do not have to be very precise. As discussed in Chapter 5, sizing a gate too large or too small by a factor of 1.5 still results in circuits within 5% of minimum delay. Therefore, the "back of the envelope" hand calculations using this technique are quite effective.

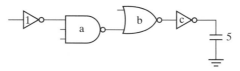

Figure 6-19   Critical path of combinational network.

Problem 6.2 Sizing an Inverter Network

Revisit Problem 5.5, but this time around use the branching-effort approach to produce the solution.

Power Consumption in CMOS Logic Gates

The sources of power consumption in a complementary CMOS inverter were discussed in detail in Chapter 5. Many of these issues apply directly to complex CMOS gates. The power dissipation is a strong function of transistor sizing (which affects physical capacitance,) input and output rise–fall times (which determine the short-circuit power,) device thresholds and temperature (which impact leakage power,) and switching activity. The dynamic power dissipation is given by $\alpha_{0 \to 1} C_L V_{DD}^2 f$. Making a gate more complex mostly affects the *switching activity* $\alpha_{0 \to 1}$, which has two components: a static component that is only a function of the topology of the logic network, and a dynamic one that results from the timing behavior of the circuit. (The latter factor is also called glitching.)

Logic Function The transition activity is a strong function of the logic function being implemented. For static CMOS gates with statistically independent inputs, the static transition probability is the probability p_0 that the output will be in the *zero* state in one cycle, multiplied by the probability p_1 that the output will be in the *one* state in the next cycle:

$$\alpha_{0 \to 1} = p_0 \cdot p_1 = p_0 \cdot (1 - p_0) \tag{6.19}$$

Assuming that the inputs are independent and uniformly distributed, any N-input static gate has a transition probability given by

$$\alpha_{0 \to 1} = \frac{N_0}{2^N} \cdot \frac{N_1}{2^N} = \frac{N_0 \cdot (2^N - N_0)}{2^{2N}} \tag{6.20}$$

where N_0 is the number of *zero* entries, and N_1 is the number of *one* entries in the output column of the truth table of the function. To illustrate, consider a static two-input NOR gate whose truth table is shown in Table 6-6. Assume that only one input transition is possible during a clock cycle and that the inputs to the NOR gate have a uniform input distribution (in other words, the four possible states for inputs A and B—00, 01, 10, 11—are equally likely).

Table 6-6 Truth table of a two-input NOR gate.

A	B	Out
0	0	1
0	1	0
1	0	0
1	1	0

From Table 6-6 and Eq. (6.20), the output transition probability of a two-input static CMOS NOR gate can be derived:

$$\alpha_{0 \to 1} = \frac{N_0 \cdot (2^N - N)}{2^{2N}} = \frac{3 \cdot (2^2 - 3)}{2^{2 \cdot 2}} = \frac{3}{16} \tag{6.21}$$

Problem 6.3 N-Input XOR Gate

Assuming the inputs to an N-input XOR gate are uncorrelated and uniformly distributed, derive the expression for the switching activity factor.

Signal Statistics   The switching activity of a logic gate is a strong function of the input signal statistics. Using a uniform input distribution to compute activity is not a good technique, since the propagation through logic gates can significantly modify the signal statistics. For example, consider once again a two-input static NOR gate, and let p_a and p_b be the probabilities that the inputs A and B are *one*. Assume further that the inputs are not correlated. The probability that the output node is 1 is given by

$$p_1 = (1 - p_a)(1 - p_b) \tag{6.22}$$

Therefore, the probability of a transition from 0 to 1 is

$$\alpha_{0 \to 1} = p_0 \, p_1 = (1 - (1 - p_a)(1 - p_b))(1 - p_a)(1 - p_b) \tag{6.23}$$

Figure 6-20 shows the transition probability as a function of p_a and p_b. Observe how this graph degrades into the simple inverter case when one of the input probabilities is set to 0. From

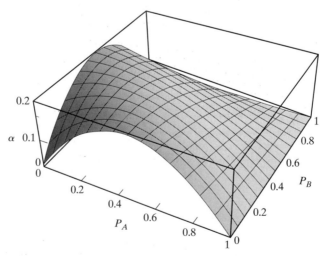

Figure 6-20   Transition activity of a two-input NOR gate as a function of the input probabilities (p_A, p_B).

this plot, it is clear that understanding the signal statistics and their impact on switching events can be used to significantly impact the power dissipation.

Problem 6.4 Power Dissipation of Basic Logic Gates

Derive the $0 \rightarrow 1$ output transition probabilities for the basic logic gates (AND, OR, XOR). The results to be obtained are given in Table 6-7.

Table 6-7 Output transition probabilities for static logic gates.

	$\alpha_{0 \rightarrow 1}$
AND	$(1 - p_A p_B) p_A p_B$
OR	$(1 - p_A)(1 - p_B)[1 - (1 - p_A)(1 - p_B)]$
XOR	$[1 - (p_A + p_B - 2p_A p_B)](p_A + p_B - 2p_A p_B)$

Intersignal Correlations The evaluation of the switching activity is further complicated by the fact that signals exhibit correlation in space and time. Even if the primary inputs to a logic network are uncorrelated, the signals become correlated or "colored," as they propagate through the logic network. This is best illustrated with a simple example. Consider first the circuit shown in Figure 6-21a, and assume that the primary inputs A and B are uncorrelated and uniformly distributed. Node C has a **1 (0)** probability of 1/2, and a $0 \rightarrow 1$ transition probability of 1/4. The probability that the node Z undergoes a power consuming transition is then determined using the AND-gate expression of Table 6-7:

$$p_{0 \rightarrow 1} = (1 - p_a p_b) \, p_a p_b = (1 - 1/2 \cdot 1/2) \, 1/2 \cdot 1/2 = 3/16 \tag{6.24}$$

The computation of the probabilities is straightforward: signal and transition probabilities are evaluated in an ordered fashion, progressing from the input to the output node. This approach, however, has two major limitations: (1) it does not deal with circuits with feedback as found in sequential circuits, and (2) it assumes that the signal probabilities at the input of each gate are independent. This is rarely the case in actual circuits, where reconvergent fan-out often causes intersignal dependencies. For instance, the inputs to the AND gate in Figure 6-21b (C and B) are interdependent because both are a function of A. The approach to computing

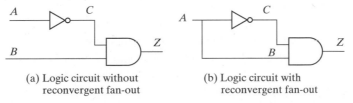

(a) Logic circuit without reconvergent fan-out

(b) Logic circuit with reconvergent fan-out

Figure 6-21 Example illustrating the effect of signal correlations.

probabilities that we presented previously fails under these circumstances. Traversing from inputs to outputs yields a transition probability of 3/16 for node Z, similar to the previous analysis. This value clearly is false, as logic transformations show that the network can be reduced to $Z = C \cdot B = A \cdot \overline{A} = 0$, and thus no transition will ever take place.

To get the precise results in the progressive analysis approach, its is essential to take signal interdependencies into account. This can be accomplished with the aid of conditional probabilities. For an AND gate, Z equals 1 if and only if B and C are equal to 1. Thus,

$$p_Z = p(Z = 1) = p(B = 1, C = 1) \tag{6.25}$$

where $p(B = 1, C = 1)$ represents the probability that B and C are equal to 1 simultaneously. If B and C are independent, $p(B = 1, C = 1)$ can be decomposed into $p(B = 1) \cdot p(C = 1)$, and this yields the expression for the AND gate derived earlier: $p_Z = p(B = 1) \cdot p(C = 1) = p_B \, p_C$. If a dependency between the two exists (as is the case in Figure 6-21b), a conditional probability has to be employed, such as the following:

$$p_Z = p(C = 1|B = 1) \cdot p(B = 1) \tag{6.26}$$

The first factor in Eq. (6.26) represents the probability that $C = 1$ given that $B = 1$. The extra condition is necessary because C is dependent upon B. Inspection of the network shows that this probability is equal to 0, since C and B are logical inversions of each other, resulting in the signal probability for Z, $p_Z = 0$.

Deriving those expressions in a structured way for large networks with reconvergent fan-out is complex, especially when the networks contain feedback loops. Computer support is therefore essential. To be meaningful, the analysis program has to process a typical sequence of input signals, because the power dissipation is a strong function of statistics of those signals.

Dynamic or Glitching Transitions   When analyzing the transition probabilities of complex, multistage logic networks in the preceding section, we ignored the fact that the gates have a non-zero propagation delay. In reality, the finite propagation delay from one logic block to the next can cause spurious transitions known as *glitches or dynamic hazards* to occur: a node can exhibit multiple transitions in a single clock cycle before settling to the correct logic level.

A typical example of the effect of glitching is shown in Figure 6-22, which displays the simulated response of a chain of NAND gates for all inputs going simultaneously from 0 to 1. Initially, all the outputs are 1 since one of the inputs was 0. For this particular transition, all the odd bits must transition to 0, while the even bits remain at the value of 1. However, due to the finite propagation delay, the even output bits at the higher bit positions start to discharge, and the voltage drops. When the correct input ripples through the network, the output goes high. The glitch on the even bits causes extra power dissipation beyond what is required to strictly implement the logic function. Although the glitches in this example are only partial (i.e., not from rail to rail), they contribute significantly to the power dissipation. Long chains of gates often occur in important structures such as adders and multipliers, and the glitching component can easily dominate the overall power consumption.

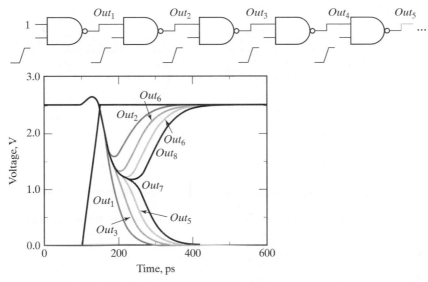

Figure 6-22 Glitching in a chain of NAND gates.

Design Techniques to Reduce Switching Activity

The dynamic power of a logic gate can be reduced by minimizing the physical capacitance and the switching activity. The physical capacitance can be minimized in a number ways, including circuit style selection, transistor sizing, placement and routing, and architectural optimizations. The switching activity, on the other hand, can be minimized at all levels of the design abstraction, and is the focus of this section. Logic structures can be optimized to minimize both the fundamental transitions required to implement a given function and the spurious transitions.

1. **Logic Restructuring** Changing the topology of a logic network may reduce its power dissipation. Consider, for example, two alternative implementations of $F = A \cdot B \cdot C \cdot D$, as shown in Figure 6-23. Ignore glitching and assume that all primary inputs (A,B,C,D) are uncorrelated and uniformly distributed (this is, $p_{1\ (a,b,c,d)} = 0.5$). Using the expressions from Table 6-7, the activity can be computed for the two topologies, as shown in Table 6-8. The results indicate that the chain implementation has an overall lower switching activity than the tree implementation for random inputs. However, as mentioned before, it is also important to consider the timing behavior to

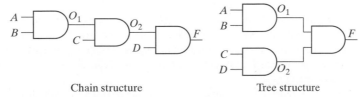

Chain structure   Tree structure

Figure 6-23 Simple example to demonstrate the influence of circuit topology on activity.

Table 6-8 Probabilities for tree and chain topologies.

	O_1	O_2	F
p_1 (chain)	1/4	1/8	1/16
$p_0 = 1\text{-}p_1$ (chain)	3/4	7/8	15/16
$p_{0\to1}$ (chain)	3/16	7/64	15/256
p_1 (tree)	1/4	1/4	1/16
$p_0 = 1\text{-}p_1$ (tree)	3/4	3/4	15/16
$p_{0\to1}$ (tree)	3/16	3/16	15/256

accurately make power trade-offs. In this example, the tree topology experiences (virtually) no glitching activity since the signal paths are balanced to all the gates.

2. **Input ordering** Consider the two static logic circuits of Figure 6-24. The probabilities that A, B, and C are equal to **1** are listed in the Figure. Since both circuits implement identical logic functionality, it is clear that the activity at the output node Z is equal in both cases. The difference is in the activity at the intermediate node. In the first circuit, this activity equals $(1 - 0.5 \times 0.2)(0.5 \times 0.2) = 0.09$. In the second case, the probability that a $0 \to 1$ transition occurs equals $(1 - 0.2 \times 0.1)(0.2 \times 0.1) = 0.0196$, a substantially lower value. From this, we learn that it is beneficial to postpone the introduction of signals with a high transition rate (i.e., signals with a signal probability close to 0.5). A simple reordering of the input signals is often sufficient to accomplish that goal.

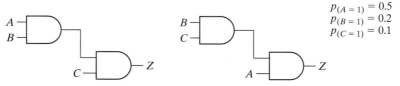

$$P_{(A = 1)} = 0.5$$
$$P_{(B = 1)} = 0.2$$
$$P_{(C = 1)} = 0.1$$

Figure 6-24 Reordering of inputs affects the circuit activity.

3. **Time-multiplexing resources** Time-multiplexing a single hardware resource—such as a logic unit or a bus—over a number of functions is a technique often used to minimize the implementation area. Unfortunately, the minimum area solution does not always result in the lowest switching activity. For example, consider the transmission of two input bits (A and B) using either dedicated resources or a time-multiplexed approach, as shown in Figure 6-25. To the first order, ignoring the multiplexer overhead, it would seem that the degree of time multiplexing should not affect the switched capacitance, since the time-multiplexed solution has half the physical capacitance switched at twice the frequency (for a fixed throughput).

If the data being transmitted are random, it will make no difference which architecture is used. However, if the data signals have some distinct properties (such as temporal correlation), the power dissipation of the time-multiplexed solution can be significantly higher. Suppose, for instance, that A is always (or mostly) 1, and B is (mostly) 0. In the parallel solution, the switched capacitance is very low since there are very few transitions on the data bits. However, in the time-multiplexed solution,

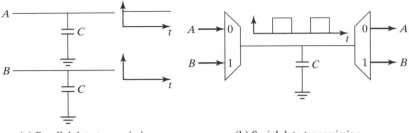

(a) Parallel data transmission   (b) Serial data transmission

Figure 6-25   Parallel versus time-multiplexed data busses.

the bus toggles between 0 and 1. Care must be taken in digital systems to avoid time-multiplexing data streams with very distinct data characteristics.

4. **Glitch Reduction by balancing signal paths**   The occurrence of glitching in a circuit is mainly due to a mismatch in the path lengths in the network. If all input signals of a gate change simultaneously, no glitching occurs. On the other hand, if input signals change at different times, a dynamic hazard might develop. Such a mismatch in signal timing is typically the result of different path lengths with respect to the primary inputs of the network. This is illustrated in Figure 6-26. Assume that the XOR gate has a unit delay. The first network (a) suffers from glitching as a result of the wide disparity between the arrival times of the input signals for a gate. For example, for gate F_3, one input settles at time 0, while the second one only arrives at time 2. Redesigning the network so that all arrival times are identical can dramatically reduce the number of superfluous transitions (network b).

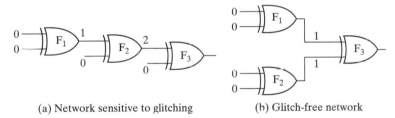

(a) Network sensitive to glitching   (b) Glitch-free network

Figure 6-26   Glitching is influenced by matching of signal path lengths.
The annotated numbers indicate the signal arrival times.

■

Summary

The CMOS logic style described in the previous section is highly robust and scalable with technology, but requires $2N$ transistors to implement an N-input logic gate. Also, the load capacitance is significant, since each gate drives two devices (a PMOS and an NMOS) per *fan-out*. This has opened the door for alternative logic families that either are simpler or faster.

6.2.2   Ratioed Logic

Concept

Ratioed logic is an attempt to reduce the number of transistors required to implement a given logic function, often at the cost of reduced robustness and extra power dissipation. The purpose

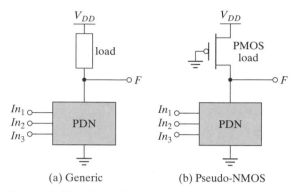

(a) Generic   (b) Pseudo-NMOS

Figure 6-27   Ratioed logic gate.

of the PUN in complementary CMOS is to provide a conditional path between V_{DD} and the output when the PDN is turned *off*. In ratioed logic, the entire PUN is replaced with a single unconditional load device that pulls up the output for a high output as in Figure 6-27a. Instead of a combination of active pull-down and pull-up networks, such a gate consists of an NMOS pull-down network that realizes the *logic function*, and a simple *load device*. Figure 6-27b shows an example of ratioed logic, which uses a grounded PMOS load and is referred to as a pseudo-NMOS gate.

The clear advantage of a pseudo-NMOS gate is the reduced number of transistors ($N + 1$, versus $2N$ for complementary CMOS). The nominal high output voltage (V_{OH}) for this gate is V_{DD} since the pull-down devices are turned *off* when the output is pulled high (assuming that V_{OL} is below V_{Tn}). On the other hand, the **nominal low output voltage is not 0 V**, since there is contention between the devices in the PDN and the grounded PMOS load device. This results in reduced noise margins and, more importantly, static power dissipation. The sizing of the load device relative to the pull-down devices can be used to trade off parameters such as *noise margin*, *propagation delay*, and *power dissipation*. Since the voltage swing on the output and the overall functionality of the gate depend on the ratio of the NMOS and PMOS sizes, the circuit is called *ratioed*. This is in contrast to the *ratioless* logic styles, such as complementary CMOS, where the low and high levels do not depend on transistor sizes.

Computing the dc-transfer characteristic of the pseudo-NMOS proceeds along paths similar to those used for its complementary CMOS counterpart. The value of V_{OL} is obtained by equating the currents through the driver and load devices for $V_{in} = V_{DD}$. At this operation point, it is reasonable to assume that the NMOS device resides in linear mode (since, ideally, the output should be close to 0V), while the PMOS load is saturated:

$$k_n\left((V_{DD} - V_{Tn})V_{OL} - \frac{V_{OL}^2}{2}\right) + k_p\left((-V_{DD} - V_{Tp}) \cdot V_{DSATp} - \frac{V_{DSATp}^2}{2}\right) = 0 \qquad (6.27)$$

Assuming that V_{OL} is small relative to the gate drive ($V_{DD} - V_T$), and that V_{Tn} is equal to V_{Tp} in magnitude, V_{OL} can be approximated as

$$V_{OL} \approx \frac{k_p(V_{DD} + V_{Tp}) \cdot V_{DSATp}}{k_n(V_{DD} - V_{Tn})} \approx \frac{\mu_p \cdot W_p}{\mu_n \cdot W_n} \cdot V_{DSATp} \tag{6.28}$$

In order to make V_{OL} as small as possible, the PMOS device should be sized much smaller than the NMOS pull-down devices. Unfortunately, this has a negative impact on the *propagation delay* for charging up the output node since the current provided by the PMOS device is limited.

A major disadvantage of the pseudo-NMOS gate is the static power that is dissipated when the output is low through the direct current path that exists between V_{DD} and GND. The static power consumption in the low-output mode is easily derived:

$$P_{low} = V_{DD}I_{low} \approx V_{DD} \cdot \left| k_p\left((-V_{DD} - V_{Tp}) \cdot V_{DSATp} - \frac{V_{DSATp}^2}{2}\right) \right| \tag{6.29}$$

Example 6.7 Pseudo-NMOS Inverter

Consider a simple pseudo-NMOS inverter (where the PDN network in Figure 6-27 degenerates to a single transistor) with an NMOS size of 0.5 μm/0.25 μm. In this example, we study the effect of sizing the PMOS device to demonstrate the impact on various parameters. The *W–L* ratio of the grounded PMOS is varied over values from 4, 2, 1, 0.5 to 0.25. Devices with a *W–L* < 1 are constructed by making the length greater than the width. The voltage transfer curve for the different sizes is plotted in Figure 6-28.

Table 6-9 summarizes the nominal output voltage (V_{OL}), static power dissipation, and the low-to-high propagation delay. The low-to-high delay is measured as the time it takes to reach 1.25 V from V_{OL} (which is not 0V for this inverter)—by definition. The trade-off between the static and dynamic properties is apparent. A larger pull-up device not only improves performance, but also increases static power dissipation and lowers noise margins by increasing V_{OL}.

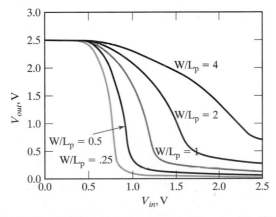

Figure 6-28   Voltage-transfer curves of the pseudo-NMOS inverter as a function of the PMOS size.

Table 6-9   Performance of a pseudo-NMOS inverter.

Size	V_{OL}	Static Power Dissipation	t_{plh}
4	0.693 V	564 μW	14 ps
2	0.273 V	298 μW	56 ps
1	0.133 V	160 μW	123 ps
0.5	0.064 V	80 μW	268 ps
0.25	0.031 V	41 μW	569 ps

Notice that the simple first-order model to predict V_{OL} is quite effective. For a PMOS $W–L$ of 4, V_{OL} is given by $(30/115)$ (4) $(0.63V) = 0.66V$.

The static power dissipation of pseudo-NMOS limits its use. When area is most important however, its reduced transistor count compared with complementary CMOS is quite attractive. Pseudo-NMOS thus still finds occasional use in large fan-in circuits. Figure 6-29 shows the schematics of pseudo-NMOS NOR and NAND gates.

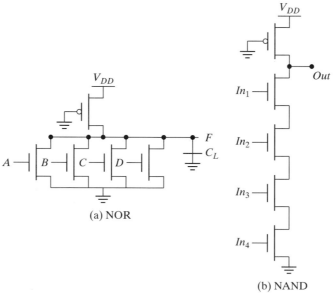

(a) NOR

(b) NAND

Figure 6-29   Four-input pseudo-NMOS NOR and NAND gates.

Problem 6.5   NAND versus NOR in Pseudo-NMOS

Given the choice between NOR or NAND logic, which one would you prefer for implementation in pseudo-NMOS?

How to Build Even Better Loads

It is possible to create a ratioed logic style that completely eliminates static currents and provides rail-to-rail swing. Such a gate combines two concepts: *differential logic* and *positive feedback*. A differential gate requires that each input is provided in complementary format, and it produces complementary outputs in turn. The feedback mechanism ensures that the load device is turned off when not needed. An example of such a logic family, called *Differential Cascode Voltage Switch Logic* (or DCVSL), is presented conceptually in Figure 6-30a [Heller84].

The pull-down networks PDN1 and PDN2 use NMOS devices and are mutually exclusive—that is, when PDN1 conducts, PDN2 is off, and when PDN1 is off, PDN2 conducts—such that the required logic function and its inverse are simultaneously implemented. Assume now that, for a given set of inputs, PDN1 conducts while PDN2 does not, and that *Out* and $\overline{Out}$ are initially high and low, respectively. Turning *on PDN1,* causes *Out* to be pulled down, although there is still contention between M_1 and PDN1. $\overline{Out}$ is in a high impedance state, as M_2 and PDN2 are both turned *off.* PDN1 must be strong enough to bring *Out* below $V_{DD} - |V_{Tp}|$, the point at which M_2 turns *on* and starts charging $\overline{Out}$ to V_{DD}, eventually turning off M_1. This in turn enables $\overline{Out}$ to discharge all the way to GND. Figure 6-30b shows an example of an XOR–XNOR gate. Notice that it is possible to share transistors among the two pull-down networks, which reduces the implementation overhead.

The resulting circuit exhibits a rail-to-rail swing, and the static power dissipation is eliminated: in steady state, none of the stacked pull-down networks and load devices are

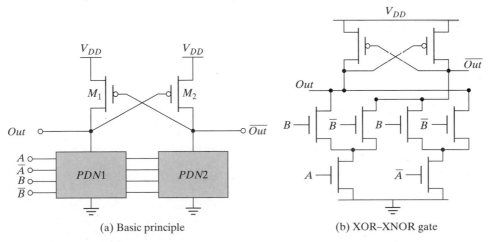

(a) Basic principle   (b) XOR–XNOR gate

Figure 6-30   DCVSL logic gate.

simultaneously conducting. However, the circuit is still ratioed since the sizing of the PMOS devices relative to the pull-down devices is critical to functionality, not just performance. In addition to the problem of increased design complexity, this circuit style has a power-dissipation problem that is due to cross-over currents. During the transition, there is a period of time when PMOS and PDN are turned on simultaneously, producing a short circuit path.

Example 6.8   DCVSL Transient Response

An example transient response is shown in Figure 6.31 for an AND/NAND gate in DCVSL. Notice that as *Out* is pulled down to $V_{DD} - |V_{Tp}|$, $\overline{Out}$ starts to charge up to V_{DD} quickly. The delay from the input to *Out* is 197 ps and to $\overline{Out}$ is 321 ps. A static CMOS AND gate (NAND followed by an inverter) has a delay of 200 ps.

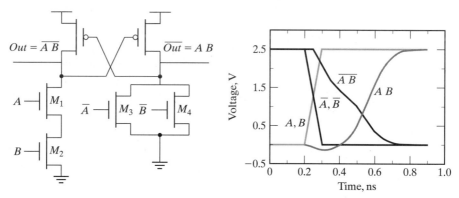

Figure 6-31   Transient response of a simple AND/NAND DCVSL gate. M_1 and M_2 1 µm/0.25 µm, M_3 and M_4 are 0.5 µm/0.25 µm and the cross-coupled PMOS devices are 1.5 µm/0.25 µm.

Design Consideration—Single-Ended versus Differential

The DCVSL gate provides *differential* (or *complementary*) *outputs*. Both the output signal (V_{out}) and its inverted value ($\overline{V}_{out}$) are simultaneously available. This is a distinct advantage, because it eliminates the need for an extra inverter to produce the complementary signal. It has been observed that a differential implementation of a complex function may reduce the number of gates required by a factor of two! The number of gates in the critical timing path is often reduced as well. Finally, the approach prevents some of the time-differential problems introduced by additional inverters. For example, in logic design, it often happens that both a signal and its complement are needed simultaneously. When the complementary signal is generated using an inverter, the inverted signal is delayed with respect to the original (Figure 6-32a). This causes timing problems, especially in very high-speed designs. Logic families with differential output capability avoid this problem to a major extent, if not completely (Figure 6-32b).

With all these positive properties, why not always use differential logic? The reason is that the differential nature virtually doubles the number of wires that have to be routed, often leading to unwieldy designs on top of the additional implementation overhead in the individual gates. The dynamic power dissipation also is high.

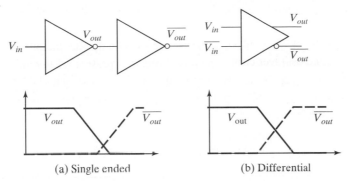

(a) Single ended   (b) Differential

Figure 6-32   Advantage of over single-ended (a) differential (b) gate.   ■

6.2.3   Pass-Transistor Logic

Pass-Transistor Basics

A popular and widely used alternative to complementary CMOS is *pass-transistor logic*, which attempts to reduce the number of transistors required to implement logic by allowing the primary inputs to drive gate terminals as well as source–drain terminals [Radhakrishnan85]. This is in contrast to logic families that we have studied so far, which only allow primary inputs to drive the gate terminals of MOSFETS.

Figure 6-33 shows an implementation of the AND function constructed that way, using only NMOS transistors. In this gate, if the *B* input is high, the top transistor is turned on and copies the input *A* to the output *F*. When *B* is low, the bottom pass-transistor is turned on and passes a 0. The switch driven by $\overline{B}$ seems to be redundant at first glance. Its presence is essential to ensure that the gate is static—a low-impedance path must exist to the supply rails under all circumstances (in this particular case, when *B* is low).

The promise of this approach is that fewer transistors are required to implement a given function. For example, the implementation of the AND gate in Figure 6-33 requires 4 transistors (including the inverter required to invert *B*), while a complementary CMOS implementation would require 6 transistors. The reduced number of devices has the additional advantage of lower capacitance.

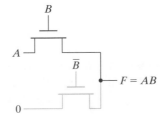

Figure 6-33   Pass-transistor implementation of an AND gate.

Unfortunately, as discussed earlier, an NMOS device is effective at passing a 0, but it is poor at pulling a node to V_{DD}. When the pass-transistor pulls a node high, the output only charges up to $V_{DD} - V_{Tn}$. In fact, the situation is worsened by the fact that the devices experience body effect, because a significant source-to-body voltage is present when pulling high. Consider the case in which the pass-transistor is charging up a node with the gate and drain terminals set at V_{DD}. Let the source of the NMOS pass-transistor be labeled x. The node x will charge up to $V_{DD} - V_{Tn}(V_x)$. We obtain

$$V_x = V_{DD} - (V_{tn0} + \gamma((\sqrt{|2\phi_f| + V_x}) - \sqrt{|2\phi_f|})) \tag{6.30}$$

Example 6.9   Voltage Swing for Pass-Transistors Circuits

The transient response of Figure 6-34 shows an NMOS charging up a capacitor. The drain voltage of the NMOS is at V_{DD}, and its gate voltage is being ramped from 0 V to V_{DD}. Assume that node x is initially at 0 V. We observe that the output initially charges up quickly, but the tail end of the transient is slow. The current drive of the transistor (gate-to-source voltage) is reduced significantly as the output approaches $V_{DD} - V_{Tn}$, and the current available to charge up node x is reduced drastically. Manual calculation using Eq. (6.30) results in an output voltage of 1.8 V, which is close to the simulated value.

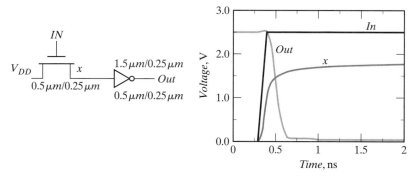

Figure 6-34   Transient response of charging up a node using an N device. Notice the slow tail after an initial quick response. $V_{DD} = 2.5$ V.

WARNING: The preceding example demonstrates that **pass-transistor gates cannot be cascaded by connecting the output of a pass gate to the gate input of another pass-transistor.** This is illustrated in Figure 6-35a, where the output of M_1 (node x) drives the gate of another MOS device. Node x can charge up to $V_{DD} - V_{Tn1}$. If node C has a rail-to-rail swing, node Y only charges up to the voltage on node $x - V_{Tn2}$, which works out to $V_{DD} - V_{Tn1} - V_{Tn2}$. Figure 6-35b, on the other hand, has the output of M_1 (x) driving the junction of M_2, and there is only one threshold drop. This is the proper way of cascading pass gates.

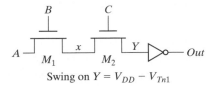

Swing on $Y = V_{DD} - V_{Tn1}$

Figure 6-35 Pass-transistor output (drain–source) terminal should not drive other gate terminals to avoid multiple threshold drops.

Example 6.10   VTC of the Pass-Transistor AND Gate

The voltage transfer curve of a pass-transistor gate shows little resemblance to complementary CMOS. Consider the AND gate shown in Figure 6-36. Similar to complementary CMOS, the VTC of pass-transistor logic is data dependent. For the case when $B = V_{DD}$, the top pass-transistor is turned *on*, while the bottom one is turned *off*. In this case, the output just follows the input A until the input is high enough to turn *off* the top pass-transistor (i.e., reaches $V_{DD} - V_{Tn}$). Next, consider the case in which $A = V_{DD}$, and B makes a transition from $0 \rightarrow 1$. Since the inverter has a threshold of $V_{DD}/2$, the bottom pass-transistor is turned *on* until then and the output remains close to zero. Once the bottom pass-transistor turns *off*, the output follows the input B minus a threshold drop. A similar behavior is observed when both inputs A and B transition from $0 \rightarrow 1$.

Observe that a pure pass-transistor gate is not regenerative. A gradual signal degradation will be observed after passing through a number of subsequent stages. This can be remedied by the occasional insertion of a CMOS inverter. With the inclusion of an inverter in the signal path, the VTC resembles one of the CMOS gates.

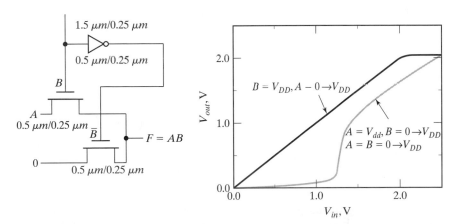

Figure 6-36 Voltage transfer characteristic for the pass-transistor AND gate of Figure 6-33.

Pass-transistors require lower switching energy to charge up a node, due to the reduced voltage swing. For the pass-transistor circuit in Figure 6-34, assume that the drain voltage is at

V_{DD} and the gate voltage transitions to V_{DD}. The output node charges from 0V to $V_{DD} - V_{Tn}$ (assuming that node x was initially at 0V), and the energy drawn from the power supply for charging the output of a pass-transistor is given by

$$E_{0 \rightarrow 1} = \int_0^T P(t)dt = V_{DD}\int_0^T i_{supply}(t)dt$$

$$= V_{DD}\int_0^{(V_{DD} - V_{Tn})} C_L dV_{out} = C_L \cdot V_{DD} \cdot (V_{DD} - V_{Tn})$$

(6.31)

While the circuit exhibits lower switching power, it may also consume static power when the output is high—the reduced voltage level may be insufficient to turn off the PMOS transistor of the subsequent CMOS inverter.

Differential Pass-Transistor Logic

For high performance design, a differential pass-transistor logic family, called *CPL* or *DPL*, is commonly used. The basic idea (similar to DCVSL) is to accept true and complementary inputs and produce true and complementary outputs. Several CPL gates (AND/NAND, OR/NOR, and XOR/NXOR) are shown in Figure 6-37. These gates possess some interesting properties:

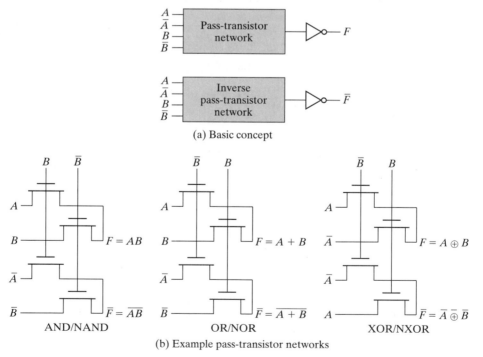

(a) Basic concept

(b) Example pass-transistor networks

Figure 6-37   Complementary pass-transistor logic (CPL).

- Since the circuits are *differential*, complementary data inputs and outputs are always available. Although generating the differential signals requires extra circuitry, the differential style has the advantage that some complex gates such as XORs and adders can be realized efficiently with a small number of transistors. Furthermore, the availability of both polarities of every signal eliminates the need for extra inverters, as is often the case in static CMOS or pseudo-NMOS.
- CPL belongs to the class of *static* gates, because the output-defining nodes are always connected to either V_{DD} or GND through a low-resistance path. This is advantageous for the noise resilience.
- The design is very modular. In effect, all gates use exactly the same topology. Only the inputs are permutated. This makes the design of a library of gates very simple. More complex gates can be built by cascading the standard pass-transistor modules.

Example 6.11   Four-Input NAND in CPL

Consider the implementation of a four-input AND/NAND gate using CPL. Based on the associativity of the boolean AND operation $[A \cdot B \cdot C \cdot D = (A \cdot B) \cdot (C \cdot D)]$, a two-stage approach has been adopted to implement the gate (Figure 6-38). The total number of transistors in the gate (including the final buffer) is 14. This is substantially higher than previously discussed gates.[3] This factor, combined with the complicated routing requirements, makes this circuit style not particularly efficient for this gate. One should, however, be aware of the fact that the structure simultaneously implements the AND and the NAND functions, which might reduce the transistor count of the overall circuit.

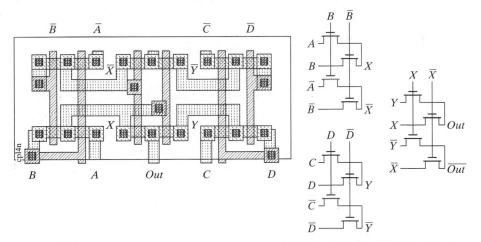

Figure 6-38   Layout and schematics of four-input NAND gate using CPL. The final inverter stage is omitted.

---

[3]This particular circuit configuration is only acceptable when zero-threshold pass-transistors are used. If not, it directly violates the concepts introduced in Figure 6-35.

In sum, CPL is a conceptually simple and modular logic style. Its applicability depends strongly on the logic function to be implemented. The availability of a simple XOR and the ease of implementing multiplexers makes it attractive for structures such as adders and multipliers. Some extremely fast and efficient implementations have been reported in that application domain [Yano90]. When considering CPL, the designer should not ignore the implicit routing overhead of the complementary signals, which is apparent in the layout of Figure 6-38.

Robust and Efficient Pass-Transistor Design

Unfortunately, differential pass-transistor logic, like single-ended pass-transistor logic, suffers from static power dissipation and reduced noise margins, since the high input to the signal-restoring inverter only charges up to $V_{DD} - V_{Tn}$. There are several solutions proposed to deal with this problem, outlined as follows:

Solution 1: Level Restoration A common solution to the voltage drop problem is the use of a *level restorer*, which is a single PMOS configured in a feedback path (see Figure 6-39). The gate of the PMOS device is connected to the output of the inverter its drain is connected to the input of the inverter and the source is connected to V_{DD}. Assume that node X is at 0V (*out* is at V_{DD} and the M_r is turned *off*) with $B = V_{DD}$ and $A = 0$. If input A makes a 0 to V_{DD} transition, M_n only charges up node X to $V_{DD} - V_{Tn}$. This is, however, enough to switch the output of the inverter low, turning on the feedback device M_r and pulling node X all the way to V_{DD}. This eliminates any static power dissipation in the inverter. Furthermore, no static current path can exist through the level restorer and the pass-transistor, since the restorer is only active when A is high. In sum, this circuit has the advantage that all voltage levels are either at GND or V_{DD}, and no static power is consumed.

While this solution is appealing in terms of eliminating static power dissipation, it adds complexity since the circuit is ratioed. The problem arises during the transition of node X from high to low (seeFigure 6-40). The pass-transistor network attempts to pull down node X, while

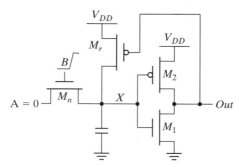

Figure 6-39 Transistor-sizing problem in level-restoring circuits.

Level restorer

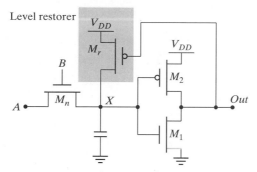

Figure 6-40 Level-restoring circuit.

the level restorer pulls X to V_{DD}. Therefore, the pull-down network, represented by M_n, must be stronger than the pull-up device Mr to switch node X (and the output). Careful transistor sizing is necessary to make the circuit function correctly. Assume the notation R_1 to denote the equivalent on-resistance of transistor M_1, R_2 for M_2, and R_r for M_r. When R_r is too small, it is impossible to bring the voltage at node X below the switching threshold of the inverter. Hence, the inverter output never switches to V_{DD}, and the gate is locked in a single state. The problem can be resolved by sizing transistors M_n and M_r such that the voltage at node X drops below the threshold of the inverter V_M, which is a function of R_1 and R_2. This condition is sufficient to guarantee the switching of the output voltage V_{out} to V_{DD} and the turning off of the level-restoring transistor.

Example 6.12   Sizing of a Level Restorer

Analyzing the circuit as a whole is nontrivial, because the restoring transistor acts as a feedback device. One way to simplify the circuit for manual analysis is to open the feedback loop and to ground the gate of the restoring transistor when determining the switching point (this is a reasonable assumption, as the feedback only becomes active once the inverter starts to switch). Hence, M_r and M_n form a configuration that resembles pseudo-NMOS with M_r the load transistor, and M_n acting as a pull-down network to GND. Assume that the inverter M_1, M_2 is sized to have its switching threshold at $V_{DD}/2$ (NMOS: 0.5 µm/0.25 µm and PMOS: 1.5 µm/0.25 µm). Therefore, node X must be pulled below $V_{DD}/2$ to switch the inverter and to shut off M_r.

This is confirmed in Figure 6-41, which shows the transient response as the size of the level restorer is varied, while keeping the size of M_n fixed (0.5 µm/0.25 µm). As the simulation indicates, for sizes above 1.5 µm/0.25 µm, node X cannot be brought below the switching threshold of the inverter, and can't switch the output.

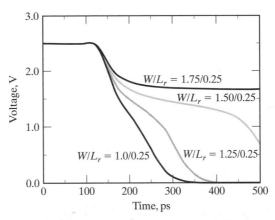

Figure 6-41   Transient response of the circuit in Figure 6-39.
A level restorer that is too large results in incorrect evaluation.

Another concern is the influence of the level restorer on the switching speed of the device. Adding the restoring device increases the capacitance at the internal node X, slowing down the gate. In addiction, the rise time of the gate is affected negatively. The level restoring transistor M_r fights the decrease in voltage at node X before being switched off. On the other hand, the level restorer reduces the fall time, since the PMOS transistor, once turned on, accelerates the pull-up action.

Problem 6.6   Device Sizing in Pass-Transistors

For the circuit shown in Figure 6-39, assume that the pull-down device consists of six pass-transistors in series each with a device size of 0.5 µm/0.25 µm (replacing transistor M_n). Determine the maximum $W–L$ size for the level restorer transistor for correct functionality.

A modification of the level restorer concept is shown in Figure 6-42. It is applicable in differential networks and is known as *swing-restored pass-transistor logic*. Instead of a simple inverter at the output of the pass-transistor network, two back-to-back inverters configured in a cross-coupled fashion are used for level restoration and performance improvement. Inputs are fed to both the gate and source–drain terminals, as in the case of conventional pass-transistor networks. Figure 6-42 shows a simple XOR/XNOR gate of three variables A, B, and C. The complementary network can be optimized by sharing transistors between the true and complementary outputs. This logic family comes with a major caveat: When cascading gates, buffers may have to be included in between the gates. If not, contention between the level-restoring devices of the cascaded gates negatively impacts the performance.

Solution 2: Multiple-Threshold Transistors   A technology solution to the voltage-drop problem associated with pass-transistor logic is the use of multiple-threshold devices. Using *zero-*

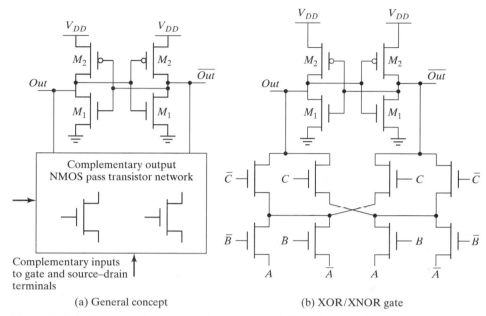

(a) General concept (b) XOR/XNOR gate

Figure 6-42 Swing-restored pass-transistor logic [Landman91, Parameswar96].

threshold devices for the NMOS pass-transistors eliminates most of the threshold drop, and passes a signal close to V_{DD}. All devices other than the pass-transistors (i.e., the inverters) are implemented using standard high-threshold devices. The use of multiple-threshold transistors is becoming more common, and involves simple modifications to existing process flows. Observe that even if the device implants were carefully calibrated to yield thresholds of exactly zero, the body effect of the device still would prevent a full swing to V_{DD}.

The use of zero-threshold transistors has some negative impact on the power consumption due to the subthreshold currents flowing through the pass-transistors, even if V_{GS} is below V_T. This is demonstrated in Figure 6-43, which points out a potential sneak dc-current path. While these leakage paths are not critical when the device is switching constantly, they do pose a significant energy overhead when the circuit is in the idle state.

Solution 3: Transmission-Gate Logic The most widely used solution to deal with the voltage-drop problem is the use of *transmission gates*.[4] This technique builds on the complementary properties of NMOS and PMOS transistors: NMOS devices pass a strong 0, but a weak 1, while PMOS transistors pass a strong 1 but a weak 0. The ideal approach is to use an NMOS to pull down and a PMOS to pull up. The transmission gate combines the best of both device flavors by placing an NMOS device in parallel with a PMOS device as in Figure 6-44a. The control

---

[4]The transmission gate is only one of the possible solutions. Other styles of pass-transistor networks that combine NMOS and PMOS transistors have been devised. Double pass-transistor logic (DPL) is an example of such [Bernstein98, pp. 84].

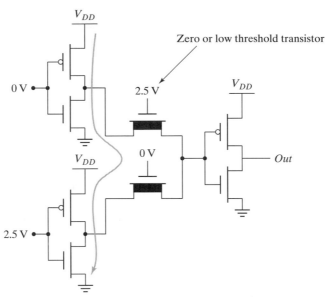

Figure 6-43 Static power consumption when using zero-threshold pass-transistors.

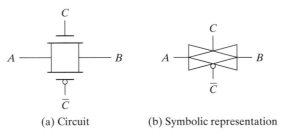

(a) Circuit   (b) Symbolic representation

Figure 6-44 CMOS transmission gate.

signals to the transmission gate (C and $\overline{C}$) are complementary. The transmission gate acts as a bidirectional switch controlled by the gate signal C. When $C = 1$, both MOSFETs are on, allowing the signal to pass through the gate. In short,

$$A = B \quad \text{if} \quad C = 1 \qquad (6.32)$$

On the other hand, $C = 0$ places both transistors in cutoff, creating an open circuit between nodes A and B. Figure 6-44b shows a commonly used transmission-gate symbol.

Consider the case of charging node B to V_{DD} for the transmission-gate circuit in Figure 6-45a. Node A is set at V_{DD}, and the transmission gate is enabled ($C = 1$ and $\overline{C} = 0$). If only the NMOS pass device were present, node B would only charge up to $V_{DD} - V_{Tn}$, at which point the NMOS device would turn off. However, since the PMOS device is present

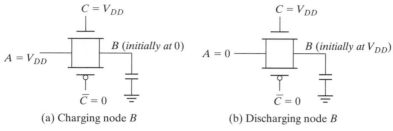

(a) Charging node B (b) Discharging node B

Figure 6-45 Transmission gates enable rail-to-rail switching.

and is "on" ($V_{GSp} = -V_{DD}$), the output charges all the way up to V_{DD}. Figure 6-45b shows the opposite case—that is, discharging node B to 0. B is initially at V_{DD} when node A is driven low. The PMOS transistor by itself can only pull-down node B to V_{Tp} at which point it turns *off*. The parallel NMOS device stays turned on, however (since its $V_{GSn} = V_{DD}$), and pulls node B all the way to GND. Although the transmission gate requires two transistors and more control signals, it enables rail-to-rail swing.

Transmission gates can be used to build some complex gates very efficiently. Figure 6-46 shows an example of a simple inverting two input multiplexer. This gate either selects input A or B on the basis of the value of the control signal S, which is equivalent to implementing the following Boolean function:

$$\bar{F} = (A \cdot S + B \cdot \bar{S}) \tag{6.33}$$

A complementary implementation of the gate requires eight transistors instead of six.

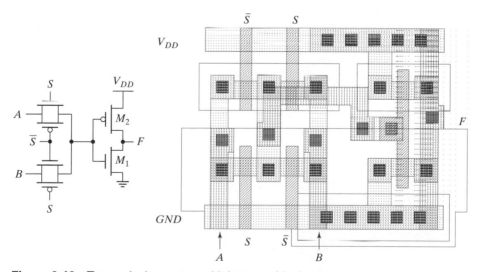

Figure 6-46 Transmission-gate multiplexer and its layout.

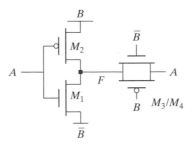

Figure 6-47   Transmission-gate XOR.

Another example of the effective use of transmission gates is the popular XOR circuit shown in Figure 6-47. The complete implementation of this gate requires only 6 transistors (including the inverter used for the generation of $\overline{B}$), compared with the 12 transistors required for a complementary implementation. To understand the operation of this circuit, we need only analyze the $B = 0$ and $B = 1$ cases separately. For $B = 1$, transistors M_1 and M_2 act as an inverter, while the transmission gate M_3/M_4 is off; hence, $F = \overline{A}B$. In the opposite case, M_1 and M_2 are disabled, and the transmission gate is operational, or $F = A\overline{B}$. The combination of both leads to the XOR function. Notice that regardless of the values of A and B, node F always has a connection to either V_{DD} or GND and thus is a low-impedance node. When designing static-pass-transistor networks, it is essential to adhere to the low-impedance rule under all circumstances. Other examples in which transmission-gate logic is effectively used are fast adder circuits and registers.

Performance of Pass-Transistor and Transmission-Gate Logic

The pass-transistor and the transmission gate are, unfortunately, not ideal switches, and they have a series resistance associated with them. To quantify the resistance, consider the circuit in Figure 6-48, which involves charging a node from 0 V to V_{DD}. In this discussion, we use the large-signal definition of resistance, which involves dividing the voltage across the switch by the drain current. The effective resistance of the switch is modeled as a parallel connection of the resistances R_n and R_p of the NMOS and PMOS devices, defined as $(V_{DD} - V_{out})/I_{Dn}$ and $(V_{DD} - V_{out})/(-I_{Dp})$, respectively. The currents through the devices obviously are dependent on the value of V_{out} and the operating mode of the transistors. During the low-to-high transition, the pass-transistors traverse through a number of operation modes. For low values of V_{out}, the NMOS device is saturated and the resistance is approximated as

$$R_p = \frac{V_{out} - V_{DD}}{I_{Dp}} = \frac{V_{out} - V_{DD}}{k_p \cdot \left((-V_{DD} - V_{Tp})(V_{out} - V_{DD}) - \dfrac{(V_{out} - V_{DD})^2}{2} \right)}$$

$$\approx \frac{1}{k_p(-V_{DD} - V_{Tp})}$$

(6.34)

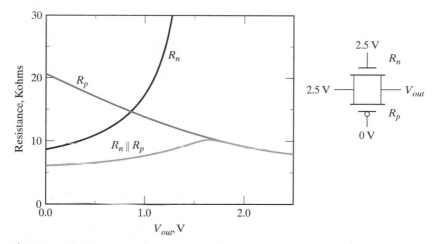

Figure 6-48 Simulated equivalent resistance of transmission gate for low-to-high transition (for $(W-L)_n = (W-L)_p = 0.5\ \mu m/0.25\ \mu m$). A similar response for overall resistance is obtained for the high-to-low transition.

The resistance goes up for increasing values of V_{out} and approaches infinity when V_{out} reaches $V_{DD} - V_{Tn}$ and the device shuts off. Similarly, we can analyze the behavior of the PMOS transistor. When V_{out} is small, the PMOS is saturated, but it enters the linear mode of operation for V_{out} approaching V_{DD}. This gives the following approximated resistance:

$$R_p = \frac{V_{out} - V_{DD}}{I_{Dp}} = \frac{V_{out} - V_{DD}}{k_p \cdot \left((-V_{DD} - V_{Tp})(V_{out} - V_{DD}) - \dfrac{(V_{out} - V_{DD})^2}{2}\right)} \tag{6.35}$$

$$\approx \frac{1}{k_p(-V_{DD} - V_{Tp})}$$

The simulated value of $R_{eq} = R_p \parallel R_n$ as a function of V_{out} is plotted in Figure 6-48. It can be observed that R_{eq} is relatively constant ($\approx 8\ k\Omega$ in this particular case). The same is true in other design instances (for example, when discharging C_L). When analyzing transmission-gate networks, the simplifying assumption that the switch has a constant resistive value is therefore acceptable.

Problem 6.7   Equivalent Resistance during Discharge

Determine the equivalent resistance by simulation for the high-to-low transition of a transmission gate. (In other words, produce a plot similar to the one presented in Figure 6-48).

An important consideration is the delay associated with a chain of transmission gates. Figure 6-49 shows a chain of n transmission gates. Such a configuration often occurs in circuits such as adders or deep multiplexors. Assume that all transmission gates are turned on and a step

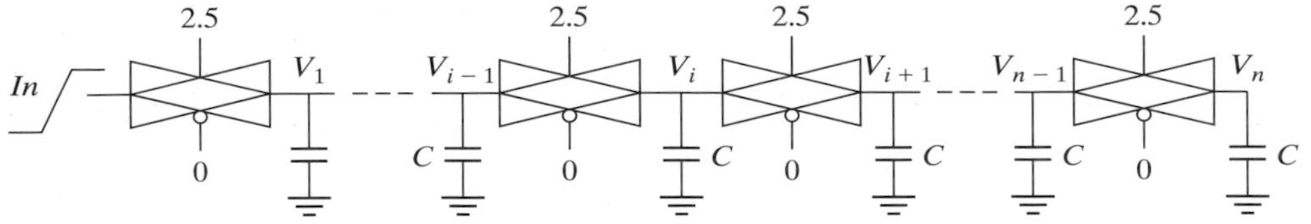

(a) A chain of transmission gates

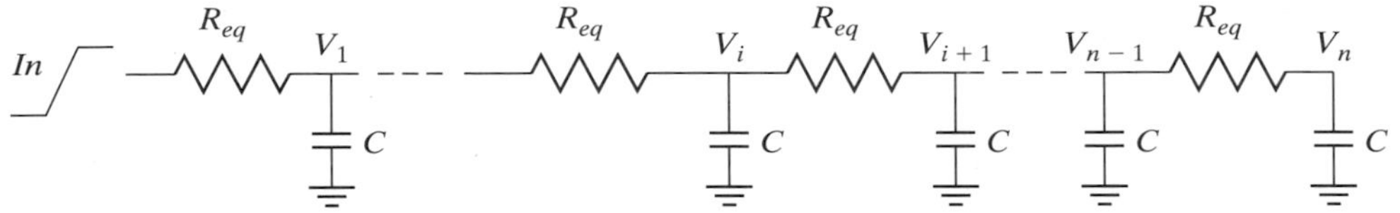

(b) Equivalent RC network

Figure 6-49 Speed optimization in transmission-gate networks.

is applied at the input. To analyze the propagation delay of this network, the transmission gates are replaced by their equivalent resistances R_{eq}. This produces the network of Figure 6-49b.

The delay of a network of n transmission gates in sequence can be estimated by using the Elmore approximation (see Chapter 4):

$$t_p(V_n) = 0.69 \sum_{k=0}^{n} CR_{eq}k = 0.69 CR_{eq}\frac{n(n+1)}{2} \qquad (6.36)$$

This means that the propagation delay is proportional to n^2 and increases rapidly with the number of switches in the chain.

Example 6.13 Delay of Transmission-Gate Chain

Consider 16 cascaded minimum-sized transmission gates, each with an average resistance of 8 kΩ. The node capacitance consists of the capacitance of two NMOS and PMOS devices (junction and gate). Since the gate inputs are assumed to be fixed, there is no Miller multiplication. The capacitance can be calculated to be approximately 3.6 fF for the low-to-high transition. The delay is given by

$$t_p = 0.69 \cdot CR_{eq}\frac{n(n+1)}{2} = 0.69 \cdot (3.6 \text{ fF})(8 \text{ K}\Omega)\left(\frac{16(16+1)}{2}\right) \approx 2.7 \text{ ns} \qquad (6.37)$$

The transient response for this particular example is shown in Figure 6-50. The simulated delay is 2.7 ns. It is remarkable that a simple RC model predicts the delay so accurately. It is also clear that the use of long pass-transistor chains causes significant delay degradation.

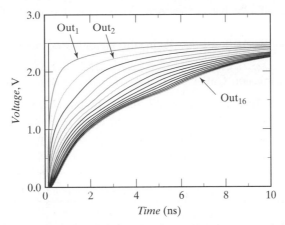

Figure 6-50 Speed optimization in transmission-gate networks.

The most common approach for dealing with the long delay is to break the chain and insert buffers every m switches (Figure 6-51). Assuming a propagation delay t_{buf} for each buffer, the overall propagation delay of the transmission gate–buffer network is then computed as follows:

$$t_p = 0.69\left[\frac{n}{m}CR_{eq}\frac{m(m+1)}{2}\right] + \left(\frac{n}{m}-1\right)t_{buf}$$

$$= 0.69\left[CR_{eq}\frac{n(m+1)}{2}\right] + \left(\frac{n}{m}-1\right)t_{buf}$$

(6.38)

The resulting delay exhibits only a linear dependence on the number of switches n, in contrast to the unbuffered circuit, which is quadratic in n. The optimal number of switches m_{opt} between buffers can be found by setting the derivative $\dfrac{\partial t_p}{\partial m}$ to 0, which yields

$$m_{opt} = 1.7\sqrt{\frac{t_{pbuf}}{CR_{eq}}}$$

(6.39)

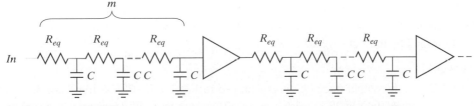

Figure 6-51 Breaking up long transmission-gate chains by inserting buffers.

Obviously, the number of switches per segment grows with increasing values of t_{buf}. In current technologies, m_{opt} typically equals 3 or 4. The presented analysis ignores that tp_{buf} itself is a function of the load m. A more accurate analysis taking this factor into account is presented in Chapter 9.

Example 6.14   Transmission-Gate Chain

Consider the same 16-transmission-gate chain. The buffers shown in Figure 6-51 can be implemented as inverters (instead of two cascaded inverters). In some cases, it might be necessary to add an extra inverter to produce the correct polarity. Assuming that each inverter is sized such that the NMOS is 0.5 µm/0.25 µm and PMOS is 0.5 µm /0.25 µm, Eq. (6.39) predicts that an inverter must be inserted every 3 transmission gates. The simulated delay when placing an inverter every two transmission gates is 154 ps; for every three transmission gates, the delay is 154 ps; and for four transmission gates, it is 164 ps. The insertion of buffering inverters reduces the delay by a factor of almost 2.

CAUTION: Although many of the circuit styles discussed in the previous sections sound very interesting, and might be superior to static CMOS in many respects, none has the *robustness and ease of design* of complementary CMOS. Therefore, use them sparingly and with caution. For designs that have no extreme area, complexity, or speed constraints, complementary CMOS is the recommended design style.

6.3 Dynamic CMOS Design

It was noted earlier that static CMOS logic with a fan-in of N requires $2N$ devices. A variety of approaches were presented to reduce the number of transistors required to implement a given logic function including pseudo-NMOS, pass-transistor logic, etc. The pseudo-NMOS logic style requires only $N + 1$ transistors to implement an N input logic gate, but unfortunately it has static power dissipation. In this section, an alternate logic style called *dynamic logic* is presented that obtains a similar result, while avoiding static power consumption. With the addition of a clock input, it uses a sequence of *precharge* and conditional *evaluation* phases.

6.3.1   Dynamic Logic: Basic Principles

The basic construction of an (*n*-type) dynamic logic gate is shown in Figure 6-52a. The PDN (pull-down network) is constructed exactly as in complementary CMOS. The operation of this circuit is divided into two major phases—*precharge* and *evaluation*—with the mode of operation determined by the *clock signal CLK*.

Precharge

When $CLK = 0$, the output node *Out* is precharged to V_{DD} by the PMOS transistor M_p. During that time, the evaluate NMOS transistor M_e is off, so that the pull-down path is disabled. The

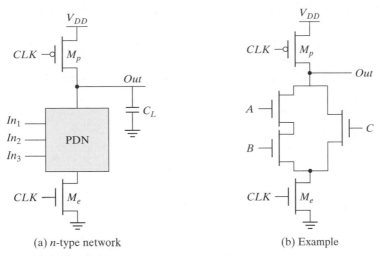

(a) *n*-type network   (b) Example

Figure 6-52   Basic concepts of a dynamic gate.

evaluation FET eliminates any static power that would be consumed during the precharge period (i.e., static current would flow between the supplies if both the pull-down and the precharge device were turned on simultaneously).

Evaluation

For $CLK = 1$, the precharge transistor M_p is off, and the evaluation transistor M_e is turned on. The output is conditionally discharged based on the input values and the pull-down topology. If the inputs are such that the PDN conducts, then a low resistance path exists between *Out* and GND, and the output is discharged to GND. If the PDN is turned off, the precharged value remains stored on the output capacitance C_L, which is a combination of junction capacitances, the wiring capacitance, and the input capacitance of the fan-out gates. During the evaluation phase, the only possible path between the output node and a supply rail is to GND. Consequently, once *Out* is discharged, it cannot be charged again until the next precharge operation. **The inputs to the gate can thus make at most one transition during evaluation**. Notice that the output can be in the *high-impedance state* during the evaluation period if the pull-down network is turned off. This behavior is fundamentally different from the static counterpart that always has a low resistance path between the output and one of the power rails.

As an example, consider the circuit shown in Figure 6-52b. During the precharge phase ($CLK = 0$), the output is precharged to V_{DD} regardless of the input values, because the evaluation device is turned off. During evaluation ($CLK = 1$), a conducting path is created between *Out* and GND if (and only if) $A \cdot B + C$ is TRUE. Otherwise, the output remains at the precharged state of V_{DD}. The following function is thus realized:

$$Out = \overline{CLK} + \overline{(A \cdot B + C)} \cdot CLK \tag{6.40}$$

A number of important properties can be derived for the dynamic logic gate:

- The logic function is implemented by the NMOS pull-down network. The construction of the PDN proceeds just as it does for static CMOS.
- The *number of transistors* (for complex gates) is substantially lower than in the static case: $N + 2$ versus $2N$.
- It is *nonratioed*. The sizing of the PMOS precharge device is not important for realizing proper functionality of the gate. The size of the precharge device can be made large to improve the low-to-high transition time (of course, at a cost to the high-to-low transition time). There is, however, a trade-off with power dissipation, since a larger precharge device directly increases clock-power dissipation.
- It only consumes *dynamic power*. Ideally, no static current path ever exists between V_{DD} and GND. The overall power dissipation, however, can be significantly higher compared with a static logic gate.
- The logic gates have *faster switching speeds*, for two main reasons. The first (obvious) reason is due to the reduced load capacitance attributed to the lower number of transistors per gate and the single-transistor load per *fan-in*. This translates in a *reduced logical effort*. For instance, the logical effort of a two-input dynamic NOR gate equals 2/3, which is substantially smaller than the 5/3 of its static CMOS counterpart. The second reason is that the dynamic gate does not have short circuit current, and all the current provided by the pull-down devices goes towards discharging the load capacitance.

The low and high output levels of V_{OL} and V_{OH} are easily identified as GND and V_{DD}, and they are not dependent on the transistor sizes. The other VTC parameters are dramatically different from static gates. Noise margins and switching thresholds have been defined as static quantities that are not a function of time. To be functional, a dynamic gate requires a periodic sequence of precharges and evaluations. Pure static analysis, therefore, does not apply. During the evaluation period, the pull-down network of a dynamic inverter starts to conduct when the input signal exceeds the threshold voltage (V_{Tn}) of the NMOS pull-down transistor. Therefore, it is reasonable to assume that the switching threshold (V_M) as well as V_{IH} and V_{IL} are equal to V_{Tn}. This translates to a low value for the NM_L.

Design Consideration

It is also possible to implement dynamic logic using the dual approach, where the output node is connected by a predischarge NMOS transistor to GND, and the evaluation PUN network is implemented in PMOS. The operation is similar: During precharge, the output node is discharged to GND; during evaluation, the output is conditionally charged to V_{DD}. This *p*-type dynamic gate has the disadvantage of being slower than the *n*-type because of the lower current drive of the PMOS transistors.   ∎

6.3 Dynamic CMOS Design   287

6.3.2 Speed and Power Dissipation of Dynamic Logic

The main advantages of dynamic logic are increased speed and reduced implementation area. Fewer devices to implement a given logic function implies that the overall load capacitance is much smaller. The analysis of the switching behavior of the gate has some interesting peculiarities to it. After the precharge phase, the output is high. For a low input signal, no additional switching occurs. As a result, $t_{pLH} = 0$! The high-to-low transition, on the other hand, requires the discharging of the output capacitance through the pull-down network. Therefore, t_{pHL} is proportional to C_L and the current-sinking capabilities of the pull-down network. The presence of the evaluation transistor slows the gate somewhat, as it presents an extra series resistance. Omitting this transistor, while functionally not forbidden, may result in static power dissipation and potentially a performance loss.

The preceding analysis is somewhat unfair because it ignores the influence of the precharge time on the switching speed of the gate. The precharge time is determined by the time it takes to charge C_L through the PMOS precharge transistor. During this time, the logic in the gate cannot be utilized. Very often, however, the overall digital system can be designed in such a way that the precharge time coincides with other system functions. For instance, the precharge of the arithmetic unit in a microprocessor could coincide with the instruction decode. The designer has to be aware of this "dead zone" in the use of dynamic logic and thus should carefully consider the pros and cons of its usage, taking the overall system requirements into account.

Example 6.15 A Four-Input Dynamic NAND Gate

Figure 6-53 shows the design of a four-input NAND example designed using the dynamic-circuit style. Due to the dynamic nature of the gate, the derivation of the voltage-transfer

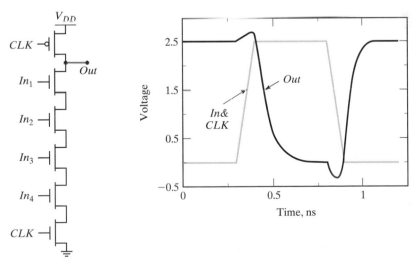

Figure 6-53   Schematic and transient response of a four-input dynamic NAND gate.

characteristic diverges from the traditional approach. As discussed earlier, we assume that the switching threshold of the gate equals the threshold of the NMOS pull-down transistor. This results in asymmetrical noise margins, as shown in Table 6-10.

Table 6-10   The dc and ac parameters of a four-input dynamic NAND.

Transistors	V_{OH}	V_{OL}	V_M	NM_H	NM_L	t_{pHL}	t_{pLH}	t_{pre}
6	2.5 V	0 V	V_{TN}	$2.5 - V_{TN}$	V_{TN}	110 ps	0 ps	83 ps

The dynamic behavior of the gate is simulated with SPICE. It is assumed that all inputs are set high when the clock goes high. On the rising edge of the clock, the output node is discharged. The resulting transient response is plotted in Figure 6-53, and the propagation delays are summarized in Table 6-10. The duration of the precharge cycle can be adjusted by changing the size of the PMOS precharge transistor. Making the PMOS too large should be avoided, however, as it both slows down the gate and increases the capacitive load on the clock line. For large designs, the latter factor might become a major design concern because the clock load can become excessive and hard to drive.

As mentioned earlier, the static gate parameters are time dependent. To illustrate this, consider a four-input NAND gate with all the partial inputs tied together, and are making a low-to-high transition. Figure 6-54 shows a transient simulation of the output voltage for three different input transitions—from 0 to 0.45 V, 0.5 V and 0.55 V, respectively. In the preceding discussion, we have defined the switching threshold of the dynamic gate as the device threshold. However, notice that the amount by which the output voltage drops is a strong function of the input voltage and the *available evaluation time*. The noise voltage needed to corrupt the signal has to be larger if the evaluation time is short. In other words, the switching threshold is truly time dependent.

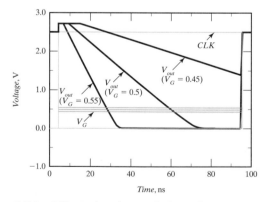

Figure 6-54   Effect of an input glitch on the output.
The switching threshold depends on the time for evaluation.
A larger glitch is acceptable if the evaluation phase is shorter.

It would appear that dynamic logic presents a significant advantage from a power perspective. There are three reasons for this. First, the physical capacitance is lower since dynamic logic uses fewer transistors to implement a given function. Also, the load seen for each fan-out is one transistor instead of two. Second, dynamic logic gates *by construction* can have at most one transition per clock cycle. Glitching (or dynamic hazards) does not occur in dynamic logic. Finally, dynamic gates do not exhibit short-circuit power since the pull-up path is not turned on when the gate is evaluating.

While these arguments generally are true, they are offset by other considerations: (1) the clock power of dynamic logic can be significant, particularly since the clock node has a guaranteed transition on every single clock cycle; (2) the number of transistors is greater than the minimal set required for implementing the logic; (3) short-circuit power may exist when leakage-combatting devices are added (as will be discussed further); and (4), most importantly, dynamic logic generally displays a higher switching activity due to the periodic *precharge* and *discharge* operations. Earlier, the transition probability for a static gate was shown to be $p_0 \, p_1 = p_0 \, (1 - p_0)$. For dynamic logic, the output transition probability does not depend on the state (history) of the inputs, but rather on the signal probabilities. For an *n*-tree dynamic gate, the output makes a $0 \rightarrow 1$ transition during the precharge phase only if the output was discharged during the preceding evaluate phase. Hence, the $0 \rightarrow 1$ transition probability for an *n*-type dynamic gate is given by

$$a_{0 \rightarrow 1} = p_0 \tag{6.41}$$

where p_0 is the probability that the output is zero. This number is always greater than or equal to $p_0 \, p_1$. For uniformly distributed inputs, the transition probability for an *N*-input gate is

$$a_{0 \rightarrow 1} = \frac{N_0}{2^N} \tag{6.42}$$

where N_0 is the number of zero entries in the truth table of the logic function.

Example 6.16   Activity Estimation in Dynamic Logic

To illustrate the increased activity for a dynamic gate, consider again a two-input NOR gate. An *n*-tree dynamic implementation is shown in Figure 6-55, along with its static counterpart. For equally probable inputs, there is a 75% probability that the output node of the dynamic gate discharges immediately after the precharge phase, implying that the activity for such a gate equals 0.75 (i.e., $P_{NOR} = 0.75 \, C_L V_{dd}^2 f_{clk}$). The corresponding activity is a lot smaller, 3/16, for a static implementation. For a dynamic NAND gate, the transition probability is 1/4 (since there is a 25% probability the output will be discharged) while it is 3/16 for a static implementation. Although these examples illustrate that the switching activity of dynamic logic is generally higher, it should be noted that dynamic logic has lower physical capacitance. Both factors must be accounted for when analyzing dynamic power dissipation.

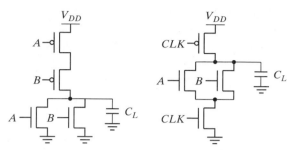

Figure 6-55   Static NOR versus *n*-type dynamic NOR.

Problem 6.8   Activity Computation

For the four-input dynamic NAND gate, compute the activity factor with the following assumption for the inputs: They are independent, and $p_{A=1} = 0.2, p_{B=1} = 0.3, p_{C=1} = 0.5$, and $p_{D=1} = 0.4$.

6.3.3   Signal Integrity Issues in Dynamic Design

Dynamic logic clearly can result in high-performance solutions compared to static circuits. However, there are several important considerations that must be taken into account if one wants dynamic circuits to function properly. These include charge leakage, charge sharing, capacitive coupling, and clock feedthrough. These issues are discussed in some detail in this section.

Charge Leakage

The operation of a dynamic gate relies on the dynamic storage of the output value on a capacitor. If the pull-down network is *off*, ideally, the output should remain at the precharged state of V_{DD} during the evaluation phase. However, this charge gradually leaks away due to leakage currents, eventually resulting in a malfunctioning of the gate. Figure 6-56a shows the sources of leakage for the basic dynamic inverter circuit.

Source 1 and 2 are the *reverse-biased diode* and *subthreshold leakage* of the NMOS pull-down device M_1, respectively. The charge stored on C_L will slowly leak away through these leakage channels, causing a degradation in the high level (Figure 6-56b). Dynamic circuits therefore require a minimal clock rate, which is typically on the order of a few kHz. This makes the usage of dynamic techniques unattractive for low-performance products such as watches, or processors that use conditional clocks (where there are no guarantees on minimum clock rates). Note that the PMOS precharge device also contributes some leakage current due to the reverse bias diode (source 3) and the subthreshold conduction (source 4). To some extent, the leakage current of the PMOS counteracts the leakage of the pull-down path. As a result, the output voltage is going to be set by the resistive divider composed of the pull-down and pull-up paths.

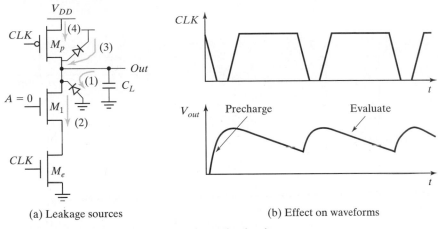

(a) Leakage sources   (b) Effect on waveforms

Figure 6-56   Leakage issues in dynamic circuits.

Example 6.17   Leakage in Dynamic Circuits

Consider the simple inverter with all devices set at $0.5\ \mu m/0.25\ \mu m$. Assume that the input is low during the evaluation period. Ideally, the output should remain at the precharged state of V_{DD}. However, as seen from Figure 6-57, the output voltage drops. Once the output drops below the switching threshold of the fan-out logic gate, the output is interpreted as a low voltage. Notice that the output settles to an intermediate voltage, due to the leakage current provided by the PMOS pull-up.

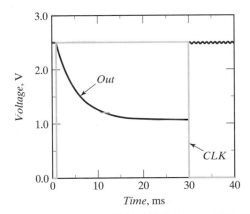

Figure 6-57   Impact of charge leakage. The output settles to an intermediate voltage determined by a resistive divider of the pull-down and pull-up devices.

Leakage is caused by the high-impedance state of the output node during the evaluate mode, when the pull-down path is turned off. The leakage problem may be counteracted by reducing the output impedance on the output node during evaluation. This often is done by

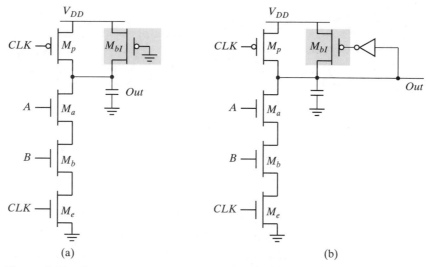

Figure 6-58   Static bleeders compensate for the charge leakage.

adding a *bleeder transistor*, as shown in Figure 6-58a. The only function of the bleeder—an NMOS style pull-up device—is to compensate for the charge lost due to the pull-down leakage paths. To avoid the ratio problems associated with this style of circuit and the associated static power consumption, the bleeder resistance is made high (in other words, the device is kept small). This allows the (strong) pull-down devices to lower the *Out* node substantially below the switching threshold of the next gate. Often, the bleeder is implemented in a feedback configuration to eliminate the static power dissipation altogether (Figure 6-58b).

Charge Sharing

Another important concern in dynamic logic is the impact of charge sharing. Consider the circuit in Figure 6-59. During the precharge phase, the output node is precharged to V_{DD}. Assume that all inputs are set to 0 during precharge, and that the capacitance C_a is discharged. Assume further that input B remains at 0 during evaluation, while input A makes a $0 \rightarrow 1$ transition, turning transistor M_a on. The charge stored originally on capacitor C_L is redistributed over C_L and C_a. This causes a drop in the output voltage, which cannot be recovered due to the dynamic nature of the circuit.

The influence on the output voltage is readily calculated. Under the assumptions given previously, the following initial conditions are valid: $V_{out}(t = 0) = V_{DD}$ and $V_X(t = 0) = 0$. As a result, two possible scenarios must be considered:

1. $\Delta V_{out} < V_{Tn}$. In this case, the final value of V_X equals $V_{DD} - V_{Tn}(V_X)$. Charge conservation then yields

$$C_L V_{DD} = C_L V_{out}(\text{final}) + C_a[V_{DD} - V_{Tn}(V_X)]$$

or

$$\Delta V_{out} = V_{out}(\text{final}) + (-V_{DD}) = -\frac{C_a}{C_L}[V_{DD} - V_{Tn}(V_X)]$$

(6.43)

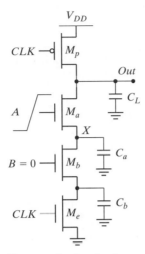

Figure 6-59 Charge sharing in dynamic networks.

2. $\Delta V_{out} > V_{Tn}$. V_{out} and V_X then reach the same value:

$$\Delta V_{out} = -V_{DD}\left(\frac{C_a}{C_a + C_L}\right) \tag{6.44}$$

We determine which of these scenarios is valid by the capacitance ratio. The boundary condition between the two cases can be determined by setting ΔV_{out} equal to V_{Tn} in Eq. (6.44), yielding

$$\frac{C_a}{C_L} = \frac{V_{Tn}}{V_{DD} - V_{Tn}} \tag{6.45}$$

Case 1 holds when the (C_a/C_L) ratio is smaller than the condition defined in Eq. (6.45). If not, Eq. (6.44) is valid. Overall, it is desirable to keep the value of ΔV_{out} below $|V_{Tp}|$. The output of the dynamic gate might be connected to a static inverter, in which case the low level of V_{out} would cause static power consumption. One major concern is a circuit malfunction if the output voltage is brought below the switching threshold of the gate it drives.

Example 6.18 Charge Sharing

Let us consider the impact of charge sharing on the dynamic logic gate shown in Figure 6-60, which implements a three-input EXOR function $y = A \oplus B \oplus C$. The first question to be resolved is what conditions cause the worst case voltage drop on node y. For simplicity, ignore the load inverter, and assume that all inputs are low during the precharge operation and that all isolated internal nodes (V_a, V_b, V_c, and V_d) are initially at 0 V.

Inspection of the truth table for this particular logic function shows that the output stays high for 4 out of 8 cases. The worst case change in output is obtained by exposing the maximum amount of internal capacitance to the output node during the evaluation

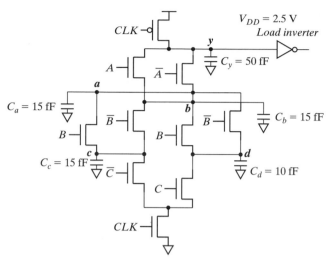

Figure 6-60   Example illustrating the charge-sharing effect in dynamic logic.

period. This happens for $\bar{A}\ B\ C$ or $A\ \bar{B}\ C$. The voltage change can then be obtained by equating the initial charge with the final charge as done with equation Eq. (6.44), yielding a worst case change of $30/(30 + 50) * 2.5\ V = 0.94\ V$. To ensure that the circuit functions correctly, the switching threshold of the connecting inverter should be placed below $2.5 − 0.94 = 1.56\ V$.

The most common and effective approach to deal with the charge redistribution is to also precharge critical internal nodes, as shown in Figure 6-61. Since the internal nodes are charged

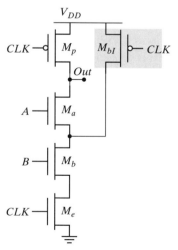

Figure 6-61   Dealing with charge sharing by precharging internal nodes. An NMOS precharge transistor may also be used, but this requires an inverted clock.

to V_{DD} during precharge, charge sharing does not occur. This solution obviously comes at the cost of increased area and capacitance.

Capacitive Coupling

The relatively high impedance of the output node makes the circuit very sensitive to crosstalk effects. A wire routed over or next to a dynamic node may couple capacitively and destroy the state of the floating node. Another equally important form of capacitive coupling is *backgate* (or *output-to-input) coupling.* Consider the circuit shown in Figure 6-62a, in which a dynamic two-input NAND gate drives a static NAND gate. A transition in the input *In* of the static gate may cause the output of the gate *(Out₂)* to go low. This output transition couples capacitively to the other input of the gate (the dynamic node Out_1) through the gate–source and gate–drain capacitances of transistor M_4. A simulation of this effect is shown in Figure 6-62b. It demonstrates how the coupling causes the output of the dynamic gate Out_1 to drop significantly. This further causes the output of the static NAND gate not to drop all the way down to 0 V and a small amount of static power to be dissipated. If the voltage drop is large enough, the circuit can evaluate incorrectly, and the NAND output may not go low. When designing and laying out dynamic circuits, special care is needed to minimize capacitive coupling.

Clock Feedthrough

A special case of capacitive coupling is clock feedthrough, an effect caused by the capacitive coupling between the clock input of the precharge device and the dynamic output node. The coupling capacitance consists of the gate-to-drain capacitance of the precharge device, and includes both the overlap and channel capacitances. This capacitive coupling causes the output of the dynamic node to rise above V_{DD} on the low-to-high transition of the clock, assuming that the pull-down network is turned off. Subsequently, the fast rising and falling edges of the clock couple onto the signal node, as is quite apparent in the simulation of Figure 6-62b.

The danger of clock feedthrough is that it may cause the normally reverse-biased junction diodes of the precharge transistor to become forward biased. This causes electron injection into the substrate, which can be collected by a nearby high-impedance node in the **1** state, eventually resulting in faulty operation. CMOS latchup might be another result of this injection. For all purposes, high-speed dynamic circuits should be carefully simulated to ensure that clock feedthrough effects stay within bounds.

All of the preceding considerations demonstrate that the design of dynamic circuits is rather tricky and requires extreme care. It should therefore be attempted only when high performance is required, or high quality design-automation tools are available.

6.3.4   Cascading Dynamic Gates

Besides the signal integrity issues, there is one major catch that complicates the design of dynamic circuits: Straightforward cascading of dynamic gates to create multilevel logic structures does not work. The problem is best illustrated with two cascaded *n*-type dynamic

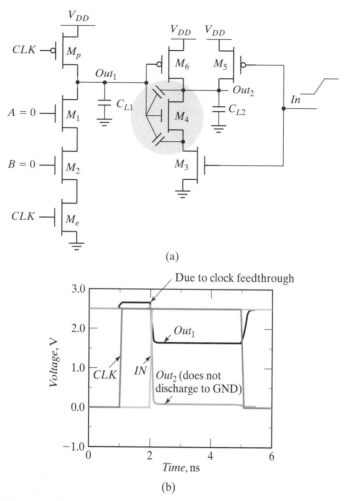

(a)

(b)

Figure 6-62   Example demonstrating the effect of backgate coupling:
(a) circuit schematics; (b) simulation results.

inverters, shown in Figure 6-63a. During the precharge phase (i.e., $CLK = 0$), the outputs of both inverters are precharged to V_{DD}. Assume that the primary input In makes a $0 \rightarrow 1$ transition (Figure 6-63b). On the rising edge of the clock, output Out_1 starts to discharge. The second output should remain in the precharged state of V_{DD} as its expected value is 1 (Out_1 transitions to 0 during evaluation). However, there is a finite propagation delay for the input to discharge Out_1 to GND. Therefore, the second output also starts to discharge. As long as Out_1 exceeds the switching threshold of the second gate, which approximately equals V_{Tn}, a conducting path exists between Out_2 and GND, and precious charge is lost at Out_2. The conducting path is only disabled once Out_1 reaches V_{Tn}, and turns off the NMOS pull-down transistor. This leaves Out_2 at an intermediate voltage level. The correct level will not be recovered, because dynamic gates rely on capacitive storage, in contrast to static gates, which have dc restoration. The charge loss leads to reduced noise margins and potential malfunctioning.

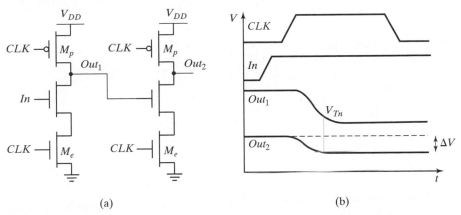

Figure 6-63 Cascade of dynamic *n*-type blocks.

The cascading problem arises because the outputs of each gate—and thus the inputs to the next stages—are precharged to 1. This may cause inadvertent discharge in the beginning of the evaluation cycle. Setting all the inputs to 0 during precharge addresses that concern. When doing so, all transistors in the pull-down network are turned off after precharge, and no inadvertent discharging of the storage capacitors can occur during evaluation. In other words, correct operation is guaranteed as long as **the inputs can only make a single $0 \rightarrow 1$ transition during the evaluation period**.[5] Transistors are turned on only when needed—and at most, once per cycle. A number of design styles complying with this rule have been conceived, but the two most important ones are discussed next.

Domino Logic

Concept A domino logic module [Krambeck82] consists of an *n*-type dynamic logic block followed by a static inverter (Figure 6-64). During precharge, the output of the *n*-type dynamic gate is charged up to V_{DD}, and the output of the inverter is set to 0. During evaluation, the dynamic gate conditionally discharges, and the output of the inverter makes a conditional transition from $0 \rightarrow 1$. If one assumes that all the inputs of a domino gate are outputs of other domino gates,[6] then it is ensured that all inputs are set to 0 at the end of the precharge phase, and that the only transitions during evaluation are $0 \rightarrow 1$ transitions. Hence, the formulated rule is obeyed. The introduction of the static inverter has the additional advantage that the fan-out of the gate is driven by a static inverter with a low-impedance output, which increases noise immunity. Also, the buffer reduces the capacitance of the dynamic output node by separating internal and load capacitances. Finally, the inverter can be used to drive a bleeder device to combat leakage and charge redistribution, as shown in the second stage of Figure 6-64.

---

[5]This ignores the impact of charge distribution and leakage effects, discussed earlier.
[6]It is required that all other inputs that do not fall under this classification (for instance, primary inputs) stay constant during evaluation.

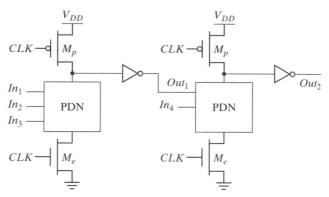

Figure 6-64 Domino CMOS logic.

Consider now the operation of a chain of domino gates. During precharge, all inputs are set to 0. During evaluation, the output of the first domino block either stays at 0 or makes a $0 \rightarrow 1$ transition, affecting the second gate. This effect might ripple through the whole chain, one after the other, similar to a line of falling dominoes—hence the name. Domino CMOS has the following properties:

- Since each dynamic gate has a static inverter, only noninverting logic can be implemented. Although there are ways to deal with this, as discussed in a subsequent section, this is a major limiting factor, and pure domino design has thus become rare.
- Very high speeds can be achieved: only a rising edge delay exists, while t_{pHL} equals zero. The inverter can be sized to match the *fan-out*, which is already much smaller than in the complimentary static CMOS case, as only a single gate capacitance has to be accounted for per fan-out gate.

Since the inputs to a domino gate are low during precharge, it is tempting to eliminate the evaluation transistor because this reduces clock load and increases pull-down drive. However, eliminating the evaluation device extends the precharge cycle—the precharge now has to ripple through the logic network as well. Consider the logic network shown in Figure 6-65, where the evaluation devices have been eliminated. If the primary input In_1 is 1 during evaluation, the output of each dynamic gate evaluates to 0, and the output of each static inverter is 1. On the falling edge of the clock, the precharge operation is started. Assume further that In_1 makes a high-to-low transition. The input to the second gate is initially high, and it takes two gate delays before In_2 is driven low. During that time, the second gate cannot precharge its output, as the pull-down network is fighting the precharge device. Similarly, the third gate has to wait until the second gate precharges before it can start precharging, etc. Therefore, the time taken to precharge the logic circuit is equal to its critical path. Another important negative is the extra power dissipation when both pull-up and pull-down devices are on. Therefore, it is good practice to always utilize evaluation devices.

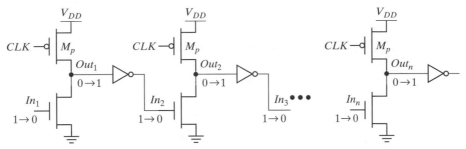

Figure 6-65 Effect of ripple precharge when the evaluation transistor is removed. The circuit also exhibits static power dissipation.

Dealing with the Noninverting Property of Domino Logic A major limitation in domino logic is that only noninverting logic can be implemented. This requirement has limited the widespread use of pure domino logic. There are several ways to deal with it, though. Figure 6-66 shows one approach to the problem—reorganizing the logic using simple boolean transforms such as De Morgan's Law. Unfortunately, this sort of optimization is not always possible, and more general schemes may have to be used.

A general (but expensive) approach to solving the problem is the use of differential logic. *Dual-rail domino* is similar in concept to the DCVSL structure discussed earlier, but it uses a precharged load instead of a static cross-coupled PMOS load. Figure 6-67 shows the circuit schematic of an AND/NAND differential logic gate. Note that all inputs come from other differential domino gates. They are low during the precharge phase, while making a conditional $0 \rightarrow 1$ transition during evaluation. Using differential domino, it is possible to implement any arbitrary function. This comes at the expense of an increased power dissipation, since a transition is guaranteed every single clock cycle regardless of the input values—either O or $\overline{O}$ must make a $0 \rightarrow 1$ transition. The function of transistors M_{f1} and M_{f2} is to keep the circuit static when the clock is high for extended periods of time (*bleeder*). Notice that this circuit is not ratioed, even in the presence of the PMOS pull-up devices! Due to its high performance, this differential approach is very popular, and is used in several commercial microprocessors.

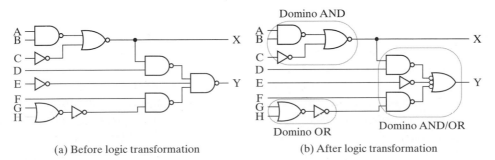

(a) Before logic transformation   (b) After logic transformation

Figure 6-66 Restructuring logic to enable implementation by using noninverting domino logic.

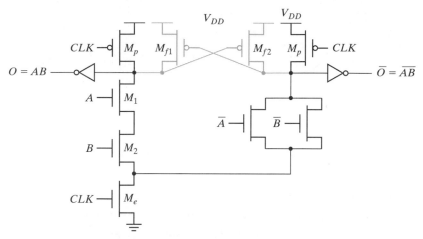

Figure 6-67 Simple dual rail (differential) domino logic gate.

Optimization of Domino Logic Gates   Several optimizations can be performed on domino logic gates. The most obvious performance optimization involves the sizing of the transistors in the static inverter. With the inclusion of the evaluation devices in domino circuits, all gates precharge in parallel, and the precharge operation takes only two gate delays—charging the output of the dynamic gate to V_{DD}, and driving the inverter output low. The critical path during evaluation goes through the pull-down path of the dynamic gate and through the PMOS pull-up transistor of the static inverter. Therefore, to speed up the circuit during evaluation, the beta ratio of the static inverter should be made high so that its switching threshold is close to V_{DD}. This can be accomplished by using a small (minimum-sized) NMOS and a large PMOS device. The minimum-sized NMOS only affects the precharge time, which is generally limited due to the parallel precharging of all gates. The only disadvantage of using a large beta ratio is a reduction in noise margin. Hence, a designer should consider reduced noise margin and performance impact simultaneously during the device sizing.

Numerous variations of domino logic have been proposed [Bernstein98]. One optimization that reduces area is *multiple-output domino logic*. The basic concept is illustrated in Figure 6-68. It exploits the fact that certain outputs are subsets of other outputs to generate a number of logical functions in a single gate. In this example, $O3 = C + D$ is used in all three outputs, and thus it is implemented at the bottom of the pull-down network. Since $O2$ equals $B \cdot O3$, it can reuse the logic for $O3$. Notice that the internal nodes have to be precharged to V_{DD} to produce the correct results. Given that the internal nodes precharge to V_{DD}, the number of devices driving precharge devices is not reduced. However, the number of evaluation transistors is drastically reduced because they are amortized over multiple outputs. Additionally, this approach results in a reduction of the fan-out factor, again due to the reuse of transistors over multiple functions.

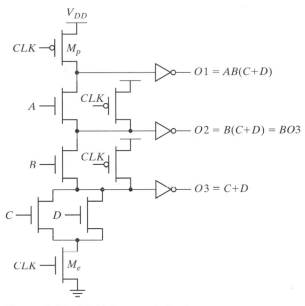

Figure 6-68   Multiple-output domino.

Compound domino (Figure 6-69) represents another optimization of the generic domino gate, once again minimizing the number of transistors. Instead of each dynamic gate driving a static inverter, it is possible to combine the outputs of multiple dynamic gates with the aid of a complex static CMOS gate, as shown in Figure 6-69. The outputs of three dynamic structures (implementing $O1 = \overline{A\,B\,C}$, $O2 = \overline{D\,E\,F}$ and $O3 = \overline{G\,H}$) are combined using a single complex CMOS static gate that implements $O - \overline{(O1 + O2)\,O3}$. The total logic function realized this way is $O = A\,B\,C\,D\,E\,F + GH$.

Compound domino is a useful tool for constructing complex dynamic logic gates. Large dynamic stacks are replaced by parallel structures with small fan-in and complex CMOS gates. For example, a large *fan in* domino AND can be implemented as a set of parallel dynamic NAND structures with lower *fan-in,* combined with a static NOR gate. One important consideration in Compound domino is the problem associated with backgate coupling. Care must be taken to ensure that the dynamic nodes are not affected by the coupling between the output of the static gates and the output of dynamic nodes.

np-CMOS

An alternative approach to cascading dynamic logic is provided by *np*-CMOS, which uses two flavors (*n*-tree and *p*-tree) of dynamic logic, and avoids the extra static inverter in the critical path that comes with domino logic. In a *p*-tree logic gate, PMOS devices are used to build a pull-up logic network, including a PMOS evaluation transistor ([Gonçalvez83, Friedman84, Lee86]).

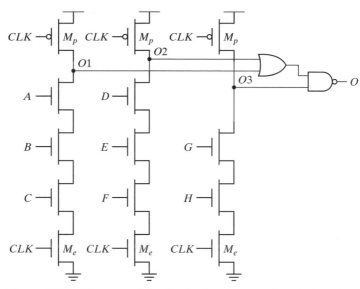

Figure 6-69   Compound domino logic uses complex static gates at the output of the dynamic gates.

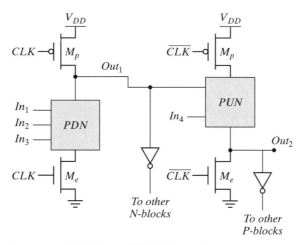

Figure 6-70   The *np*-CMOS logic circuit style.

(see Figure 6-70). The NMOS predischarge transistor drives the output low during precharge The output conditionally makes a $0 \to 1$ transition during evaluation depending on its inputs.

   np-CMOS logic exploits the duality between *n*-tree and *p*-tree logic gates to eliminate the cascading problem. If the *n*-tree gates are controlled by *CLK*, and *p*-tree gates are controlled using $\overline{CLK}$, *n*-tree gates can directly drive *p*-tree gates, and vice versa. Similar to domino, *n*-tree outputs must go through an inverter when connecting to another *n*-tree gate. During the precharge phase ($CLK = 0$), the output of the n-tree gate, Out_1, is charged to V_{DD}, while the output of the p-tree gate, Out_2, is predischarged to 0 V. Since the n-tree gate connects PMOS pull-up devices, the PUN of the p-tree is turned off at that time. During evaluation, the output of the n-tree gate can only make a $1 \rightarrow 0$ transition, conditionally turning on some transistors in the p-tree. This ensures that no accidental discharge of Out_2 can occur. Similarly, n-tree blocks can follow p-tree gates without any problems, because the inputs to the n-gate are precharged to 0. A disadvantage of the np-CMOS logic style is that the p-tree blocks are slower than the n-tree modules, due to the lower current drive of the PMOS transistors in the logic network. Equalizing the propagation delays requires extra area. Also, the lack of buffers requires that dynamic nodes are routed between gates.

6.4 Perspectives

6.4.1 How to Choose a Logic Style?

In the preceding sections, we have discussed several gate-implementation approaches using the CMOS technology. Each of the circuit styles has its advantages and disadvantages. Which one to select depends upon the primary requirement: ease of design, robustness, area, speed, or power dissipation. No single style optimizes all these measures at the same time. Even more, the approach of choice may vary from logic function to logic function.

The static approach has the advantage of being robust in the presence of noise. This makes the design process rather trouble free and amenable to a high degree of automation. It is clearly the best general-purpose logic design style. This ease of design does come at a cost: For complex gates with a large fan-in, complementary CMOS becomes expensive in terms of area and performance. Alternative static logic styles have therefore been devised. Pseudo-NMOS is simple and fast at the expense of a reduced noise margin and static power dissipation. Pass-transistor logic is attractive for the implementation of a number of specific circuits, such as multiplexers and XOR-dominated logic like adders.

Dynamic logic, on the other hand, makes it possible to implement fast and small complex gates. This comes at a price, however. Parasitic effects such as charge sharing make the design process a precarious job. Charge leakage forces a periodic refresh, which puts a lower bound on the operating frequency of the circuit.

The current trend is towards an increased use of complementary static CMOS. This tendency is inspired by the increased use of design-automation tools at the logic design level. These tools emphasize optimization at the logic level, rather than at the circuit level, and they put a premium on robustness. Another argument is that static CMOS is more amenable to voltage scaling than some of the other approaches discussed in this chapter.

6.4.2 Designing Logic for Reduced Supply Voltages

In Chapter 3, we projected that the supply voltage for CMOS processes will continue to drop over the coming decade, and may go as low as 0.6 V by 2010. To maintain performance under

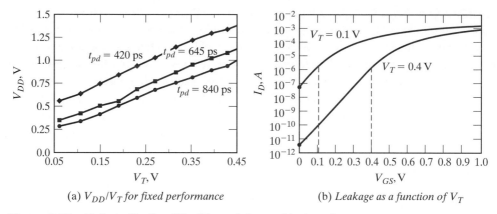

(a) V_{DD}/V_T for fixed performance   (b) Leakage as a function of V_T

Figure 6-71   Voltage Scaling (V_{DD}/V_T on delay and leakage).

those conditions, it is essential that the device thresholds scale as well. Figure 6-71a shows a plot of the (V_T, V_{DD}) ratio required to maintain a given performance level (assuming that other device characteristics remain identical).

This trade-off is not without penalty. Reducing the threshold voltage increases the subthreshold leakage current exponentially, as we derived in Eq. (3.39) (repeated here for the sake of clarity):

$$I_{leakage} = I_S 10^{\frac{V_{GS}-V_{th}}{S}} \left(1 - 10^{-\frac{nV_{DS}}{S}}\right) \tag{6.46}$$

In Eq. (6.46), S is the *slope factor* of the device. The subthreshold leakage of an inverter is the current of the NMOS for $V_{in} = 0$ V and $V_{out} = V_{DD}$ (or the PMOS current for $V_{in} = V_{DD}$ and $V_{out} = 0$). The exponential increase in inverter leakage for decreasing thresholds is illustrated in Figure 6-71b.

These leakage currents are a concern particularly for designs that feature intermittent computational activity separated by long periods of inactivity. For example, the processor in a cellular phone remains in idle mode for a majority of the time. While the processor is in idle mode, ideally, the system should consume zero or near-zero power. This is only possible if leakage is low—that is, the devices have a high threshold voltage. This is in contrast to the scaling scenario that we just depicted, where high performance under low supply voltage means reduced thresholds. To satisfy the contradicting requirements of high performance during active periods and low leakage during standby, several process modifications or leakage-control techniques have been introduced in CMOS processes. Most processes with feature sizes at or below 0.18 μm CMOS support devices with different thresholds—typically a device with low threshold for high-performance circuits, and a transistor with high threshold for leakage control. Another approach gaining popularity is the dynamic control of the threshold voltage of a device by exploiting the body effect of the transistor. Use of this approach to control individual devices requires a dual-well process (see Figure 2-2).

Clever circuit design can also help reduce the leakage current, which is a function of the circuit topology and the value of the inputs applied to the gate. Since V_T depends on body bias (V_{BS}), the subthreshold leakage of an MOS transistor depends not only on the gate drive (V_{GS}), but also on the body bias. In an inverter with $In = 0$, the subthreshold leakage of the inverter is set by the NMOS transistor with its $V_{GS} = V_{BS} = 0$ V. In more complex CMOS gates, such as the two-input NAND gate of Figure 6-72, the leakage current depends on the input vector. The subthreshold leakage current of this gate is the least when $A = B = 0$. Under these conditions, the intermediate node X settles to

$$V_X \approx V_{th}\ln(1 + n) \qquad\qquad (6.47)$$

The leakage current of the gate is then determined by the topmost NMOS transistor with $V_{GS} = V_{BS} = -V_X$. Clearly, the subthreshold leakage under this condition is smaller than that of the inverter. This reduction due to stacked transistors is called the *stack effect*. The Table in Figure 6-72 analyzes the leakage components for the two-input NAND gate under different input conditions.

The reality is even better. In short-channel MOS transistors, the subthreshold leakage current depends not only on the gate drive (V_{GS}) and the body bias (V_{BS}), but also on the drain voltage (V_{DS}). The threshold voltage of a short-channel MOS transistor decreases with increasing V_{DS} due to *drain-induced barrier lowering* (DIBL). Typical values for DIBL can range from a 20- to a 150-mV change in V_T per voltage change in V_{DS}. Because of this, the impact of the stack effect is even more significant for short-channel transistors. The intermediate voltage reduces the drain–source voltage of the topmost device, increases its threshold, and thus lowers its leakage.

Example 6.19 Stack Effect in Two-Input NAND Gate

Consider again the two-input NAND gate of Figure 6-72a, when both N_1 and N_2 are *off* ($A = B = 0$). From the simulated load lines shown in Figure 6-72c, we see that V_X settles to approximately 100 mV in steady state. The steady-state subthreshold leakage in the gate is therefore due to $V_{GS} = V_{BS} = -100$ mV and $V_{DS} = V_{DD} - 100$ mV, which is 20 times smaller than the leakage of a stand-alone NMOS transistor with $V_{GS} = V_{BS} = 0$ mV and $V_{DS} = V_{DD}$ [Ye98].

In sum, the subthreshold leakage in complex stacked circuits can be significantly lower than in individual devices. Observe that the maximum leakage reduction occurs when all the transistors in the stack are *off*, and the intermediate node voltage reaches its steady–state value. Exploiting this effect requires a careful selection of the input signals to every gate during standby or sleep mode.

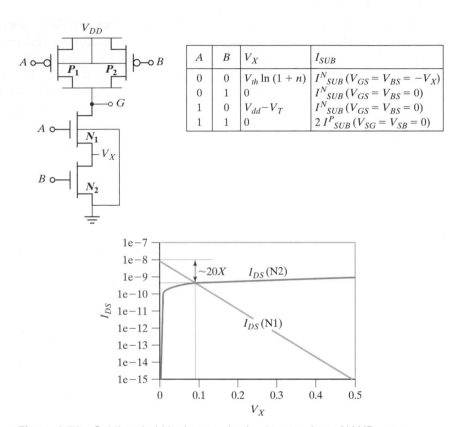

Figure 6-72   Subthreshold leakage reduction in a two-input NAND gate (a) due to stack effect for different input conditions (b). Figure (c) plots the simulated load lines of the gate for $A = B = 0$.

Problem 6.9   Computing V_X

Equation (6.47) calculates the intermediate node voltage for a two-input NAND with less than 10% error, for $A = B = 0$. Derive Eq. (6.47) assuming (1) V_T and I_S of N_1 and N_2 are approximately equal, (2) NMOS transistors are identically sized, and (3) $n < 1.5$.

6.5 Summary

In this chapter, we have extensively analyzed the behavior and performance of combinational CMOS digital circuits with regard to area, speed, and power. We summarize the major points as follows:

- *Static complementary* CMOS combines dual pull-down and pull-up networks, only one of which is enabled at any time.

- The performance of a CMOS gate is a strong function of the *fan-in*. Techniques to deal with fan-in include transistor sizing, input reordering, and partitioning. The speed is also a linear function of the fan-out. Extra buffering is needed for large fan-outs.
- The *ratioed logic* style consists of an active pull-down (-up) network connected to a load device. This results in a substantial reduction in gate complexity at the expense of static power consumption and an asymmetrical response. Careful transistor sizing is necessary to maintain sufficient noise margins. The most popular approaches in this class are the pseudo-NMOS techniques and differential DCVSL, which require complementary signals.
- *Pass-transistor logic* implements a logic gate as a simple switch network. This results in very simple implementations for some logic functions. Long cascades of switches are to be avoided due to a quadratic increase in delay with respect to the number of elements in the chain. NMOS-only pass-transistor logic produces even simpler structures, but might suffer from static power consumption and reduced noise margins. This problem can be addressed by adding a level-restoring transistor.
- The operation of *dynamic logic* is based on the storage of charge on a capacitive node and the conditional discharging of that node as a function of the inputs. This calls for a two-phase scheme, consisting of a precharge followed by an evaluation step. Dynamic logic trades off noise margin for performance. It is sensitive to parasitic effects such as leakage, charge redistribution, and clock feedthrough. Cascading dynamic gates can cause problems and thus should be addressed carefully.
- The *power consumption* of a logic network is strongly related to the switching activity of the network. This activity is a function of the input statistics, the network topology, and the logic style. Sources of power consumption such as glitches and short-circuit currents can be minimized by careful circuit design and transistor sizing.
- Threshold voltage scaling is required for *low-voltage operation*. Leakage control is critical for low-voltage operation.

6.6 To Probe Further

The topic of (C)MOS logic styles is treated extensively in the literature. Numerous texts have been devoted to the issue. Some of the most comprehensive treatments can be found in [Weste93] and [Chandrakasan01]. Regarding the intricacies of high-performance design, [Shoji96] and [Bernstein98] offer the most in-depth discussion of the optimization and analysis of digital MOS circuits. The topic of power minimization is relatively new, but comprehensive reference works are available in [Chandrakasan95], [Rabaey95], and [Pedram02].

Innovations in the MOS logic area are typically published in the proceedings of the ISSCC Conference and the VLSI circuits symposium, as well as the *IEEE Journal of Solid State Circuits* (especially the November issue).

References

[Bernstein98] K. Bernstein et al., *High-Speed CMOS Design Styles*, Kluwer Academic Publishers, 1998.

[Chandrakasan95] A. Chandrakasan and R. Brodersen, *Low Power Digital CMOS Design*, Kluwer Academic Publishers, 1995.

[Chandrakasan01] A. Chandrakasan, W. Bowhill, and F. Fox, ed., *Design of High-Performance Microprocessor Circuits*, IEEE Press, 2001.

[Gonçalvez83] N. Gonçalvez and H. De Man, "NORA: A Racefree Dynamic CMOS Technique for Pipelined Logic Structures," *IEEE Journal of Solid State Circuits*, vol. SC-18, no. 3, pp. 261–266, June 1983.

[Heller84] L. Heller et al., "Cascade Voltage Switch Logic: A Differential CMOS Logic Family," *Proc. IEEE ISSCC Conference*, pp. 16–17, February 1984.

[Krambeck82] R. Krambeck et al., "High-Speed Compact Circuits with CMOS," *IEEE Journal of Solid State Circuits*, vol. SC-17, no. 3, pp. 614–619, June 1982.

[Landman91] P. Landman and J. Rabaey, "Design for Low Power with Applications to Speech Coding," *Proc. International Micro-Electronics Conference*, Cairo, December 1991.

[Parameswar96] A. Parameswar, H. Hara, and T. Sakurai, "A Swing Restored Pass-Transistor Logic-Based Multiply and Accumulate Circuit for Multimedia Applications," *IEEE Journal of Solid State Circuits*, vol. SC-31, no. 6, pp. 805–809, June 1996.

[Pedram02] M. Pedram and J. Rabaey, ed., *Power-Aware Design Methodologies*, Kluwer, 2002.

[Rabaey95] J. Rabaey and M. Pedram, ed., *Low Power Design Methodologies,* Kluwer, 1995.

[Radhakrishnan85] D. Radhakrishnan, S. Whittaker, and G. Maki, "Formal Design Procedures for Pass-Transistor Switching Circuits," *IEEE Journal of Solid State Circuits*, vol. SC-20, no. 2, pp. 531–536, April 1985.

[Shoji88] M. Shoji, *CMOS Digital Circuit Technology*, Prentice Hall, 1988.

[Shoji96] M. Shoji, *High-Speed Digital Circuits*, Addison-Wesley, 1996.

[Sutherland99] I. Sutherland, B. Sproull, and D. Harris, *Logical Effort*, Morgan Kaufmann, 1999.

[Weste93] N. Weste and K. Eshragian, *Principles of CMOS VLSI Design: A Systems Perspective*, Addison-Wesley, 1993.

[Yano90] K. Yano et al., "A 3.8 ns CMOS 16×16 b Multiplier Using Complimentary Pass-Transistor Logic," *IEEE Journal of Solid State Circuits,* vol. SC-25, no. 2, pp. 388–395, April 1990.

[Ye98] Y. Ye, S. Borkar, and V. De, "A New Technique for Standby Leakage Reduction in High-Performance Circuits," *Symposium on VLSI Circuits*, pp. 40–41, 1998.

Exercises

For the latest problem sets and design challenges in CMOS digital logic, log in to **http://bwrc.eecs.berkeley.edu/ IcBook**.

C

How to Simulate Complex Logic Circuits

Timing- and Switch-Level Simulation

Logic and Functional Simulation

Behavioral Simulation

Register-Transfer Languages

While circuit simulation in the SPICE style proves to be an extremely valuable element of the designers tool box, it has one major deficiency. By taking into account all the peculiarities and second-order effects of the semiconductor devices, it tends to be time consuming. It rapidly becomes unwieldy when designing complex circuits, unless one is willing to spend days of computer time. Even though computers are always getting faster and simulators are getting better, circuits are getting complex even faster. The designer can address the complexity issue by giving up modeling accuracy and resorting to higher representation levels. A discussion of the different abstraction levels available to the designer and their impact on simulation accuracy is the topic of this insert.

The best way of differentiating among the myriad of simulation approaches and abstraction levels is to identify how the data and time variables are represented—as analog, continuous variables, as discrete signals, or as abstract data models.

C.1 Representing Digital Data as a Continuous Entity

Circuit Simulation and Derivatives

In Design Methodology Insert B, we established that a circuit simulator is "digitallyagnostic," meaning that it is, in essence, an analog simulator. Voltage, current, and time are treated as analog variables. This accurate modeling, combined with the nonlinearity of most of the devices leads to a high overhead in simulation time.

Substantial effort has been invested to decrease the computation time at the expense of generality. Consider an MOS digital circuit. Due to the excellent isolation property of the MOS gate, it is often possible to partition the circuit into a number of sections that have limited interaction. A possible approach is to solve each of these partitions individually over a given period, assuming that the inputs from other sections are known or constant. The resulting waveforms can then be iteratively refined. This *relaxation-based* approach has the advantage of being computationally more effective than the traditional technique by avoiding expensive matrix inversions, but it is restricted to MOS circuits [White87]. When the circuit contains feedback paths, the partitions can become large, and simulation performance degrades.

Another approach is to reduce the complexity of the transistor models used. For example, linearization of the model leads to a dramatic reduction in the computational complexity. Yet another approach is to employ a simplified table-lookup model. While this approach, by necessity, leads to a decreased accuracy of the waveforms, it still allows for a good estimation of timing parameters such as propagation delay and rise and fall times. This explains why these tools often are called *timing simulators*. The big advantage is in the execution speed, which can be one or two orders of magnitude higher than that of SPICE-like tools. Another advantage is that, in contrast to the tools that are discussed next, timing simulators still can incorporate second-order effects such as leakage, threshold drops, and signal glitches. Examples of an offering in this class is the NanoSim (formerly TimeMill/PowerMill) tool set from Synopsys [TimeMill]. As a point of reference, simulators in this class typically give up 5 to 10% in accuracy on timing parameters, with respect to full-blown circuit simulators.

C.2 Representing Data as a Discrete Entity

In digital circuits, we generally are not interested in the actual value of the voltage variable, but only in the digital value it represents. Therefore, it is possible to envision a simulator in which data signals are either in the 0 or 1 range. Signals that do no comply with either condition are denoted as X, or undefined.

This tertiary representation $\{0, 1, X\}$ is used extensively in simulators at both the device and gate level. By augmenting this set of allowable data values, we can obtain more detailed information, while retaining the capability of handling complex designs. Possible extensions are the Z-value for a tristate node in the high-impedance state, and R- and F-values for the rising and falling transients. Some commercially offered simulators provide as many as a dozen possible signal states.

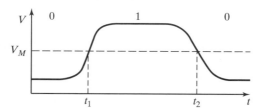

Figure C-1 Discretizing the time variable.

While substantial performance improvement is obtained by making the data representation space discrete, similar benefits can be obtained by making time a discrete variable as well. Consider the voltage waveform of Figure C-1, which represents the signal at the input of an inverter with a switching threshold V_M. It is reasonable to assume that the inverter output changes its value one propagation delay after its input crossed V_M. When one is not strictly interested in the exact shape of the signal waveforms, it is sufficient to evaluate the circuit only at the interesting time points, t_1 and t_2.

Similarly, the interesting points of the output waveform are situated at $t_1 + t_{pHL}$ and $t_2 + t_{pLH}$. A simulator that only evaluates a gate at the time an event happens at one of its inputs is called an *event-driven* simulator. The evaluation order is determined by putting projected events on a time queue and processing them in a time-ordered fashion. Suppose that the waveform of Figure C-1 acts as the input waveform to a gate. An event is scheduled to occur at time t_1. Upon processing that event, a new event is scheduled for the fan-out nodes at $t_1 + t_{pHL}$ and is put on the time queue. This event-driven approach is evidently more efficient than the time-step-driven approach of the circuit simulators. To take the impact of fan-out into account, the propagation delay of a circuit can be expressed in terms of an intrinsic delay (t_{in}) and a load-dependent factor (t_l), and it can differ over edge transitions:

$$t_{pLH} = t_{inLH} + t_{lLH} \times C_L \tag{C.1}$$

The load C_L can be entered in absolute terms (in pF) or as a function of the number of fan-out gates. Observe how closely this equation resembles the *logical-effort* model we introduced in the preceding chapter.

While offering a substantial performance benefit, the preceding approach still has the disadvantage that events can happen any time. Another simplification could be to make the time even more discrete and allow events to happen only at integer multiples of a *unit time* variable. An example of such an approach is the *unit-delay* model, where each circuit has a single delay of one unit. Finally, the simplest model is the *zero-delay model*, in which gates are assumed to be free of delay. Under this paradigm, time proceeds from one clock event to the next, and all events are assumed to occur instantaneously upon arrival of a clocking event. These concepts can be applied on a number of abstraction levels, resulting in the simulation approaches discussed next.

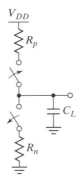

Figure C-2   Switch-level model of CMOS inverter.

Switch-Level Simulation

The nonlinear nature of semiconductor devices is one of the major impediments to higher simulation speeds. The switch-level model [Bryant81] overcomes this hurdle by approximating the transistor behavior with a linear resistance whose value is a function of the operating conditions. In the off-mode, the resistance is set to infinity, while in the on-mode, it is set to the average "on" resistance of the device (Figure C-2). The resulting network is a time-variant, linear network of resistors and capacitors that can be more efficiently analyzed. Evaluation of the resistor network determines the steady-state values of the signals and typically employs a $\{0, 1, X\}$ model. For instance, if the total resistance between a node and GND is substantially smaller than the resistance to V_{DD}, the node is set to the 0-state. The timing of the events can be resolved by analyzing the RC network. Simpler timing models such as the unit-delay model are also employed.

Example C.1   Switch versus Circuit-Level Simulation

A four-bit adder is simulated using the switch-level simulator IRSIM ([Salz89]). The simulation results are plotted in Figure C-3. Initially, all inputs (*IN*1 and *IN*2) and the carry-input *CIN* are set to 0. After 10 nsec, all inputs *IN*2 as well as *CIN* are set to 1. The display window plots the input signals, the output vector *OUT*[0–3], and the most significant output bits *OUT*[2] and *OUT*[3]. The output converges to the correct value 0000 after a transition period. Notice how the data assumes only 0 and 1 levels. The glitches in the output signals go rail to rail, although in reality they might represent only partial excursions. During transients, the signal is marked *X*, which means "undefined." To put this result in perspective, Figure C-3 plots the SPICE results for the same input vectors. Notice the partial glitches. Also, it shows that the IRSIM timing, which is based on an RC model, is relatively accurate and sufficient to get a first-order impression.

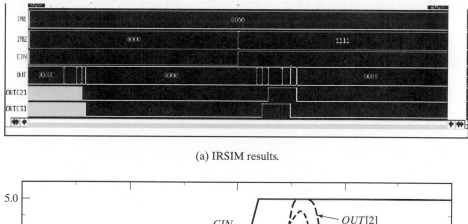

(a) IRSIM results.

(b) SPICE results.

Figure C-3   Comparison between circuit and switch-level simulations.

Gate-Level (or Logic) Simulation

Gate-level simulators use the same signal values as the switch-level tools, but the simulation primitives are gates instead of transistors. This approach enables the simulation of more complex circuits at the expense of detail and generality. For example, some common VLSI structures such as tristate busses and pass transistors are hard to deal with at this level. Since gate level is the preferred entry level for many designers, this simulation approach remained extremely popular until the introduction of logic synthesis tools, which moved the focus to the functional or behavioral abstraction layer. The interest in logic simulation was so great that special and expensive hardware accelerators were developed to expedite the simulation process (e.g., [Agrawal90]).

Functional Simulation

Functional simulation can be considered as a simple extension of logic simulation. The primitive elements of the input description can be of an arbitrary complexity. For instance, a simulation element can be a NAND gate, a multiplier, or an SRAM memory. The functionality of one of these complex units can be described using a modern programming language or a dedicated hardware description language. For instance, the THOR simulator uses the C programming language to determine the output values of a module as a function of its inputs [Thor88].

The *SystemC* language [SystemC] uses most of the syntax and semantics of C, but adds a number of constructs and data types to deal with the peculiarities of hardware design—such as the presence of concurrency. On the other hand, *VHDL (VHSIC Hardware Description Language)* [VHDL88] is a specially developed language for the description of hardware designs.

In the *structural mode*, VHDL describes a design as a connection of functional modules. Such a description often is called a *netlist*. For example, Figure C-4 shows a description of a 16-bit accumulator consisting of a register and adder.

The adder and register can in turn be described as a composition of components such as full-adder or register cells. An alternative approach is to use the *behavioral mode* of the language that describes the functionality of the module as a set of input/output relations regardless of the chosen implementation. As an example, Figure C-5 describes how the output of the adder is the two's-complement sum of its inputs.

```
entity accumulator is
    port ( -- definition of input and output terminals
        DI: in bit_vector(15 downto 0) -- a vector of 16 bit wide
        DO: inout bit_vector(15 downto 0);
        CLK: in bit
    );
end accumulator;

architecture structure of accumulator is
    component reg -- definition of register ports
        port (
            DI : in bit_vector(15 downto 0);
            DO : out bit_vector(15 downto 0);
            CLK : in bit
        );
    end component;
    component add -- definition of adder ports
        port (
            IN0 : in bit_vector(15 downto 0);
            IN1 : in bit_vector(15 downto 0);
            OUT0 : out bit_vector(15 downto 0)
        );
    end component;
-- definition of accumulator structure
signal X : bit_vector(15 downto 0);
begin
    add1 : add
        port map (DI, DO, X); -- defines port connectivity
    reg1 : reg
        port map (X, DO, CLK);
end structure;
```

Figure C-4   Functional description of an accumulator in VHDL.

```
entity add is
    port (
        IN0 : in bit_vector(15 downto 0);
        IN1 : in bit_vector(15 downto 0);
        OUT0 : out bit_vector(15 downto 0)
    );
end add;

architecture behavior of add is
begin
    process(IN0, IN1)
        variable C : bit_vector(16 downto 0);
        variable S : bit_vector(15 downto 0);
    begin
        loop1:
        for i in 0 to 15 loop
            S(i) := IN0(i) xor IN1(i) xor C(i);
            C(i+1):= IN0(i) and IN1(i) or C(i) and (IN0(i) or IN1(i));
        end loop loop1;
        OUT0 <= S;
    end process;
end behavior;
```

Figure C-5   Behavioral description of 16-bit adder.

The signal levels of the functional simulator are similar to the switch and logic levels. A variety of timing models can be used—for example, the designer can describe the delay between input and output signals as part of the behavioral description of a module. Most often the zero-delay model is employed, since it yields the highest simulation speed.

C.3 Using Higher-Level Data Models

When conceiving a digital system such as a compact disk player or an embedded microcontroller, the designer rarely thinks in terms of bits. Instead, she envisions data moving over busses as integer or floating-point words, and patterns transmitted over the instruction bus as members of an enumerated set of instruction words (such as {*ACC, RD, WR,* or *CLR*}). Modeling a discrete design at this level of abstraction has the distinct advantage of being more understandable, and it also results in a substantial benefit in simulation speed. Since a 64-bit bus is now handled as a single object, analyzing its value requires only one action instead of the 64 evaluations it formerly took to determine the current state of the bus at the logic level. The disadvantage of this approach is another sacrifice of timing accuracy. Since a bus is now considered to be a single entity, only one global delay can be annotated to it, while the delay of bus elements can vary from bit to bit at the logic level.

It also is common to distinguish between *functional* (or *structural*) and *behavioral* descriptions. In a functional-level specification, the description mirrors the intended hardware structure. Behavioral-level specifications only mimic the input/output functionality of a design. Hardware delay loses its meaning, and simulations are normally performed on a per clock-cycle (or higher) basis. For instance, the behavioral models of a microprocessor that are used to verify the completeness and the correctness of the instruction set are performed on a per instruction basis.

The most popular languages at this level of abstraction are the VHDL and VERILOG hardware-description languages. VHDL allows for the introduction of user-defined data types such as 16-bit, two's-complement words or enumerated instruction sets. Many designers tend to use traditional programming approaches such as C or C++ for their first-order behavioral models. This approach has the advantage of offering more flexibility, but it requires the user to define all data types and to essentially write the complete simulation scenario.

Example C.2   Behavioral-Level VHDL Description

To contrast the functional and behavioral description modes and the use of higher level data models, consider again the example of the accumulator (see Figure C-6). In this case, we use a fully behavioral description that employs integer data types to describe the module operation.

Figure C-7 shows the results of a simulation performed at this level of abstraction. Even for this small example, the simulation performance in terms of CPU time is three times better than what is obtained with the structural description of Figure C-4.

```
entity accumulator is
    port (
        DI : in integer;
        DO : inout integer := 0;
        CLK : in bit
    );
end accumulator;

architecture behavior of accumulator is
begin
    process(CLK)
    variable X : integer := 0; -- intermediate variable
    begin
        if CLK = '1' then
            X <= DO + D1;
            DO <= X;
        end if;
    end process;
end behavior;
```

Figure C-6 Accumulator for Example C.2.

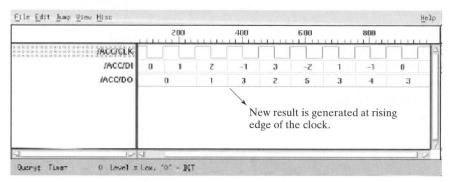

Figure C-7   Display of simulation results for accumulator example as obtained at the behavioral level. The WAVES display tool (and VHDL simulator) are part of the Synopsis VHDL tool suite (Courtesy of Synopsys.).

References

[Agrawal90] P. Agrawal and W. Dally, "A Hardware Logic Simulation System," *IEEE Trans. Computer-Aided Design*, CAD-9, no. 9, pp. 19–29, January 1990.

[Bryant81] R. Bryant, *A Switch-Level Simulation Model for Integrated Logic Circuits*, Ph. D. diss., MIT Laboratory for Computer Science, report MIT/LCS/TR-259, March 1981.

[Salz89] A. Salz and M. Horowitz, "IRSIM: An Incremental MOS Switch-Level Simulator," *Proceedings of the 26th Design Automation Conference*, pp. 173–178, 1989.

[SystemC] *Everything You Wanted to Know about SystemC*, *http://www.systemc.org/*

[Thor88] R. Alverson et al., "THOR User's Manual," Technical Report CSL-TR-88-348 and 349, Stanford University, January 1988.

[TimeMill] *http://www.synopsys.com/products/mixedsignal/mixedsignal.html*

[VHDL88] VHDL Standards Committee, IEEE Standard VHDL Language Reference Manual, IEEE standard 1076–1077, 1978.

[White87] J. White and A. Sangiovanni-Vincentelli, *Relaxation Techniques for the Simulation of VLSI Circuits*, Kluwer Academic, 1987.

D

Layout Techniques for Complex Gates

Weinberger and standard-cell layout techniques

Euler graph approach

In Chapter 6, we discussed in detail how to construct the schematics of complex gates and how to size the transistors. The last step in the design process is to derive a layout for the gate or cell; in other words, we must determine the exact shape of the various polygons composing the gate layout. The composition of a layout is strongly influenced by the *interconnect approach*. How does the cell fit in the overall chip layout, and how does it communicate with neighboring cells? Keeping these questions in mind from the start results in denser designs with less parasitic capacitance.

Weinberger and Standard-Cell Layout Techniques

We now examine two important layout approaches, although many others can be envisioned. In the *Weinberger approach* [Weinberger67], the data wires (inputs and outputs) are routed (in metal) parallel to the supply rails and perpendicular to the diffusion areas, as illustrated in Figure D-1. Transistors are formed at the cross points of the polysilicon signal wires (connected to the horizontal metal wires) and the diffusion zones. The "over-the-cell" wiring approach makes the Weinberger technique particularly suited for bit-sliced datapaths. While it is still used on an occasional base, the Weinberger technique has lost its appeal over the years in favor of the standard-cell style.

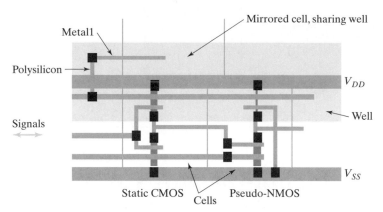

Figure D-1 The Weinberger approach for complex gate layout (using a single metal layer).

In the *standard-cell technique*, signals are routed in polysilicon perpendicular to the power distribution (Figure D-2). This approach tends to result in a dense layout for static CMOS gates, as the vertical polysilicon wire can serve as the input to both the NMOS and the PMOS transistors. An example of a cell implemented using the standard-cell approach is shown in Figure 6-12. Interconnections between cells generally are established in so-called routing channels, as demonstrated in Figure D-2. The standard-cell approach is very popular at present due to its high degree of automation. (For a detailed description of the design automation tools supporting the standard-cell approach, see Chapter 8.)

Layout Planning using the Euler Path Approach

The common use of this layout strategy makes it worth analyzing how a complex Boolean function can be mapped efficiently onto such a structure. For density reasons, it is desirable to realize the NMOS and PMOS transistors as an unbroken row of devices with abutting source–drain con-

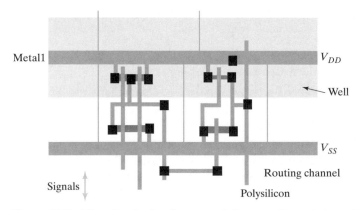

Figure D-2 The standard-cell approach for complex gate layout.

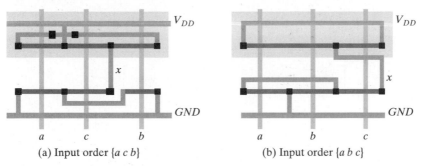

(a) Input order {a c b}   (b) Input order {a b c}

Figure D-3   Stick Diagram for $x = (a + b) \cdot c$.

nections, and with the gate connections of the corresponding NMOS and PMOS transistors aligned. This approach requires only a single strip of diffusion in both wells. To achieve this goal, a careful ordering of the input terminals is necessary. This is illustrated in Figure D-3, where the logical function $\overline{x} = (a + b) \cdot c$ is implemented. In the first version, the order {a c b} is adopted. It can easily be seen that no solution will be found using only a single diffusion strip. A reordering of the terminals (for instance, using {a b c}), generates a feasible solution, as shown in Figure D-3b. Observe that the "layouts" in Figure D-3 do not represent actual mask geometries, but are rather symbolic diagrams of the gate topologies. Wires and transistors are represented as dimensionless objects, and positioning is relative, not absolute. Such conceptual representations are called *stick diagrams*, and often are used at the conception time of the gate, before determining the actual dimensions. We use stick diagrams whenever we want to discuss gate topologies or layout strategies.

Fortunately, a systematic approach has been developed to derive the permutation of the input terminals so that complex functions can be realized by uninterrupted diffusion strips that minimize the area [Uehara81]. The systematic nature of the technique also has the advantage that it is easily automated. It consists of the following two steps:

1. **Construction of *logic graph*.**   The logic graph of a transistor network (or a switching function) is the graph of which the vertices are the nodes (signals) of the network, and the edges represent the transistors. Each edge is named for the signal controlling the corresponding transistor. Since the PUN and PDN networks of a static CMOS gate are dual, their corresponding graphs are dual as well—that is, a parallel connection is replaced by a series one and vice versa. This is demonstrated in Figure D-4, where the logic graphs for the PDN and PUN networks of the Boolean function $\overline{x} = (a + b) \cdot c$ are overlaid (notice that this approach can be used to derive dual networks).

2. **Identification of Euler paths.**   An Euler path in a graph is defined as a path through all nodes in the graph such that each edge in the graph is only visited once. Identification of such a path is important, because an ordering of the inputs leading to an uninterrupted diffusion strip of NMOS (PMOS) transistors is possible only if there exists an Euler path in

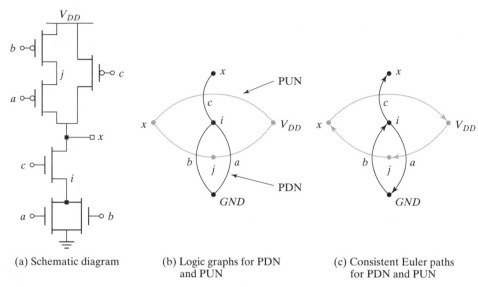

(a) Schematic diagram   (b) Logic graphs for PDN   (c) Consistent Euler paths
                                 and PUN   for PDN and PUN

Figure D-4   Schematic diagram, logic graph, and Euler paths for $x = (a + b) \cdot c$.

the logic graph of the PDN (PUN) network. The reasoning behind this finding is as follows:

To form an interrupted strip of diffusion, all transistors must be visited in sequence; that is, the drain of one device is the source of the next one. This is equivalent to traversing the logic graph along an Euler path. Be aware that Euler paths are not unique: many different solutions may exist.

The sequence of edges in the Euler path equals the ordering of the inputs in the gate layout. To obtain the same ordering in both the PUN and PDN networks, as is necessary if we want to use a single poly strip for every input signal, the Euler paths must be *consistent*—that is, they must have the same sequence.

Consistent Euler paths for the example of Figure D-4a are shown in Figure D-4c. The layout associated with this solution is shown in Figure D-3b. An inspection of the logic diagram of the function shows that {a c b} is an Euler path for the PUN, but not for the PDN. A single-diffusion-strip solution is, hence, nonexistent (Figure D-3a).

Example D.1   Derivation of Layout Topology of Complex Logic Gate

As an example, let us derive the layout topology of the following logical function:

$$x = \overline{ab + cd}$$

The logical function and one consistent Euler path are shown in Figure D-5a and Figure D-5b. The corresponding layout is shown in Figure D-5c.

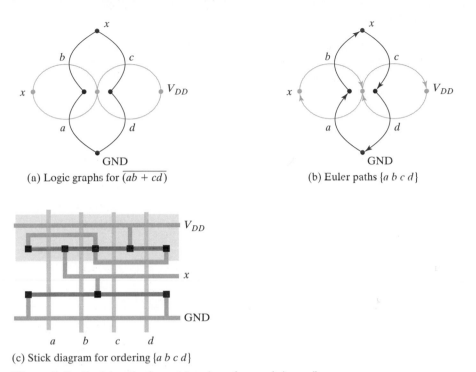

(a) Logic graphs for $\overline{(ab + cd)}$

(b) Euler paths $\{a\ b\ c\ d\}$

(c) Stick diagram for ordering $\{a\ b\ c\ d\}$

Figure D-5   Deriving the layout topology for $x = \overline{(ab + cd)}$.

The reader should be aware that the existence of consistent Euler paths depends on the way the Boolean expressions (and the corresponding logic graphs) are constructed. For example, no consistent Euler paths can be found for $\bar{x} = a + b \cdot c + d \cdot e$, but the function $\bar{x} = b \cdot c + a + d \cdot e$ has a simple solution (confirm that this is true by preserving the ordering of the function when constructing the logic graphs). A restructuring of the function is sometimes necessary before a set of consistent paths can be identified. This could lead to an exhaustive search over all possible path combinations. Fortunately, a simple algorithm to avoid this plight has been proposed [Uehara81]. A discussion of this is beyond our scope, however, and we refer the interested reader to that text.

Finally, it is worth mentioning that the layout strategies presented are not the only possibilities. For example, sometimes it might be more effective to provide multiple diffusion strips stacked vertically. In this case, a single polysilicon input line can serve as the input for multiple transistors. This might be beneficial for certain gate structures, such as the NXOR gate, and therefore, case-by-case analysis is recommended.

To Probe Further

A good overview of cell-generation techniques can be found in [Rubin87, pp. 116–128]. Some of the landmark papers in this area include the following:

[Clein00] D. Clein, *CMOS IC Layout*, Newnes, 2000.

[Rubin87] S. Rubin, *Computer Aids for VLSI Design*, Addison-Wesley, 1987.

[Uehara81] T. Uehara and W. Van Cleemput, "Optimal Layout of CMOS Functional Arrays," *IEEE Trans. on Computers*, vol. C-30, no. 5, pp. 305–311, May 1981.

[Weinberger67] A. Weinberger, "Large Scale Integration of MOS Complex Logic: A Layout Method," *IEEE Journal of Solid State Circuits*, vol. 2, no. 4, pp. 182–190, 1967.

Designing Sequential Logic Circuits

Implementation techniques for registers, latches, flip-flops, oscillators, pulse generators, and Schmitt triggers

Static versus dynamic realization

Choosing clocking strategies

7.1 Introduction
    7.1.1 Timing Metrics for Sequential Circuits
    7.1.2 Classification of Memory Elements

7.2 Static Latches and Registers
    7.2.1 The Bistability Principle
    7.2.2 Multiplexer-Based Latches
    7.2.3 Master–Slave Edge-Triggered Register
    7.2.4 Low-Voltage Static Latches
    7.2.5 Static SR Flip-Flops—Writing Data by Pure Force

7.3 Dynamic Latches and Registers
    7.3.1 Dynamic Transmission-Gate Edge-Triggered Registers
    7.3.2 C^2MOS—A Clock-Skew Insensitive Approach
    7.3.3 True Single-Phase Clocked Register (TSPCR)

7.4 Alternative Register Styles*
    7.4.1 Pulse Registers
    7.4.2 Sense-Amplifier-Based Registers

7.5 Pipelining: An Approach to Optimize Sequential Circuits
    7.5.1 Latch- versus Register-Based Pipelines
    7.5.2 NORA–CMOS—A Logic Style for Pipelined Structures

7.6　　Nonbistable Sequential Circuits
　　　7.6.1　　The Schmitt Trigger
　　　7.6.2　　Monostable Sequential Circuits
　　　7.6.3　　Astable Circuits
7.7　　Perspective: Choosing a Clocking Strategy
7.8　　Summary
7.9　　To Probe Further

7.1 Introduction

As described earlier, combinational logic circuits have the property that the output of a logic block is only a function of the *current* input values, assuming that enough time has elapsed for the logic gates to settle. Still, virtually all useful systems require storage of state information, leading to another class of circuits called *sequential logic* circuits. In these circuits, the output depends not only on the *current* values of the inputs, but also on *preceding* input values. In other words, a sequential circuit remembers some of the past history of the system—it has memory.

　　Figure 7-1 shows a block diagram of a generic *finite-state machine* (FSM) that consists of combinational logic and registers, which hold the system state. The system depicted here belongs to the class of *synchronous* sequential systems, in which all registers are under control of a single global clock. The outputs of the FSM are a function of the current *Inputs* and the *Current State*. The *Next State* is determined based on the *Current State* and the current *Inputs* and is fed to the inputs of registers. On the rising edge of the clock, the *Next State* bits are copied to the outputs of the registers (after some propagation delay), and a new cycle begins. The register then ignores changes in the input signals until the next rising edge. In general, registers can be *positive edge triggered* (where the input data is copied on the rising edge of the clock) or *negative edge triggered* (where the input data is copied on the falling edge, as indicated by a small circle at the clock input).

　　This chapter discusses the CMOS implementation of the most important sequential building blocks. A variety of choices in sequential primitives and clocking methodologies exist; making the correct selection is getting increasingly important in modern digital circuits, and can

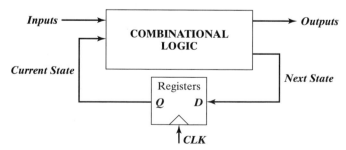

Figure 7-1　Block diagram of a finite-state machine, using *positive edge-triggered* registers.

have a great impact of performance, power, and/or design complexity. Before embarking on a detailed discussion of the various design options, a review of the relevant design metrics and a classification of the sequential elements is necessary.

7.1.1 Timing Metrics for Sequential Circuits

There are three important timing parameters associated with a register. They are shown in Figure 7-2. The *setup time* (t_{su}) is the time that the data inputs (D) must be valid before the clock transition (i.e., the $0 \rightarrow 1$ transition for a *positive edge-triggered* register). The *hold time* (t_{hold}) is the time the data input must remain valid after the clock edge. Assuming that the *setup* and *hold* times are met, the data at the D input is copied to the Q output after a worst case *propagation delay* (with reference to the clock edge) denoted by t_{c-q}.

Once we know the timing information for the registers and the combinational logic blocks, we can derive the system-level timing constraints (see Figure 7-1 for a simple system view). In synchronous sequential circuits, switching events take place concurrently in response to a clock stimulus. Results of operations await the next clock transitions before progressing to the next stage. In other words, the next cycle cannot begin unless all current computations have completed and the system has come to rest. The *clock period T,* at which the sequential circuit operates, must thus accommodate the longest delay of any stage in the network. Assume that the worst case propagation delay of the logic equals t_{plogic}, while its minimum delay—also called the *contamination delay*—is t_{cd}. The minimum clock period T required for proper operation of the sequential circuit is given by

$$T \geq t_{c-q} + t_{plogic} + t_{su} \tag{7.1}$$

The *hold time* of the register imposes an extra constraint for proper operation, namely

$$t_{cdregister} + t_{cdlogic} \geq t_{hold} \tag{7.2}$$

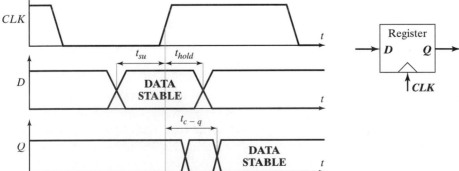

Figure 7-2 Definition of *setup time, hold time,* and *propagation delay* of a synchronous register.

where $t_{cdregister}$ is the minimum *propagation delay* (or *contamination delay*) of the register. This constraint ensures that the input data of the sequential elements is held long enough after the clock edge and is not modified too soon by the new wave of data coming in.

As seen from Eq. (7.1), it is important to minimize the values of the timing parameters associated with the register, as these directly affect the rate at which a sequential circuit can be clocked. In fact, modern high-performance systems are characterized by a very low logic depth, and the register *propagation delay* and *setup* times account for a significant portion of the clock period. For example, the DEC Alpha EV6 microprocessor [Gieseke97] has a maximum logic depth of 12 gates, and the register overhead stands for approximately 15% of the clock period. In general, the requirement of Eq. (7.2) is not difficult to meet, although it becomes an issue when there is little or no logic between registers.[1]

7.1.2   Classification of Memory Elements

Foreground versus Background Memory

At a high level, memory is classified into background and foreground memory. Memory that is embedded into logic is *foreground memory* and is most often organized as individual registers or register banks. Large amounts of centralized memory core are referred to as *background memory*. Background memory, discussed in Chapter 12, achieves higher area densities through efficient use of array structures and by trading off performance and robustness for size. In this chapter, we focus on foreground memories.

Static versus Dynamic Memory

Memories can be either static or dynamic. Static memories preserve the state as long as the power is turned on. They are built by using *positive feedback* or regeneration, where the circuit topology consists of intentional connections between the output and the input of a combinational circuit. Static memories are most useful when the register will not be updated for extended periods of time. Configuration data, loaded at power-up time, is a good example of static data. This condition also holds for most processors that use conditional clocking (i.e., gated clocks) where the clock is turned off for unused modules. In that case, there are no guarantees on how frequently the registers will be clocked, and static memories are needed to preserve the state information. Memory based on positive feedback falls under the class of elements called *multivibrator circuits*. The *bistable* element is its most popular representative, but other elements such as *monostable* and *astable* circuits also are frequently used.

Dynamic memories store data for a short period of time, perhaps milliseconds. They are based on the principle of temporary *charge storage* on parasitic capacitors associated with MOS devices. As with dynamic logic, discussed earlier, the capacitors have to be refreshed periodically to compensate for charge leakage. Dynamic memories tend to be simpler, resulting in significantly higher performance and lower power dissipation. They are most useful in datapath

---

[1] Or when the clocks at different registers are somewhat out of phase due to clock skew. We discuss this topic in Chapter 10.

circuits that require high performance levels and are periodically clocked. It is possible to use dynamic circuitry even when circuits are conditionally clocked, if the state can be discarded when a module goes into idle mode.

Latches versus Registers

A latch is an essential component in the construction of an *edge-triggered* register. It is a *level-sensitive* circuit that passes the D input to the Q output when the clock signal is high. This latch is said to be in *transparent* mode. When the clock is low, the input data sampled on the falling edge of the clock is held stable at the output for the entire phase, and the latch is in *hold* mode. The inputs must be stable for a short period around the falling edge of the clock to meet setup and hold requirements. A latch operating under these conditions is a *positive latch*. Similarly, a *negative latch* passes the D input to the Q output when the clock signal is low. Positive and negative latches are also called *transparent high* or *transparent low*, respectively. The signal waveforms for a positive and negative latch are shown in Figure 7-3. A wide variety of static and dynamic implementations exists for the realization of latches.

Contrary to *level-sensitive* latches, *edge-triggered* registers only sample the input on a clock transition—that is, $0 \rightarrow 1$ for a *positive edge-triggered* register, and $1 \rightarrow 0$ for a *negative edge-triggered* register. They are typically built to use the latch primitives of Figure 7-3. An often-recurring configuration is the *master–slave* structure, that cascades a positive and negative latch. Registers also can be constructed by using one-shot generators of the clock signal ("glitch" registers), or by using other specialized structures. Examples of these are shown later in this chapter.

The literature on sequential circuits has been plagued by ambiguous definitions for the different types of storage elements (i.e., register, flip-flop, and latch). To avoid confusion, we adhere strictly to the following set of definitions in this book:

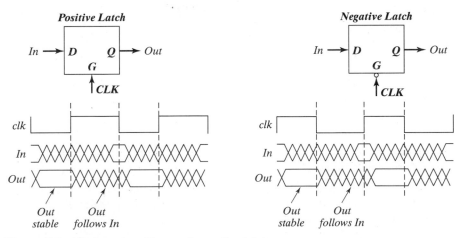

Figure 7-3 Timing of positive and negative latches.

- An *edge-triggered* storage element is called a *register*;
- A *latch* is a *level-sensitive* device;
- and any *bistable* component, formed by the cross coupling of gates, is called a *flip-flop*.[2]

7.2 Static Latches and Registers

7.2.1 The Bistability Principle

Static memories use positive feedback to create a *bistable circuit*—a circuit having two stable states that represent 0 and 1. The basic idea is shown in Figure 7-4a, which shows two inverters connected in cascade along with a voltage-transfer characteristic typical of such a circuit. Also plotted are the VTCs of the first inverter—that is, V_{o1} versus V_{i1}—and the second inverter (V_{o2} versus V_{o1}). The latter plot is rotated to accentuate that $V_{i2} = V_{o1}$. Assume now that the output of the second inverter V_{o2} is connected to the input of the first V_{i1}, as shown by the dotted lines in Figure 7-4a. The resulting circuit has only three possible operation points (*A, B,* and *C*), as demonstrated on the combined VTC. It is easy to prove the validity of the following important conjecture:

> **When the gain of the inverter in the transient region is larger than 1, *A* and *B* are the only stable operation points, and *C* is a metastable operation point.**

Suppose that the cross-coupled inverter pair is biased at point *C*. A small deviation from this bias point, possibly caused by noise, is amplified and *regenerated* around the circuit loop.

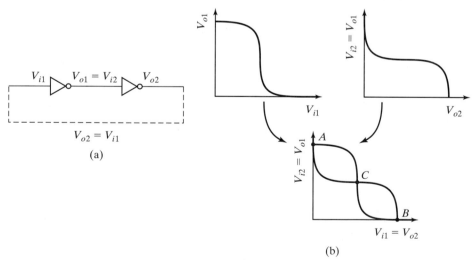

(a)

(b)

Figure 7-4   Two cascaded inverters (a) and their VTCs (b).

---

[2]An edge-triggered register is often referred to as a flip-flop as well. In this text, flip-flop is used to **uniquely** mean bistable element.

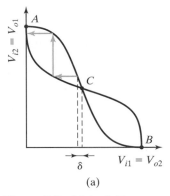

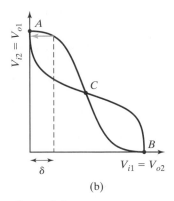

Figure 7-5 Metastable versus stable operation points.

This is a result of the gain around the loop being larger than 1. The effect is demonstrated in Figure 7-5a. A small deviation δ is applied to V_{i1} (biased in C). This deviation is amplified by the gain of the inverter. The enlarged divergence is applied to the second inverter and amplified once more. The bias point moves away from C until one of the operation points A or B is reached. In conclusion, C is an unstable operation point. Every deviation (even the smallest one) causes the operation point to run away from its original bias. The chance is indeed very small that the cross-coupled inverter pair is biased at C and stays there. Operation points with this property are termed *metastable*.

On the other hand, A and B are stable operation points, as demonstrated in Figure 7-5b. In these points, the **loop gain is much smaller than unity**. Even a rather large deviation from the operation point reduces in sizesand disappears.

Hence, the cross coupling of two inverters results in a *bistable* circuit—that is, a circuit with two stable states, each corresponding to a logic state. The circuit serves as a memory, storing either a 1 or a 0 (corresponding to positions A and B).

In order to change the stored value, we must be able to bring the circuit from state A to B and vice versa. Since the precondition for stability is that the loop gain G is smaller than unity, we can achieve this by making A (or B) temporarily unstable by increasing G to a value larger than 1. This is generally done by applying a trigger pulse at V_{i1} or V_{i2}. For example, assume that the system is in position A ($V_{i1} = 0$, $V_{i2} = 1$). Forcing V_{i1} to 1 causes both inverters to be on simultaneously for a short time and the loop gain G to be larger than 1. The positive feedback regenerates the effect of the trigger pulse, and the circuit moves to the other state (B, in this case). The width of the trigger pulse need be only a little larger than the total propagation delay around the circuit loop, which is twice the average propagation delay of the inverters.

In summary, a bistable circuit has two stable states. In absence of any triggering, the circuit remains in a single state (assuming that the power supply remains applied to the circuit) and thus remembers a value. Another common name for a bistable circuit is *flip-flop*. A flip-flop is useful only if there also exists a means to bring it from one state to the other one. In general, two different approaches may be used to accomplish the following:

- **Cutting the feedback loop.** Once the feedback loop is open, a new value can easily be written into *Out* (or *Q*). Such a latch is called *multiplexer based*, as it realizes that the logic expression for a synchronous latch is identical to the multiplexer equation:

$$Q = \overline{Clk} \cdot Q + Clk \cdot In \qquad (7.3)$$

This approach is the most popular in today's latches, and thus forms the bulk of this section.

- **Overpowering the feedback loop.** By applying a trigger signal at the input of the flip-flop, a new value is forced into the cell by overpowering the stored value. A careful sizing of the transistors in the feedback loop and the input circuitry is necessary to make this possible. A weak trigger network may not succeed in overruling a strong feedback loop. This approach used to be in vogue in the earlier days of digital design, but has gradually fallen out of favor. It is, however, the dominant approach to the implementation of static background memories (which we discuss more fully in Chapter 12). A short introduction will be given later in the chapter.

7.2.2 Multiplexer-Based Latches

The most robust and common technique to build a latch involves the use of transmission-gate multiplexers. Figure 7-6 shows an implementation of positive and negative static latches based on multiplexers. For a negative latch, input 0 of the multiplexer is selected when the clock is low, and the *D* input is passed to the output. When the clock signal is high, input 1 of the multiplexer, which connects to the output of the latch, is selected. The feedback ensures a stable output as long as the clock is high. Similarly in the positive latch, the *D* input is selected when the clock signal is high, and the output is held (using feedback) when the clock signal is low.

A transistor-level implementation of a positive latch based on multiplexers is shown in Figure 7-7. When *CLK* is high, the bottom transmission gate is on and the latch is transparent—that is, the *D* input is copied to the *Q* output. During this phase, the feedback loop is open, since the top transmission gate is off. Sizing of the transistors therefore is not critical for realizing correct functionality. The number of transistors that the clock drives is an important metric from a power perspective, because the clock has an *activity factor* of 1. This particular latch implemen-

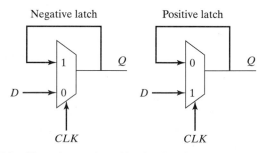

Figure 7-6   Negative and positive latches based on multiplexers.

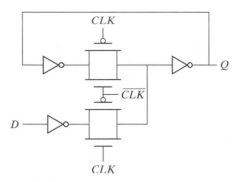

Figure 7-7   Transistor-level implementation of a positive latch built by using transmission gates.

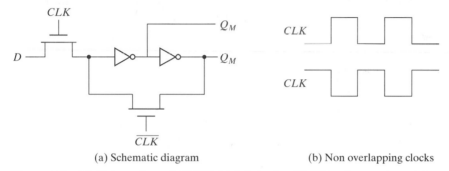

(a) Schematic diagram   (b) Non overlapping clocks

Figure 7-8   Multiplexer-based NMOS latch by using NMOS-only pass transistors for multiplexers.

tation is not very efficient from this perspective: It presents a load of four transistors to the *CLK* signal.

It is possible to reduce the clock load to two transistors by implementing multiplexers that use as NMOS-only pass transistors, as shown in Figure 7-8. When *CLK* is high, the latch samples the *D* input, while a low clock signal enables the feedback loop, and puts the latch in the hold mode. While attractive for its simplicity, the use of NMOS-only pass transistors results in the passing of a degraded high voltage of $V_{DD} - V_{Tn}$ to the input of the first inverter. This impacts both noise margin and the switching performance, especially in the case of low values of V_{DD} and high values of V_{Tn}. It also causes static power dissipation in the first inverter, because the maximum input voltage to the inverter equals $V_{DD} - V_{Tn}$, and the PMOS device of the inverter is never fully turned off.

7.2.3   Master–Slave Edge-Triggered Register

The most common approach for constructing an *edge-triggered* register is to use a *master–slave* configuration, as shown in Figure 7-9. The register consists of cascading a negative latch (master stage) with a positive one (slave stage). A multiplexer-based latch is used in this particular

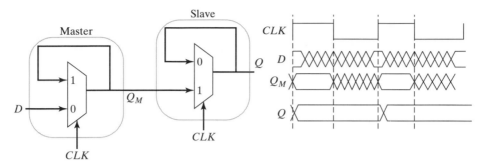

Figure 7-9   Positive edge-triggered register based on a master–slave configuration.

implementation, although any latch could be used. On the low phase of the clock, the master stage is transparent, and the D input is passed to the master stage output, Q_M. During this period, the slave stage is in the hold mode, keeping its previous value by using feedback. On the rising edge of the clock, the master stage stops sampling the input, and the slave stage starts sampling. During the high phase of the clock, the slave stage samples the output of the master stage (Q_M), while the master stage remains in a hold mode. Since Q_M is constant during the high phase of the clock, the output Q makes only one transition per cycle. The value of Q is the value of D right before the rising edge of the clock, achieving the *positive edge-triggered* effect. A negative edge-triggered register can be constructed by using the same principle by simply switching the order of the positive and negative latches (i.e., placing the positive latch first).

A complete transistor-level implementation of the master–slave positive edge-triggered register is shown in Figure 7-10. The multiplexer is implemented by using transmission gates as discussed in the previous section. When the clock is low ($\overline{CLK} = 1$), T_1 is on and T_2 is off, and the D input is sampled onto node Q_M. During this period, T_3 and T_4 are off and on, respectively. The cross-coupled inverters (I_5, I_6) hold the state of the slave latch. When the clock goes high, the master stage stops sampling the input and goes into a hold mode. T_1 is off and T_2 is on, and the cross-coupled inverters I_2 and I_3 hold the state of Q_M. Also, T_3 is *on* and T_4 is off, and Q_M is copied to the output Q.

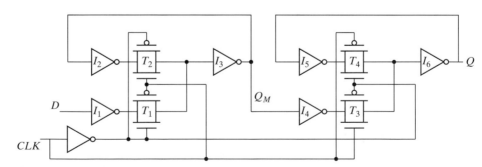

Figure 7-10   Master–slave positive edge-triggered register, using multiplexers.

Problem 7.1   Optimization of the Master–Slave Register

It is possible to remove the inverters I_1 and I_4 from Figure 7-10 without loss of functionality. Is there any advantage to including these inverters in the implementation?

Timing Properties of Multiplexer-Based Master–Slave Registers

Registers are characterized by three important timing parameters: the setup time, the hold time and the propagation delay. It is important to understand the factors that affect these timing parameters and develop the intuition to manually estimate them. Assume that the propagation delay of each inverter is t_{pd_inv}, and the propagation delay of the transmission gate is t_{pd_tx}. Also assume that the contamination delay is 0, and that inverter, deriving $\overline{CLK}$ from CLK, has a delay of 0 as well.

The setup time is the time before the rising edge of the clock that the input data D must be valid. This is similar to asking the question, how long before the rising edge of the clock must the D input be stable such that Q_M samples the value reliably? For the transmission gate multiplexer-based register, the input D has to propagate through I_1, T_1, I_3, and I_2 before the rising edge of the clock. This ensures that the node voltages on both terminals of the transmission gate T_2 are at the same value. Otherwise, it is possible for the cross-coupled pair I_2 and I_3 to settle to an incorrect value. The setup time is therefore equal to $3 \times t_{pd\_inv} + t_{pd\_tx}$.

The propagation delay is the time it takes for the value of Q_M to propagate to the output Q. Note that, since we included the delay of I_2 in the setup time, the output of I_4 is valid before the rising edge of the clock. Therefore, the delay t_{c-q} is simply the delay through T_3 and I_6 ($t_{c-q} = t_{pd\_tx} + t_{pd\_inv}$).

The *hold time* represents the time that the input must be held stable after the rising edge of the clock. In this case, the transmission gate T_1 turns off when the clock goes high. Since both the D input and the CLK pass through inverters before reaching T_1, any changes in the input after the clock goes high do not affect the output. Therefore, the hold time is 0.

Example 7.1   Timing Analysis, Using SPICE

To obtain the setup time of the register while using SPICE, we progressively skew the input with respect to the clock edge until the circuit fails. Figure 7-11 shows the setup-time simulation assuming a skew of 210 ps and 200 ps. For the 210 ps case, the correct value of input D is sampled (in this case, the Q output remains at the value of V_{DD}). For a skew of 200 ps, an incorrect value propagates to the output, as the Q output transitions to 0. Node Q_M starts to go high, and the output of I_2 (the input to transmission gate T_2) starts to fall. However, the clock is enabled before the two nodes across the transmission gate T_2 settle to the same value. This results in an incorrect value being written into the master latch. The setup time for this register is 210 ps.

In a similar fashion, the hold time can be simulated. The D-input edge is once again skewed relative to the clock signal until the circuit stops functioning. For this design, the

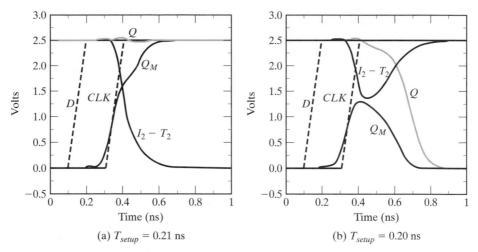

Figure 7-11   Setup time simulation.

hold time is 0 (i.e., the inputs can be changed on the clock edge). Finally, for the propagation delay, the inputs transition at least one setup time before the rising edge of the clock, and the delay is measured from the 50% point of the *CLK* edge to the 50% point of the *Q* output. From this simulation (Figure 7-12), $t_{c-q(lh)}$ was 160 ps, and $t_{c-q(hl)}$ was 180 ps.

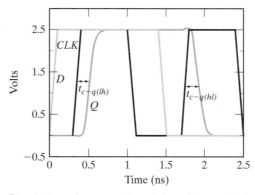

Figure 7-12   Simulation of propagation delay of transmission gate register.

The drawback of the transmission-gate register is the high capacitive load presented to the clock signal. The clock load per register is important, since it directly impacts the power dissipation of the clock network. Ignoring the overhead required to invert the clock signal—since the inverter overhead can be amortized over multiple register bits—each register has a clock load of eight transistors. One approach to reduce the clock load at the cost of robustness is to make the circuit

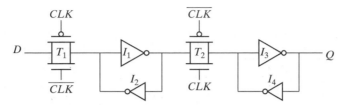

Figure 7-13 Reduced load clock load static master–slave register.

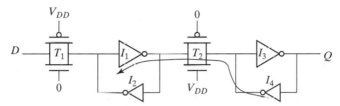

Figure 7-14 Reverse conduction possible in the transmission gate.

ratioed. Figure 7-13 shows that the feedback transmission gate can be eliminated by directly cross-coupling the inverters.

The penalty paid for the reduced in clock load is an increased design complexity. The transmission gate (T_1) and its source driver must overpower the feedback inverter (I_2) to switch the state of the cross-coupled inverter. The sizing requirements for the transmission gates can be derived by using an analysis similar to the one used for the sizing of the level-restoring device in Chapter 6. The input to the inverter I_1 must be brought below its switching threshold in order to make a transition. If minimum-sized devices are to be used in the transmission gates, it is essential that the transistors of inverter I_2 should be made even weaker. This can be accomplished by making their channel lengths larger than minimum. Using minimum or close-to-minimum size devices in the transmission gates is desirable to reduce the power dissipation in the latches and the clock distribution network.

Another problem with this scheme is *reverse conduction*—the second stage can affect the state of the first latch. When the slave stage is on (Figure 7-14), it is possible for the combination of T_2 and I_4 to influence the data stored in the I_1–I_2 latch. As long as I_4 is a weak device, this fortunately not a major problem.

Non-Ideal Clock Signals

So far, we have assumed that $\overline{CLK}$ is a perfect inversion of *CLK*, or in other words, that the delay of the generating inverter is zero. Even if this were possible, this still would not be a good assumption. Variations can exist in the wires used to route the two clock signals, or the load capacitances can vary based on data stored in the connecting latches. This effect, known as *clock skew*, is a major problem, causing the two clock signals to overlap, as shown in Figure 7-15b. *Clock overlap* can cause two types of failures, which we illustrate for the NMOS-only negative master–slave register of Figure 7-15a.

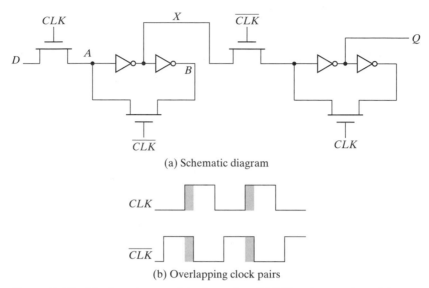

(a) Schematic diagram

(b) Overlapping clock pairs

Figure 7-15   Master–slave register based on NMOS-only pass transistors.

1. When the clock goes high, the slave stage should stop sampling the master stage output and go into a hold mode. However, since CLK and $\overline{CLK}$ are both high for a short period of time (the *overlap period*), both sampling pass transistors conduct, and there is a direct path from the D input to the Q output. As a result, data at the output can change on the rising edge of the clock, which is undesired for a negative edge-triggered register. This is known as a *race* condition in which the value of the output Q is a function of whether the input D arrives at node X before or after the falling edge of $\overline{CLK}$. If node X is sampled in the meta-stable state, the output will switch to a value determined by noise in the system.

2. The primary advantage of the multiplexer-based register is that the feedback loop is open during the sampling period, and therefore the sizing of the devices is not critical to func-tionality. However, if there is clock overlap between CLK and $\overline{CLK}$, node A can be driven by both D and B, resulting in an undefined state.

These problems can be avoided by using two *nonoverlapping clocks* instead, PHI_1 and PHI_2 (Figure 7-16), and by keeping the nonoverlap time $t_{non\_overlap}$ between the clocks large enough so that no overlap occurs even in the presence of clock-routing delays. During the non-overlap time, the FF is in the high-impedance state—the feedback loop is open, the loop gain is zero, and the input is disconnected. Leakage will destroy the state if this condition holds for too long—hence the name *pseudostatic*: The register employs a combination of static and dynamic storage approaches, depending upon the state of the clock.

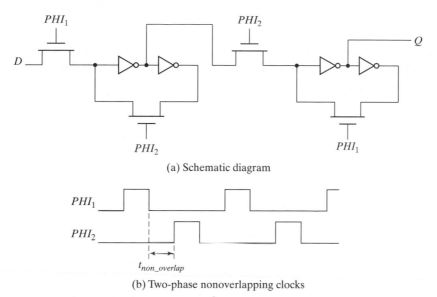

(a) Schematic diagram

(b) Two-phase nonoverlapping clocks

Figure 7-16 Pseudostatic two-phase *D* register.

Problem 7.2   Generating Nonoverlapping Clocks

Figure 7-17 shows one possible implementation of the clock generation circuitry for generating a two-phase nonoverlapping clock. Assuming that each gate has a unit gate delay, derive the timing relationship between the input clock and the two output clocks. What is the nonoverlap period? How can this period be increased if needed?

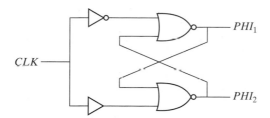

Figure 7-17 Circuitry for generating a two-phase nonoverlapping clock.

7.2.4   Low-Voltage Static Latches

The scaling of supply voltages is critical for low-power operation. Unfortunately, certain latch structures do not function at reduced supply voltages. For example, without the scaling of device thresholds, NMOS-only pass transistors (e.g., Figure 7-16) don't scale well with supply voltage due to its inherent threshold drop. At very low power supply voltages, the input to the inverter cannot be raised above the switching threshold, resulting in incorrect evaluation. Even with the use of transmission gates, performance degrades significantly at reduced supply voltages.

Scaling to low supply voltages thus requires the use of reduced threshold devices. However, this has the negative effect of exponentially increasing the subthreshold leakage power (as discussed in Chapter 6). When the registers are constantly accessed, the leakage energy typically is insignificant compared with the switching power. However, with the use of conditional clocks, it is possible that registers are idle for extended periods, and the leakage energy expended by registers can be quite significant.

Many solutions are being explored to address the problem of high leakage during idle periods. One approach involves the use of Multiple Threshold devices, as shown in Figure 7-18 [Mutoh95]. Only the negative latch is shown. The shaded inverters and transmission gates are implemented in low-threshold devices. The low-threshold inverters are gated by using high-threshold devices to eliminate leakage.

During the normal mode of operation, the sleep devices are turned on. When the clock is low, the D input is sampled and propagates to the output. The latch is in the hold mode when the clock is high. The feedback transmission gate conducts and the cross-coupled feedback is enabled. An extra inverter, in parallel with the low-threshold one, is added to store the state when the latch is in *idle* (or *sleep*) mode. Then, the high-threshold devices in series with the low-threshold inverter are turned off (the *SLEEP* signal is high), eliminating leakage. It is assumed that clock

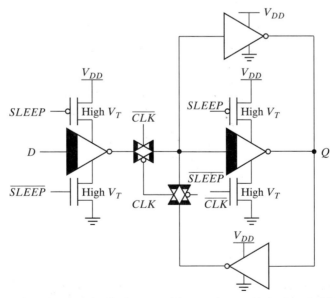

Figure 7-18   Solving the leakage problem, using multiple-threshold CMOS.

is held high when the latch is in the sleep state. The feedback low-threshold transmission gate is turned on and the cross-coupled high-threshold devices maintain the state of the latch.

Problem 7.3 Transistor Minimization in the MTCMOS Register

Unlike combinational logic, both NMOS and PMOS high-threshold devices are required to eliminate the leakage in low-threshold latches. Explain why this is the case.

Hint: Eliminate the high-V_T NMOS or high-V_T PMOS of the low-threshold inverter on the right of Figure 7-18, and investigate potential leakage paths.

7.2.5 Static SR Flip-Flops—Writing Data by Pure Force

The traditional way of causing a bistable element to change state is to overpower the feedback loop. The simplest incarnation accomplishing this is the well-known *SR*, or *set-reset, flip-flop*, an implementation of which is shown in Figure 7-19a. This circuit is similar to the cross-coupled inverter pair with NOR gates replacing the inverters. The second input of the NOR gates is connected to the trigger inputs (*S* and *R*) that make it possible to force the outputs Q and $\overline{Q}$ to a given state. These outputs are complimentary (except for the *SR* = 11 state). When both *S* and *R* are 0, the flip-flop is in a quiescent state and both outputs retain their values. (A NOR gate with one of its inputs being 0 looks like an inverter, and the structure looks like a cross-coupled inverter.) If a positive (or 1) pulse is applied to the *S* input, the Q output is forced into the 1 state (with $\overline{Q}$ going to 0) and vice versa: A 1-pulse on *R* resets the flip-flop, and the Q output goes to 0.

These results are summarized in the *characteristic table* of the flip-flop, shown in Figure 7-19c. The characteristic table is the truth table of the gate and lists the output states as functions of all possible input conditions. When both *S* and *R* are high, both Q and $\overline{Q}$ are forced to zero. Since this does not correspond with our constraint that Q and $\overline{Q}$ must be complementary, this input mode is considered forbidden. An additional problem with this condition is that when the input triggers return to their zero levels, the resulting state of the latch is unpredictable, and depends on whatever input is last to go low. Finally, Figure 7-19b shows the schematic symbol of the *SR* flip-flop.

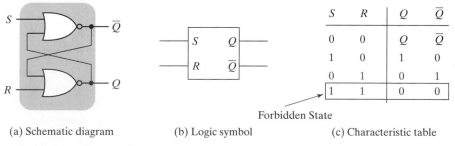

S	R	Q	$\overline{Q}$
0	0	Q	$\overline{Q}$
1	0	1	0
0	1	0	1
1	1	0	0

Forbidden State

(a) Schematic diagram   (b) Logic symbol   (c) Characteristic table

Figure 7-19 NOR-based *SR* flip-flop.

Problem 7.4   SR Flip-Flop, Using NAND Gates

An SR flip-flop can also be implemented by using a cross-coupled NAND structure, as shown in Figure 7-20. Derive the truth table for a such an implementation.

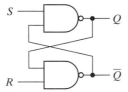

Figure 7-20   NAND-based SR flip-flop.

The SR flip-flop shown so far is purely asynchronous, which does not match well with the synchronous design methodology, the preferred strategy for more than 99% of today's integrated circuits. A clocked version of the latch is shown in Figure 7-21. It consists of a cross-coupled inverter pair, plus four extra transistors to drive the flip-flop from one state to another, and to provide synchronization. In steady state, one inverter resides in the high state, while the other one is low. No static paths between V_{DD} and GND exist. Transistor sizing is, however, essential to ensure that the flip-flop can transition from one state to the other when requested.

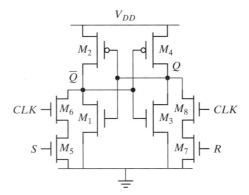

Figure 7-21   Ratioed CMOS *SR* latch.

Example 7.2   Transistor Sizing of Clocked SR Latch

Consider the case in which Q is high and an R pulse is applied. In order to make the latch switch, we must succeed in bringing Q below the switching threshold of the inverter M_1–M_2. Once this is achieved, the positive feedback causes the flip-flop to invert states. This requirement forces us to increase the sizes of transistors M_5, M_6, M_7, and M_8. The combination of transistors M_4, M_7, and M_8 forms a ratioed inverter. Assume that the cross-coupled inverter pair is designed such that the inverter threshold V_M is located at $V_{DD}/2$. For a 0.25-µm CMOS technology, the following transistor sizes were selected: $(W/L)_{M_1} = (W/L)_{M_3} = (0.5 \text{ μm}/0.25 \text{ μm})$, and $(W/L)_{M_2} = (W/L)_{M_4} = (1.5 \text{ μm}/0.25 \text{ μm})$. Assuming $Q = 0$, we determine the minimum sizes of M_5, M_6, M_7, and M_8 to make the device switchable.

To switch the latch from the $Q = 0$ to the $Q = 1$ state, it is essential that the low level of the ratioed, pseudo-NMOS inverter $(M_5-M_6)-M_2$ be below the switching threshold of the inverter M_3-M_4 that equals $V_{DD}/2$. It is reasonable to assume that as long as $V_{\overline{Q}} > V_M$, V_Q equals 0 and the gate of transistor M_2 is grounded. The boundary conditions on the transistor sizes can be derived by equating the currents in the inverter for $V_{\overline{Q}} = V_{DD}/2$, as given in Eq. (7.4) (this ignores channel-length modulation). The currents are determined by the saturation current, since $V_S = V_{DD} = 2.5$ V and $V_M = 1.25$ V. We assume that M_5 and M_6 have identical sizes and that W/L_{5-6} is the effective ratio of the series-connected devices. Under this condition, the pull-down network can be modeled by a single transistor M_{5-6}, whose length is twice the length of the individual devices:

$$k'_n\left(\frac{W}{L}\right)_{5-6}\left((V_{DD} - V_{Tn})V_{DSATn} - \frac{V^2_{DSATn}}{2}\right)$$

$$= -k'_p\left(\frac{W}{L}\right)_2\left((-V_{DD} - V_{Tp})V_{DSATp} - \frac{V^2_{DSATp}}{2}\right)$$

(7.4)

Using the parameters for the 0.25-μm process, Eq. (7.4) results in the constraint that the effective $(W/L)_{M_{5-6}} \geq 2.26$. This implies that the individual device ratio for M_5 or M_6 must be larger than approximately 4.5. Figure 7-22a shows the DC plot of $V_{\overline{Q}}$ as a function of the individual device sizes of M_5 and M_6. We notice that the individual device ratio of greater than 3 is sufficient to bring the $\overline{Q}$ voltage to the inverter switching threshold. The difference between the manual analysis and simulation arises from second-order effects

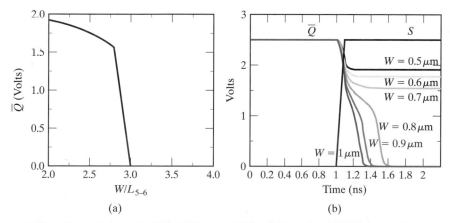

(a) (b)

Figure 7-22 Sizing issues for *SR* flip-flop. (a) DC output voltage versus pull-down device size M_{5-6} (with $W/L_2 = 1.5$ μm/0.25 μm). (b) Transient response showing that M_5 and M_6 must each have a W/L larger than 3 to switch the *SR* flip-flop.

such as channel length modulation and DIBL. Figure 7-22b plots the transient response for different device sizes and confirms that an individual *W/L* ratio of greater than 3 is required to overpower the feedback and switch the state of the latch.

7.3 Dynamic Latches and Registers

Storage in a static sequential circuit relies on the concept that a cross-coupled inverter pair produces a bistable element and can thus be used to memorize binary values. This approach has the useful property that a stored value remains valid as long as the supply voltage is applied to the circuit—hence the name *static*. The major disadvantage of the static gate, however, is its complexity. When registers are used in computational structures that are constantly clocked (such as a pipelined datapath), the requirement that the memory should hold state for extended periods of time can be significantly relaxed.

This results in a class of circuits based on temporary storage of charge on parasitic capacitors. The principle is exactly identical to the one used in dynamic logic—charge stored on a capacitor can be used to represent a logic signal. The absence of charge denotes a 0, while its presence stands for a stored 1. No capacitor is ideal, unfortunately, and some charge leakage is always present. A stored value can thus only be kept for a limited amount of time, typically in the range of milliseconds. If one wants to preserve signal integrity, a periodic *refresh* of the value is necessary; hence, the name *dynamic* storage. Reading the value of the stored signal from a capacitor without disrupting the charge requires the availability of a device with a high-input impedance.

7.3.1 Dynamic Transmission-Gate Edge-Triggered Registers

A fully dynamic positive edge-triggered register based on the master–slave concept is shown in Figure 7-23. When $CLK = 0$, the input data is sampled on storage node 1, which has an equivalent capacitance of C_1, consisting of the gate capacitance of I_1, the junction capacitance of T_1, and the overlap gate capacitance of T_1. During this period, the slave stage is in a hold mode, with node 2 in a high-impedance (floating) state. On the rising edge of clock, the transmission gate T_2 turns on, and the value sampled on node 1 right before the rising edge propagates to the output Q (note that node 1 is stable during the high phase of the clock, since the first transmission gate is

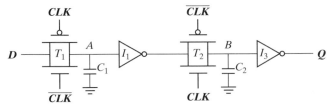

Figure 7-23   Dynamic edge-triggered register.

turned off). Node 2 now stores the inverted version of node 1. This implementation of an edge-triggered register is very efficient because it requires only eight transistors. The sampling switches can be implemented using NMOS-only pass transistors, resulting in an even simpler six transistor implementation. The reduced transistor count is attractive for high-performance and low-power systems.

The setup time of this circuit is simply the delay of the transmission gate, and it corresponds to the time it takes node 1 to sample the D input. The hold time is approximately zero, since the transmission gate is turned off on the clock edge and further inputs changes are ignored. The propagation delay (t_{c-q}) is equal to two inverter delays plus the delay of the transmission gate T_2.

One important consideration for such a dynamic register is that the storage nodes (i.e., the state) have to be refreshed at periodic intervals to prevent
losses due to charge leakage, diode leakage, or subthreshold currents. In datapath circuits, the refresh rate is not an issue, since the registers are periodically clocked, and the storage nodes are constantly updated.

Clock overlap is an important concern for this register. Consider the clock waveforms shown in Figure 7-24. During the 0–0 overlap period, the NMOS of T_1 and the PMOS of T_2 are simultaneously on, creating a direct path for data to flow from the D input of the register to the Q output. In other words, a *race condition* occurs. The output Q can change on the falling edge if the overlap period is large—obviously an undesirable effect for a positive edge-triggered register. The same is true for the 1–1 overlap region, where an input-output path exists through the PMOS of T_1 and the NMOS of T_2. The latter case is taken care of by enforcing a *hold* time constraint. That is, the data must be stable during the high-overlap period. The former situation (0–0 overlap) can be addressed by making sure that there is enough delay between the D input and node B, ensuring that new data sampled by the master stage does not propagate through to the slave stage. Generally, the built-in single inverter delay should be sufficient. The overlap period constraint is given by

$$t_{overlap0-0} < t_{T1} + t_{I1} + t_{T2} \qquad (7.5)$$

Similarly, the constraint for the 1–1 overlap is given as:

$$t_{hold} > t_{overlap1-1} \qquad (7.6)$$

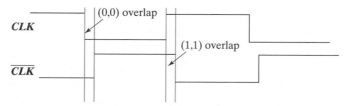

Figure 7-24 Impact of nonoverlapping clocks.

WARNING: The dynamic circuits shown in this section are very appealing from the perspective of complexity, performance, and power. Unfortunately, robustness considerations limit their use. In a fully dynamic circuit like that shown in Figure 7-23, a signal net that is capacitively coupled to the internal storage node can inject significant noise and destroy the state. This is especially important in ASIC flows, where there is little control over coupling between signal nets and internal dynamic nodes. Leakage currents cause another problem: Most modern processors require that the clock can be slowed down or completely halted, to conserve power in low-activity periods. Finally, the internal dynamic nodes do not track variations in power supply voltage. For example, when *CLK* is high for the circuit in Figure 7-23, node *A* holds its state, but it does not track variations in the power supply seen by I_1. This results in reduced noise margins.

Most of these problems can be adequately addressed by adding a weak feedback inverter and making the circuit *pseudostatic* (Figure 7-25). While this comes at a slight cost in delay, it improves the noise immunity significantly. Unless registers are used in a highly-controlled environment (for instance, a custom-designed high-performance datapath), they should be made pseudostatic or static. This holds for all latches and registers discussed in this section.

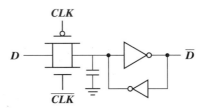

Figure 7-25   Making a dynamic latch pseudostatic.

7.3.2   C²MOS—A Clock-Skew Insensitive Approach

The C²MOS Register

Figure 7-26 shows an ingenious positive edge-triggered register that is based on a master–slave concept insensitive to clock overlap. This circuit is called the *C²MOS* (Clocked CMOS) *register* [Suzuki73], and operates in two phases:

1. $CLK = 0$ ($\overline{CLK} = 1$): The first tristate driver is turned on, and the master stage acts as an inverter sampling the inverted version of *D* on the internal node *X*. The master stage is in the evaluation mode. Meanwhile, the slave section is in a high-impedance mode, or in a hold mode. Both transistors M_7 and M_8 are off, decoupling the output from the input. The output *Q* retains its previous value stored on the output capacitor C_{L2}.
2. The roles are reversed when $CLK = 1$: The master stage section is in hold mode (M_3-M_4 off), while the second section evaluates (M_7–M_8 on). The value stored on C_{L1} propagates to the output node through the slave stage, which acts as an inverter.

The overall circuit operates as a positive edge-triggered master–slave register very similar to the transmission-gate-based register presented earlier. However, there is an important difference:

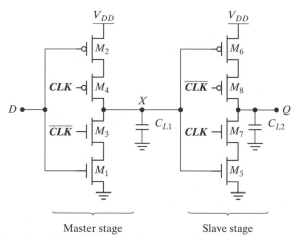

Figure 7-26 C²MOS master–slave positive edge-triggered register.

A C²MOS register with *CLK*–$\overline{CLK}$ clocking is insensitive to overlap, as long as the rise and fall times of the clock edges are sufficiently small.

To prove this statement, we examine both the (0–0) and (1–1) overlap cases (see Figure 7-24). In the (0–0) overlap case, the circuit simplifies to the network shown in Figure 7-27a in which both PMOS devices are *on* during this period. To operate correctly, none of the new data sampled during the overlap window should propagate to the output Q, since data should not change on the negative edge of a positive edge-triggered register. Indeed, new data is sampled on node X through the series PMOS devices M_2–M_4, and node X can make a 0-to-1 transition during the overlap period. However, this data cannot propagate to the output since the NMOS device M_7 is turned off. At the end of the overlap period, $\overline{CLK} = 1$ and both M_7 and M_8 turn off, putting the slave stage in the hold mode. Therefore, any new data sampled on the falling clock edge is not seen at the slave output Q, since the slave state is off till the next rising edge of the clock. As the circuit consists of a cascade of inverters, signal propagation requires one pull-up followed by a pull-down, or vice versa, which is not feasible in the situation presented.

The (1–1) overlap case where both NMOS devices M_3 and M_7 are turned on, is somewhat more contentious (see Figure 7-27b). The question is again if new data sampled during the overlap period (right after clock goes high) propagates to the Q output. A positive edge-triggered register may only pass data that is presented at the input before the rising edge. If the D input changes during the overlap period, node X can make a 1-to-0 transition, but cannot propagate further. However, as soon as the overlap period is over, the PMOS M_8 turns on and the 0 propagates tooutput, which is not desirable. The problem is fixed by imposing a hold-time constraint on the input data, D; or, in other words, the data D should be stable during the overlap period.

In sum, it can be stated that the C²MOS latch is insensitive to clock overlaps because those overlaps activate either the pull-up or the pull-down networks of the latches, but never both of them simultaneously. If the rise and fall times of the clock are sufficiently slow, however, there

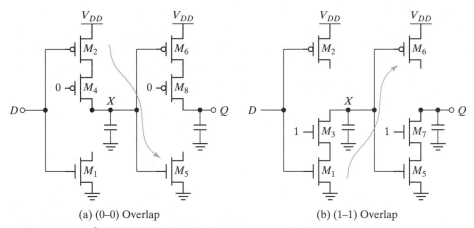

(a) (0–0) Overlap   (b) (1–1) Overlap

Figure 7-27   C^{2}MOS *D* FF during overlap periods. No feasible signal path can exist between *In* and *D,* as illustrated by the arrows.

exists a time slot where both the NMOS and PMOS transistors are conducting. This creates a path between input and output that can destroy the state of the circuit. Simulations have shown that the circuit operates correctly as long as the clock rise time (or fall time) is smaller than approximately five times the propagation delay of the register. This criterion is not too stringent, and it is easily met in practical designs. The impact of the rise and fall times is illustrated in Figure 7-28, which plots the simulated transient response of a C^{2}MOS *D* FF for clock slopes of, respectively, 0.1 and 3 ns. For slow clocks, the potential for a *race condition* exists.

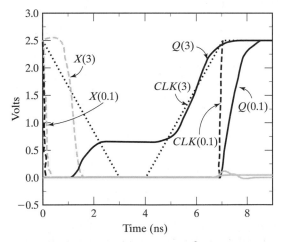

Figure 7-28   Transient response of C^{2}MOS FF for 0.1-ns and 3-ns clock rise/fall times, assuming *In* = 1.

Dual-Edge Registers

So far, we have focused on edge-triggered registers that sample the input data on only one of the clock edges (rising or falling). It also is possible to design sequential circuits that sample the input on both edges. The advantage of this scheme is that a lower frequency clock—half the original rate—is distributed for the same functional throughput, resulting in power savings in the clock distribution network. Figure 7-29 shows a modification of the C^{2}MOS register enabling sampling on both edges. It consists of two parallel master–slave edge-triggered registers, whose outputs are multiplexed by using tristate drivers.

When clock is high, the positive latch composed of transistors M_1–M_4 is sampling the inverted D input on node X. Node Y is held stable, since devices M_9 and M_{10} are turned off. On the falling edge of the clock, the top slave latch M_5–M_8 turns on, and drives the inverted value of X to the Q output. During the low phase, the bottom master latch (M_1, M_4, M_9, M_{10}) is turned on, sampling the inverted D input on node Y. Note that the devices M_1 and M_4 are reused, reducing the load on the D input. On the rising edge, the bottom slave latch conducts and drives the inverted version of Y on node Q. Data thus changes on both edges. Note that the slave latches operate in a complementary fashion—that is, only one of them is turned on during each phase of the clock.

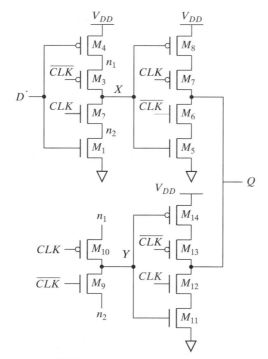

Figure 7-29 C^{2}MOS-based dual-edge triggered register.

Problem 7.5   Dual-Edge Registers

Determine how the adoption of dual-edge registers influences the power dissipation in the clock-distribution network.

7.3.3   True Single-Phase Clocked Register (TSPCR)

In the two-phase clocking schemes described earlier, care must be taken in routing the two clock signals to ensure that overlap is minimized. While the C²MOS provides a skew-tolerant solution, it is possible to design registers that only use a single phase clock. The *True Single-Phase Clocked Register* (TSPCR), proposed by Yuan and Svensson, uses a **single clock** [Yuan89]. The basic single-phase positive and negative latches are shown in Figure 7-30. For the positive latch, when *CLK* is high, the latch is in the transparent mode and corresponds to two cascaded inverters; the latch is noninverting, and propagates the input to the output. On the other hand, when *CLK* = 0, both inverters are disabled, and the latch is in hold mode. Only the pull-up networks are still active, while the pull-down circuits are deactivated. As a result of the dual-stage approach, no signal can ever propagate from the input of the latch to the output in this mode. A register can be constructed by cascading positive and negative latches. The clock load is similar to a conventional transmission gate register, or C²MOS register. The main advantage is the use of a single clock phase. The disadvantage is the slight increase in the number of transistors—12 transistors are now required.

As a reminder, note that a dynamic circuit in the style of Figure 7-30 must be used with caution. When the clock is low (for the positive latch), the output node may be floating, and it is exposed to coupling from other signals. Also, charge sharing can occur if the output node drives transmission gates. Dynamic nodes should be isolated with the aid of static inverters, or made pseudostatic for improved noise immunity.

As with many other latch families, TSPC offers an additional advantage that we have not explored so far: The possibility of embedding logic functionality into the latches. This reduces

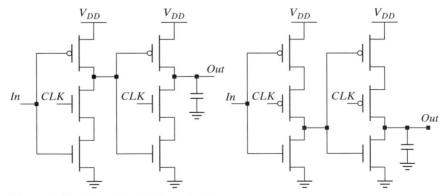

Figure 7-30   True Single-Phase Latches.

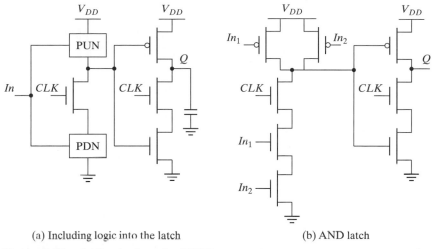

(a) Including logic into the latch   (b) AND latch

Figure 7-31 Adding logic to the TSPC approach.

the delay overhead associated with the latches. Figure 7-31a outlines the basic approach for embedding logic, while Figure 7-31b shows an example of a positive latch that implements the AND of In_1 and In_2 in addition to performing the latching function. While the setup time of this latch has increased over the one shown in Figure 7-30, the overall performance of the digital circuit (that is, the clock period of a sequential circuit) has improved: The increase in setup time typically is smaller than the delay of an AND gate. This approach of embedding logic into latches has been used extensively in the design of the EV4 DEC Alpha microprocessor [Dobberpuhl92] and many other high-performance processors.

Example 7.3   Impact of Embedding Logic into Latches on Performance

Consider embedding an AND gate into the TSPC latch, as shown in Figure 7-31b. In a 0.25-μm technology, the setup time of such a circuit, using minimum-size devices is 140 ps. A conventional approach, composed of an AND gate followed by a positive latch, has an effective setup time of 600 ps (we treat the AND plus latch as a black box that performs both functions). The embedded logic approach thus results in significant performance improvements.

The TSPC latch circuits can be further reduced in complexity, as illustrated in Figure 7-32, where only the first inverter is controlled by the clock. Besides the reduced number of transistors, these circuits have the advantage that the clock load is reduced by half. On the other hand, not all node voltages in the latch experience the full logic swing. For instance, the voltage at node A (for $V_{in} = 0$ V) for the positive latch maximally equals $V_{DD} - V_{Tn}$, which results in a reduced drive for the output NMOS transistor and a loss in performance. Similarly, the voltage

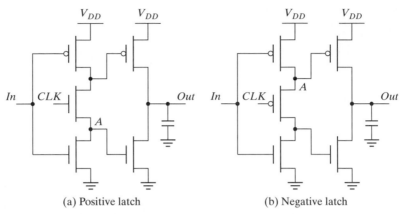

(a) Positive latch   (b) Negative latch

Figure 7-32   Simplified TSPC latch (also called split output).

on node A (for $V_{in} = V_{DD}$) for the negative latch is only driven down to $|V_{TP}|$. This also limits the amount of V_{DD} scaling possible on the latch.

Figure 7-33 shows the design of a specialized single-phase edge-triggered register. When $CLK = 0$, the input inverter is sampling the inverted D input on node X. The second (dynamic) inverter is in the precharge mode, with M_6 charging up node Y to V_{DD}. The third inverter is in the hold mode, since M_8 and M_9 are off. Therefore, during the low phase of the clock, the input to the final (static) inverter is holding its previous value and the output Q is stable. On the rising edge of the clock, the dynamic inverter M_4–M_6 evaluates. If X is high on the rising edge, node Y discharges. The third inverter M_7–M_9 is on during the high phase, and the node value on Y is passed to the output Q. On the positive phase of the clock, note that node X transitions to a low if the D input transitions to a high level. Therefore, the input must be kept stable until the value on node X before the rising edge of the clock propagates to Y. This represents the hold time of the register (note that the hold time is less than 1 inverter delay, since it takes 1 delay for the input to affect node X). The propagation delay of the register is essentially three inverters, because the

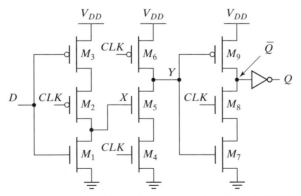

Figure 7-33   Positive edge-triggered register in TSPC.

value on node X must propagate to the output Q. Finally, the setup time is the time for node X to be valid, which is one inverter delay.

WARNING: Similar to the C²MOS latch, the TSPC latch malfunctions when the *slope of the clock* is not sufficiently steep. Slow clocks cause both the NMOS and PMOS clocked transistors to be on simultaneously, resulting in undefined values of the states and race conditions. The clock slopes should therefore be carefully controlled. If necessary, local buffers must be introduced to ensure the quality of the clock signals.

Example 7.4   TSPC Edge-Triggered Register

Transistor sizing is critical for achieving correct functionality in the TSPC register. With improper sizing, glitches may occur at the output due to a *race condition* when the clock transitions from low to high. Consider the case where D is low and $\overline{Q} = 1$ ($Q = 0$). While *CLK* is low, Y is precharged high turning on M_7. When *CLK* transitions from low to high, nodes Y and $\overline{Q}$ start to discharge simultaneously (through M_4–M_5 and M_7–M_8, respectively). Once Y is sufficiently low, the trend on $\overline{Q}$ is reversed and the node is pulled high again through M_9. In a sense, this sequence of events is comparable to what happens when we chain dynamic logic gates. Figure 7-34 shows the transient response of the circuit of Figure 7-33 for different sizes of devices in the final two stages.

This glitch may be the cause of fatal errors, because it may create unwanted events (for instance, when the output of the latch is used as a clock signal input to another register). It also reduces the contamination delay of the register. The problem can be corrected by resizing the relative strengths of the pull-down paths through M_4–M_5 and M_7–M_8, so that Y discharges much faster than $\overline{Q}$. This is accomplished by reducing the strength of the M_7–M_8 pull-down path, and by speeding up the M_4–M_5 pull-down path.

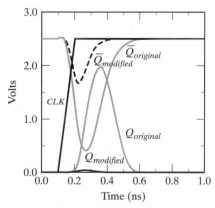

	M_4, M_5	M_7, M_8
Original Width	$0.5\,\mu m$	$2\,\mu m$
Modified Width	$1\,\mu m$	$1\,\mu m$

Figure 7-34   Transistor sizing issues in TSPC (for the register of Figure 7-33).

7.4 Alternative Register Styles*

7.4.1   Pulse Registers

Until now, we have used the master–slave configuration to create an edge-triggered register. A fundamentally different approach for constructing a register uses *pulse signals*. The idea is to construct a short pulse around the rising (or falling) edge of the clock. This pulse acts as the clock input to a latch (for example, Figure 7-35a), sampling the input only in a short window. Race conditions are thus avoided by keeping the opening time (i.e, the transparent period) of the latch very short. The combination of the glitch-generation circuitry and the latch results in a positive edge-triggered register.

Figure 7-35b shows an example circuit for constructing a short intentional glitch on each rising edge of the clock [Kozu96]. When $CLK = 0$, node X is charged up to V_{DD} (M_N is off since $CLKG$ is low). On the rising edge of the clock, there is a short period of time when both inputs of the AND gate are high, causing $CLKG$ to go high. This in turn activates M_N, pulling X and eventually $CLKG$ low (Figure 7-35c). The length of the pulse is controlled by the delay of the AND gate and the two inverters. Note that there exists also a delay between the rising edges of the input clock (CLK) and the glitch clock ($CLKG$), which also is equal to the delay of the AND gate and the two inverters. If every register on the chip uses the same clock generation mechanism, this sampling delay does not matter. However, process variations and load variations may

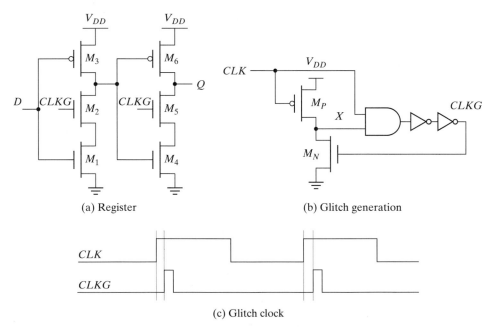

(a) Register   (b) Glitch generation

(c) Glitch clock

Figure 7-35   TSPC-based glitch latch-timing generation and register.

cause the delays through the glitch clock circuitry to be different. This must be taken into account when performing timing verification and clock skew analysis (the topics of Chapter 10).

If the setup time and hold time are measured in reference to the rising edge of the glitch clock, the setup time is essentially zero, the hold time is essentially equal to the length of the pulse, and the propagation delay (t_{c-q}) equals two gate delays. The advantage of the approach is the **reduced clock load and the small number of transistors** required. The glitch-generation circuitry can be amortized over multiple register bits. The disadvantage is a substantial increase in verification complexity. For this circuit to function properly, simulations must be performed across all corners to ensure that the clock pulse always exists (i.e., that the glitch-generation circuit works reliably). Despite the increased complexity, such registers do provide an alternate approach to conventional schemes, and they have been adopted in a number of high-performance processors (e.g., [Kozu96]).

Another version of the pulsed register is shown in Figure 7-36 (as used in the AMD-K6 processor [Partovi96]). When the clock is low, M_3 and M_6 are off, and device P_1 is turned on. Node X is precharged to V_{DD}, the output node (Q) is decoupled from X and is held at its previous state. $\overline{CLKD}$ is a delay-inverted version of CLK. On the rising edge of the clock, M_3 and M_6 turn on while devices M_1 and M_4 stay on for a short period, determined by the delay of the three inverters. During this interval, the circuit is transparent and the input data D is sampled by the latch. Once $\overline{CLKD}$ goes low, node X is decoupled from the D input and is either held or starts to precharge to V_{DD} through PMOS device P_2. On the falling edge of the clock, node X is held at V_{DD} and the output is held stable by the cross-coupled inverters.

Note that this circuit also uses a *pulse generator*, but it is integrated into the register. The transparency period also determines the hold time of the register. The window must be wide enough for the input data to propagate to the Q output. In this particular circuit, the setup time can be negative. This is the case if the transparency window is longer than the delay from input to output. This is attractive, as data can arrive at the register even after the clock goes high, which means that time is borrowed from the previous cycle.

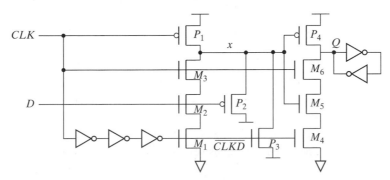

Figure 7-36   Flow-through positive edge-triggered register.

Example 7.5   Setup Time of Glitch Register

The glitch register of Figure 7-36 is transparent during the (1–1) overlap of *CLK* and $\overline{CLKD}$. As a result, the input data can actually change after the rising edge of the clock, resulting in a negative setup time (Figure 7-37). The *D*-input transitions to low after the rising edge of the clock, and transitions to high before the falling edge of $\overline{CLKD}$ (i.e., during the transparency period). Observe how the output follows the input. The output *Q* does go to the correct value of V_{DD} as long as the input *D* is set up correctly some time before the falling edge of $\overline{CLKD}$. When the negative setup time is exploited, there can be no guarantees on the monotonic behavior of the output. That is, the output can have multiple transitions around the rising edge, and therefore, the output of the register should not be used for driving dynamic logic or as a clock as a clock to other registers.

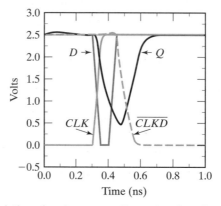

Figure 7-37   Simulation showing a negative setup time for the glitch register.

Problem 7.6   Converting a Glitch Register to a Conditional Glitch Register

Modify the circuit in Figure 7-36 so that it takes an additional *Enable* input. The goal is to convert the register to a conditional register which latches only when the enable signal is asserted.

7.4.2   Sense-Amplifier-Based Registers

In addition to the *master–slave* and the *glitch* approaches to implement an edge-triggered register, a third technique based on *sense amplifiers* can be used, as introduced in Figure 7-38 [Montanaro96].[3] Sense-amplifier circuits accept small input signals and amplify them to generate rail-to-rail swings. They are used extensively in memory cores and in low-swing bus drivers to either improve performance or reduce power dissipation. There are many techniques to construct these amplifiers. A common approach is to use feedback—for instance, through a set of

---

[3]In a sense, these sense-amplifier-based registers are similar in operation to the glitch registers—that is, the first stage generates the pulse, and the second latches it.

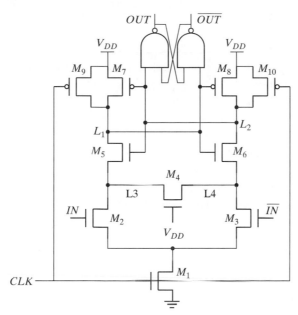

Figure 7-38 Positive edge-triggered register based on sense amplifier.

cross-coupled inverters. The circuit shown in Figure 7-38 uses a precharged front-end amplifier that samples the differential input signal on the rising edge of the clock signal. The outputs of front end are fed into a NAND cross-coupled SR *flip-flop* that holds the data and guarantees that the differential outputs switch only once per clock cycle. The differential inputs in this implementation don't have to have rail-to-rail swing.

The core of the front end consists of a cross-coupled inverter (M_5–M_8), whose outputs (L_1 and L_2) are precharged by using devices M_9 and M_{10} during the low phase of the clock. As a result, PMOS transistors M_7 and M_8 are turned off and the NAND flip-flop is holding its previous state. Transistor M_1 is similar to an evaluate switch in dynamic circuits and is turned off to ensure that the differential inputs do not affect the output during the low phase of the clock. On the rising edge of the clock, the evaluate transistor turns on and the differential input pair (M_2 and M_3) is enabled, and the difference between the input signals is amplified on the output nodes on L_1 and L_2. The cross-coupled inverter pair flips to one of its stable states based on the value of the inputs. For example, if *IN* is 1, L_1 is pulled to 0, and L_2 remains at V_{DD}. Due to the amplifying properties of the input stage, it is not necessary for the input to swing all the way up to V_{DD}, which enables the use of low-swing signaling on the input wires.

The shorting transistor, M_4, is used to provide a DC-leakage path from either node L_3, or L_4, to ground. This is necessary to accommodate the case in which the inputs change their value after the positive edge of *CLK* has occurred, resulting in either L_3 or L_4 being left in a high-impedance state with a logical low-voltage level stored on the node. Without the leakage path, that node would be susceptible to charging by leakage currents. The latch could then actually change state prior to the next rising edge of *CLK*! This is best illustrated graphically, as in Figure 7-39.

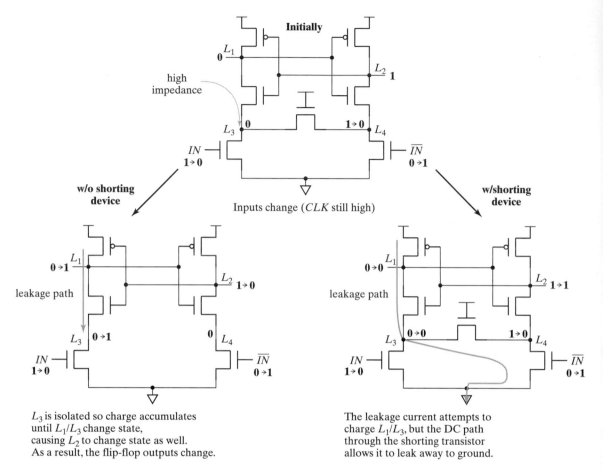

L_3 is isolated so charge accumulates until L_1/L_3 change state, causing L_2 to change state as well. As a result, the flip-flop outputs change.

The leakage current attempts to charge L_1/L_3, but the DC path through the shorting transistor allows it to leak away to ground.

Figure 7-39   The need for the shorting transistor M_4.

7.5 Pipelining: An Approach to Optimize Sequential Circuits

Pipelining is a popular design technique often used to accelerate the operation of datapaths in digital processors. The concept is explained with the example of Figure 7-40a. The goal of the presented circuit is to compute $\log(|a + b|)$, where both a and b represent streams of numbers (i.e., the computation must be performed on a large set of input values). The minimal clock period T_{min} necessary to ensure correct evaluation is given as

$$T_{min} = t_{c-q} + t_{pd,logic} + t_{su} \tag{7.7}$$

where t_{c-q} and t_{su} are the propagation delay and the setup time of the register, respectively. We assume that the registers are edge-triggered D registers. The term $t_{pd,logic}$ stands for the worst case delay path through the combinational network, which consists of the adder, absolute value, and logarithm functions. In conventional systems (that don't push the edge of technology), the

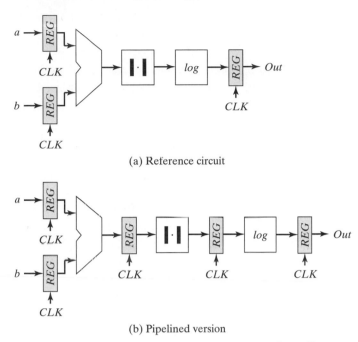

(a) Reference circuit

(b) Pipelined version

Figure 7-40 Datapath for the computation of $\log(|a + b|)$.

latter delay is generally much larger than the delays associated with the registers and dominates the circuit performance. Assume that each logic module has an equal propagation delay. We note that each logic module is then active for only one-third of the clock period (if the delay of the register is ignored). For example, the adder unit is active during the first third of the period and remains idle (no useful computation) during the other two-thirds of the period. Pipelining is a technique to improve the resource utilization, and increase the functional through-put. Assume that we introduce registers between the logic blocks, as shown in Figure 7-40b. This causes the computation for one set of input data to spread over a number of clock-periods, as shown in Table 7-1. The result for the data set (a_1, b_1) only appears at the output after three clock periods.

Table 7-1 Example of pipelined computations.

Clock Period	Adder	Absolute Value	Logarithm
1	$a_1 + b_1$		
2	$a_2 + b_2$	$|a_1 + b_1|$	
3	$a_3 + b_3$	$|a_2 + b_2|$	$\log(|a_1 + b_1|)$
4	$a_4 + b_4$	$|a_3 + b_3|$	$\log(|a_2 + b_2|)$
5	$a_5 + b_5$	$|a_4 + b_4|$	$\log(|a_3 + b_3|)$

At that time, the circuit has already performed parts of the computations for the next data sets, (a_2, b_2) and (a_3, b_3). The computation is performed in an assembly-line fashion—hence the name *pipeline*.

The advantage of pipelined operation becomes apparent when examining the minimum clock period of the modified circuit. The combinational circuit block has been partitioned into three sections, each of which has a smaller propagation delay than the original function. This effectively reduces the value of the minimum allowable clock period:

$$T_{min,pipe} = t_{c-q} + \max(t_{pd,add}, t_{pd,abs}, t_{pd,log}) + t_{su} \qquad (7.8)$$

Suppose that all logic blocks have approximately the same propagation delay, and that the register overhead is small with respect to the logic delays. The pipelined network outperforms the original circuit by a factor of three under these assumptions (i.e., $T_{min,pipe} = T_{min}/3$). The increased performance comes at the relatively small cost of two additional registers and an increased latency.[4] This explains why pipelining is popular in the implementation of very high-performance datapaths.

7.5.1 Latch- versus Register-Based Pipelines

Pipelined circuits can be constructed by using level-sensitive latches instead of edge-triggered registers. Consider the pipelined circuit of Figure 7-41. The pipeline system is implemented using pass-transistor-based positive and negative latches instead of edge-triggered registers. That is, logic is introduced between the master and slave latches of a master–slave system. In the following discussion, we use the CLK–$\overline{CLK}$ notation to denote a two-phase clock system without loss of generality. Latch-based systems give significantly more flexibility in implementing a pipelined system, and they often offer higher performance. When the CLK and $\overline{CLK}$ clocks are nonoverlapping, correct pipeline operation is obtained. Input data is sampled on C_1 at the negative edge of CLK and the computation of logic block F starts; the result of the logic block F is stored on C_2 on the falling edge of $\overline{CLK}$, and the computation of logic block G starts. The non-overlapping of the clocks ensures correct operation. The value stored on C_2 at the end of the CLK low phase is the result of passing the previous input (stored on the falling edge of CLK on C_1) through the logic function F. When overlap exists between CLK and $\overline{CLK}$, the next input is already being applied to F, and its effect might propagate to C_2 before $\overline{CLK}$ goes low (assuming that the contamination delay of F is small). In other words, a *race* develops between the previous input and the current one. Which value wins depends upon the logic and is often a function of the applied inputs. The latter factor makes the detection and elimination of race conditions non-trival in nature.

---

[4]*Latency* is defined here as the number of clock cycles it takes for the data to propagate from the input to the output. For the example at hand, pipelining increases the latency from 1 to 3. An increased latency is generally acceptable, but it can cause a global performance degradation if not treated with care.

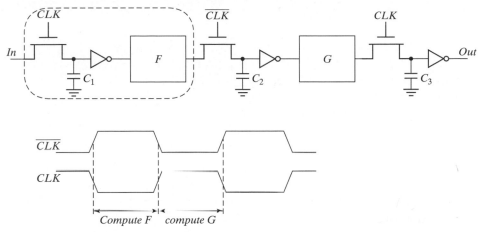

Figure 7-41 Operation of two-phase pipelined circuit, using dynamic registers.

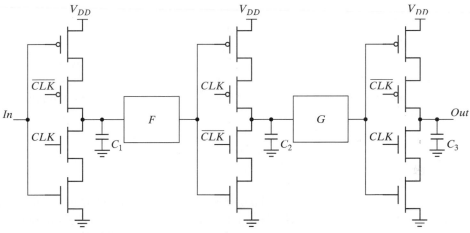

Figure 7-42 Pipelined datapath, using C²MOS latches.

7.5.2 NORA–CMOS—A Logic Style for Pipelined Structures

The latch-based pipeline circuit can also be implemented by using C²MOS latches, as shown in Figure 7-42. The operation is similar to the one discussed in Section 7.5.1. This topology has one additional important property:

> **A C²MOS-based pipelined circuit is race free as long as all the logic functions *F* (implemented by using static logic) between the latches are noninverting.**

The reasoning for the preceding argument is similar to the argument made in the construction of a C²MOS register. During a (0–0) overlap between *CLK* and $\overline{CLK}$, all C²MOS latches simplify to pure pull-up networks (see Figure 7-27). The only way a signal can race

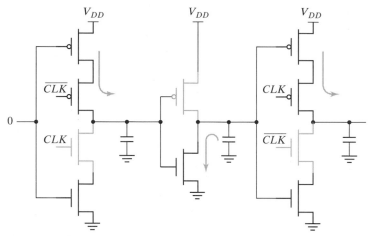

Figure 7-43   Potential race condition during (0–0) overlap in C^{2}MOS-based design.

from stage to stage under this condition is when the logic function F is inverting, as illustrated in Figure 7-43, where F is replaced by a single, static CMOS inverter. Similar considerations are valid for the (1–1) overlap.

Based on this concept, a logic circuit style called NORA–CMOS was conceived [Gonçalves83]. It combines C^{2}MOS pipeline registers and NORA dynamic logic function blocks. Each module consists of a block of combinational logic that can be a mixture of static and dynamic logic, followed by a C^{2}MOS latch. Logic and latch are clocked in such a way that both are simultaneously in either evaluation, or hold (precharge) mode. A block that is in evaluation during $CLK = 1$ is called a *CLK module*, while the inverse is called a $\overline{CLK}$ *module*. Examples of both classes are shown in Figure 7-44a and 7-44b, respectively. The operation modes of the modules are summarized in Table 7-2.

A NORA datapath consists of a chain of alternating CLK and $\overline{CLK}$ modules. While one class of modules is precharging with its output latch in hold mode, preserving the previous output value, the other class is evaluating. Data is passed in a pipelined fashion from module to module. NORA offers designers a wide range of design choices. Dynamic and static logic

Table 7-2   Operation modes for NORA logic modules.

	CLK block		$\overline{CLK}$ block	
	Logic	**Latch**	**Logic**	**Latch**
$CLK = 0$	Precharge	Hold	Evaluate	Evaluate
$CLK = 1$	Evaluate	Evaluate	Precharge	Hold

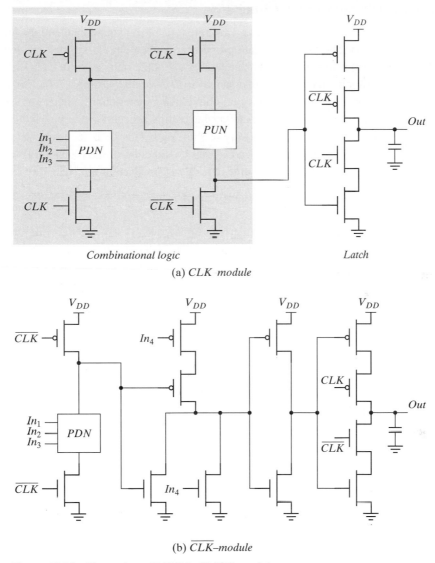

(a) *CLK module*

(b) $\overline{CLK}$*–module*

Figure 7-44   Examples of NORA–CMOS modules.

can be mixed freely, and both CLK_p and CLK_n dynamic blocks can be used in cascaded or in pipelined form. Although this style of logic avoids the extra inverter required in domino CMOS, there are many rules that must be followed to achieve reliable and race-free operation. As a result of this added complexity, the use of NORA has been limited to high-performance applications.

7.6 Nonbistable Sequential Circuits

In the preceding sections, we have focused on a single type of sequential element: the latch (and its sibling, the register). The most important property of such a circuit is that it has two stable states—hence, the term *bistable*. The bistable element is not the only sequential circuit of interest. Other regenerative circuits can be catalogued as *astable* and *monostable*. The former act as oscillators and can, for instance, be used for on-chip clock generation. The latter serve as pulse generators, also called *one-shot circuits*. Another interesting regenerative circuit is the *Schmitt trigger*. This component has the useful property of showing hysteresis in its dc characteristics—its switching threshold is variable and depends upon the direction of the transition (low to high or high to low). This peculiar feature can come in handy in noisy environments.

7.6.1   The Schmitt Trigger

Definition

A *Schmitt trigger* [Schmitt38] is a device with two important properties:

1. It responds to a slowly changing input waveform with a **fast transition time at the output**.
2. The voltage-transfer characteristic of the device displays *different switching thresholds for positive- and negative-going input signals*. This is demonstrated in Figure 7-45, where a typical voltage-transfer characteristic of the Schmitt trigger is shown (and its schematics symbol). The switching thresholds for the low-to-high and high-to-low transitions are called V_{M+} and V_{M-}, respectively. The *hysteresis voltage* is defined as the difference between the two.

One of the main uses of the Schmitt trigger is to turn a noisy or slowly varying input signal into a clean digital output signal. This is illustrated in Figure 7-46. Notice how the hysteresis suppresses the ringing on the signal. At the same time, the fast low-to-high (and high-to-low) transi-

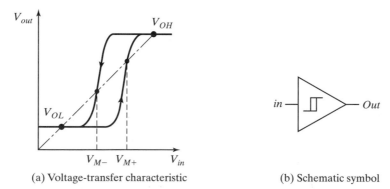

(a) Voltage-transfer characteristic   (b) Schematic symbol

Figure 7-45   Noninverting Schmitt trigger.

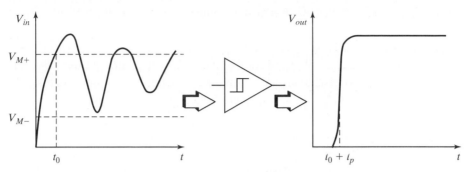

Figure 7-46 Noise suppression, using a Schmitt trigger.

tions of the output signal should be observed. Steep signal slopes are beneficial in general, for instance for reducing power consumption by suppressing direct-path currents. The "secret" behind the Schmitt trigger concept is the use of positive feedback.

CMOS Implementation

One possible CMOS implementation of the Schmitt trigger is shown in Figure 7-47. The idea behind this circuit is that the switching threshold of a CMOS inverter is determined by the (k_n/k_p) ratio between the PMOS and NMOS transistors. Increasing the ratio raises the threshold, while decreasing it lowers V_M. Adapting the ratio depending upon the direction of the transition results in a shift in the switching threshold and a hysteresis effect. This adaptation is achieved with the aid of feedback.

Suppose that V_{in} is initially equal to 0, so that $V_{out} = 0$ as well. The feedback loop biases the PMOS transistor M_4 in the conductive mode, while M_3 is off. The input signal effectively connects to an inverter consisting of two PMOS transistors in parallel (M_2 and M_4) as a pull-up network, and a single NMOS transistor (M_1) in the pull-down chain. This modifies the effective transistor ratio of the inverter to $k_{M1}/(k_{M2} + k_{M4})$, which moves the switching threshold upwards.

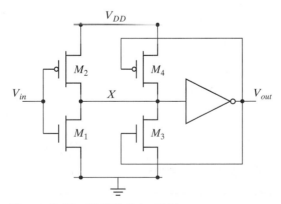

Figure 7-47 CMOS Schmitt trigger.

Once the inverter switches, the feedback loop turns off M_4, and the NMOS device M_3 is activated. This extra pull-down device speeds up the transition and produces a clean output signal with steep slopes.

A similar behavior can be observed for the high-to-low transition. In this case, the pull-down network originally consists of M_1 and M_3 in parallel, while the pull-up network is formed by M_2. This reduces the value of the switching threshold to V_{M-}.

Example 7.6   CMOS Schmitt Trigger

Consider the Schmitt trigger of Figure 7-47, with M_1 and M_2 sized at 1 μm/0.25 μm, and 3 μm/0.25 μm, respectively. The inverter is designed such that the switching threshold is around $V_{DD}/2$ (= 1.25 V). Figure 7-48a shows the simulation of the Schmitt trigger assuming that devices M_3 and M_4 are 0.5 μm/0.25 μm and 1.5 μm/0.25 μm, respectively. As apparent from the plot, the circuit exhibits hysteresis. The high-to-low switching point (V_{M-} = 0.9 V) is lower than $V_{DD}/2$, while the low-to-high switching threshold (V_{M+} = 1.6 V) is larger than $V_{DD}/2$.

It is possible to shift the switching point by changing the sizes of M_3 and M_4. For example, to modify the low-to-high transition, we need to vary the PMOS device. The high-to-low threshold is kept constant by keeping the device width of M_3 at 0.5 μm. The device width of M_4 is varied as $k \times 0.5$ μm. Figure 7-48b demonstrates how the switching threshold increases with raising values of k.

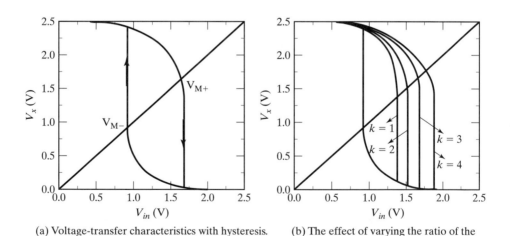

(a) Voltage-transfer characteristics with hysteresis.

(b) The effect of varying the ratio of the PMOS device M_4. The width is $k \times 0.5$ μm.

Figure 7-48   Schmitt trigger simulations.

Problem 7.7 An Alternative CMOS Schmitt Trigger

Another CMOS Schmitt trigger is shown in Figure 7-49. Discuss the operation of the gate, and derive expressions for V_{M-} and V_{M+}.

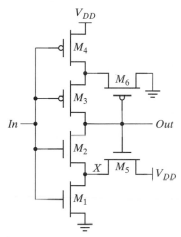

Figure 7-49 Alternate CMOS Schmitt trigger.

7.6.2 Monostable Sequential Circuits

A monostable element is a circuit that generates a pulse of a predetermined width every time the quiescent circuit is triggered by a pulse or transition event. It is called *monostable* because it has only one stable state (the quiescent one). A trigger event, which is either a signal transition or a pulse, causes the circuit to go temporarily into another quasi-stable state. This means that it eventually returns to its original state after a time period determined by the circuit parameters. This circuit, also called a *one-shot*, is useful in generating pulses of a known length. This functionality is required in a wide range of applications. We have already seen the use of a one-shot in the construction of glitch registers. Another well-known example is the *address transition detection* (ATD) circuit, used for the timing generation in static memories. This circuit detects a change in a signal or group of signals, such as the address or data bus, and produces a pulse to initialize the subsequent circuitry.

The most common approach to the implementation of one-shots is the use of a simple delay element to control the duration of the pulse. The concept is illustrated in Figure 7-50. In the quiescent state, both inputs to the XOR are identical, and the output is low. A transition on the input causes the XOR inputs to differ temporarily and the output to go high. After a delay t_d (of the delay element), this disruption is removed, and the output goes low again. A pulse of length t_d is created. The delay circuit can be realized in many different ways, such as an *RC*-network or a chain of basic gates.

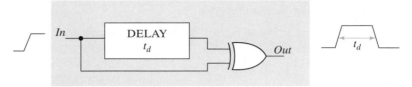

Figure 7-50   Transition-triggered one shot.

7.6.3   Astable Circuits

An astable circuit has no stable states. The output oscillates back and forth between two quasi-stable states, with a period determined by the circuit topology and parameters (delay, power supply, etc.). One of the main applications of oscillators is the on-chip generation of clock signals. (This application is discussed in detail in a later chapter on timing.)

The ring oscillator is a simple example of an astable circuit. It consists of an odd number of inverters connected in a circular chain. Due to the odd number of inversions, no stable operation point exists, and the circuit oscillates with a period equal to $2 \times t_p \times N$, where N is the number of inverters in the chain and t_p is the propagation delay of each inverter.

Example 7.7   Ring Oscillator

The simulated response of a ring oscillator with five stages is shown in Figure 7-51 (all gates use minimum-size devices). The observed oscillation period approximately equals 0.5 ns, which corresponds to a gate propagation delay of 50 ps. By tapping the chain at various points, different phases of the oscillating waveform are obtained. (Phases 1, 3, and 5 are displayed in the plot.) A wide range of clock signals with different duty-cycles and phases can be derived from those elementary signals, using simple logic operations.

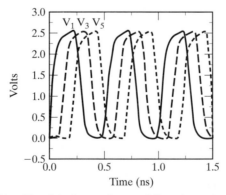

Figure 7-51   Simulated waveforms of five-stage ring oscillator. The outputs of stages 1, 3, and 5 are shown.

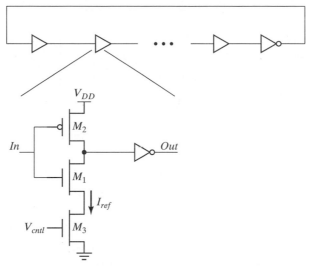

Figure 7-52 Voltage-controlled oscillator based on current-starved inverters.

The ring oscillator composed of cascaded inverters produces a waveform with a fixed oscillating frequency determined by the delay of an inverter in the CMOS process. In many applications, it is necessary to control the frequency of the oscillator. An example of such a circuit is the *voltage-controlled oscillator (VCO)*, whose oscillation frequency is a function (typically, nonlinear) of a control voltage. The standard ring oscillator can be modified into a *VCO* by replacing the standard inverter with a *current-starved inverter* like the one shown in Figure 7-52 [Jeong87]. The mechanism for controlling the delay of each inverter is to limit the current available to discharge the load capacitance of the gate.

In this modified inverter circuit, the maximal discharge current of the inverter is limited by adding an extra series device. Note that the low-to-high transition on the inverter can also be controlled by adding a PMOS device in series with M_2. The added NMOS transistor M_3, is controlled by an analog control voltage V_{cntl}, which determines the available discharge current. Lowering V_{cntl} reduces the discharge current and, hence, increases t_{pHL}. The ability to alter the propagation delay per stage allows us to control the frequency of the ring structure. The control voltage is generally set by using feedback techniques. Under low-operating current levels, the current-starved inverter suffers from slow fall times at its output. This can result in significant short-circuit current. We solve this problem by feeding its output into a CMOS inverter or, better yet, a Schmitt trigger. An extra inverter is needed at the end to ensure that the structure oscillates.

Example 7.8 Current-Starved Inverter Simulation

Figure 7-53 shows the simulated delay of the current-starved inverter as a function of the control voltage V_{cntl}. The delay of the inverter can be varied over a large range. When the control voltage is smaller than the threshold, the device enters the subthreshold region.

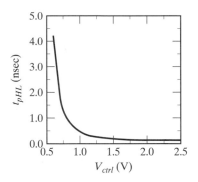

Figure 7-53 t_{pHL} of current-starved inverter as a function of the control voltage.

This results in large variations of the propagation delay, as the drive current is exponentially dependent on the drive voltage. When operating in this region, the delay is very sensitive to variations in the control voltage and hence to noise.

Another approach to implement the delay cell is to use a differential element as shown in Figure 7-54a. Since the delay cell provides both inverting and noninverting outputs, an oscillator with an even number of stages can be implemented. Figure 7-54b shows a two-stage differential *VCO*, where the feedback loop provides 180° phase shift through two gate delays, one noninverting and the other inverting, therefore forming an oscillation. The simulated waveforms of this two-stage *VCO* are shown in Figure 7-54c. The in-phase and quadrature phase outputs are available simultaneously. The differential-type *VCO* has better immunity to common mode noise (for example, supply noise) compared with the common ring oscillator. However, it consumes more power due to its increased complexity and its static current.

7.7 Perspective: Choosing a Clocking Strategy

A crucial decision that must be made in the earliest phases of chip design is to select the appropriate clocking methodology. The reliable synchronization of the various operations occurring in a complex circuit is one of the most intriguing challenges facing the digital designer of the next decade. Choosing the right clocking scheme affects the functionality, speed, and power of a circuit.

A number of widely used clocking schemes were introduced in this chapter. The most robust and conceptually simple scheme is the two-phase master–slave design. The predominant approach is to use the multiplexer-based register, and to generate the two clock phases locally by simply inverting the clock. More exotic schemes such as the glitch register are also used in practice. However, these schemes require significant fine-tuning and must only be used in specific situations. An example of such is the need for a negative setup time to cope with clock skew.

The general trend in high-performance CMOS VLSI design is therefore to **use simple clocking schemes**, even at the expense of performance. Most automated design methodologies

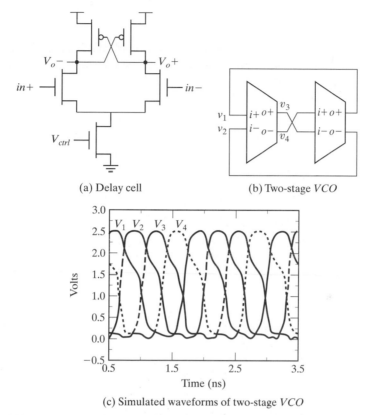

(a) Delay cell (b) Two-stage *VCO*

(c) Simulated waveforms of two-stage *VCO*

Figure 7-54 Differential delay element and *VCO* topology.

such as standard cell employ a single-phase, edge-triggered approach, based on static flip-flops. Nevertheless, the tendency towards simpler clocking approaches also is apparent in high-performance designs such as microprocessors. The use of latches between logic to improve circuit performance is common as well.

7.8 Summary

This chapter has explored the subject of sequential digital circuits. The following topics were discussed:

- The cross coupling of two inverters creates a *bistable* circuit, also known as a *flip-flop*. A third potential operation point turns out to be metastable; that is, any diversion from this bias point causes the flip-flop to converge to one of the stable states.
- A latch is a *level-sensitive* memory element that samples data on one phase and holds data on the other phase. A register, on the other hand, samples the data on the rising or falling edge. A register has three important parameters: *the setup time, the hold time, and the*

propagation delay. These parameters must be carefully optimized, because they may account for a significant portion of the clock period.

- Registers can be *static* or *dynamic*. A static register holds state as long as the power supply is turned on. It is ideal for memory that is accessed infrequently (e.g., reconfiguration registers or control information). Static registers use either multiplexers or overpowering to enable the writing of data.

- Dynamic memory is based on temporary charge storage on capacitors. The primary advantage is reduced complexity, higher performance, and lower power consumption. However, charge on a dynamic node leaks away with time, and dynamic circuits thus have a minimum clock frequency. Pure dynamic memory is hardly used anymore. Register circuits are made pseudostatic to provide immunity against capacitive coupling and other sources of circuit induced noise.

- Registers can also be constructed by using the *pulse or glitch concept.* An intentional pulse (using a one-shot circuit) is used to sample the input around an edge. Sense-amplifier-based schemes also are used to construct registers; they should be used as well when high-performance or low-signal-swing signalling is required.

- Choice of *clocking style* is an important consideration. Two-phase design can result in race problems. Circuit techniques such as C^2MOS can be used to eliminate race conditions in two-phase clocking. Another option is to use true single-phase clocking. However, the rise time of clocks must be carefully optimized to eliminate races.

- The combination of dynamic logic with dynamic latches can produce extremely fast computational structures. An example of such an approach, the NORA logic style, is very effective in *pipelined datapaths*.

- *Monostable structures* have only one stable state; thus, they are useful as pulse generators.

- *Astable multivibrators*, or oscillators, possess no stable state. The ring oscillator is the best-known example of a circuit of this class.

- *Schmitt triggers* display hysteresis in their dc characteristic and fast transitions in their transient response. They are mainly used to suppress noise.

7.9 To Probe Further

The basic concepts of sequential gates can be found in many logic design textbooks (e.g., [Mano82] and [Hill74]). The design of sequential circuits is amply documented in most of the traditional digital circuit handbooks. [Partovi01] and [Bernstein98] provide in-depth overviews of the issues and solutions in the design of high-performance sequential elements.

References

[Bernstein98] K. Bernstein et al., *High-Speed CMOS Design Styles*, Kluwer Academic Publishers, 1998.

[Dopperpuhl92] D. Dopperpuhl et al., "A 200 MHz 64-b Dual Issue CMOS Microprocessor," *IEEE Journal of Solid-State Circuits*, vol. 27, no. 11, Nov. 1992, pp. 1555–1567.

[Gieseke97] B. Gieseke et al., "A 600 MHz Superscalar RISC Microprocessor with Out-of-Order Execution," *IEEE International Solid-State Circuits Conference*, pp. 176–177, Feb. 1997.

[Gonçalves83] N. Gonçalves and H. De Man, "NORA: a racefree dynamic CMOS technique for pipelined logic structures," *IEEE Journal of Solid-State Circuits*, vol. SC-18, no. 3, June 1983, pp. 261–266.

[Hill74] F. Hill and G. Peterson, *Introduction to Switching Theory and Logical Design*, Wiley, 1974.

[Jeong87] D. Jeong et al., "Design of PLL-based clock generation circuits," *IEEE Journal of Solid-State Circuits*, vol. SC-22, no. 2, April 1987, pp. 255–261.

[Kozu96] S. Kozu et al., "A 100 MHz 0.4 W RISC Processor with 200 MHz Multiply-Adder, using Pulse-Register Technique," *IEEE ISSCC*, pp. 140–141, February 1996.

[Mano82] M. Mano, *Computer System Architecture*, Prentice-Hall, 1982.

[Montanaro96] J. Montanaro et al., "A 160-MHz, 32-b, 0.5-W CMOS RISC Microprocessor," *IEEE Journal of Solid-State Circuits*, pp. 1703–1714, November 1996.

[Mutoh95] S. Mutoh et al., "1-V Power Supply High-Speed Digital Circuit Technology with Multithreshold-Voltage CMOS," *IEEE Journal of Solid State Circuits*, pp. 847–854, August 1995.

[Partovi96] H. Partovi, "Flow-Through Latch and *Edge-Triggered* Flip-Flop Hybrid Elements," *IEEE ISSCC*, pp. 138–139, February 1996.

[Partovi01] H. Partovi, "Clocked Storage Elements," in *Design of High-Performance Microprocessor Circuits*, Chandakasan et al., ed., Chapter 11, pp. 207–233, 2001.

[Schmitt38] O. H. Schmitt, "A Thermionic Trigger," *Journal of Scientific Instruments*, vol. 15, January 1938, pp. 24–26.

[Suzuki73] Y. Suzuki, K. Odagawa, and T. Abe, "Clocked CMOS calculator circuitry," *IEEE Journal of Solid State Circuits*, vol. SC-8, December 1973, pp. 462–469.

[Yuan89] J. Yuan and Svensson C., "High-Speed CMOS Circuit Technique," *IEEE JSSC*, vol. 24, no. 1, February 1989, pp. 62–70.

A System Perspective

"Art, it seems to me, should simplify. That, indeed, is very nearly the whole of the higher artistic process; finding what conventions of form and what of detail one can do without and yet preserve the spirit of the whole."

Willa Sibert Cather,
On the Art of Fiction (1920).

"Simplicity and repose are the qualities that measure the true value of any work of art"

Frank Lloyd Wright.

CHAPTER

8

Implementation Strategies
for Digital ICs

Semicustom and structured design methodologies

ASIC and system-on-a-chip design flows

Configurable hardware

8.1   Introduction

8.2   From Custom to Semicustom and Structured-Array Design Approaches

8.3   Custom Circuit Design

8.4   Cell-Based Design Methodology
    8.4.1   Standard Cell
    8.4.2   Compiled Cells
    8.4.3   Macrocells, Megacells, and Intellectual Property
    8.4.4   Semicustom Design Flow

8.5   Array-Based Implementation Approaches
    8.5.1   Prediffused (or Mask-Programmable) Arrays
    8.5.2   Prewired Arrays

8.6   Perspective—The Implementation Platform of the Future

8.7   Summary

8.8   To Probe Further

8.1 Introduction

The dramatic increase in complexity of contemporary integrated circuits poses an enormous design challenge. Designing a multimillion-transistor circuit and ensuring that it operates correctly when the first silicon returns is a daunting task that is virtually impossible without the help of computer aids and well-established design methodologies. In fact, it has often been suggested that technology advancements might be outpacing the absorption bandwidth of the design community. This is articulated in Figure 8-1, which shows how IC complexity (in logic transistors) is growing faster than the productivity of a design engineer, creating a "design gap." One way to address this gap is to increase steadily the size of the design teams working on a single project. We observe this trend in the high-performance processor world, where teams of more than 500 people are no longer a surprise.

Obviously, this approach cannot be sustained in the long term—just imagine all the design engineers in the world working on a single design. Fortunately, about once in a decade we witness the introduction of a novel design methodology that creates a step function in design productivity, helping to bridge the gap temporarily. Looking back over the past four decades, we can identify a number of these productivity leaps. Pure custom design was the norm in the early integrated circuits of the 1970s. Since then, programmable logic arrays (PLAs), standard cells, macrocells, module compilers, gate arrays, and reconfigurable hardware have steadily helped to ease the time and cost of mapping a function onto silicon. In this chapter, we provide a description of some commonly used design implementation approaches. Due to the extensive nature of the field, we cannot be comprehensive—doing so would require a textbook of its own. Instead, we present *a user perspective* that provides a basic perception and insight into what is offered and can be expected from the different design methodologies.

The preferred approach to mapping a function onto silicon depends largely upon the function itself. Consider, for instance, the simple digital processor of Figure 8-2. Such a processor

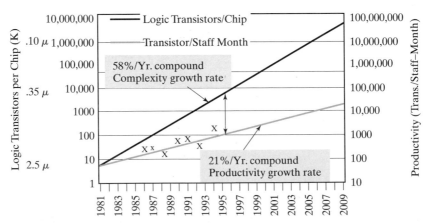

Figure 8-1   The design productivity gap. Technology (in logic transistors/chip) outpaces the design productivity (in transistors designed by a single design engineer per month). Source: SIA [SIA97].

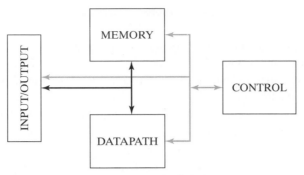

Figure 8-2   Composition of a generic digital processor. The arrows represent the possible interconnections.

could be the brain of a personal computer (PC), or the heart of a compact-disc player or cellular phone. It is composed of a number of building blocks that occur in one form or another in almost every digital processor:

- **The datapath** is the core of the processor; it is where all computations are performed. The other blocks in the processor are support units that either store the results produced by the datapath or help to determine what will happen in the next cycle. A typical datapath consists of an interconnection of basic combinational functions, such as logic (AND, OR, EXOR) or arithmetic operators (addition, multiplication, comparison, shift). Intermediate results are stored in registers. Different strategies exist for the implementation of datapaths—structured custom cells versus automated standard cells, or fixed hard-wired versus flexible field-programmable fabric. The choice of the implementation platform is mostly influenced by the trade-off between different design metrics such as area, speed, energy, design time, and reusability.
- **The control module** determines what actions happen in the processor at any given point in time. A controller can be viewed as a finite state machine (FSM). It consists of registers and logic, and thus is a sequential circuit. The logic can be implemented in different ways—either as an interconnection of basic logic gates (standard cells), or in a more structured fashion using programmable logic arrays (PLAs) and instruction memories.
- **The memory module** serves as the centralized data storage area. A broad range of different memory classes exist. The main difference between those classes is in the way data can be accessed, such as "read only" versus "read–write," sequential versus random access, or single-ported versus multiported access. Another way of differentiating between memories is related to their data-retention capabilities. Dynamic memory structures must be refreshed periodically to keep their data, while static memories keep their data as long as the power source is turned on. Finally, nonvolatile memories such as flash memories conserve the stored data even when the supply voltage is removed. A single processor might combine different memory classes. For example, random access memory can be used to store data, and read-only memory may store instructions.

• **The interconnect** network joins the different processor modules to one another, while the **input/output circuitry** connects to the outside world. For a long time, interconnections were an afterthought in the design process. Unfortunately, the wires composing the interconnect network are less than ideal and present a capacitive, resistive, and inductive load to the driving circuitry. As die sizes grow larger, the length of the interconnect wires also tends to grow, resulting in increasing values for these parasitics. Today, automated or structured design methodologies are being introduced that ease the deployment of these interconnect structures. Examples include *on-chip busses*, *mesh interconnect* structures, and even complete *networks on a chip*. Some components of the interconnect network typically are abstracted away on schematic block diagrams, such as the one shown in Figure 8-2, yet are of critical importance to the well-being of the design. These include the power- and clock-distribution networks. Early planning of these "service" networks can go a long way toward ensuring the correct operation of the integrated circuit.

The structure of Figure 8-2 may be repeated many times on a single die. Figure 8-3 shows an example of a *system on a chip*, which combines all the functions needed for the realization of a complete high-definition digital TV set. It combines two processors, memory units, specialized accelerators for functions such as MPEG (de)coding and data filtering, as well as a range of

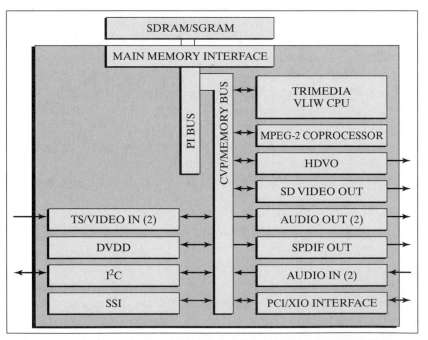

Figure 8-3   The "Nexperia" system on a chip [Philips99]. This single chip combines a general-purpose microprocessor core, a VLIW (very large instruction word) signal processor, a memory system, an MPEG coprocessor, multiple accelerator units, and input/output peripherals, as well as two system busses.

peripheral units. Other applications such as wireless transceivers or hard-disk read/write units may even include some sizable analog modules.

Choosing an effective implementation approach strongly depends upon the function of the modules under consideration. For example, memory units tend to be very regular and structured. A module compiler that stacks cells in an arraylike fashion is thus the preferred implementation approach. Controllers, on the other hand, tend to be unstructured, and other implementation approaches are desirable. The choice of the implementation strategy can have a tremendous effect on the quality of the final product. The challenge for the designer is to pick the style that meets the product specifications and constraints. What works well for one design may well be a disaster for another one.

Example 8.1 Trading Off Energy Efficiency and Flexibility

A design that embraces flexibility (or programmability) is very attractive from an application perspective. It allows for "late binding," in which the application can still be changed after the chip has gone to fabrication. Flexibility makes it possible to reuse a single design for multiple applications, or to upgrade the firmware of a component in the field, reducing the risk for the manufacturer. In contrast, a hard-wired component is totally fixed at manufacturing time and cannot be modified afterwards.

So, why not use flexible or programmable components for every possible design? As always, there is no free lunch. Flexibility comes at a price in both performance and energy efficiency. Providing programmability means adding overhead to implementation. For example, a programmable processor uses stored instructions and an instruction decoder to make a single datapath perform multiple functions. Most designers are not aware of the large cost of flexibility. The impact is illustrated in Figure 8-4, which compares the *energy*

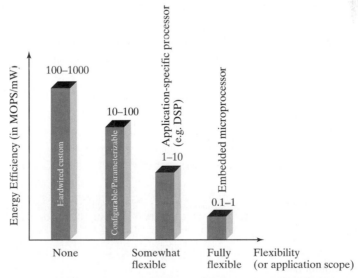

Figure 8-4 Trading off flexibility versus energy efficiency (in MOPS/mW or millions of operations per mJ of energy) for different implementation styles. The numbers were collected for a 0.25 μm CMOS process [Rabaey00].

efficiency—the number of operations that can be performed for a given amount of energy—of various implementation styles versus their *flexibility*—that is, the range of applications that can be mapped onto them. A staggering **three orders of magnitude** in variation can be observed. This clearly demonstrates that hard-wired or implementation styles with limited flexibility (such as configurable or parameterizable modules) are preferable when energy efficiency is a must.

In this and the following three chapters, we discuss, respectively, implementation techniques for random logic and controllers (this chapter), interconnect (Chapter 9), datapaths (Chapter 11), and memories (Chapter 12). Observe that the choice of the implementation approach can have a tremendous effect on the quality of the final product. Important aspects in the design of complex systems consisting of multiple blocks and thus deserving special attention are synchronization and timing (Chapter 10) and the power distribution network (Chapter 9). The distribution of clock signals and supply current has become one of the dominant problems in the design of state-of-the-art processors. A number of Design Methodology Inserts, interspersed between the chapters, address the design challenge posed by these complex components, and introduce the advanced design automation tools that are available to the designer. Inserts F, G, and H discuss design synthesis, verification, and test, respectively.

8.2 From Custom to Semicustom and Structured-Array Design Approaches

The viability of a microelectronics design depends on a number of (often) conflicting factors, such as performance in terms of speed or power consumption, cost, and production volume. For example, to be competitive in the market, a microprocessor has to excel in performance at a low cost to the customer. Achieving both goals simultaneously is only possible through large sales volumes. The high development cost associated with high-performance design is then amortized over many parts. Applications such as supercomputing and some defense applications present another scenario. With ultimate performance as the primary design goal, high-performance custom design techniques often are desirable. The production volume is small, but the cost of electronic parts is only a fraction of the overall system costs and thus not much of an issue. Finally, reducing the system size through integration, not performance, is the major objective in most consumer applications. Under these circumstances, the design cost can be reduced substantially by using advanced design-automation techniques, which compromise performance, but minimize design time. As noted in Chapter 1, the cost of a semiconductor device is the sum of two components:

- The *nonrecurring expense* (NRE), which is incurred only once for a design and includes the cost of designing the part.
- The *production cost per part*, which is a function of the process complexity, design area, and process yield.

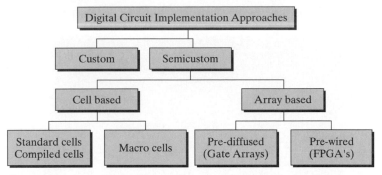

Figure 8-5 Overview of implementation approaches for digital integrated circuits (after [DeMicheli94]).

These economic considerations have spurred the development of a number of distinct implementation approaches ranging from high-performance, handcrafted design to fully programmable, medium-to-low performance designs. Figure 8-5 provides an overview of the different methodologies. In the sections that follow, we discuss first the custom design methodology, followed by the semicustom and array-based approaches.

8.3 Custom Circuit Design

When performance or design density is of primary importance, handcrafting the circuit topology and physical design seems to be the only option. Indeed, this approach was the only option in the early days of digital microelectronics, as is adequately demonstrated in the design of the Intel 4004 microprocessor (see Figure 8-5a). The labor-intensive nature of custom design translates into a high cost and a long *time to market*. Therefore, it can only be justified economically under the following conditions:

- The custom block can be reused many times (for example, as a library cell).
- The cost can be amortized over a large volume. Microprocessors and semiconductor memories are examples of applications in this class.
- Cost is not the prime design criterion, as it is in supercomputers or hypersupercomputers.

With continuous progress in the design-automation arena, the share of custom design reduces from year to year. Even in the most advanced high-performance microprocessors, such as the Intel Pentium® 4 processor (see Figure 8-6), virtually all portions are designed automatically using semicustom design approaches. Only the most performance-critical modules such as the phase locked-loops and the clock buffers are designed manually. In fact, library cell design is the only area where custom design still thrives today.

The amount of design automation in the custom-design process is minimal, yet some design tools have proven indispensable. In concert with a wide range of verification, simulation, extraction and modeling tools, layout editors, design-rule and electrical-rule checkers—as

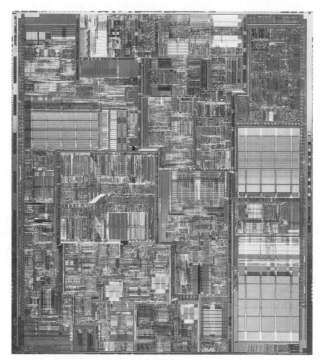

Figure 8-6 Chip microphotograph of Intel Pentium® 4 processor. It contains 42 million transistors, designed in a 0.18-µm CMOS technology. Its first generation runs at a clock speed of 1.5 GHz (Courtesy Intel Corp.).

described earlier in Design Methodology Insert A—are at the core of every custom-design environment. A excellent discussion of the opportunities and challenges of custom design can be found in [Grundman97].

8.4 Cell-Based Design Methodology

Since the custom-design approach proves to be prohibitively expensive, a wide variety of design approaches have been introduced over the years to shorten and automate the design process. This automation comes at the price of reduced integration density and/or performance. The following rule tends to hold: **the shorter the design time, the larger is the penalty incurred.** In this section, we discuss a number of design approaches that still require a full run through the manufacturing process for every new design. The *array-based design* approach discussed in the next section cuts the design time and cost even further by requiring only a limited set of extra processing steps or by eliminating processing completely.

The idea behind cell-based design is to reduce the implementation effort by *reusing* a limited library of cells. The advantage of this approach is that the cells only need to be designed and verified once for a given technology, and they can be reused many times, thus amortizing the design cost. The disadvantage is that the constrained nature of the library reduces the possibility of fine-tuning the design. Cell-based approaches can be partitioned into a number of classes depending on the granularity of the library elements.

8.4.1 Standard Cell

The standard-cell approach standardizes the design entry level at the logic gate. A library containing a wide selection of logic gates over a range of fan-in and fan-out counts is provided. Besides the basic logic functions, such as inverter, AND/NAND, OR/NOR, XOR/XNOR, and flip-flops, a typical library also contains more complex functions, such as AND-OR-INVERT, MUX, full adder, comparator, counter, decoders, and encoders. A design is captured as a schematic containing only cells available in the library, or is generated automatically from a higher level description language. The layout is then automatically generated. This high degree of automation is made possible by placing strong restrictions on the layout options. In the standard-cell philosophy, cells are placed in rows that are separated by routing channels, as illustrated in Figure 8-7. To be effective, this requires that all cells in the library have identical heights. The width of the cell can vary to accommodate for the variation in complexity between the cells. As illustrated in the drawing, the standard-cell technique can be intermixed with other layout approaches to allow for the introduction of modules such as memories and multipliers that do not adapt easily or efficiently to the logic-cell paradigm.

An example of a design implemented in an early standard-cell design style is shown in Figure 8-8a. A substantial fraction of the area is devoted to signal routing. The minimization of the interconnect overhead is the most important goal of the standard-cell placement and routing tools. One approach to minimizing the wire length is to introduce feed-through cells (Figure 8-7) that make it possible to connect between cells in different rows without having to route around a complete row. A far more important reduction in wiring overhead is obtained by adding more

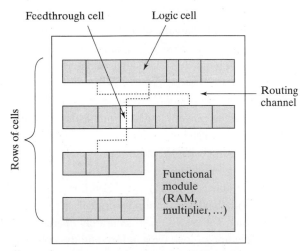

Figure 8-7 Standard-cell layout methodology.

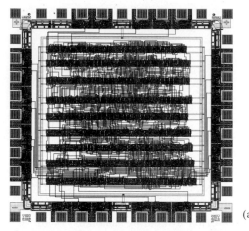
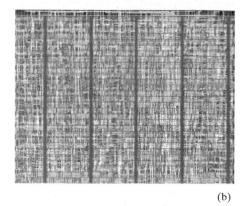

(a)   (b)

Figure 8-8   The evolution of standard-cell design. (a) Design in a three-layer metal technology. Wiring channels represent a substantial amount of the chip area. (b) Design in a seven-layer metal technology. Routing channels have virtually disappeared, and all interconnection is laid on top of the logic cells.

interconnect layers. The seven or more metal layers that are available in contemporary CMOS processes make it possible to all but eliminate the need for routing channels. Virtually all signals can be routed on top of the cells, creating a truly three-dimensional design. Figure 8-8b shows a fraction of a standard-cell design, implemented by using seven metal layers. The design achieves more than 90% density, which means that virtually all of the chip area is covered by logic cells, and that only a limited amount of the area is wasted for interconnect.

The design of a standard-cell library is a time-intensive undertaking that, fortunately, can be amortized over a large number of designs. Determining the composition of the library is a nontrivial task. A pertinent question is, Are we better off with a small library in which most cells have a limited fan-in, or is it more beneficial to have a large library with many versions of every gate (e.g., containing two-, three-, and four-input NAND gates, and different sizes for each of these gates)? Since the fan-out and load capacitance due to wiring are not known in advance, it used to be common practice to ensure that each gate had large current-driving capabilities, (i.e., employs large output transistors). While this simplifies the design procedure, it has a detrimental effect on area and power consumption. Today's libraries employ many versions of each cell, sized for different driving strengths, as well as performance and power consumption levels. It is left to the synthesis tool to select the correct cells, given speed and area requirements.

To make the library-based approach work, a detailed documentation of the cell library is an absolute necessity. The information should not only contain the layout, a description of functionality and terminal positioning, but it also must accurately characterize the delay and power consumption of the cell as a function of load capacitance and the input rise and fall times. Generating this information accounts for a large portion of the library generation effort. How to characterize logic and sequential cells is the topic of "Design Methodology Insert E."

Example 8.2   A Three-Input NAND-Gate Cell

To illustrate some of the preceding observations, the design of a three-input NAND standard-cell gate, implemented in a 0.18 μm CMOS technology, is depicted in Figure 8-9. The library actually contains five versions of the cell, supporting capacitive loads from 0.18 pF up to 0.72 pF and ranging in area from 16.4 μm^2 to 32.8 μm^2. The cell shown represents the low-performance, energy-efficient design corner, and uses high-threshold transistors to reduce leakage. The NMOS and PMOS transistors in the pull-down (-up) networks are both sized at a (W/L) ratio of approximately 8.

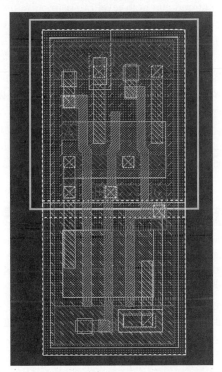

Figure 8-9   Three-input NAND standard cell (Courtesy ST Microelectronics).

Observe how the layout strategy follows the approach outlined in Figure D-2. Supply lines are distributed horizontally and shared between cells in the same row. Input signals are wired vertically using polysilicon. The input/output terminals are located throughout the cell body (as exemplified by the *pin* terminal in the layout drawing), in line with the over-the-cell wiring approach of today's standard-cell methodology.

The standard-cell approach has become immensely popular, and is used for the implementation of virtually all logic elements in today's integrated circuits. The only exceptions are when extreme high performance or low energy consumption is needed, or when the structure of the targeted function is very regular (such as a memory or a multiplier). The success of the standard-cell approach can be attributed to a number of developments, including the following:

- The **increased quality of the automatic cell placement and routing tools** in conjunction with the availability of multiple routing layers. In fact, it has been shown in a number of studies that the automated approach of today rivals if not surpasses manual design for complex, irregular logic circuits. This is a major departure from a couple of years ago, when automated layout carried a large overhead.

- **The advent of sophisticated *logic-synthesis* tools**. The logic-synthesis approach allows for the design to be entered at a high level of abstraction using Boolean equations, state machines, or register-transfer languages such as VHDL or Verilog. The synthesis tools automatically translate this specification into a gate netlist, minimizing a specific cost function such as area, delay, or power. Early synthesis tools—such as those used in the first half of the 1980s—focused mostly on two-level logic minimization. While this enabled automatic design mapping for the first time, it limited the area efficiency and the performance of the generated circuits. It is only with the arrival of *multilevel logic synthesis* in the late 1980s that automated design generation has really taken off. Today, virtually no designer uses the standard-cell approach without resorting to automatic synthesis. A more detailed description of the design synthesis process can be found in "Design Methodology Insert F" which follows this chapter.

Historical Perspective: The Programmable Logic Array

In the early days of MOS integrated circuit design, logic design and optimization was a manual and labor-intensive task. Karnaugh maps and Quine–McCluskey tables were the techniques of choice at that time. In the late 1970s, a first approach toward automating the tedious process of designing logic circuits emerged, triggered by two important developments:

- Rather than using the ad hoc approach to laying out logic circuits, a regular structured design approach was adopted called the *Programmable Logic Array* or PLA. This methodology enabled the automatic layout generation of two-level logic circuits, and, more importantly, it did so in a predictable fashion in terms of area and performance.

- The emergence of automated logic synthesis tools for two-level logic [Brayton84] made it possible to translate any possible Boolean expression into an optimized two-level (sum-of-products or product-of-sums) logic structure. Tools for the synthesis of sequential circuits followed shortly thereafter.

The idea of structured logic design gained a rapid foothold, and already in the mid-1980s it was adopted by major microprocessor design companies such as Intel and DEC. While PLAs are only sparingly used in today's semicustom logic design, the topic deserves some discussion (especially since PLAs might be poised for a come-back).

The concept is best explained with the aid of an example. Consider the following logic functions, for which we have transformed the equations into the sum-of-products format by using logic manipulations:

$$f_0 = x_0 x_1 + \overline{x}_2$$

$$f_1 = x_o x_1 x_2 + \overline{x}_2 + \overline{x_0 x_1}$$

$$(8.1)$$

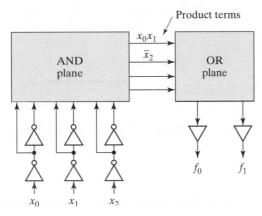

Figure 8-10 Regular two-level implementation of Boolean functions.

One important advantage of this representation is that a regular realization is easily conceived, as illustrated in Figure 8-10. A first layer of gates implements the AND operations—also called *product terms* or *minterms*—while a second layer realizes the OR functions, called the *sumterms*. Hence, a PLA is a rectangular macrocell, consisting of an array of transistors aligned to form rows in correspondence with product terms, and columns in correspondence with inputs and outputs. The input and output columns partition the array into two subarrays, called AND and OR planes, respectively.

The schematic of Figure 8-10 is not directly realizable since single-layer logic functions in CMOS are always inverting. With a few simple Boolean manipulations, Eq. (8.1) can be rewritten into a NOR–NOR format:

$$\overline{f}_0 = \overline{\overline{(\overline{x}_0 + \overline{x}_1)} + \overline{x}_2}$$

$$\overline{f}_1 = \overline{\overline{(\overline{x}_0 + \overline{x}_1 + \overline{x}_2)} + \overline{x}_2 + \overline{(x_0 + \overline{x}_1)}}$$

$$(8.2)$$

Problem 8.1 Two-Level Logic Representations

It is equally conceivable to represent Eq. (8.1) in a NAND–NAND format. In general, the NOR–NOR representation is preferred due to the prohibitively slow speed of large fan-in NAND gates. The NAND–NAND configuration is very dense, however, and thus can help to reduce power consumption. Derive the NAND–NAND representation for the example of Eq. (8.2).

Translating a set of two-level logic functions into a physical design now boils down to a "programming" task—that is, deciding where to place transistors in both the AND and the OR planes. This task is easily automated—hence, the early success of PLAs. An automatically generated PLA implementation of the logic functions described by Eq. (8.2) is shown in Figure 8-11. Unfortunately, the regular structure, while predictable,

V_{DD} **AND plane**   **OR plane**   GND

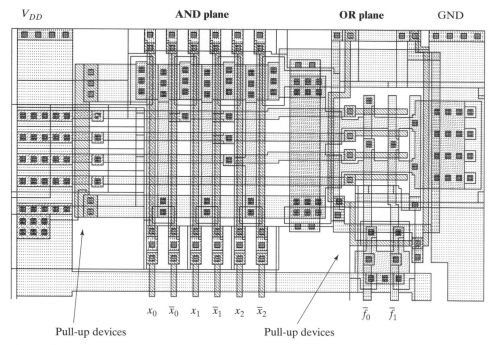

x_0 $\bar{x}_0$ x_1 $\bar{x}_1$ x_2 $\bar{x}_2$ $\bar{f}_0$ $\bar{f}_1$

Pull-up devices   Pull-up devices

Figure 8-11   PLA layout implementing Eq. (8.2).

brings with it a lot of overhead in area and delay (as is quite visible in the layout), which was its ultimate demise in the semicustom design world. Those who are curious on how these AND and OR planes are actually implemented must wait until we get to Chapter 12, where we discuss the transistor-level implementation of PLAs. ∎

8.4.2 Compiled Cells

The cost of implementing and characterizing a library of cells should not be underestimated. Today's libraries contain from several hundred to more than a thousand cells. These cells have to be redesigned with every migration to a new technology. Moreover, changes happen during the development of a single technology generation. For example, minimum metal widths or contact rules often are changed to improve yield. As a result, the complete library has to be laid out and characterized again. In addition, even an extensive library has the disadvantage of being discrete, which means that the number of design options is limited. When targeting performance or power, customized cells with optimized transistor sizes are attractive. With the increased impact of interconnect load, providing cells with adjusted driver sizes is an absolute necessity from both a performance and a power perspective [Sylvester98]—hence, the quest for automated (or compiled) cell generation.

A number of automated approaches have been devised that generate cell layouts on the fly, given the transistor netlists, but high-quality automatic cell layout has remained elusive. Earlier approaches relied on fixed topologies. Later approaches allowed for more flexibility in the transistor placement (e.g., [Hill85]). Layout densities close to what can be accomplished by a human designer are now within reach, and a number of cell-generation tools are commercially available—for example [Cadabra01, Prolific01]:

Example 8.3 Automatic Cell Generation

The flow of a typical cell-generation process is illustrated with the example of a simple inverter (using the Abracad tool [Cadabra01]).

- The cell schematics are developed first. The Spice netlist is the starting point for the automatic layout generation. The generator examines the netlist and starts with transistor geometries. In case of a CMOS inverter, the cell contains just two transistors (see Figure 8-12a) .
- The tool proceeds along the same lines that a designer would follow. The transistors are placed in a cell architecture with predefined topology rules (Figure 8-12b). This architecture is common for all the cells in the library, including the cell height, power rails, pin placements, routing and contact styles.
- The cell is routed symbolically (Figure 8-12c).
- The routing is rearranged, and the cell is compacted to meet design rules and library preferences (Figure 8-12d).
- The final step cleans the cell of any remaining design rule errors and produces the final layout (Figure 8-12e).

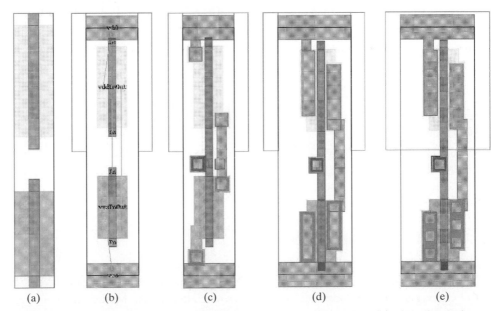

(a) (b) (c) (d) (e)

Figure 8-12 Automatic cell layout (a) initial transistor geometries, (b) placed transistors with flylines indicating intended interconnections, (c) initially routed cell, and (d) compacted cell, (e) finished cell.

8.4.3   Macrocells, Megacells and Intellectual Property

Standardizing at the logic-gate level is attractive for random logic functions, but it turns out to be inefficient for more complex structures such as multipliers, data paths, memories, and embedded microprocessors and DSPs. By capturing the specific nature of these blocks, implementations can be obtained that outperform the results of the standard ASIC design process by a wide margin. Cells that contain a complexity that surpasses what is found in a typical standard-cell library are called *macrocells* (or, sometimes, *megacells*). Two types of macrocells can be identified:

The Hard Macro   represents a module with a given functionality and a predetermined physical design. The relative location of the transistors and the wiring within the module is fixed. In essence, a hard macro represents a custom design of the requested function. In some cases, the macro is parameterized, which means that versions with slightly different properties are available or can be generated. Multipliers and memories are examples: A hard multiplier macro may not only generate a 32×16 multiplier, but also an 8×8 one.

The advantage of the hard macro is that it brings with it all the good properties of custom design: dense layout, and optimized and predictable performance and power dissipation. By encapsulating the function into a macromodule, it can be reused over and over in different designs. This reuse helps to offset the initial design cost. The disadvantage of the hard macro is that it is hard to port the design to other technologies or to other manufacturers. For every new technology, a major redesign of the block is necessary. For this reason, hard macros are used less and less, and are employed mainly when the automated generation approach is far inferior or even impossible. Embedded memories and microprocessors are good examples of hard macros. They typically are provided by the IC manufacturer (who also provides the standard cell library), or the semiconductor vendor who has a particularly desirable function to offer (such as a standard microprocessor or DSP).

In the case of a macro that can be parameterized, a generator called the *module compiler* is used to create the actual physical layout. Regular structures such as PLAs, memories, and multipliers are easily constructed by abutting predesigned leaf cells in a two-dimensional array topology. All interconnections are made by abutment, and no or little extra routing is needed if the cells are designed correctly, which minimizes the parasitic capacitance. The PLA of Figure 8-11 is an example of such a configuration. The whole array can be constructed with a minimal number of cells. The generator itself is a simple software program that determines the relative positioning of the various leaf cells in the array.

Example 8.4   A Memory Macromodule

Figure 8-13 shows an example of a "hard" memory macrocell. The 256×32 SRAM block is generated by a parameterizable module generator. Besides creating the layout, the generator also provides accurate timing and power information. Modern memory generators also include an amount of redundancy to deal with defects.

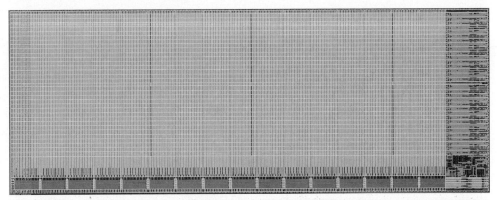

Figure 8-13 Parameterizable memory "hard" macrocell. This particular instance stores 256 × 32 (or 8192) bits. The decoders are located on the bottom. All eight address bits, as well as the 32 data input and output ports are placed on the right side of the cell. The total area of the memory module, implemented in a 0.18-μm CMOS technology, equals a mere 0.094 mm^2 (courtesy ST Microelectronics).

A Soft Macro represents a module with a given functionality, but without a specific physical implementation. The placement and the wiring of a soft macro may vary from instance to instance. This means that the timing data can only be determined after the final synthesis and placement and routing steps—in other words, the process is unpredictable. Yet, through intrinsic knowledge of the internal structure of the module, and by imposing precise timing and placement constraints on the physical generation process, soft macros most often succeed in offering well-defined timing guarantees. While stepping away from the advantages of the custom design process and relying on the semicustom physical design process, soft macros have the major advantage that they can be ported over a wide range of technologies and processes. This amortizes the design effort and cost over a wide set of designs.

Soft-macrocell generators come in different styles depending on the type of function they target. Virtually all of them can be classified as *structural generators*: Given the desired function and values for the requested parameters, the generator produces a netlist, which is an enumeration of the standard cells used and their interconnections. It also provides a set of timing constraints that the placement and routing tools should meet. The advantage of this approach is that the generator exploits its knowledge of the function under consideration to come up with clever structures that are more efficient than what logic synthesis would produce. For example, the design of fast and area-efficient multipliers has been the topic of decades of research.[1] The multiplier generator just incorporates the best of what the multiplier literature has to offer into an automated generation tool.

---

[1]Multiplier design is explained more thoroughly in Chapter 11, which discusses the design of arithmetic structures.

Example 8.5   Multiplier Macromodule

Two instances of an 8×8 multiplier module with different aspect ratios are shown in Figure 8-14. The modules are generated using the ModuleCompiler tool from Synopsys [ModuleCompiler01]. As can observed from the layout, a common standard-cell methodology is used to generate the physical artwork. The contribution of the macrocell generator is to translate the compact input description into an optimized connection of standard cells that meets the timing constraints. This "soft" approach has the advantage that modules with different aspect ratios can easily be generated. Also, porting between different manufacturing technologies is relatively easy.

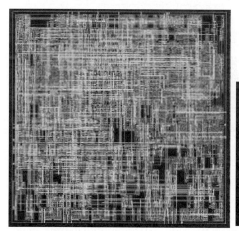
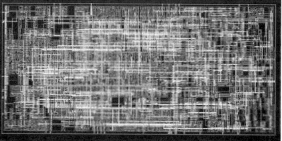

```
string mat = "booth";
directive (multtype = mat);
output signed [16] Z = A * B;
```

Figure 8-14   Multiplier "soft" macro modules. Both layouts implement an 8×8 booth multiplier, but with different aspects ratios. The compact input description to the compiler is shown in the gray box on top.

The availability of macromodules has substantially changed the semicustom design landscape in the 21st century. With the complexity of ICs going up exponentially, the idea of building every new IC from scratch becomes an uneconomic and nonplausible proposition. More and more, circuits are being built from reusable building blocks of increasing complexity and functionality. Typically, these modules are acquired from third-party vendors, who make the functions available through royalty or licensing agreements. Macromodels distributed in this style are called *intellectual property* (or IP) modules. This approach is somewhat comparable to the software world, where a large programming project typically makes intensive use of reusable software libraries. Good examples of commonly available intellectual property modules are embedded microprocessors and microcontrollers, DSP processors, bus interfaces such as PCI, and several special-purpose functional modules such as FFT and filter modules for DSP applications, error-correction coders for wireless communications, and MPEG decoding and encoding for video. Obviously, for an IP module to be useful, it has to not only deliver the hardware, but it also has to come with the appropriate software tools (such as compilers and debuggers for embedded processors), prediction models, and test benches. The latter are quite important because they represent the only means for the end user to verify that the module delivers the promised functionality and performance.

The design of a system on a chip is rapidly becoming an exercise in reuse at different levels of granularity. At the lowest level, we have the standard cell library; at a level higher, we have the functional modules such as multipliers, datapaths and memories; next, we have the embedded processors; and finally, the application-specific megacells. With more and more of the system functionality migrating onto a single die, it is not surprising to see that a typical ASIC consists of a blend of design styles and modules, embedding a number of hard or soft macrocells within a sea of standard cells.

Example 8.6   A Processor for Wireless Communications

Figure 8-15 shows an integrated circuit implementing the protocol stack for a wireless indoor communication system [Silva01]. The majority of the area is occupied by the embedded microprocessor (the Tensilica Xtensa processor [Xtensa01]) and its memory system. This processor allows for a flexible implementation of the higher levels of the protocol stack (Application/Network), and enables changes in the functionality of the chip, even after fabrication. The memory modules are generated using module compilers provided by the process vendor. The processor core itself is automatically generated from a higher level description in Verilog, and uses standard cells for its physical implementation. The advantage of using the "soft-core" approach is that the processor instruction set can be

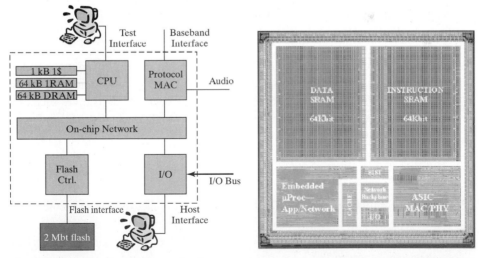

Figure 8-15 Wireless communications processor—an example of a hybrid ASIC design methodology. The processor combines an embedded microprocessor and its memory system with dedicated hardware accelerators and I/O modules. Observe also the on-chip network module [Silva01].

tailored to the application, and that the processor itself can easily be ported to different technologies and fabrication processes.

Implementing the computation-intensive parts of the protocol (MAC/PHY) on the microprocessor would require very high clock speeds and would unnecessarily increase the power dissipation of the chip. Fortunately, these functions are fixed and typically do not require a flexible implementation. Hence, they are implemented as an accelerator module in standard cells. The hard-wired implementation accomplishes the task of implementing a huge number of computations at a relatively low power level and clock frequency. The designer of a system on a chip is continuously faced with the challenge of deciding what is more desirable—after-the-fabrication flexibility versus higher performance at lower power levels. Fortunately, tools are emerging that help the designer to explore the overall design space and analyze the trade-offs in an informed fashion [Silva01]. Observe also that the chip contains a set of I/O interfaces, as well as an embedded network module, which helps to orchestrate the traffic between processor and the various accelerator and I/O modules.

The generation process of a macro module depends on the hard or soft nature of the block, as well as the level of design entry. In the following sections, we briefly discuss some commonly: used approaches.

8.4.4   Semicustom Design Flow

So far, we have defined the components that make up the cell-based design methodology. In this section, we discuss how it all comes together. Figure 8-16 details the traditional sequence of steps to design a semicustom circuit. The steps of what we call the design flow are enumerated in the figure, with a brief description of each:

1. **Design Capture** enters the design into the ASIC design system. A variety of methods can be used to do so, including schematics and block diagrams; hardware description languages (HDLs) such as VHDL, Verilog, and, more recently, C-derivatives such as SystemC; behavioral description languages followed by high-level synthesis; and imported intellectual property modules.
2. **Logic Synthesis** tools translate modules described using an HDL language into a *netlist*. Netlists of reused or generated macros can then be inserted to form the complete netlist of the design.
3. **Prelayout Simulation and Verification**. The design is checked for correctness. Performance analysis is performed based on estimated parasitics and layout parameters. If the design is found to be nonfunctional, extra iterations over the design capture or the logic synthesis are necessary.
4. **Floor Planning**. Based on estimated module sizes, the overall outlay of the chip is created. The global-power and clock-distribution networks also are conceived at that time.

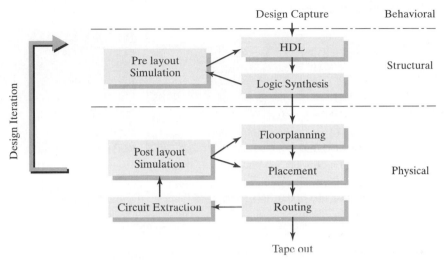

Figure 8-16 The Semicustom (or ASIC) design flow.

5. **Placement**. The precise positioning of the cells is decided.
6. **Routing**. The interconnections between the cells and blocks are wired.
7. **Extraction**. A model of the chip is generated from the actual physical layout, including the precise device sizes, devices parasitics, and the capacitance and resistance of the wires.
8. **Postlayout Simulation and Verification**. The functionality and performance of the chip is verified in the presence of the layout parasitics. If the design is found to be lacking, iterations on the floorplanning, placement, and routing might be necessary. Very often, this might not solve the problem, and another round of the structural design phase might be necessary.
9. **Tape Out**. Once the design is found to be meeting all design goals and functions, a binary file is generated containing all the information needed for mask generation. This file is then sent out to the ASIC vendor or foundry. This important moment in the life of a chip is called *tape out*.

While the design flow just described has served us well for many years, it was found to be severely lacking once technology reached the 0.25-μm CMOS boundary. With design technology proceeding into the deep submicron region, layout parasitics—especially from the interconnect—are playing an increasingly important role. The prediction models used by the logic and structural synthesis tools have a hard time providing accurate estimates for these parasitics. The chances that the generated design meets the timing constraints at the first try are thus very small (Figure 8-17a). The designer (or design team) is then forced to go through a number of costly iterations of synthesis followed by layout generation until an acceptable artwork that meets the timing constraints is obtained (Figure 8-17b and c). Each of these iterations may take several days—just routing a complex chip can take a week on the most advanced computers! The

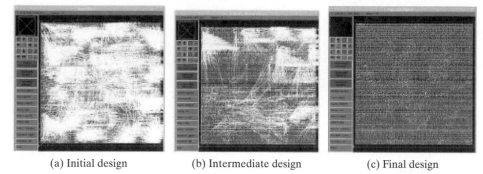

(a) Initial design   (b) Intermediate design   (c) Final design

Figure 8-17 The timing closure process. The white lines indicate nets with timing violations. In each iteration of the design process, timing errors are removed by optimizing the logic, by insertion of buffers, by constraining the placement, or by streamlining the routing until an error-free design is obtained [Avanti01].

number of needed iterations continues to grow with the scaling of technology. This problem, called *timing closure*, made it obvious that new solutions and a change in design methodology were required.

The common answer is to create a tighter integration between the logical and physical design processes. If the logic synthesis tool, for example, also performs some part of the placement—or directs the placement—more precise estimates of the layout parameters can be obtained. Figure 8-18 shows an example of a design environment that merges RTL synthesis with first-order placement and routing. The resulting netlist is then fed into an optimization tool that performs the detailed placement and routing, while guaranteeing the timing constraints are met. While this approach has shown to be quite successful in reducing the number of design iter-

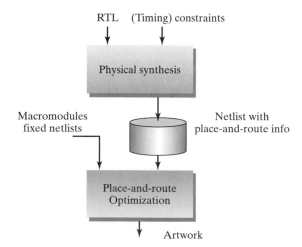

Figure 8-18 Integrated synthesis place-and-route reduces the number of iterations to reach timing closure in deep submicron.

ations, it throws quite a challenge at the design-tool developers. With the number of parasitic effects increasing with every round of technology scaling, the design optimization process that must take all this into account becomes exponentially complex as well. As a result, other approaches might be required as well. In the coming chapters, we will highlight "design solutions" that can help to alleviate some of these problems. An example is the use of regular and predictable structures, both at the logical and the physical level.

8.5 Array-Based Implementation Approaches

While design automation can help reduce the design time, it does not address the time spent in the manufacturing process. All of the design methodologies discussed thus far require a complete run through the fabrication process.This can take from three weeks to several months, and it can substantially delay the introduction of a product. Additionally, with ever-increasing mask costs, a dedicated process run is expensive, and product economics must determine if this is a viable route.

Consequently, a number of alternative implementation approaches have been devised that do not require a complete run through the manufacturing process, or they avoid dedicated processing completely. These approaches have the advantage of having a lower NRE (nonrecurring expense) and are, therefore, more attractive for small series. This comes at the expense of lower performance, lower integration density, or higher power dissipation.

8.5.1 Prediffused (or Mask-Programmable) Arrays

In this approach, batches of wafers containing arrays of primitive cells or transistors are manufactured by the vendors and stored. All the fabrication steps needed to make transistors are standardized and executed without regard to the final application.

To transform these uncommitted wafers into an actual design, only the desired interconnections have to be added, determining the overall function of the chip with only a few metallization steps. These layers can be designed and applied to the premanufactured wafers much more rapidly, reducing the turnaround time to a week or less.

This approach is often called the *gate-array* or the *sea-of-gates* approach, depending on the style of the prediffused wafer. To illustrate the concept, consider the gate-array primitive cell shown in Figure 8-19a. It comprises four NMOS and four PMOS transistors, polysilicon gate connections, and a power and ground rail. There are two possible contact points per diffusion area and two potential connection points for the polysilicon strips. We can turn this cell, which does not implement any logic function so far, into a real circuit by adding some extra wires on the metal layer and contact holes. This is illustrated in Figure 8-19b, where the cell is turned into a four-input NOR gate.

The original *gate-array* approach[2] places the cells in rows separated by wiring channels, as shown in Figure 8-20a. The overall look is similar to the traditional standard-cell technique. With the advent of extra metallization layers, the routing channels can be eliminated, and routing can

---

[2]This approach is often called the *channeled* gate array.

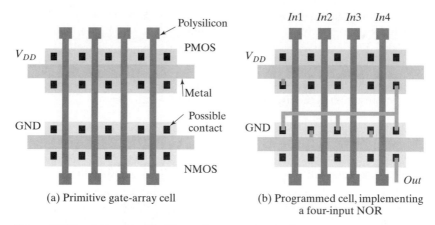

(a) Primitive gate-array cell

(b) Programmed cell, implementing a four-input NOR

Figure 8-19   An example of the gate-array approach.

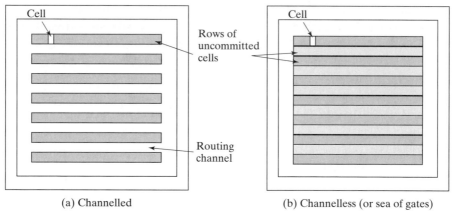

(a) Channelled

(b) Channelless (or sea of gates)

Figure 8-20   Gate-array architectures.

be performed on top of the primitive cells—occasionally leaving a cell unused. This channelless architecture, also called *sea of gates* (Figure 8-20b), yields an increased density, and makes it possible to achieve integration levels of millions of gates on a single die. Another advantage of the sea-of-gates approach is that it customizes the contact layer between metal-1 and diffusion and/or polysilicon, in contrast to the standard gate-array approach where the contacts are predefined (see Figure 8-19a). This extra flexibility leads to a further reduction in cell size.

The primary challenge when designing a gate-array (or sea-of-gates) template is to determine the composition of the primitive cell and the size of the individual transistors. A sufficient number of wiring tracks must be provided to minimize the number of cells wasted to interconnect. The cell should be chosen so that the prefabricated transistors can be utilized to a maximal extent over a wide range of designs. For example, the configuration of Figure 8-19 is well suited for the realization of four-input gates, but wastes devices when implementing two-input gates.

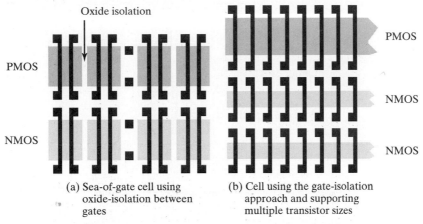

(a) Sea-of-gate cell using oxide-isolation between gates

(b) Cell using the gate-isolation approach and supporting multiple transistor sizes

Figure 8-21 Examples of sea-of-gates primitive cells (from [Veendrick92]).

Multiple cells are needed when implementing a flip-flop. A number of alternative cell structures are pictured in Figure 8-21 in a simplified format. In one approach, each cell contains a limited number of transistors (four to eight). The gates are isolated by means of *oxide isolation* (also called *geometry isolation*). The "dog-bone" terminations on the poly gates allow for denser routing. A second approach provides long rows of transistors, all sharing the same diffusion area. In this architecture, it is necessary to electrically turn off some devices to provide isolation between neighboring gates by tying NMOS and PMOS transistors to *GND* and V_{DD}, respectively. This technique is called *gate isolation*. This approach wastes a number of transistors to provide the isolation, but provides an overall higher transistor density.

Figure 8-22 shows the base cell for a gate-isolated gate array (from [Smith97]). The cell is one routing track wide, and contains one *p*-channel and one *n*-channel transistor. Also shown is a base cell containing all possible contact positions. There is room for 21 contacts in the vertical direction, which means that the cell has a height of 21 tracks.

It is worth observing that the cell in Figure 8-21b provides two rows of smaller NMOS transistors that can be connected in parallel if needed. Smaller transistors come in handy when implementing pass-transistor logic or memory cells. Sizing the transistors in the cells is a clear challenge. Due to the interconnect-oriented nature of the array-based design methodology, the propagation delay is generally dominated by the interconnect capacitance. This seems to favor larger device sizes that cause a larger area loss when unused. On the other hand, it is possible to construct larger transistors by putting several smaller devices in parallel.

Mapping a logic design onto an array of cells is a largely automated process, involving logic synthesis followed by placement and routing. The quality of these tools has an enormous impact on the final density and performance of a sea-of-gates implementation. Utilization factors in sea-of-gates structures are a strong function of the type of application being implemented. Utilization factors of nearly 100% can be obtained for regular structures such as memories. For

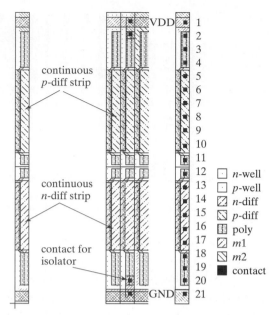

Figure 8-22   Base cell of gate-isolated gate array (from [Smith97]).

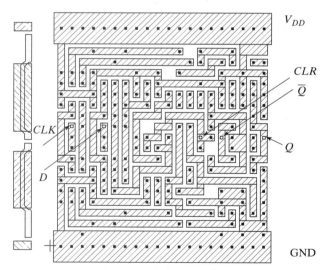

Figure 8-23   Flip-flop implemented in a gate-isolated gate-array library.
The base cell is shown on the left (from [Smith97]).

other applications, utilization factors can be substantially lower (< 75%), due largely to wiring restrictions. Figure 8-23 shows an example of a flip-flop macrocell, implemented in a gate-isolated, gate-array library.

Similar to the scenarios unfolding in the standard-cell arena, designers of sea-of-gate arrays discovered that a design with a large number of gates also has large memory needs. Implementing these memory cells on top of the gate-array base-cells is possible, but not very efficient. A more efficient approach is to set aside some area for dedicated memory modules. The mixing of gate arrays with fixed macros is called the *embedded gate-array* approach. Other modules such as microprocessor and microcontrollers are also ideal candidates for embedding.

Example 8.7 Sea-of-Gates

An example of a sea-of-gates implementation is shown in Figure 8-24. The array has a maximum capacity of 300 K gates and is implemented in a 0.6-μm CMOS technology. The upper left part of the array implements a memory subsystem, which results in a regular modular layout. The rest of the array implements random logic.

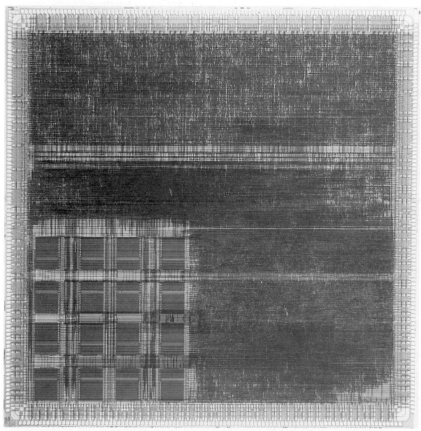

Figure 8-24 Gate-array die microphotograph (LEA300K) (Courtesy of LSI Logic.)

Design Consideration—Gate Arrays versus Standard Cells

In the 1980s and 1990s, when the majority of the chips were less than 50,000 gates, design cycles often could be measured in weeks or a few months. The two- or three-week savings in turnaround time for a gate-array design was then a significant portion of the total design cycle, more than enough to offset the additional die size. With today's deep-submicron processes and multimillion-gate complexities come longer design times, and the small reduction in turnaround time is no longer much of an issue. Furthermore, metallization has become the most time-consuming and yield-impacting part of the semiconductor manufacturing process, reducing further the advantage that gate arrays had to offer. Consequently, gate arrays have lost a lot of their luster. Another alternative for rapid prototyping—the prewired arrays discussed in the next section—has arisen, and it has taken a large portion out of the gate-array market.

Still, beware of dismissing the idea of the mask-programmable logic module as a thing of the past. A regular and fixed layout style has the advantage that load factors, wiring parasitics, and cross-coupling noise are easily and accurately estimated. This is in contrast to the standard-cell approach, where these values are ultimately only known after placement, routing, and extraction. One may consider populating sections of a large chip with a regular logic array consisting of uncommitted (prediffused) logic cells superimposed by a wiring grid. The actual programming of the module is performed by placing vias at predefined positions. As shown in Figure 8-25, the use of a via-programmable cross-point switch makes it possible to overlay a wide variety of wiring patterns on a regular repetitive wiring grid. It is the opinion of the authors that prediffused arrays have quite some life left into them.

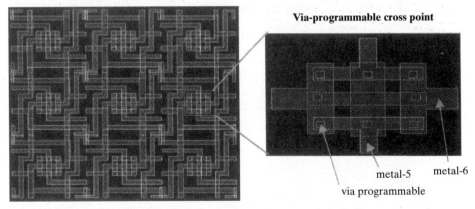

Via-programmable cross point

metal-5   metal-6

via programmable

Figure 8-25 Via-programmable gate array. Vias are used to dedicate a generic wiring grid to a specific wiring pattern, resulting in predictable arrays [Pileggi02].

■

8.5.2 Prewired Arrays

While the prediffused arrays offer a fast road to implementation, it would be even more efficient if dedicated manufacturing steps could be avoided altogether. This leads to the concept of the preprocessed die that can be programmed in the field (i.e., outside the semiconductor foundry) to implement a set of given Boolean functions. Such a programmable, prewired array of cells is called a *field-programmable gate array (FPGA)*. The advantage of this approach is that the manufacturing process is completely separated from the implementation phase and can be amortized over a large number of designs. The implementation itself can be performed at the user site with negligible turnaround time. The major drawback of this technique is a loss in performance and design density, compared with the more customized approaches.

Two main issues have to be addressed when attempting to implement a set of Boolean functions on top of a regular array of cells without requiring any processing steps:

1. How do we implement "programmable" logic—that is, logic that can committed to perform any possible Boolean function?
2. How and where do we store the *program*—also called the configuration—that dedicates the programmable array to a certain logic function?

The answer to the second question depends on the memory technology used. Since memory technology is the topic of a later chapter, we limit ourselves here to a high-level overview. In general, three different techniques can be identified:

- **The write-once or fuse-based FPGA.** The logic array is committed to a particular function by blowing "fuses" or by short-circuiting "antifuses." A fuse is a connection element that is short-circuited by default. A large current causes it to blow, and then it becomes an open circuit. The antifuse has the opposite behavior. An example of an antifuse implementation is shown in Figure 8-26 [El-Ayat89]. The advantage of the write-once approach is that the area overhead of the program memory (i.e., the fuses) is very small. But it has the important disadvantage of being *one-time programmable*. Circuit corrections or extensions are not possible, and new components are required for every design change.
- **The nonvolatile FPGA.** The program is stored in nonvolatile memory, which is memory that retains its value even when the supply voltage is turned off. Examples include EEPROM (*Electrically Erasable Programmable Read-Only Memory*) or Flash memories. Once programmed, the logic remains functional and fixed until a new programming round. The disadvantage of this approach is that nonvolatile memories require special steps in the manufacturing process, such as the deposition of ultrathin oxides. Also, high voltages

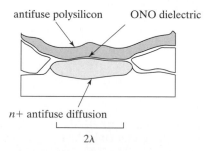

Figure 8-26   Example of antifuse. A 10-nm-thin layer (< 10 nm) of ONO (oxide–nitride–oxide) dielectric is deposited between conducting polysilicon and diffusion layers. The circuit is open by default, unless a large programming current is forced through it. This causes the dielectric to melt, and a permanent connection with fixed resistance is formed (from [Smith97]).

(> 10 V) are needed for the programming and erasure of the memory cells. Generating these high voltages and distributing them through the logic array adds extra complexity to the design.

- **The Volatile or RAM-Based FPGA.** This popular approach to programming the logic array employs volatile static RAM (random-access memory) cells for the storage of the program. Since these memories lose their stored contents when the FPGA is powered down, a reloading of the configuration from an external permanent memory is necessary every time the part is turned on. To program the component at start-up time, programming data is shifted serially into the part over a single line (or pin). For all practical purposes, one can consider the FPGA RAM cells to be configured as a giant shift register during that period. Once all memories are loaded, normal execution is started. The configuration time is proportional to the number of programmable elements. This can become excessive for today's larger FPGAs, which often feature more than one million gates. Recent parts therefore rely more and more on a parallel programming interface, allowing multiple cells to be programmed at the same time.

  In contrast to their nonvolatile counterparts, volatile FPGAs do not have special manufacturing process requirements, and can be implemented in a regular CMOS process. In addition, designers can reuse chips during prototyping. Logic can be modified and upgraded once deployed in the field—a customer can be sent a new configuration file to upgrade the chip, instead of sending a new chip. In addition, logic can be dynamically modified on the fly during execution. The latter approach is called *reconfiguration*, and it became quite popular in the late 1990s. In some sense, this brings a paradigm that was extremely successful in the world of programming (as embodied by the microprocessor) to the domain of logic design.

As for the first question, the answer is somewhat more extensive. Implementing a complex circuit in a programmable fashion requires that both the logic functions as well as the interconnect between them are realized in a configurable fashion. In the coming sections, we first discuss different ways of implementing programmable logic, followed by an overview of programmable interconnection. Finally, we detail a number of specific ways of putting the two together.

Programmable Logic

Similar to the situation in semicustom design, two fundamentally different approaches towards programmable logic are currently in vogue: array based and cell based.

Array-Based Programmable Logic   Earlier we discussed how a *programmable logic array* (PLA) implements arbitrary Boolean logic functions in a regular fashion (see page 388). A similar approach can be applied to field-programmable devices as well. Consider, for example, the logic structure of Figure 8-27. A circle (o) at an intersection indicates a programmable connection—that is, an interconnect point that is either enabled or not. An inspection of the diagram reveals that it is equivalent to a PLA, where both the AND and OR planes can be programmed by selectively enabling connections. This approach allows for the implementation of arbitrary

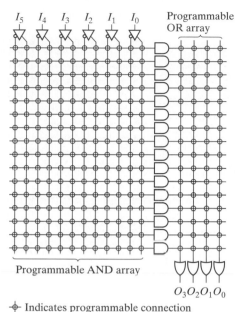

Figure 8-27   Fuse-programmable logic array (PLA).

logic functions in a two-level *sum-of-products* format. The AND plane creates the required min-terms, while the OR plane takes the sum of a selected set of products to form the outputs. To include a given input variable (for instance, I_1) in a specific minterm, just close the switch at the intersection of the input signal and the minterm. Similarly, a minterm is included into an output by closing the appropriate connection in the OR plane. The functionality of PLA is restricted by the number of inputs, outputs, and minterms.

We can envision variations on this theme, some of which are represented in Figure 8-28. The dot (•) at the intersection of two lines represents a nonfusible, hard-wired link. The first structure represents the PROM architecture, in which the AND plane is fixed and enumerates all possible minterms. The second structure, called a *programmable array logic device* (PAL), is located at the other end of the spectrum, where the OR plane is fixed, and the AND plane is programmable. The PLA architecture is the most generic one for the implementation of arbitrary logic functions. The PROM and PAL structures, on the other hand, trade off flexibility for density and performance. Which structure to select depends strongly on the nature of the Boolean functions to be implemented. All these approaches are generally classified under the common term of *programmable logic devices* (or PLDs).

The single-array architecture of the PLA, PROM, and PAL structures in Figure 8-27 and Figure 8-28 becomes less attractive in the era of higher integration density. First of all, implementing very complex logic functions on a single, large array results in a loss of programming density and performance. Secondly, the arrays shown implement only combinational logic. To realize complete, sequential subdesigns, the presence of registers and/or flip-flops is an absolute requirement. These deficiencies can be addressed as follows:

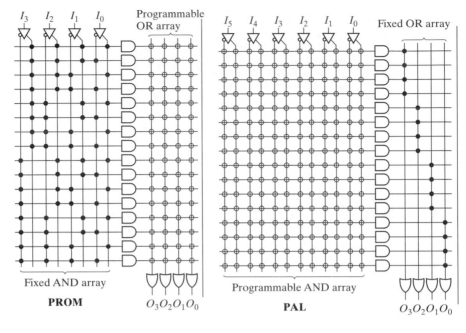

Figure 8-28   Alternative fuse-based programmable logic devices (or PLDs).

1. Partition the array into a number of smaller sections, often called macrocells.
2. Introduce flip-flops and provide a potential feedback from output signals to the inputs.

One example of how this can be accomplished is shown in Figure 8-29. The PAL consists of k macrocells, each of which can select from i inputs and features, at most, j product terms. Each macrocell contains a single register, which also is programmable—it can be configured as a D, T, J-K, or a clocked S-R flip-flop. The k output signals are fed back to the input bus, and thus form a subset of the i input signals.

The PLA approach to configurable logic has two distinct advantages:

• The structure is very regular, which makes the estimation of the parasitics quite easy, and enables accurate predictions of area, speed, and power dissipation.
• It provides an efficient implementation for logic functions that map well into a two-level logic description. Functions with a large fan-in fall into that category. Examples of such are finite-state machines used in controllers and sequencers.

On the other hand, the array structure has the disadvantage of higher overhead. Every intermediate node has a sizable capacitance, which negatively affects performance and power. This is especially true when parts of the array are underutilized—that is, if only some of the minterms are actively used.

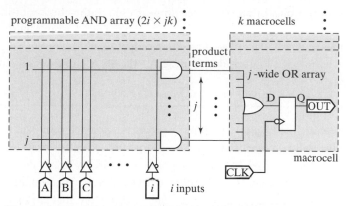

Figure 8-29   Schematic diagram of a PAL with *i* inputs,
j minterms/macrocell and *k* macrocells (or outputs) [Smith97].

Example 8.8   Example of Programmed Macrocell

Figure 8-30 shows an example of how to program a PROM module. The structure is programmed to realize the logical functions used earlier during the discussion on PLAs (Eq. (8.1)):

$$f_0 = x_0 x_1 + \overline{x_2}$$

$$f_1 = x_0 x_1 x_2 + \overline{x_2} + \overline{x_0} x_1$$

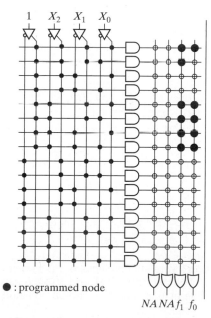

● : programmed node

$NA \, NA \, f_1 \, f_0$

Figure 8-30   Programming a PROM.

Observe that only a fraction of the array is used as the number of input (3) and output (2) variables are smaller than the dimensions of the 4×4 array. Unused input variables are tied either to 0 or 1. The large dots in the output planes represent programmed nodes. The reader is invited to repeat the exercise for the PLA and PAL modules presented in Figure 8-28 and Figure 8-29.

Cell-Based Programmable Logic   The sum-of-products approach results in regular structures, and is very effective for logic functions that have a large fan-in such as finite-state machines. On the other hand, it performs rather poorly for logic that features a large fan-out, or that benefits from a multilevel logic implementation. (Arithmetic operations such as addition and multiplication are an example of such). Other approaches can be conceived that are more in line with the multilevel approach favored in the standard-cell and sea-of-gate approaches.

There are many ways to design a logic block that can be configured to perform a wide range of logic functions. One approach is to use *multiplexers as function generators*. Consider the two-input multiplexer of Figure 8-31, which implements the logic function F:

$$F = A \cdot \bar{S} + B \cdot S \qquad (8.3)$$

By carefully choosing the connections between the variables X and Y and the input ports A, B, and S of the multiplexer, we can program it to perform ten useful logic operations on one or more of those inputs (see Figure 8-31).

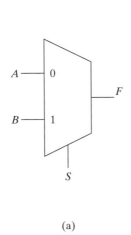

Configuration			
A	B	S	$F =$
0	0	0	0
0	X	1	X
0	Y	1	Y
0	Y	X	XY
X	0	Y	$X\bar{Y}$
Y	0	X	$\bar{X}Y$
Y	1	X	$X + Y$
1	0	X	$\bar{X}$
1	0	Y	$\bar{Y}$
1	1	1	1

(a)   (b)

Figure 8-31   Using a two-input multiplexer (a) as a configurable logic block. By properly connecting the inputs A, B, and S to the input variables X or Y, or to 0 or 1, 10 different logic functions can be obtained (b).

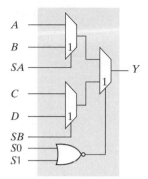

Figure 8-32 Logic cell as used in the Actel fuse-based FPGA.

A number of multiplexers can be combined to form more complex logic gates. Consider, for example, the logic cell of Figure 8-32, which is used in the Actel ACT family of FPGAs. It consists of three two-input multiplexers and a two-input NOR gate. The cell can be programmed to realize any two- and three-input logic functions, some four-input Boolean functions, and a latch.

Example 8.9   Programmable Logic Cell

It can be verified that the logic cell of Figure 8-32 acts as a two-input XOR under the programming conditions that follow. Assume the multiplexers select the bottom input signal when the control signal is high. We have the following:

$$A = 1; B = 0; C = 0; D = 1; SA = SB = In1; S0 = S1 = In2$$

As an exercise, determine the programming required for the two-input XNOR function. A three-input AND gate can be realized as follows:

$$A = 0; B = In1; C = 0; D = 0; SA = In2; SB = 0; S0 = S1 = In3$$

Finally, the largest function that can be realized is the four-input multiplexer. A, B, C, and D act as inputs, while SA, SB, and $(S0 + S1)$ are control signals.

The "multiplexer-as-functional-block" approach provides configurability through programmable interconnections. The *lookup table* (LUT) method employs a vastly different strategy. To configure a fully programmable module with fan-in of i for a specific function, a two-bit large memory, called the lookup table, is programmed to capture the truth table of that function. The input variables serve as control inputs to a multiplexer, which picks the appropriate value from the memory. The idea is illustrated in Figure 8-33 for a two-input cell. To implement an EXOR function, the lookup table is loaded with the output column of the EXOR truth table, this is "0 1 1 0". For an input value of "0 0", the multiplexer selects the first value in the table ("0"), etc. With this approach, any logic function of two inputs can be realized by a simple (re)programming of the memory.

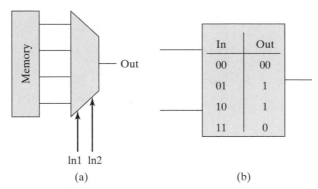

In	Out
00	00
01	1
10	1
11	0

In1 In2

(a)   (b)

Figure 8-33   Configurable logic cell based on lookup table. (a) cell schematic;
(b) programming the cell to implement an EXOR function.

As in the case of the multiplexer-based approach, more complex gates can be constructed.
This is accomplished by either combining a number of LUTs, or by increasing the LUT sizes, or
a combination of both. Additional functionality is provided by incorporating flip-flops.

Example 8.10   LUT-Based Programmable Logic Cell

Figure 8-34 shows the basic cell, called a *Configurable Logic Block* or CLB, used in the
Xilinx 4000 FPGA series [Xilinx4000]. It combines two four-input LUTs feeding a three-

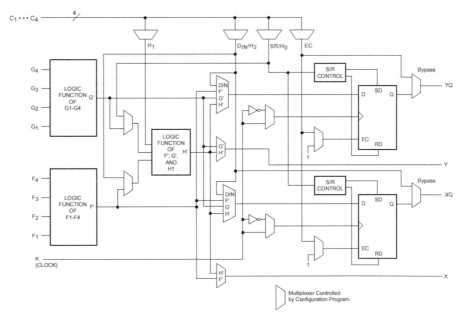

Figure 8-34   Simplified block diagram of XC4000 Series CLB (RAM and Carry-logic
functions not shown) [Xilinx4000].

input LUT. The cell features two flip-flops, whose inputs can be any one of the LUT outputs F, G, or H, or an extra external input D_{in}, and whose outputs are available at the XQ and YQ output pins. The X and Y outputs export the outputs of the LUTs and make it possible to build more complex combinational functions. The cell has four extra inputs ($C1...C4$) that either can be used as inputs or as set/reset and clock-enable signals for the flip-flops.

Programmable Interconnect

So far, we have discussed in some depth how to make logic programmable. A compelling question is how to make interconnections between those gates changeable or programmable as well. To fully utilize the available logic cells, the interconnect network must be flexible and routing bottlenecks must be avoided. Speed is another prerequisite, since interconnect delay tends to dominate the performance in this style of design. At the same time, the reader should be aware that programmable interconnect comes at a substantial cost in performance in area, performance, and power. In fact, most of the power dissipation in field-programmable architectures is attributable to the interconnect network [George01].

Once again, we can differentiate between mask-programmable, one-time programmable and reprogrammable approaches. It also is worth differentiating between local cell-to-cell interconnections and global signals, such as clocks, that have to be distributed over the complete chip with low delay. In the local-area class, programmable wiring can be classified into two major groupings: array and switchbox routers.

Array-Based Programmable Wiring   In this approach, wiring is grouped into routing channels, each of which contains a complete grid of horizontal and vertical wires. An interconnect wire can then be programmed into the structure by short-circuiting some of the intersections between horizontal and vertical wires (see Figure 8-35). This can be accomplished by providing a pass transistor at each of the cross points. Closing the interconnection means raising the control signal—by programming a "1" into the connected memory cell M (see Figure 8-36). This approach is prohibitive and expensive because it requires a large number of transistors and control signals. Also, the large number of transistors connected to each wire leads to a high fan-out, translating into delay and power consumption. A fuse is a more effective programmable connector. In this approach, each routing channel as a fully connected grid of horizontal and vertical interconnect wires, and a fuse is blown whenever a connection is not needed. Unfortunately, interconnect networks tend to be sparsely populated, which requires the interruption of an excessive number of switches and results in prohibitively long programming times.

To circumvent this problem, an *antifuse* can be used (as in Figure 8-26). Antifuses only need to be enabled when a connection is required in the routing channel. This represents a small fraction of the overall grid. Notice in Figure 8-35 how only two antifuses are needed to set up a connection. Be aware that this figure hides the programming circuitry. This operation is a one-time event and cannot be undone. The array-based wiring approach has thus been most successful in the write-once class of FPGAs. Circuit corrections or extensions are not possible, and new

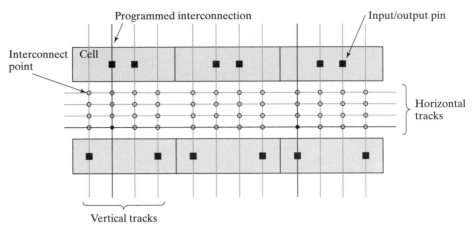

Figure 8-35 Array-based programmable wiring.

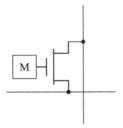

Figure 8-36 Programmable interconnect point. The memory cell controls the interconnection. A stored 0 and 1 mean an open or a closed circuit, respectively. The memory cell can be nonvolatile (EEPROM) or volatile (SRAM).

components are required for every design change. Providing true field (re)programmability requires a more efficient routing strategy.

Switch-Box-Based Programmable Wiring It's easy to imagine more efficient programmable-routing approach once we realize that the fully connected wiring grid represents major overkill. By restricting the number of routing resources and interconnect points, we can still manage to wire the desired interconnections, while drastically reducing the overhead. The disadvantage of this approach is that occasionally an interconnection cannot be routed. Most often, this can be addressed by remapping the design—for instance, by choosing another group of logic cells for a given function.

A large number of local interconnections can be accounted for by providing a mesh-like interconnection between neighboring cells. For instance, the outputs of each logic cell (LC) can be distributed to its neighbors to the north, east, south, and west. To account for interconnections between disjoint cells or to provide global interconnections, routing channels are placed between the cells containing a fixed number of uncommitted vertical and horizontal routing wires (Figure 8-37). At the junctions of the horizontal and vertical wires, RAM-programmable switching matrices (S-boxes) are

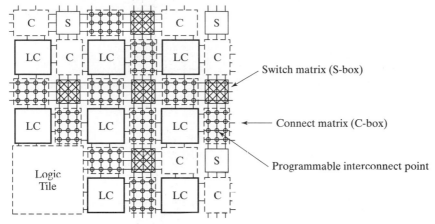

Figure 8-37 Programmable mesh-based interconnect network (Courtesy Andre Dehon and John Wawrzyniek.).

provided that direct the routing of the data. Cell inputs and outputs are connected to the global interconnect network by RAM-programmable interconnect points (C-box). Figure 8-38 provides a more detailed view, showing the transistor implementation of the switch and interconnect boxes. Be aware that the single pass-transistor implementation of the switches comes with a threshold-voltage drop. While advantageous from a power perspective, this reduced signal swing has a negative impact on the performance. Special design techniques such as zero-threshold devices, level restorers, or boosted control signals might be required.

The mesh architecture provides a flexible and scalable means for connecting a large number of components. It is quite efficient for local connections, as the number of switches traversed by a single interconnection is small and the fan-out is small. However, the mesh network does not lend itself well to global interconnections. The delay caused by the combination of the many

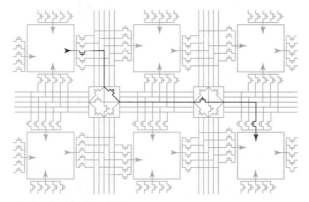

Figure 8-38 Transistor-level schematic diagram of mesh-based programmable routing network (Courtesy Andre Dehon and John Wawrzyniek.).

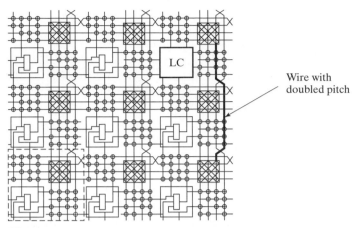

Figure 8-39   Programmable mesh-based interconnect architecture
with overlaid 2 x 2 grid (Courtesy Andre Dehon and John Wawrzyniek.).

switches and the large capacitive load becomes excessive. Most mesh-based FPGA architectures therefore offer alternative wiring resources that allow for effective global wiring. One approach for accomplishing this task is shown in Figure 8-39. In addition to the standard S-box-to-S-box wiring, the network also includes wires connecting S-boxes that are two steps away from each other. Eliminating one S-box from an interconnection decreases the resistance. Similarly, we can include long wires that connect every 4th, 8th, or 16th S-box. What we are creating, in fact, is a number of overlaying meshes with different granularity (single pitch, double pitch, etc.). Long wires are, by preference, mapped on the wiring meshes with the larger pitch.

Putting It All Together

A complete field-programmable gate array can now be assembled by joining logic-cell and interconnect approaches. Many alternative architectures can be (and have been) conceived. The most important decision to make at the start is the configuration style (write once, nonvolatile, volatile). This puts some constraints on the types of cells and interconnects that can be used. Giving a complete overview is out of the scope of this textbook, so we limit ourselves to two popular architectures, which are illustrative for the field. The interested reader can find more information in [Trimberger94], [Smith97], [Betz99], and [George01].

The Altera MAX Series [Altera01]   The MAX family of devices (Figure 8-40) belongs to the class of nonvolatile FPGAs (often called EPLDs, or *Electrically Programmable Logic Devices*). It uses a PAL module, (as introduced in Figure 8-29) as the basic logic module. The module (called the *Logic Array Block* or LAB in Altera language) varies little over the members of the family: a wide programmable AND array followed by a narrow fixed OR array and programmable inversion. A LAB typically contains 16 macrocells.

   The major differentiation lies in the interconnect architecture between the LABs. The smaller devices (MAX5000, MAX7000) use an array-based routing architecture. The back-

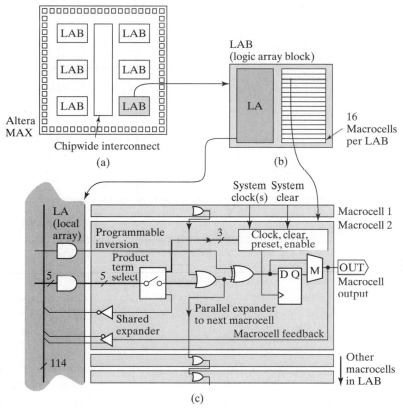

Figure 8-40 The Altera MAX Architecture. (a) Organization of logic and interconnect; (b) LAB module; (c) a MAX family macrocell. The expanders increase the number of products available by taking another pass through the logic array (from [Smith97]).

bone of the routing channel is formed by the outputs of all the macrocells, complemented with the direct chip inputs. These can be connected to the inputs of the LABs through programmable interconnect points. The advantage of this architecture, called the *Programmable Interconnect Array* or *PIA*, is that it is simple, and the routing delay between the blocks is totally predictable and fixed (see Figure 8-41). The disadvantage is that it does not scale very well. This is why the larger members of the series (MAX9000) have to resort to another scheme. With the number of macrocells reaching up to 560, the single-channel approach runs out of steam, and becomes slow. A mesh-based routing architecture has been opted for instead. Individual macrocells can connect to both row and column channels, which are quite wide (48 to 96 wires).

The EPLD approach delivers up to 15,000 logic gates, and typically is used when high performance is a necessity. Other architectures become desirable when more complex functions have to be implemented.

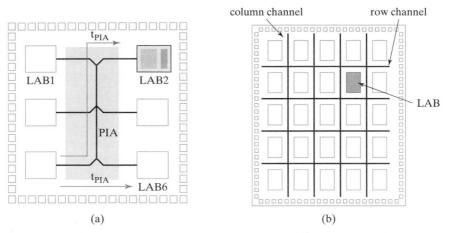

Figure 8-41   Interconnect architectures used in the Altera MAX series. (a) Array-based architecture used in MAX 3000-7000; (b) Mesh architecture of the MAX9000.

The Xilinx XC40xx Series   This popular RAM-programmable device family combines the lookup table approach for the implementation of the logic cells, with a mesh-based interconnect network. The largest part in the series (XC4085) supports almost 100,000 gates using a 56×56 CLB array. The architecture of the CLB was shown in Figure 8-34. An interesting feature is that the CLB can also be configured as an array of Read/Write memory cells, using the memory lookup tables in the F' and G' blocks. Depending on the selected mode, a single CLB can be configured as either a 16×2, 32×1, or 16×1 bit array. This feature comes in handy, because it is typical for large modules of logic to need comparable amounts of storage.

The interconnect architecture is also quite rich, and combines a wide variety of wiring resources, as shown in Figure 8-42. The overlaid meshes consist of wire segments of lengths 1, 2, and 4. Some components also support direct connections, which link adjacent CLBs without using general wiring resources. Signals routed on the direct interconnect experience minimum wiring delay, as the fan-out is small. These *Directs* are especially effective in the implementation of fast arithmetic modules, which feature many critical local connections. To address global wiring, *long lines* are provided that form a grid of metal interconnect segments that run the entire length or width of the array. These are intended for high fan-out, time-critical signal nets, or nets that are distributed over long distances (such as buses). In addition, special wires are provided for the routing of the clocks.

One topic we have ignored so far in our discussion of configurable array structures is the input/output architecture. For maximum usability, it is crucial that the I/O pins of the component are flexible, and that they provide a wide range of options in terms of direction, logic levels, and drive strengths. One style of input/output block (IOB), used in the XC4000 series, is shown in Figure 8-43. It can be programmed to act as an input, output, or bidirectional port. It includes a flip-flop that can be programmed to be either edge triggered or level sensitive. The slew-rate

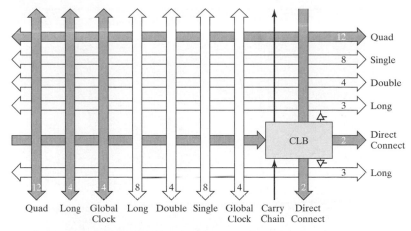

Figure 8-42 Interconnect architecture of the Xilinx XC4000 series. The numbers annotated on the diagram indicate the amount of each of the resources available.

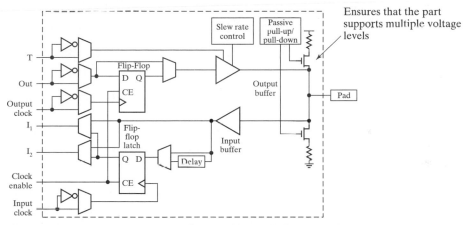

Figure 8-43 Programmable input/output Block of XC4000 series.

control provides variable drive strengths and allows for a reduction in the rise–fall time for non-critical signals.

Example 8.11 FPGA Complexity and Performance

To get an impression of what can be achieved with the volatile field-programmable components, consider the Xilinx 4025. It contains approximately 1000 CLBs organized in a 32×32 array. This translates into a maximum equivalent gate count of 25,000 gates. The chip contains 422 Kbits of RAM, used mostly for programming. A single CLB is specified to operate at 250 MHz. When taking into account the interconnect network and attempting more complex logic configurations such as adders, clock speeds between 20 and 50 MHz

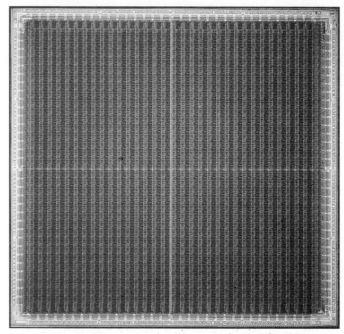

Figure 8-44   Chip microphotograph of XC4025 volatile FPGA (Courtesy of Xilinx, Inc.).

are attainable. To put the integration complexity in perspective, a 32-bit adder requires approximately 62 CLBs. A chip microphotograph of the XC4025 part is shown in Figure 8-44. The horizontal and vertical routing channels are easily recognizable.

Prewired logic arrays have rapidly claimed a significant part of the logic component market. Their arrival has effectively ended the era of logic design using discrete components represented by the TTL logic family. It is generally believed that the impact of these components is increasing with a further scaling of the technology. To make this approach successful, however, advanced software support in terms of cell placement, signal routing, and synthesis are required. Also, one should not ignore the overhead that flexibility brings with it. Programmable logic is at least 10 times less efficient in terms of energy and performance with respect to ASIC solutions. Hence, its scope has been mostly restricted to prototyping and small-volume applications so far. Yet, flexibility and reuse are alluring. Field-programmable components are bound to see a substantial growth in the years to come.

8.6 Perspective—The Implementation Platform of the Future

The designer of today's advanced systems-on-a-chip is offered a broad range of implementation choices. What approach is ultimately chosen is determined by a broad range of factors:

- performance, power and cost constraints
- design complexity
- testability
- time to market, or more precisely, time to revenue
- uncertainty of the market, or late changes in the design
- application range to be covered by the design
- prior experiences of the design team

A number of these factors seem to imply a trend towards more flexible, programmable components that can be reused and that can be modified even after manufacturing. At the same time, solutions that offer the best "bang for the buck" most often end up the winners. Too much flexibility often results in ineffective and expensive solutions, which rapidly end up on the dust heap. Finding the balance between the two extremes is the ultimate challenge of the chip architects of today.

On the basis of these observations, it seems logical to assume that the implementation platform of the future will be a combination of the strategies we have discussed in this chapter, providing implementation efficiency and flexibility when and where needed. The system on a chip is becoming a combination of embedded microprocessors with their memory subsystems, DSPs, fixed ASIC-style hardware accelerators, parameterizable modules, and flexible logic implemented in FPGA style. How these components are balanced is a function of the application requirements and the intended market.

Example 8.12   Examples of Hybrid Implementation Platforms

Figure 8-45 shows two contrasting implementation platforms for wireless applications. The first device, the Virtex-II Pro from Xilinx [XilinxVirtex01] is centered around a large FPGA array. A PowerPC microprocessor is embedded in the center of the array. The processor provides an effective implementation approach for application-level functionality and system-level control. To provide higher performance for signal processing applications, an array of embedded 18×18 multipliers is added. These dedicated components offer a significant performance, power, and area advantage over a pure FPGA implementation of the same function. Finally, a number of very fast 3.125-Gbps transceivers are provided, allowing for high-speed serial communication off chip.

A somewhat contrasting approach is offered in the design of Figure 8-45b [Zhang00]. The center of this device is an ARM-7 embedded microprocessor, acting as the overall chip manager. Functions that need high performance and energy efficiency are offloaded to a configurable array of functional units such as multipliers, ALUs, memories, and address generators. These components can be combined dynamically into application-specific processors. The chip also provides an embedded FPGA array for functions that need bit-level granularity.

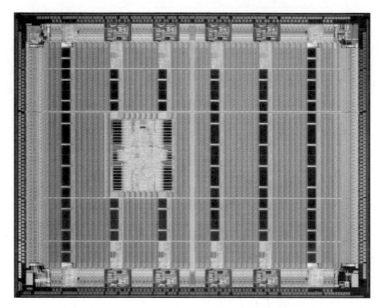

(a) The Xilinx Virtex-II Pro embeds a PowerPC microprocessor into an FPGA fabric (Courtesy Xilinx, Inc.).

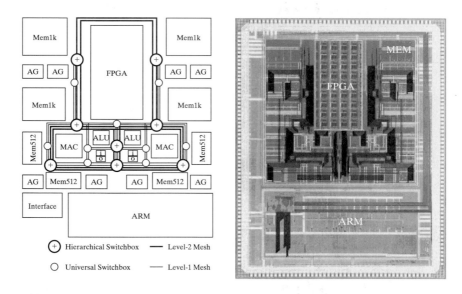

(b) The Maia chip combines embedded microprocessor, configurable accelerators, and FPGA [Zhang00].

Figure 8-45   Examples of hybrid implementation platforms.

8.7 Summary

In this chapter, we have briefly scanned the complex world of design implementation strategies for digital integrated circuits. New implementation styles have rapidly emerged over the last few decades, presenting the designer with a wide variety of options. These design techniques and the accompanying tools are having a major impact on the way design is performed today, and make possible the exciting and impressive processors and application-specific circuits to which we have become so accustomed. We have touched on the following issues in this chapter:

- *Custom design*, where each transistor is individually handcrafted, offers the implementation from an area and performance perspective. This approach has become prohibitively expensive, and should be reserved for the design of the few critical modules in which extreme performance is required, or for often-reused library cells.
- The *semicustom* approach, based on the standard-cell methodology, is the workhorse of today's digital design industry. The advantage is the high degree of automation. The challenge is to deal with the impact of deep-submicron technologies.
- To deal with the increasing complexity of integrated circuits, designers increasingly rely on the availability of large *macrocells* such as memories, multipliers, and microprocessors. These modules are often provided by third-party vendors, and they have spurned a new industry focused on "*intellectual property.*"
- Starting a new design for every new emerging application has become prohibitively expensive. The majority of the semiconductor market now focuses on flexible solutions that allow a single component to be used for a variety of applications, either through software programming or reconfiguration. *Configurable hardware* delays the time when the required function is actually committed to the hardware. Different approaches toward late binding also have been discussed. Delaying the binding time comes with an efficiency penalty: The more flexibility that is provided, the larger the impact on performance and power dissipation.

Undoubtedly, new design styles will come on the scene in the near future. Becoming familiar with the available options is an essential part of the learning experience of the beginning digital designer. We hope this chapter, although compressed, entices the reader to further explore the numerous possibilities. One final observation is as follows: Even with the increasing automation of the digital circuit design process, new challenges are continuously emerging—challenges that require the profound insight and intuition offered only by a human designer.

8.8 To Probe Further

The literature on design methodologies and automation for digital integrated circuits has exploded in the last few decades. Several reference works are worth mentioning:

- ASIC and FPGA design methodologies: [Smith97]
- FPGA architectures: [Trimberger94], [George01]

• System on a Chip: [Chang99]
• Design methodology and technology: [Bryant01]
• Design synthesis: [DeMicheli94]

State-of-the-art developments in the design automation domain are generally reported in the *IEEE Transactions on CAD*, the *IEEE Transactions on VLSI Systems,* and the *IEEE Design and Test Magazine.* Premier conferences are, among others, the Design Automation Conference (DAC) and the International Conference on CAD (ICCAD). The web sites of the major Electronic Design Automation Companies (Cadence, Synopsys, Mentor, etc.) provide a treasure of information as well.

References

[Altera01] Altera Device Index, *http://www.altera.com/products/devices/dev-index.html*, 2001.

[Avanti01] Saturn Efficient and Concurrent Logical and Physical Optimization of SoC Timing, Area and Power, *http://www.synopsis.com/product/avmrg/saturn_ds.html.*

[Betz99] V. Betz, J. Rose, and A. Marquardt, *Architecture and CAD for Deep-Submicron FPGAs*, Kluwer International Series in Engineering and Computer Science, Kluwer Academic Publishers, 1999.

[Brayton84] R. Brayton et al., *Logic Minimization Algorithms for VLSI Synthesis*, Kluwer Academic Publishers, 1984.

[Bryant01] R. Bryant, T. Cheng, A. Kahng, K. Keutzer, W. Maly, R. Newton, L. Pileggi, J. Rabaey, and A. Sangiovanni-Vincentelli, "Limitations and Challenges of Computer-Aided Design Technology for CMOS VLSI," *IEEE Proceedings*, pp. 341–365, March 2001.

[Cadabra01] AbraCAD Automated Layout Creation, *http://www.cadabratech.com/?id=145products*, Cadabra Design Automation.

[Chang99] H. Chang et al., "Surviving the SOC Revolution: A Guide to Platform-Based Design," Kluwer Academic Publishers, 1999.

[DeMicheli94] G. De Micheli, *Synthesis and Optimization of Digital Circuits*, McGraw-Hill, 1994.

[El-Ayat89] K. El-Ayat, "A CMOS Electrically Configurable Gate Array," *IEEE Journal of Solid State Circuits,* vol. SC-24, no. 3, pp. 752–762, June 1989.

[George01] V. George and J. Rabaey, *Low-Energy FPGAs*, Kluwer Academic Publishers, 2001.

[Grundman97] W. Grundmann, D. Dobberpuhl, R. Allmon, and N. Rethman "Designing High-Performance CMOS Processors Using Full Custom Techniques," *Proceedings Design Automation Conference*, pp. 722–727, Anaheim, June 1997.

[Hill85] D. Hill, "S2C—A Hybrid Automatic Layout System," *Proc. ICCAD-85*, pp. 172–174, November 1985.

[ModuleCompiler01] Synopsys Module Compiler Datasheet, *http://www.synopsys.com/products/datapath/datapath.html*, Synopsys, Inc.

[Philips99] The Nexperia System Silicon Implementation Platform, *http://www.semiconductors.philips.com/platforms/nexperia/*, Philips Semiconductors.

[Pileggi02] Pileggi, Schmit et al., *"Via Patterned Gate Array,"* CMU Center for Silicon System Implementation Technical Report Series, no. CSSI 02-15, April 2002.

[Prolific01] The ProGenesis Cell Compiler, *http://www.prolificinc.com/progenesis.html*, Prolific, Inc.

[Rabaey00] J. Rabaey, "Low-Power Silicon Architectures for Wireless Applications," *Proceedings ASPDAC Conference*, Yokohama, January 2000.

[Silva01] J. L. da Silva Jr., J. Shamberger, M. J. Ammer, C. Guo, S. Li, R. Shah, T. Tuan, M. Sheets, J. M. Rabaey, B. Nikolic, A. Sangiovanni-Vincentelli, P. Wright, "Design Methodology for PicoRadio Networks," *Proc. DATE Conference*, Munich, March 2000.

[Smith97] M. Smith, *Application-Specific Integrated Circuits*, Addison-Wesley, 1997.

[Sylvester98] D. Sylvester and K. Keutzer, "Getting to the Bottom of Deep Submicron," *Proc. ICCAD Conference*, pp. 203, San Jose, November 1998.

[Trimberger94] S. Trimberger, *Field-Programmable Gate Array Technology*, Kluwer Academic Publishers, 1994.

[Veendrick92] H. Veendrick, *MOS IC's: From Basics to ASICS*, Wiley-VCH, 1992.

[Xilinx4000] The Xilinx-4000 Product Series, *http://www.xilinx.com/apps/4000.htm*, Xilinx, Inc.

[XilinxVirtex01] Virtex-II Pro Platform FPGAs, *http://www.xilinx.com/xlnx/xil_prodcat_landing page.jsp?title=Virtex-II+Pro+FPGAs*, Xilinx, Inc.

[Xtensa01] Xtensa Configurable Embedded Processor Core, *http://www.tensilica.com/technology.html*, Tensilica.

[Zhang00] H. Zhang, V. Prabhu, V. George, M. Wan, M. Benes, A. Abnous, and J. Rabaey, "A 1 V Heterogeneous Reconfigurable Processor IC for Baseband Wireless Applications," *Proc. ISSCC*, pp. 68–69, February 2000.

Exercises

For problems and exercises on design methodology, please check **http://bwrc.eecs.berkeley.edu/IcBook**.

E

Characterizing Logic
and Sequential Cells

The challenge of library characterization

Characterization methods for logic cells and registers

Cell parameters

The Importance and Challenge of Library Characterization

The quality of the results produced by a logic synthesis tool is a strong function of the level of detail and accuracy with which the individual cells were characterized. To estimate the delay of a complex module, a logic synthesis program must rely on higher level delay models of the individual cells—falling back to a full circuit- or switch-level timing model for each delay estimation is simply not possible because it takes too much compute time. Hence, an important component in the development process of a standard-cell library is the generation of the delay models. In previous chapters, we learned that the delay of a complex gate is a function of the fan-out (consisting of connected gates and wires), and the rise and fall times of the input signals. Furthermore, the delay of a cell can vary between manufacturing runs as a result of process variations.

In this insert, we first discuss the models and characterization methods that are commonly used for logic cells. Sequential registers require extra timing parameters and thus deserve a separate discussion.

Characterization of Logic Cells

Unfortunately, no common delay model for standard cells has been adopted. Every vendor has his own favored methods of cell characterization. Even within a single tool, various delay models often can be used, trading off accuracy for performance. The basic concepts are, however, quite similar, and they are closely related to the ones we introduced in Chapters 5 and 6. We therefore opt to concentrate on a single set of models in this section—more precisely, those used in the Synopsys Design Compiler [DesignCompiler01], one of the most popular synthesis tools. Once a model has been adopted, it has to be adopted for all the cells in the block; in other words, it cannot be changed from cell to cell.

The total delay consists of four components, as illustrated in Figure E-1:

$$D_{total} = D_I + D_T + D_S + D_C. \tag{E.1}$$

D_I represents the *intrinsic delay*, which is the delay with no output loading. D_T is the transition component, or the part of the delay caused by the output load. D_S is the fraction of the delay due to the *input slope*. Finally, D_C is the *delay of the wire* following the gate. All delays are characterized for both rising and falling transitions.

The simplest model for the transition delay is the linear delay model of Chapter 5. We have

$$D_T = R_{driver}(\Sigma C_{gate} + C_{wire}), \tag{E.2}$$

where ΣC_{gate} is the sum of all input pin capacitances of gates connected to the output of this gate, and C_{wire} is the estimated wire capacitance. The slope delay D_S is approximated as a linear function of the transition delay D_T of the previous gate, written as

$$D_S = S_S D_{Tprev} \tag{E.3}$$

where S_S is the *slope-sensitivity factor,* and D_{Tprev} is the transition delay of the previous stage.

The characterization of a library cell must therefore provide the following components, each of them for both rising and falling transitions, and with respect to each of the input pins:

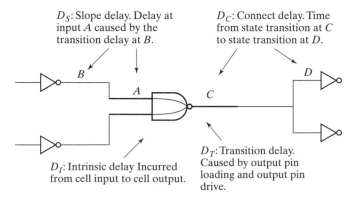

D_S: Slope delay. Delay at input A caused by the transition delay at B.

D_C: Connect delay. Time from state transition at C to state transition at D.

D_I: Intrinsic delay Incurred from cell input to cell output.

D_T: Transition delay. Caused by output pin loading and output pin drive.

Figure E-1　　Delay components of a combinational gate [DesignCompiler01].

- intrinsic delay
- input pin capacitance
- equivalent output driving resistance
- slope sensitivity

In addition to the cell models, the synthesis tools also must have access to a wire model. Since the length of the wires is unknown before the placement of the cells, estimates of C_{wire} and R_{wire} are made on the basis of the size of the block and the fan-out of the gate. The length of a wire is most often proportional to the number of destinations it has to connect.

Example E.1   Three-Input NAND Gate Cell

The characterization of the three-input NAND standard cell gate, presented earlier in Example 8.2, is given in Table E-1. The table characterizes the performance of the cell as a function of the load capacitance and the input-rise (fall) time for two different supply voltages and operating temperatures. The cell is designed in a 0.18-μm CMOS technology.

Table E-1   Delay characterization of a three-input NAND gate (in ns) as a function of the input node for two operation corners (supply-voltage–temperature pairs of 1.2 V–125°C, and 1.6 V–40°C). The parameters are the load capacitance C and the input rise (fall) time T. (*Courtesy ST Microelectronics.*)

Path	1.2 V–125°C	1.6 V–40°C
$In1-t_{pLH}$	$0.073 + 7.98C + 0.317T$	$0.020 + 2.73C + 0.253T$
$In1-t_{pHL}$	$0.069 + 8.43C + 0.364T$	$0.018 + 2.14C + 0.292T$
$In2-t_{pLH}$	$0.101 + 7.97C + 0.318T$	$0.026 + 2.38C + 0.255T$
$In2-t_{pHL}$	$0.097 + 8.42C + 0.325T$	$0.023 + 2.14C + 0.269T$
$In3-t_{pLH}$	$0.120 + 8.00C + 0.318T$	$0.031 + 2.37C + 0.258T$
$In3-t_{pHL}$	$0.110 + 8.41C + 0.280T$	$0.027 + 2.15C + 0.223T$

While linear delay models offer good first-order estimates, more precise models are often used in synthesis, especially when the real wire lengths are back annotated onto the design. Under those circumstances, nonlinear models have to be adopted. The most common approach is to capture the nonlinear relations as lookup tables for each of these parameters. To increase computational efficiency and minimize storage and characterization requirements, only a limited set of loads and slopes are captured, and linear interpolation is used to determine the missing values.

Example E.2   Delay Models Using Lookup Tables

A (partial) characterization of a two-input AND cell (AND2), designed in a 0.25-μm CMOS technology (*Courtesy ST Microelectronics*) follows. The delays are captured for output capacitances of 7 fF, 35 fF, 70 fF, and 140 fF, and input slopes of 40 ps, 200 ps, 800 ps, and 1.6 ns, respectively.

```
cell(AND) {
 area : 36 ;
 pin(Z) {
  direction : output ;
  function : "A*B";
  max_capacitance : 0.14000 ;

  timing() {
    related_pin : "A" ; /* delay between input pin A and output pin Z */
cell_rise {
    values( "0.10810, 0.17304, 0.24763, 0.39554", \
        "0.14881, 0.21326, 0.28778, 0.43607", \
        "0.25149, 0.31643, 0.39060, 0.53805", \
        "0.35255, 0.42044, 0.49596, 0.64469" ); }
    rise_transition {
     values( "0.08068, 0.23844, 0.43925, 0.84497", \
        "0.08447, 0.24008, 0.43926, 0.84814", \
        "0.10291, 0.25230, 0.44753, 0.85182", \
        "0.12614, 0.27258, 0.46551, 0.86338" );}
    cell_fall(table_1) {
     values( "0.11655, 0.18476, 0.26212, 0.41496", \
        "0.15270, 0.22015, 0.29735, 0.45039", \
        "0.25893, 0.32845, 0.40535, 0.55701", \
        "0.36788, 0.44198, 0.52075, 0.67283" );}
    fall_transition(table_1) {
     values( "0.06850, 0.18148, 0.32692, 0.62442", \
        "0.07183, 0.18247, 0.32693, 0.62443", \
        "0.09608, 0.19935, 0.33744, 0.62677", \
        "0.12424, 0.22408, 0.35705, 0.63818" );}
    intrinsic_rise : 0.13305 ; /* unloaded delays */
    intrinsic_fall : 0.13536 ;
    }
  timing() {
    related_pin : "B" ; /* delay between input pin A and output pin Z */
    ...
```

```
    intrinsic_rise : 0.12426 ;
        intrinsic_fall : 0.14802 ;
      }
    }
    pin(A) {
      direction : input ;
      capacitance : 0.00485 ; /* gate capacitance */
    }
    pin(B) {
      direction : input ;
      capacitance : 0.00519 ;
    }
  }
```

Characterization of Registers

In Chapter 7, we identified the three important timing parameters of a register. The *setup time* (t_{su}) is the time that the data inputs (*D* input) must be valid before the clock transition (in other words, the 0 to 1 transition for a *positive edge-triggered* register). The *hold time* (t_{hold}) is the time the data input must remain valid after the clock edge. Finally, the propagation delay (t_{c-q}) equals the time it takes for the data to be copied to the *Q* output after a clock event. The latter parameter is illustrated in Figure E-2a.

Latches have a bit more complex behavior, and thus require an extra timing parameter. While t_{C-Q}, corresponds to the delay of relaunching of data that arrived to a closed latch, t_{D-Q} equals the delay between *D* and *Q* terminals when the latch is in transparent mode (Figure E-2b).

The characterization of the t_{C-Q} (t_{D-Q}) delay is fairly straightforward. It consists of a delay measurement between the 50% transitions of *Clk* (*D*) and *Q*, for different values of the input slopes and the output loads, not unlike the case of combinational logic cells.

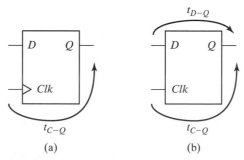

(a)   (b)

Figure E-2 Propagation delay definitions for sequential components: (a) register; (b) latch.

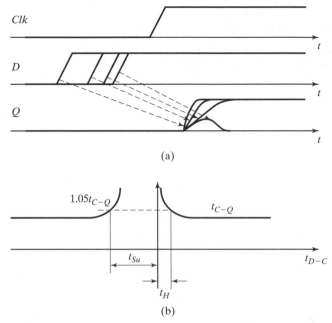

Figure E-3   Characterization of sequential elements: (a) determining the setup time of a register; (b) definition of setup and hold times.

The characterization of setup and hold times is more elaborate, and depends on what is defined as "valid" in the definitions of both setup and hold times. Consider the case of the setup time. Narrowing the time interval between the arrival of the data at the D input and the Clk event does not lead to instantaneous failure (as assumed in the first-order analysis in Chapter 7), but rather to a gradual degradation in the delay of the register. This is documented in Figure E-3a, which illustrates the behavior of a register when the data arrives close to the setup time. If D changes long before the clock edge, the t_{C-Q} delay has a constant value. Moving the data transition closer to the clock edge causes t_{C-Q} to increase. Finally, if the data changes too close to the clock edge, the register fails to register the transition altogether.

Clearly, a more precise definition of the "setup time" concept is necessary. An unambiguous specification can be obtained by plotting the t_{C-Q} delay against the data-to-clock offset, as shown in Figure E-3b. The degradation of the delay for smaller values of the offset can be observed. The actual definition of the setup time is rather precarious. If it were defined as the minimum D-Clk offset that causes the flip-flop to fail, the logic following the register would suffer from excessive delay if the offset is close to, but larger than, that point of failure. Another option is to place it at the operation point of the register that minimizes the sum of the data-clock offset and the t_{C-Q} delay. This point, which minimizes the overall flip-flop overhead, is reached when the slope of the delay curve in Figure E-3b equals 45 degrees [Stojanovic99].

While custom design can take advantage of driving flip-flops close to their point of failure—and take all the risk that comes with it—semicustom design must take a more conservative approach. For the characterization of registers in a standard cell library, both setup and hold times are commonly defined as data-clock offsets that correspond to **some fixed percentage increase in** t_{C-Q}, typically set **at 5%**, as indicated in Figure E-3b. Note that these curves are different for 0–1 and 1–0 transitions, resulting in different setup (an hold) times for 0 and 1 values. As with clock-to-output delays, setup times also are dependent on clock and data slopes, and they are represented as a two-dimensional table in nonlinear delay models. Identical definitions hold for latches.

Example E.3 Register Setup and Hold Times

In this example, we examine setup and hold behavior of the transmission gate master-slave register introduced in Chapter 7. (See Figure 7.18.) The register is loaded with a 100-fF capacitor, and its setup and hold times are examined for clock and data slopes of 100 ps. The simulation results are plotted in Figure E-4. When data settles a "long time" before the clock edge, the clock-to-output delay equals 193 ps. Moving the data transition closer to the clock edge causes the t_{C-Q} delay to increase. This becomes noticeable at an offset between data and clock of about 150 ps. The register completely fails to latch the data when data precedes the clock by 77 ps. The sum of D-Q offset and the t_{C-Q} is minimal at 93 ps. A 5% increase in t_{C-Q} is observed at 125 ps, and this time is entered in the library as the setup time for this particular slope of data and clock. This characterization of setup time adds a margin to the design of about 30 ps. From these simulations, we also can determine that this register has a hold time of –15 ps.

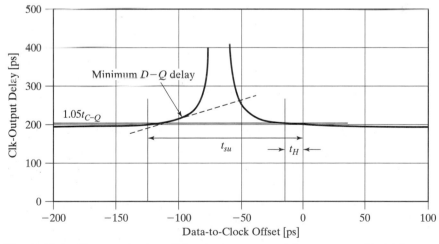

Figure E-4 Characterization of the clock-to-output delay, setup and hold times of a transmission-gate latch pair.

References

[DesignCompiler01] Design Compiler, Product Information, *http://www.synopsys.com/products/logic/ design_compiler.html*, Synopsys, Inc.

[Stojanovic99] V. Stojanovic, V.G. Oklobdzija, "Comparative analysis of master-slave latches and flip-flops for high-performance and low-power systems," *IEEE Journal of Solid-State Circuits*, vol. 34, no. 4, April 1999.

F

Design Synthesis

Circuit, Logic, and Architectural Synthesis

One of the most enticing proposals one can make to a designer who has to generate a circuit with tough specifications in a short time is to offer him a tool that automatically translates his specifications into a working circuit that meets all the requirements. One of the main reasons that semiconductor circuits have reached the mind-boggling complexity they have today, is that such synthesis tools actually exist—at least to a certain extent. Synthesis can be defined as the transformation between two different design views. Typically, it represents a translation from a *behavioral* specification of a design entity into a *structural* description. In simple terms, it translates a description of the function a module should perform (the behavior) into a composition— that is, an interconnection of elements (the structure). Synthesis approaches can be defined at each level of abstraction: circuit, logic, and architecture. An overview of the various synthesis levels and their impact is given in Figure F-1. The synthesis procedures may differ depending on the targeted implementation style. For example, logic synthesis translates a logic description given by a set of Boolean equations into an interconnection of gates. The techniques involved in this process strongly depend on the choice of either a two-level (PLA) or a multilevel (standard-cell or gate-array) implementation style. We briefly describe the synthesis tasks at each of the different modeling levels. Refer to [DeMicheli94] for more information and a deeper insight into design synthesis.

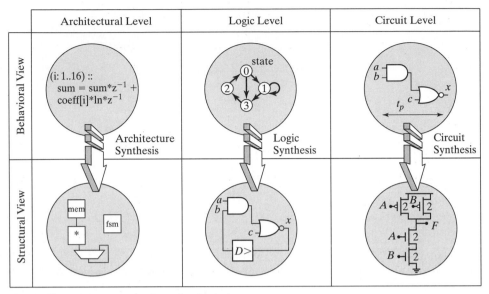

Figure F-1   A taxonomy of synthesis tasks.

Circuit Synthesis

The task of circuit synthesis is to translate a logic description of a circuit into a network of transistors that meets a set of timing constraints. This process can be divided into two stages:

1. Derivation of *transistor schematics* from the logic equations. This requires the selection of a circuit style (complementary static, pass transistors, dynamic, DCVSL, etc.) and the construction of the logic network. The former task is usually up to the designer, while the latter depends upon the chosen style. For instance, the logic graph technique introduced in Design Methodology Insert D can be used to derive the complementary pull-down and pull-up networks of a static CMOS gate. Similarly, automated techniques have been developed to generate the pull-down trees for the DCVSL logic style so that the number of required transistors is minimized [Chu86].

2. *Transistor sizing* to meet performance constraints. This has been a recurring subject throughout this book. The choice of the transistor dimensions has a major impact on the area, performance, and power dissipation of a circuit. We have also learned that this is a subtle process. For instance, the performance of a gate is sensitive to a number of layout parasitics, such as the size of the diffusion area, fan-out, and wiring capacitances. Notwithstanding these daunting challenges, some powerful transistor-sizing tools have been developed [e.g., Fishburn85, AMPS99, Northrop01]. The key to the success of these tools is the accurate modeling of the performance of the circuit using *RC* equivalent circuits and a detailed knowledge of the subsequent layout-generation process. The latter allows for an accurate estimation of the values of the parasitic capacitances.

While circuit synthesis has proven to be a powerful tool, it has not penetrated the design world as much as we might expect. One of the main reasons for this is that the quality of the cell library has a strong influence on the complete design, and designers are reluctant to pass this important task to automatic tools that might produce inferior results. Yet, the need for ever-larger libraries and the impact of transistor-sizing on circuit performance and power dissipation is providing a strong push for a more pervasive introduction of circuit-synthesis tools.

Logic Synthesis

Logic synthesis is the task of generating a structural view of a logic-level model. This model can be specified in many different ways, such as state-transition diagrams, state charts, schematic diagrams, Boolean equations, truth tables, or HDL (Hardware Description Language) descriptions.

The synthesis techniques differ according to the nature of the circuit (combinational or sequential) or the intended implementation architecture (multilevel logic, PLA, or FPGA). The synthesis process consists of a sequence of optimization steps, the order and nature of which depend on the chosen cost function—area, speed, power, or a combination of these. Typically, logic optimization systems divide the task into two stages:

1. A *technology-independent phase*, where the logic is optimized using a number of Boolean or algebraic manipulation techniques.
2. A *technology-mapping phase*, which takes into account the peculiarities and properties of the intended implementation architecture. The technology-independent description resulting from the first phase is translated into a gate netlist or a PLA description.

The *two-level minimization* tools were the first logic-synthesis techniques to become widely available. The Espresso program developed at the University of California at Berkeley [Brayton84] is an example of a popular two-level minimization program. For some time, the wide availability of these tools made regular, array-based architectures like PLAs and PALs the prime choice for the implementation of random logic functions.

At the same time, the groundwork was laid for sequential or state-machine synthesis. Tasks involved include the *state minimization* that aims at reducing the number of machine states, and the *state encoding* that assigns a binary encoding to the states of a finite state machine [DeMicheli94].

The emergence of *multilevel logic synthesis* environments such as the Berkeley MIS tool [Brayton87] swung the pendulum towards the standard-cell and FPGA implementations that offer higher performance or integration density for a majority of random-logic functions.

The combination of these techniques with sequential synthesis has opened the road to complete register-transfer (RTL) synthesis environments that take as an input an HDL description (in VHDL or Verilog—see Design Methodology Insert C) of a sequential circuit and produce a gate netlist [Carlson91, Kurup97]. Saying the logic synthesis has fundamentally altered the digital circuit design landscape is by no means an understatement. It also is fair to say that the tool set that made this major paradigm change in design methodology ultimately happen is the *Design Compiler* environment from Synopsys. Even after being in place for almost two decades, Design Compiler continues to dominate the market and represents the synthesis tool of choice for the majority of the digital ASIC designers. Built around a core of Boolean optimization and technology mapping, Design Compiler incorporates advanced techniques such as timing, area and power optimization, cell-based sizing, and test insertion [Kurup97, DesignCompiler].

Example F.1   Logic Synthesis

To demonstrate the difference between two-level and multilevel logic synthesis, both approaches were applied to the following full-adder equations, which will be treated in substantial detail in Chapter 11.

$$S = (A \oplus B) \oplus C_i$$
$$C_o = A \cdot B + A \cdot C_i + B \cdot C_i \tag{F.1}$$

The MIS-II logic synthesis environment was employed for both the two-level and multilevel synthesis. The minimized truth table representing the PLA implementation is shown in Table F-1. It can be verified that the resulting network corresponds to the preceding full-adder equations. The PLA counts three inputs, seven product terms, and two outputs. Observe that no product terms can be shared between the sum and carry outputs. A NOR-NOR implementation requires 26 transistors in the PLA array (17 and 9 in the OR plane and AND planes, respectively). This count does not include the input and output buffers.

Table F-1   Minimized PLA truth table for full adder. The dashes (–) mean that the corresponding input does not appear in the product term.

A	B	C_i	S	C_o
1	1	1	1	–
0	0	1	1	–
0	1	0	1	–
1	0	0	1	–
1	1	–	–	1
1	–	1	–	1
–	1	1	–	1

Figure F-2 shows the multilevel implementation as generated by MIS-II. In the technology-mapping phase, a generic standard-cell library was targeted. Implementation of the adder requires only six standard cells. This corresponds to 34 (!) transistors in a static CMOS implementation.[1] Observe the usage of complex logic gates such as EXOR and OR-AND-INVERT. For this case study, minimization of the area was selected as the prime optimization target. Other implementations can be obtained by targeting performance instead. For instance, the critical timing path from C_i to C_o can be reduced by signal reordering. This requires the designer to identify this path as the most critical, a fact that is not obvious from a simple inspection of the full-adder equations.

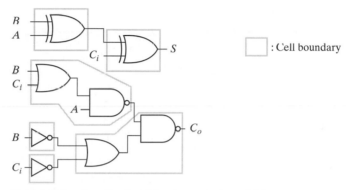

: Cell boundary

Figure F-2 Standard-cell implementation of full adder, as generated by multilevel logic synthesis.

Architecture Synthesis

Architecture synthesis is the latest development in the synthesis area. It is also referred to as *behavioral* or *high-level synthesis*. Its task is to generate a structural view of an architecture design, given a behavioral description of the task to be executed, and a set of performance, area, and/or power constraints. This corresponds to determining what architectural resources are needed to perform the task (execution units, memories, busses, and controllers), binding the behavioral operations to hardware resources, and determining the execution order of the operations on the produced architecture. In synthesis jargon, these functions are called *allocation, assignment,* and *scheduling* [Gajski92, DeMicheli94]. While these operations represent the core of architecture synthesis, other steps can have a dramatic impact on the quality of the solution. For example, optimizing transformations manipulate the initial behavioral description so that a superior solution can be obtained in terms of area or speed. *Pipelining* is a typical example of such a transformation. In a sense, this component of the synthesis process is similar to the use of optimizing transformations in software compilers.

---

[1] How to implement a static complementary CMOS EXOR gate with only nine transistors is left as an exercise for the reader.

Example F.2   Architecture Synthesis

To illustrate the concept and capabilities of architecture synthesis, consider the simple computational flowgraph of Figure F-3. It describes a program that inputs three numbers *a, b,* and *c* from off-chip and produces their sum *x* at the output.

Two possible implementations, as generated by the HYPER synthesis system ([Rabaey91]), are shown in Figure F-4. The first instance requires four clock cycles and

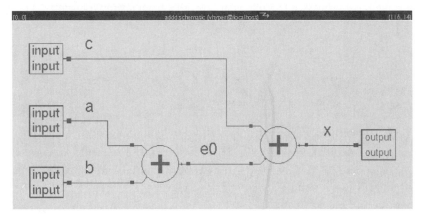

Figure F-3   Simple program performing the sum of three numbers.

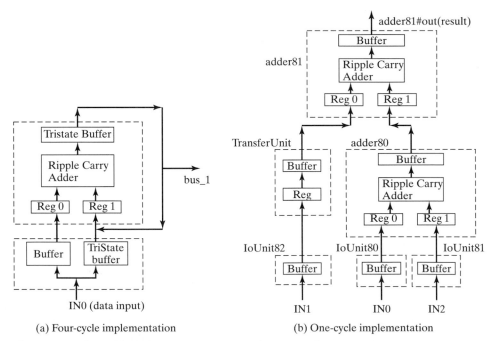

(a) Four-cycle implementation   (b) One-cycle implementation

Figure F-4   Two alternative architectures implementing the sum program.

time-shares the input bus as well as the adder. The second architecture performs the program in a single clock cycle. To achieve this performance, it was necessary to pipeline the algorithm; that is, multiple iterations of the computation overlap. The increased speed translates as expected to a higher hardware cost—one extra adder, extra registers, and a more dedicated bus architecture, including three input ports. Both architectures were produced automatically, given the behavioral description and the clock-cycle constraint. This includes the pipelining transformation.

While architecture compilers have been extensively researched in the academic community (e.g., [DeMan86], [Rabacy91]), their overall impact has remained limited. Commercial introductions have been largely unsuccessful. A number of reasons for this slow penetration can be enumerated:

- Behavioral synthesis assumes the availability of an established synthesis approach at the register-transfer level. This has only recently come to a widespread acceptance. In addition, the discussion about the appropriate input language at the behavioral level has created a lot of confusion. The emergence of widely accepted input languages such as SystemC can change the momentum.
- For a long time, architecture synthesis has concentrated on a limited aspect of the overall design process. The impact of interconnect on the overall design cost, for example, was long ignored. Also, limitations on the architectural scope resulted in inferior solutions apparent to every experienced designer.
- Most importantly, the revolutionary advent of the system-on-a-chip has outstripped the evolutionary progress of the synthesis world. The hybrid nature of embedded system architectures that combine embedded processors with ASIC accelerators ultimately limits the usability of architectural synthesis. Current logic and sequential synthesis tools probably suffice for the accelerators. The challenge has shifted to the synthesis of the software that runs on the embedded processors, chip-level operation systems, driver generators, interconnect network synthesis, and architectural exploration.

Notwithstanding these observations, architectural synthesis has proven to be very successful in a number of application-specific areas. The design of high-performance accelerator units in areas such as wireless communications, storage, imaging, and consumer electronics has benefited greatly from compilers that translate high-level algorithmic functions into hard-wired dedicated solutions.

Example F.3   Architectural Synthesis of Wireless-Communications Processor [Silva01]

An advanced baseband processor for a wireless modem is generated automatically from a high-level description in the Simulink environment ([Mathworks01]). Simulink and Mathlab are tools used extensively in the world of communications design. Capturing the

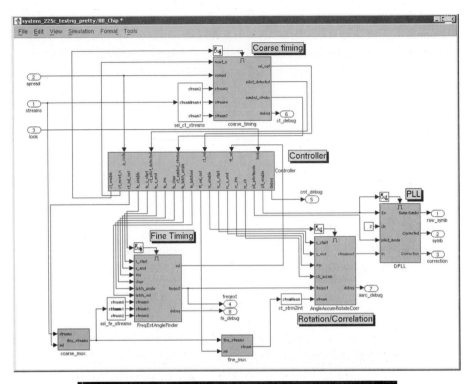

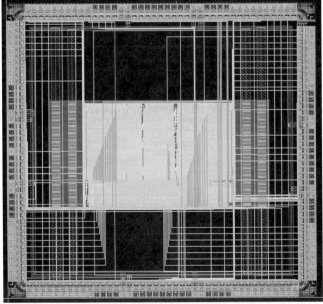

Figure F-5   Architectural synthesis of wireless baseband processor from Simulink (a) to silicon (b). The core area of the chip, which is pad limited, measures only 2 mm² in a 0.18 µm CMOS technology, and counts 600,000 transistors. The high transistor density (0.3 transistor/µm²) demonstrates the effectiveness of today's physical design tools.

design specifications in that environment is a major help in bridging the chasm between systems and implementation engineer. The translation process from Simulink to implementation is managed by the *"Chip-in-a-day"* design environment [Davis01]. This tool manages the synthesis of the individual blocks from behavior to gate level, introduces the chip floorplan, performs the clock tree generation, and oversees the execution of the physical synthesis. The overall generation and verification process takes little more than 24 hours. A similar approach has also proven to be very successful in the mapping of high-level signal-processing functions on rapid-prototyping platforms such as FPGAs. The *System Generator* tool from Xilinx, Inc, for instance, maps modules such as filters, modulators, and correlators, described in the Mathworks Simulink environment, directly onto an FPGA module [SystemGenerator].

To Probe Further

For an in-depth overview of design synthesis, please refer to [DeMicheli94].

References

[AMPS99] AMPS, Intelligent Design Optimization, *http://www.synopsys.com/products/analysis/amps_ds.html*, Synopsys, Inc.

[Brayton84] R. Brayton et al., *Logic Minimization Algorithms for VLSI Synthesis*, Kluwer Academic Publishers, 1984.

[Brayton87] R. Brayton, R. Rudell, A. Sangiovanni-Vincentelli, and A. Wang, "MIS: A Multilevel Logic Optimization System," *IEEE Trans. on CAD*, CAD-6, pp. 1062–81, November 1987.

[Carlson91] S. Carlson, *Introduction to HDL-Based Design Using VHDL*, Synopsys, Inc. 1991.

[Chu86] K. Chu and D. Pulfrey, "Design Procedures for Differential Cascode Logic," *IEEE Journal of Solid State Circuits*, vol. SC-21, no. 6, Dec. 1986, pp. 1082–1087.

[Davis01] W.R. Davis, N. Zhang, K. Camera, F. Chen, D. Markovic, N. Chan, B. Nikolic, R.W. Brodersen, "A Design Environment for High Throughput, Low Power, Dedicated Signal Processing Systems," *Proceedings CICC 2001*, San Diego, 2001.

[DesignCompiler] Design Compiler Technical Datasheet, *http://www.synopsys.com/products/logic/ design_compiler.html*, Synopsys, Inc.

[DeMan86] H. De Man, J. Rabaey, P. Six, and L. Claesen, "Cathedral-II: A Silicon Compiler for Digital Signal Processing," *IEEE Design and Test*, vol. 3, no. 6, pp. 13 25, December 1986.

[DeMicheli94] G. De Micheli, *Synthesis and Optimization of Digital Circuits*, McGraw-Hill, 1994.

[Fishburn85] J. Fishburn and A. Dunlop, "TILOS: A Polynomial Programming Approach to Transistor Sizing," *Proceedings ICCAD-85*, pp. 326–328, Santa Clara, 1985.

[Gajski92] D. Gajski, N. Dutt, A. Wu, and S. Lin, *High-Level Synthesis—Introduction to Chip and System Design*, Kluwer Academic Publishers, 1992.

[Kurup97] P. Kurub and T. Abassi, *Logic Synthesis using Synopsys*, Kluwer Academic Publishers, 1997.

[Mathworks01] Matlab and Simulink, *http://www.mathworks.com*, The Mathworks

[Northrop01] G. Northrop, P. Lu, "A Semicustom Design Flow in High-Performance Microprocessor Design," *Proceedings 38th Design Automation Conference*, Las Vegas, June 2001.

[Rabaey91] J. Rabaey, C. Chu, P. Hoang and M. Potkonjak, "Fast Prototyping of Datapath-Intensive Architectures," *IEEE Design and Test*, vol. 8, pp. 40–51, 1991.

[Silva01] J.L. da Silva Jr., J. Shamberger, M.J. Ammer, C. Guo, S. Li, R. Shah, T. Tuan, M. Sheets, J.M. Rabaey, B. Nikolic, A. Sangiovanni-Vincentelli, P. Wright, "Design Methodology for PicoRadio Networks," *Proceedings DATE Conference*, Munich, March 2000.

[SystemGenerator] The Xilinx System Generator for DSP,
http://www.xilinx.com/xlnx/xil_prodcat_product.jsp?title=system_generator, Xilinx, Inc.

Coping with Interconnect

Driving large capacitors

Dealing with transmission line effects in wires

Signal integrity in the presence of interconnect parasitics

Noise in supply networks

9.1   Introduction

9.2   Capacitive Parasitics
    9.2.1   Capacitance and Reliability—Cross Talk
    9.2.2   Capacitance and Performance in CMOS

9.3   Resistive Parasitics
    9.3.1   Resistance and Reliability—Ohmic Voltage Drop
    9.3.2   Electromigration
    9.3.3   Resistance and Performance—*RC* Delay

9.4   Inductive Parasitics*
    9.4.1   Inductance and Reliability—Voltage Drop
    9.4.2   Inductance and Performance—Transmission-Line Effects

9.5   Advanced Interconnect Techniques
    9.5.1   Reduced-Swing Circuits
    9.5.2   Current-Mode Transmission Techniques

9.6   Perspective: Networks on a Chip

9.7   Summary

9.8   To Probe Further

9.1 Introduction

In previous chapters, we have pointed out the growing impact of interconnect parasitics on all design metrics of digital integrated circuits. Interconnect introduces three types of parasitic effects—capacitive, resistive, and inductive—all of which influence the signal integrity and degrade the performance of the circuit. While so far we have concentrated on the modeling aspects of the wire, we now analyze how interconnect affects the circuit operation, and we present a collection of design techniques to cope with these effects. The discussion addresses each parasitic in sequence.

9.2 Capacitive Parasitics

9.2.1 Capacitance and Reliability—Cross Talk

An unwanted coupling from a neighboring signal wire to a network node introduces an interference that is generally called *cross talk*. The resulting disturbance acts as a noise source and can lead to hard-to-trace intermittent errors, since the injected noise depends upon the transient value of the other signals routed in the neighborhood. In integrated circuits, this intersignal coupling can be both capacitive and inductive (as shown earlier in Figure 1-10). Capacitive cross talk is the dominant effect at current switching speeds. Inductive coupling is already a major concern in the design of the input-output circuitry of mixed-signal circuits, but has not been an issue in digital designs so far.

The potential impact of capacitive cross talk is influenced by the impedance of the line under examination. If the line is floating, the disturbance caused by the coupling persists and may be worsened by subsequent switching on adjacent wires. If the wire is driven, on the other hand, the signal returns to its original level.

Floating Lines

Let us consider the circuit configuration of Figure 9-1. Line X is coupled to wire Y by a parasitic capacitance C_{XY}. Line Y sees a total capacitance to ground equal to C_Y. Assume that the voltage at node X experiences a step change equal to ΔV_X. This step appears on node Y attenuated by the capacitive voltage divider.

$$\Delta V_Y = \frac{C_{XY}}{C_Y + C_{XY}}\Delta V_X \tag{9.1}$$

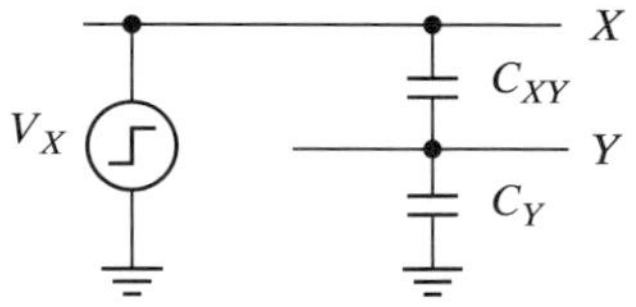

Figure 9-1 Capacitive coupling to a floating line.

Circuits that are particularly susceptive to capacitive cross talk are networks with low-swing precharged nodes, located in adjacency to full-swing wires (with $\Delta V_X = V_{DD}$). Examples include dynamic memories, low-swing on-chip busses, and some dynamic logic families. To address the cross-talk issue, *level-restoring devices* (or keepers) are a must in dynamic logic today!

Example 9.1   Interwire Capacitance and Cross Talk

Consider the dynamic logic circuit of Figure 9-2. The storage capacitance C_Y of the dynamic node Y is composed of the diffusion capacitances of the precharge and discharge transistors, the gate capacitance of the connecting inverter, and the wire capacitance. A nonrelated signal Y is routed as an Al-1 (first layer of aluminum) wire over the polysilicon gate of one of the transistors in the inverter. This creates a parasitic capacitance C_{XY} with respect to node Y. Suppose now that node Y is precharged to 2.5 V and that signal X undergoes a transition from 2.5 V to 0 V. The charge redistribution causes a voltage drop ΔV_Y on node Y, as given by Eq. (9.1).

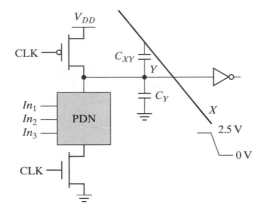

Figure 9-2 Cross talk in dynamic circuits.

Assume that C_Y equals 6 fF. An overlap of $3 \times 1 \ \mu m^2$ between Al-1 and polysilicon results in a parasitic coupling capacitance of 0.5 fF ($3 \times 1 \times 0.057 + 2 \times 3 \times 0.054$), as obtained from Table 4-2. The computation of the fringing effect in the case of overlapping wires is complex and depends upon the relative orientation of the wires. We assumed in this analysis that two sides of the cross section contribute. A 2.5-V transition on X thus causes a voltage disturbance of 0.19 V on the dynamic node (or 7.5%). Combined with other parasitic effects, such as charge redistribution and clock feed-through, this might lead to circuit malfunctioning.

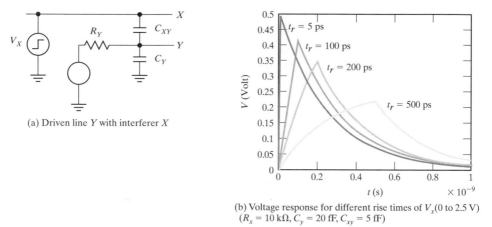

(a) Driven line Y with interferer X

(b) Voltage response for different rise times of V_x(0 to 2.5 V)
(R_x = 10 kΩ, C_y = 20 fF, C_{xy} = 5 fF)

Figure 9-3   Capacitive coupling to a driven line.

Driven Lines

If the line Y is driven with a resistance R_Y, a step on line X results in a transient on line Y (see Figure 9-3a). The transient decays with a time constant $\tau_{XY} = R_Y(C_{XY} + C_Y)$. The actual impact on the "victim line" is a strong function of the rise (fall) time of the interfering signal. If the rise time is comparable or larger than the time constant, the peak value of disturbance is diminished. This is illustrated by the simulated waveforms in Figure 9-3b. Obviously, keeping the driving impedance of a wire—and hence τ_{XY}—low goes a long way toward reducing the impact of capacitive cross talk. The keeper transistor, added to a dynamic gate or precharged wire, is an excellent example of how impedance reduction helps to control noise.

In summary, the impact of cross talk on the signal integrity of driven nodes is rather limited. The resulting glitches may cause malfunctioning of connecting sequential elements, and should therefore be carefully monitored. However, the most important effect is an increase in delay, which we discuss later.

Design Techniques—Dealing with Capacitive Cross Talk

Cross talk is a proportional noise source. This means that scaling the signal levels to increase noise margins does not help since the noise sources scale in a similar way. The only options in addressing the problem is to control the circuit geometry, or to adopt signaling conventions that are less sensitive to coupled energy. The following ground rules, as advocated in [Dally98], can be established:

1. If at all possible, avoid floating nodes. Nodes sensitive to cross-talk problems, such as precharged busses, should be equipped with keeper devices to reduce the impedance.
2. Sensitive nodes should be well separated from full-swing signals.
3. Make the rise (fall) time as large as possible, subject to timing constraints. Be aware, however, of the impact this might have on the short-circuit power.
4. Use differential signaling in sensitive low-swing wiring networks. This turns the cross-talk signal into a common-mode noise source that does not impact the operation of the circuit.
5. In order to keep cross talk to a minimum, do not allow the capacitance between two signal wires to grow too large. For example, it is bad practice to have two wires on the same layer run parallel for a

long distance, although one is often tempted to do just that when distributing the two clocks in a two-phase system or when routing a bus. Parallel wires on the same layer should be *spaced sufficiently*. Increasing the wire pitch (e.g., for a bus) reduces the cross talk and increases the performance. Wires on adjacent layers should run perpendicular to each other.

6. If necessary, provide a *shielding* wire—*GND* or V_{DD}—between the two signals (see Figure 9-4). This effectively turns the interwire capacitance into a capacitance-to-ground and eliminates interference. An adverse effect of shielding is the increased capacitive load.

7. The interwire capacitance between signals on different layers can be further reduced by the addition of extra routing layers. When four or more routing layers are available, we can fall back to an approach often used in printed circuit board design: interleave every signal layer with a GND or V_{DD} metal plane (see Figure 9-4).

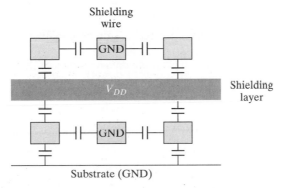

Figure 9-4 Cross section of routing layers, illustrating the use of shielding to reduce capacitive cross talk.

9.2.2 Capacitance and Performance in CMOS

Cross Talk and Performance

The previous section discussed the impact of capacitive cross talk on the signal integrity. Even when cross talk does not result in fatal breakdown of the circuit, it still should be monitored carefully because it also affects the performance of the gate. The circuit schematic of Figure 9-5 is illustrative of how capacitive cross talk may result in a data-dependent variation of the propagation delay. Assume that the inputs to the three parallel wires X, Y, and Z experience simultaneous transitions. Wire Y (called the victim wire) switches in a direction that is opposite to the transitions of its neighboring signals, X and Z. The coupling capacitances experience a voltage swing that is double the signal swing, and thus represent an effective capacitive load that is twice as large as C_c. In other words, it experiences the *Miller effect* (as introduced in Chapter 5). Since the coupling capacitance represents a large fraction of the overall capacitance in the deep-submicron dense wire structures, this increase in capacitance is substantial, and it has a major impact on the propagation delay of the circuit. Note that this is a worst case scenario. If all inputs experience a simultaneous transition in the same direction, the voltage over the coupling capacitances remains constant, resulting in a zero contribution to the effective load capacitance. The total load

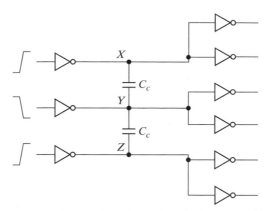

Figure 9-5   Impact of cross talk on propagation delay.

capacitance C_L of gate Y therefore depends on the data activities of the neighboring signals, and it varies between the bounds given by

$$C_{\text{GND}} \leq C_L \leq C_{\text{GND}} + 4C_c \qquad\qquad (9.2)$$

where C_{GND} is the capacitance of node Y to ground, including the diffusion and fan-out capacitances. In [Sylvester98], it was established that for a 0.25-μm technology, the wire delay for a noise is potentially 80% larger than the wire delay without noise (as shown in Figure 9-5, with wire length = 100 μm, and fan-out = 2).

Complicating the analysis of this problem is the fact that the capacitance not only depends upon the values of the surrounding wires, but also upon the exact timing of the transitions. The simultaneity of transitions can only be detected from detailed timing simulations, making the timing verification process substantially more complex. Hence, the ensuing explosion in verification cost caused by the unpredictability of the actual delay is a source of major concern. Assuming worst case conditions for the capacitances—in other words, assuming the Miller effect—leads to an overly pessimistic estimation, and ultimately overkill in the circuit design.

Example 9.2   Impact of Cross Talk on Performance

To illustrate the effects of cross talk, let us consider a bus of N bits, where the wires (of length L) are routed in parallel, evenly spaced, and driven independently. Assume that all the N inputs transition at the same time. Due to the interwire capacitance, the delay of the kth wire is a function of the transitions on the neighboring wires, $k-1$ and $k+1$. Based on Elmore, we obtain a good approximation of this delay with

$$t_{p,k} = gC_W(0.38R_W + 0.69R_D) \qquad\qquad (9.3)$$

where $C_W = c_w L$ and $R_W = r_w L$. c_w and r_w stand for the capacitance to ground and the resistance of the wire per unit length, respectively, while R_D is the equivalent resistance of the driver. The correction factor g introduces the cross-talk effect and is a function of the ratio

$r = c_i/c_w$, and the activities on the wires. c_i represents the interwire capacitance per unit length. The value of g for a set of representative scenarios is given in Table 9-1. When all three wires ($k - 1$, k, $k + 1$) transition in the same direction, the interwire capacitance does not come into the picture and $g = 1$. The worst case occurs when both neighbors transition in the opposite direction of wire k, leading to $g = 1 + 4r$. For the plausible case of $c_i = c_w$, $g = 5$. Hence, the wire delay may vary over 500% between the worst and best case, purely as a function of the direction of the transitions on the wires!

Table 9-1 Delay-ratio factor g for bus wire as a function of simultaneous transitions on neighboring lines. The table only contains representative examples, and it is easily expanded to contain a complete set of all possible transition scenarios. The symbols —, ↑, and ↓ stand for no, positive, and negative transitions, respectively.

bit $k - 1$	bit k	bit $k + 1$	Delay factor g
↑	↑	↑	1
↑	↑	—	$1 + r$
↑	↑	↓	$1 + 2r$
—	↑	—	$1 + 2r$
—	↑	↓	$1 + 3r$
↓	↑	↓	$1 + 4r$

Design Techniques—Circuit Fabrics with Predictable Wire Delay

With cross talk making wire delay more and more unpredictable, a designer can choose between a number of different methodology options to address the issue, including the following:

1. **Evaluate and improve.** After detailed extraction and simulation, the bottlenecks in delay are identified, and the circuit is appropriately modified.
2. **Constructive layout generation.** Wire routing programs take into account the effects of the adjacent wires, ensuring that the performance requirements are met.
3. **Predictable structures.** By using predefined, known, or conservative wiring structures, the designer is ensured that the circuit will meet his specifications and that cross talk will not be a show stopper.
4. **Avoid worst case patterns.** Eliminate or avoid wire transitions that cause the worst case delays.

The first approach is the most often used. It has the disadvantage of requiring many iterations through the complete design generation process, and hence, it is very slow. The second technique is appealing. The complexity of the required tool set, however, makes it doubtful that this ambitious goal is actually achievable in a reliable (and affordable) way in the foreseeable future (although some major inroads already have been made; see [Apollo02] for example). The third approach may be the only one that really is workable, at least in the short run. Just as regular structures helped to tame the complexity of transistor layout in the early 1980s, regular and predictable wiring topologies may be the way to address the

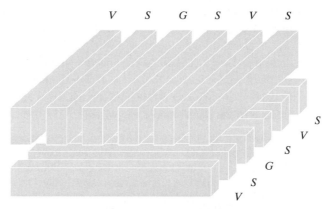

Figure 9-6   Dense Wire Fabric (DWF) [Khatri01].
S, *V*, and *G* stand for Signal, V_{DD}, and GND, respectively.

capacitive cross talk problem. For example, preserving bit slices in datapath structures keeps the wires organized and makes cross-talk manageable. Field-programmable gate arrays (FPGAs), with their well-characterized interconnect grid, provide another representative example. The availability of multiple layers of metal interconnect makes this approach viable for semicustom (and custom) design as well. The *dense wire fabric* [Khatri01] shown in Figure 9-6 presents such a solution. Minimum-width wires are prewired on a minimum pitch distance. Wires on adjacent layers are routed orthogonally, minimizing the cross talk. Signals on the same layer are separated by V_{DD} or GND shields. The wiring structure is personalized by providing vias at the appropriate places. The advantage of this approach is that cross talk is virtually eliminated, and delay variation is reduced to no more than 2%. This comes at a cost in area and a capacitance increase of 5%—with a concomitant increase in delay and power consumption. For most designs, the reduction in design and verification time easily compensates for this performance penalty. The VPGA approach introduced in Chapter 8 represents an example of the Dense Wire Fabric approach.

   Approach 4—avoiding worst case patterns—presents an intriguing possibility. In the case of a bus, it is possible to encode the data in such a way that the transitions that "victimize" delay are eliminated. This requires that the bus interface units include encoder and decoder functions (see Figure 9-7). While these represent hardware and delay overhead, they may help to reduce the delay by as much as a factor of 2 for large busses [Sotiriadis01]. Encoding data before transmitting it on a bus can also help to reduce power dis-

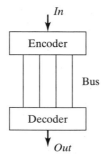

Figure 9-7   Encoding data to eliminate worst case conditions
can speed up a bus.

sipation by minimizing the number of transitions [Stan95]. The latter approach in itself does not address cross talk, however. ∎

Capacitive Load and Performance

The increasing values of the interconnect capacitances, especially those of the global wires, emphasize the need for effective driver circuits that can (dis)charge capacitances with sufficient speed. This need is further highlighted by the fact that in complex designs, a single gate often has to drive a large fan-out and hence has a large capacitive load. Typical examples of large on-chip loads are busses, clock networks, and control wires. The latter include, for instance, reset and set signals. These signals control the operation of a large number of gates, so fan-out normally is high. Other examples of large fan-outs are encountered in memories, in which a large number of storage cells is connected to a small set of control and data wires. The capacitance of these nodes is easily in the multi-picofarad range. The worst case occurs when signals go off chip. In this case, the load consists of the package wiring, the printed circuit board wiring, and the input capacitance of the connected ICs or components. Off-chip loads can be as large as 50 pF, which is many thousands of times larger than a standard on-chip load. Driving those nodes with sufficient speed becomes one of the most crucial design problems.

The main secrets to the efficient driving of large capacitive loads were already revealed in Chapter 5. They boil down to two dominant concepts:

- Adequate **transistor sizing** is instrumental when dealing with large loads.
- **Partitioning drivers** into chains of gradually increasing buffers helps to deal with large fan-out factors.

Some other important conclusions from that analysis are worth repeating for the sake of clarity:

- When optimizing for performance, the delay of a multistage driver should be **divided equally over all stages**.
- If the number of stages can be chosen freely, a **fan-out (sizing) factor of approximately 4 per stage** leads to the minimum delay for contemporary semiconductor processes. Choosing factors that are somewhat larger does not affect the performance that much, while yielding substantial area benefits.

In the following sections, we elaborate further on this topic. We focus especially on the driving of very large capacitances, such as those encountered when going off chip. We also introduce some special and quite useful driver circuits.

Driving off-Chip Capacitances—A Case Study As established in Chapter 2, the increase in complexity of today's integrated circuits translates to an explosive need for input/output pins. Packages with over 1000 pins have become a necessity. This puts tough requirements on the bonding-pad design in terms of noise immunity. The simultaneous switching of a lot of pads, each driving a large capacitor, causes large transient currents and creates voltage fluctuations on the power and ground lines. This reduces the noise margins and affects the signal integrity, as will become apparent later in this chapter.

At the same time, technology scaling reduces the internal capacitances on the chip, while off-chip capacitances remain approximately constant—typically 10 to 20 pF. As a result, the overall effective fan-out factor F of an output-pin driver experiences a net increase with technology scaling. In a consistent system-design approach, we expect the off-chip propagation delays to scale in the same way as the on-chip delays. This puts an even tougher burden on the bonding-pad buffer design. The challenges associated with a bonding-pad driver design are illustrated with the following simple case study.

Example 9.3   Output Buffer Design

Consider the case where an on-chip minimum-size inverter has to drive an off-chip capacitor C_L of 20 pF. C_i equals approximately 2.5 fF for a standard gate in a 0.25-μm CMOS process. This corresponds to a t_{p0} of approximately 30 ps. The overall effective fan-out F (the ratio between C_L and C_i) equals 8000, which definitely calls for a multistage buffer design. From Eq. 5-36, and assuming that $\gamma = 1$, we learn that seven stages result in a near-optimal design, with a scaling factor f of 3.6 and an overall propagation delay of 0.76 ns. With a PMOS over NMOS ratio of 1.9—the optimum ratio derived for our standard process parameters in Example 5.6—and a minimum dimension of 0.25 μm, we can derive the widths for the NMOS and PMOS transistors in the consecutive inverter stages, as shown in Table 9-2.

Table 9-2   Transistor sizes for optimally sized cascaded buffers.

Stage	1	2	3	4	5	6	7
W_n (μm)	0.375	1.35	4.86	17.5	63	226.8	816.5
W_p (μm)	0.71	2.56	9.2	33.1	119.2	429.3	1545.5

This solution obviously requires some extremely large transistors with gate widths of up to 1.5 mm! The overall size of this buffer equals several thousand minimum-size inverters, which is quite prohibitive because a complex chip requires a large number of those drivers. This argument clearly illustrates the enormous cost associated with attempting to achieve optimum delay.

Trading Off Performance for Area and Energy Reduction   Fortunately, it is not necessary to achieve the optimal buffer delay in most cases. Off-chip communications can often be performed at a fraction of the on-chip clocking speeds. Relaxing the delay requirements still allows for off-chip clock speeds in excess of 100 MHz, while substantially reducing the buffering requirements. This redefines the buffer design problem, as follows:

> **Given a maximum propagation delay time $t_{p,max}$, determine the number of buffer stages N and the required scaling factor f, such that the overall area is minimized. This is equivalent to finding a solution that sets t_p as close as possible to $t_{p,max}$.**

The optimization problem is now reformulated into the finding of the minimum integer value of N that obeys the constraints of Eq. (9.4):

$$\frac{t_{p,max}}{t_{p0}} \geq \ln(F)\frac{f}{\ln(f)} = N \times F^{1/N} \tag{9.4}$$

This transcendental optimization problem can be solved by using a small computer program or by using mathematical packages such as MATLAB [Etter93]. Figure 9-8 plots the right-hand side of this equation as a function of N for some values of F. For a given value of the overall effective fan-out F, and a maximum value of the delay $(t_{p,max}/t_{p0})$, the minimum number of buffers N is derived from inspection of the appropriate curve.[1]

Using the total width of the transistors as a measure, we can project the area savings of the larger scaling factor. Assuming that the area of the minimum inverter equals A_{min}, and that scaling the transistors with a factor f results in a similar area increase, we can derive the area of the driver as a function of f:

$$A_{driver} = (1 + f + f^2 + \dots + f^{N-1})A_{min}$$

$$\left(\frac{f^N - 1}{f - 1}\right)A_{min} = \frac{F - 1}{f - 1}A_{min} \tag{9.5}$$

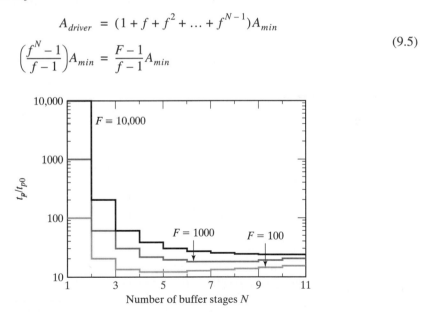

Figure 9-8 t_p/t_{p0} as a function of the number of buffer stages for various values of the overall fan-out F.

---

[1] When minimizing the buffer area under timing constraints, keeping the fan-out factor f constant actually is not the optimal solution. A gradual increase in the fan-out factor from stage to stage provides a better solution [Ma94]. We will briefly re-address this issue in Chapter 11.

In short, the area of the driver is approximately inversely proportional to the tapering factor f. Choosing larger values of f can help to reduce the area cost substantially.

It also is worth reflecting on the energy dissipation of the exponential buffer. While it takes an amount of energy equal to $C_L V_{DD}^2$ simply to switch the load capacitance, the driver itself consumes energy to switch its internal capacitances (ignoring short-circuit energy). This overhead energy is approximated by the following expression:

$$E_{driver} = (1 + f + f^2 + \dots + f^{N-1})C_i V_{DD}^2$$

$$= \left(\frac{F-1}{f-1}\right)C_i V_{DD}^2 \approx \frac{C_L}{f-1}V_{DD}^2 \qquad (9.6)$$

This means that driving a large capacitance fast—that is, using the optimal tapering factor of 3.6—consumes 40% more energy than the unbuffered case. For large load capacitances, this is a substantial overhead. Backing off somewhat from the optimum tapering factor can help to reduce the extra dissipation.

Example 9.4   Output Driver Design—Revisited

Applying these results to the bonding-pad driver problem and setting $t_{p,max}$ to 2 ns results in the following solution: $N = 3$, $f = 20$, and $t_p = 1.8$ ns. The required transistor sizes are summarized in Table 9-3.

Table 9-3   Transistor sizes of redesigned cascaded buffer.

Stage	1	2	3
W_n (µm)	0.375	7.5	150
W_p (µm)	0.71	14.4	284

The overall area of this solution is approximately 7.5 times smaller than the optimum solution, while its speed is reduced by a factor of less than 2.5. The extra energy dissipation per switching transition due to the buffer intrinsic capacitances is reduced to an almost negligible amount. The overall dissipation of the combined buffer and load is reduced by 24%, which is a hefty amount given the size of the load capacitor. Clearly, then, it pays to **design a circuit for the right speed, not for the maximum speed!**

Design Consideration—Implementing Wide Transistors

Even this redesigned buffer requires wide transistors, and one must be careful when designing those devices, since the large value of W translates into very long gate connections. Long polysilicon wires tend to be highly resistive, which degrades the switching performance. This problem can be addressed in the fol-

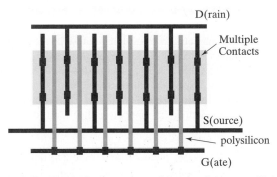

Figure 9-9 To implement very wide transistors, multiple transistors (or "fingers") are placed in parallel.

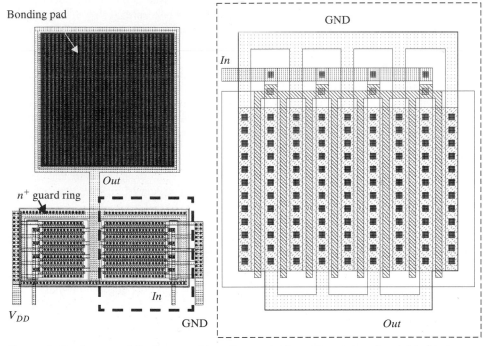

Figure 9-10 Layout of final stage of bonding-pad driver. The plot on the right side is a magnification of the NMOS transistor connected between GND and *Out*.

lowing way: A wide transistor can be constructed by connecting many smaller transistors (also called "fingers") in parallel (Figure 9-9).

The resistance of the gate is reduced with the aid of a low-resistance metal bypass connecting the shorter polysilicon sections. An example of a pad driver designed using these techniques is shown in Figure 9-10. When trying to decide if a very wide transistor should be partitioned into smaller devices, use the following rule of thumb: The *RC* delay of the polysilicon gate should always be substantially smaller than the switching delay of the overall digital gate. ■

Design Challenge—Designing Reliable Output and Input Pads

The design of bonding-pad drivers is obviously a critical and nontrivial task, which is further complicated by noise and reliability considerations. For instance, the large transient currents resulting from the switching of the huge output capacitance can cause latch-up to occur. Multiple well and substrate contacts supplemented with guard rings help avoid the onset of this destructive effect.

Guard rings are grounded p+ diffusions in a *p*-well and supply-connected n+ diffusions in an *n*-well that are used to collect injected minority carriers before they reach the base of the parasitic bipolar transistors. These rings should surround the NMOS and PMOS transistors in the final stage of the output pad driver.

The designer of an input pad faces some different challenges. The input of the first stage of the input buffer is directly connected to external circuitry, and thus is sensitive to any voltage excursions on the connected input pins. A human walking over a synthetic carpet in a relative air humidity of 80% or more can accumulate a voltage potential of 1.5 kV—you have probably experienced the sparks that jump from your hand when touching a metal object under those circumstances. The same is true for the assembly machinery. The gate connection of an MOS transistor has a very high input resistance (10^{12} to 10^{13} Ω). The voltage at which the gate oxide punctures and breaks down is about 10 to 20 V, and is getting smaller with reducing oxide thicknesses. Therefore, either human or a machine charged up to a high static potential can easily cause a fatal breakdown of the input transistors when brought in contact with the input pin. This phenomenon, known as called *electrostatic discharge* (ESD), has proven to be fatal to many circuits during manufacturing and assembly.

A combination of resistance and diode clamps are used to defray and limit this potentially destructive voltage. A typical *electrostatic protection circuit* is shown in Figure 9-11a. The protection diodes D_1 and D_2 turn on when the voltage at node X rises below V_{DD} or goes below ground. The resistor R is used to limit the peak current that flows in the diodes in the event of an unusual voltage excursion. Currently, designers tend to use tub resistors (*p*-diffusion in an *n*-well, and *n*-diffusion for a *p*-well) to implement the resistors, and values can be anywhere from 200 Ω to 3 kΩ. The designer should be aware that the resulting *RC* time constant can be a performance limiting factor in high-speed circuits. Figure 9-11b shows the layout of a typical input pad.   ∎

The task of the pad designer seems to become more difficult with every new technology generation. Fortunately, a number of novel packaging technologies that reduce off-chip capacitance are emerging. For instance, advanced techniques such as *ball grid arrays* and chip-on-board go a long way in helping to reduce the off-chip driving requirements. (Recall Chapter 2.)

Some Interesting Driver Circuits   So far, most of our driver circuits have been simple inverters. Sometimes, other functions are necessary. The *tristate* buffer is an example of such a variant. Busses are essential in most digital systems—a *bus* is a shared bundle of wires that connect a set of sender and receiver devices, such as processors, memories, disks, and input/output devices. When one device is sending information through the bus, all other transmitting devices should be disconnected. This can be achieved by putting the output buffers of those devices in a high-impedance state Z that effectively disconnects the gate from the output wire. Such a buffer has three possible states—0, 1, and Z—and is therefore called a *tristate* buffer.

Implementing a tri-state inverter is straightforward in CMOS. Simultaneously turning off the NMOS and PMOS transistor produces a floating output node that is disconnected from its

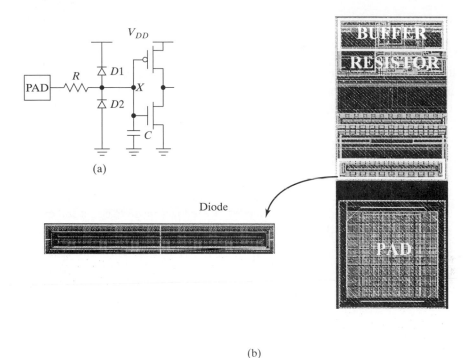

Figure 9-11 Input protection circuit (a), and layout of input pad (b).

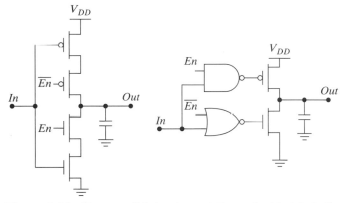

Figure 9-12 Two possible implementations of a tri-state buffer. *En* = 1 enables the buffer.

input. Two possible implementations of a tri-state buffer are shown in Figure 9-12. While the first one is simple,[2] the second is more appropriate for driving large capacitances. Having stacked transistors in the output stage should be avoided due to the huge area overhead.

---

[2]Observe how this tristate buffer resembles a C²MOS latch.

9.3 Resistive Parasitics

9.3.1   Resistance and Reliability—Ohmic Voltage Drop

Current flowing through a resistive wire results in an ohmic voltage drop that degrades the signal levels. This is especially important in the power-distribution network, where current levels can easily reach amperes, as illustrated in Figure 9-13 for the Compaq (formerly Digital Equipment Corporation) Alpha processor family.

Consider now a 2-cm-long V_{DD} or GND wire with a current of 1 mA per μm width. This current is about the maximum that can be sustained by an aluminum wire due to *electromigration*, which is discussed in the subsequent section. Assuming a sheet resistance of 0.05 Ω/□, the resistance of this wire (per μm width) equals 1 kΩ. A current of 1 mA/μm would result in a voltage drop of 1 V. The altered value of the voltage supply reduces noise margins and changes the logic levels as a function of the distance from the supply terminals. This is demonstrated by the circuit in Figure 9-14, where an inverter placed far from the power and ground pins connects to a device closer to the supply. The difference in logic levels caused by the *IR* voltage drop over the supply rails might partially turn on transistor M_1. This can result in an accidental discharging of

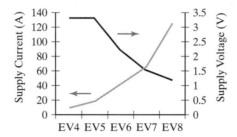

Figure 9-13   Evolution of power-supply current and supply voltage for different generations of the high-performance Alpha microprocessor family from Compaq. Even with the reductions in supply voltage, a total supply current of over 100 A has to be supported for the newer generations [Herrick00].

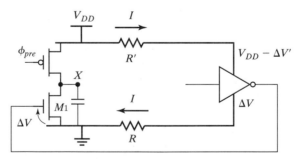

Figure 9-14   Ohmic voltage drop on the supply rails reduces the noise margins.

the precharged, dynamic node X, or it can cause static power consumption if the connecting gate is static. In short, the current pulses from the on-chip logic and memories, and the I/O pins cause voltage drops over the power-distribution network and are the major source for on-chip power-supply noise. In addition to causing a reliability risk, *IR* drops on the supply network also impact the performance of the system, as a small drop in the supply voltage may cause a significant increase in delay.

The most obvious solution to this problem is to reduce the maximum distance between the supply pins and the circuit supply connections. This is most easily accomplished through a structured layout of the power distribution network. A number of on-chip power-distribution networks with peripheral bonding are shown in Figure 9-15. In all solutions, power and ground are brought onto the chip via bonding pads located on the four sides of the chip. Which approach to use depends on the number of coarse metal layers (that is, thick, high-pitch, topmost metal layers) that one is allowed to allocate to power distribution. In the first approach (a), power and ground are routed vertically (or horizontally) on the same layer. Power is brought in from two sides of the chip. Local power strips are strapped to this upper grid and then further routed on the

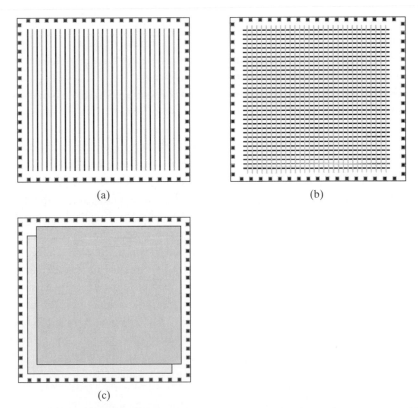

(a)

(b)

(c)

Figure 9-15   On-chip power-distribution networks. (a) Single layer power grid; (b) dual layer grid; (c) dual power plane.

lower metal levels. Method (b) uses two coarse metal layers for the power distribution, and the power is brought in from the four sides of the die. This approach was used in the EV5 generation of the Compaq Alpha processor [Herrick00], in which the combination of power and clock distribution occupied more than 90% of the third and fourth Al layers. A more aggressive step to take is to use two solid metal planes for the distribution of V_{dd} and GND (c). This approach has the advantage of drastically reducing the resistance of the network. The metal planes also act as shields between data signalling layers—hence, reducing cross talk. In addition, they help to reduce the on-chip inductance. Obviously, this approach is only feasible when sufficient metal layers are available.

The sizing of the power network is a nontrivial task. The power grid can be modeled as a network of resistors (wires) and current sources (logic), containing hundreds of millions of elements. Very often, many paths exist between the power pins and the supply connections of a chip module or gate. While, in general, the current follows the path of the lowest resistance, the exact flow depends on factors such as the current draw from neighboring modules that share the same network. The analysis is complicated by the fact that peak currents drawn by the connected modules are distributed over time. *IR* drop is a dynamic phenomenon! The largest stress on the power grid can primarily be attributed to simultaneous switching events, such as caused by clocks and bus drivers. The demand for current from the power network usually peaks after clock transients, or when large drivers switch. At the same time, a worst-case analysis adding all the peak currents most often leads to a gross over dimensioning of the wires.

Therefore, it is clear that computer-aided design tools are a necessary companion of the power-distribution network designer. Given the complexity of today's integrated circuits, transistor-level analysis of the complete power-network requirements is not feasible. At the same time, partitioning of the problem over subsections may not result in an accurate picture, either. Changes to the power grid in one section tend to have a global impact. This is illustrated in Figure 9-16, which shows the simulated *IR* voltage drop of a complex digital circuit. A first implementation (see Figure 9-16a) suffers from a more than acceptable drop in the upper right module of the design, because only the top portion of the power grid feeds the large drivers at the top. The lower portions of the module are not directly connected to the grid. Adding just one single strap to the network largely resolves this problem, as shown in Figure 9-16b.

Therefore, an accurate picture of the *IR* risk cannot be obtained unless the entire chip is verified as a single entity. Any tool used for this purpose must have the capacity to analyze millions of resistor grids. Fortunately, a number of efficient and quite accurate power-grid analysis tools are currently available [Cadence-Power, RailMill]. They combine a dynamic analysis of the current requirements of the circuit modules with a detailed modeling of the power network. Such tools are quite indispensable in the design flows of today.

9.3.2   Electromigration

The current density (current per unit area) in a metal wire is limited due to an effect called *electromigration*. A *direct* current in a metal wire running over a substantial time period causes a

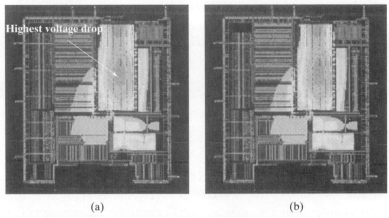

(a)   (b)

Figure 9-16 Simulated *IR* voltage drop in power-distribution network of two versions of a complex digital integrated circuit. The gray scale is used to indicate the severeness of the voltage drop. Adding an extra strap to the upper layers of the power network (b) helps to address the large voltage drop in the upper right module in the initial design (a) [Cadence-Power]. See also Colorplate 8.

transport of the metal ions. Eventually, this causes the wire to break or to short-circuit to another wire. This type of failure will only occur after the device has been in use for some time. Some examples of failure caused by migration are shown in Figure 9-17. Notice how the first photo clearly shows hillock formation in the direction of the electron current flow.

The rate of the electromigration depends on the temperature, the crystal structure, and the average current density. The latter is the only factor that can be effectively controlled by the circuit designer. Keeping the current below 0.5 to 1 mA/μm normally prevents migration. This parameter can be used to determine the minimal wire width of the power and ground network. Signal wires normally carry an ac current and are less susceptible to migration. The bidirectional

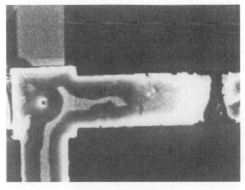
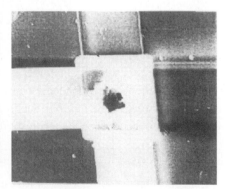

(a) Line-open failure   (b) Open failure in contact plug

Figure 9-17 Electromigration-related failure modes. (Courtesy of N. Cheung and A. Tao, U.C. Berkeley.)

flow of the electrons tends to anneal any damage done to the crystal structure. Most companies impose a number of strict wire-sizing guidelines on their designers, based on measurements and past experience. Research results have shown that many of these rules tend to be overly conservative [Tao94].

Design Rule

Electromigration effects are proportional to the average current flow through the wire, while *IR* voltage drops are a function of the peak current. ∎

At the technology level, a number of precautions can be taken to reduce the migration risk. One option is to add alloying elements (such as Cu or Tu) to the aluminum to prevent the movement of the Al ions. Another approach is to control the granularity of the ions. The introduction of new interconnect materials is a big help as well. For instance, the expected lifetime of a wire is increased by a factor of 100 when using copper interconnect instead of aluminum.

9.3.3   Resistance and Performance—*RC* Delay

In Chapter 4, we established that the delay of a wire grows quadratically with its length. Doubling the length of a wire increases its delay by a factor of 4! The signal delay of long wires therefore tends to be dominated by the *RC effect*. This is becoming an ever larger problem in modern technologies, which feature an increasing average length of the global wires (Figure 4-27)—at the same time that the average delay of the individual gates is going down. This leads to the rather strange situation that it may take multiple clock cycles to get a signal from one side of a chip to its opposite end [Dally01]. Providing accurate synchronization and correct operation becomes a major challenge under these circumstances. In this section, we discuss a number of design techniques that can help you cope with the delay imposed by the resistance of a wire.

Better Interconnect Materials

A first option for reducing *RC* delays is to use better interconnect materials when they are available and appropriate. The introduction of silicides and copper has helped to reduce the resistance of polysilicon and metal wires, respectively, while the adoption of dielectric materials with a lower permittivity lowers the capacitance. Both copper and low-permittivity dielectrics have become common in advanced CMOS technologies (starting from the 0.18-μm technology generation). Yet, the designer should be aware that these new materials only provide a temporary respite of one or two generations, and they do not solve the fundamental problem of the delay of long wires. Innovative design techniques are often the only way of coping with the latter.

Example 9.5   Impact of Advanced Interconnect Materials

Copper offers a resistivity that is 1.6 times lower than that of aluminum (Table 4-4) from a pure material perspective. Cladding and other manufacturing artifacts increase the effective resistivity of on-chip Cu wires to approximately $2.2 \cdot 10^{-8}$ Ω-m.

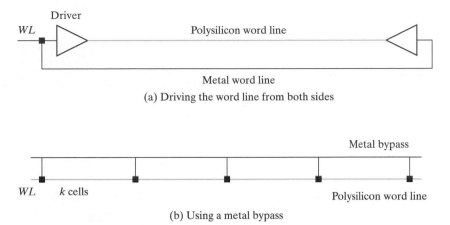

(a) Driving the word line from both sides

(b) Using a metal bypass

Figure 9-18 Approaches to reduce the address-line delay.

It is sometimes hard to avoid the use of long polysilicon wires. A typical example is the address lines in a memory, which must connect to a large number of transistor gates. Keeping the wires in polysilicon increases the memory density substantially by avoiding the overhead of the extra metal contacts. The polysilicon-only option unfortunately leads to an excessive propagation delay. One possible solution is to drive the address line from both ends, as shown in Figure 9-18a. This effectively reduces the worst-case delay by a factor of four. Another option is to provide an extra metal wire, called a *bypass*, which runs parallel to the polysilicon one and connects to it every k cell (see Figure 9-18b). The delay is now dominated by the much shorter polysilicon segments between the contacts, and is proportional to $(k/2)^2$. Providing contacts only every k cells helps to preserve the implementation density. For example, providing a connection to the bypass line every 16 cells in a address line of 1024 cells reduces the delay by approximately 4000.

Better Interconnect Strategies

With the length of the wire being a prime factor in both the delay and the energy consumption of an interconnect wire, any approach that helps to reduce the wire length is bound to have an essential impact. It was pointed out earlier that the addition of interconnect layers tends to reduce the average wire length, as routing congestion is reduced and interconnections can pretty much follow a direct path between source and destination. Yet, the "Manhattan-style" wiring approach that is typical in today's routing tools brings with it a substantial amount of overhead that is often overlooked. In the Manhattan-style routing, interconnections are routed first along one of two preferred directions, followed by a connection in the other direction (see Figure 9-19a). It seems obvious that routing along the diagonal direction would yield a sizable reduction in wire length—up to 29% in the best case. Ironically, 45° lines were very popular in the early days of integrated circuits designs, but they got of out of vogue because of complexity issues, impact on tools, and mask-making concerns. Recently, it has

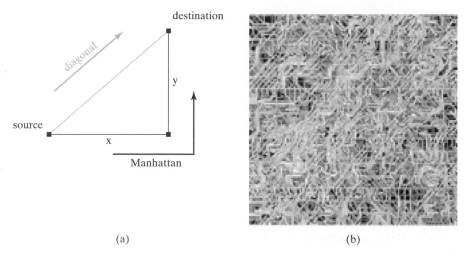

(a)   (b)

Figure 9-19   Manhattan versus diagonal routing. (a) Manhattan routing uses preferential routing along orthogonal axis, while diagonal routing allows for 45° lines; (b) example of layout using 45° lines.

been demonstrated that these concerns can be addressed adequately, and that 45° lines are perfectly feasible [Cadence-X]. The impact on wiring is quite tangible: a reduction of 20% in wire length! This, in turn, results in higher performance, lower power dissipation, and smaller chip area. The density of wiring structures that go beyond pure Manhattan directions is demonstrated clearly in the sample layout shown in Figure 9-19b.

Introducing Repeaters

The most popular design approach to reducing the propagation delay of long wires is to introduce intermediate buffers, (also called *repeaters*), in the interconnect line (see Figure 9-20). Making an interconnect line m times shorter reduces its propagation delay quadratically, and is sufficient to offset the extra delay of the repeaters when the wire is sufficiently long [Cong99, Adler00]. Assuming that the repeaters have a fixed delay t_{pbuf}, we can derive the optimum number of repeaters m_{opt} to insert using an analysis similar to the one used for the optimization of a chain of transmission gates (see page 283 in Chapter 6). This yields

$$m_{opt} = L \sqrt{\frac{0.38rc}{t_{pbuf}}} = \sqrt{\frac{t_{pwire(unbuffered)}}{t_{pbuf}}} \tag{9.7}$$

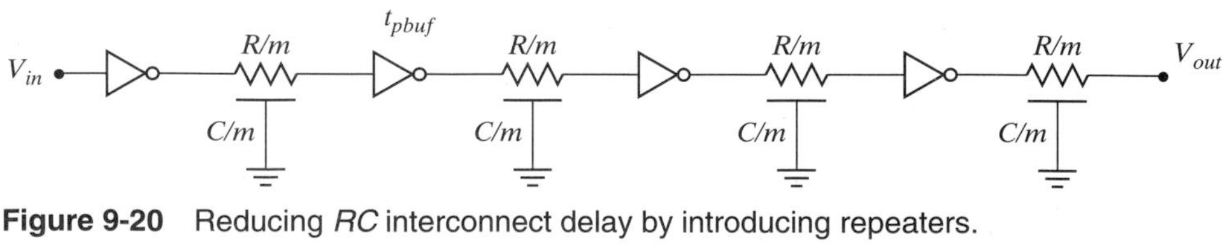

Figure 9-20   Reducing *RC* interconnect delay by introducing repeaters.

and a corresponding minimum delay for the wire

$$t_{p,\,opt} = 2\sqrt{t_{pwire(unbuffered)}t_{pbuf}}.\tag{9.8}$$

The optimum is obtained when the delay of the individual wire segments is made equal to that of a repeater.

Example 9.6   Reducing the Wire Delay through Repeaters

In Example 4.8, we derived the propagation delay of a 10-cm long, 1-µm wide Al-1 wire to be 31.4 ns. Eq. (9.7) indicates that a partitioning of the wire into 18 sections would minimize its delay, assuming a fixed t_{pbuf} of 0.1 ns. This results in an overall delay time of 3.5 ns, an important improvement. Similarly, the delays of equal-length Polysilicon and Al-5 wires would be reduced to 212 ns (from 112 µs) and 1.3 ns (from 4.2 ns) by introducing 1058 and 6 stages, respectively.

The preceding analysis is simplified and optimistic in the sense that t_{pbuf} is a function of the load capacitance. Sizing the repeaters is essential to reducing the delay. A more precise expression of the delay of the interconnect chain is obtained by modeling the repeater as an *RC* network, and by using the Elmore delay approach. With R_d and C_d the resistance and input capacitance, respectively, of a minimum-sized repeater, the following expression is obtained:

$$t_p = m\left(0.69\frac{R_d}{s}\left(s\gamma C_d + \frac{cL}{m} + sC_d\right) + 0.69\left(\frac{rL}{m}\right)(sC_d) + 0.38rc\left(\frac{L}{m}\right)^2\right)\tag{9.9}$$

Here, γ represents the ratio between intrinsic output and input capacitances of the repeater, as defined in Eq. (5.29). Setting $\partial t_p/\partial m$ and $\partial t_p/\partial s$ to 0 yields optimal values for m, s and $t_{p,\,min}$.

$$m_{opt} = L\sqrt{\frac{0.38rc}{0.69R_dC_d(\gamma+1)}} = \sqrt{\frac{t_{pwire(unbuffered)}}{t_{p1}}}$$

$$s_{opt} = \sqrt{\frac{R_dc}{rC_d}}\tag{9.10}$$

$$t_{p,\,min} = (1.38 + 1.02\sqrt{1+\gamma})L\sqrt{R_dC_drc}$$

Here, $t_{p1} = 0.69R_dC_d(\gamma+1) = t_{p0}(1 + 1/\gamma)$ represents the delay of an inverter for a fan-out of 1 ($f = 1$). Equation (9.10) clearly demonstrates how the insertion of repeaters linearizes the delay of a wire. Also, observe that **for a given technology and a given interconnect layer, there exists an optimal length of the wire segments between repeaters**. This *critical length* is given by the following expression:

$$L_{crit} = \frac{L}{m_{opt}} = \sqrt{\frac{t_{p1}}{0.38rc}}\tag{9.11}$$

It can be derived that the **delay of a segment of critical length** is always given by

$$t_{p, crit} = \frac{t_{p, min}}{m_{opt}} = 2\left(1 + \sqrt{\frac{0.69}{0.38(1 + \gamma)}}\right)t_{p1} \qquad (9.12)$$

and **is independent of the routing layer.** For the typical value of $\gamma = 1$, $t_{p, crit} = 3.9t_{p1} = 7.8t_{p0}$. Inserting repeaters to reduce the delay of a wire only makes sense when the wire is at least twice as long as the critical length.

Example 9.7 Minimizing the Wire Delay (Revised)

In Chapter 5 (Example 5.5), we determined that the t_{p1} of a minimum-sized inverter in our 0.25-μm CMOS process was 32.5 ps with (average) values of R_d and C_d of 7.8 kΩ and 3 fF, respectively, and $\gamma = 1$. For an Al-1 wire ($c = 110$ aF/μm; $r = 0.075$ Ω/μm), this yields an optimum repeater sizing factor $s_{opt} = 62$. Inserting 31 of these sized-up inverters sets the minimum delay of the 10-cm wire of Example 9.6 to a more realistic 3.9 ns. This is in contrast to inserting 31 minimum-sized repeaters, which would actually increase the delay to 61 ns due to the poor driving capability of the repeater.

The critical length for Al-1 evaluates to 3.2 mm, while it equals 54 μm and 8.8 mm for polysilicon and Al-5, respectively.

Repeater insertion has proven to be such an efficient and essential tool in combatting long wire delays that modern design automation tools perform this task automatically.

Optimizing the Interconnect Architecture

Even with buffer insertion, the delay of a resistive wire cannot be reduced below the minimum dictated by Eq. (9.10). Hence, long wires often exhibit a delay that is longer than the clock period of the design. For example, the 10-cm long Al-1 wire of Example 9.7 comes with a minimum delay of 3.9 ns—even after optimal buffer insertion and sizing—while the 0.25-μm CMOS process featured in this text can sustain clock speeds in excess of 1 GHz (i.e., clock periods below 1 ns). The wire delay all by itself becomes the limiting factor on the performance achievable by the integrated circuit. The only way to address this bottleneck is to tackle it at the system architecture level.

Wire pipelining is a popular performance-improvement technique in this category. The concept of pipelining was introduced in Chapter 7 as a means of improving the through-put performance of logic modules with long critical paths. A similar approach can be used to increase the through-put of a wire, as is illustrated in Figure 9-21. The wire is partitioned into k segments by inserting registers or latches. While this does not reduce the delay through the wire segment—it takes k clock cycles for a signal to proceed through the wire—it helps to increase its through-put because the wire is handling k signals simultaneously at any point in time. The delay of the individual wire segments can be further optimized by repeater insertion, and should be less than a single clock period.

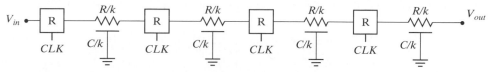

Figure 9-21 Wire pipelining improves the throughput of a wire.

This is only one example of the many techniques that the chip architect has at her disposal to deal with the wire delay problem. The most important message that emerges from this discussion is that wires have to be considered early on in the design process, and can no longer be treated as an afterthought as was most often the case in the past.

9.4 Inductive Parasitics*

Besides having a parasitic resistance and capacitance, interconnect wires also exhibit an inductive parasitic. An important source of parasitic inductance is introduced by the bonding wires and chip packages. Even for intermediate-speed CMOS designs, the current through the input/output connections can experience fast transitions that cause voltage drops as well as ringing and overshooting, phenomena not found in RC circuits. At higher switching speeds, wave propagation and transmission line effects can come into the picture. Both effects are analyzed in this section, and design solutions are proposed.

9.4.1 Inductance and Reliability—$L\dfrac{di}{dt}$ Voltage Drop

During each switching action, a transient current is sourced from (or sunk into) the supply rails to charge (or discharge) the circuit capacitances, as modeled in Figure 9-22. Both V_{DD} and V_{SS} connections are routed to the external supplies through bonding wires and package pins and possess a nonignorable series inductance. Hence, a change in the transient current creates a voltage difference between the external and internal (V'_{DD}, GND′) supply voltages. This situation is especially severe at the output pads, where the driving of the large external capacitances generates large current surges. The deviations on the internal supply voltages affect the logic levels and result in reduced noise margins.

Example 9.8   Noise Induced by Inductive Bonding Wires and Package Pins

Assume that the circuit in Figure 9-22 is the last stage of an output pad driver driving a load capacitance of 10 pF over a voltage swing of 2.5 V. The inverter has been dimensioned so that the 10–90% rise and fall times of the output signal (t_r, t_f) equal 1 ns. Since the power and ground connections are connected to the external supplies through the supply pins, both connections have a series inductance L. For a traditional through-hole packaging approach, an inductance of around 2.5 nH is typical. To simplify the analysis, assume first that the inverter acts as a current source with a constant current (dis)charging the load capacitance. An average current of 20 mA is required to achieve the 1 ns output rise and fall times:

$$I_{avg} = (10 \text{ pF} \times (0.9 - 0.1) \times 2.5 \text{ V}) / 1 \text{ ns} = 20 \text{ mA}$$

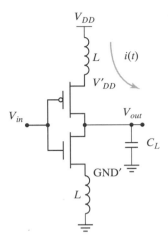

Figure 9-22   Inductive coupling between external
and internal supply voltages.

This scenario occurs when the buffer is driven by a steep step function at its input. The left side of Figure 9-23 plots the simulated evolution of output voltage, inductor current, and inductor voltage over time for t_f = 50 ps. The abrupt current change causes a steep voltage spike of up to 0.95 V over the inductor. In fact, the voltage drop would have been larger, if the drop itself did not slow down the transients and reduce the demand for current. Nevertheless, a supply voltage with that much variation is unacceptable.

In reality, the charging current is rarely a constant. A better approximation assumes that the current rises linearly to a maximum after which it drops back to zero, again in a linear fashion. The current distribution over time models as a triangle. This situation occurs when the input signal to the buffer displays slow rise and fall times. Under this model, the (estimated) voltage drop over the inductor is given by

$$v_L = L\, di_L/dt = (2.5\ \text{nH} \times 40\ \text{mA})/((1.25/2) \times 1\ \text{ns}) = 133\ \text{mV}.$$

The peak current of (approximately) 40 mA results from the fact that the total delivered charge—or the integrated area under I_L—is fixed, which means that the peak of the triangular current distribution is double the value of the rectangular one. The denominator factor estimates the time it takes to go from zero to peak current. Simulated results of this scenario are shown on the right side of Figure 9-23 for t_f = 800 ps. It can be observed that the current distribution indeed becomes triangular for this slow input slope. The simulated voltage drop over the inductor is below 100 mV. This indicates that avoiding fast-changing signals at the inputs of the drivers for large capacitors goes a long way in reducing the $L\, di/dt$ effects.

In an actual circuit, a single supply pin serves a large number of gates or output drivers. A simultaneous switching of those drivers causes even worse current transients and voltage drops.

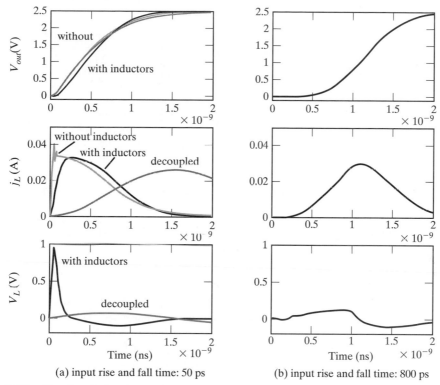

(a) input rise and fall time: 50 ps   (b) input rise and fall time: 800 ps

Figure 9-23 Simulated signal waveforms for output driver connected to bonding pads for input rise/fall times of 50 ps (a) and 800 ps (b), respectively. The waveforms in (a) show the results when (1) no inductors are present in the power distribution network (ideal case); (b) in the presence of inductors; and (c) with added decoupling capacitance of 200 pF.

As a result, the internal supply voltages deviate in a substantial way from the external ones. For instance, the simultaneous switching of the 16 output drivers of an output bus would cause a voltage drop of at least 1.1 V if the supply connections of the buffers were connected to the same pin on the package.

Improvements in packaging technologies are leading to ever-increasing numbers of pins per package. Packages with up to 1000 pins are currently available. Simultaneous switching of a substantial number of those pins results in huge spikes on the supply rails that are bound to disturb the operation of the internal circuits as well as other external components connected to the same supplies.

Design Techniques

A number of approaches are available to the designer to address the $L(di/dt)$ problem:

1. **Separate power pins for I/O pads and chip core.** Since the I/O drivers require the largest switching currents, they also cause the largest current changes. Therefore, it is wise to isolate the core of

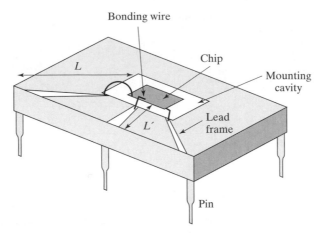

Figure 9-24 The inductance of a bonding-wire/pin combination depends upon the pin position.

the chip—where most of the logic action occurs—from the drivers by providing different power and ground pins.

2. **Multiple power and ground pins**. In order to reduce the *di/dt* per supply pin, we can restrict the number of I/O drivers connected to a single supply pin. Typical numbers are 5 to 10 drivers per supply pin. Be aware that this number strongly depends upon the switching characteristics of the drivers, such as the number of simultaneously switching gates and the rise and fall times.

3. **Careful selection of the positions of the power and ground pins on the package**. The inductance of leads and bond-wires associated with pins located at the corners of the package is substantially higher (see Figure 9-24).

4. **Increase the rise and fall times** of the off-chip signals to the maximum extent allowable, and distributed all over the chip, especially under the data busses. The preceding example demonstrates that output drivers with overdesigned rise and fall times are not only expensive in terms of area, but also affect the circuit operation and reliability. We concluded in Section 9.2.2 that the better driver in terms of area is the one that achieves a specified delay, not the one with the fastest delay. When noise is considered, the best driver is the one that achieves a specified delay with the maximum allowable rise and fall times at the output.

Problem 9.1   Design of Output Driver with Reduced Rise/Fall Times

Given a cascaded output driver designed for a given delay that produces excessive noise on the supply lines, determine the best approach to address the noise problem: (a) Scale down all stages of the buffer; (b) scale down the last stage only; (c) scale down all stages except the last one.

---

5. **Schedule current-consuming transitions** so that they do not occur simultaneously. It is, for example, possible to slightly stagger the switching of a set of output drivers by skewing their control inputs.

6. **Use advanced packaging technologies** such as surface-mount or hybrids that come with a substantially reduced capacitance and inductance per pin. For instance, we can see from Table 2-2 that the bonding inductance of a chip mounted in flip-chip style on a substrate by using the solder-bump techniques is reduced to 0.1nH, which is 50 to 100 times smaller than for standard packages.

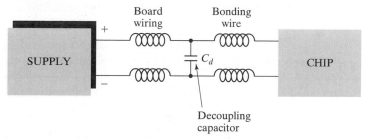

Figure 9-25 Decoupling capacitors isolate the board inductance from the bonding wire and pin inductance.

7. **Add decoupling capacitances on the board.** These capacitances, which should be added for every supply pin, act as local supplies and stabilize the supply voltage seen by the chip. They separate the bonding-wire inductance from the inductance of the board interconnect (see Figure 9-25). The bypass capacitor, combined with the inductance, actually acts as a low-pass network that filters away the high-frequency components of the transient voltage spikes on the supply lines.

Example 9.9   Impact of Decoupling Capacitances

A decoupling capacitance of 200 pF was added between the supply connections of the buffer circuit examined in Example 9.8. Its impact is shown in the simulations of Figure 9-23. For an input signal rise time of 50 ps, the spike on the supply rail is reduced from 0.95 V—without decoupling capacitance—to approximately 70 mV. The waveforms also demonstrate how the current for charging the output capacitance is drawn from the decoupling capacitor initially. Later in the transition, current is gradually being drawn from the supply network through the bonding-wire inductors.

---

8. **Add decoupling capacitances on the chip.** In high-performance circuits with high switching speeds and steep signal transitions, it is becoming common practice to integrate decoupling capacitances on the chip, which ensures cleaner supply voltages. On-chip bypass capacitors reduce the peak current demand to the average value. To limit the voltage ripple to 0.25 V, a capacitance of around 12.5 nF must be provided for every 50-K gate module in a 0.25-μm CMOS process [Dally98]. This capacitance is typically implemented using the thin gate oxide. A thin oxide capacitor is essentially an MOS transistor with its drain and source tied together. Diffusion capacitors can be used for that purpose as well.

Example 9.10   On-Chip Decoupling Capacitances in Compaq's Alpha Processor Family [Herrick00]

The Alpha processor family, mentioned earlier in this chapter, has been at the leading edge in high-performance microprocessors for the last decade. Pushing the clock speed to the extreme seriously challenges the power-distribution network. The Alpha processors are therefore at the forefront when it comes to on-chip decoupling capacitance. Table 9-4 enumerates some characteristics of consecutive processors in the family.

Table 9-4 On-chip decoupling capacitances in Alpha processor family.

Processor	Technology	Clock Frequency	Total Switching Capacitance	On-chip Decoupling Capacitance
EV4	0.75-μm CMOS	200 MHz	12.5 nF	128 nF
EV5	0.5-μm CMOS	350 MHz	13.9 nF	160 nF
EV6	0.35-μm CMOS	575 MHz	34 nF	320 nF

Given a gate capacitance of 4 fF/μm^2 (t_{ox} = 9.0 nm), the 320-nF decoupling capacitance of the EV6 requires 80 mm^2 of die area, which is 20% of the total chip. To minimize the impact, the capacitance was placed under the major busses. Even with this large amount of area dedicated to decoupling, the designers were running out of capacitance, and they had to resort to some innovative techniques to provide more near-chip decoupling. The problem was solved by wire-bonding a 2-μF, 2-cm^2 capacitor to the chip, and connecting the power grid to the capacitance through 160 V_{dd}/GND wire-bond pairs. A diagram of the *wire-bond-attached chip capacitor* (WACC) approach is shown in Figure 9-26. This example serves as a clear illustration of the problems that power-grid designers are facing in the high-performance arena. Other solutions that are in use today include the use of advanced flip-chip packaging approaches that allow the power network to tap into decoupling capacitors that are integrated in the package. For example, this approach is used in the Intel Pentium 4™ [Pentium02].

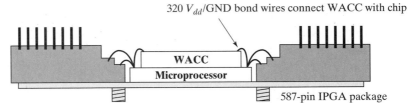

Figure 9-26 The wire-bond-attached chip capacitor (WACC) approach provides extra decoupling capacitance close to the chip for the Compaq EV6 microprocessor [Gieseke97]. The microprocessor is connected to the outside world via 320 signal pins and 198 supply/ground pins.

Finally, be aware that the mutual inductance between neighboring wires also introduces cross talk. This effect is not yet a major concern in CMOS, but definitely is emerging as an issue at the highest switching speeds [Johnson93].

WARNING: The combination of capacitance (wire and decoupling) and inductance in the power-distribution network makes it a resonant circuit that might oscillate. Of the various resonances, the most significant is the one arising from the package inductance L combined with the decoupling capacitance C_d with a resonance frequency of $f = 1/(2\pi\sqrt{L \cdot C_d})$. If the power grid resonates with the system clock, dangerous fluctuations in the power supply might occur. In the past, the resonance frequency of the power network was much higher than the clock frequency. Over the years, the resonant frequency has steadily decreased—as C_d has increased from generation to generation for an approximately constant L—while processor clock frequencies continue to increase. It is now common for the resonant frequency of the power network to lie well below the clock frequency. Hence, oscillations in the power network are a severe problem.

Proper damping of these oscillations is an absolute necessity. In fact, this is only one of the few places in digital design where resistance comes in handy. Decoupling capacitances often are designed in series with a controlled amount of resistance. However, this results in an increased *IR* drop. A careful trade-off must be made between *IR* voltage drop and drop due to resonance. In addition, the correct choice and distribution of the decoupling capacitances can help to move the resonant frequency to regions that are never triggered or excited.

9.4.2 Inductance and Performance—Transmission-Line Effects

When an interconnection wire becomes sufficiently long or when the circuits become sufficiently fast, the inductance of the wire starts to dominate the delay behavior, and transmission-line effects must be considered. This is more precisely the case when the rise and fall times of the signal become comparable to the time of flight of the signal waveform across the line as determined by the speed of light. Until recently, this was only the case for the fastest digital designs implemented on a board level, or in exotic technologies such as GaAs and SiGe. As advancing technology increases line lengths and switching speeds, this situation is gradually becoming common in the fastest CMOS circuits as well, and transmission-line effects are bound to become a concern of the CMOS designer.

In this section, we discuss some techniques to minimize or mitigate the impact of the transmission-line behavior. A first and foremost technique is to use appropriate termination. Termination is, however, only effective if the current-return path is well behaved. Shielding of wires that are prone to transmission-line effects will prove to be a necessity.[3]

Termination

In our discussion of transmission lines in Chapter 4, it became apparent that appropriate termination is the most effective way of minimizing delay. Matching the load impedance to the characteristic impedance of the line results in the fastest response. This leads to the following design rule:

---

[3] A number of insightful animations of transmission-line effects and the impact of termination can be found at the web site of P. Restle of IBM [Restle01].

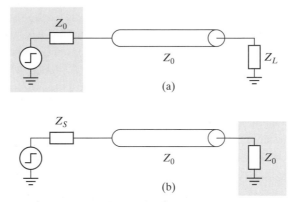

Figure 9-27 Matched termination scenarios for wires behaving as transmission lines: (a) series termination at the source; (b) parallel termination at the destination.

> **To avoid the negative effects of transmission-line behavior such as ringing or slow propagation delays, the line should be terminated, either at the source (series termination), or at the destination (parallel termination), with a resistance matched to its characteristic impedance Z_0.**

The two scenarios—series and parallel termination—are depicted in Figure 9-27. Series termination requires that the impedance of the signal source is matched to the connecting wire. This approach is appropriate for many CMOS designs, where the destination load is purely capacitive. The impedance of the driver inverter can be matched to the line by careful transistor sizing. To drive a 50-Ω line, for example, requires a 53-μm-wide NFET and a 135-μm-wide PFET (in our 0.25-μm CMOS technology) to give a nominal output impedance of 50 Ω.

It is important that the impedance of the driver be closely matched to the line, typically to within 10% or better, if excessive reflections of travelling waves are to be avoided. Unfortunately, the on resistance of a FET may vary by 100% across process, voltage, and temperature variations. This can be compensated for by making the resistance of the driver transistors electrically tunable, as in Figure 9-28. Each of the driver transistors is replaced by a segmented driver, with the segments being switched in and out by control lines c_1 to c_n, to match the driver impedance to the line impedance as closely as possible. Each of the segments has a different resistance, shaped by the shape factors s_i (typically ratioed with factors of 2). A fixed element (s_0) is added in parallel with the adjustable ones, so that the range of adjustment is limited and a more precise adjustment can be made with fewer bits. The control lines usually are driven by a feedback control circuit that compares an on-chip reference driver transistor with a fixed external reference transistor [Dally98].

Similar considerations are valid when the termination is provided at the destination end, called *parallel termination*. Be aware that this approach results in a standby current, and that keeping this current flowing continuously might result in unacceptably large power dissipation. Parallel termination is very popular for high-speed interchip, board-level interconnections,

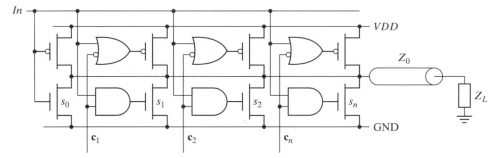

Figure 9-28   Tunable segmented driver providing matched series-termination to a transmission line load.

where the termination often is implemented by inserting a grounded resistor next to the input pin (of the package) at the end of the wire. This off-chip termination does not take into account the package parasitics and the internal circuitry, and it may introduce unacceptably large reflections on the signal line. A more effective approach is to include the terminating resistor inside the package after the package parasitics.

The CMOS fabrication process does not provide us with means to make precise, temperature-insensitive transistors. Interconnect materials or FETs can be used as resistors, but static or dynamic tuning is a necessity to overcome their temperature, supply-voltage and process dependence. Digital trimming such as that proposed earlier in this section can be used to accomplish this goal. Another issue when using MOSFETs as resistors is ensuring that the device exhibits a linear behavior over the operation region of interest. Since PMOS transistors typically display a larger region of linear operation than their NMOS counterparts, they are the preferred way of implementing a terminating resistance. Assume that the triode-connected PMOS transistor of Figure 9-29a is used as a 50-Ω matched termination, connected to V_{DD}. From simulations (Figure 9-29d), we observe that the resistance is fairly constant for small values of V_R, but increases rapidly once the transistor saturates. We can extend the linear region by increasing the bias voltage of the PMOS transistor (Figure 9-29b). This requires an extra supply voltage, however, which is not practical in most situations. A better approach is to add a diode-connected NMOS transistor in parallel with the PMOS device (Figure 9-29c). The combination of the two devices gives a near-constant resistance over the complete voltage range (as we saw in Chapter 6 when discussing transmission gates). Many other clever schemes have been devised, but they are beyond the scope of this book. We refer the reader to [Dally98] for a more detailed perspective.

Shielding

A transmission line is, in essence, a two-port (four-terminal) network. While we focus mostly on the signal path, the *signal return path* should not be ignored. Kirchhoff's law tells us that when we inject a current i into a signal conductor, a net current of $-i$ must flow in the return. The characteristics of the transmission line (such as its characteristic impedance) are a strong function of the signal return. Hence, if we want to control the behavior of a wire behaving as a transmission

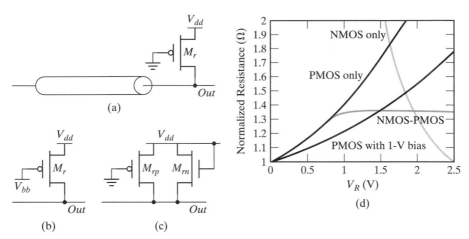

Figure 9-29   Parallel termination of transmission line using transistors as resistors:
(a) grounded PMOS; (b) PMOS with negative gate bias; (c) PMOS-NMOS combination;
(d) simulation.

line, we should carefully plan and manage how the return current flows. A good example of a
well-defined transmission line is the coaxial cable, where the signal wire is surrounded by a
cylindrical ground plane. To accomplish similar effects on a board or on a chip, designers often
surround the signal wire with ground (supply) planes and shielding wires. While expensive in
terms of space, adding shielding makes the behavior and the delay of an interconnection a lot
more predictable. Even with these precautions, however, powerful extraction and simulation
tools will be needed in the future for the high-performance circuit designer.

Example 9.11   Design of an Output Driver—Revisited

To conclude to this section, we simulate an output driver that includes the majority of the
parasitic effects introduced in this chapter. The schematics of the driver are shown in
Figure 9-30a. The inductance of the power, ground, and signal pins (estimated at 2.5 nH),
as well as the transmission line behavior of the board, have been included (assuming a
wire length of 15 cm, which is typical for board traces). The driver is initially designed to
produce rise and fall times of 0.33 ns for a total load capacitance of 10 pF. Producing the
required average current of 60 mA requires transistor widths of 120 μm and 275 μm for
the NMOS and PMOS transistors in the final stage of the driver. The first graph of
Figure 9-30b shows the simulated waveforms for this configuration. Severe ringing is
observed. It is quite obvious, therefore, that the driver resistance is underdamped—hence,
the large overshoots and the large settling time.

   The circuit was redesigned to eliminate some of these effects. First of all, the sizes
of the driver transistors were reduced so that their impedances now match the characteristic impedance of the transmission line (to 65 μm and 155 μm for NMOS and PMOS,

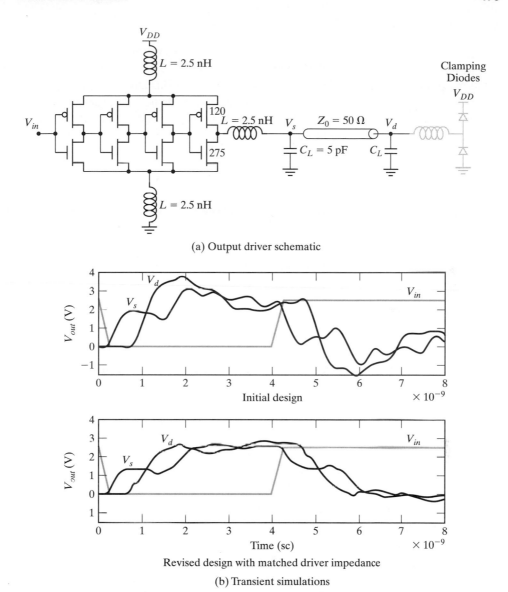

(a) Output driver schematic

Initial design

Revised design with matched driver impedance

(b) Transient simulations

Figure 9-30 Simulation of output driver for various terminations.

respectively). A decoupling capacitance of 200 pF was added to the supplies of the drivers. Finally, the input-protection diodes of thc fan-out device were added to the model to create a more realistic perspective. The simulation shows that these modifications are quite effective, giving a circuit that is both faster and better behaved.

9.5 Advanced Interconnect Techniques

So far, we discussed a number of techniques on how to cope with the capacitive, resistive and inductive parasitics that come with interconnect wires. To offer some perspective on the future, in this section we discuss a number of more advanced circuits that have emerged in recent years. More specifically, we discuss how reducing the signal swing can help to reduce the delay and the power dissipation when driving wires with large capacitances, and we introduce a number of techniques to realize these reduced swings.

9.5.1 Reduced-Swing Circuits

Increasing the size of the driver transistor—and thus increasing the average current I_{av} during switching—is only one way of coping with the delay caused by a large load capacitance. Another approach is the reduction of the signal swing at the output of the driver and over the load capacitance. Intuitively, we can understand that reducing the charge that has to be displaced may be beneficial to the performance of the gate. This is best understood by revisiting Eq. (5.16), the propagation-delay equation:

$$t_p = \int_{v_1}^{v_2} \frac{C_L(v)}{i(v)} dv \tag{9.13}$$

Assuming that the load capacitance is a constant, we can simplify the expression for the delay, writing it as

$$t_p = \frac{C_L V_{swing}}{I_{av}} \tag{9.14}$$

where $V_{swing} = v_2 - v_1$ is the signal swing at the output and I_{av} is the average (dis)charge current. This expression clearly demonstrates that the delay reduces linearly with the signal swing, under the condition that the (dis)charge current is not affected by the reduction in voltage swing. Simply lowering the overall supply voltage does not work: While reducing the swing, it also lowers the current by a similar ratio. This was demonstrated earlier in Figure 3-28, which showed that the equivalent resistance of a (dis)charge transistor is approximately constant over a wide range of supply voltages. While potentially offering an increased performance, lower signal swings further carry the major benefit of lowering the dynamic power consumption, which can be substantial when the load capacitance is large.

On the other hand, reduced signal swings result in smaller noise margins, and the signal integrity and reliability is therefore negatively affected. Furthermore, CMOS gates are not particularly effective in detecting and reacting to small signal changes, because of the relatively small transconductance of the MOS device. In order to work properly and to achieve high performance, reduced-swing circuits require amplifier circuits, whose task it is to restore the signal to its full swing in a minimum amount of time and with a minimum amount of extra energy consumption. The overhead resulting from adding extra amplifiers is only justifiable for network

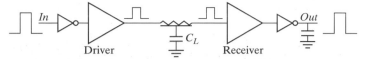

Figure 9-31 Reduced-swing interconnect circuit. The driver circuit reduces the normal voltage swing to a lower value, while the receiver detects the signal and restores it to the normal swing value.

nodes with a large fan-in, in which the circuits can be shared over many input gates. Typical examples of such nodes are the data or address buses of a microprocessor or the data lines in a memory array. In the former case, the amplifier is most often called a *receiver*; in the latter, it is called a *sense amplifier*.[4]

A typical diagram of a reduced-swing network, consisting of a driver, a large capacitance/ resistance interconnect wire, and a receiver circuit, is shown in Figure 9-31. Many different designs for both drivers and receivers have been devised [Zhang00], but we will only discuss a few. In general, the reduced-swing circuits fall into two major categories, *static* and *dynamic* (or precharged). Another differentiating factor between circuits is the signaling technique. Most receiver circuits use a *single-ended* approach, where the receiver detects an absolute change in voltage on a single wire. Other circuits use *differential* or *double-ended* signaling techniques, where both the signal and its complement are transmitted, and the receiver detects a relative change in voltage between the two wires. While the latter approach requires substantially more wiring space, it has the advantage of being more robust in the presence of noise.

Static Reduced-Swing Networks

The task of designing a reduced-swing network simplifies substantially when a second lower supply rail V_{DDL} is available. A simple and robust single-ended driver circuit using that second supply is shown in Figure 9-32. The challenge is in the receiver design. Just using an inverter does not work very well: The small swing at the input of the NMOS transistor results in a small pull-down current and thus a slow high-to-low transition at the output; furthermore, the low value of V_{DDL} is not sufficient to turn off the PMOS transistor, which deteriorates the performance even more, and causes static power dissipation to occur in addition.

A more refined receiver circuit, inspired by the DCVSL gate discussed in Chapter 6, is shown in Figure 9-32. A low-voltage inverter is used in the receiver to generate a local complement of the input signal. The receiver now acts as a differential amplifier. The cross-coupled load transistors ensure that the output is restored to V_{DD}, and that no static power is consumed in steady-state mode. The positive feedback helps to accelerate the transitions. The disadvantage of the circuit is that it becomes unacceptably slow for low swing values.

Figure 9.33 shows a circuit that avoids the use of a second supply rail. By reversing the positions of the NMOS and the PMOS transistors in the final stage of the driver, we have limited the signal swing on the interconnect from $|V_{Tp}|$ to $V_{DD}-V_{Tn}$, or approximately two threshold

---

[4]More details on sense amplifier circuits can be found in Chapter 12 on semiconductor memories.

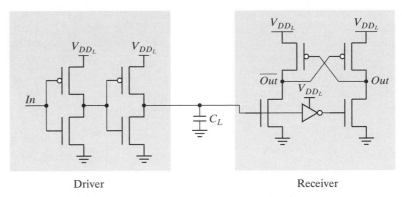

Driver   Receiver

Figure 9-32   Single-ended, static reduced-swing driver and receiver.

values below V_{DD}.[5] The symmetrical receiver/level converter consists of two cross-coupled transistor pairs (P1–P2 and N1–N2), and two transistors (N3–P3) that isolate the reduced-swing interconnect wire from the full-swing output signal.

To understand the operation of the receiver, assume that node *In2* goes from low to high, or from $|V_{Tp}|$ to V_{DD}-V_{Tn}. Initially, nodes A and B are positioned at $|V_{Tp}|$ and GND, respectively. During the transition period, with both N3 and P3 conducting, A and B rise to V_{DD}–V_{Tn}, as shown in Figure 9-33b. Consequently, N2 turns on, and *Out* goes to low. The feedback transistor P1 pulls A further up to V_{DD} to turn P2 completely off. *In2* and B stay at V_{DD}–V_{Tn}. There is no standby current path from V_{DD} to GND through N3, although the gate-source voltage of N3 is almost V_{Tn}. Since the circuit is symmetric, a similar explanation holds for the high-to-low transition. Attentive readers will notice that transistors P1 and N1 act as level restorers—as we saw earlier in pass-transistor logic. Hence, they can be very weak, which tends to minimize their contention with the driver. The sensing delay of the receiver is as small as two inverter delays.

Problem 9.2   Energy Consumption of Reduced-Swing Receiver

Assuming that the capacitance is dominated by the load capacitance of the wire, derive the energy reduction per signal transition that the circuit of Figure 9-33 offers over the full-swing version.

The interconnect systems shown so far are all single ended. Figure 9-34 shows a differential scheme. The driver generates two complementary reduced-swing signals using a second supply rail. The receiver is nothing less than a sense-amplifier-based register, introduced earlier in Section 7.4.2. The differential approach offers a high rejection of common-mode noise signals such as supply-rail noise and cross-talk interference. The signal swing can therefore be reduced to very low levels—operation with swings as low as 200 mV has been demonstrated. The driver

---

[5]This approach has the drawback that the voltage swing on the interconnect line depends on technology parameters, such as the threshold voltage and the body effect, and thus varies from die to die.

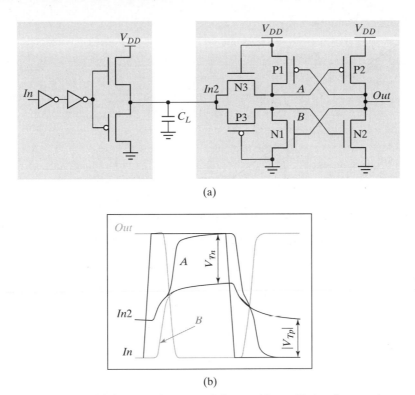

(a)

(b)

Figure 9-33 (a) Symmetric source-follower driver with level converter; (b) simulated waveforms.

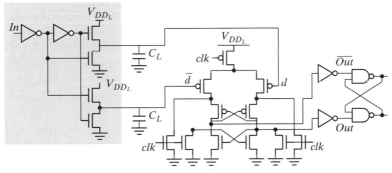

Figure 9-34 Differential reduced-swing interconnect system using an additional supply rail at the driver. The receiver is a clocked differential flip-flop [Burd00].

uses NMOS transistors for both pull-up and pull-down networks. The main disadvantage of the differential approach is the doubling of the number of wires, which presents a major concern in many designs. The extra clock signal adds further to the overhead.

Dynamic Reduced-Swing Networks

Another approach to speeding up the response of large fan-in circuits such as busses is to make use of precharging, an example of which is shown in Figure 9-35. During $\phi = 0$, the bus wire is precharged to V_{DD} through transistor M_2. Because this device is shared by all input gates, it can be made large enough to ensure a fast precharging time. During $\phi = 1$, the bus capacitance is conditionally discharged by one of the pull-down transistors. This operation is slow because the large capacitance C_{bus} must be discharged through the small pull-down device M_1.

A speedup at the expense of noise margin can be obtained by observing that all transitions on the bus are from high to low during evaluation. A faster response can be obtained by moving the switching threshold of the subsequent inverter upwards, which results in an asymmetrical gate. In a traditional inverter design, M_3 and M_4 are sized so that t_{pHL} and t_{pLH} are identical, and the switching threshold (V_M) of the inverter is situated around $0.5\ V_{DD}$. This means that the bus voltage V_{bus} has to drop over $V_{DD}/2$ before the output inverter switches. This can be avoided by making the PMOS device larger, moving V_M upwards, which causes the output buffer to start switching earlier.

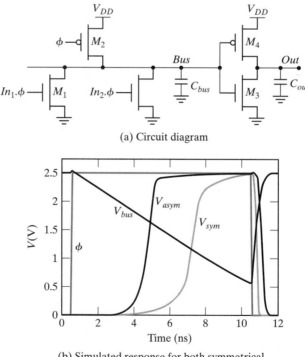

(a) Circuit diagram

(b) Simulated response for both symmetrical and asymmetrical read-out inverters. M_4 is made fifteen times wider in the asymmetrical case. The load capacitance equals 1 pF.

Figure 9-35   Precharged bus.

The precharged approach can result in a substantial speedup for the driving of large capacitive lines. However, it also suffers from all the disadvantages of dynamic circuit techniques—charge sharing, leakage, and inadvertent charge loss resulting from cross talk and noise. Cross talk between neighboring wires is an issue especially in densely wired bus networks. The reduced noise margin NM_H of the asymmetrical receiver makes this circuit particularly sensitive to parasitic effects. Therefore, extreme caution and extensive simulation are required when designing large precharged networks. Making the network pseudostatic through the addition of a small level-restoring device goes a long way towards making the circuit more resilient.

The simulated response of a precharged bus network is shown in Figure 9-35b. The output signal is plotted for both symmetrical and asymmetrical output inverters. Skewing the switching threshold upwards reduces the propagation delay by more than 2.5 ns. This allows for a further reduction in the evaluation period, reducing the voltage swing of the bus by 0.6 V and the energy consumption by 18%. Careful timing of the precharge/evaluate signals is required to realize the maximum benefits of this scheme.

A variant on the dynamic theme is offered by the pulse-controlled driver scheme of Figure 9-36. The idea is to control the (dis)charging time of the drivers so that a desired swing is obtained on the interconnect. The interconnect wire is precharged to a reference voltage REF, typically situated at $V_{DD}/2$. The receiver consists of a (pseudo-)differential sense amplifier, which compares the voltage on the interconnect wire to REF. Since the amplifier consumes static power, it should only be enabled for a short while. The advantage of this circuit is that the pulse width can be fine-tuned to realize a very low swing, while no extra supply voltage is needed. This concept has been widely applied in memory designs. However, it only works well when the capacitive loads are known beforehand. Furthermore, the wire is floating when the driver is disabled, making it more susceptible to noise.

The circuits just described present only a small subgroup of the many reduced-swing circuits that have been devised. For example, the concept of charge-sharing can be exploited to create buses that operate at a reduced voltage swing while recycling the charge from wire to wire. We refer the interested reader to in-depth references such as [Zhang00]. When venturing into the area of reduced-swing interconnect circuits, you should always be aware of the trade-offs involved: power and performance versus signal integrity and reliability.

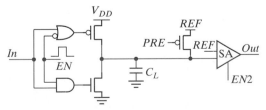

Figure 9-36   Pulse-controlled driver with sense amplifier.

9.5.2 Current-Mode Transmission Techniques

All of these approaches assume that the data to be transmitted over the wire is represented by a set of voltage levels referenced to the supply rails. While this approach conforms with the logical levels typically used in digital logic, it does not necessarily represent the best solution from a performance, power, and reliability perspective. Consider the problem of transmitting a bit over a lengthy transmission line with a characteristic impedance of 50 Ω. The traditional voltage-mode approach is depicted in Figure 9-37a. The driver switches the line between the two supply lines, which represent a **1** and a **0**, respectively, and it has an impedance of R_{out}. The receiver is a CMOS inverter that compares the incoming voltage with a supply-referenced threshold voltage, typically centered in the middle of the two supply rails. The signal swing is lower bounded by noise considerations. More specifically, power supply noise has a large impact because it affects both the signal levels and the switching threshold of the receiver. The latter is also a strong function of manufacturing process variations.

An alternative approach is to use a low-swing current-mode transmission system like the one shown in Figure 9-37b. The driver injects a current I_{in} into the line for a **1** and a reversed current of $-I_{in}$ for a **0**. This induces a voltage wave of $2 \times I_{in} \times Z_0$ into the transmission line, which ultimately gets absorbed into the parallel termination resistance R_T. The convergence time depends on how well the termination resistance is matched to the characteristic impedance of the transmission line. A differential amplifier is used to detect the voltage changes over R_T. Observe that the signal and its return path are isolated from the supply rails and the associated noise, making all supply noise common mode to the differential receiver. Analog designers will testify that this type of noise is easily suppressed in any decent differential amplifier design.

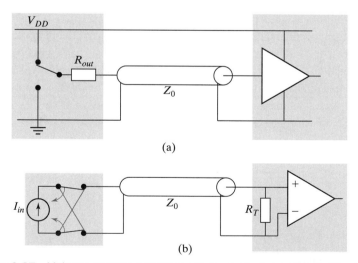

Figure 9-37 Voltage- versus current-mode transmission systems. The latter has the advantage of higher noise immunity with respect to supply noise.

While both approaches can be tuned to accomplish approximately the same performance levels, the current-mode approach holds a definite edge in terms of (dynamic) power dissipation. Due to its immunity to power-supply noise, it can operate at a much lower noise margin than the voltage-mode network, and thus at a much lower swing as well. The value of the voltage wave propagating over the transmission line can be as low as 100 mV. The main challenge in designing an efficient current-mode circuit is the static power consumption. This is not an issue in ultra high-speed networks, where dynamic power tends to play the major role.

Current-mode CMOS transmission systems have become quite popular in the area of high-speed off-chip interconnections. The ferocious appetite for I/O bandwidth in high performance processors and memories has made this a topic of considerable importance. Readers interested in high-speed interchip signaling are referred to [Dally98] and [Sidiropoulos01].

9.6 Perspective: Networks-on-a-Chip

With chip sizes and clock rates continuing to increase, and feature sizes getting smaller at a constant rate, it is not hard to see that interconnect problems will be with us for quite some time to come. Physics constraints such as the speed of light and thermal noise determine how fast and how reliably communication can be established over "long" distances. As communication bottlenecks have been the major damper on hyper-supercomputer performance, similar constraints are starting to hamper integrated systems ("systems-on-a-chip"). New solutions conceived at the technology and circuit level only help to postpone the problems temporarily. Ultimately, the solution is to address interconnections on a chip as a communication problem and to apply techniques and approaches that have made large-scale communications and networking systems around the world operate reliably and correctly for many years. For example, it is somewhat amazing that the Internet is working correctly given its span and number of connection points. The secret to its success is a well-thought-out protocol stack, which isolates and orthogonalizes the various functionality, performance, and reliability concerns. Hence, rather than considering on-chip interconnections as point-to-point wires, they should be abstracted as communication channels over which we want to transmit data within a "quality-of-service" setting, putting constraints on through-put, latency, and correctness [Sgroi01].

As an example, today's on-chip interconnect signalling techniques are designed with noise margins large enough to ensure that a bit will always be transmitted correctly. It is probably safe to say that future designs might take a more extreme tack—they may give up integrity entirely for the sake of energy/performance, and thus allow errors to occur on the transmitted signals. These errors can be corrected by other circuitry that is designed to provide error-correcting capabilities and/or retransmit corrupted data.

On a higher level of abstraction, we are already witnessing the emergence of complete "networks" on a chip, as shown in Figure 9-38. Rather than being statically wired from source-to-destination, data is injected as packets into a complete network of wires, switches, and routers, and it is the network that dynamically decides how and when to route these packets through

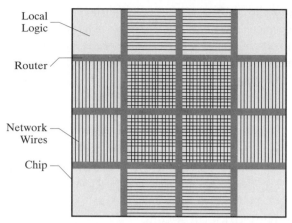

Figure 9-38   The network-on-a-chip combines multiple processors and their interconnect network on the same die.

its segments [Dally01]. Ultimately, this is the only approach that can work reliably when the discrepancy between device sizes and on-chip distances becomes macroscopic.

9.7 Summary

This chapter has introduced a number of techniques to cope with the impact of interconnect on the performance and reliability of digital integrated circuits. The parasitics introduced by the interconnect have a dual effect on circuit operation: (1) They introduce noise and (2) they increase the propagation delay and power dissipation. In summary,

- *Capacitive cross talk* in dense wire networks affects the reliability of the system and therefore its performance. Careful design using well-behaved regular structures or using advanced design-automation tools is a necessity. Providing the necessary shielding is important for wires such as busses or clock signals.
- Driving large capacitances rapidly in CMOS requires the introduction of a *cascade of buffer stages* that must be carefully sized. More advanced techniques include the lowering of the signal-swing on long wires, and the use of current-mode signalling.
- Resistivity affects the reliability of the circuit by introducing *IR drops*. This is especially important for the *supply network*, where wire sizing is important.
- The extra delay introduced by the *rc* effects can be minimized by *repeater insertion* and by using a better interconnect technology.
- The *inductance of the interconnect* becomes important at higher switching speeds. The chip package is currently one of the most important contributors of inductance. Novel packaging techniques are gaining importance with faster technologies.
- *Ground bounce* introduced by the *L di/dt* voltage drop over the supply wires is one of the most important sources of noise in current integrated circuits. Ground bounce can be

reduced by providing sufficient supply pins and by controlling the slopes of the off-chip signals.

- *Transmission-line effects* are rapidly becoming an issue in super-GHz designs. Providing the *correct termination* is the only means of dealing with the transmission-line delay.
- Ultimately, we should choose for a proactive approach in dealing with interconnect problems. Constructive fabrics combined with architectural and system-level solutions will go a long way.

9.8 To Probe Further

Excellent overviews of the issues involved in dealing with interconnect in digital designs can be found in [Bakoglu90], [Dally98], and [Chandrakasan01].

References

[Adler00] V. Adler and E. Friedman, "Uniform Repeater Insertion in RC Trees," *IEEE Transactions on Circuits and Systems—I: Fundamental Theory and Applications*, vol. 47, no. 10, Oct. 2000.

[Apollo] Apollo-II, High-Performance VDSM Place and Route System for Soc Designs, *http://www.synopsys.com/products/avmrg/apolloII_ds.html*.

[Bakoglu90] H. Bakoglu, *Circuits, Interconnections and Packaging for VLSI*, Addison-Wesley, 1990.

[Burd00] T. Burd, T. Pering, A. Stratakos, and R. Brodersen, "A Dynamic Voltage Scaled Microprocessor System," *IEEE ISSCC Dig. Tech. Papers*, pp. 294–5, Feb. 2000.

[Cadence-Power] R. Saleh, M. Benoit, and P. McCrorie, "Power Distribution Planning," *http://www.cadence.com/whitepapers/powerdistplan.html*, Cadence Design, 2001.

[Cadence-X Initiative] The X Initiative, *http://www.cadence.com/industry/x2.html*, Cadence Design, 2001.

[Chandrakasan01] A. Chandrakasan, W. Bowhill, and F. Fox, Ed., *Design of High-Performance Microprocessor Circuits*, IEEE Press, 2001.

[Cong99] J. Cong and D.Z. Pan "Interconnect Estimation and Planning for Deep Submicron Designs," *Proc. 36th ACM/IEEE Design Automation Conf.*, New Orleans, LA., pp. 507–10, June, 1999.

[Dally98] B. Dally, *Digital Systems Engineering,* Cambridge University Press, 1998.

[Dally01] W. Dally, "Route Packets, Not Wires: On-Chip Interconnection Networks," *Proceedings Design Automation Conference*, pp. 684–9, Las Vegas, June 2001.

[Etter93] D. Etter, "Engineering Problem Solving with Matlab," Prentice Hall, 1993.

[Gieseke97] B. Gieseke et al., "A 600 MHz superscalar RISC microprocessor with out-of-order execution," *IEEE ISSCC Digest of Technical Papers*, pp. 176-177, Febr. 1997.

[Herrick00] B. Herrick, "Design Challenges in Multi-GHz Microprocessors," *Proceedings ASPDAC 2000*, Yokohama, January 2000.

[Johnson93] H. Johnson and M. Graham, *High-Speed Digital Design—A Handbook of Black Magic*, Prentice Hall, 1993.

[Khatri01] S. Khatri, R. K. Brayton, A. L. Sangiovanni-Vincentelli, *Cross Talk Immune VLSI Design Using Regular Layout Fabrics*, Kluwer Academic Publishers, June 2001.

[Ma94] S. Ma and P. Franzon, "Energy Control and Accurate Delay Estimation in the Design of CMOS Buffers," *IEEE Journal of Solid-State Circuits*, vol. 29, pp. 1150–1153, Sept. 1994.

[Pentium02] Intel Pentium 4 Processor Home Page, *http://www.intel.com/products/desk_lap/processors/desktop/pentium4*.

[RailMill] "RailMill Datasheet—Power Network Analysis to Assure IC Performance," *http://www.synopsys.com/products/phy_syn/railmill_ds.html*, Synopsys, Inc.

[Restle01] P. Restle Home Page, *http://www.research.ibm.com/people/r/restle*, IBM Research, 2001.

[Sgroi01] M. Sgroi, M. Sheets, A. Mihal, K. Keutzer, S. Malik, J. Rabaey, A. Sangiovanni-Vincentelli, "Addressing the System-on-a-Chip Interconnect Woes Through Communication-Based Design," *Proceedings Design Automation Conference*, pp. 678–83, Las Vegas, June 2001.

[Sidiropoulos01] S. Sidiropoulos, C. Yang, and M. Horowitz, "High Speed Inter-Chip Signaling," Chapter 19, pp. 397–425, in [Chandrakasan01].

[Sotiriadis01] P. Sotiriadis and A. Chandrakasan, "Reducing Bus Delay in Submicron Technology Using Coding," *Proceedings ASPDAC Conference 2001*, Yokohama, January 2001.

[Stan95] M. Stan and W. Burleson, "Bus-Invert Coding for Low-Power I/O," *IEEE Transactions on VLSI*, pp. 49–58, March 1995.

[Sylvester98] D. Sylvester and K. Keutzer, "Getting to the Bottom of Deep Submicron," *Proceedings ICCAD Conference*, pp. 203, San Jose, November 1998.

[Tao94] J. Tao, N. Cheung, and C. Hu, "An Electromigration Failure Model for Interconnects under Pulsed and Bidirectional Current Stressing," *IEEE Trans. on Devices*, vol. 41, no. 4, pp. 539–45, April 1994.

[Zhang00] H. Zhang, V. George, J. Rabaey, "Low-swing on-chip Signaling Techniques: Effectiveness and Robustness," *IEEE Transactions on VLSI Systems*, vol. 8, no. 3, pp. 264–272, June 2000.

Exercises and Design Problems

New problem sets, design problems, and exercises can always be found at **http://bwrc.eecs.berkeley.edu/IcBook**.

Timing Issues in Digital Circuits

Impact of clock skew and jitter on performance and functionality

Alternative timing methodologies

Synchronization issues in digital IC and board design

Clock generation

10.1   Introduction

10.2   Timing Classification of Digital Systems

   10.2.1   Synchronous Interconnect

   10.2.2   Mesochronous Interconnect

   10.2.3   Plesiochronous Interconnect

   10.2.4   Asynchronous Interconnect

10.3   Synchronous Design—An In-Depth Perspective

   10.3.1   Synchronous Timing Basics

   10.3.2   Sources of Skew and Jitter

   10.3.3   Clock-Distribution Techniques

   10.3.4   Latch-Based Clocking*

10.4   Self-Timed Circuit Design*

   10.4.1   Self-Timed Logic—An Asynchronous Technique

   10.4.2   Completion-Signal Generation

   10.4.3   Self-Timed Signaling

   10.4.4   Practical Examples of Self-Timed Logic

10.5   Synchronizers and Arbiters*

   10.5.1   Synchronizers—Concept and Implementation

   10.5.2   Arbiters

10.6   Clock Synthesis and Synchronization Using a Phase-Locked Loop*
        10.6.1   Basic Concept
        10.6.2   Building Blocks of a PLL
10.7   Future Directions and Perspectives
        10.7.1   Distributed Clocking Using DLLs
        10.7.2   Optical Clock Distribution
        10.7.3   Synchronous versus Asynchronous Design
10.8   Summary
10.9   To Probe Further

10.1 Introduction

All sequential circuits have one property in common—a well-defined ordering of the switching events must be imposed if the circuit is to operate correctly. If this were not the case, wrong data might be written into the memory elements, resulting in a functional failure. The *synchronous* system approach, in which all memory elements in the system are simultaneously updated using a globally distributed periodic synchronization signal (that is, a global clock signal), represents an effective and popular way to enforce this ordering. Functionality is ensured by imposing some strict constraints on the generation of the clock signals and their distribution to the memory elements distributed over the chip; noncompliance often leads to malfunction.

This chapter starts with an overview of the different timing methodologies. The majority of the text is devoted to the popular *synchronous approach*. We analyze the impact of spatial variations of the clock signal, called *clock skew*, and temporal variations of the clock signal, called *clock jitter*, and introduce techniques to cope with both. These variations fundamentally limit the performance that can be achieved using a conventional design methodology.

At the other end of the spectrum is an approach called *asynchronous design*, which avoids the problem of clock uncertainty altogether by eliminating the need for globally distributed clocks. After discussing the basics of asynchronous design approach, we analyze the associated overhead and identify some practical applications. The important issue of synchronization between different *clock domains* and *interfacing* between asynchronous and synchronous systems also deserve in-depth treatment. Finally, the fundamentals of on-chip clock generation using feedback are introduced, along with trends in timing.

10.2 Timing Classification of Digital Systems

In digital systems, signals can be classified depending on how they are related to a local clock [Messerschmitt90][Dally98]. Signals that transition only at predetermined periods in time can be classified as *synchronous*, *mesochronous*, or *plesiochronous* with respect to a system clock. A signal that can transition at arbitrary times, on the other hand, is considered *asynchronous*.

10.2.1 Synchronous Interconnect

A synchronous signal is one that has the exact same frequency as the local clock and maintains a known fixed phase offset to that clock. In such a timing framework, the signal is "synchronized" with the clock, and the data can be sampled directly without any uncertainty. In digital logic

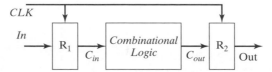

Figure 10-1 Synchronous interconnect methodology.

design, synchronous systems are the most straightforward type of interconnect. The flow of data in such a circuit proceeds in lockstep with the system clock, as illustrated in Figure 10-1.

Here, the input data signal *In* is sampled with register R_1 to produce signal C_{in}, which is synchronous with the system clock, and then it is passed along to the combinational logic block. After a suitable setting period, the output C_{out} becomes valid. Its value is sampled by R_2 which synchronizes the output with the clock. In a sense, the *certainty period* of signal C_{out}—the period during which data are valid—is synchronized with the system clock. This allows register R_2 to sample the data with complete confidence. The **length of the uncertainty period**, or the period during which data are not valid, **places an upper bound on how fast a synchronous system can be clocked**.

10.2.2 Mesochronous Interconnect

A *mesochronous* signal—*meso* is Greek for "middle"—is a signal that not only has the same frequency as the local clock, but also has an unknown phase offset with respect to that clock. For example, if data are being passed between two different clock domains, the data signal transmitted from the first module can have an unknown phase relationship to the clock of the receiving module. In such a system, it is not possible to directly sample the output at the receiving module because of the uncertainty in the phase offset. A (mesochronous) synchronizer can be used to synchronize the data signal with the receiving clock, as shown in Figure 10.2. The synchronizer serves to adjust the phase of the received signal to ensure proper sampling.

In Figure 10-2, signal D_1 is synchronous with respect to Clk_A. However, D_1 and D_2 are mesochronous with Clk_B because of the unknown phase difference between Clk_A and Clk_B and the unknown interconnect delay in the path between Block A and Block B. The role of the synchronizer is to adjust the variable delay line such that the data signal D_3 (a delayed version of D_2) is aligned properly with the system clock of Block B. In this example, the variable delay element is adjusted by measuring the phase difference between the received signal and the local clock. Register R_2 samples the incoming data during the certainty period, after which the signal D_4 becomes synchronous with Clk_B.

10.2.3 Plesiochronous Interconnect

A *plesiochronous* signal is one that has a frequency that is nominally the same as that of the local clock, yet is slightly different. (In Greek, *plesio* means "near.") This causes the phase difference to drift in time. This scenario can easily arise when two interacting modules have independent clocks generated from separate crystal oscillators. Since the transmitted signal can arrive at the receiving module at a different rate than the local clock, one needs to utilize a buffering scheme to ensure that all data are received. Typically, plesiochronous interconnect occurs only in distributed

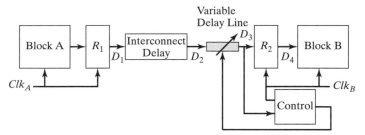

Figure 10-2   Mesochronous communication approach using variable delay line.

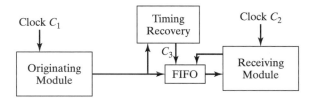

Figure 10-3   Plesiochronous communications by using a FIFO.

systems that contain long-distance communications, since chip- or even board-level circuits typically utilize a common oscillator to derive local clocks. A possible framework for plesiochronous interconnect is shown in Figure 10-3.

In this digital communications framework, the originating module issues data at some unknown rate C_1, which is plesiochronous with respect to C_2. The timing recovery unit is responsible for deriving clock C_3 from the data sequence and buffering the data in a FIFO. As a result, C_3 will be synchronous with the data at the input of the FIFO and will be mesochronous with C_1. Since the clock frequencies from the originating and receiving modules are mismatched, data might have to be dropped if the transmit frequency is faster, or data can be duplicated if the transmit frequency is slower than the receive frequency. However, by making the FIFO large enough, as well as periodically resetting the system whenever an overflow condition occurs, robust communication can be achieved.

10.2.4 Asynchronous Interconnect

Asynchronous signals can transition arbitrarily at any time, and they are not slaved to any local clock. As a result, it is not straightforward to map these arbitrary transitions into a synchronized data stream. It is possible to synchronize asynchronous signals by detecting events and by introducing latencies into the data stream synchronized to a local clock. A more natural way to handle asynchronous signals, however, is simply to eliminate the use of local clocks and utilize a self-timed asynchronous design approach. In such an approach, communication between modules is controlled through a handshaking protocol that ensures the proper ordering of operations.

When a logic block completes an operation (Figure 10-4), it will generate a completion signal DV to indicate that output data are valid. The handshaking signals then initiate a data transfer to the next block, which latches in the new data and begins a new computation by asserting the initialization signal I. Asynchronous designs are advantageous because computations are

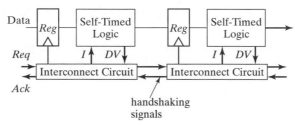

Figure 10-4 Asynchronous design methodology for simple pipeline interconnect.

performed at the native speed of the logic, and block computations occur whenever data become available. There is no need to manage clock *skew*, and the design methodology leads to a very modular approach in which interaction between blocks simply occurs through a handshaking procedure. However, these protocols result in increased complexity and overhead in communication, which impacts performance.

10.3 Synchronous Design—An In-Depth Perspective

10.3.1 Synchronous Timing Basics

Virtually all systems designed today use a periodic *synchronization* signal or clock. The generation and distribution of a clock has a significant impact on the performance and power dissipation of the system. For the time being, let us assume a positive *edge-triggered* system, in which the rising edge of the clock denotes the beginning and completion of a clock cycle. In an ideal world, the phase of the clock (i.e., the position of the clock edge relative to the reference) at various points in the system is exactly equal, assuming that the clock paths from the central distribution point to each register are perfectly balanced. Figure 10-5 shows the basic structure of a synchronous pipelined datapath. In the ideal scenario, the clocks at registers 1 and 2 have the same period and transition at the exact same time.

Assume that the following timing parameters of the sequential circuit are available:

- The contamination or minimum delay ($t_{c-q,cd}$) and the maximum propagation delay of the register (t_{c-q}).
- The setup (t_{su}) and hold times (t_{hold}) for the registers.
- The contamination delay ($t_{logic,cd}$) and the maximum delay (t_{logic}) of the combinational logic.
- The positions of the rising edges of the clocks CLK_1 and CLK_2 (t_{clk1} and t_{clk2}, respectively), relative to a global reference.

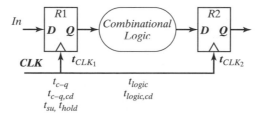

Figure 10-5 Pipelined datapath circuit and timing parameters.

Under the ideal condition that $t_{clk1} = t_{clk2}$, the minimum clock period required for this sequential circuit is determined solely by the worst case propagation delays. The period must be long enough for the data to propagate through the registers and logic and to be set up at the destination register before the next rising edge of the clock. As we saw in Chapter 7, this constraint is given by the following expression:

$$T > t_{c-q} + t_{logic} + t_{su} \tag{10.1}$$

At the same time, the hold time of the destination register must be shorter than the minimum propagation delay through the logic network:

$$t_{hold} < t_{c-q,cd} + t_{logic,cd} \tag{10.2}$$

Unfortunately, the preceding analysis is somewhat simplistic, since the clock is never ideal. The different clock events turn out to be neither perfectly periodic nor perfectly simultaneous. As a result of process and environmental variations, the clock signal can have both *spatial* and *temporal* variations, which lead to performance degradation and/or circuit malfunction.

Clock Skew

The spatial variation in arrival time of a clock transition on an integrated circuit is commonly referred to as *clock skew*. The *clock skew* between two points i and j on an IC is given by $\delta(i, j) = t_i - t_j$, where t_i and t_j are the positions of the rising edge of the clock with respect to the reference. Consider the transfer of data between registers $R1$ and $R2$ in Figure 10-5. The clock skew can be positive or negative depending upon the routing direction and position of the clock source. The timing diagram for the case with positive skew is shown in Figure 10-6. As the figure illustrates, the rising clock edge is delayed by a positive δ at the second register.

    Clock skew is caused by static mismatches in the clock paths and differences in the clock load. By definition, skew is constant from cycle to cycle. That is, if in one cycle CLK_2 lagged CLK_1 by δ, then on the next cycle, it will lag it by the same amount. It is important to note that clock skew does not result in clock period variation, but only in phase shift.

    The clock-skew phenomenon has strong implications for both the performance and the functionality of sequential systems. First, consider the impact of clock skew on performance. We can see from Figure 10-6 that a new input *In* sampled by $R1$ at edge ① will propagate through the com-

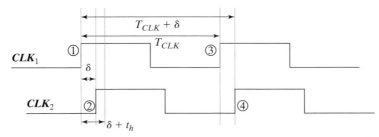

Figure 10-6 Timing diagram to study the impact of clock skew on performance and functionality. In this sample timing diagram, $\delta > 0$.

binational logic and be sampled by $R2$ on edge ④. If the clock skew is positive, the time available for a signal to propagate from $R1$ to $R2$ is increased by the skew δ. The output of the combinational logic must be valid one setup time before the rising edge of CLK_2 (point ④). The constraint on the minimum clock period can then be derived as follows:

$$T + \delta \geq t_{c-q} + t_{logic} + t_{su} \quad \text{or} \quad T \geq t_{c-q} + t_{logic} + t_{su} - \delta \tag{10.3}$$

This equation suggests that clock skew actually has the potential to improve the performance of the circuit. That is, the minimum clock period required to operate the circuit reliably reduces with increasing clock skew! This is indeed correct, but unfortunately, increasing skew makes the circuit more susceptible to race conditions, which may harm the correct operation of sequential systems.

This can be illustrated by the following example: Assume again that input In is sampled on the rising edge of CLK_1 at edge ① into $R1$. The new value at the output of $R1$ propagates through the combinational logic and should be valid before edge ④ at CLK_2. However, if the minimum delay of the combinational logic block is *small*, the inputs to $R2$ may change before the clock edge ②, resulting in incorrect evaluation. To avoid races, we must ensure that the minimum propagation delay through the register and logic is long enough that the inputs to $R2$ are valid for a hold time after edge ②. The constraint can be formally stated as

$$\delta + t_{hold} < t_{(c-q, cd)} + t_{(logic, cd)}$$

or $\tag{10.4}$

$$\delta < t_{(c-q, cd)} + t_{(logic, cd)} - t_{hold}$$

Figure 10-7 shows the timing diagram for the case in which $\delta < 0$. For this case, the rising edge of CLK_2 happens before the rising edge of CLK_1. On the rising edge of CLK_1, a new input is sampled by $R1$. The new data propagate through the combinational logic, and they are sampled by $R2$ on the rising edge of CLK_2, which corresponds to edge ④. As Figure 10-7 and Eq. (10.3) clearly show, a negative skew adversely impacts the performance of a sequential system. However, assuming $t_{hold} + \delta < t_{(c-q, cd)} + t_{(logic, cd)}$, a negative skew implies that the system never fails, since edge ② happens before edge ①!

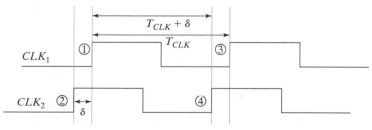

Figure 10-7   Timing diagram for the case when $\delta < 0$. The rising edge of CLK_2 arrives earlier than the edge of CLK_1.

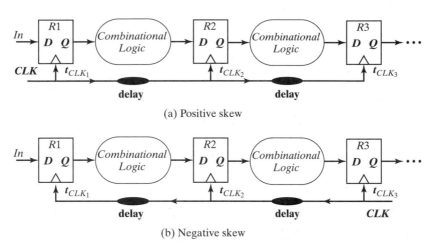

(a) Positive skew

(b) Negative skew

Figure 10-8   Positive and negative clock skew scenarios.

Example scenarios for positive and negative clock skew are shown in Figure 10-8.

- $\delta > 0$—This corresponds to a clock routed in the same direction as the flow of the data through the pipeline (Figure 10-8a). In this case, the skew has to be strictly controlled and satisfy Eq. . If the constraint is not met, the circuit malfunctions **independently of the clock period**. Reducing the clock frequency of an edge-triggered circuit does not help getting around skew problems! It is therefore necessary to satisfy the hold-time constraints at design time. On the other hand, positive skew increases the through put of the circuit as expressed by Eq. (10.3). The clock period can be shortened by δ. The extent of this improvement is limited, as large values of δ soon provoke violations of Eq. .
- $\delta < 0$—When the clock is routed in the opposite direction of the data (Figure 10-8b), the skew is negative and provides significant immunity to races; if the hold time is zero or negative, races are eliminated because Eq. is unconditionally met! The skew reduces the time available for actual computation so that the clock period has to be increased by $|\delta|$. In summary, routing the clock in the opposite direction of the data avoids disasters, but hampers the circuit performance.

Unfortunately, since a general logic circuit can have data flowing in both directions (for example, circuits with feedback), this solution to eliminate races does not always work. Figure 10-9 shows that the skew can assume both positive and negative values, depending on the direction of the data transfer. Under these circumstances, the designer has to account for the worst case skew condition. In general, routing the clock so that only negative skew occurs is not feasible. Therefore, the design of a low-skew clock network is essential.

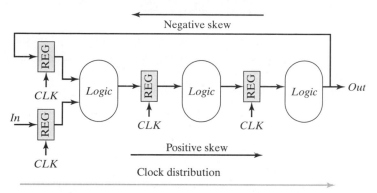

Figure 10-9 Datapath structure with feedback.

Example 10.1 Propagation and Contamination Delay Estimation

Consider the logic network shown in Figure 10-10. Determine the contamination and propagation delays of the network, given a worst case gate delay of t_{gate}. We also assume that the maximum and minimum delays of the gates are identical.

The contamination delay is easily found; it equals $2t_{gate}$, and is the delay through OR_1 and OR_2. On the other hand, computation of the worst case propagation delay is not as simple. At first glance, it would appear that the worst case corresponds to path ①, and its delay is $5t_{gate}$. However, when analyzing the data dependencies, it becomes obvious that path ① can never be exercised. Path ① is called a *false path*. If $A = 1$, the critical path goes through OR_1 and OR_2. If $A = 0$ and $B = 0$, the critical path is through I_1, OR_1 and OR_2 (corresponding to a delay of $3t_{gate}$). For the case in which $A = 0$ and $B = 1$, the longest path goes through I_1, OR_1, AND_3 and OR_2. In other words, for this simple (but contrived) network, the output does not even depend on inputs C and D (that is, there is **redundancy**). Therefore, the actual propagation delay is $4t_{gate}$. Given the propagation and contamination delay, the minimum and maximum allowable skew can be easily computed.

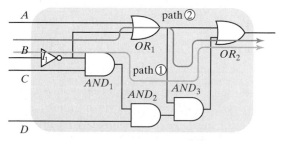

Figure 10-10 Logic network for computation of performance.

WARNING: The computation of the worst case propagation delay for combinational logic, due to the existence of *false paths*, cannot be obtained simply by adding the propagation delays of individual logic gates. The critical path is strongly dependent on circuit topology and data dependencies.

Clock Jitter

Clock jitter refers to the temporal variation of the clock period at a given point on the chip—that is, the clock period can reduce or expand on a cycle-by-cycle basis. It is strictly a temporal uncertainty measure, and it is often specified at a given point. **Jitter** can be measured and characterized in a number of ways and **is a zero-mean random variable**. The *absolute jitter* (t_{jitter}) refers to the worst case variation (absolute value) of a clock edge at a given location with respect to an ideally periodic reference clock edge. The *cycle-to-cycle jitter* (T_{jitter}) typically refers to the time-varying deviations of a single clock period relative to an ideal reference clock. For a given spatial location i, it is given as $T^i_{jitter}(n) = t^i_{clk,n+1} - t^i_{clk,n} - T_{CLK}$, where $t^i_{clk,n+1}$ and $t^i_{clk,n}$ represent the arrival time of the $n + 1^{\text{th}}$ and the nth clock edges at node i, respectively, and T_{CLK} is the nominal clock period. Under the worst case conditions, the magnitude of the cycle-to-cycle jitter equals twice the absolute jitter ($2t^i_{jitter}$).

Jitter directly impacts the performance of a sequential system. Figure 10-11 shows the nominal clock period, as well as the variation in period. Ideally, the clock period starts at edge ② and ends at edge ⑤, with a nominal clock period of T_{CLK}. However, the worst case scenario happens when the leading edge of the current clock period is delayed by jitter (edge ③), while jitter causes the leading edge of the next clock period to occur early (edge ④). As a result, the total time available to complete the operation is reduced by $2t_{jiiter}$ in the worst case and is given by

$$T_{CLK} - 2t_{jitter} \geq t_{c-q} + t_{logic} + t_{su} \quad \text{or} \quad T \geq t_{c-q} + t_{logic} + t_{su} + 2t_{jitter} \tag{10.5}$$

Equation (10.5) illustrates that jitter directly reduces the performance of a sequential circuit. Keeping it within strict bounds is essential if one is concerned about performance.

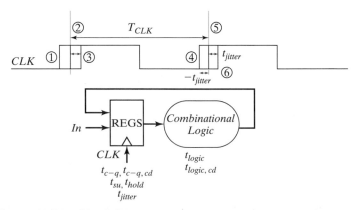

Figure 10-11   Circuit for studying the impact of jitter on performance.

The Combined Impact of Skew and Jitter

In this section, the combined impact of skew and jitter is studied for conventional edge-triggered clocking. Consider the sequential circuit shown in Figure 10-14.

Assume that as a result of the clock distribution, there is a static skew δ between the clock signals at the two registers (assume that $\delta > 0$). Furthermore, the two clocks experience a jitter of t_{jitter}. To determine the constraint on the minimum clock period, we must look at the minimum available time to perform the required computation. The worst case occurs when the leading edge of the current clock period on CLK_1 happens late (edge ③) and the leading edge of the next cycle of CLK_2 happens early (edge ⑩). This results in the following constraint:

$$T_{CLK} + \delta - 2t_{jitter} \geq t_{c-q} + t_{logic} + t_{su}$$

or (10.6)

$$T \geq t_{c-q} + t_{logic} + t_{su} - \delta + 2t_{jitter}$$

This equation illustrates that positive skew can provide a performance advantage. On the other hand, *jitter* always has a negative impact on the minimum clock period.[1]

To formulate the minimum delay constraint, consider the case in which the leading edge of the CLK_1 cycle arrives early (edge ①), and the leading edges the current cycle of CLK_2 arrives late (edge ⑥). The separation between edges ① and ⑥ should be smaller than the minimum delay through the network. This results in

$$\delta + t_{hold} + 2t_{jitter} < t_{(c-q, cd)} + t_{(logic, cd)}$$

or (10.7)

$$\delta < t_{(c-q, cd)} + t_{(logic, cd)} - t_{hold} - 2t_{jitter}$$

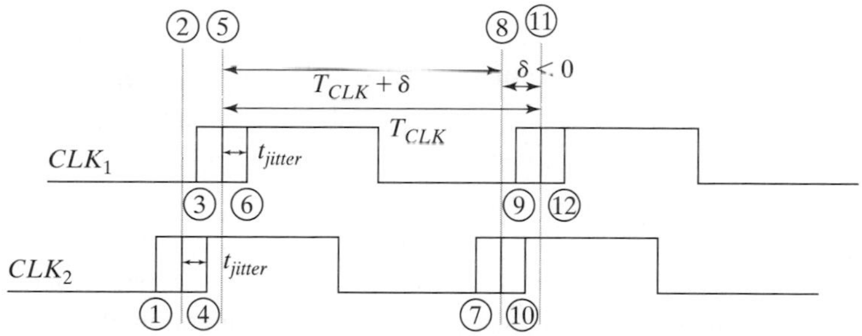

Figure 10-12 Sequence circuit with a negative clock skew (δ). The skew is assumed to be larger than the *jitter*.

---

[1] This analysis is definitely for the worst case. It assumes that the jitter values at the source and the destination nodes are independent statistical variables. In reality, the clock edges involved in the hold-time analysis are derived from the same clock edge and are statistically dependent. Taking this dependence into account reduces the timing constraints substantially.

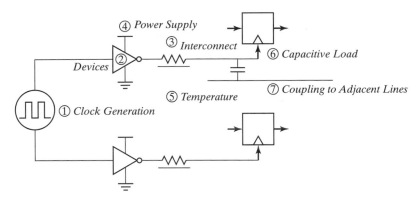

Figure 10-13   Skew and jitter sources in synchronous clock distribution.

This relation indicates that the acceptable skew is reduced by the *jitter* of the two signals.

Now consider the case in which the skew is negative ($\delta < 0$), as shown in Figure 10-12. Assume that $|\delta| > t_{jitter}$. It can be verified that the worst case timing is exactly the same as in the previous analysis, with δ taking a negative value. That is, negative skew reduces performance.

10.3.2 Sources of Skew and Jitter

A perfect *clock* is defined as a periodic signal that simultaneously triggers various memory elements on the chip. However, due to a variety of process and environmental variations, clocks are not ideal. To illustrate the sources of skew and jitter, consider a simplistic view of a typical clock generation and distribution network, as shown in Figure 10-13. A high-frequency clock is either provided from off chip or generated on chip. From a central point, the clock is distributed using multiple *matched* paths to low-level sequential elements. In this picture, two paths are shown. The clock paths include the wiring and the associated distributed buffers required to drive interconnect and loads. A key point to realize in clock distribution is that the **absolute delay through a clock distribution path is not important**; what matters is the relative arrival time at the register points at the end of each path. It is perfectly acceptable for the clock signal to take multiple cycles to get from a central distribution point to a low-level register as long as all clocks arrive at the same time at all the registers on the chip.

There are many reasons why the two parallel paths don't result in exactly the same delay. The sources of clock uncertainty can be classified in several ways. First, errors can be divided into two categories: *systematic* and *random*. *Systematic* errors are nominally identical from chip to chip and are predictable (for instance, variation in total load capacitance of each clock path). In principle, such errors can be modeled and corrected at design time, given sufficiently good models and simulators. Short of that, systematic errors can be deduced from measurements over a set of chips, and the design can be adjusted to compensate. *Random* errors are due to manufacturing variations that are difficult to model and eliminate (for instance, dopant fluctuations that result in threshold variations).

Mismatches may also be characterized as *static* or *time varying*. In practice, a continuum exists between changes that are slower than the time constant of interest and those that are faster. For example, temperature gradients on a chip vary on a millisecond time scale. A clock network

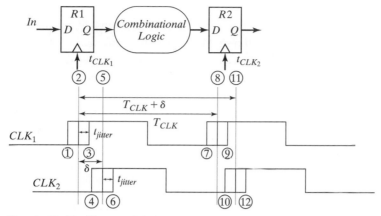

Figure 10-14 Sequential circuit to study the impact of skew and jitter on *edge-triggered* systems. In this example, a positive *skew* (δ) is assumed.

tuned by a one-time calibration is vulnerable to the time-varying mismatch caused by the varying thermal gradients. On the other hand, thermal changes appear essentially static to a feedback network with a bandwidth of several megahertz. Another example is fielded by power-supply noise. The clock net is usually by far the largest signal net on the chip, and simultaneous transitions on the clock drivers induce noise in the power supply. This high-speed effect does not create a time-varying mismatch, because it is the same at every clock cycle and affects each rising clock edge the same way. Of course, this power-supply glitch may still cause static mismatch if it is not the same throughout the chip. The various sources of skew and jitter introduced in Figure 10-13 are described and characterized in detail in the sections that follow.

Clock-Signal Generation (1)

The generation of the clock signal itself causes **jitter**. A typical on-chip clock generator, as described at the end of this chapter, takes a low-frequency reference clock signal and produces a high-frequency global reference for the processor. The core of such a generator is a *voltage-controlled oscillator* (VCO). This is an analog circuit, sensitive to intrinsic device noise and power-supply variations. A major problem is the coupling from the surrounding noisy digital circuitry through the substrate. This is especially a problem in modern fabrication processes that use a lightly doped epitaxy on the heavily doped substrate (to combat latch up). This causes substrate noise to travel over large distances on the chip [Maneatis00]. These noise sources cause temporal variations in the clock signal that propagate unfiltered through the clock drivers to the flip-flops, and result in *cycle-to-cycle* clock-period variations.

Manufacturing Device Variations (2)

Distributed buffers are integral components of the clock distribution networks. They are required to drive both the register loads and the global and local interconnects. The matching of devices in the buffers along multiple clock paths is critical to minimizing timing uncertainty. Unfortunately, as a result of process variations, device parameters in the buffers vary along different paths, resulting in *static skew*. There are many sources of variations that contribute, such as oxide variations (which affect the gain and threshold), dopant variations, and lateral dimension (width and length) variations. The doping variations can affect the depth of junction and dopant profiles and cause electrical parameters (such as device threshold and parasitic capacitances) to vary.

The orientation of polysilicon also can have some impact on the device parameters. Keeping the orientation the same across the chip for the clock drivers is therefore critical. Variation in the polysilicon critical dimension is particularly important, because it translates directly into MOS transistor channel length, impacting the drive current and switching characteristics. Spatial variation usually consists of a wafer-level (or within-wafer) variation and a die-level (or within-die) variation. At least part of this variation is systematic and therefore can be modeled and compensated for. The random variations, however, ultimately limit the matching and lower bound of the skew that can be achieved.

Interconnect Variations (3)

Vertical and lateral dimension variations cause the interconnect capacitance and resistance to vary across a chip. Since this variation is static, it causes **skew** between different paths. One important source of interconnect variation is the *Inter-layer Dielectric (ILD)* thickness variation. In the formation of aluminum interconnect, layers of silicon dioxide are interposed between layers of patterned metallization. Oxide is deposited over a layer of patterned metal features, generally resulting in some remaining step height or surface topography. *Chemical–mechanical polishing* (CMP) is used to "planarize" the surface and remove the topography resulting from deposition and etch (as described in Chapter 3 and shown in Figure 10-15a). While at the feature scale (i.e., over an individual metal line), CMP can achieve excellent planarity, there are limitations on it over a global range. This is due primarily to variations in the polish rate, which is a function of the circuit layout density and pattern effects. Figure 10-15b shows this effect— the polish rate is higher for the lower-spatial-density region, resulting in a smaller dielectric thickness and higher capacitance.

The assessment and control of variation is of critical importance in semiconductor process development and manufacturing. Significant advances have been made to develop analytical

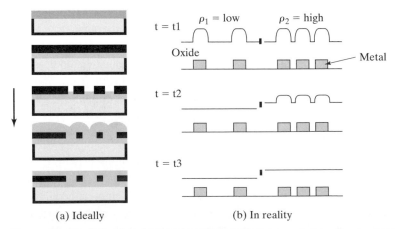

(a) Ideally   (b) In reality

Figure 10-15   Inter-level Dielectric (ILD) thickness variation due to density (Courtesy of Duane Boning.).

models for estimating the ILD thickness variations, based on spatial density. Since this component is often predictable from the layout, it is possible to actually correct for the systematic component at design time (e.g., by adding appropriate delays or making the density uniform by adding "dummy fills"). Figure 10-16 shows the spatial pattern density and ILD thickness for a

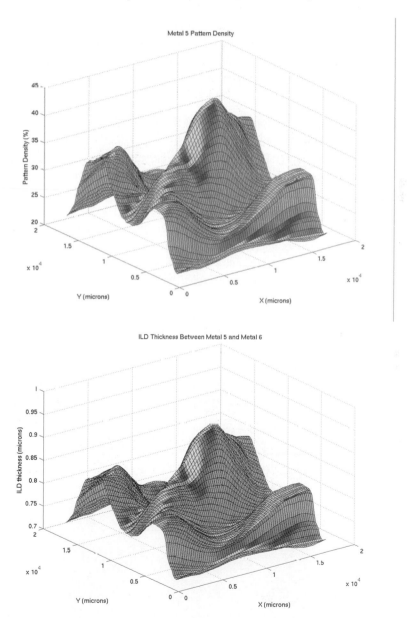

Figure 10-16 Pattern density and ILD thickness variation for a high-performance microprocessor. (Courtesy of Duane Boning)

high-performance microprocessor. The graphs show a clear correlation between the density and the thickness of the dielectric. Hence, clock distribution networks must exploit such information in order to reduce clock skew.

Other interconnect variations include deviations in the width of the wires and line spacing, which result from photolithography and etch dependencies. At the lower levels of the metallization hierarchy, lithographic effects are more important, while etch effects that depend on width and layout are dominant at the higher levels. The width is a critical parameter because it directly impacts the resistance of the line, and the wire spacing affects the wire-to-wire capacitance. A detailed review of device and interconnect variations is presented in [Boning00]. Recent processors use copper interconnects, in which line thickness variations are also seen to be highly pattern dependent due to CMP dishing and erosion effects [Park00].

Environmental Variations (4 and 5)

Environmental variations probably are the most significant contributors to **skew and jitter**. The two major sources of environmental variations are *temperature* and *power supply*. Temperature gradients across the chip result from variations in power dissipation across the die. These gradients can be quite large, as shown in Figure 10-17, which displays a snapshot of the surface temperature of the DEC 21064 microprocessor. Temperature variation has become an important issue with *clock gating*, where some parts of the chip may be idle, while other parts of the chip are fully active. Clock gating has become popular in recent years as a means to minimize power dissipation in idle modules (as described in a later section). Shutting off parts of the chip leads to large temperature variations. Since the device parameters (such as threshold and mobility) depend strongly on temperature, the buffer delay for a clock distribution network can vary drastically from path to path. More importantly, this component is time varying, since the temperature changes as the logic activity of the circuit varies. Hence, it is not sufficient to simulate the clock networks at worst case corners of temperature; instead, the worst case variation in temperature must be simulated. An interesting question is whether temperature variation contributes to skew or to jitter. Clearly, the difference in temperature is time varying, but the changes are rela-

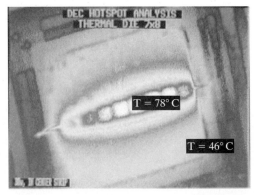

Figure 10-17   Temperature variation (snapshot) over DEC 21064 microprocessor. The highest temperature occurs at the central clock driver [Herrick00].

tively slow (typical time constants for temperature changes are on the order of milliseconds). Therefore, it is usually considered as a skew component and the worst case conditions are used. Fortunately, by using feedback, it is possible to calibrate the temperature and to compensate for this effect.

Power-supply variations, on the other hand, are the major source of **jitter** in clock distribution networks. The delay through buffers is a very strong function of power supply, as it directly affects the drive of the transistors. As with temperature, the power-supply voltage is a strong function of the switching activity. Therefore, the buffer delay varies strongly from path to path. Power-supply variations can be classified into slow- (or static) and high-frequency variations. Static power-supply variations may result from fixed currents drawn from various modules, while high-frequency variations result from instantaneous *IR* drops along the power grid due to fluctuations in switching activity. Inductive effects on the power supply also are a major concern since they cause voltage fluctuations. Again, clock gating has exacerbated this problem, because the logic transitions between the idle and active states can cause major changes in current drawn from the supply. Since the power supply can change rapidly, the period of the clock signal is modulated on a cycle-by-cycle basis, resulting in jitter. The jitter on two different clock points may be correlated or uncorrelated, depending on how the power network is configured and the profile of switching patterns. Unfortunately, high-frequency power-supply changes are difficult to compensate for, even with feedback techniques. Consequently, **power-supply noise fundamentally limits the performance of clock networks**. To minimize power-supply variations, high-performance designs add decoupling capacitance around major clock drivers.

Capacitive Coupling (6 and 7)

Changes in capacitive load also contribute to timing uncertainty. There are two major sources of capacitive-load variations: coupling between the clock lines and adjacent signal wires, and variation in gate capacitance. The clock network includes both the interconnect and the gate capacitance of latches and registers. Any coupling between the clock wire and adjacent signal results in timing uncertainty. Since the adjacent signal can transition in arbitrary directions and at arbitrary times, the exact coupling to the clock network is not fixed from cycle to cycle, causing **jitter**. Another major source of clock uncertainty is the variation in the gate capacitance contributed by the connecting sequential elements. The load capacitance is highly nonlinear and depends on the applied voltage. For many latches and registers, the clock load is a function of the current state of the latch/register (i.e., the values stored on the internal nodes of the circuit), as well as the next state. This causes the delay through the clock buffers to vary from cycle to cycle, which causes jitter.

Example 10.2   Data-Dependent Clock Jitter

Consider the circuit shown in Figure 10-18, where a minimum-sized local clock buffer drives a register. (Actually, each clock buffer drives four registers, though only one is shown here.) The simulation shows *CKb*, the output of the first inverter for four possible

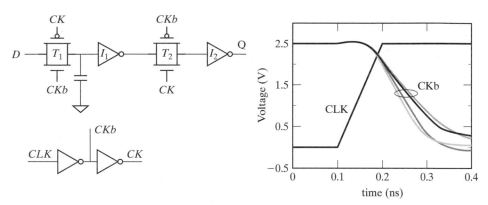

Figure 10-18 Impact of data-dependent clock load on clock jitter for transmission-gate register.

transitions $(0 \to 0, 0 \to 1, 1 \to 0$ and $1 \to 1)$. The jitter on the clock based on data-dependent capacitance is illustrated. In general, the only way to deal with this problem is to use registers that do not exhibit a large variation in load as a function of data—for example, the differential sense-amplifier register shown in Chapter 7.

10.3.3 Clock-Distribution Techniques

It is clear from the previous discussion that clock skew and jitter are major issues in digital circuits, and can fundamentally limit the performance of a digital system. It is therefore necessary to design a clock network that minimizes both. While designing that clock network, a close eye should be kept on the associated power dissipation. In most high-speed digital processors, a majority of the power is dissipated in the clock network. To reduce power dissipation, clock networks must support clock conditioning—that is, the ability to shut down parts of the clock network. Unfortunately, clock gating results in additional clock uncertainty (as described earlier).

In this section, an overview of basic constructs in high-performance clock distribution techniques is presented, along with a case study of clock distribution in the Alpha microprocessors. There are many degrees of freedom in the design of a clock network, including the type of material used for wires, the basic topology and hierarchy, the sizing of wires and buffers, the rise and fall times, and the partitioning of load capacitances.

Fabrics for Clocking

Clock networks typically include a network that is used to distribute a **global reference** to various parts of the chip, and a final stage that is responsible for **local distribution** of the clock while considering the local load variations. Most clock distribution schemes exploit the fact that the absolute delay from a central clock source to the clocking elements is irrelevant—only the relative phase between two clocking points is important. Therefore, one common approach to distributing a clock is to use balanced paths (called *trees*).

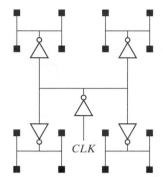

Figure 10-19 Example of an *H*-tree clock-distribution network for 16 leaf nodes.

The most common type of clock distribution scheme is the H-*tree network*, which is illustrated in Figure 10-19 for a 4 × 4 processor array. The clock is first routed to a central point on the chip. Balanced paths that include both matched interconnect and buffers then distribute the reference to the various leaf nodes. Ideally, if each path is perfectly balanced, the clock skew is zero. Although it might take multiple clock cycles for a signal to propagate from the central point to each leaf node, the arrival times are identical at every leaf node. However, in reality, process and environmental variations cause clock skew and jitter to occur.

The *H*-tree configuration is particularly useful for regular array networks in which all elements are identical and the clock can be distributed as a binary tree (for example, arrays of identical tiled processors). The concept can be generalized to a more generic setting. The more general approach, referred to as *matched* RC *trees*, represents a floor plan that distributes the clock signal so that the interconnections carrying the clock signals to the functional subblocks are of equal length. That is, the general approach does not rely on a regular physical structure. An example of a matched *RC* is shown in Figure 10-20. The chip is partitioned into 10 balanced load segments (tiles). The global clock driver distributes the clock to the tile drivers located at the dots in the figure. A lower level *RC*-matched tree is used to drive 580 additional drivers inside each tile. A 3D visualization of the clock delay in a tree network is shown in Figure 10-21.

An alternative clock distribution approach is the grid structure of Figure 10-22 [Bailey00]. Grids typically are used in the final stage of a clock network to distribute the clock to the clocking element loads. This approach is fundamentally different from the balanced *RC* approach. The main difference is that the delay from the final driver to each load is not matched. Rather, the absolute delay is minimized, assuming that the grid size is small. A major advantage of such a grid structure is that it allows for late design changes, since the clock is easily accessible at various points on the die. Unfortunately, the penalty is a relatively large power dissipation since the structure has a lot of "excess" interconnect. In addition to the schemes described earlier, other approaches have been devised for clock distribution. The length-matched serpentine approach is just one of them [Young97].

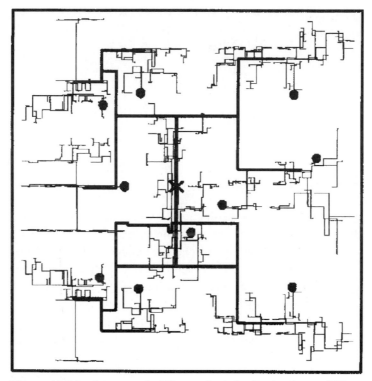

Figure 10-20   An example RC-matched distribution for an IBM microprocessor [Restle98].

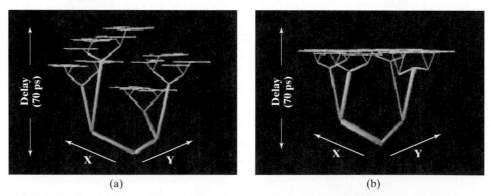

Figure 10-21   Visualization of clock delay in a tree network driving different loads. The *X*- and *Y*-axes represent the die, while the *Z*-axis represents the clock delay. The width of the lines is proportional to the designed wire width. The unbalanced load creates a large skew, as is clear in (a). By careful tuning of the wire widths, the load is balanced, minimizing the skew, as shown in (b) [Restle01]. See also Colorplate 9.

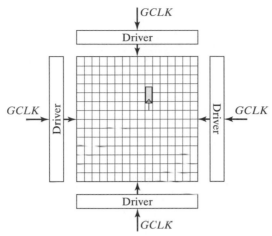

Figure 10-22 Grid structures allow a low skew distribution and physical design flexibility at the cost of power dissipation [Bailey00].

It is essential to consider clock distribution in the earlier phases of the design of a complex circuit, since it might influence the shape and form of the chip's floor plan. It is tempting for a designer to ignore the clock network in the early phases of a project and consider it only at the end of the design cycle, when most of the chip layout is already frozen. This results in unwieldy clock networks and multiple timing constraints that hamper the performance and operation of the final circuit. With careful planning, a designer can avoid many of these problems, and clock distribution becomes a manageable operation.

Clock Distribution Case Study—The Digital Alpha Microprocessors

In this section, the clock distribution strategies for three generations of the Alpha microprocessor are discussed in detail. These processors have always been at the cutting edge of the technology and therefore represent an interesting perspective on the evolution of clock distribution [Herrick00].

The Alpha 21064 Processor—The first-generation Alpha microprocessor (21064 or EV4) from Digital Equipment Corporation used a single global clock driver [Dobberpuhl92]. The distribution of clock load capacitance among various functional blocks is shown in Figure 10-23. The total clock load equals 3.25 nF! The processor uses a single-phase clock methodology, and the 200-MHz clock is fed to a binary tree with five levels of buffering. The inputs to the clock drivers are shorted to smooth out the asymmetry in the incoming signals. The final output stage, residing in the middle of the chip, drives the clock net. The clock driver and the associated pre-drivers account for 40% of the effective switched capacitance (12.5 nF), resulting in significant power dissipation. The overall width of the clock driver was on the order of 35 cm in a 0.75-μm technology. A detailed clock skew simulation with process variations indicates that a clock uncertainty of less than 200 ps (< 10%) was achieved.

The Alpha 21164 Processor—The second-generation Alpha microprocessor (EV5) operates at a clock frequency of 300 Mhz while using 9.3 million transistors on a 16.5 mm. × 18.1 mm die in a 0.5-μm CMOS technology [Bowhill95]. A single-phase clocking methodology was selected, and the design made extensive use of dynamic logic, resulting in a substantial clock load of 3.75 nF. The clock distribution system consumes 20 W, which is 40% of the total dissipation of the processor.

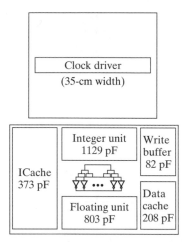

Figure 10-23   Distribution of clock load capacitance for the 21064 Alpha processor.

The incoming clock signal is first routed through a single six-stage buffer placed at the center of the chip. The resulting signal is distributed in metal-3 to the left and right banks of final clock drivers, positioned between the secondary cache memory and the outside edge of the execution unit (see Figure 10-24a). The produced clock signal is driven onto a grid of metal-3 and metal-4 wires. The equivalent transistor width of the final driver inverter equals 58 cm! To ensure the integrity of the clock grid across the chip, the grid was extracted from the layout, and the resulting *RC*-network was simulated. A

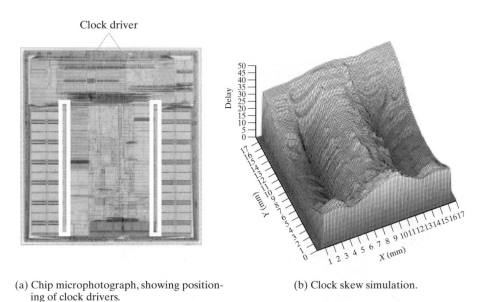

(a) Chip microphotograph, showing position-   (b) Clock skew simulation.
ing of clock drivers.

Figure 10-24   Clock distribution and skew in a 300-MHz microprocessor. (Courtesy of Digital Equipment Corporation.)

three-dimensional representation of the simulation results is plotted in Figure 10-24b. As evident from the plot, the skew is zero at the output of the left and right drivers. The maximum value of the absolute skew is smaller than 90 ps. The critical instruction and execution units all see the clock within 65 ps.

Clock skew and race problems were addressed using a "mix-and-match" approach. The clock skew problems were eliminated by either routing the clock in the opposite direction of the data (at a small cost in terms of performance) or by ensuring that the data could not overtake the clock. A standardized library of level-sensitive transmission-gate latches was used for the complete chip. To avoid race-through conditions, the following design guidelines were used:

- Careful sizing of the local clock buffers were carefully sized so that their skew was minimal.
- At least one gate had to be inserted between connecting latches. This gate, which can be part of the logic function or just a simple inverter, ensures was a minimum contamination delay. Special design verification tools were developed to guarantee that this rule was obeyed over the complete chip.

To improve the interlayer dielectric uniformity, filler polygons were inserted between widely spaced lines, as shown in Figure 10-25. Though this may increase the capacitance to nearby signal lines, the improved uniformity results in lower variation and clock uncertainty. The dummy fills are automatically inserted, and tied to one of the power rails (V_{DD} or GND). Dummy insertion is required today to equalize the etch-away of CMP (Chapter 3). This technique is used in many of today's processes for controlling the clock skew.

This example demonstrates that managing clock skew and clock distribution for large, high-performance synchronous designs is a feasible task. However, making such a circuit work in a reliable way requires careful planning and intensive analysis.

The Alpha 21264 Processor—A hierarchical clocking scheme is used in the 600-Mhz Alpha 21264 (EV6) processor (in 0.35-μm CMOS), shown in Figure 10-26 [Bailey98]. The choice of a hierarchical clocking scheme for this processor is a major departure from the preceding processors, which did not have

Figure 10-25   Dummy fills reduce the ILD variation and improve clock skew.

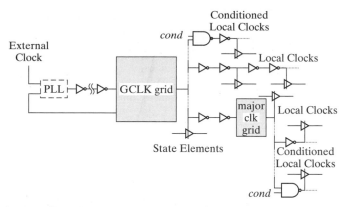

Figure 10-26   Clock hierarchy for the Alpha 21264 Processor.

a hierarchy of clocks beyond the global clock grid. Using a hierarchical clocking approach enables trade-offs between power and skew management. Power is reduced, because the clocking networks for individual blocks can be gated. As seen in previous-generation microprocessors, the clock power contributes to a large fraction of overall power consumption. Also, the flexibility of having local clocks provides the designers with more freedom with respect to circuit styles at the module level. The drawback of using a hierarchical clock network is that skew reduction becomes more difficult. Clocks to various local registers may go through very different paths, which may contribute to the skew. However, by using timing verification tools, the skew can be managed by tweaking of the clock drivers.

The clock hierarchy consists of a *global clock grid*, called *GCLK*, that covers the entire die. State elements and clocking points exist from zero to height levels past *GCLK*. The on-chip generated clock is routed to the center of the die and distributed using tree structures to 16 distributed clock drivers (see Figure 10-27). The global clock distribution network utilizes a windowpane configuration, which achieves low skew by dividing the clock into four regions, which reduces the distance from the drivers to the loads. Each grid pane is driven from four sides, reducing the dependence on process variations. This also helps the power supply and thermal problems, as the drivers are distributed through the chip.

The use of a gridded clock has the advantage of reducing the clock skew, while providing universal availability of the clock signals. The drawback is the increased capacitance of *GCLK*, compared with a tree-distribution approach. In addition to the *GCLK*, at the next level of clock hierarchy, there is a major clock grid. The major clocks were introduced to reduce power: They have localized loads, and they can be sized appropriately to meet the skew and edge requirements for the local loading conditions at timing-critical units.

The lowest level in this hierarchy is formed by the *local clocks*, which are generated as needed from any other clock. Typically, they can be customized to meet local timing constraints. The local clocks provide great flexibility in the design of the logic blocks, but at the same time make it significantly more difficult to manage skew. Furthermore, the local clocks are more susceptible to coupling from data lines, because they are not shielded like the global gridded clocks. As a result, the local clock distribution is highly dependent on its local interconnection and thus has to be designed very carefully.

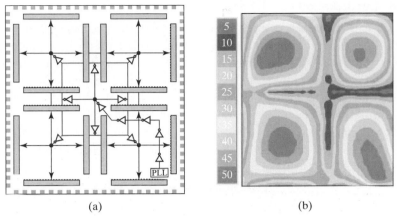

(a)   (b)

Figure 10-27 Global clock distribution network in a windowpane structure. Pane structure (a); clock skew distribution over die (b).

■

Design Techniques—Dealing with Clock Skew and Jitter

From the preceding discussions, some useful guidelines for reducing clock skew and jitter can be derived:

1. To minimize skew, *balance clock paths* from a central distribution source to individual clocking elements, using H-*tree* structures or more generally routed matched-tree structures. When using routed clock trees, the effective clock load of each path that includes wiring as well as transistor loads must be equalized.

2. The use of *local clock grids* (instead of routed trees) can reduce skew at the cost of increased capacitive load and power dissipation.

3. If data-dependent clock load variations cause significant jitter, *differential registers* that have a data-independent clock load should be used. The use of gated clocks to save power results in a data-dependent clock load and increased jitter. In clock networks where the fixed load is large (e.g., in clock grids), the data-dependent variation might not be significant.

4. If data flow in one direction, route the *data and the clock in opposite directions*. This eliminates races at the cost of performance.

5. Avoid data-dependent noise by *shielding clock wires* from adjacent signal wires. By placing power lines (V_{DD} or GND) next to the clock wires, coupling from neighboring signal nets can be minimized or avoided.

6. Variations in interconnect capacitance due to interlayer dielectric thickness variation can be greatly reduced through the use of *dummy fills*. Dummy fills are very common and reduce skew by increasing uniformity. Systematic variations should be modeled and compensated for.

7. Variation in chip temperature across the die causes variations in clock buffer delay. The use of *feedback circuits based on delay-locked loops*, discussed later in this chapter, can compensate for temperature variations.

8. Power-supply variation is a significant component of jitter, as it impacts the cycle-to-cycle delay through clock buffers. High-frequency power-supply variation can be reduced by adding *on-chip decoupling capacitors*. Unfortunately, decoupling capacitors require a significant amount of area, and efficient packaging solutions must be leveraged in order to reduce chip area.   ■

10.3.4 Latch-Based Clocking*

The use of registers in a sequential circuit translates into a robust, reliable design methodology. Yet there are significant performance advantages to be made through the use of a latch-based design style, in which combinational logic is separated by transparent latches. In an edge-triggered system, the worst case logic path between two registers determines the minimum clock period for the entire system. If a logic block finishes before the end of the clock period, it has to sit idle until the next clock edge. The use of a latch-based methodology enables more flexible timing by allowing one stage to pass slack or to borrow time from other stages. The added flexibility also increases overall performance. Note that the latch-based methodology can be thought of as adding logic between latches of master–slave flip-flops.

Consider the latch-based system of Figure 10-28, which uses a two-phase clocking scheme. Assume that the clocks are ideal and that the two clocks are inverted versions of each other (for the sake of simplicity). In this configuration, a stable input is available to the combinational logic block A (CLB_A) on the falling edge of CLK_1 (at edge ②). It has a maximum time equal to the $T_{CLK}/2$ to evaluate—that is, the entire low phase of CLK_1. On the falling edge of CLK_2 (at edge ③), the output CLB_A is latched, and the computation of CLK_B is launched. CLB_B computes on the low phase of CLK_2, and the output is available on the falling edge of CLK_1 (at edge ④). From a timing perspective, this scenario appears to be equivalent to an edge-triggered system, where CLB_A and CLB_B are cascaded between two edge-triggered registers (see Figure 10-29). In both cases, it appears that the time available to perform the combination of CLB_A and CLB_B is T_{CLK}.

However, there is an important performance-related difference. In a latch-based system, it is possible for a logic block to utilize time that is left over from the previous logic block. This phenomenon, which results from the transparency of a latch during its on time, is referred to as *slack borrowing* [Bernstein98]. It requires no explicit design changes, as the passing of slack

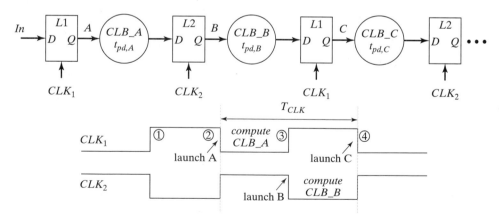

Figure 10-28 Latch-based design in which transparent latches are separated by combinational logic.

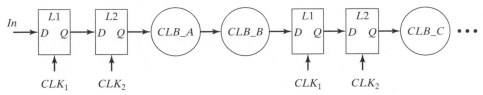

Figure 10-29 Edge-triggered pipeline (back-to-back latches for edge-triggered registers) of the logic in Figure 10-28.

from one block to the next is automatic. The key advantage of slack borrowing is that it allows logic between cycle boundaries to use more than one clock cycle while satisfying the overall cycle time constraint. Stated in another way, if the sequential system works at a particular clock rate, and the total logic delay for a complete cycle is larger than the clock period, then unused time, or *slack,* has been implicitly borrowed from preceding stages. Formally stated, slack passing has taken place if $T_{CLK} < t_{pd, A} + t_{pd, B}$, and the logic functions correctly (for simplicity, the delay associated with latches are ignored). This implies that the clock rate can be higher than the worst case critical path!

As mentioned earlier, slack passing results from the level-sensitive nature of the latches. In Figure 10-28, the input to *CLB_A* must be valid by the falling edge of CLK_1 (edge ②). What happens if the combinational logic block of the previous stage finishes early and has a valid input data for *CLB_A* before edge ②? Since the latch is transparent during the entire high phase of the clock, the input data for *CLB_A* are valid as soon as the previous stage has finished computing. This implies that the maximum time available for *CLB_A* is its phase time (i.e., the low phase of CLK_1) plus any time left from the previous computation.

Consider the latch-based system of Figure 10-30. In this example, signal *a* (input to *CLB_A*) is valid well before edge ②. This implies that the previous block did not use its entire allotment, producing slack time as denoted by the shaded area. By construction, *CLB_A* can start computing as soon as signal *a* becomes valid. It uses this slack time to finish well before its allocated time (edge ③). Since L_2 is a transparent latch, valid on the high phase of CLK_2, *CLB_B* starts to compute using the slack provided by *CLB_A*. Again, *CLB_B* completes before its allocated time (edge ④), and it passes a small amount of slack to the next cycle. As this picture indicates, the total cycle delay—that is, the sum of the delay for *CLB_A* and *CLB_B*—is larger than the clock period. Since the pipeline behaves correctly, slack passing has taken place and a higher through-put has been achieved.

An important question related to slack passing involves to the maximum possible slack that can be passed across cycle boundaries. In Figure 10-30, it is easy to see that the earliest time that *CLB_A* can start computing is ①. This happens if the previous logic block did not use any of its allocated time (CLK_1 high phase) or if it finished by using slack from previous stages. Therefore, the maximum time that can be borrowed from the previous stage is half of a cycle or $T_{CLK}/2$. Similarly, *CLB_B* must finish its operation by edge ④. This implies that

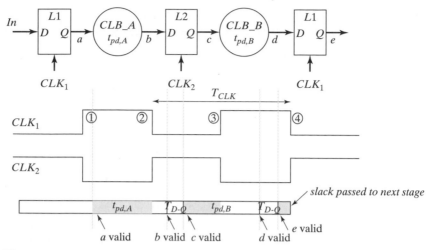

Figure 10-30   Example of slack borrowing.

the maximum logic cycle delay is equal to $1.5 \times T_{CLK}$. However, note that for an n-stage pipeline, the overall logic delay cannot exceed the time available of $n \times T_{CLK}$.

Example 10.3   Slack-Passing Example

First, consider the negative edge-triggered pipeline of Figure 10-31. Assume that the primary input *In* is valid slightly after the rising edge of the clock. We can derive that the minimum clock period required is 125 ns. The latency is two clock cycles. (Actually, the output is valid 2.5 cycles after the input settles.) Note that for the first pipeline stage, one-half cycle is wasted, as the input data are available only to CL_1 after the falling edge of the clock. This time can be exploited in a latch-based system.

Figure 10-32 shows a latch-based version of the same circuit. As the timing indicates, exactly the same timing can be achieved with a clock period of 100 ns. This is enabled by slack borrowing between the logical partitions.

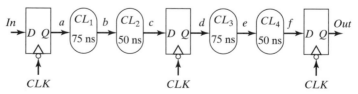

Figure 10-31   Conventional edge-triggered pipeline.

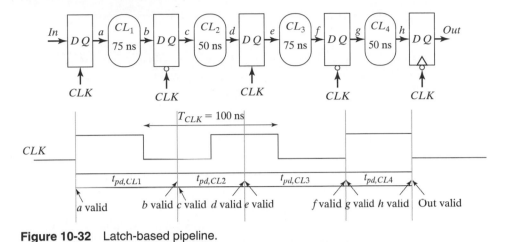

Figure 10-32 Latch-based pipeline.

If slack passing or borrowing intrigues you, we refer you to [Bernstein98], which presents an excellent quantitative analysis of this attractive, yet challenging, approach.

10.4 Self-Timed Circuit Design*

10.4.1 Self-Timed Logic—An Asynchronous Technique

The synchronous design approach advocated in the previous sections assumes that all circuit events are orchestrated by a central clock. Those clocks have a dual function:

- They ensure that the *physical timing constraints* are met. The next clock cycle can start only when all logic transitions have settled and the system has come to a steady state. This ensures that only legal logical values are applied in the next round of computation. In short, clocks account for the worst case delays of logic gates, sequential logic elements, and the wiring.
- Clock events serve as a *logical ordering mechanism* for the global system events. A clock provides a time base that determines what will happen and when. On every clock transition, a number of operations are initiated that change the state of the sequential network.

Consider the pipelined datapath of Figure 10-33. In this circuit, the data transitions through logic stages under the command of the clock. The important point is that the clock period is chosen to be larger than the worst case delay of each pipeline stage, or $T > \max(t_{pd1}, t_{pd2}, t_{pd3}) + t_{pd,reg}$. This will ensure satisfaction of the physical constraint. At each clock transition, a new set of inputs is sampled, and computation is started again. The throughput of the system—which is equivalent to the number of data samples processed per second—is equivalent to the clock rate. The time at which to sample a new input and the availability of an output depend upon the *logical ordering* of the system events and clearly are orchestrated by the clock in this example.

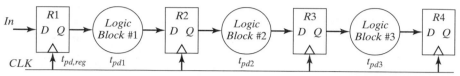

Figure 10-33   Pipelined, synchronous datapath.

The synchronous design methodology has some clear advantages. It presents a structured, deterministic approach to the problem of choreographing the myriad of events that take place in digital designs. The approach taken is to equalize the delays of all operations by making them as bad as the worst of the set. The approach is robust and easy to adhere to, which explains its enormous popularity; however, it does have several pitfalls:

- It assumes that all clock events or timing references happen simultaneously over the complete circuit. This is not the case in reality, because of effects such as clock skew and jitter.
- As all the clocks in a circuit transition at the same time, significant current flows over a very short period of time (due to the large capacitance load). This causes significant noise problems due to package inductance and power-supply grid resistance.
- The linking of physical and logical constraints has some obvious effects on the performance. For instance, the throughput rate of the pipelined system of Figure 10-33 is directly linked to the worst case delay of the slowest element in the pipeline. On average, the delay of each pipeline stage is smaller. The same pipeline could support an average throughput rate that is substantially higher than the synchronous one. For example, the propagation delay of a 16-bit adder is highly data dependent: Adding two 4-bit numbers requires a much shorter time compared with adding two 16-bit numbers.

One way to avoid these problems is to opt for an *asynchronous* design approach and to eliminate all the clocks. Designing a purely asynchronous circuit is a nontrivial and potentially hazardous task. Ensuring a correct circuit operation that avoids all potential race conditions under any operation condition and input sequence requires a careful timing analysis of the network. In fact, the logical ordering of the events is dictated by the structure of the transistor network and the relative delays of the signals. Enforcing timing constraints by manipulating the logic structure and the lengths of the signal paths requires an extensive use of CAD tools and is recommended only when strictly necessary.

A more reliable and robust technique is the self-timed approach, which presents a local solution to the timing problem [Seitz80]. Figure 10-34 uses a pipelined datapath to illustrate how this can be accomplished. It is assumed that each combinational function has a means of indicating that it has completed a computation for a particular piece of data. The computation of a logic block is initiated by asserting a *Start* signal. The combinational logic block computes on the input data, and in a data-dependent fashion (taking the physical constraints into account), generates a *Done* flag once the computation is finished. In addition, the operators must signal each other that either they are ready to receive a next input word or they have a legal data word at

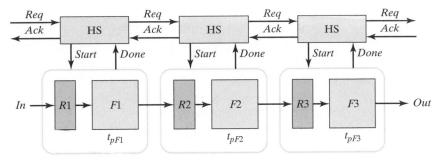

Figure 10-34 Self-timed, pipelined datapath.

their outputs that is ready for consumption. This signaling ensures the logical ordering of the events and can be achieved with the aid of extra *Ack(nowledge)* and *Req(uest)* signals. In the case of the pipelined datapath, the scenario could proceed as follows:

1. An input word arrives, and a *Req(uest)* to the block *F*1 is raised. If *F*1 is inactive at the time, it transfers the data and acknowledges this fact to the input buffer, which can go ahead and fetch the next word.
2. *F*1 is enabled by raising the *Start* signal. After a certain amount of time, which is dependent upon the data values and operating conditions, the *Done* signal goes high, indicating the completion of the computation.
3. A *Req(uest)* is issued to the *F*2 module. If this function is free, an *Ack(nowledge)* is raised, the output value is transferred, and *F*1 can go ahead with its next computation.

The self-timed approach effectively separates the physical and logical ordering functions implied in circuit timing. The completion signal *Done* ensures that the physical timing constraints are met and that the circuit is in steady state before accepting a new input. The logical ordering of the operations is ensured by the acknowledge–request scheme, often called *a handshaking protocol*. Both interested parties synchronize with each other by mutual agreement (or, if you want, by "shaking hands"). The ordering protocol described previously and implemented in the module HS is only one of many that are possible. The choice of protocol is important, since it has a profound effect on the circuit performance and robustness.

When compared with the synchronous approach, self-timed circuits display some alluring properties:

- In contrast to the global centralized approach of the synchronous methodology, timing signals are generated *locally*, which avoids all problems and overheads associated with distributing high-speed clocks.
- Separating the physical and logical ordering mechanisms results in a potential increase in performance. In synchronous systems, the period of the clock has to be stretched to accommodate the slowest path over all possible input sequences. In self-timed systems, a

completed data word does not have to wait for the arrival of the next clock edge in order to proceed to the subsequent processing stages. Since circuit delays often are dependent on the actual data value, a self-timed circuit proceeds at the *average speed* of the hardware, in contrast to the *worst case* model of synchronous logic. For a ripple-carry adder, the average length of carry propagation is $O(\log (N))$. This is a fact that can be exploited in self-timed circuits, while in a synchronous methodology, a worst case performance that varies linearly with the number of bits ($O(N)$) must be assumed.

- The automatic shutdown of blocks that are not in use can result in power savings. In addition, the power consumption overhead of generating and distributing high-speed clocks can be partially avoided. As discussed earlier, this overhead can be substantial. The use of gated clocks in synchronous design yields similar results.

- Self-timed circuits are, by nature, robust regarding variations in manufacturing and operating conditions such as temperature. Synchronous systems are limited by their performance at the extremes of the operating conditions. The performance of a self-timed system is determined by the actual operating conditions.

Unfortunately, these general properties are not without cost—they come at the expense of a substantial circuit-level overhead, which is caused by the need to generate completion signals and the need for handshaking logic that acts as a local traffic agent to order the circuit events (see block HS in Figure 10-34). Both of these topics are treated in more detail in the subsequent sections.

10.4.2 Completion-Signal Generation

A necessary component of self-timed logic is the circuitry to indicate when a particular piece of circuitry has completed its operation for the current piece of data. There are two common and reliable ways to generate the completion signal.

Dual-Rail Coding

One common approach to completion-signal generation is the use of *dual-rail coding*. It actually requires the introduction of redundancy in the data representation in order to signal that a particular bit is in either a transition or a steady-state mode. Consider the redundant data model presented in Table 10-1. Two bits ($B0$ and $B1$) are used to represent a single data bit B. For the data

Table 10-1   Redundant signal representation to include transition state.

B	$B0$	$B1$
in transition (or reset)	0	0
0	0	1
1	1	0
illegal	1	1

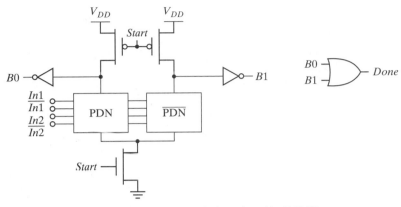

Figure 10-35 Generation of a completion signal in DCVSL.

to be valid or the computation to be completed, the circuit must be in a legal 0 ($B0 = 0$, $B1 = 1$) or 1 ($B0 = 1$, $B1 = 0$) state. The ($B0 = 0$, $B1 = 0$) condition signals that the data are nonvalid and the circuit is in either a reset or a transition mode. The ($B0 = 1$, $B1 = 1$) state is illegal and should never occur in an actual circuit.

A circuit that actually implements such a redundant representation is shown in Figure 10-35, which is a dynamic version of the DCVSL logic style where the clock is replaced by the *Start* signal [Heller84]. DCVSL uses a redundant data representation by nature of its differential dual-rail structure. When the *Start* signal is low, the circuit is precharged by the PMOS transistors, and the output ($B0$, $B1$) goes in the *Reset–Transition* state (0, 0). When the *Start* signal goes high, signaling the initiation of a computation, the NMOS pull-down network evaluates, and one of the precharged nodes is lowered. Either $B0$ or $B1$— but never both—goes high, which raises *Done* and signals the completion of the computation.

DCVSL is more expensive in terms of area than a nonredundant circuit due to its differential nature. The completion generation is performed in series with the logic evaluation, and its delay adds directly to the total delay of the logic block. The completion signals of all the individual bits must be combined to generate the completion for an N-bit data word. Completion generation thus comes at the expense of both area and speed. The benefits of the dynamic timing generation often justify this overhead.

Redundant signal representations other than the one presented in Table 10-1 can also be envisioned. One essential element is the presence of a *transition state* denoting that the circuit is in evaluation mode and the output data are not valid.

Example 10.4   Self-Timed Adder Circuit

An efficient implementation of a self-timed adder circuit is shown in Figure 10-36 [Abnous93]. A Manchester-carry scheme is used to boost the circuit performance. The proposed approach is based on the observation that the circuit delay of an adder is dominated by the carry-propagation path. It is therefore sufficient to use the differential

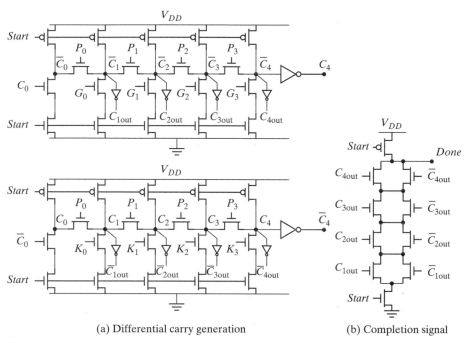

(a) Differential carry generation (b) Completion signal

Figure 10-36 Manchester-carry scheme with differential signal representation.

signaling in the carry path only (Figure 10-36a). The completion signal is efficiently derived by combining the carry signals of the different stages (Figure 10-36b). This safely assumes that the sum generation, which depends upon the arrival of the carry signal, is faster than the completion generation. The benefit of this approach is that the completion generation starts earlier and proceeds in parallel with sum generation, which reduces the critical timing path. All other signals, such as *P(ropagate)*, *G(enerate)*, *K(ill)*, and *S(um)*, do not require completion generation and can be implemented in single-ended logic. As shown in the circuit schematics, the differential carry paths are virtually identical. The only difference is that the *G(enerate)* signal is replaced by a *K(ill)*. A simple logic analysis demonstrates that this indeed results in an inverted carry signal and, hence, a differential signaling.

Replica Delay

While the dual-rail coding just described allows tracking of the signal statistics, it comes at the cost of power dissipation. Every single gate must transition for every new input vector, regardless of the value of the data vector. A way to reduce the overhead of completion detection is to use a *critical-path replica* configured as a delay element, as shown in Figure 10-37. To start a computation, the *Start* signal is raised and the computation of the logic network is initiated. At the same time, the start signal is fed into the replica delay line, which tracks the critical path of

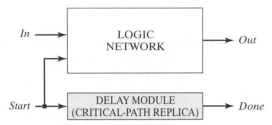

Figure 10-37   Completion-signal generation using delay module.

the logic network. It is important that the replica is structured such that no glitching transitions occur. When the output of the delay line makes a transition, it indicates that the logic is complete as the delay line mimics the critical path. In general, it is important to add extra padding in the delay line in order to compensate for possible random process variations.

The advantage of this approach is that the logic can be implemented using a standard non-redundant circuit style, such as complementary CMOS. Also, if multiple logic units are computing in parallel, it is possible to amortize the overhead of the delay line over multiple blocks. Note that this approach generates the completion signal after a time equal to the worst case delay through the network. As a result, it does not exploit the statistical properties of the incoming data. However, it can track the local effects of process variations and environmental variations (e.g., temperature or power-supply variations). This approach is widely used to generate the internal timing of semiconductor memories where self-timing is a commonly used technique.

Example 10.5   An Alternate Completion Detection Circuit Using Current Sensing

Ideally, logic should be implemented using nonredundant CMOS (e.g., static CMOS), and the completion signal circuitry should track data dependencies. Figure 10-38 shows an approach that attempts to realize this principle [Dean94]. A current sensor is inserted in series with the combinational logic, and it monitors the current flowing through the logic. The current sensor outputs a low value when no current flows through the logic (i.e., the logic is idle) and a high value when the combinational logic is switching. This signal

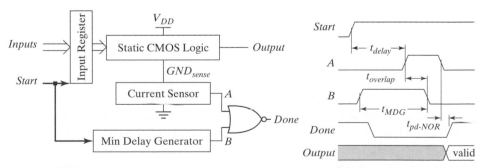

Figure 10-38   Completion-signal generation using current sensing.

effectively determines when the logic has completed its cycle. Note that this approach tracks data-dependent computation times—if only the lower order bits of an adder switch, current will stop flowing once the lower order bits switch to the final value. If the input data vector does not provoke any change in the logic from one cycle to the next, no current is drawn from the supply for static CMOS logic. In this case, a delay element that tracks the minimum delay through the logic and current sensor is used for signal completion. The outputs of the current sensor and minimum delay element are then combined.

This approach is interesting, but requires careful analog design. Ensuring reliability while keeping the overhead circuitry small is the main challenge. These concerns have kept the applicability of the approach very limited, despite its obvious potential.

10.4.3 Self-Timed Signaling

Besides the generation of the completion signals, a self-timed approach also requires a handshaking protocol to logically order the circuit events in order to avoid races and hazards. The functionality of the signaling (or handshaking) logic is illustrated by the example of Figure 10-39, which shows a *sender module* transmitting data to a *receiver* [Sutherland89]. The sender places the data value on the data bus ① and produces an event on the *Req* control signal by changing the polarity of the signal ②. In some cases, the request event is a rising transition; at other times, it is a falling one—the protocol described here does not distinguish between them. Upon receiving the request, the receiver accepts the data when possible and produces an event on the *Ack* signal to indicate that the data have been accepted ③. If the receiver is busy or its input buffer is full, no *Ack* event is generated, and the transmitter is stalled until the receiver becomes available by, for instance, freeing space in the input buffer. Once the *Ack* event is produced, the transmitter goes ahead and produces the next data word ①. The four events—*data change*, *request*, *data acceptance*, and *acknowledge*—proceed in a cyclic order. Successive cycles may take different amounts of time, depending on the time it takes to produce or consume the data.

This is called the *two-phase protocol*, since only two phases of operation can be distinguished for each data transmission—the active cycle of the sender and the active cycle of the receiver. Both phases are terminated by certain events. The *Req* event terminates the active cycle of the sender, while the receiver's cycle is completed by the *Ack* event. The sender is free to change the data during its active cycle. Once the *Req* event is generated, it has to keep the data constant as long as the receiver is active. The receiver can only accept data during its active cycle.

The correct operation of the sender–receiver system requires a strict ordering of the signaling events, as indicated by the arrows in Figure 10-39. Imposing this order is the task of the handshaking logic, which, in a sense, performs logic manipulations on events. An essential component of virtually any handshaking module is the *Muller C-element*. This gate, whose schematic symbol and truth table are given in Figure 10-40, performs an AND-operation on events. The output of the C-element is a copy of its inputs when both inputs are identical. When the inputs differ, the output retains its previous value. Phrased in a different way, events must occur

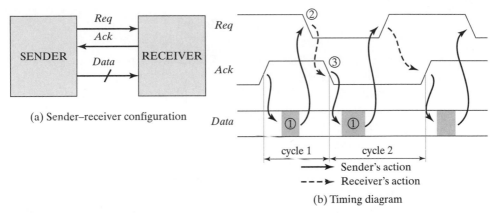

(a) Sender–receiver configuration

(b) Timing diagram

Figure 10-39   Two-phase handshaking protocol.

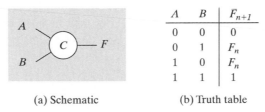

(a) Schematic

(b) Truth table

Figure 10-40   Muller *C*-element.

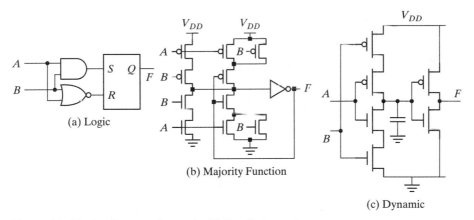

(a) Logic

(b) Majority Function

(c) Dynamic

Figure 10-41   Implementations of a Muller *C*-element.

at both inputs of a Muller *C*-element for its output to change state and to create an output event. As long as this does not happen, the output remains unchanged and no output event is generated. The implementation of a *C*-element is centered around a latch, which should not be a surprise, given the similarities in their truth tables. Figure 10-41 presents a static and a dynamic circuit realization of the function.

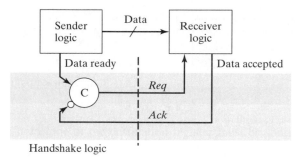

Figure 10-42 A Muller *C*-element implements a two-phase handshake protocol. The circle at the lower input of the Muller *C*-element stands for inversion.

Figure 10-42 shows how to use this component to enforce the two-phase handshaking protocol for the example of the sender–receiver. Assume that *Req*, *Ack*, and *Data Ready* are initially 0. When the sender wants to transmit the next word, the *Data Ready* signal is set to 1, which triggers the *C*-element, because both its inputs are at 1. *Req* goes high—this is commonly denoted as *Req*↑. The sender now resides in the wait mode, and control is passed to the receiver. The *C*-element is blocked, and no new data are sent to the data bus (*Req* stays high) as long as the transmitted data are not processed by the receiver. Once this happens, the *Data accepted* signal is raised. This can be the result of many different actions, possibly involving other *C*-elements communicating with subsequent blocks. An *Ack*↑ ensues, which unblocks the *C*-element and passes the control back to the sender. A *Data ready*↓ event, which might already have happened before *Ack*↑, produces a *Req*↓, and the cycle is repeated.

Problem 10.1 Two-Phase Self-Timed FIFO

Figure 10-43 shows a two-phase, self-timed implementation of a FIFO (first-in first-out) buffer with three registers. Assuming that the registers accept a data word on both positive- and negative-going transitions of the *En* signals, and that the *Done* signal is simply a delayed version of *En*, examine the operation of the FIFO by plotting the timing behavior of all signals of interest. How can you observe that the FIFO is completely empty? Full? (Hint: Determine the necessary conditions on the *Ack* and *Req* signals.)

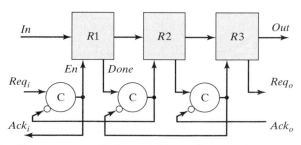

Figure 10-43 Three-stage, self-timed FIFO, using a two-phase signaling protocol.

The two-phase protocol has the advantage of being simple and fast. However, this protocol requires the detection of transitions that may occur in either direction. Most logic devices in the MOS technology tend to be sensitive to levels or to transitions in one particular direction. Event-triggered logic, as required in the two-phase protocol, requires extra logic as well as state information in the registers and the computational elements. Since the transition direction is important, initializing all the Muller *C*-elements in the appropriate state is essential. If this is not done, the circuit might become deadlocked, which means that all elements are permanently blocked and nothing will ever happen. A detailed study on how to implement event-triggered logic can be found in [Sutherland89], the defining text on micropipelines.

The only alternative is to adopt a different signaling approach, such as *four phase signaling*, or *return-to-zero* (*RTZ*). This class of signaling requires that all controlling signals be brought back to their initial values before the next cycle can be initiated. Once again, this is illustrated with the example of the sender–receiver. The four-phase protocol for this example is shown in Figure 10-44.

The protocol presented is initially the same as the two-phase one. Both the *Req* and the *Ack* are initially in the zero state, however. Once a new data word is put on the bus ①, the *Req* is raised (*Req*↑ or ②), and control is passed to the receiver. When ready, the receiver accepts the data and raises *Ack* (*Ack*↑ or ③). So far, nothing new. The protocol proceeds by bringing both *Req* (*Req*↓ or ④) and *Ack* (*Ack*↓ or ⑤) back to their initial state in sequence. Only when that state is reached is the sender allowed to put new data on the bus ①. This protocol is called *four-phase* signalling because four distinct time zones can be recognized per cycle: two for the sender; two for the receiver. The first two phases are identical to the two-phase protocol, while the last two are devoted to resetting of the state. An implementation of the protocol, based on Muller *C*-elements, is shown in Figure 10-45. It is interesting to note that the four-phase protocol requires two *C*-elements in series (since four states must be represented). The *Data ready* and *Data accepted* signals must be pulses instead of single transitions.

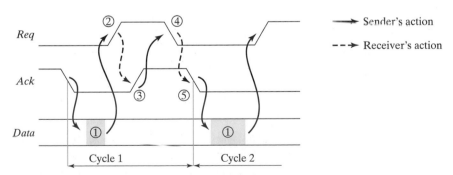

Figure 10-44 Four-phase handshaking protocol.

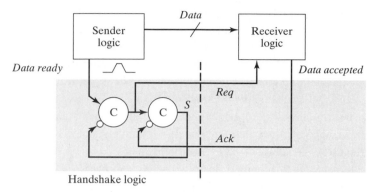

Figure 10-45   Implementation of four-phase handshake protocol using Muller *C*-elements.

Problem 10.2   Four-Phase Protocol

Derive the timing diagram for the signals shown in Figure 10-45. Assume that the *Data Ready* signal is a pulse and that the *Data Accepted* signal is a delayed version of *Req*.

The four-phase protocol has the disadvantage of being more complex and slower, since two events on *Req* and *Ack* are needed per transmission. On the other hand, it has the advantage of being robust. The logic in the sender and receiver modules does not have to deal with transitions, which can go either way; it has to consider only rising (or falling) transition events or signal levels. This is readily accomplished with traditional logic circuits. For this reason, four-phase handshakes are the preferred implementation approach for most of the current self-timed circuits. The two-phase protocol is mainly selected when the sender and receiver are far apart and the delays on the control wires (*Ack* and *Req*) are substantial.

Example 10.6   The Pipelined Datapath—Revisited

We have introduced both the signaling conventions and the concepts of the completion-signal generation. Now it is time to bring them all together. We do this with the example of the pipelined datapath, which was presented earlier. A view of the self-timed datapath, including the timing control, is offered in Figure 10-46. The logic functions $F1$ and $F2$ are implemented using dual-rail, differential logic.

To understand the operation of this circuit, assume that all *Req* and *Ack* signals, including the internal ones, are set to 0, which means there is no activity in the data path. All *Start* signals are low so that all logic circuits are in precharge condition. An input request ($Req_i\uparrow$) triggers the first *C*-element. The enable signal *En* of $R1$ is raised, effectively latching the input data into the register, assuming a positive edge-triggered or a level-sensitive implementation. $Ack_i\uparrow$ acknowledges the acceptance of the data. The second *C*-element is triggered as well, since Ack_{int} is low. This raises the *Start* signal and starts the evaluation of $F1$. At its completion, the output data are placed on the bus, and a

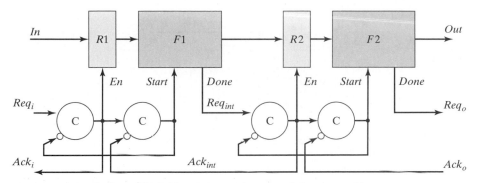

Figure 10-46   Self-timed pipelined datapath—complete composition.

request is initiated to the second stage ($Req_{int}\uparrow$), which acknowledges its acceptance by raising Ack_{int}.

At this point, stage 1 is still blocked for further computations. However, the input buffer can respond to the $Ack_i\uparrow$ event by resetting Req_i to its zero state ($Req_i\downarrow$). In turn, this lowers En and Ack_i. Upon receipt of $Ack_{int}\uparrow$, $Start$ goes low, the precharge phase starts, and $F1$ is ready for new data. Note that this sequence corresponds to the four-phase handshake mechanism described earlier. The dependencies among the events are presented in a more pictorial fashion in the *state transition diagram* (*STG*) shown in Figure 10-47. These STGs can become very complex. Computer tools are often used to derive STGs that ensure proper operation and optimize the performance.

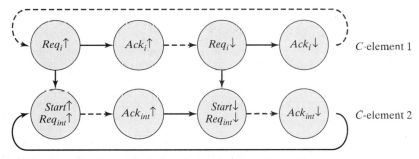

Figure 10-47   State transition diagram for pipeline stage 1. The nodes represent the signaling events, while the arrows express dependencies. Arrows in dashed lines express actions in either the preceding or the following stage.

10.4.4   Practical Examples of Self-Timed Logic

Self-timed circuits can provide a significant performance advantage. Unfortunately, the overhead circuitry precludes a widespread application in general-purpose digital computing. Nevertheless, the key concepts of self-timed circuits are being exploited in a number of practical applications. We present a few examples that illustrate the use of self-timed concepts for either power savings or performance enhancement.

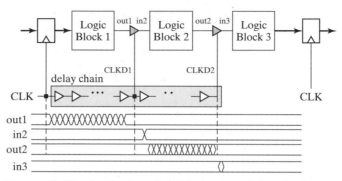

Figure 10-48   Application of self-timing for glitch reduction.

Glitch Reduction Using Self-Timing

A major source of unnecessary switched capacitance in large datapaths, such as bit-sliced adders and multipliers, is due to spurious transitions caused by *glitch propagation*. Imbalances in a logic network cause inputs of a logic gate or block to arrive at different times, resulting in glitching transitions. In large combinational blocks such as multipliers, transitions happen in waves as the primary input changes ripple through the logic. Enabling a logic block only when all the inputs have settled helps reduce or eliminate the glitching transitions. One approach for doing so is the use of a self-timed gating approach, which partitions each computational logic block into smaller blocks and distinct phases. Tristate buffers are inserted between each of these phases in order to prevent glitches from propagating through the datapath, as shown in Figure 10-48. Assuming an arbitrary logic network, it is fair to assume that the outputs of logic block 1 will not be synchronized. When the tristate buffers at the output of logic block 1 are enabled, the computation of logic block 2 is allowed to proceed. To reduce glitching transitions, the tristate buffer should be enabled only when it is ensured that the outputs of logic block 1 are stable and valid. The control of the tristate buffer can be performed through the use of a self-timed enable signal, generated by passing the system clock through a delay chain that models the critical path of the processor. The delay chain is then tapped at the points corresponding to the locations of the buffers, and the resulting signals are distributed throughout the chip. This technique succeeds in substantially reducing the switching capacitance of combinational logic blocks such as multipliers, even when taking into account the overhead of the delay chain, gating signal distribution, and buffers [Goodman98].

Post-Charge Logic

An interesting form of self-timed logic is self-resetting CMOS. This structure uses a control structure different from the conventional self-timed pipeline. Instead of waiting for all the logic stages to complete their operation before transitioning to the reset stage, the idea is to precharge a logic block as soon as it completes its operation. Since the precharge happens after the operation instead of before evaluation, it is often termed *postcharge logic*. A block diagram of post-charge logic is shown in Figure 10-49 [Williams00]. As can be seen from this block diagram, the

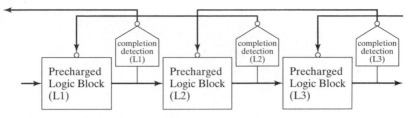

Figure 10-49 Self-resetting logic.

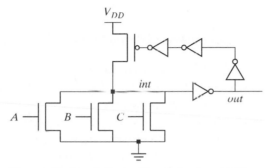

Figure 10-50 Self-resetting three-input OR.

precharging of L1 happens when its successor stage has completed and does not need its input anymore. It is possible to precharge a block based on the completion of its own output, but care must be taken to ensure that the stage that follows has properly evaluated and the input has become obsolete.

It should be noted that, unlike with other logic styles, the signals are represented as pulses and are valid only for a limited duration. The pulse width must be shorter than the reset delay or else there is a contention between the precharge device and the evaluate switches. While this logic style offers potential speed advantages, special care must be taken to ensure correct timing. Also, circuitry that converts level signals to pulses, and vice versa is required. An example of self-resetting logic is shown in Figure 10-50 [Bernstein98]. Assume that all inputs are low and that *int* is initially precharged. If *A* goes high, *int* will fall, causing *out* to go high. This causes the gate to precharge. When the PMOS precharge device is active, the inputs must be in a reset state in order to avoid contention.

Clock-Delayed Domino

One interesting application of self-timed circuits that uses the delay-matching concept is *clock-delayed (CD) domino* [Yee96][Bernstein00]. This is a style of dynamic logic in which there is no global clock signal. Instead, the clock for one stage is derived from that for the previous stage. A simple example of a CD domino stage is shown in Figure 10-51. The two inverter delays plus the transmission gate along the clock path emulate the worst-case delay through the dynamic logic gate. The transmission gate is always turned on, and the delay of the clock path is adjusted

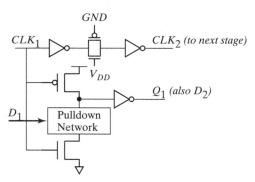

Figure 10-51   Clock-delayed domino logic: a self-clocked logic style.

through the sizing of the devices. Clock-delayed domino was used in IBM's 1-GHz micro-processor, and it is used widely in high-speed domino logic. Clock-delayed domino can provide both inverting and noninverting functionality. This alleviates a major limitation of conventional domino techniques, which are capable only of noninverting logic. The inverter after the pull-down network is not essential because the clock arrives at the next stage only after the current stage has evaluated. The clock-evaluation edge thus arrives only after the inputs are stable. On the other hand, adding the inverter allows for the elimination of the foot switch (see Chapter 6).

10.5 Synchronizers and Arbiters*

10.5.1 Synchronizers—Concept and Implementation

Even though a complete system may be designed in a synchronous fashion, it must still commu-nicate with the outside world, which is generally asynchronous. An asynchronous input can change value at any time related to the clock edges of the synchronous system, as is illustrated in Figure 10-52.

Consider a typical personal computer. All operations within the system are strictly orches-trated by a central clock that provides a time reference. This reference determines what happens within the computer system at any point in time. This synchronous computer has to communicate with a human through the mouse or the keyboard, even though the human does not have knowl-edge of this time reference and may decide to press a keyboard key at any point in time. The way a synchronous system deals with such an asynchronous signal is to sample or poll it at regular intervals and to check its value. If the sampling rate is high enough, no transitions will be missed—this is known as the *Nyquist criterion* in the communication community. However, it might happen that the signal is polled in the middle of a transition. The resulting value is neither low nor high, but undefined. At that point, it is not clear whether the key was pressed or not. Feed-ing the undefined signal into the computer could be the source of all kinds of trouble, especially when it is routed to different functions or gates that might interpret it differently. For instance, one function might decide that the key is pushed and start a certain action, while another function might lean the other way and issue a competing command. This results in a conflict and thus a

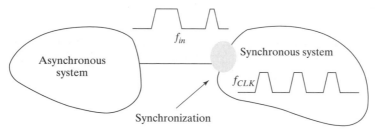

Figure 10-52 Asynchronous–synchronous interface.

potential crash. Therefore, the undefined state must be resolved in one way or another before it is interpreted further. It does not really matter what decision is made, as long as a unique result is available. For instance, it is either decided that the key is not yet pressed, a situation that will be corrected in the next poll of the keyboard, or it is concluded that the key is already pressed.

Thus, an asynchronous signal must be resolved to be either in the high or low state before it is fed into the synchronous environment. A circuit that implements such a decision-making function is called a *synchronizer*. Now comes the bad news: **Building a perfect synchronizer that always delivers a legal answer is impossible** [Chaney73] [Glasser85]! A synchronizer needs some time to come to a decision, and in certain cases this time might be arbitrarily long. An asynchronous–synchronous interface is thus always prone to errors, known as *synchronization failures*. The designer's task is to ensure that the probability of such a failure is small enough that it is not likely to disturb the normal system behavior. Typically, this probability can be reduced in an exponential fashion by waiting longer before making a decision. This is not too troublesome in the keyboard example, but in general, waiting affects system performance and should be avoided as much as possible.

To illustrate why waiting helps reduce the failure rate of a synchronizer, consider a synchronizer as shown in Figure 10-53. This circuit is a latch that is transparent during the low phase of the clock and samples the input on the rising edge of the clock *CLK*. However, since the sampled signal is not synchronized to the clock signal, there is a finite probability that the setup time or hold time of the latch is violated. (The probability is a strong function of the transition frequencies of the input and the clock.) As a result, once the clock goes high, there is a chance that the output of the latch resides somewhere in the undefined transition zone. The sampled signal eventually evolves into a legal 0 or 1, even in the latter case, as the latch has only two stable states.

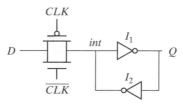

Figure 10-53 A simple synchronizer.

Example 10.7   Synchronizer Trajectories

Figure 10-54 shows the simulated trajectories of the output nodes of a cross-coupled static CMOS inverter pair for an initial state close to the metastable point. The inverters are composed of minimum-size devices.

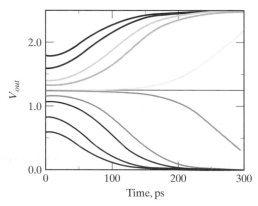

Figure 10-54   Simulated trajectory for a simple synchronizer.

If the input is sampled such that the cross-coupled inverter pair starts at the metastable point, the voltage will remain at the metastable state forever in the absence of noise. If the data are sampled with a small offset (positive or negative), the differential voltage will evolve in an exponential way, with a time constant that is dependent on the strength of the transistors as well as the parasitic capacitances. The time it takes to reach the acceptable signal zones depends upon the initial distance of the sampled signal from the metastable point.

In order to determine the required waiting period, let us build a mathematical model of the behavior of the bistable element and use the results to determine the probability of synchronization failure as a function of the waiting period [Veendrick80].

The transient behavior around the metastable point of the cross-coupled inverter pair is accurately modeled by a system with a single dominant pole. Under this assumption, the transient response of the bistable element after the sampling clock has been turned off can be modeled as

$$v(t) = V_{MS} + (v(0) - V_{MS})e^{t/\tau} .$$

(10.8)

where V_{MS} is the metastable voltage of the latch, $v(0)$ is the initial voltage after the sampling clock is turned off, and τ is the time constant of the system. Simulations indicate that this first-order model is quite precise.

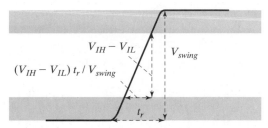

Figure 10-55　Linear approximation of signal slope.

The model can now be used to compute the range of values for $v(0)$ that causes an error or a voltage in the undefined range, after a waiting period T. A signal is called undefined if its value is situated between V_{IH} and V_{IL}:

$$V_{MS} - (V_{MS} - V_{IL})e^{-T/\tau} \le v(0) \le V_{MS} + (V_{IH} - V_{MS})e^{-T/\tau} \tag{10.9}$$

Equation (10.9) conveys an important message: **The range of input voltages that cause a synchronization error decreases exponentially with the waiting period T.** Increasing the waiting period from 2τ to 4τ decreases the interval and the chances of an error by a factor of 7.4.

Some information about the asynchronous signal is required in order to compute the probability of an error. Assume that V_{in} is a periodical waveform with an average period T_{signal} between transitions and with identical rise and fall times t_r. Assume also that the slopes of the waveform in the undefined region can be approximated by a linear function, as shown in Figure 10-55. Using this model, we can estimate the probability P_{init} that $v(0)$, the value of V_{in} at the sampling time, resides in the undefined region:

$$P_{init} = \frac{\left(\dfrac{V_{IH} - V_{IL}}{V_{swing}}\right)t_r}{T_{signal}} \tag{10.10}$$

The chances for a synchronization error to occur depend upon the frequency of the synchronizing clock, ϕ. The greater the number of sampling events, the higher the chance of running into an error. This means that the average number of synchronization errors per second, $N_{sync}(0)$, equals Eq. (10.11) if no synchronizer is used. Therefore, we can write

$$N_{sync}(0) = \frac{P_{init}}{T_\phi} \tag{10.11}$$

where T_ϕ is the sampling period.

From Eq. (10.9), we learned that waiting a period T before observing the output reduces exponentially the probability that the signal is still undefined:

$$N_{sync}(T) = \frac{P_{init}e^{-T/\tau}}{T_\phi} = \frac{(V_{IH} - V_{IL})e^{-T/\tau}}{V_{swing}}\frac{t_r}{T_{signal}T_\phi} \tag{10.12}$$

The robustness of an asynchronous–synchronous interface is thus determined by the following parameters: signal switching rate and rise time, sampling frequency, and waiting time T.

Example 10.8 Synchronizers and Mean Time to Failure

Consider the following design example: $T_\phi = 5$ ns, which corresponds to a 200-Mhz clock; $T = T_\phi = 5$ ns; $T_{signal} = 50$ ns; $t_r = 0.5$ ns; and $\tau = 150$ ps. From the VTC of a typical CMOS inverter, it can be derived that $V_{IH} - V_{IL}$ approximately equals 0.5 V for a voltage swing of 2.5 V. Evaluation of Eq. (10.12) yields an error probability of 1.38×10^{-9} errors/s. The inverse of N_{sync} is called the *mean time to failure*, or the MTF, and equals 7×10^8 s, or 23 years. If no synchronizer were used, the MTF would have been only 2.5 μs!

Design Consideration

When designing a synchronous–asynchronous interface, we must keep in mind the following observations:

- The acceptable failure rate of a system depends upon many economic and social factors and is a strong function of the system's application area.
- The exponential relation in Eq. (10.12) makes the failure rate extremely sensitive to the value of τ. Defining a precise value of τ is not easy in the first place, because τ varies from chip to chip and is a function of temperature as well. The probability of an error occurring can thus fluctuate over large ranges even for the same design. A worst case design scenario is definitely advocated here. If the worst case failure rate exceeds a certain criterion, it can be reduced by increasing the value of T. A problem occurs when T exceeds the sampling period T_ϕ. This can be avoided by cascading (or pipelining) a number of synchronizers, as shown in Figure 10-56. Each of those synchronizers has a waiting period equal to T_ϕ. Notice that this arrangement requires the ϕ pulse to be short enough to avoid race conditions. The global waiting period equals the sum of the Ts of all the individual synchronizers. The increase in MTF comes at the expense of an increased latency.

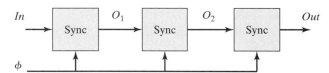

Figure 10-56 Cascading (edge-triggered) synchronizers reduces the main time to failure.

- Synchronization errors are very hard to trace, due to their random nature. Making the mean time to failure very large does not preclude errors. The **number of synchronizers in a system should therefore be severely restricted**. A maximum of one or two per system is advocated. ∎

10.5.2 Arbiters

Finally, a sibling of the synchronizer called the *arbiter,* interlock element, or mutual-exclusion circuit, should be mentioned. An arbiter is an element that decides which of two events has occurred first. For example, such components allow multiple processors to access a single

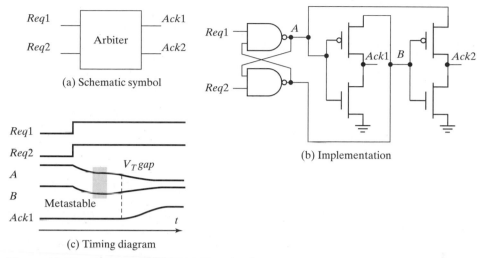

(a) Schematic symbol

(b) Implementation

(c) Timing diagram

Figure 10-57   Mutual-exclusion element (or arbiter).

resource, such as a large shared memory. A synchronizer is actually a special case of an arbiter, since it determines whether a signal transition happened before or after a clock event. A synchronizer is thus an arbiter with one of its inputs tied to the clock.

   An example of a mutual-exclusion circuit is shown in Figure 10-57. It operates on two input-request signals that operate on a four-phase signaling protocol; that is, the *Req(uest)* signal has to go back to the reset state before a new *Req(uest)* can be issued. The output consists of two *Ack(nowledge)* signals that should be mutually exclusive. While *Requests* may occur concurrently, only one of the *Acknowledges* is allowed to go high. The operation is most easily visualized starting with both inputs low—neither device issuing a request—nodes *A* and *B* high, and both *Acknowledges* low. An event on one of the inputs (e.g., *Req1*↑) causes the flip-flop to switch, node *A* to go low, and *Ack1*↑. Concurrent events on both inputs force the flip-flop into the metastable state, and the signals *A* and *B* might be undefined for a while. The cross-coupled output structure keeps the output values low until one of the NAND outputs differs from the other by more than a threshold value V_T. This approach eliminates glitches at the output.

10.6   Clock Synthesis and Synchronization Using a Phase-Locked Loop*

There are numerous digital applications that require the on-chip generation of a periodic signal. Synchronous circuits need a global periodic clock reference to orchestrate the events. Current microprocessors and high-performance digital circuits require clock frequencies in the gigahertz range. Crystal oscillators, by contrast, generate only accurate, low-jitter clocks over a frequency range from 10's of MHz to approximately 200 MHz. To generate the higher frequency required by digital circuits, a *phase-locked loop* (PLL) structure typically is used. A PLL takes an external low-frequency reference crystal frequency signal and multiplies its frequency by a rational number N (see the left side of Figure 10-58).

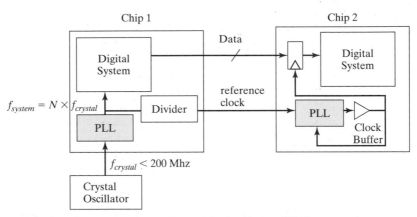

Figure 10-58   Applications of phase locked loops (PLL).

Another and equally important function of a PLL is to synchronize communications between chips. As shown in Figure 10-58, a reference clock is sent along with the parallel data being communicated. (In this example only the transmit path from chip 1 to chip 2 is shown.) Since chip-to-chip communication most often occurs at a lower rate than the on-chip clock rate, the reference clock is divided, but kept in phase with the system clock. In chip 2, the reference clock is used to synchronize all the input flip-flop, which can present a significant clock load in the case of wide data busses. Unfortunately, implementing clock buffers to deal with this problem introduces skew between the data and the sample clock. A PLL aligns (i.e., de-skews) the output of the clock buffer with respect to the data. In addition, the PLL can multiply the frequency of the incoming reference clock, allowing the core of the second chip to operate at a higher frequency than the input reference clock.

10.6.1 Basic Concept

A set of two or more periodic signals of the same frequency is well defined if we know one of them and its *phase* with respect to the other signals, as shown in Figure 10-59, where ϕ_1 and ϕ_2 represent the phases of the two signals. The *relative phase* is the difference between the two.

A PLL is a complex, nonlinear feedback circuit. Its basic operation is best understood with the aid of Figure 10-60 [Jeong87]. The *voltage-controlled oscillator* (VCO) takes an analog control input and generates a clock signal of the desired frequency. In general, there is a nonlinear relationship between the control voltage (v_{cont}) and the output frequency. To synthesize a system clock of a particular frequency, it is necessary to set the control voltage to the appropriate value. This is a function of the rest of the blocks and the feedback loop in the PLL. The feedback loop is critical to the tracking process and environmental variations. The feedback also allows for frequency multiplication.

The reference clock is, in general, generated off chip from an accurate crystal reference. It is compared with a divided version of the system clock (i.e., the local clock), using a phase detector, which determines the phase difference between the signals and produces an

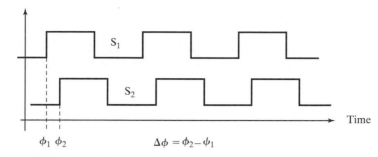

Figure 10-59   Relative and absolute phase of two periodic signals.

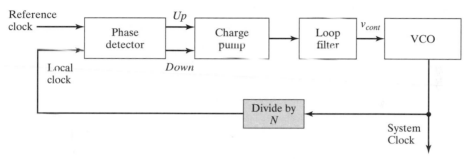

Figure 10-60   Functional composition of a phase-locked loop (PLL).

Up or *Down* signal when the local clock lags or leads the reference signal. It detects which of the two input signals arrives earlier and produces the appropriate output signal. The *Up* and *Down* signals are fed into a charge pump, which translates the digital encoded control information into an analog voltage [Gardner80]. An *Up* signal increases the value of the con trol voltage and speeds up the VCO, which causes the local signal to catch up with the reference clock. A *Down* signal, on the other hand, slows down the oscillator and eliminates the phase lead of the local clock.

Passing the output of the charge pump directly into the VCO creates a jittery clock signal. The edge of the local clock jumps back and forth instantaneously and oscillates around the targeted position. As discussed earlier, clock jitter is highly undesirable, since it reduces the time available for logic computation, and therefore should be kept within a given percentage of the clock period. This is partially accomplished by the introduction of the *loop filter*. This low-pass filter removes the high-frequency components from the VCO control voltage and smooths out its response, which results in a reduction of the jitter. Note that the PLL structure is a feedback structure. The addition of extra phase shifts, as introduced by a high-order filter, may result in instability. Important properties of a PLL are its *lock range*, the range of input frequencies over which the loop can maintain functionality; the *lock time*, the time it takes for the PLL to lock onto a given input signal; and the jitter. When in lock, the system clock is N times the reference clock frequency.

A PLL is an analog circuit and is inherently sensitive to noise and interference. This is especially true for the loop filter and VCO, for which induced noise has a direct effect on the resulting clock jitter. A major source of interference is the noise coupling through the supply rails and the substrate. This is a particular concern in digital environments, where noise is introduced by a variety of sources. Analog circuits with a high supply rejection, such as differential VCOs, are therefore desirable [Kim90]. In summary, integrating a highly sensitive component into a hostile digital environment is nontrivial and requires expert analog design. More detailed descriptions of various components of a PLL are given in the following subsections.

10.6.2 Building Blocks of a PLL

Voltage Controlled Oscillator (VCO)

A VCO generates a periodic signal with a frequency that is a linear function of the input control voltage v_{cont}. In other words, the VCO frequency can be expressed as

$$\omega = \omega_0 + K_{vco} \cdot v_{cont} \tag{10.13}$$

Since the phase is the time integral of the frequency, the output phase of the VCO block is given by

$$\phi_{out} = \omega_0 t + K_{vco} \cdot \int_{-\infty}^{t} v_{cont} dt \tag{10.14}$$

where K_{vco} is the gain of the VCO given in rad/s/V, ω_0 is a fixed frequency offset, and v_{cont} controls a frequency centered around ω_0. The output signal has the form

$$x(t) = A \cdot \cos\left(\omega_0 t + K_{vco} \cdot \int_{-\infty}^{t} v_{cont} dt\right) \tag{10.15}$$

Various implementation strategies for VCOs were discussed in Chapter 7. Differential ring oscillators are the method of choice at present.

Phase Detectors

The phase detector determines the relative phase difference between two incoming signals and outputs a signal that is proportional to this phase difference. One of the inputs to the phase detector is a reference clock that typically is generated off chip, while the other clock input is a divided version of the VCO. Two basic types of phase detectors are commonly used—the *XOR gate* and the *phase frequency detector* (PFD).

XOR Phase Detector   An XOR gate, the simplest of the two detectors, is useful as a phase detector because of the following observation: the relative phase difference between the inputs is reflected by the time over which the two inputs are different. Figure 10-61 shows the XOR-function of two waveforms. Feeding this output in a low-pass filter results into a dc-signal that is proportional to the phase difference, as shown in Figure 10-61(c).

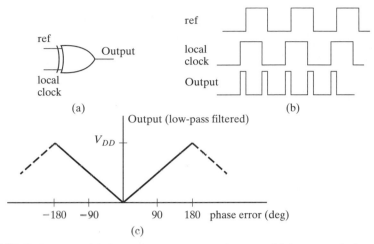

Figure 10-61   (a) The XOR as a phase detector; (b) its output before low-pass filtering; and (c) transfer characteristics.

For this detector, a deviation in a positive or negative direction from the perfect in-phase condition (i.e., a phase error of zero) produces the same change in duty factor, resulting in the same average voltage. Thus the linear phase range is only 180 degrees. In steady state, a PLL locks into a quadrature phase relationship ($\frac{1}{4}$ cycle offset) between the two inputs.

Phase Frequency Detector   The phase frequency detector (PFD) is the most commonly used form of phase detector, and it solves several of the shortcomings of the detector discussed previously [Dally98]. As the name implies, the output of the PFD is dependent on both the phase and frequency difference between the applied signals. Accordingly, it cannot lock to an incorrect multiple of the frequency. The PFD takes two clock inputs and produces two outputs, *UP* and *DN*, as shown in Figure 10-62.

The PFD is a state machine with three states. Assume that both *UP* and *DN* outputs are initially low. When input *A* leads *B*, the *UP* output is asserted on the rising edge of input *A*. The *UP* signal remains in this state until a low-to-high transition occurs on input *B*. At that time, the *DN* output is asserted, causing both flip-flops to reset through the asynchronous reset signal. Notice that there is a small pulse on the *DN* output, whose duration is equal to the delay through the AND gate and register *Reset*-to-*Q* delay. The pulse width of the *UP* pulse is equal to the phase error between the two signals. The roles are reversed for the case in which input *A* lags *B*. A pulse proportional to the phase error is generated on the *DN* output. If the loop is in lock, short pulses will be generated on the *UP* and *DN* outputs.

The circuit also acts as a frequency detector, providing a measure of the frequency error (see Figure 10-63). For the case in which *A* is at a higher frequency than *B*, the PFD generates a lot more *UP* pulses—with the average pulse proportional to the frequency difference—while the number of *DN* pulses are close to zero, on average. Exactly the opposite is true for the case in

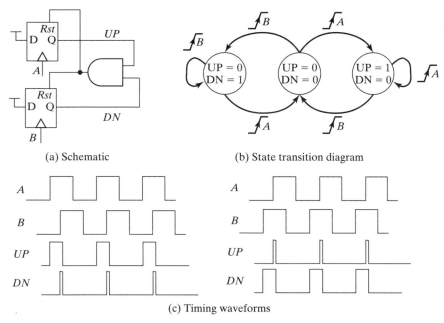

(a) Schematic   (b) State transition diagram

(c) Timing waveforms

Figure 10-62   Phase frequency detector.

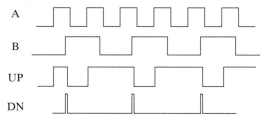

Figure 10-63   Timing of the PFD measuring frequency error.

which B has a frequency larger than that of A—many more pulses are generated on the DN output than on the UP output.

The phase characteristics of the phase detector are shown in Figure 10-64. Notice that the linear range has been expanded to 4π.

Charge Pump

The UP and DN pulses must be converted into an analog voltage that controls the VCO. One possible implementation is shown in Figure 10-65. A pulse on the UP signal adds an amount of charge to the capacitor proportional to the width of the UP pulse, while a pulse on the DN signal removes charge proportional to the DN pulse width. If the width of the UP pulse is larger than that of the DN pulse, there is a net increase in the control voltage. This effectively increases the frequency of the VCO.

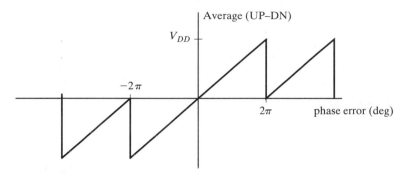

Figure 10-64   PFD phase characteristics.

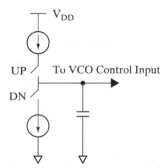

Figure 10-65   Charge pump.

Example 10.9   Transient Behavior of Phase-Locked Loop

The settling time of a PLL is the time it takes to reach steady-state behavior. The length of the startup transient is strongly dependent on the bandwidth of the loop filter. Figure 10-66 shows a SPICE level simulation of a PLL implemented in a 0.25-μm CMOS. An ideal VCO is used to speed up the simulation. In this example, a reference frequency of 100 MHz is chosen, and the PLL multiplies this frequency by 8 to 800 MHz. The Figure 10-66a illustrates the transient response and the settling process of the control voltage input of the VCO. Once the control voltage reaches its final value, the output frequency (the clock of the digital system) settles to steady state. The simulation on the right shows the reference frequency, the output of the divider, and the output frequency in steady state. Figure 10-66b shows the PLL in lock at f_{out} = 8 × f_{ref}, while Figure 10-66c shows the waveforms during the locking process when the output is not in phase with the input, and the frequencies are not related yet by the divide ratio.

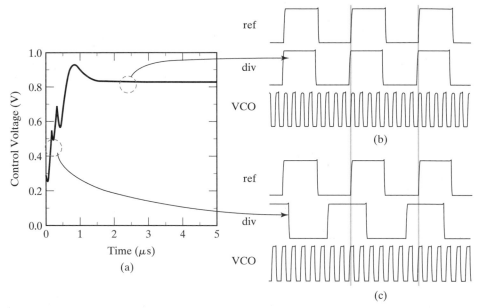

Figure 10-66 SPICE simulation of a PLL. The control voltage of the VCO is shown along with the waveforms before lock and after lock.

Summary

In a short span of time, phase-locked loops have become an essential component of any high-performance digital design. Their design requires considerable skill, integrating analog circuitry into a hostile digital environment. Yet experience has demonstrated that this combination is perfectly feasible, and it leads to new and better solutions.

10.7 Future Directions and Perspectives

This section highlights some of the trends in timing optimization for combining high performance and low power dissipation.

10.7.1 Distributed Clocking Using DLLs

A recent trend in high-performance clocking is the use of *delay locked loop* (DLL), a variation of the PLL structure. The schematic of a DLL is shown in Figure 10-67a [Maneatis00]. The key component of a DLL is a *voltage-controlled delay line* (VCDL). It consists of a cascade of adjustable delay elements (for instance, a current-starved inverter). The idea is to **delay the output clock such that it lines up perfectly with the reference**. Unlike the case of a VCO, there is no clock generator. The reference frequency is fed into the input of the VCDL. Similar to the case of a PLL structure, a phase detector compares the reference frequency with the output of the delay line (F_O) and generates an *UP–DN* error signal. Note that only a phase and not a phase-frequency detection, is required. When in lock, there is no error between the two clocks. The

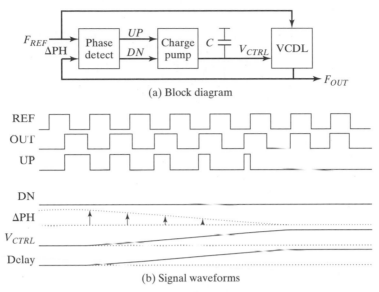

(a) Block diagram

(b) Signal waveforms

Figure 10-67   Delay-locked loop [Maneatis00].

function of the feedback is to adjust the delay through the VCDL such that the rising edge of the input reference clock (f_{REF}) and the output clock (f_O) are aligned.

A qualitative sketch of the signals in the DLL is shown in Figure 10-67b and c. Initially, the DLL is out of lock. Since the first edge of the output arrives before the reference edge, an *UP* pulse of width equal to the error between the two signals occurs. The role of the charge pump is to generate a charge packet proportional to the error, increasing the VCDL control voltage. This causes the edge of the output signal to be delayed in the next cycle. (This implementation of the VCDL assumes that a larger voltage results in larger delay.) After many cycles, the phase error is corrected, and the two signals are in lock. Note that a DLL does not alter the frequency of the input reference, but rather adjusts its phase.

Figure 10-68 shows the utilization of a DLL structure in a clock distribution network. The chip is partitioned into many small regions (or tiles). A global clock is distributed to each tile in a low-skew manner; this could be done through the package or by using low-skew, on-chip routing schemes. For the purpose of simplicity, the figure shows a two-tile chip, but this is easily extended to many regions. Inside each tile, the global clock is buffered before driving the digital load. In front of each buffer is a VCDL. The goal of the clock network is to deliver a signal to the digital circuit with close-to-zero skew and jitter. As we observed earlier, the static and dynamic variations of the buffers cause the phase error between the buffered clocks to be nonzero and time varying. The feedback inside each tile adjusts the control voltage of VCDL such that the buffered output is locked in phase to the global input clock. The feedback loop compensates for both static process variations and for slow dynamic variations (e.g., temperature). Such configurations have become common in high-performance digital microprocessors, as well as in media

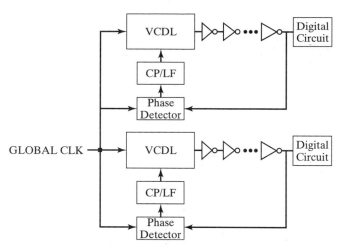

Figure 10-68   DLL approach to clock distribution.

processors. The preceding approach of clock distribution using multiple DLLs can be extended to the use of multiple distributed PLLs on a chip [Gutnik00]. Distributed PLLs offer potential performance advantages, but careful system analysis is required for stable operation.

10.7.2 Optical Clock Distribution

By now, the reader should have become aware of the fact that future high-performance multi-GHz systems face some fundamental synchronization problems. The performance of a digital design is fundamentally limited by process and environmental variations. Even with aggressive active clock management schemes, such as the use of DLLs and PLLs, the variations in power supply and clock load result in unacceptable clock uncertainty. Researchers are furiously searching for innovative solutions. An alternative approach that has received a lot of attention is the use of optics for systemwide synchronization. An excellent review of the rationale and trade-offs in optical interconnects versus electrical interconnects is given in [Miller00].

The potential advantages of optical technology for clock distribution are that the delay of optical signals is not sensitive to temperature and that the clock edges do not degrade over long distances (i.e., tens of meters). In fact, it is possible to deliver an optical clock signal with, at most, 10–100 ps of uncertainty over tens of meters. Optical clocks can be distributed on-chip via waveguides or through free space. Figure 10-69 shows the plot of an optical clock architecture, using waveguides. The off-chip optical source is brought to the chip, distributed through waveguides, and converted through receiver circuitry to a local electrical clock distribution network. The chip is divided into small sections (or tiles), and the global clock is distributed from the photon source through waveguides with splitters and bends to each of the sections. Notice that an *H*-tree is used in distributing the optical clock. At the end of the waveguide is a photodetector, which can be implemented using silicon or germanium. Upon reaching the detector in each section, the optical pulses that represent the global clock are converted into current pulses.

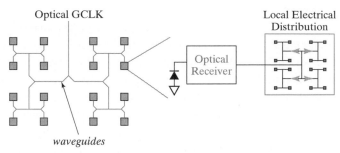

Figure 10-69 Architecture for optical clock distribution (courtesy of Lionel Kimerling).

These have a very small magnitude (tens of μA), and are fed into a amplifier that amplifies the signal into voltage signals appropriate for digital processing. The electrical clock is then distributed to the local load using conventional techniques [Kimerling00].

The optical approach has the advantage that the skew of the global clock to the photodectors is virtually zero. There are some variations in the arrival time of the optical signal, but they are minimal. For instance, variations at the waveguide bends may cause the energy losses to differ from path to path. Optics has the additional advantage that many of the difficulties associated with electromagnetic wave propagation—such as cross talk and inductive coupling—are avoided.

On the other hand, the challenge lies in the design of the optical receiver. To amplify and transform the small current pulses into reasonable voltage signals requires multiple amplifier stages. These are susceptible to process and environmental variations, and thus also to skew. The problem becomes very similar to the conventional electrical approach. The advantage, however, is that the problem is confined to these receivers, which makes it more tractable.

Optical clocking may have an important future in high-performance systems. The challenges of dealing with process variations in the opto-electronic circuitry must be addressed first for this to become a reality.

10.7.3 Synchronous versus Asynchronous Design

The self-timed approach offers a potential solution to the growing clock-distribution problem. It translates the global clock signal into a number of local synchronization problems. Independence from physical timing constraints is achieved with the aid of completion signals. Handshaking logic is needed to ensure the logical ordering of the circuit events and to avoid race conditions. This requires adherence to a certain protocol, which normally consists of either two or four phases.

Despite all its advantages, self-timing has only been used in a number of isolated cases. Examples of self-timed circuits can be found in signal processing [Jacobs90], fast arithmetic units (such as self-timed dividers) [Williams87], simple microprocessors [Martin89] and memory (static RAM, FIFOs). In general, synchronous logic is both faster and simpler, since the overhead of completion-signal generation and handshaking logic is avoided. The design of a foolproof network of handshaking units that is robust with respect to races, live lock, and dead lock is nontrivial and requires the availability of dedicated design-automation tools.

On the other hand, distributing a clock at high speed becomes exceedingly difficult. Skew management requires extensive modeling and analysis, as well as careful design. It will not be easy to extend this methodology into the next generation of designs. With the increasing timing uncertainty in synchronous circuits, self-timing is bound to become more attractive in the years to come [Sutherland02]. This observation is already reflected in the fact that the routing network for the latest generation of massively parallel supercomputers is completely implemented using self-timing [Seitz92]. For self-timing to become a mainstream design technique, however (if it ever will), further innovations in circuit and signaling techniques and design methodologies are needed.

Other alternative timing approaches might emerge as well. Possible candidates are fully asynchronous designs or islands of synchronous units connected by an asynchronous network. The latter method, called the *globally–asynchronous locally–synchronous approach*, is quite attractive. It avoids the pitfalls of self-timed design at the local circuit level, while eliminating the need for strict phase synchronization between blocks that are quite distant from each other on the die. In fact, the pure phase-synchronicity requirement between large modules in a system-on-a-chip is most often too large a constraint. Mesochronous or plesiochronous communication would work just as well. It is our conjecture that these styles of synchronization will become much more common in the coming decade.

10.8 Summary

This chapter has explored the timing of sequential digital circuits:

- An in-depth analysis of the synchronous digital circuits and clocking approaches was presented. Clock *skew* and *jitter* substantially impact the functionality and performance of a system. Important parameters are the clocking scheme used and the nature of the clock generation and distribution network.
- Alternative timing approaches, such as *self-timed design*, are becoming attractive for dealing with clock distribution problems. Self-timed design uses completion signals and handshaking logic to isolate physical timing constraints from event ordering.
- The connection of synchronous and asynchronous components introduces the risk of synchronization failure. The introduction of *synchronizers* helps to reduce that risk, but can never eliminate it.
- *Phase-locked loops* are becoming an important element of the digital designer's toolbox. They are used to generate high-speed clock signals on a chip. The analog nature of the PLL makes its design a real challenge.
- Important trend in clock distribution include the use of *delay-locked loops* to actively adjust delays on a chip.
- The key message of this chapter is that synchronization and timing are among the most intriguing challenges facing the digital designer of the next decade.

10.9 To Probe Further

While system timing is an important topic, a comprehensive reference work is not available in this area. One of the best discussions so far is the chapter by Chuck Seitz in [Mead80, Chapter 7]. Other in-depth overviews are given in [Bakoglu90], [Johnson93, Chapter 11], [Bernstein98, Chapter 7], and [Chandrakasan00, Chapters 12 and 13]. A collection of papers on clock distribution networks is presented in [Friedman95]. Numerous other publications are available on this topic in the leading journals, some of which are mentioned in the following list of references.

References

[Abnous93] A. Abnous and A. Behzad, "A High-Performance Self-Timed CMOS Adder," in *EE241 Final Class Project Reports*, by J. Rabaey, Univ. of California–Berkeley, May 1993.

[Bailey98] D. Bailey and B. Benschneider, "Clocking Design and Analysis for a 600-MHz Alpha Microprocessor," *IEEE Journal of Solid-State Circuits*, vol. 33, pp. 1627–1633, November 1998.

[Bailey00] D. Bailey, "Clock Distribution," in [Chandrakasan00].

[Bakoglu90] H. Bakoglu, *Circuits, Interconnections and Packaging for VLSI*, Addison-Wesley, pp. 338–393, 1980.

[Boning00] D. Boning, S. Nassif, "Models of Process Variations in Device and Interconnect," in [Chandrakasan00].

[Bowhill95] W. Bowhill et al., "A 300 MHz Quad-Issue CMOS RISC Microprocessor," *Technical Digest of the 1995 International Solid State Circuits Conference (ISSCC)*, San Francisco, pp. 182–183, February 1995.

[Bernstein98] K. Bernstein et al., *High Speed CMOS Design Styles*, Kluwer Academic Publishers, 1998.

[Bernstein00] K. Bernstein, "Basic Logic Families," in [Chandrakasan00].

[Chandrakasan00] A. Chandrakasan, W. Bowhill, and F. Fox, *Design of High-Performance Microprocessor Circuits*, IEEE Press, 2000.

[Chaney73] T. Chaney and F. Rosenberger, "Anomalous Behavior of Synchronizer and Arbiter Circuits," *IEEE Trans. on Computers*, vol. C-22, pp. 421–422, April 1973.

[Dally98] W. Dally and J. Poulton, *Digital Systems Engineering*, Cambridge University Press, 1998.

[Dean94] M. Dean, D. Dill, and M. Horowitz, "Self-Timed Logic Using Current-Sensing Completion Detection," *Journal of VLSI Signal Processing*, pp. 7–16, February 1994.

[Dobberpuhl92] D. Dopperpuhl et al., "A 200 MHz 64-b Dual Issue CMOS Microprocessor," *IEEE Journal on Solid State Circuits*, vol. 27, no. 11, pp. 1555–1567, November 1992.

[Friedman95] E. Friedman, ed., *Clock Distribution Networks in VLSI Circuits and Systems*, IEEE Press, 1995.

[Gardner80] F. Gardner, "Charge-Pump Phase-Locked Loops," *IEEE Trans. on Communications*, vol. COM-28, pp. 1849–58, November 1980.

[Glasser85] L. Glasser and D. Dopperpuhl, *The Design and Analysis of VLSI Circuits*, Addison-Wesley, pp. 360–365, 1985.

[Goodman98] J. Goodman, A. P. Dancy, A. P. Chandrakasan, "An Energy/Security Scalable Encryption Processor Using an Embedded Variable Voltage DC/DC Converter," *IEEE Journal of Solid State Circuits*, pp. 1799–1809, 1998.

[Gutnik00] Gutnik, V., A. P. Chandrakasan, "Active GHz Clock Network Using Distributed PLLs," *IEEE Journal of Solid State Circuits*, pp. 1553–1560, November 2000.

[Heller84] L. Heller et al., "Cascade Voltage Switch Logic: A Differential CMOS Logic Family," *IEEE International Solid State Conference Digest*, San Francisco, pp. 16–17, February 1984.

[Herrick00] B. Herrick, "Design Challenges in Multi-GHz Microprocessors," *Proceedings ASPDAC 2000*, Yokohama, January 2000.

[Jacobs90] G. Jacobs and R. Brodersen, "A Fully Asynchronous Digital Signal Processor," *IEEE Journal on Solid State Circuits*, vol. 25, no. 6, pp. 1526–1537, December 1990.

[Jeong87] D. Jeong et al., "Design of PLL-Based Clock Generation Circuits," *IEEE Journal on Solid State Circuits*, vol. SC-22, no. 2, pp. 255–261, April 1987.

[Johnson93] H. Johnson and M. Graham, *High-Speed Digital Design—A Handbook of Black Magic*, Prentice Hall, 1993.

[Kim90] B. Kim, D. Helman, and P. Gray, "A 30 MHz Hybrid Analog/Digital Clock Recovery Circuit in 2 µm CMOS," *IEEE Journal on Solid State Circuits*, vol. SC-25, no. 6, pp. 1385–1394, December 1990.

[Kimerling00] L. Kimerling, "Photons to the Rescue: Microelectronics becomes Microphotonics," *Interface*, vol. 9, no. 2, 2000.

[Martin89] A. Martin et al., "The First Asynchronous Microprocessor: Test Results," *Computer Architecture News,* vol. 17, no. 4, pp. 95–110, June 1989.

[Maneatis00] J. Maneatis, "Design of High-Speed CMOS PLLs and DLLs," in [Chandrakasan00].

[Mead80] C. Mead and L. Conway, *Introduction to VLSI Design*, Addison-Wesley, 1980.

[Messerschmitt90] D. Messerschmitt, "Synchronization in Digital Systems," *IEEE Journal on Selected Areas in Communications*, vol. 8, no. 8, pp. 1404–1419, October 1990.

[Miller00] D. Miller, "Rationale and Challenges for Optical Interconnects to Electronic Chips," *Proceedings of the IEEE*, pp. 728–749, June 2000.

[Park00] T. Park, T. Tugbawa, and D. Boning, "An Overview of Methods for Characterization of Pattern Dependencies in Copper CMP," *Chemical Mechanical Polish for ULSI Multilevel Interconnection Conference* (CMP-MIC 2000), pp. 196–205, March 2000.

[Restle98] P. Restle, K. Jenkins, A. Deutsch, P. Cook, "Measurement and Modeling of On-Chip Transmission Line Effects in a 400 MHz Microprocessor," *IEEE Journal of Solid State Circuits*, pp. 662–665, April 1998.

[Restle01] P. Restle, "Technical Visualizations in VLSI Design," *Proceedings 38th Design Automation Conference*, Las Vegas, June 2001.

[Seitz80] C. Seitz, "System Timing," in [Mead80], pp. 218–262, 1980.

[Seitz92] C. Seitz, "Mosaic C: An Experimental Fine-Grain Multicomputer," in *Future Tendencies in Computer Science, Control and Applied Mathematics*, Proceedings International Conference on the 25th Anniversary of INRIA, Springer-Verlag, pp. 69–85, 1992.

[Sutherland89] I. Sutherland, "Micropipelines," *Communications of the ACM*, pp. 720–738, June 1989.

[Sutherland02] I. Sutherland and J. Ebergen, "Computers Without Clocks," *Scientific American*, vol. 287, no. 2, pp. 62–69, August 2002.

[Veendrick80] H. Veendrick, "The Behavior of Flip Flops Used as Synchronizers and Prediction of Their Failure Rates," *IEEE Journal of Solid State Circuits*, vol. SC-15, no. 2, pp. 169–176, April 1980.

[Williams87] T. Williams et al., "A Self-Timed Chip for Division," in *Proceedings of Advanced Research in VLSI 1987*, Stanford Conference, pp. 75–96, March 1987.

[Williams00] T. Williams, "Self-Timed Pipelines," in [Chandrakasan00].

[Yee96] G. Yee et al., "Clock-Delayed Domino for Adder and Combinational Logic Design," *Proceedings IEEE International Conference on Computer Design (ICCD)*, pp. 332–337, 1996.

[Young97] I. Young, M. Mar, and B. Brushan, "A 0.35 µm CMOS 3-880 MHz PLL N/2 Clock Multiplier and Distribution Network with Low Jitter for Microprocessors," *IEEE International Solid State Conference Digest*, pp. 330–331, 1997.

Exercises and Design Problem

Visit the book's web site **http://bwrc.eecs.berkeley.edu/IcBook** for challenging exercises and design problems.

G

Design Verification

Simulation versus Verification

Electrical and Timing Verification

Formal Verification

Up to this point, we have relied heavily on simulation as the preferred method for ensuring the correctness of a design and for extracting critical parameters such as speed and power dissipation. Simulation is, however, only one of the techniques the designer has in his toolbox to accomplish those objectives. In general, it is useful to differentiate between *simulation* and *verification*.

In the *simulation* approach, the value of a design parameter such as noise margin, propagation delay, or dissipated energy is determined by applying a set of excitation vectors to the circuit model of choice and extracting parameters from the obtained signal waveforms. While this approach is very flexible, it has the disadvantage that the results depend strongly upon the choice of the excitations. For example, a charge-redistribution condition in a dynamic logic gate is not detected by a simulation if the exact sequence of input patterns that causes the charge sharing is not applied. Similarly, the delay of an adder varies widely depending on the input signal. Identification of the worst-case delay requires a careful choice of the excitation vector so that the complete carry path is exercised. In other words, the designer must have a good understanding of the intrinsics of the circuit module and its operation. Failing to do so produces meaningless results.

Verification, on the other hand, attempts to extract the system parameters directly from the circuit description. For instance, the critical path of an adder can be recognized from an inspection of the circuit diagram or a model of it. This approach has the advantage that the result is independent of the choice of excitation vectors and is supposedly foolproof. On the other hand, it relies on a number of implicit assumptions regarding design techniques and methodologies. For the example of the adder, determination of the propagation delay requires an understanding of the logic operation of the composing circuitry—for example, dynamic or static logic—and a definition of the term *propagation delay*. As a second example, it is necessary to identify the register elements first before the maximum clock speed of a synchronous circuit can be determined. The resulting tools are therefore restricted in scope and handle only a limited class of circuit styles, such as single-phase synchronous design. In this insert, we analyze a number of the popular verification techniques.

Electrical Verification

Given the transistor schematics of a digital design, it is possible to verify whether a number of basic rules are satisfied. Some examples of typical rules help illustrate this concept, and many others can be derived from previous chapters:

- The number of inversions between two C^2MOS gates should be even.
- In a pseudo-NMOS gate, a well-defined ratio between the PMOS pull-up and NMOS pull-down devices is necessary to guarantee a sufficient noise-margin low NM_L.
- To ensure that rise and fall times of the signal waveforms stay within limits, minimum bounds can be set on the sizes of the driver transistors as a function of the fan-out.
- The maximum amount of charge sharing in a dynamic design should be such that the noise-margin high is not violated.

Pure common sense can help define a large set of rules to which a design should always adhere. Applying them requires an in-depth understanding of the circuit structure. An electrical verifier, therefore, starts with the identification of well-known substructures in the overall circuit schematic. Typical templates are simple logic gates, pass transistors, and registers. The verifier traverses the resulting network on a rule-per-rule basis. Since electrical rules tend to be specific to a particular design style, they should be able to be easily modified. For example, *rule-based expert systems* allow for an easy updating of the rule base [DeMan85]. The individual rules can be complex and even invoke a circuit simulator for a small subsection of the network to verify whether a given condition is met. In summary, electrical verification is a helpful tool, and it can dramatically reduce the risk of malfunction.

Timing Verification

As circuits become more complex, it is increasingly difficult to define exactly which paths through the network are critical with respect to timing. One solution is to run extensive SPICE simulations that may take a long time to finish. Even then, this does not guarantee that the iden-

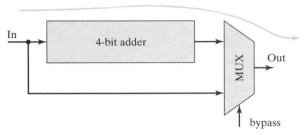

Figure G-1   Example of false path in timing verification.

tified critical path is the worst case, since the delay path is a function of the applied signal patterns. A timing verifier traverses the electrical network and rank-orders the various paths, based on delay. This delay can be determined in a number of ways. One approach is to build an *RC*-model of the network and compute bounds on the delay of the resulting passive network. To obtain more accurate results, many timing verifiers first extract the details of the longest path(s), based on the *RC*-model, and perform circuit simulation on the reduced circuit in order to obtain a better estimation. Examples of early timing verifiers are the Crystal [Ousterhout83] and TV [Jouppi84] systems.

One problem that hampered many earlier systems is that they identified *false paths*—that is, critical paths that can never be exercised during normal circuit operation. For example, a false path exists in the carry-bypass adder discussed in Chapter 11 (also shown in simplified form in Figure G-1). From a simple analysis of only the circuit topology, one would surmise that the critical path of the circuit passes through the adder and multiplexer modules as illustrated by the arrow. A closer look at the circuit operation reveals that such a path is not feasible. All individual adder bits must be in the propagate mode for *In* to propagate through the complete adder. However, the *bypass* signal is asserted under these conditions, and the bottom path through the multiplexer is selected instead. The actual critical path is thus shorter than what would be predicted from the first-order analysis. Detecting false timing paths is not easy, since it requires an understanding of the logic functionality of the network. Newer timing verifiers are remarkably successful in accomplishing this and have become one of the more important design aids for the high-performance circuit designer [Devadas91].

Example G.1   Example of Timing Verification

The output of a static timing verifier, the PathMill tool from Synopsys [PathMill], is shown in Figure G-2. The input to the verification process is a transistor netlist, but gate- and block-level models can be included as well. The analysis considers capacitive and resistive parasitics that are obtained from the transistor schematic or layout extraction.

The output of the timing analysis is an ordered list of critical timing paths. For the example of Figure G-2, the longest path extends from node *B* (falling edge) to node *Y* (falling edge) over nodes S1, S2, and X3. The predicted delay is 1.14 ns. Other paths close to

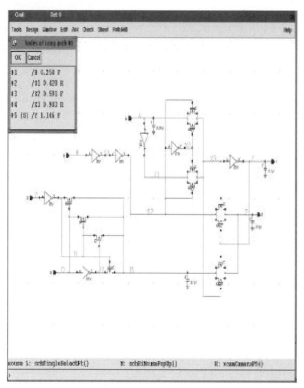

Figure G-2 Example of static timing verifier response as generated by the PathMill tool (*Courtesy of Synopsys, Inc.*). The results are displayed using the *Cadence DFII* tool. The critical timing path runs from node *B* to node *Y* via nodes *S1*, *S2*, and *X3*.

the critical one are in order of decreasing delay: B $(F) \rightarrow Y$ (R) (1.11 ns), B $(R) \rightarrow Y$ (F) (1.04 ns), C $(R) \rightarrow Z$ (R) (1.04 ns), and D $(R) \rightarrow Y$ (R) (1.00 ns). Observe that the circuit contains multiple chained pass transistors that make the analysis complicated and increase the chances for false paths to occur. An example of the latter is the path $A(F) \rightarrow X1 \rightarrow X3 \rightarrow Z(R)$ for $B = 0$. For a low value on B, node Z can only make a falling transition.

Functional (or Formal) Verification

Each component (transistor, gate, or functional block) of a circuit can be described behaviorally as a function of its inputs and internal state. By combining these component descriptions, an overall circuit model can be generated that symbolically describes the behavior of the complete circuit. Formal verification compares this derived behavior with the designer's initial specification. Although not identical, the two descriptions need to be mathematically equivalent for the circuit to be correct.

Formal verification is the designer's ultimate dream of what design automation should be able to accomplish—*proof that the circuit will work as specified*. Unfortunately, no general and widely accepted verifier has yet been realized. The complexity of the problem is illustrated by the following argument: One way to prove that two circuit descriptions are identical is to compare the outputs while enumerating all possible input patterns and sequences thereof. This is an intractable approach with a computation time exponential in the number of inputs and states.

This does not mean that formal verification is a fantasy, however. Techniques have been proposed and successfully implemented for certain classes of circuits. For instance, assuming that a circuit is synchronous helps minimize the search space. Proving the equivalence between state machines has been one of the main research targets and has led to some remarkable advances (e.g., [Coudert90]). In general, formal verification techniques fall into two major classes:

- *Equivalence Proving*—Design often follows a refinement approach. An initial high-level description is refined into a more detailed one in a step-by-step fashion. Equivalence-proving tools establish that the obtained result is functionally equivalent to the original description. As an example, such a tool could establish that the transistor schematic of an N-input static CMOS NAND gate truly implements the intended NAND function.
- *Property Proving*—A designer often wants to know if her creation possesses some well-defined properties. A possible requirement could be, "My circuit should never experience a race condition." A property-proving tool analyzes whether this goal is met; if it is not met, the tool also analyzes the circumstances.

One of the prime necessities for formal verification to work is that the design is described in a manner that is unambiguous and that has a clearly defined meaning (or semantics). When these conditions are met, formal-verification techniques have yielded some remarkable results and are very effective. For instance, finite-state machines (FSMs) represent an area where formal verification has made some major advances. The secret to the success is a well-defined mathematical description model.

Summary

Although not yet in the mainstream of the design-automation process, functional verification might become one of the important assets in the designer's toolbox. For this to happen, important progress must be made in the fields of design specification and interpretation.

References

[Coudert90] O. Coudert and J. Madre, *Proceedings ICCAD 1990,* pp. 126–129, November 1990.

[DeMan85] H. De Man, I. Bolsens, E. Vandenmeersch, and J. Van Cleynenbreugel, "DIALOG: An Expert Debugging System for MOS VLSI Design," *IEEE Trans. on CAD*, vol. 4, no. 3, pp. 301–311, July 1985.

[Devadas91] S. Devadas, K. Keutzer, and S. Malik, "Delay Computations in Combinational Logic Circuits: Theory and Algorithms," *Proc. ICCAD-91,* pp. 176–179, Santa Clara, 1991.

[Jouppi84] N. Jouppi, *Timing Verification and Performance Improvement of MOS VLSI Designs*, Ph. D. diss., Stanford University, 1984.

[Ousterhout83] J. Ousterhout, "Crystal: A Timing Analyzer for nMOS VLSI Circuits," *Proc. 3rd Caltech Conf. on VLSI* (Bryant ed.), Computer Science Press, pp. 57–69, March 1983.

[PathMill] PathMill: Transistor-Level Static Timing Analysis, *http://www.synopsys.com/products/analysis/pathmill_ds.html*, Synopsys, Inc.

CHAPTER

11

Designing Arithmetic Building Blocks

Designing adders, multipliers, and shifters
for performance, area, or power

Logic and system optimizations for datapath modules

Power-delay trade-offs in datapaths

11.1   Introduction

11.2   Datapaths in Digital Processor Architectures

11.3   The Adder
    11.3.1   The Binary Adder: Definitions
    11.3.2   The Full Adder: Circuit Design Considerations
    11.3.3   The Binary Adder: Logic Design Considerations

11.4   The Multiplier
    11.4.1   The Multiplier: Definitions
    11.4.2   Partial-Product Generation
    11.4.3   Partial-Product Accumulation
    11.4.4   Final Addition

11.5   The Shifter
    11.5.1   Barrel Shifter
    11.5.2   Logarithmic Shifter

11.6   Other Arithmetic Operators

11.7   Power and Speed Trade-Offs in Datapath Structures*
    11.7.1   Design-Time Power-Reduction Techniques
    11.7.2   Run-Time Power Management
    11.7.3   Reducing the Power in Standby (or Sleep) Mode

11.8   Perspective: Design as a Trade-off
11.9   Summary
11.10 To Probe Further

11.1 Introduction

After the in-depth study of the design and optimization of the basic digital gates, it is time to test our acquired skills on a somewhat larger scale and put them in a more system-oriented perspective.

We will apply the techniques of the previous chapters to design a number of circuits often used in the datapaths of microprocessors and signal processors. More specifically, we discuss the design of a representative set of modules such as adders, multipliers, and shifters. The speed and power of these elements often dominates the overall system performance. Hence, a careful design optimization is required. It rapidly becomes obvious that the design task is not straightforward. For each module, multiple equivalent logic and circuit topologies exist, each of which has its own positives and negatives in terms of area, speed, or power.

Although far from complete, the analysis presented helps focus on the essential trade-offs that must be made in the course of the digital design process. You will see that optimization at only one design level—for instance, through transistor sizing only—leads to inferior designs. A global picture is therefore of crucial importance. Digital designers focus their attention on the gates, circuits, or transistors that have the largest impact on their goal function. The noncritical parts of the circuit can be developed routinely. We also may develop first-order performance models that foster understanding of the fundamental behavior of a module. The discussion also clarifies which computer aids can help to simplify and automate this phase of the design process.

Before analyzing the design of the arithmetic modules, a short discussion of the role of the datapath in the digital-processor picture is appropriate. This not only highlights the specific design requirements for the datapath, but also puts the rest of this book in perspective. Other processor modules—the input/output, controller, and memory modules—have different requirements and were discussed in Chapter 8. After an analysis of the area-power-delay trade-offs in the design of adders, multipliers, and shifters, we will use the same structures to illustrate some of the power-minimization approaches introduced in Chapter 6. The chapter concludes with a short perspective on datapath design and its trade-offs.

11.2 Datapaths in Digital Processor Architectures

We introduced the concept of a digital processor in Chapter 8. Its components consist of the datapath, memory, control, and input/output blocks. The datapath is the core of the processor—this is where all computations are performed. The other blocks in the processor are support units that either store the results produced by the datapath or help to determine what will happen in the next cycle. A typical datapath consists of an interconnection of basic combinational functions, such as arithmetic operators (addition, multiplication, comparison, and shift) or logic (AND, OR, and XOR). The design of the arithmetic operators is the topic of this chapter. The

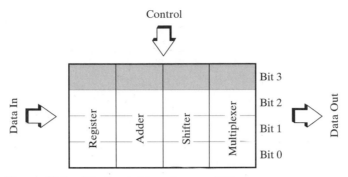

Figure 11-1 Bit-sliced datapath organization.

intended application sets constraints on the datapath design. In some cases, such as in personal computers, processing speed is everything. In most other applications, there is a maximum amount of power that is allowed to be dissipated, or there is maximum energy available for computation while maintaining the desired throughput.

Datapaths often are arranged in a *bit-sliced* organization, as shown in Figure 11-1. Instead of operating on single-bit digital signals, the data in a processor are arranged in a *word*-based fashion. Typical microprocessor datapaths are 32 or 64 bits wide, while the dedicated signal processing datapaths, such as those in DSL modems, magnetic disk drives, or compact-disc players are of arbitrary width, typically 5 to 24 bits. For instance, a 32-bit processor operates on data words that are 32 bits wide. This is reflected in the organization of the datapath. Since the same operation frequently has to be performed on each bit of the data word, the datapath consists of 32 bit slices, each operating on a single bit—hence the term *bit sliced*. Bit slices are either identical or resemble a similar structure for all bits. The datapath designer can concentrate on the design of a single slice that is repeated 32 times.

11.3 The Adder

Addition is the most commonly used arithmetic operation. It often is the speed-limiting element as well. Therefore, careful optimization of the adder is of the utmost importance. This optimization can proceed either at the logic or circuit level. Typical logic-level optimizations try to rearrange the Boolean equations so that a faster or smaller circuit is obtained. An example of such a logic optimization is the *carry lookahead adder* discussed later in the chapter. Circuit optimizations, on the other hand, manipulate transistor sizes and circuit topology to optimize the speed. Before considering both optimization processes, we provide a short summary of the basic definitions of an adder circuit (as defined in any book on logic design [e.g., Katz94]).

11.3.1 The Binary Adder: Definitions

Table 11.1 shows the truth table of a binary full adder. A and B are the adder inputs, C_i is the carry input, S is the sum output, and C_o is the carry output. The Boolean expressions for S and C_o are given in Eq. (11.1).

Table 11-1 Truth table for full adder.

A	B	C_i	S	C_o	Carry Status
0	0	0	0	0	delete
0	0	1	1	0	delete
0	1	0	1	0	propagate
0	1	1	0	1	propagate
1	0	0	1	0	propagate
1	0	1	0	1	propagate
1	1	0	0	1	generate/propagate
1	1	1	1	1	generate/propagate

$$S = A \oplus B \oplus C_i$$
$$= A\bar{B}\bar{C_i} + \bar{A}B\bar{C_i} + \bar{A}\bar{B}C_i + ABC_i \qquad (11.1)$$
$$C_o = AB + BC_i + AC_i$$

It is often useful from an implementation perspective to define S and C_o as functions of some intermediate signals G (generate), D (delete), and P (propagate).[1] $G = 1$ $(D = 1)$ ensures that a carry bit will be *generated* (*deleted*) at C_o independent of C_i, while $P = 1$ guarantees that an incoming carry will propagate to C_o. Expressions for these signals can be derived from inspection of the truth table:

$$G = AB$$
$$D = \bar{A}\bar{B} \qquad (11.2)$$
$$P = A \oplus B$$

We can rewrite S and C_o as functions of P and G (or D):

$$C_o(G, P) = G + PC_i$$
$$S(G, P) = P \oplus C_i \qquad (11.3)$$

Notice that G and P are only functions of A and B and are not dependent upon C_i. In a similar way, we can also derive expressions for $S(D, P)$ and $C_o(D, P)$.

---

[1]Note that the propagate signal sometimes is also defined as the OR function of the inputs A and B—this condition guarantees that the input carry propagates to the output when $A = B = 1$, too. We will provide appropriate warning whenever this definition is used.

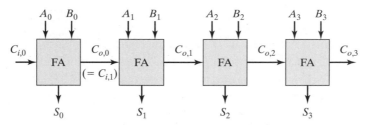

Figure 11-2   Four-bit ripple-carry adder: topology.

An N-bit adder can be constructed by cascading N full-adder (FA) circuits in series, connecting $C_{o,k-1}$ to $C_{i,k}$ for $k = 1$ to $N-1$, and the first carry-in $C_{i,0}$ to 0 (Figure 11-2). This configuration is called a *ripple-carry adder*, since the carry bit "ripples" from one stage to the other. The delay through the circuit depends upon the number of logic stages that must be traversed and is a function of the applied input signals. For some input signals, no rippling effect occurs at all, while for others, the carry has to ripple all the way from the *least significant bit (lsb)* to the *most significant bit (msb)*. The propagation delay of such a structure (also called the *critical path*) is defined as the *worst case delay over all possible input patterns*.

In the case of the ripple-carry adder, the worst case delay happens when a carry generated at the least significant bit position propagates all the way to the most significant bit position. This carry is finally consumed in the last stage to produce the sum. The delay is then proportional to the number of bits in the input words N and is approximated by

$$t_{adder} \approx (N - 1)t_{carry} + t_{sum} \tag{11.4}$$

where t_{carry} and t_{sum} equal the propagation delays from C_i to C_o and S, respectively.[2]

Example 11.1   Propagation Delay of Ripple-Carry Adder

Derive the values of A_k and B_k ($k = 0...N - 1$) so that the worst case delay is obtained for the ripple-carry adder.

The worst case condition requires that a carry be generated at the *lsb* position. Since the input carry of the first full adder C_{i0} is always 0, this both A_0 and B_0 must equal 1. All the other stages must be in propagate mode. Hence, either A_i or B_i must be high. Finally, we would like to physically measure the delay of a transition on the *msb* sum bit. Assuming an initial value of 0 for S_{N-1}, we must arrange a $0 \rightarrow 1$ transition. This is achieved by setting both A_{N-1} and B_{N-1} to 0 (or 1), which yields a high sum bit given the incoming carry of 1.

For example, the following values for A and B trigger the worst case delay for an 8-bit addition:

$$A: 0000001; B: 01111111$$

---

[2]Equation (11.4) assumes that both the delay from the input signals A_0 (or B_0) to $C_{o,0}$ for the *lsb*, and the C_i-to-C_o delay for all other bits equal to t_{carry}.

To set-up the worst case delay transition, all the inputs can be kept constant with A_0 undergoing a $0 \rightarrow 1$ transition.

The left-most bit represents the *msb* in this binary representation. Observe that this is only one of the many worst case patterns. This case exercises the $0 \rightarrow 1$ delay of the final sum. Derive several other cases that exercise the $0 \rightarrow 1$ and $1 \rightarrow 0$ transitions.

Two important conclusions can be drawn from Eq. (11.4):

- The propagation delay of the ripple-carry adder is *linearly* proportional to N. This property becomes increasingly important when designing adders for the wide data paths ($N = 16...128$) that are desirable in current and future computers.
- When designing the full-adder cell for a fast ripple-carry adder, it is far more important to optimize t_{carry} than t_{sum}, since the latter has only a minor influence on the total value of t_{adder}.

Before starting an in-depth discussion on the circuit design of full-adder cells, the following additional logic property of the full adder is worth mentioning:

Inverting all inputs to a full adder results in inverted values for all outputs.

This property, also called the *inverting property*, is expressed in the pair of equations

$$\bar{S}(A, B, C_i) = S(\bar{A}, \bar{B}, \bar{C_i})$$

$$\overline{C}_o(A, B, C_i) = C_o(\bar{A}, \bar{B}, \bar{C_i})$$

$$(11.5)$$

and will be extremely useful when optimizing the speed of the ripple-carry adder. It states that the circuits of Figure 11-3 are identical.

11.3.2 The Full Adder: Circuit Design Considerations

Static Adder Circuit

One way to implement the full-adder circuit is to take the logic equations of Eq. (11.1) and translate them directly into complementary CMOS circuitry. Some logic manipulations can help to reduce the transistor count. For instance, it is advantageous to share some logic between the

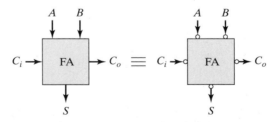

Figure 11-3   Inverting property of the full adder. The circles in the schematics represent inverters.

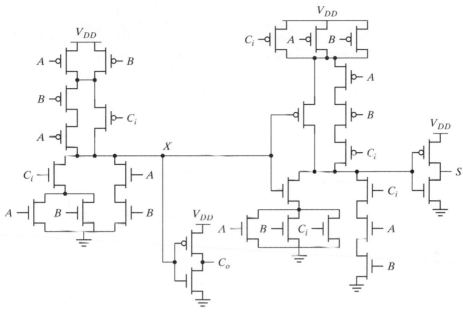

Figure 11-4 Complementary static CMOS implementation of full adder.

sum- and carry-generation subcircuits, as long as this does not slow down the carry generation, which is the most critical part, as stated previously. The following is an example of such a reorganized equation set:

$$C_o = AB + BC_i + AC_i$$

and   (11.6)

$$S = ABC_i + \overline{C_o}(A + B + C_i)$$

The equivalence with the original equations is easily verified. The corresponding adder design, using complementary static CMOS, is shown in Figure 11-4 and requires 28 transistors. In addition to consuming a large area, this circuit is slow:

- Tall PMOS transistor stacks are present in both *carry-* and *sum*-generation circuits.
- The intrinsic load capacitance of the C_o signal is large and consists of two diffusion and six gate capacitances, plus the wiring capacitance.
- The signal propagates through two inverting stages in the carry-generation circuit. As mentioned earlier, minimizing the carry-path delay is the prime goal of the designer of high-speed adder circuits. Given the small load (fan-out) at the output of the carry chain, having two logic stages is too high a number, and leads to extra delay.
- The sum generation requires one extra logic stage, but that is not that important, since a factor appears only once in the propagation delay of the ripple-carry adder of Eq. (11.4).

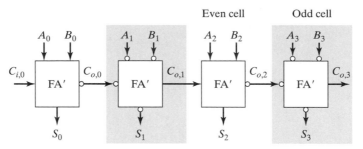

Figure 11-5   Inverter elimination in carry path. FA′ stands for a full adder without the inverter in the carry path.

Although slow, the circuit includes some smart design tricks. Notice that the first gate of the carry-generation circuit is designed with the C_i signal on the smaller PMOS stack, lowering its logical effort to 2. Also, the NMOS and PMOS transistors connected to C_i are placed as close as possible to the output of the gate. This is a direct application of a circuit-optimization technique discussed in Section 4.2—transistors on the critical path should be placed as close as possible to the output of the gate. For instance, in stage k of the adder, signals A_k and B_k are available and stable long before $C_{i,k}$ ($= C_{o,k-1}$) arrives after rippling through the previous stages. In this way, the capacitances of the internal nodes in the transistor chain are precharged or discharged in advance. On arrival of $C_{i,k}$, only the capacitance of node X has to be (dis)charged. Putting the $C_{i,k}$ transistors closer to V_{DD} and GND would require not only the (dis)charging of the capacitance of node X, but also of the internal capacitances.

The speed of this circuit can now be improved gradually by using some of the adder properties discussed in the previous section. First, the number of inverting stages in the carry path can be reduced by exploiting the inverting property—inverting all the inputs of a full-adder cell also inverts all the outputs. This rule allows us to eliminate an inverter in a carry chain, as demonstrated in Figure 11-5.

Mirror Adder Design

An improved adder circuit, also called the *mirror adder*, is shown in Figure 11-6 [Weste93]. Its operation is based on Eq. (11.3). The carry-generation circuitry is worth analyzing. First, the carry-inverting gate is eliminated, as suggested in the previous section. Secondly, the PDN and PUN networks of the gate are not dual. Instead, they form a clever implementation of the propagate/generate/delete function—when either D or G is high, $\overline{C}_o$ is set to V_{DD} or GND, respectively. When the conditions for a *Propagate* are valid (or P is 1),[3] the incoming carry is propagated (in inverted format) to $\overline{C}_o$. This results in a considerable reduction in both area and delay. The analysis of the sum circuitry is left to the reader. The following observations are worth considering:

  • This full-adder cell requires only 24 transistors.

---

[3]The $P = A + B$ definition of the *propagate* signal is used here.

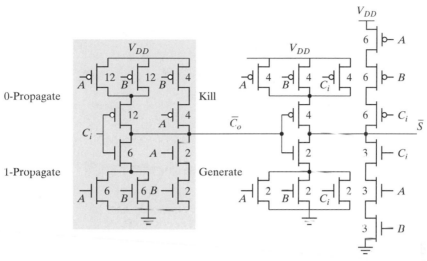

Figure 11-6 Mirror adder—circuit schematics.

- The NMOS and PMOS chains are completely symmetrical, which still yields correct operation due to *self-duality* of both the sum and carry functions. As a result, a maximum of two series transistors can be found in the carry-generation circuitry.
- The transistors connected to C_i are placed closest to the output of the gate.
- Only the transistors in the carry stage have to be optimized for speed. All transistors in the sum stage can be of minimum size. When laying out the cell, the most critical issue is the minimization of the capacitance at node $\overline{C}_o$. Shared diffusions reduce the stack node capacitances.
- In the adder cell of Figure 11-4, the inverter can be sized independently to drive the C_i input of the adder stage that follows. If the carry circuit in Figure 11-6 is symmetrically sized, each of its inputs has a logical effort of 2. This means that the optimal fan-out, sized for minimum delay, should be (4/2) = 2. However, the output of this stage drives two internal gate capacitances and six gate capacitances in the connecting adder cell. A clever solution to keep the transistor sizes the same in each stage is to increase the size of the carry stage to about three to four times the size of the sum stage. This maintains the optimal fan-out of 2. The resulting transistor sizes are annotated on Figure 11-6, where a PMOS/NMOS ratio of 2 is assumed.

Transmission-Gate-Based Adder

A full adder can be designed to use multiplexers and XORs. While this is impractical in a complementary CMOS implementation, it becomes attractive when the multiplexers and XORs are implemented as transmission gates. A full-adder implementation based on this approach is shown in Figure 11-7 and uses 24 transistors. It is based on the *propagate–generate* model, introduced in Eq. (11.3). The propagate signal, which is the XOR of inputs A and B, is used to

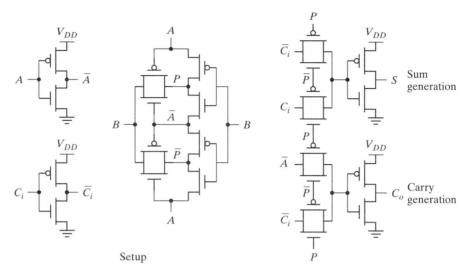

Figure 11-7   Transmission-gate-based full-adder cell with sum and carry delays of similar value (after [Weste93]).

select the true or complementary value of the input carry as the new sum output. Based on the propagate signal, the output carry is either set to the input carry, or either one of inputs A or B. One of interesting features of such an adder is that it has similar delays for both sum and carry outputs.

Manchester Carry-Chain Adder

The carry-propagation circuitry in Figure 11-7 can be simplified by adding generate and delete signals, as shown in Figure 11-8a. The propagate path is unchanged, and it passes C_i to the C_o output if the propagate signal ($A_i \oplus B_i$) is true. If the propagate condition is not satisfied, the output is either pulled low by the D_i signal or pulled up by $\overline{G}_i$. The dynamic implementation (Figure 11-8b), makes even further simplification possible. Since the transitions in a dynamic circuit are monotonic, the transmission gates can be replaced by NMOS-only pass transistors. Precharging the output eliminates the need for the kill signal (for the case in which the carry chain propagates the complementary values of the carry signals).

    A *Manchester carry-chain adder* uses a cascade of pass transistors to implement the carry chain [Kilburn60]. An example, based on the dynamic circuit version introduced in Figure 11-8, is shown in Figure 11-9. During the precharge phase ($\phi = 0$), all intermediate nodes of the pass-transistor carry chain are precharged to V_{DD}. During evaluation, the A_k node is discharged when there is an incoming carry and the propagate signal P_k is high, or when the generate signal for stage k (G_k) is high.

    Figure 11-10 shows an example layout of the Manchester carry chain in stick-diagram format. The datapath layout consists of three rows of cells organized in bit-sliced style: The top row

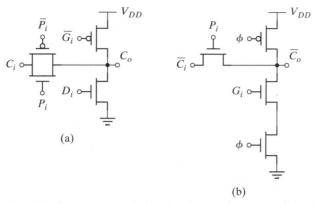

Figure 11-8 Manchester carry gates. (a) Static, using propagate, generate, and kill, (b) dynamic implementation, using only propagate and generate signals.

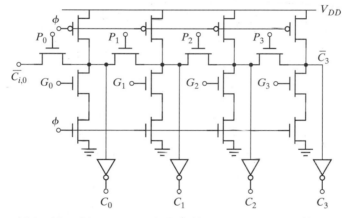

Figure 11-9 Manchester carry-chain adder in dynamic logic (four-bit section).

of cells computes the propagate and generate signals, the middle row propagates the carry from left to right, and the bottom row generates the final sums.

The worst case delay of the carry chain of the adder in Figure 11-9 is modeled by the linearized RC network of Figure 11-11. As derived in Chapter 4, the propagation delay of such a network equals

$$t_p = 0.69 \sum_{i=1}^{N} C_i \left(\sum_{j=1}^{i} R_j \right) = 0.69 \frac{N(N+1)}{2} RC \tag{11.7}$$

when all $C_i = C$ and $R_j = R$.

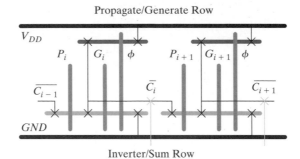

Figure 11-10   Stick diagram of two bits of a Manchester carry chain.

Carry in or clock

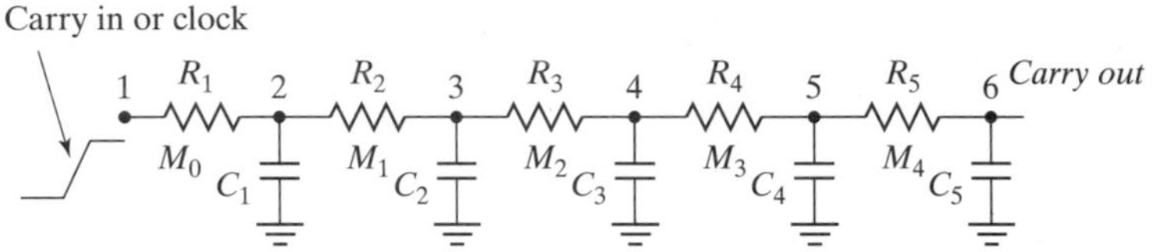

Figure 11-11   Equivalent network to determine propagation delay of a carry chain.

Example 11.2   Sizing of Manchester Carry Chain

The capacitance per node on the carry chain equals four diffusion capacitances, one inverter input capacitance, and the wiring capacitance proportional to the size of the cell. The inverter and the PMOS precharging transistor can be kept at unit size. Together with the wire capacitance, the fixed capacitance can be estimated as 15 fF (for our technology). If a unit-sized transistor with width W_0 has a resistance of 10 kΩ and a diffusion capacitance of 2 fF, then the RC time constant for a chain of transistors of width W is

$$RC = \left(6 \text{ fF} \cdot \frac{W}{W_0} + 15 \text{ fF} \right) \cdot 10 \text{ k}\Omega \cdot \frac{W_0}{W}$$

Increasing the transistor width reduces this time constant, but it also loads the gates in the previous stage. Therefore, the transistor size is limited by the input loading capacitance.

Unfortunately, the distributed RC-nature of the carry chain results in a propagation delay that is quadratic in the number of bits N. To avoid this, it is necessary to insert signal-buffering inverters. The optimum number of stages per buffer depends on the equivalent resistance of the inverter and the resistance and capacitance of the pass transistors, as was discussed in Chapter 9. In our technology, and in most other practical cases, this number is between 3 and 4. Adding the inverter makes the overall propagation delay a linear function of N, as is the case with ripple-carry adders.

11.3.3 The Binary Adder: Logic Design Considerations

The ripple-carry adder is only practical for the implementation of additions with a relatively small word length. Most desktop computers use word lengths of 32 bits, while servers require 64; very fast computers, such as mainframes, supercomputers, or multimedia processors (e.g., the Sony PlayStation2) [Suzuoki99], require word lengths of up to 128 bits. The linear dependence of the adder speed on the number of bits makes the usage of ripple adders rather impractical. Logic optimizations are therefore necessary, resulting in adders with $t_p < O(N)$. We briefly discuss a number of those in the sections that follow. We concentrate on the circuit design implications, since most of the presented structures are well known from the traditional logic design literature.

The Carry-Bypass Adder

Consider the four-bit adder block of Figure 11-12a. Suppose that the values of A_k and B_k ($k = 0...3$) are such that all propagate signals P_k ($k = 0...3$) are high. An incoming carry $C_{i,0} = 1$ propagates under those conditions through the complete adder chain and causes an outgoing carry $C_{o,3} = 1$. In other words,

$$\text{if } (P_0P_1P_2P_3 = 1) \text{ then } C_{o,3} = C_{i,0}$$
$$\text{else either DELETE or GENERATE occurred} \tag{11.8}$$

This information can be used to speed up the operation of the adder, as shown in Figure 11-12b. When $BP = P_0P_1P_2P_3 = 1$, the incoming carry is forwarded immediately to the next block through the bypass transistor M_b—hence the name *carry-bypass adder* or *carry-skip adder* [Lehman62]. If this is not the case, the carry is obtained by way of the normal route.

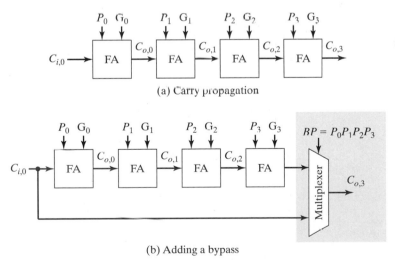

(a) Carry propagation

(b) Adding a bypass

Figure 11-12 Carry-bypass structure—basic concept.

Example 11.3   Carry Bypass in Manchester Carry-Chain Adder

Figure 11-13 shows the possible carry-propagation paths when the full-adder circuit is implemented in Manchester-carry style. This picture demonstrates how the bypass speeds up the addition: The carry propagates either through the bypass path, or a carry is generated somewhere in the chain. In both cases, the delay is smaller than the normal ripple configuration. The area overhead incurred by adding the bypass path is small and typically ranges between 10 and 20%. However, adding the bypass path breaks the regular bit-slice structure (as was present in Figure 11-10).

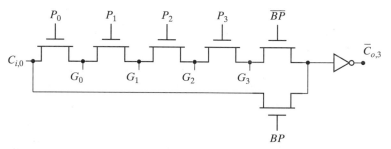

Figure 11-13   Manchester carry-chain implementation of bypass adder.

Let us now compute the delay of an N-bit adder. At first, we assume that the total adder is divided in (N/M) equal-length bypass stages, each of which contains M bits. An approximating expression for the total propagation time can be derived from Figure 11-14a and is given in Eq. (11.9). Namely,

$$t_p = t_{setup} + Mt_{carry} + \left(\frac{N}{M} - 1\right)t_{bypass} + (M-1)t_{carry} + t_{sum} \qquad (11.9)$$

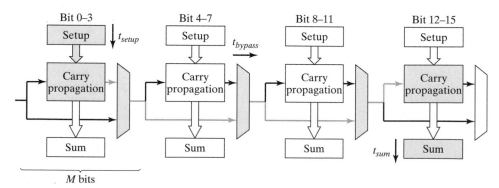

Figure 11-14   ($N = 16$) carry-bypass adder: composition. The worst case delay path is shaded in gray.

with the composing parameters defined as follows:

- t_{setup}: the fixed overhead time to create the generate and propagate signals.
- t_{carry}: the propagation delay through a single bit. The worst case carry-propagation delay through a single stage of M bits is approximately M times larger.
- t_{bypass}: the propagation delay through the bypass multiplexer of a single stage.
- t_{sum}: the time to generate the sum of the final stage.

The critical path is shaded in gray on the block diagram of Figure 11-14. From Eq. (11.9), it follows that t_p is still linear in the number of bits N, since in the worst case, the carry is generated at the first bit position, ripples through the first block, skips around ($N/M - 2$) bypass stages, and is consumed at the last bit position without generating an output carry. The optimal number of bits per skip block is determined by technological parameters such as the extra delay of the bypass-selecting multiplexer, the buffering requirements in the carry chain, and the ratio of the delay through the ripple and the bypass paths.

Although still linear, the slope of the delay function increases in a more gradual fashion than for the ripple-carry adder, as pictured in Figure 11-15. This difference is substantial for larger adders. Notice that the ripple adder is actually faster for small values of N, for which the overhead of the extra bypass multiplexer makes the bypass structure not interesting. The cross-over point depends upon technology considerations and is normally situated between four and eight bits.

Problem 11.1   Delay of Carry-Skip Adder

Determine an input pattern that triggers the worst case delay in a 16-bit (4×4) carry-bypass adder. Assuming that $t_{carry} = t_{setup} = t_{skip} = t_{sum} = 1$, determine the delay and compare it with that of a normal ripple adder.

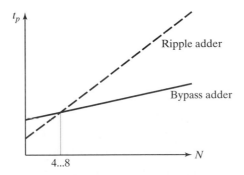

Figure 11-15   Propagation delay of ripple-carry versus carry-bypass adder.

The Linear Carry-Select Adder

In a ripple-carry adder, every full-adder cell has to wait for the incoming carry before an outgoing carry can be generated. One way to get around this linear dependency is to anticipate both possible values of the carry input and evaluate the result for both possibilities in advance. Once the real value of the incoming carry is known, the correct result is easily selected with a simple multiplexer stage. An implementation of this idea, appropriately called the *carry-select adder* [Bedrij62], is demonstrated in Figure 11-16. Consider the block of adders, which is adding bits k to $k + 3$. Instead of waiting on the arrival of the output carry of bit $k - 1$, both the 0 and 1 possibilities are analyzed. From a circuit point of view, this means that two carry paths are implemented. When $C_{o,k-1}$ finally settles, either the result of the 0 or the 1 path is selected by the multiplexer, which can be performed with a minimal delay. As is evident from Figure 11-16, the hardware overhead of the carry-select adder is restricted to an additional carry path and a multiplexer, and equals about 30% with respect to a ripple-carry structure.

A full carry-select adder is now constructed by chaining a number of equal-length adder stages, as in the carry-bypass approach (see Figure 11-17). The critical path is shaded in gray. From inspection of the circuit, we can derive a first-order model of the worst case propagation delay of the module, written as

$$t_{add} = t_{setup} + Mt_{carry} + \left(\frac{N}{M}\right)t_{mux} + t_{sum} \tag{11.10}$$

where t_{setup}, t_{sum}, and t_{mux} are fixed delays and N and M represent the total number of bits, and the number of bits per stage, respectively. t_{carry} is the delay of the carry through a single full-adder cell. The carry delay through a single block is proportional to the length of that stage or equals $M\,t_{carry}$.

The propagation delay of the adder is, again, linearly proportional to N (Eq. (11.10)). The reason for this linear behavior is that the *block-select* signal that selects between the 0 and 1 solutions still has to ripple through all stages in the worst case.

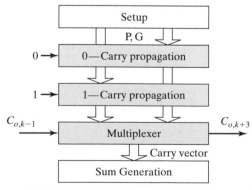

Figure 11-16   Four-bit carry-select module—topology.

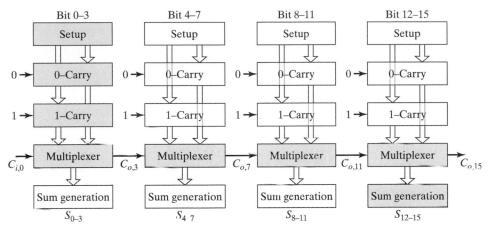

Figure 11-17   Sixteen-bit, linear carry-select adder. The critical path is shaded in gray.

Problem 11.2   Linear Carry-Select Delay

Determine the delay of a 16-bit linear carry-select adder by using unit delays for all cells. Compare the result with that of Problem 11.1. Compare various block configurations as well.

The Square-Root Carry-Select Adder

The next structure illustrates how an alert designer can make a major impact. To optimize a design, it is essential to locate the critical timing path first. Consider the case of a 16-bit linear carry-select adder. To simplify the discussion, assume that the full-adder and multiplexer cells have identical propagation delays equal to a normalized value of 1. The worst case arrival times of the signals at the different network nodes with respect to the time the input is applied are marked and annotated on Figure 11-18a. This analysis demonstrates that the critical path of the adder ripples through the multiplexer networks of the subsequent stages.

One striking opportunity is readily apparent. Consider the multiplexer gate in the last adder stage. The inputs to this multiplexer are the two carry chains of the block and the block-multiplexer signal from the previous stage. A major mismatch between the arrival times of the signals can be observed. The results of the carry chains are stable long before the multiplexer signal arrives. It makes sense to equalize the delay through both paths. This can be achieved by progressively adding more bits to the subsequent stages in the adder, requiring more time for the generation of the carry signals. For example, the first stage can add 2 bits, the second contains 3, the third has 4, and so forth, as demonstrated in Figure 11-18b. The annotated arrival times show that this adder topology is faster than the linear organization, even though an extra stage is needed. In fact, the same propagation delay is also valid for a 20-bit adder. Observe that the discrepancy in arrival times at the multiplexer nodes has been eliminated.

In effect, the simple trick of making the adder stages progressively longer results in an adder structure with sublinear delay characteristics. This is illustrated by the following analysis:

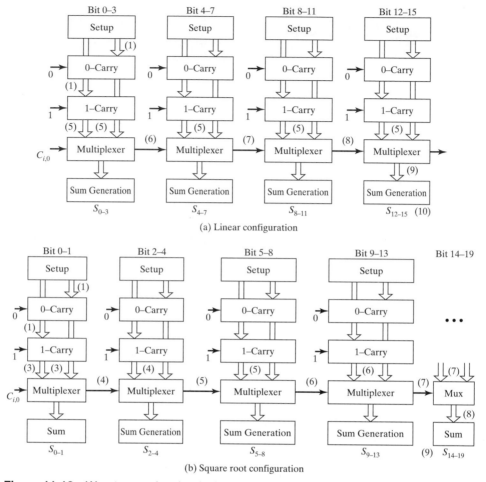

Figure 11-18 Worst case signal arrival times in carry-select adders. The signal arrival times are marked in parentheses.

Assume that an N-bit adder contains P stages, and the first stage adds M bits. An additional bit is added to each subsequent stage. The following relation then holds:

$$N = M + (M + 1) + (M + 2) + (M + 3) + \ldots + (M + P - 1)$$

$$= \dot{M}P + \frac{P(P-1)}{2} = \frac{P^2}{2} + P\left(M - \frac{1}{2}\right) \tag{11.11}$$

If $M \ll N$ (e.g., $M = 2$, and $N = 64$), the first term dominates, and Eq. (11.11) can be simplified to

$$N \approx \frac{P^2}{2} \tag{11.12}$$

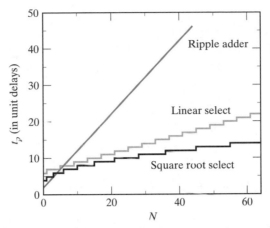

Figure 11-19 Propagation delay of square-root carry-select adder versus linear ripple and select adders. The unit delay model is used to model the cell delays.

or

$$P \approx \sqrt{2N} \tag{11.13}$$

Equation (11.13) can be used to express t_{add} as a function of N by rewriting Eq. (11.10):

$$t_{add} = t_{setup} + Mt_{carry} + (\sqrt{2N})t_{mux} + t_{sum}. \tag{11.14}$$

The delay is proportional to $\sqrt{N}$ for large adders ($N \gg M$), or $t_{add} = O(\sqrt{N})$. This square-root relation has a major impact, which is illustrated in Figure 11-19, where the delays of both the linear and square-root select adders are plotted as a function of N. It can be observed that for large values of N, t_{add} becomes almost a constant.

Problem 11.3   Unequal Bypass Groups in Carry-Bypass Adder

A careful reader might be interested in applying the previous technique to carry-bypass adders. We saw earlier that their delay is a linear function of a number of bits. Can they be modified to achieve better than linear delay by using variable group sizes?

It does make sense to make the consecutive groups gradually larger. However, the technique used in carry-select adders does not directly apply to this case, and a progressive increase in stage sizes eventually increases the delay. Consider a carry-bypass adder in which the last stage is the largest: The carry signal that propagates through that stage and gets consumed at the *msb* position (with no chance of bypassing it) is on the critical path for the sum generation. Increasing the size of the last group does not help the problem.

Based on this discussion and assuming constant delays for carry and bypass gates, sketch the profile of the carry bypass network that achieves a delay that is better than linear.

The Carry-Lookahead Adder*

The Monolithic Lookahead Adder   When designing even faster adders, it is essential to get around the rippling effect of the carry that is still present in one form or another in both the carry-bypass and carry-select adders. The *carry-lookahead* principle offers a possible way to do so [Weinberger56, MacSorley61]. As stated before, the following relation holds for each bit position in an N-bit adder:

$$C_{o,k} = f(A_k, B_k, C_{o,k-1}) = G_k + P_k C_{o,k-1} \tag{11.15}$$

The dependency between $C_{o,k}$ and $C_{o,k-1}$ can be eliminated by expanding $C_{o,k-1}$:

$$C_{o,k} = G_k + P_k(G_{k-1} + P_{k-1} C_{o,k-2}) \tag{11.16}$$

In a fully expanded form,

$$C_{o,k} = G_k + P_k(G_{k-1} + P_{k-1}(\ldots + P_1(G_0 + P_0 C_{i,0}))) \tag{11.17}$$

with $C_{i,0}$ typically equal to 0.

This expanded relationship can be used to implement an N-bit adder. For every bit, the carry and sum outputs are independent of the previous bits. The ripple effect has thus been effectively eliminated, and the addition time should be independent of the number of bits. A block diagram of the overall composition of a carry-lookahead adder is shown in Figure 11-20.

Such a high-level model contains some hidden dependencies. When we study the detailed schematics of the adder, it becomes obvious that the constant addition time is wishful thinking and that the real delay is at least increasing linearly with the number of bits. This is illustrated in Figure 11-21, where a possible circuit implementation of Eq. (11.17) is shown for $N = 4$. Note that the circuit exploits the self-duality and the recursivity of the carry-lookahead equation to build a mirror structure, similar in style to the single-bit full adder of Figure 11-6.[4] The large fan-in of the circuit makes it prohibitively slow for larger values of N. Implementing it with simpler gates requires multiple logic levels. In both cases, the propagation delay increases. Further-

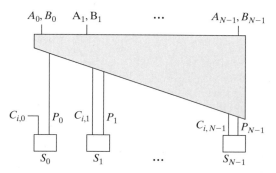

Figure 11-20   Conceptual diagram of a carry-lookahead adder.

---

[4]Similar to the mirror adder, this circuit requires that the *Propagate* signal be defined as $P = A + B$.

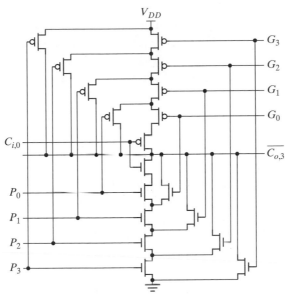

Figure 11-21 Schematic diagram of mirror implementation of four-bit lookahead adder (from [Weste85]).

more, the fan-out on some of the signals tends to grow excessively, slowing down the adder even more. For instance, the signals G_0 and P_0 appear in the expression for every one of the subsequent bits. Hence, the capacitance on these lines is substantial. Finally, the area of the implementation grows progressively with N. Therefore, the lookahead structure suggested by Eq. (11.16) is only useful for small values of N (≤ 4).

The Logarithmic Lookahead Adder—Concept For a carry-lookahead group of N bits, the transistor implementation has $N + 1$ parallel branches with up to $N + 1$ transistors in the stack. Since wide gates and large stacks display poor performance, the carry-lookahead computation has to be limited to up to two or four bits in practice. In order to build very fast adders, it is necessary to organize carry propagation and generation into recursive trees. A more effective implementation is obtained by hierarchically decomposing the carry propagation into subgroups of N bits:

$$
\begin{aligned}
C_{o,0} &= G_0 + P_0 C_{i,0} \\
C_{o,1} &= G_1 + P_1 G_0 + P_1 P_0 C_{i,0} = (G_1 + P_1 G_0) + (P_1 P_0) C_{i,0} = G_{1:0} + P_{1:0} C_{i,0} \\
C_{o,2} &= G_2 + P_2 G_1 + P_2 P_1 G_0 + P_2 P_1 P_0 C_{i,0} = G_2 + P_2 C_{o,1} \\
C_{o,3} &= G_3 + P_3 G_2 + P_3 P_2 G_1 + P_3 P_2 P_1 G_0 + P_3 P_2 P_1 P_0 C_{i,0} \\
&= (G_3 + P_3 G_2) + (P_3 P_2) C_{o,1} = G_{3:2} + P_{3:2} C_{o,1}
\end{aligned}
\tag{11.18}
$$

In Eq. (11.18), the carry-propagation process is decomposed into subgroups of two bits. $G_{i:j}$ and $P_{i:j}$ denote the generate and propagate functions, respectively, for *a group of bits* (from bit positions i to j). Therefore, we call them *block generate* and *propagate* signals. $G_{i:j}$ equals 1 if the group generates a carry, independent of the incoming carry. The block propagate $P_{i:j}$ is true if an incoming carry propagates through the complete group. This condition is equivalent to the carry bypass, discussed earlier. For example, $G_{3:2}$ is equal to 1 when a carry either is generated at bit position 3 or is generated at position 2 and propagated through position 3, or $G_{3:2} = G_3 + P_3 G_2$. $P_{3:2}$ is true when an incoming carry propagates through both bit positions, or $P_{3:2} = P_3 P_2$.

Note that the format of the new expression for the carry is equivalent to the original one, except that the generate and propagate signals are replaced with block generate and propagate signals. The notation $G_{i:j}$ and $P_{i:j}$ generalizes the original carry equations, since $G_i = G_{i:i}$ and $P_i = P_{i:i}$. Another generalization is possible by treating the generate and propagate functions as a pair $(G_{i:j}, P_{i:j})$, rather than considering them as separate functions. A new Boolean operator, called the *dot* operator ($\cdot$), can be introduced. This operator on the pairs and allows for the combination and manipulation of blocks of bits:

$$(G, P) \cdot (G', P') = (G + PG', PP') \qquad (11.19)$$

Using this operator we can now decompose $(G_{3:2}, P_{3:2}) = (G_3, P_3) \cdot (G_2, P_2)$. The dot operator obeys the associative property, but it is not commutative.

Example 11.4   Ripple-Carry Adder Expressed by Using the Dot Operator

With the dot operator, a four-bit ripple carry adder can be re-written as

$$(C_{o,3}, 0) = [(G_3, P_3) \cdot (G_2, P_2) \cdot (G_1, P_1) \cdot (G_0, P_0)] \cdot (C_{i,0}, 0)$$

The associative property allows us to rewrite this function and express $C_{o,3}$ as a function of 2 group carries:

$$(G_{3:0}, P_{3:0}) = [(G_3, P_3) \cdot (G_2, P_2)] \cdot [(G_1, P_1) \cdot (G_0, P_0)]$$
$$= (G_{3:2}, P_{3:2}) \cdot (G_{1:0}, P_{1:0})$$

By exploiting the associative property of the dot operator, a tree can be constructed that effectively computes the carries at all $2^i - 1$ positions (that is, 1, 3, 7, 15, etc.) for $i = 1...\log_2(N)$. The crucial advantage is that the computation of the carry at position $2^i - 1$ takes only $\log_2(N)$ steps. In other words, the output carry of an N-bit adder can be computed in $\log_2(N)$ time. This is a major improvement over the previously described adders. For example, for an adder of 64 bits, the propagation delay of a linear adder is proportional to 64. For a square-root select adder, it is reduced to 8, while, for a logarithmic adder, the proportionality constant is 6. This is illustrated in Figure 11-22, which shows the block diagram of a 16-bit logarithmic adder. The carry at position 15 is computed by combining the results of blocks (0:7) and (8:15). Each of these, in turn, is

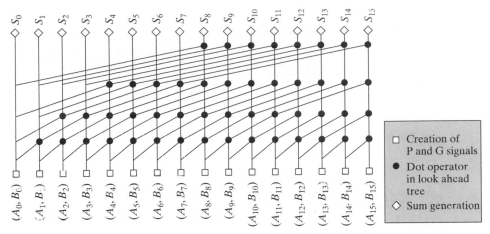

Figure 11-22: Schematic diagram for Kogge–Stone 16-bit lookahead logarithmic adder.

composed hierarchically. For instance, (0:7) is the composition of (0:3) and (4:7), while (0:3) consists of (0:1) and (2:3), etc.

Computing the carries at just the $2^i - 1$ positions is obviously not sufficient. It is necessary to derive the carry signals at the intermediate positions as well. One way to accomplish this is by replicating the tree at every bit position, as illustrated in Figure 11-22 for $N = 16$. For instance, the carry at position 6 is computed by combining the results of blocks (6:3) and (2:0). This complete structure, which frequently is referred to as a *Kogge–Stone tree* [Kogge73], is a member of the *radix-2* class of trees. Radix-2 means that the tree is binary: It combines two carry words at a time at each level of hierarchy. The total adder requires 49 complex logic gates each to implement the dot operator. In addition, 16 logic modules are needed for the generation of the propagate and generate signals at the first level (P_i and G_i), as well as 16 sum-generation gates.

Design Example—Implementing a Lookahead Adder in Dynamic Logic

The combination of carry-lookahead (CLA) techniques and dynamic logic seems to be ideal when very high performance is the ultimate goal. It is therefore useful to walk through the complete design of a dynamic CLA.

The first module generates the propagate and generate signal, as shown in Figure 11-23. The addition of a separate inverter to drive the keeper represents a small twist. This approach is beneficial in gates that drive a sizable fan-out. By decoupling its driver from the fan-out it allows for a quick disengagement of the keeper after the transition starts. The inverter that is driving successive logic gates, on the other hand, is optimized to drive a fan-out of two (for G outputs) or three (for P outputs) NMOS pull-down networks.

Each of the black dots in Figure 11-22 represents two gates that compute the block-level propagate and generate signals, as shown in Figure 11-24. Since these gates are not located at the beginning of the pipeline, the evaluation transistor (also called the *foot switch*) is optional, as discussed in Chapter 6. This approach is commonly used in dynamic datapaths. During the precharge phase, all the outputs of the domino gates are guaranteed to be low, turning off any discharge path in the succeeding domino stages. Elimination of the foot switch in any stage other than the first lowers the logical effort of the gates and speeds up

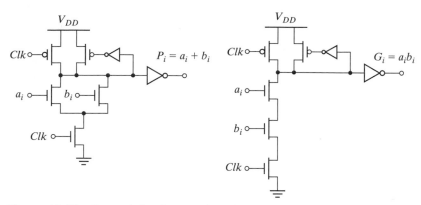

Figure 11-23 Dynamic Implementation of propagate and generate signals.

the evaluation. For example, the propagate gate in Figure 11-24 has a logical effort of two-thirds instead of unity (assuming that the inverter is symmetrical). However, there is a drawback to this method. During the precharge phase, a short-circuit current exists in all gates without a foot switch until their inputs get pre-charged. To avoid the short-circuit current, the clock at each stage Clk_k is delayed from the previous stage. This approach is called *clock-delayed domino* as was introduced in Chapter 10. Note that the clock is delayed by the same amount for all bit-slices per stage, thus simplifying the implementation.

By putting together seven stages of logic in a bit-sliced fashion—P–G generation, followed by six dot operators—a 64-bit adder can be constructed. The only logic stage that is missing to complete the dynamic adder design is the final sum generation. The sum generation requires an XOR function, which is not easily built in domino logic. Static XOR gates could be used, but these produce nonmonotonic transitions and thus cannot be used to drive other domino gates. This might not be a problem per se, since the sum generation typically is the final stage of the addition operation. However, the latch that follows the sum generation cannot be transparent, because this could cause a violation of the transition rule for the succeeding domino gates.

One way of implementing the sum in domino logic is through *sum selection*, in which both care for the sum are computed as $S_i^0 = \overline{a_i \oplus b_i}$ and $S_i^1 = a_i \oplus b_i$. The dynamic gate of Figure 11-25 is then used to select one of these possibilities, based on the incoming carry.

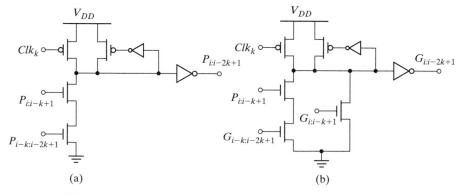

Figure 11-24 Implementation of the dot operator in dynamic logic: (a) propagate and (b) generate logic at stage *k* and bit position *i* (see Figure 11-22).

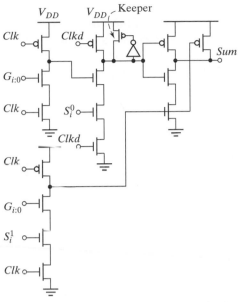

Figure 11-25 Sum select in dynamic logic.

The implementation of the multiplexer gate requires three logic levels, because no complementary carry is available in domino logic. As usual, keepers should be placed at all dynamic nodes, but the one shown in the figure is absolutely critical. The first two dynamic stages (on top) violate the dynamic logic design rules: two dynamic gates are cascaded without introducing an inverter. A glitch might happen at the output of the second gate if both gates are evaluated with the same clock. Delaying the clock of the second gate (*Clkd*) helps to address this issue, although the delayed clock presents a *hard-timing edge*—all the inputs to the second gate must have finished their transition before the clock raises.

It should be emphasized that when a designer chooses to use these design techniques, circuit robustness cannot be compromised. All clocks are subject to random skew, and the delayed clock must have enough margin to absorb the worst case skew. If *Clkd* arrives early, the adder will malfunction. However, proper sizing of the keeper may add a kind of safety net to the design. It can suppress the glitch at the output of the second stage, in case the clock arrives early. The keeper must be sized large enough to minimize this glitch, but not so large that it would compromise performance.

Another option is to implement the adder in differential domino logic, where both signal polarities are available, and the inversion problem is avoided. However, the overall design would be quite power hungry.   ■

Problem 11.4   Static Adder Tree Design

Design a 16-bit carry-lookahead adder tree in static complementary CMOS. Design the lookahead tree by using inverted logic gates, avoiding the addition of inverters. Highlight the critical path of this adder. What are the logical efforts of gates along the critical path? How would you size it for minimal delay?

Logarithmic Lookahead Adder—Alternatives Designers of fast adders sometimes revert to other styles of tree structures as they trade off for area, power, or performance. We briefly discuss the Brent–Kung adder and the radix-4 adder, two of the more common alternative structures.

The Kogge–Stone tree of Figure 11-22 has some interesting properties. First, its interconnect structure is regular, which makes implementation quite easy. Furthermore, the fan-out throughout the tree is fairly constant, especially on the critical paths. The task of sizing the transistors for optimal performance is therefore simplified. At the same time, however, the replication of the carry trees to generate the intermediate carries comes at a large cost in terms of both area and power. Designers sometimes trade off some delay for area and power by choosing less complex trees. A simpler tree structure computes only the carries to the powers-of-two bit positions [Brent82], as illustrated in Figure 11-26 for $N = 16$.

The forward binary tree realizes the carry signals only at positions $2^N - 1$:

$$(C_{o,0}, 0) = (G_0, P_0) \cdot (C_{i,0}, 0)$$

$$(C_{o,1}, 0) = [(G_1, P_1) \cdot (G_0, P_0)] \cdot (C_{i,0}, 0) = (G_{1:0}, P_{1:0}) \cdot (C_{i,0}, 0)$$

$$(C_{o,3}, 0) = [(G_{3:2}, P_{3:2}) \cdot (G_{1:0}, P_{1:0})] \cdot (C_{i,0}, 0) = (G_{3:0}, P_{3:0}) \cdot (C_{i,0}, 0) \qquad (11.20)$$

$$(C_{o,7}, 0) = [(G_{7:4}, P_{7:4}) \cdot (G_{3:0}, P_{3:0})] \cdot (C_{i,0}, 0) = (G_{7:0}, P_{7:0}) \cdot (C_{i,0}, 0)$$

$$\cdots$$

The forward binary-tree structure is not sufficient to generate the complete set of carry bits. An *inverse binary tree* is needed to realize the other carry bits (shown in gray lines in Figure 11-26). This structure combines intermediate results to produce the remaining carry bits. It is left for the reader to verify that this structure produces the correct expressions for all carry

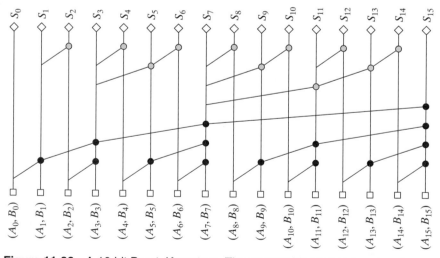

Figure 11-26 A 16-bit Brent–Kung tree. The reverse binary tree is colored gray.

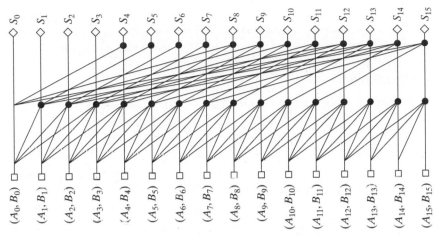

Figure 11-27 Radix-4 Kogge–Stone tree for 16-bit operands.

bits. The resulting structure, commonly called the *Brent–Kung adder*, uses 27 dot gates, or almost half of the 49 needed for a full radix-2 tree, and it needs fewer wires as well. The wiring structure is less regular, however, and fan-out varies from gate to gate, making performance optimization more difficult. Especially the fan-out of the middle node $(C_{0,7})$, which equals one sum and five dot operations for this example, is of major concern. This observation makes the Brent–Kung adder rather unsuited for very large adders (> 32 bits).

An option to reduce the depth of the tree is to combine four signals at a time at each level of the hierarchy. The resulting tree is now of class *radix-4*, because it uses building blocks of order 4 as shown in Figure 11-27. A 16-bit addition needs only two stages of carry logic. Be aware that each gate is more complex and that having less logic stages may not always result in faster operation (as we saw in Chapter 5).

Problem 11.5   Radix-4 Dot Operator in Dynamic Logic

Design the radix-4 dot operators in dynamic logic. Use the radix-2 circuits as reference. How is the sum select implemented for radix-4? Using the method of logical effort, compare the delays of radix-2 and radix-4 designs for 16-bit adders. Which one is faster?

On average, a lookahead adder is several times larger than a ripple adder, but has dramatic speed advantages for larger operands. The logarithmic behavior makes it preferable over bypass or select adders for larger values of N. The exact value of the cross point depends heavily on technology and circuit design factors.

The discussion of adders is by no means complete. Due to its impact on the performance of computational structures, the design of fast adder circuits has been the subject of many publications. It is even possible to construct adder structures with a propagation delay that is *independent of the number of bits*. Examples of those are the carry-save structures and the redundant binary arithmetic structures [Swartzlander90]. These adders require number-encoding and - decoding steps, whose delay is a function of N. Therefore, they are only interesting when embedded in larger structures such as multipliers or high-speed signal processors.

11.4 The Multiplier

Multiplications are expensive and slow operations. The performance of many computational problems often is dominated by the speed at which a multiplication operation can be executed. This observation has, for instance, prompted the integration of complete multiplication units in state-of-the-art digital signal processors and microprocessors.

Multipliers are, in effect, complex adder arrays. Therefore, the majority of the topics discussed in the preceding section are of value in this context as well. The analysis of the multiplier gives us some further insight into how to optimize the performance (or the area) of complex circuit topologies. After a short discussion of the multiply operation, we discuss the basic array multiplier. We also discuss different approaches to partial product generation, accumulation and their final summation.

11.4.1 The Multiplier: Definitions

Consider two *unsigned* binary numbers X and Y that are M and N bits wide, respectively. To introduce the multiplication operation, it is useful to express X and Y in the binary representation

$$ X = \sum_{i=0}^{M-1} X_i 2^i \qquad Y = \sum_{j=0}^{N-1} Y_j 2^j \qquad (11.21) $$

with $X_i, Y_j \in \{0, 1\}$. The multiplication operation is then defined as follows:

$$ Z = X \times Y = \sum_{k=0}^{M+N-1} Z_k 2^k $$
$$ = \left(\sum_{i=0}^{M-1} X_i 2^i \right) \left(\sum_{j=0}^{N-1} Y_j 2^j \right) = \sum_{i=0}^{M-1} \left(\sum_{j=0}^{N-1} X_i Y_j 2^{i+j} \right) \qquad (11.22) $$

The simplest way to perform a multiplication is to use a single two-input adder. For inputs that are M and N bits wide, the multiplication takes M cycles, using an N-bit adder. This *shift-and-add* algorithm for multiplication adds together M *partial products*. Each partial product is generated by multiplying the multiplicand with a bit of the multiplier—which, essentially, is an AND operation—and by shifting the result on the basis of the multiplier bit's position.

A faster way to implement multiplication is to resort to an approach similar to manually computing a multiplication. All the partial products are generated at the same time and organized in an array. A multioperand addition is applied to compute the final product. The approach is illustrated in Figure 11-28. This set of operations can be mapped directly into hardware. The

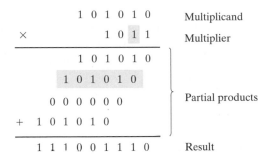

Figure 11-28   Binary multiplication—an example.

resulting structure is called an *array multiplier* and combines the following three functions: *partial-product generation, partial-product accumulation,* and *final addition.*

Problem 11.6   Multiplication by a Constant

It is often necessary to implement multiplication of an operand by a constant. Describe how you would proceed with this task.

11.4.2   Partial-Product Generation

Partial products result from the logical AND of multiplicand X with a multiplier bit Y_i (see Figure 11-29). Each row in the partial-product array is either a copy of the multiplicand or a row of zeroes. Careful optimization of the partial-product generation can lead to some substantial delay and area reductions. Note that in most cases the partial-product array has many zero rows that have no impact on the result and thus represent a waste of effort when added. In the case of a multiplier consisting of all ones, all the partial products exist, while in the case of all zeros, there is none. This observation allows us to reduce the number of generated partial products by half.

Assume, for example, an eight-bit multiplier of the form 01111110, which produces six nonzero partial-product rows. One can substantially reduce the number of nonzero rows by recoding this number ($2^7 + 2^6 + 2^5 + 2^4 + 2^3 + 2^2$) into a different format. The reader can verify that the form $1000000\overline{1}0$, with $\overline{1}$ a shorthand notation for -1, represents the same number. Using

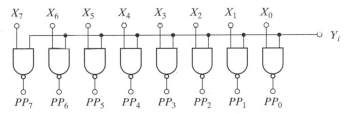

Figure 11-29   Partial-product generation logic.

this format, we have to add only two partial products, but the final adder has to be able to perform subtraction as well. This type of transformation is called *Booth's recoding* [Booth51], and it reduces the number of partial products to, at most, one half. It ensures that for every two consecutive bits, at most one bit will be 1 or −1. Reducing the number of partial products is equivalent to reducing the number of additions, which leads to a speedup as well as an area reduction. Formally, this transformation is equivalent to formatting the multiplier word into a base-4 scheme, instead of the usual binary format:

$$Y = \sum_{j=0}^{(N-1)/2} Y_j 4^j \text{ with } (Y_j \in \{-2, -1, 0, 1, 2\}) \tag{11.23}$$

Note that 1010...10 represents the worst case multiplier input because it generates the most partial products (one half). While the multiplication with {0, 1} is equivalent to an AND operation, multiplying with {−2, −1, 0, 1, 2} requires a combination of inversion and shift logic. The encoding can be performed on the fly and requires some simple logic gates.

Having a variable-size partial-product array is not practical for multiplier design, and a *modified Booth's recoding* is most often used instead [MacSorley61]. The multiplier is partitioned into three-bit groups that overlap by one bit. Each group of three is recoded, as shown in Table 11-2, and forms one partial product. The resulting number of partial products equals half of the multiplier width. The input bits to the recoding process are the two current bits, combined with the upper bit from the next group, moving from *msb* to *lsb*.

Table 11-2   Modified Booth's recoding.

Partial Product Selection Table	
Multiplier Bits	**Recoded Bits**
000	0
001	+ Multiplicand
010	+ Multiplicand
011	+ 2 × Multiplicand
100	− 2 × Multiplicand
101	− Multiplicand
110	− Multiplicand
111	0

In simple terms, the modified Booth's recoding essentially examines the multiplier for strings of ones from *msb* to *lsb* and replaces them with a leading 1, and a -1 at the end of the string. For example, 011 is understood as the beginning of a string of ones and is therefore replaced by a leading 1 (or 100), while 110 is seen as the end of a string and is replaced by a -1 at the least significant position (or $0\bar{1}0$).

Example 11.5 Modified Booth's Recoding

Consider the eight-bit binary number 01111110 shown earlier. This can be divided into four overlapping groups of three bits, going from *msb* to *lsb*: 00 (1), 11 (1), 11 (1), 10 (0). Recoding by using Table 11-2 yields: 10 (2 ×), 00 (0 ×), 00 (0 ×), $\bar{1}0$ (−2 ×), or, in combined format, $1000000\bar{1}0$. This is equivalent to the result we obtained before.

Problem 11.7 Booth's Recoder

Design the combinational logic that implements a modified Booth's recoding for a parallel multiplier, using Table 11-2. Compare its implementation in complementary and pass-transistor CMOS.

11.4.3 Partial-Product Accumulation

After the partial products are generated, they must be summed. This accumulation is essentially a multioperand addition. A straightforward way to accumulate partial products is by using a number of adders that will form an array—hence, the name, *array multiplier*. A more sophisticated procedure performs the addition in a tree format.

The Array Multiplier

The composition of an array multiplier is shown in Figure 11-30. There is a one-to-one topological correspondence between this hardware structure and the manual multiplication shown in Figure 11-28. The generation of N partial products requires $N \times M$ two-bit AND gates (in the style of Figure 11-29).[5] Most of the area of the multiplier is devoted to the adding of the N partial products, which requires $N - 1$ M-bit adders. The shifting of the partial products for their proper alignment is performed by simple routing and does not require any logic. The overall structure can easily be compacted into a rectangle, resulting in a very efficient layout.

Due to the array organization, determining the propagation delay of this circuit is not straightforward. Consider the implementation of Figure 11-30. The partial sum adders are implemented as ripple-carry structures. Performance optimization requires that the critical timing path be identified first. This turns out to be nontrivial. In fact, a large number of paths of almost identical length can be identified. Two of those are highlighted in Figure 11-31. A closer

---

[5]This particular implementation does not employ Booth's recoding. Adding recoding does not substantially change the implementation. The number of adders is reduced by half, and the partial-product generation is slightly more complex.

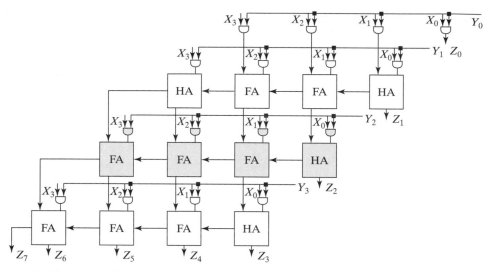

Figure 11-30 A 4 × 4 bit-array multiplier for unsigned numbers—composition. HA stands for a half adder, or an adder cell with only two inputs. The hardware for the generation and addition of one partial product is shaded in gray.

look at those critical paths yields an approximate expression for the propagation delay (derived here for critical path 2). We write this as

$$t_{mult} \approx [(M-1)+(N-2)]t_{carry} + (N-1)t_{sum} + t_{and} \qquad (11.24)$$

where t_{carry} is the propagation delay between input and output carry, t_{sum} is the delay between the input carry and sum bit of the full adder, and t_{and} is the delay of the AND gate.

Since all critical paths have the same length, speeding up just one of them—for instance, by replacing one adder by a faster one such as a carry-select adder—does not make much sense from a design standpoint. All critical paths have to be attacked at the same time. From Eq. (11.24), it can be deduced that the minimization of t_{mult} requires the minimization of both t_{carry}

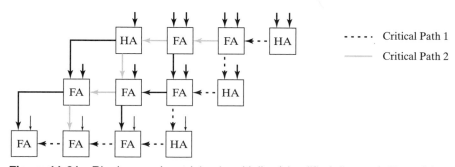

Figure 11-31 Ripple-carry based 4 × 4 multiplier (simplified diagram). Two of the possible critical paths are highlighted.

and t_{sum}. In this case, it could be beneficial for t_{carry} to equal t_{sum}. This contrasts with the requirements for adder cells discussed before, where a minimal t_{carry} was of prime importance. An example of a full-adder circuit with comparable t_{sum} and t_{carry} delays was shown in Figure 11-7.

Problem 11.8   Signed-Binary Multiplier

The multiplier presented in Figure 11-30 only handles unsigned numbers. Adjust the structure so that two's-complement numbers are also accepted.

Carry-Save Multiplier

Due to the large number of almost identical critical paths, increasing the performance of the structure of Figure 11-31 through transistor sizing yields marginal benefits. A more efficient realization can be obtained by noticing that the multiplication result does not change when the output carry bits are passed diagonally downwards instead of only to the right, as shown in Figure 11-32. We include an extra adder called a *vector-merging* adder to generate the final result. The resulting multiplier is called a *carry-save multiplier* [Wallace64], because the carry bits are not immediately added, but rather are "saved" for the next adder stage. In the final stage, carries and sums are merged in a fast carry-propagate (e.g., carry-lookahead) adder stage. While this structure has a slightly increased area cost (one extra adder), it has the advantage that its worst case critical path is shorter and uniquely defined, as highlighted in Figure 11-32 and is expressed as

$$t_{mult} = t_{and} + (N-1)t_{carry} + t_{merge} \tag{11.25}$$

still assuming that $t_{add} = t_{carry}$.

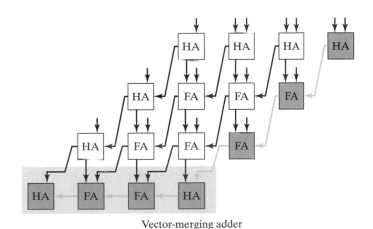

Vector-merging adder

Figure 11-32   A 4 × 4 carry-save multiplier. The critical path is highlighted in gray.

Example 11.6   Carry-Save Multiplier

When mapping the carry-save multiplier of Figure 11-32 onto silicon, one has to take into account some other topological considerations. To ease the integration of the multiplier into the rest of the chip, it is advisable to make the outline of the module approximately rectangular. A floor plan for the carry-save multiplier that achieves this goal is shown in Figure 11-33. Observe the regularity of the topology. This makes the generation of the structure amenable to automation.

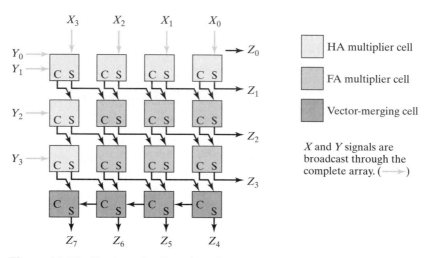

Figure 11-33   Rectangular floorplan of carry-save multiplier. Different cells are differentiated by shades of gray. X and Y signals are AND'ed before being added. The leftmost column of cells is redundant and can be eliminated.

The Tree Multiplier

The partial-sum adders can also be rearranged in a treelike fashion, reducing both the critical path and the number of adder cells needed. Consider the simple example of four partial products each of which is four bits wide, as shown in Figure 11-34a. The number of full adders needed for this operation can be reduced by observing that only column 3 in the array has to add four bits. All other columns are somewhat less complex. This is illustrated in Figure 11-34b, where the original matrix of partial products is reorganized into a tree shape to visually illustrate its varying depth. The challenge is to realize the complete matrix with a minimum depth and a minimum number of adder elements. The first type of operator that can be used to cover the array is a full adder, which takes three inputs and produces two outputs: the sum, located in the same column and the carry, located in the next one. For this reason, the FA is called a *3-2 compressor.* It is denoted by a circle covering three bits. The other operator is the half-adder, which takes two input bits in a column and produces two outputs. The HA is denoted by a circle covering two bits.

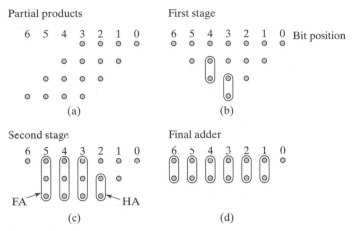

Figure 11-34 Transforming a partial-product tree (a) into a Wallace tree (b,c,d), using an iterative covering process. The example shown is for a four-bit operand.

To arrive at the minimal implementation, we iteratively cover the tree with FAs and HAs, starting from its densest part. In a first step, we introduce HAs in columns 4 and 3 (Figure 11-34b). The reduced tree is shown in Figure 11-34c. A second round of reductions creates a tree of depth 2 (Figure 11-34d). Only three FAs and three HAs are used for the reduction process, compared with six FAs and six HAs in the carry-save multiplier of Figure 11-32! The final stage consists of a simple two-input adder, for which any type of adder can be used (as discussed in the next section, "Final Addition").

The presented structure is called the *Wallace tree multiplier* [Wallace64], and its implementation is shown in Figure 11-35. The tree multiplier realizes substantial hardware savings for larger multipliers. The propagation delay is reduced as well. In fact, it can be shown that the propagation delay through the tree is equal to $O(\log_{3/2}(N))$. While substantially faster than the carry-save structure for large multiplier word lengths, the Wallace multiplier has the disadvantage of being very irregular, which complicates the task of coming up with an efficient layout. This irregularity is visible even in the four-bit implementation of Figure 11-35.

There are numerous other ways to accumulate the partial-product tree. A number of compression circuits has been proposed in the literature, a detailed discussion of which is beyond the scope of this book. They are all based on the concept that when full adders are used as 3:2 compressors, the number of partial products is reduced by two-thirds per multiplier stage. One can even go a step further and devise a 4–2 (or higher order) compressor, such as in [Weinberger81, Santoro89]. In fact, many of today's high-performance multipliers do just that.

11.4.4 Final Addition

The final step for completing the multiplication is to combine the result in the final adder. Performance of this "vector-merging" operation is of key importance. The choice of the adder style depends on the structure of the accumulation array. A carry-lookahead adder is the preferable

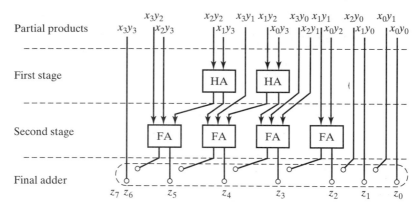

Figure 11-35 Wallace tree for four-bit multiplier.

option if all input bits to the adder arrive at the same time, as it yields the smallest possible delay. This is the case if a pipeline stage is placed right before the final addition. Pipelining is a technique frequently used in high-performance multipliers. In nonpipelined multipliers, the arrival-time profile of the inputs to the final adder is quite uneven due to the varying logic depths of the multiplier tree. Under these circumstances, other adder topologies, such as carry select, often yield performance numbers similar to lookahead at a substantially reduced hardware cost [Oklobdzija01].

11.4.5 Multiplier Summary

All the presented techniques can be combined to yield multipliers with extremely high performance. For instance, a 54×54 multiplier can achieve a propagation delay of only 4.4 ns in a 0.25-μm CMOS technology by combining Booth encoding with a Wallace tree by using 4–2 compression in pass-transistor implementation and by using a mixed carry-select, carry-lookahead topology for the final adder [Ohkubo95]. More information on these multipliers (and others) can be found in the references [e.g., Swartzlander90, Oklobdzija01].

11.5 The Shifter

The shift operation is another essential arithmetic operation that requires adequate hardware support. It is used extensively in floating-point units, scalers, and multiplications by constant numbers. The latter can be implemented as a combination of add and shift operations. Shifting a data word left or right over a constant amount is a trivial hardware operation and is implemented by the appropriate signal wiring. A programmable shifter, on the other hand, is more complex and requires active circuitry. In essence, such a shifter is nothing less than an intricate multiplexer circuit. A simple one-bit left–right shifter is shown in Figure 11-36. Depending on the control signals, the input word is either shifted left or right, or else it remains unchanged. Multi-bit shifters can be built by cascading a number of these units. This approach rapidly becomes complex, unwieldy, and ultimately too slow for larger shift values. Therefore, a more structured

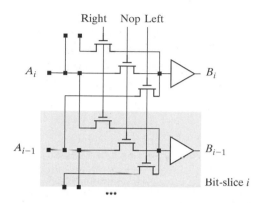

Figure 11-36 One–bit (left–right) programmable shifter. The data passes undisturbed under the nop condition.

approach is advisable. Next, we discuss two commonly used shift structures, the *barrel shifter* and the *logarithmic shifter*.

11.5.1 Barrel Shifter

The structure of a barrel shifter is shown in Figure 11-37. It consists of an array of transistors, in which the number of rows equals the word length of the data, and the number of columns equals the maximum shift width. In this case, both are set equal to four. The control wires are routed diagonally through the array. A major advantage of this shifter is that the signal has to pass through at most one transmission gate. In other words, the propagation delay is theoretically constant and independent of the shift value or shifter size. This is not true in reality, however, because the capacitance at the input of the buffers rises linearly with the maximum shift width.

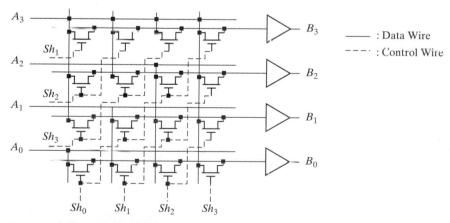

Figure 11-37 Barrel shifter with a programmable shift width from zero to three bits to the right. The structure supports automatic repetition of the sign bit (A_3), also called *sign-bit extension*.

An important property of this circuit is that the layout size is not dominated by the active transistors as in the case of all other arithmetic circuits, but by the number of wires running through the cell. More specifically, the size of the cell is bounded by the pitch of the metal wires!

Another important consideration when selecting a shifter is the format in which the shift value must be presented. From the schematic diagram of Figure 11-37, we see that the barrel shifter needs a control wire for every shift bit. For example, a four-bit shifter needs four control signals. To shift over three bits, the signals $Sh_3:Sh_0$ take on the value 1000. Only one of the signals is high. In a processor, the required shift value normally comes in an encoded binary format, which is substantially more compact. For instance, the encoded control word needs only two control signals and is represented as 11 for a shift over three bits. To translate the latter representation into the former (with only one bit high), an extra module called a *decoder* is required. (Decoders are treated in detail in Chapter 12.)

Problem 11.9   Two's Complement Shifter

Explain why the shifter shown in Figure 11-37 implements a two's complement shift.

11.5.2 Logarithmic Shifter

While the barrel shifter implements the whole shifter as a single array of pass transistors, the logarithmic shifter uses a staged approach. The total shift value is decomposed into shifts over powers of two. A shifter with a maximum shift width of M consists of a $\log_2 M$ stages, where the ith stage either shifts over 2^i or passes the data unchanged. An example of a shifter with a maximum shift value of seven bits is shown in Figure 11-38. For instance, to shift over five bits, the first stage is set to shift mode, the second to pass mode, and the last stage again to shift. Notice that the control word for this shifter is already encoded, and no separate decoder is required.

The speed of the logarithmic shifter depends on the shift width in a logarithmic way, since an M-bit shifter requires $\log_2 M$ stages. Furthermore, the series connection of pass transistors slows the shifter down for larger shift values. A careful introduction of intermediate buffers is therefore necessary, as discussed in Chapter 6.

In general, we conclude that a barrel shifter is appropriate for smaller shifters. For larger shift values, the logarithmic shifter becomes more effective, in terms of both area and speed. Furthermore, the logarithmic shifter is easily parameterized, allowing for automatic generation. The most important concept of this section is that the exploitation of regularity in an arithmetic operator can lead to dense and high-speed circuit implementations.

11.6 Other Arithmetic Operators

In the previous sections, we only discussed a subset of the large number of arithmetic circuits required in the design of microprocessors and signal processors. Besides adders, multipliers, and shifters, others operators such as comparators, dividers, counters, and goniometric operators (sine, cosine, tangent) are often needed. A full analysis of these circuits is beyond the scope of

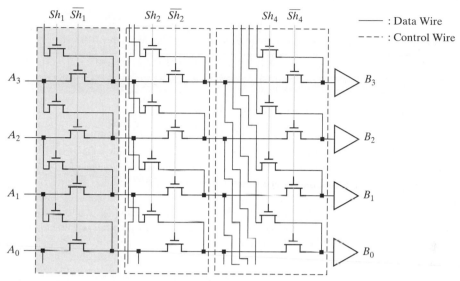

Figure 11-38   Logarithmic shifter with maximal shift width of seven bits to the right.
(Only the four least significant bits are shown.)

this book. We refer the interested reader to some of the excellent references available on the
topic (e.g., [Swartzlander90], [Oklobdzija01].

The reader should be aware that most of the design ideas introduced in this chapter apply
to these other operators as well. For instance, comparators can be devised with a linear, square
root, and logarithmic dependence on the number of bits. In fact, some operators are simple
derivatives of the adder or multiplier structures presented earlier. For example, Figure 11-39
shows how a two's complement subtractor can be realized by combining a two's complement
adder with an extra inversion stage, or how a subtractor can be used to implement $A \geq B$.

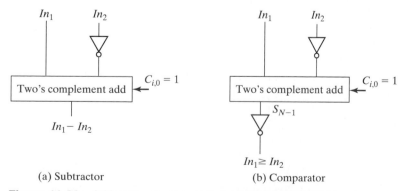

(a) Subtractor   (b) Comparator

Figure 11-39   Arithmetic structures derived from a full adder.

Problem 11.10 Comparator

Derive a logic diagram for a comparator that implements the following logic functions: ≥, =, and ≤.

Case Study—Design of an Arithmetic-Logic Unit (ALU)

The core of any microprocessor or microcontroller is the arithmetic-logic unit (or ALU). The ALU combines addition and subtraction with other operations, such as shifting and bitwise logic operations (AND, OR, and XOR).

An ALU taken from a 64-bit high-end microprocessor is shown in Figure 11-40 [Mathew01.] It consists of two levels of wide multiplexers, a 64-bit adder, a logic unit, an operator-merging multiplexer and the write-back bus driver. The first pair of 9:1 multiplexers selects the from nine different input sources, three of which are from register files and caches. The other six are bypass paths that come directly from the six ALUs—the microprocessor in question can issue six integer instructions in one cycle. One of the bypass loops actually feeds the output of the ALU back to its own input, creating a single cycle loop that defines the critical path. The second 5:1 multiplexer on the *A* input performs a partial shifting of the operand, while the 2:1 multiplexer on the *B* input simply inverts the operand to implement subtraction. The adder executes two's complement addition or subtraction on the operands *A* and *B*. Two's complement subtraction is performed by inverting all the bits of the operand *B* (using the 2:1 MUX) and by setting the carry-in to one. The sum-selection block not only implements the sum selection, but also merges the outputs of arithmetic and logic units. Finally, a strong buffer drives the loop-back bus, which presents a large load to the ALU. For example, for a six-issue Intel/HP Itanium processor [Fetzer02], the bus has to connect to all six units, translating into more than 2 mm of wire length in a 0.18-μm technology. When added to the load from the register file, it presents over 0.5 pF of capacitive load.

An ALU represents a typical example of a bit-sliced design. Each slice in each block is pitch matched, which minimizes the vertical routing. A floor plan of the 64-bit ALU is shown in Figure 11-41. Intercell routing is done horizontally in metal-3, except for the long horizontal wires between adder stages 1 and 2 and 2 and 3, which are laid out in a combination of metal-3 and metal-4. For example, if the adder is implemented as a radix-4 CLA, the longest wire after the second PG stage crosses 48 cells.

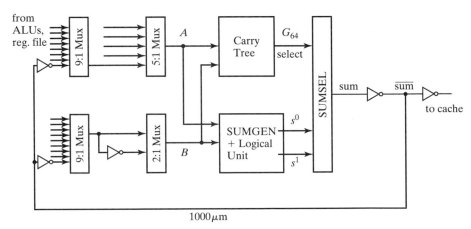

Figure 11-40   A 64-bit arithmetic-logic unit.

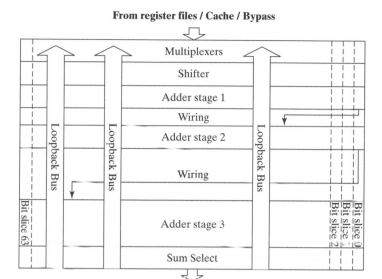

Figure 11-41 Floor plan of a 64-bit ALU.

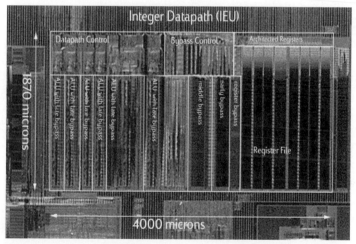

Figure 11-42 Microphotograph of the Itanium processor, showing bit-sliced design of six ALUs together with bypasses, register file, and control logic. (Courtesy of Intel.)

The loop-back wiring is routed in the top-level metal (e.g., metal-5). In case of a multiple-issue microprocessor, these busses span across all ALUs, as can be seen in Figure 11-42. The number of loop-back busses frequently determines the bit-pitch in multiissue ALUs. In our example, nine 64-bit-wide busses have to cross back over the ALU. To avoid needing to add extra wiring space, the bit width is set to nine bus-wire pitches. Since these wires are long; they are designed with double width and at least double spacing to reduce their resistance and capacitance.   ∎

11.7 Power and Speed Trade-Offs in Datapath Structures*

In the preceding discussion on adders, multipliers, and shifters, we mainly explored the trade-off between speed and area and ignored power considerations. In this section, we briefly analyze the third dimension of the design exploration space. Since most of the approaches to minimize power were already introduced in Chapter 6, the discussion that follows serves mostly as an illustration of the concepts advanced there.

Typical digital designs are either *latency* or *throughput* constrained. A latency-constrained design has to finish computation by a given deadline. Interactive communication and gaming are examples of such. Throughput-constrained designs, on the other hand, must maintain a required data throughput. A 1000BaseT Gigabit Ethernet connection has to maintain a constant throughput of 1 Gigabit/s. The architectural optimization techniques that are available to the designer differ based on the design constraints. Pipelining or parallelization, for instance, works effectively for the throughput-constrained scenario, but it may not be applicable to the latency-constrained case. For example, a throughput of 1 Gb/s over copper wires is achieved in Gigabit Ethernet by processing four 250 Mb/s streams of data in parallel.

With a fixed architecture of the datapath, speed, area, and power can be traded off through the choice of the supply voltage(s), transistor thresholds, and device sizes. This opens the door for a large variety of power minimization techniques, which are summarized in Table 11-3. They are classified as follows:

- **Enable Time**—Some design techniques are implemented (or enabled) at *design time*. Transistor widths and lengths, for instance, are fixed at the time of design. Supply voltage and transistor thresholds, on the other hand, can be either assigned statically during the design phase, or changed dynamically at *run time*. Other techniques primarily address the time that a function or module in a digital design is in idle mode (or standby). It is only logical to require that the power dissipation of a module in *sleep mode* should be absolutely minimal.

Table 11-3   Power minimization techniques.

| | Constant Throughput/Latency | | Variable Throughput/Latency |
	Design Time	Sleep Mode	Run Time
Active	Lower V_{DD}, Multi-V_{DD}, Transistor Sizing Logic Optimizations	Clock gating	Dynamic voltage scaling
Leakage	Multi-V_{Th} + Active Techniques	Sleep transistors, Variable V_{Th}	Variable V_{Th} + Active Techniques

- **Targeted Dissipation Source**—Another classification of the power-management techniques concerns the source of power dissipation they address: active (dynamic) power or leakage (static) power. Lowering the supply voltage, for example, is a very attractive technique: It not only reduces the energy consumed per transition in a quadratic way—albeit at the expense of performance—but also reduces the leakage current. On the other hand, increasing the threshold voltage mainly impacts the leakage component.

The sleep mode of operation deserves some special attention. If a digital block still receives a clock while in idle mode, its clock distribution network, together with the attached flip-flops, continues to consume energy, even while no computation is performed. Recall that, typically, one third of total energy of a digital system is consumed in clock distribution network. A common method to reduce power in idle mode is the *clock gating* technique introduced in Chapter 10. In this approach, the main clock connection to a module is turned off (or *gated*) whenever the block is idle. However, clock gating does not reduce the leakage power of the idle block. More complicated schemes to lower the standby power have to be used, such as increasing the transistor thresholds or switching off the power rails.

In the following sections, we discuss each of these design-time and run-time techniques in detail.

11.7.1   Design-Time Power-Reduction Techniques

Reducing the Supply Voltage

A reduction in supply voltage results in quadratic power savings and thus is the most attractive approach. On the negative side, the delay of CMOS gates increases inversely with supply voltage. At the datapath level, this loss of performance can be compensated for by other means, such as logical and architectural optimizations. For example, a ripple-carry adder can be replaced by a faster structure, such as a lookahead adder. The latter implementation is larger and more complex, which translates into a larger physical and switching capacitance. This is more than offset by the fact that the faster adder can run at a lower supply voltage for the same performance.

The trade-off between ripple and lookahead operates at the logical level. Similarly, and even more effectively, architectural optimizations can be employed to compensate for the effect of a reduced V_{DD}, as illustrated in Example 11.7.

Example 11.7   Minimizing the Power Consumption by Using Parallelism

To illustrate how architectural techniques can be used to compensate for reduced speed, a simple eight-bit datapath consisting of an adder and a comparator is analyzed, assuming a 2–µm CMOS technology [Chandrakasan92]. As shown in Figure 11-43, inputs A and B are added, and the result is compared with input C. Assume that the worst case delay through the adder, comparator, and latch is approximately 25 ns at a supply voltage of 5 V. At best, the system can be clocked with a clock period of $T = 25$ ns. When required to run at this maximum possible throughput, it is clear that the operating

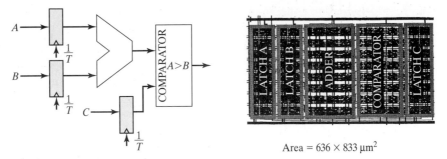

Figure 11-43   A simple datapath with corresponding layout.

voltage cannot be reduced any further because no extra delay can be tolerated. We use this topology as the reference datapath for our study and present power improvement numbers with respect to this reference. The average power consumed by the reference datapath is given by

$$P_{ref} = C_{ref} \; V_{ref}^2 \; f_{ref} \tag{11.26}$$

where C_{ref} is the total effective capacitance being switched per clock cycle. The effective capacitance can be determined by averaging the energy over a sequence of random input patterns with a uniform distribution.

One way to maintain throughput while reducing the supply voltage is to utilize a parallel architecture. As shown in Figure 11-44, two identical adder-comparator datapaths connect in parallel, allowing each unit to work at half the original rate while maintaining the original throughput. Since the speed requirements for the adder, comparator, and latch have decreased from 25 ns to 50 ns, the voltage can be dropped from 5 V to 2.9 V—the voltage that doubles the delay. While the datapath capacitance has increased by a factor of two, the operating frequency has correspondingly decreased by a factor of two. Unfortunately, there also is a slight increase in the total effective capacitance due to the extra routing and data multiplexing. This results in an increased capacitance by a factor of 2.15. The power for the parallel datapath is thus given by

$$P_{par} = C_{par}V^2{}_{par}f_{par} = (2.15C_{ref})(0.58V_{ref})^2\left(\frac{f_{ref}}{2}\right) \approx 0.36P_{ref} \tag{11.27}$$

The approach presented *trades off area for power*, as the resulting area is approximately 3.4 times larger than the original design. This technique is only applicable when the design is not area constrained. Furthermore, parallelism introduces extra routing overhead, which might cause additional dissipation. Careful optimization is needed to minimize this overhead.

Parallelism is not the only way to compensate for the loss in performance. Other architectural approaches, such as the use of pipelining, can accomplish the same goal. The most impor-

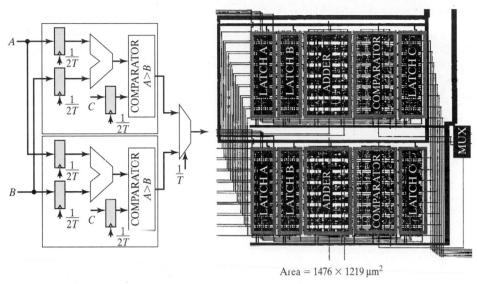

Figure 11-44   Parallel implementation of the simple datapath.

tant message in the preceding analysis is that if power dissipation is the prime concern, dropping the supply voltage is the most effective means to achieve that goal. The subsequent loss in performance can be compensated for, if necessary, by an increase in area. Within certain bounds, this is acceptable, since area is no longer the compelling issue it used to be due to the dramatic increase in integration levels of the last decade.

Problem 11.11   Reducing the Supply Voltage, Using Pipelining

A pipeline stage is introduced between the adder and the comparator of the reference datapath of Figure 11-44. You may assume that this roughly halves the propagation delay of the logic, while it increases the capacitance by 15%. Obviously, an extra pipeline register is needed on input C as well. Determine how much power can be saved by this approach, given that the throughput has to remain constant compared with the reference datapath. Comment on the area overhead.

Using Multiple Supply Voltages

Reducing the supply voltage is fairly straightforward, but it may not be optimal. Reduced supply evenly lowers the power dissipation of all logic gates, while evenly increasing their delay. A better approach is to selectively decrease the supply voltage on some of the gates:

- those which correspond to fast paths and finish the computation early, and
- those with gates that drive large capacitances, have the largest benefit for the same delay increment.

This approach, however requires the use of more than one supply voltage. Multiple supply voltages are already employed frequently in today's ICs. A separate supply voltage is provided for the I/O for compatibility reasons, where the I/O ring is designed with transistors with thicker gate oxides to sustain higher voltages. The logic core is powered from lower voltage supplies and uses transistors with thinner oxides. This approach can be extended to lower the power dissipation of a circuit. For instance, every module could select the most appropriate voltage from two (or more) supply options. Even more extreme, multiple voltages can be assigned on a gate-by-gate basis.

Module-Level Voltage Selection Consider, for instance, the digital system shown in Figure 11-45, which consists of a datapath block with a critical path of 10 ns and a control block with a much shorter critical path of 4 ns, operating from the same supply voltage of 2.5 V. Also, assume that the datapath block has a fixed latency and throughput and that no architectural transformation can be applied to lower its supply. Since the control block finishes early (in other words, it has *timing slack*), its supply voltage can be lowered. A reduction to $V_{DDL} = 1$ V increases its critical path delay to 10 ns, and lowers its power dissipation by more than five times.[6] We effectively exploit the discrepancy in the critical-path length of the various modules (called the *slack*) to selectively lower the power consumption of the modules with the larger slack.

When combining multiple supply voltages on a die, *level converters* are required whenever a module at the lower supply has to drive a gate at the higher voltage. If a gate supplied by V_{DDL} drives a gate at V_{DDH}, the PMOS transistor never turns off, resulting in static current and reduced output swing as illustrates in Figure 11-46. A level conversion performed at the boundaries of supply voltage domains prevents these problems. An asynchronous level converter, based on the DVSL template (Chapter 6) and similar to the low-swing signalling gate of Figure 9-32, is shown in Figure 11-46. The cross-coupled PMOS transistors perform the level conversion by using positive feedback. The delay of this level converter is quite sensitive to transistor-sizing and supply-voltage issues. The NMOS transistors operate with a reduced overdrive, $V_{DDL}-V_{Th}$, compared with the PMOS devices. They have to be made large to be able to overpower the positive feedback. For a low value of V_{DDL}, the delay can become very long. Due to

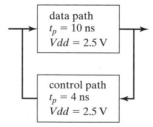

Figure 11-45   Design with diverging critical-path lengths.

---

[6]This uses the simplifying assumption that the propagation delay is inversely proportional to the supply voltage.

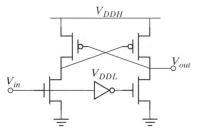

Figure 11-46   Level converter.

the reduced overdrive, the circuit is also very sensitive to variations in the supply voltage. Note that level converters are not needed for a step-down change in voltage.

The overhead of the level conversion can be somewhat mitigated by observing that most conversions are performed at a register boundary. For instance, the control inputs to the datapath block of Figure 11-45 are commonly sampled in a register. A practical way to perform the level conversion is, then, to embed it inside the register. A level-converting register is shown in Figure 11-47. It is a conventional transmission gate implementation of a master–slave latch pair, where a cross-coupled PMOS pair is embedded in the slave latch to perform level conversion.

Multiple Supplies Inside a Block   The same approach can be applied at much smaller granularity by individually setting the supply voltage for each cell inside a block [Usami95, Hamada01]. Examining the histogram of the critical-path delays for a typical digital block reveals that only a few paths are critical or near critical and that many paths have much shorter delays. The shorter paths essentially waste energy, as there is no reward for finishing early. For each of these paths, the supply voltage could be lowered to the optimum level. Minimum energy consumption would be achieved if all paths become critical. However, this is not easily achievable, as many logic gates are shared between different paths, and it is impractical to generate and

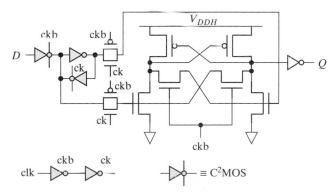

Figure 11-47   Level-converting register. Shaded gates are supplied from V_{DDL} [Usami95].

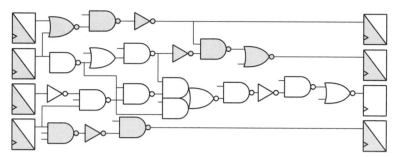

Figure 11-48 Dual-supply design using clustered voltage scaling. Each path starts with a high supply and switches to a low supply (gray logic gates) if there is a delay slack. Level conversion is performed in the registers.

distribute a wide range of supply voltages. A more practical implementation, employing only two supplies, is shown in Figure 11-48. Using a *clustered voltage-scaling* technique, each path starts with the high supply voltage and switches to the low supply when delay slack is available. Level conversion is performed in the registers at the end of the paths, using circuits such as the one introduced in Figure 11-47. Note that the level conversion is necessary only if the logic section that follows cannot run entirely off the low supply.

The impact of this approach is illustrated in Figure 11-49, which plots the path-length distribution of a typical logic block for single and dual supplies. When a single supply is used, a large fraction of the paths is substantially shorter than critical. Adding the second supply shifts the delay distribution closer to the target delay, making many more paths critical. If the design has very few critical paths and most of paths finish early, energy savings would be large. On the other hand, if most of the paths inside a block are critical, introducing the second supply does not yield any energy savings. You may wonder if adding a third or even a fourth supply voltage may yield even greater benefits to equalize the larger delay spreads. Unfortunately, a number of studies have shown that for typical path-delay distributions, adding more supplies only yields marginal savings [Hamada01, Augsburger02].

Similarly, one may ask whether this dual-supply technique yields a larger benefit, compared with a uniform reduction the supply. In using clustered voltage scaling, the dual-supply approach is **more effective when large switched capacitances are concentrated toward the**

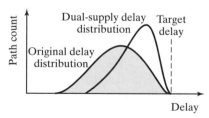

Figure 11-49 Typical path-delay distribution before and after applying the second supply.

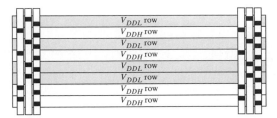

Figure 11-50   Layout of a dual-supply logic block where all
cells at different supplies are placed in different rows.

end of the block, such as in buffer chains [Stojanovic02]. This observation is in accordance with
the concept of low-swing busses, introduced in Chapter 9. A bus typically presents the largest
capacitance, and it is more advantageous to lower the supply voltage on its driver than that on
any other logic gate. Since it has the largest capacitance, the payoff in power savings is the largest for the same increase in delay.

Distributing multiple supply voltages on a die complicates the design of the power-distribution network and tends to tax the traditional place-and-route tools. As we saw many times
before, a structured approach can help to minimize the impact. One simple option is to place
cells with different supplies into different rows of a standard cell layout, as shown in
Figure 11-50. The second supply can be brought in from outside the chip, or it can be generated on the die, using an internal DC–DC converter. Step-down switching DC–DC converters
have a conversion efficiency of well over 90% and yield only a minimal overhead. Still the
V_{DDL} distribution network has to meet all necessary design requirements, such as decoupling to
minimize the voltage variations and immunity to electromigration.

Using Multiple Device Thresholds

The use of devices with multiple thresholds offers another way of trading off speed for power.
Most sub-0.25–μm CMOS technologies offer two types of n- and p-type transistors, with thresholds differing by about 100 mV. This higher threshold device features a leakage current that is
about one order of magnitude lower than that of the lower threshold transistor, at the expense of
a ~30% reduction in active current. Therefore, the low-threshold devices are preferably used in
timing-critical paths, while the high thresholds are used anywhere else. The assignment can be
done on a per-cell basis, rather than per individual transistor [Wei99, Kato00]. Note that the use
of multiple thresholds does not require level converters or any other special circuits, as shown
Figure 11-51. Clustering of the logic is not required either; as many gates as possible should be
converted to high threshold until the timing slack is completely consumed. A careful assignment
of the thresholds can reduce the leakage power by as much as 80%–90%.

While the multiple-threshold approach primarily addresses the leakage current, it yields a
small reduction in active power as an additional benefit. This is primarily due to a reduced gate-
channel capacitance in the off state and a small reduction in signal swing on the internal nodes of
a gate ($V_{DD}-V_{ThH}$). Such a reduction is partially offset by increased source and drain junction
sidewall capacitances. The overall active power reduction turns out to be only about 4%.

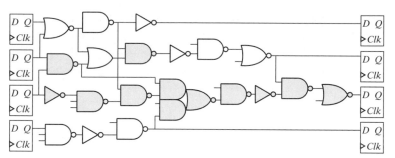

Figure 11-51   Use of dual-threshold gates. Gray-shaded gates use low thresholds and are employed only in critical paths.

Reducing Switching Capacitance through Transistor Sizing

The input capacitance of a complementary CMOS gate is directly proportional to its size and, therefore, also to its speed. In Chapter 6, we examined the optimal sizing of gates and found that each gate in a logic path should be sized to have an effective fan-out of approximately 4 to achieve the minimum delay for that path. An interesting question is how to size a circuit for minimum energy when the allowable delay is longer than minimal?

We know that for an inverter chain with a given load and number of stages, the minimum delay is achieved when the size of each inverter in the chain is the geometric mean of its neighbors:

$$s_i^2 = s_{i+1}s_{i-1} \tag{11.28}$$

When the minimum delay that is achieved by this sizing is below the desired delay, an optimization problem can be formulated that minimizes the switching capacitance under delay constraints. One option is to reduce the number of stages and increase the tapering factor, as suggested in Chapter 9. Even better, an analytical solution exists for this problem [Ma94], which establishes that the optimal approach is to adjust the tapering factor at each stage according to the following equation:

$$s_i^2 = \frac{s_{i+1}s_{i-1}}{1 + \mu s_{i-1}} \tag{11.29}$$

The parameter μ is a nonnegative number that depends on the amount of timing slack available (with respect to the chain sized for the minimum delay). This solution is intuitively clear—the last stages in the inverter chain are the largest; hence, they are the prime candidates for downsizing. Since downsizing any of the inverters in the chain by a given percentage causes the same delay increment. Hence, we rather do it at the stage where we get the largest impact, which is the largest one.

The same principle applies to a logic path. When it is optimized for minimum delay, the delay of each stage is the same (while the intrinsic delays of gates may differ). Therefore, the delay of the largest energy consuming gates should be increased first. This idea can be extrapolated to the energy–delay optimization of a general combinational logic block. However, applying it in practice is not trivial. Downsizing one path affects the delay of all paths that share logic gates with it, making it difficult to isolate and optimize one particular path. In addition, many paths in a general combinational block are reconvergent.

Example 11.8　Energy-Delay Trade-Offs in Adder Design

Let us examine the energy-delay optimization of the 64-bit Kogge–Stone tree adder. There are many paths through an adder, and not all of these paths are balanced. A crucial question is how to identify the paths for resizing or initial sizing. One option is to select all the paths in the adder equal to the critical path. Since the paths through an adder roughly correspond to different bit slices, we allocate each gate in the adder to a bit slice. There are 64 bit slices, and a total of nine stages of logic. This partition works well for Figure 11-52a shows the resulting energy map for the minimum delay. It can be seen that the adder consumes the largest energy along the longest carry paths. Figure 11-52b shows the energy profile of the same adder, this time resized, allowing for a 10% delay increase. The energy dissipation is reduced by 54%. In contrast, a dual-supply solution saves only 27%, while a single reduced supply yields 22% savings [Stojanovic02]. In summary, **sizing is an effective power-reduction method for datapaths, where the majority of energy is consumed inside the block, rather than in driving the external load**.

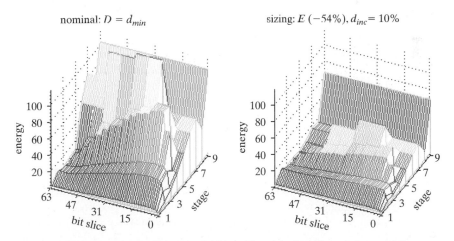

Figure 11-52 Energy profiles of a 64-bit adder: (a) sized for minimum delay, (b) sized for 10% delay increment and minimized energy.

Reducing Switching Capacitance through Logic and Architecture Optimizations

As the effective capacitance is the product of the physical capacitance and the switching activity, minimization of both factors is recommended. A wide variety of logic and architectural optimizations exist that reduce the activity at no expense in performance. Some of those were already introduced in Chapter 6. For the sake of brevity, we only show a small number of representative optimizations. (Refer to [Rabaey96] and [Chandrakasan98] for an in-depth overview.)

Reducing Switching Activity by Resource Allocation   Multiplexing multiple operations on a single hardware unit has a detrimental effect on the power consumption. Besides increasing the physical capacitance, it can also increase the switching activity. This is illustrated with a simple experiment in Figure 11-53, which compares the power consumption of two counters running simultaneously. In the first case, both counters run on separate hardware, while in the second case, they are multiplexed on the same unit. Figure 11-53b plots the number of switching events as a function of the skew between the two counters. The nonmultiplexed case is always superior, except when both counters run in a completely synchronous fashion. The multiplexing tends to randomize the data signals presented to the operational unit, which results in increased switching activity. When power consumption is a concern, it is often beneficial to avoid the excessive reuse of resources. Observe that CMOS hardware units consume only negligible amounts of power when idle. Providing dedicated, specialized operators only presents an extra cost in area, while being generally beneficial in terms of speed and power.

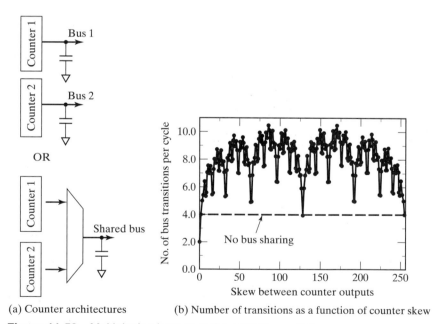

(a) Counter architectures   (b) Number of transitions as a function of counter skew

Figure 11-53   Multiplexing increases the switching activity.

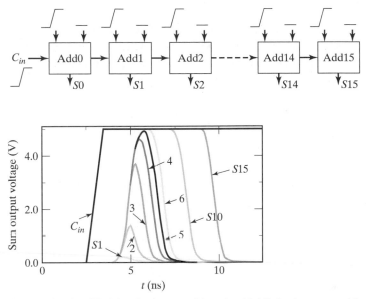

Figure 11-54 Glitching in the sum bits of a 16-bit ripple-carry adder.

Reducing Glitching through Path Balancing Dynamic hazards, or *glitching,* are a major contribution to the dissipation in complex structures such as adders and multipliers. The wide discrepancy in the lengths of the signal paths in some of those structures can be the cause of spurious transients. This is demonstrated in Figure 11-54, which displays the simulated response of a 16-bit ripple adder for all inputs going simultaneously from 0 to 1. A number of the sum bits are shown as a function of time. The sum signals should be zero for all bits. Unfortunately, a 1 appears briefly at all of the outputs, since the carry takes a significant amount of time to propagate from the first bit to the last. Notice how the glitch becomes more pronounced as it travels down the chain.

A dramatic reduction in glitching activity can be obtained by selecting structures with balanced signal paths. The tree lookahead adder structures (such as Kogge–Stone) and the tree multipliers have this property; therefore, they should be more attractive from a power point of view, even in the presence of a larger physical capacitance. An inspection of the lookahead structure of Figure 11-22 reveals that the timing paths to the inputs of dot operators are of a similar length, although some deviations may occur due to differences in loading and fan-out.

11.7.2 Run-Time Power Management

Dynamic Supply Voltage Scaling (DVS)

A static reduction of the supply voltage, as discussed in the preceding paragraphs, lowers the energy per operation and extends the battery life at the expense of performance. This performance penalty is often not acceptable, especially in applications that are latency constrained.

Substantial power reductions are still obtainable, based on the realization that the peak performance is not continuously required. Consider, for instance, a general-purpose processor to be used in portable applications, such as notebooks, electronic organizers, and cellular phones. The computational functions to be executed on such a processor fall into three major categories: compute-intensive tasks, low-speed functions, and idle-mode operation. Compute-intensive and short-latency tasks need the full computational throughput of the processor to achieve real-time constraints. MPEG video and audio decompression are examples of such. Low-throughput and long-latency tasks, such as text processing, data entry, and memory backups, operate under far more relaxed completion deadlines and require only a fraction of the maximum throughput of the microprocessor. There is no reward for finishing the computation early, and if a task is completed early, it can be considered a waste of energy. Finally, portable processors spend a large fraction of their time on idle, waiting for a user action or an external wake-up event. In sum, the computational throughput and latency expected from a mobile processor vary drastically over time.

Even compute-intensive operations, such as MPEG decoding, show variable computational requirements while processing a typical stream of data. For example, the number of times an MPEG decoder computes an inverse discrete cosine transform (IDCT) per video frame varies widely, depending upon the amount of motion in the video scenes. This is illustrated in Figure 11-55, which plots the distribution of the number of IDCTs/frame for a typical video sequence. The processor that is executing this algorithm experiences a different computational workload from frame to frame.

Lowering the clock frequency when executing the reduced workloads reduces the power, but does not save on energy—every operation is still executed at the high voltage level. However, if both supply voltage and frequency are lowered simultaneously, the energy is reduced. In order to maintain the required throughput for high workloads and minimize energy for low workloads, both supply and frequency must be dynamically varied according to the requirements application that is currently being executed. This technique is called *dynamic voltage scaling* (DVS). The concept is illustrated in Figure 11-56 [Burd00, Gutnik97]. It operates under

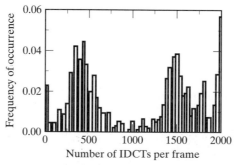

Figure 11-55   Typical IDCT histogram for MPEG decoding.

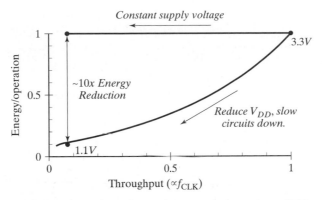

Figure 11-56   Energy/operation versus throughput (1/T) for constant and variable supply voltage operation.

the guideline that a function should always be **operated at the lowest supply voltage that meets the timing constraints**.

The DVS concept is enabled by the observation that the delay of most CMOS circuits and functions track each other well over a range of supply voltages, which is a necessity for system operation under varying supply conditions. Figure 11-57 shows the delays of a number of representative CMOS blocks (such as NAND gates, ring oscillators, register files, and SRAM), over a supply voltage range from 1 V to 4 V [Burd00]. Excellent performance tracking can be observed. Note that some circuit families, such as NMOS-only pass-transistor logic do not follow this behavior over the complete range of supplies.

A practical implementation of a dynamic-voltage scaling system now consists of the following components:

• a processor that can operate under a wide variety of supply voltages,

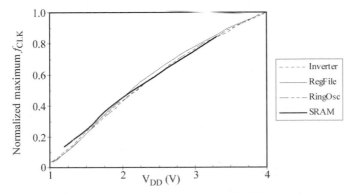

Figure 11-57   Delay of representative CMOS functions as a function of supply voltage for a 0.6-μm CMOS technology.

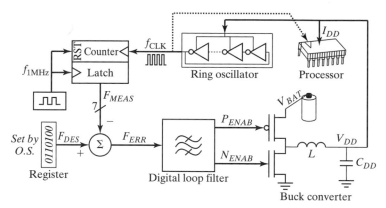

Figure 11-58   Block diagram of a dynamic voltage-scaled system.

- a supply-regulation loop that sets the minimum voltage necessary for operation at a desired frequency, and
- an operating system that calculates the required frequencies to meet requested throughputs and task completion deadlines.

   One possible implementation is shown in Figure 11-58. The core of the DVS system is a a ring oscillator, whose oscillating frequency matches the microprocessor critical path. When included inside the power-supply control loop, this ring oscillator provides the translation between the supply voltage and the clock frequency. The operating system digitally sets the desired frequency (F_{DES}). The current value of ring oscillator frequency is measured and compared with the desired frequency. The difference is used as a feedback error. By adjusting the supply voltage, the supply-voltage loop changes the ring oscillator frequency to set this error value to 0.

   The task of the scheduler (or real-time operating system) is to determine dynamically the optimal frequency (or voltage) as a function of the combined computational requirements of all active tasks in the system. In the more complex case of a general-purpose processor, each task should supply a completion deadline (e.g., video frame rate) or a desired execution frequency. The voltage scheduler then estimates the number of processor cycles necessary for completing each of the tasks and computes the optimal processor frequency [Pering99]. In the case of a single task with varying performance requirements, a queue can be used to determine the computational load and to adjust the voltage accordingly. This is illustrated in Figure 11-59, where the depth of the input queue is used to set the supply voltage (and frequency) of a stream-based signal processor [Gutnik97].

Dynamic Threshold Scaling (DTS)

In analogy to the dynamic variation of the supply voltage, it is attractive to adjust the threshold voltage of the transistors dynamically. For low-latency computation, the threshold should be lowered to its minimal value; for low speed computation, it can be increased; and in the standby

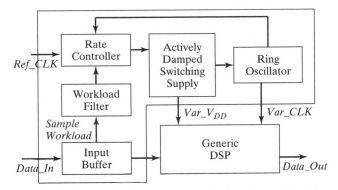

Figure 11-59 Using a queue to determine the workload in a signal processor [Gutnik97].

mode, it should be set to the highest possible value to minimize the leakage current. Substrate bias is the control knob that allows us to vary the threshold voltages dynamically. In order to do so, we have to operate the transistors as four-terminal devices. This is only possible in a triple-well process, as independent control of all four terminals of both n- and p-type devices is required, as shown in Figure 11-60. Substrate biasing can be implemented for a complete chip, on a block-by-block or a cell-by-cell basis. Per cell granularity of substrate biasing, however, has a large layout cost.

Similar to dynamic supply-voltage scaling scheme, the variable threshold voltage scheme is based on a feedback loop, which can be set to accomplish a variety of goals:

- It can lower the leakage in the standby mode
- It can compensate for threshold variations across the chip during normal operation of the circuit
- It can throttle the throughput of the circuit to lower both the active and leakage power based on performance requirements.

Since the current flow into the substrate is much smaller than into the supply lines, DTS has a smaller circuit overhead than DVS. A feedback system, designed to control the leakage in

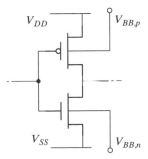

Figure 11-60 An inverter with body terminals used for threshold control.

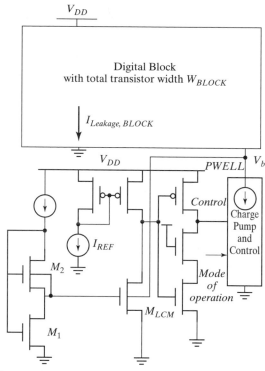

Figure 11-61   Variable threshold control scheme
implemented with a leakage current monitor [Kuroda96].

a digital module, is shown in Figure 11-61 [Kuroda96 and Kuroda01]. It consists of a leakage control monitor and a substrate bias charge pump, added to the digital block of interest.

The leakage current monitor is crucial for implementing this scheme. Transistors M_1 and M_2 are biased to operate in the subthreshold region. When an NMOS transistor is in the subthreshold, its current is given by

$$I_D = I_S / W_0 \cdot W \cdot 10^{V_{GS}/S} \tag{11.30}$$

where S is the subthreshold slope. The output of the bias generator V_b equals

$$V_b = S \cdot \log(W_2 / W_1) \tag{11.31}$$

From the total transistor width of all transistors in the block that is to be monitored, W_{BLOCK}, the total current scaling factor can be found:

$$\frac{I_{LCM}}{I_{BLOCK}} = \frac{W_{LCM}}{W_{BLOCK}} \cdot 10^{\frac{V_b}{S}} = \frac{W_{LCM}}{W_{BLOCK}} \cdot \frac{W_2}{W_1} \tag{11.32}$$

To minimize the power penalty of the monitoring, the leakage monitor should be made as small as possible. However, if it is too small, the overall leakage monitoring response gets slower and the substrate biasing does not track variations closely. The optimal value can be set very low, with the monitor transistor being as small as 0.001% of the total transistor width on the chip.

A conventional charge pump siphons current in and out of the substrate. The control circuit monitors the leakage current. If it is above the preset value—corresponding to the mode of operation, set externally by the user or the operating system—the charge pump increases the negative back bias by pumping current out of the substrate. The charge pump shuts off when the leakage current reaches the target value. Junction leakages and impact ionization in the circuit will eventually raise the substrate bias voltage again, which activates the feedback loop new. Since it does not have to provide large supply currents on a continuous basis, this scheme is simpler to implement than dynamic voltage scaling.

Unfortunately, the effectiveness of adaptive body biasing is decreasing with further technology scaling. This is due to inherently lower body-effect factors and increased junction leakage attributable to band-to-band tunneling.

11.7.3 Reducing the Power in Standby (or Sleep) Mode

The idle mode represents an extreme corner of the dynamic power-management space. As no active switching occurs, all power dissipation is due to leakage—assuming that appropriate clock and input gating is in place. One option to reduce the leakage during standby is the DTS technique presented in the previous section. A simpler power-down scheme utilizes large sleep transistors to switch off the power supply rails when the circuit is in the sleep mode. This straightforward approach significantly reduces the leakage current, but increases the design complexity. It can be implemented by using a power switch on the supply rail only, or, even better, on both supply and ground rails, as shown Figure 11-62.

In normal operation mode, the *SLEEP* signal is high, and the sleep transistors must present as small a resistance as possible. The finite resistance of these transistors results in noise on supply rails, attributable to changes in supply current drawn by the logic. The sizing and the selection of the thresholds of the sleep transistors is subject to a trade-off process. To minimize fluctuations in the supply voltage, the sleep transistors should have a very low on-resistance, and therefore be very wide. However, increasing their size brings with it a major layout penalty. When transistors with a higher threshold are available, a better leakage suppression can be achieved. However, high-threshold devices must be even larger to yield the same resistance as low-threshold devices. The principles of leakage reduction are a direct extension of principles introduced in Section 6.4.2. Adding the sleep transistor effectively increases the transistor stack height, resulting in leakage reductions of the order of tens (for low threshold switches) to a thousand (for high threshold switches) times.

As opposed to simple clock gating, switching off the power supplies erases the state of the registers inside the block. In some applications, this is acceptable, such as when a completely

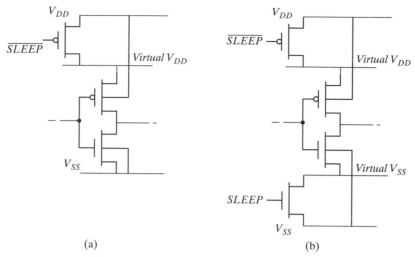

Figure 11-62 Sleep transistors used (a) only on the supply rail, (b) on both supply and ground.

new task is executed after the block wakes up again. However, additional effort is required when the state needs to be preserved. One option is to connect all the registers to the nongated supply rails, V_{DD} and V_{SS}, as discussed in Chapter 7 and shown in Figure 7-18. An alternative is to use the operating system to save the state of all registers to a nonvolatile memory.

11.8 Perspective: Design as a Trade-off

The analysis of the adder and multiplier circuits makes it clear again that digital circuit design is a trade-off between area, speed, and power requirements. This is demonstrated in Figure 11-63, which plots the normalized area and speed for some of the adders discussed earlier as a function of the number of bits.[7] The overall project goals and constraints determine which factor is dominant.

The die area has a strong impact on the cost of an integrated circuit. A larger chip size means that fewer parts fit on a single wafer as discussed in Chapter 1. Reducing the area can help the viability of a product. Ultimate performance is what makes the newest microprocessor sell, and the lowest possible power consumption is a great marketing argument for a cellular phone. Understanding the market of a product is therefore essential when deciding on how to play the trade-off game. One should be aware that all design constraints—speed, power, and area—contribute to the feasibility or market success of a design.

In this context, it is worth summarizing some of the important design concepts that have been introduced in the course of this chapter:

---

[7]Be aware that these results are for a particular implementation in a particular technology. Extrapolation to other technologies should be done with care.

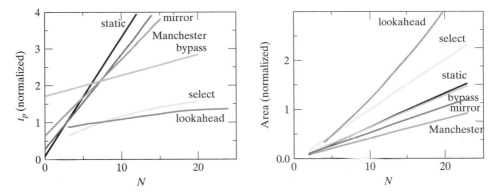

Figure 11-63 Area and propagation delay of various adder structures as a function of the number of bits *N*. Based on results from [Vermassen86].

1. The most important rule is to select the *right structure* before starting an elaborate circuit optimization. Going for the optimal performance of a complex structure by rigorously optimizing transistor sizes and topologies probably will not give you the best result. Optimizations at higher levels of abstraction, such as the logic or architectural level, can often generate more dramatic results. Simple first-order calculations can help give a global picture on the pros and cons of a proposed structure.

2. Determine the *critical timing path* through the circuit, and focus most of your optimization efforts on that part of the circuit. In addition to hand analysis, computer-aided design tools are available to help determine the critical paths and size the transistors appropriately. Be aware that some noncritical paths can be downsized to reduce power consumption.

3. Circuit size is not only determined by the number and size of transistors, but also by other factors such as *wiring and the number of vias and contacts*. These factors are becoming even more important with shrinking dimensions or when extreme performance is a goal.

4. Although an obscure optimization can sometimes help to get a better result, be wary if this results in an irregular and convoluted topology. *Regularity and modularity* are a designer's best friend.

5. Power and speed can be traded off through a choice of circuit sizing, supply voltages, and transistor thresholds.

11.9 Summary

In this chapter, we have studied the implementation of arithmetic datapath operators from a performance, area, and power perspective. Special attention was devoted to the development of *first-order performance models* that allow for a fast analysis and comparison of various logic structures before diving into the tedious transistor-level optimizations.

- A datapath is best implemented in a *bit-sliced* fashion. A single layout slice is used repetitively for every bit in the data word. This regular approach eases the design effort and results in fast and dense layouts.
- A *ripple-carry* adder has a performance that is linearly proportional to the number of bits. Circuit optimizations concentrate on reducing the delay of the carry path. A number of circuit topologies were examined, showing how careful optimization of the circuit topology and the transistor sizes helps to reduce the capacitance on the carry bit.
- Other adder structures use logic optimizations to increase the performance. The performance of *carry-bypass, carry-select and carry-lookahead* adders depends on the number of bits in square root and logarithmic fashion, respectively. This increase in performance comes at a cost in area, however.
- A *multiplier* is nothing more than a collection of cascaded adders. Its critical path is far more complex, and performance optimizations proceed along vastly different routes. The carry–save technique relies on a logic manipulation to turn the adder array into a regular structure with a well-defined critical timing path that can easily be optimized. Booth recoding and partial product accumulation in a tree reduces the complexity and delay of larger multipliers.
- The performance and area of a programmable shifter are dominated by the *wiring*. The exploitation of regularity can help to minimize the impact of the interconnect wires. This is exemplified in the barrel and the logarithmic shifter structures.
- *Power consumption* can be reduced substantially by the proper choice of circuit, logical, or architectural structure. This might come at the expense of area, but area might not be that critical in the age of submicron devices.
- A wide range of design-time and run-time techniques are at the disposition of a designer to minimize the power consumption. At *design time*, power and delay can be traded off through the choice of supply voltages and thresholds in addition to transistor sizing and logic optimization. The use of parallelism and pipelining can help to *reduce the supply voltage*, while maintaining the same throughput. The *effective capacitance* can be reduced by avoiding waste, as introduced by excessive multiplexing, for example.
- Some applications operate under variable throughput or latency conditions. Using *variable supplies and transistor thresholds* can lower the active or leakage power in such systems. Minimization of the standby energy consumption is essential for portable battery-operated devices.

11.10 To Probe Further

The literature on arithmetic and computer elements is vast. Important sources for newer developments are the *Proceedings of the IEEE Symposium on Computer Arithmetic*, the *IEEE Transactions on Computers* and the *IEEE Journal of Solid-State Circuits* (for integrated circuit implementation). An excellent collection of the most significant papers in the area can be found in some IEEE Press reprint volumes [Swartzlander90]. A number of other references, such as [Omondi94], [Koren98], and [Oklobdzija01], are provided for further reading.

References

[Augsburger02] S. Augsburger, "Using Dual-Supply, Dual-Threshold and Transistor Sizing to Reduce Power in Digital Integrated Circuits," M.S. Project Report, University of California, Berkeley, April 2002.

[Bedrij62] O. Bedrij, "Carry Select Adder," *IRE Trans. on Electronic Computers*, vol. EC-11, pp. 340–346, 1962.

[Booth51] A. Booth, "A Signed Binary Multiplication Technique," *Quart. J. Mech. Appl. Math.*, vol. 4., part 2, 1951.

[Brent82] R. Brent and H.T. Kung, "A Regular Layout for Parallel Adders," *IEEE Trans. on Computers,* vol. C-31, no. 3, pp. 260–264, March 1982.

[Burd00] T.D. Burd, T.A. Pering, A.J. Stratakos, R.W. Brodersen, "A Dynamic Voltage Scaled Microprocessor System," *IEEE Journal of Solid-State Circuits*, vol. 35, no. 11, November 2000.

[Chandrakasan92] A. Chandrakasan, S. Sheng, and R. Brodersen, "Low Power CMOS Digital Design," *IEEE Journal of Solid State Circuits,* vol. SC-27, no. 4, pp. 1082–1087, April 1992.

[Chandrakasan98] A. Chandrakasan and R. Brodersen, Ed., *Low-Power CMOS Design*, IEEE Press, 1998.

[Fetzer02] E.S. Fetzer, J.T. Orton, "A Fully-Bypassed 6-Issue Integer Datapath and Register File on an Itanium Microprocessor," 2002 *IEEE International Solid-State Circuits Conference, Digest of Technical Papers*, San Francisco, CA, pp. 420–421, February 2002.

[Gutnik97] V. Gutnik, and A. P. Chandrakasan, "Embedded Power Supply for Low-Power DSP," *IEEE Transactions on Very Large Integration (VLSI) Systems*, vol. 5, no. 4, pp. 425–435, December 1997.

[Hamada01] M. Hamada, Y. Ootaguro, and T. Kuroda, "Utilizing Surplus Timing for Power Reduction," *Proceedings of the IEEE 2001 Custom Integrated Circuits Conference*, San Diego, CA, pp. 89–92, May 2001.

[Kato00] N. Kato *et al.*, "Random Modulation: Multi-Threshold-Voltage Design Methodology in Sub-2-V Power Supply CMOS," *IEICE Transactions on Electronics*, vol. E83-C, no. 11, p. 1747–1754, November 2000.

[Katz94] R.H. Katz, *Contemporary Logic Design*, Benjamin Cummings, 1994.

[Kilburn60] T. Kilburn, D.B.G. Edwards and D. Aspinall, "A Parallel Arithmetic Unit Using a Saturated-Transistor Fast-Carry Circuit," *Proceedings of the IEE*, Pt. B, vol. 107, pp. 573–584, November 1960.

[Kogge73] P.M. Kogge and H.S. Stone, "A Parallel Algorithm for the Efficient Solution of a General Class of Recurrence Equations," *IEEE Transactions on Computers*, vol. C-22, pp. 786–793, August 1973.

[Koren98] I. Koren, *Computer Arithmetic Algorithms*, Brookside Court Publishers, 1998.

[Kuroda96] T. Kuroda and T. Sakurai, "Threshold Voltage Control Schemes Through Substrate Bias for Low-Power High-Speed CMOS LSI Design," *Journal on VLSI Signal Processing,* vol. 13, no. 2/3, pp. 191–201, 1996.

[Kuroda01] T. Kuroda, and T. Sakurai, "Low-Voltage Technologies," in *Design of High-Performance Microprocessor Circuits*, Chandrakasan, Bowhil, Fox (eds.), IEEE Press, 2001.

[Lehman62] M. Lehman and N. Burla, "Skip Techniques for High-Speed Carry Propagation in Binary Arithmetic Units," *IRE Transactions on Electronic Computers*, vol. EC-10, pp. 691–698, December 1962.

[Ma94] S. Ma, P. Franzon, "Energy Control and Accurate Delay Estimation in the Design of CMOS Buffers," *IEEE Journal of Solid-State Circuits*, vol. 29, no. 9, pp. 1150–1153, September 1994.

[MacSorley61] O. MacSorley, "High Speed Arithmetic in Binary Computers," *IRE Proceedings*, vol. 49, pp. 67–91, 1961.

[Mathew01] S. K. Mathew *et al.*, "Sub-500-ps 64-b ALUs in 0.18-μm SOI/Bulk CMOS: Design and Scaling Trends," *IEEE Journal of Solid-State Circuits*, vol. 36, no. 11, pp. 1636–1646, November 2001.

[Ohkubo95] N. Ohkubo *et al.*, "A 4.4 ns CMOS 54*54-b Multiplier Using Pass-Transistor Multiplexer," *IEEE Journal of Solid-State Circuits*, vol. 30, no. 3, pp. 251–257, March 1995.

[Oklobdzija01] V. G. Oklobdzija, "High-Speed VLSI Arithmetic Units: Adders and Multipliers," in *Design of High-Performance Microprocessor Circuits*, Chandrakasan, Bowhil, Fox (eds.), IEEE Press, 2001.

[Omondi94] A. Omondi, *Computer Arithmetic Systems*, Prentice Hall, 1994.

[Pering00] T. A. Pering, *Energy-Efficient Operating System Techniques*, Ph.D. Dissertation, University of California, Berkeley, 2000.

[Rabaey96] J. Rabaey, and M. Pedram, Ed., "Low-Power Design Methodologies," Kluwer Academic Publishers, 1996.

[Stojanovic02] V. Stojanovic, D. Markovic, B. Nikolic, M. A. Horowitz, and R. W. Brodersen, "Energy-Delay Tradeoffs in Combinational Logic using Gate Sizing and Supply Voltage Optimization," *European Solid-State Circuits Conference, ESSCIRC'2002*, Florence, Italy, September 24–26, 2002.

[Suzuoki99] M. Suzuoki *et al.*, "A Microprocessor with a 128-bit CPU, Ten Floating-Point MAC's, Four Floating-Point Dividers, and an MPEG-2 Decoder," *IEEE Journal of Solid-State Circuits*, vol. 34, no. 11, pp. 1608–1618, November 1999.

[Swartzlander90] E. Swartzlander, ed., *Computer Arithmetic—Part I and II*, IEEE Computer Society Press, 1990.

[Usami95] K. Usami, and M. Horowitz, "Clustered Voltage Scaling Technique for Low-Power Design," *Proceedings, 1995 International Symposium on Low Power Design*, Dana Point, CA, pp. 3–8, April 1995.

[Vermassen86] H. Vermassen, "Mathematical Models for the Complexity of VLSI," (in dutch), Master's Thesis, Katholieke Universiteit, Leuven, Belgium, 1986.

[Wallace64] C. Wallace, "A Suggestion for a Fast Multiplier," *IEEE Trans. on Electronic Computers*, EC-13, pp. 14–17, 1964.

[Weinberger56] A. Weinberger and J. L. Smith, "A One-Microsecond Adder Using One-Megacycle Circuitry," *IRE Transactions on Electronic Computers*, vol. 5, pp. 65–73, 1956.

[Wei99] L. Wei, Z. Chen, K. Roy, M.C. Johnson, Y. Ye, and V. K. De, "Design and Optimization of Dual-Threshold Circuits for Low-Power Applications," *IEEE Transactions on Very Large Scale Integration (VLSI) Systems*, vol. 7, no. 1, pp. 16–24, March 1999.

[Weinberger81] A. Weinberger, "A 4:2 Carry-Save Adder Module," *IBM Technical Disclosure Bulletin*, vol. 23, January 1981.

[Weste85&93] N. Weste and K. Eshragian, *Principles of CMOS VLSI Design: A Systems Perspective*, 2d ed., Addison-Wesley, 1985 and 1993.

Exercises

For the latest problem sets, design challenges and design projects related to arithmetic modules, log in to **http://bwrc.eecs.berkeley.edu/IcBook.**

CHAPTER

12

Designing Memory and Array Structures

Memory classification and architecture

Data-storage cells for read-only, nonvolatile, and read–write memories

Peripheral circuits—sense amplifiers, decoders, drivers, and timing generators

Power consumption and reliability issues in memory design

12.1   Introduction
    12.1.1 Memory Classification
    12.1.2 Memory Architectures and Building Blocks

12.2   The Memory Core
    12.2.1 Read-Only Memories
    12.2.2 Nonvolatile Read–Write Memories
    12.2.3 Read–Write Memories (RAM)
    12.2.4 Contents-Addressable or Associative Memory (CAM)

12.3   Memory Peripheral Circuitry*
    12.3.1 The Address Decoders
    12.3.2 Sense Amplifiers
    12.3.3 Voltage References
    12.3.4 Drivers/Buffers
    12.3.5 Timing and Control

12.4   Memory Reliability and Yield*
    12.4.1 Signal-to-Noise Ratio
    12.4.2 Memory Yield

12.5   Power Dissipation in Memories*
    12.5.1 Sources of Power Dissipation in Memories

12.5.2 Partitioning of the Memory
12.5.3 Addressing the Active Power Dissipation
12.5.4 Data-Retention Dissipation
12.5.5 Summary
12.6   Case Studies in Memory Design
12.6.1 The Programmable Logic Array (PLA)
12.6.2 A 4-Mbit SRAM
12.6.3 A 1-Gbit NAND Flash Memory
12.7   Perspective: Semiconductor Memory Trends and Evolutions
12.8   Summary
12.9   To Probe Further

12.1 Introduction

A large portion of the silicon area of many contemporary digital designs is dedicated to the storage of data values and program instructions. More than half of the transistors in today's high-performance microprocessors are devoted to cache memories, and this ratio is expected to further increase. The situation is even more dramatic at the system level. High-performance workstations and computers contain several Gbytes of semiconductor memory, a number that is continuously rising. With the introduction of semiconductor audio (MP3) and video players (MPEG4), the demand for nonvolatile storage has skyrocketed. Obviously, dense data-storage circuitry is and will be one of the primary concerns of a digital circuit or system designer. The total market share for semiconductor memories is expected to be over $45 billion in 2003, which is twice as large as it was in 1998.

In Chapter 7, we introduced means of storing Boolean values based on either positive feedback or capacitive storage. While semiconductor memories are built on the same concepts, the simple use of a register cell as a means for mass storage leads to excessive area requirements. Memory cells are therefore combined into large arrays, which minimizes the overhead caused by peripheral circuitry and increases the storage density. The sheer size and complexity of these array structures introduces a variety of design problems, some of which are discussed in this chapter.

We first introduce the basic memory architectures and their essential building blocks. Next, we analyze the different memory cells and their properties. The cell structure and topology is mainly driven by the available technology, and is somewhat out of the control of the digital designer. On the other hand, the peripheral circuitry has a tremendous impact on the robustness, performance, and power consumption of the memory unit. Therefore, a careful analysis of the options and considerations of the periphery design is appropriate. Reliability and power dissipation are two other large concerns of the semiconductor memory designer, and they are discussed in separate sections.

An interesting aspect of Chapter 12 is that it applies a large number of the circuit techniques introduced in the earlier chapters. In a sense, one can consider memory design as a case study of high-performance, high-density, low-power circuit design. This becomes quite clear from the two case studies that conclude the chapter.

12.1.1 Memory Classification

Electronic memories come in many different formats and styles. The type of memory unit that is preferable for a given application is a function of the required memory size, the time it takes to access the stored data, the access patterns, the application, and the system requirements.

Size Depending upon the level of abstraction, different means are used to express the size of a memory unit. The *circuit* designer tends to define the size of the memory in terms of *bits* that are equivalent to the number of individual cells (flip-flops or registers) needed to store the data. The *chip* designer expresses the memory size in *bytes* (groups of 8 or 9 bits) or its multiples—kilobytes (Kbyte), megabytes (Mbyte), gigabytes (Gbyte), and ultimately terabytes (Tbyte). The *system* designer likes to quote the storage requirement in terms of *words*, which represent a basic computational entity. For instance, a group of 32 bits represents a word in a computer that operates on 32-bit data.

Timing Parameters The timing properties of a memory are illustrated in Figure 12-1. The time it takes to retrieve (*read*) from the memory is called the *read-access time*, which is equal to the delay between the read request and the moment the data is available at the output. This time is different from the *write-access time*, which is the time elapsed between a write request and the final writing of the input data into the memory. Finally, another important parameter is the (read or write) *cycle time* of the memory, which is the minimum time required between successive reads or writes. This time is normally greater than the access time for reasons that become apparent later in the chapter. Read and write cycles do not necessarily have the same length, but their lengths are considered equal for simplicity of system design.

Function Semiconductor memories are most often classified on the basis of memory functionality, access patterns, and the nature of the storage mechanism. A distinction is made between *read-only* (ROM) and *read–write* (RWM) memories. The RWM structures have the advantage of

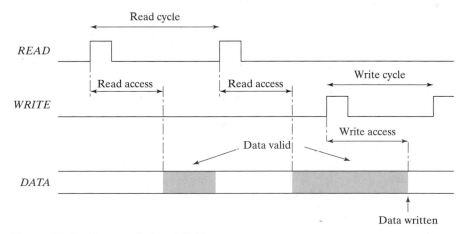

Figure 12-1 Memory-timing definitions.

offering both read and write functionality with comparable access times and are the most flexible memories. Data are stored either in flip-flops or as a charge on a capacitor. As in the classification introduced in the discussion on sequential circuitry, these memory cells are called *static* and *dynamic,* respectively. The former retain their data as long as the supply voltage is retained, while the latter need periodic refreshing to compensate for the charge loss caused by leakage. Since RWM memories use active circuitry to store the information, they belong to the class called *volatile* memories, in which the data is lost when the supply voltage is turned off.

Read-only memories, on the other hand, encode the information into the circuit topology—for example, by adding or removing transistors. Since this topology is hard wired, the data cannot be modified; it can only be read. Furthermore, ROM structures belong to the class of the *nonvolatile* memories. Disconnection of the supply voltage does not result in a loss of the stored data.

The most recent entry in the field are memory modules that can be classified as nonvolatile, yet offer both read and write functionality. Typically, their write operation takes substantially more time than the read. We call them *nonvolatile read-write* (NVRWM) memories. Members of this family are the *EPROM* (erasable programmable read-only memory), E^2PROM (electrically erasable programmable read-only memory), and *flash* memories. The emergence of novel, cheap, and dense nonvolatile technologies over the last decade has made this approach to storage the fastest growing in the memory arena.

Access Pattern   A second memory classification is based on the order in which data can be accessed. Most memories belong to the *random-access* class, which means memory locations can be read or written in a random order. One would expect memories of this class to be called *RAM* modules (random-access memory). For historical reasons, this name has been reserved for the random-access RWM memories, probably because the RAM acronym is more easily pronounced than the awkward RWM. Be aware that most ROM or NVRWM units also provide random access, but the acronym RAM should not be used for them.

Some memory types restrict the order of access, which results in either faster access times, smaller area, or a memory with a special functionality. Examples of such are the *serial memories*: the FIFO (*first-in first-out*), LIFO (*last-in first-out*, most often used as a stack), and the *shift register. Video memories* are an important member of this class. In video processing, data is acquired and outputted serially. and random access is not required. *Contents-addressable memories* (CAM) represent another important class of nonrandom access memories. Instead of using an address to locate the data, a CAM (also called an *associative memory*), uses a word of data itself as input in a query-style format. When the input data matches a data word stored in the memory array, a MATCH flag is raised. The MATCH signal remains low if no data stored in the memory corresponds to the input word. Associative memories are an important component of the cache architecture of many microprocessors.

An overview of the memory classes, as introduced earlier, is given in Figure 12-2. Implementations for each of the mentioned memory structures are discussed in subsequent sections. It

RWM		NVRWM	ROM
Random Access	**Non-Random Access**	EPROM E^{2}PROM FLASH	Mask-programmed programmable (PROM)
SRAM DRAM	FIFO LIFO Shift register CAM		

Figure 12-2 Semiconductor memory classification.

will be demonstrated how the nature of the memory affects not only the choice of the basic storage cell, but also the composition of the peripheral units.

Input/Output Architecture A final classification of semiconductor memories is based on the *number of data input and output ports*. While a majority of the memory units presents only a single port that is shared between input and output, memories with higher bandwidth requirements often have multiple input and output ports—and thus are called *multiport memories*. Examples of the latter are the register files used in RISC (reduced instruction set computer) microprocessors. Adding more ports tends to complicate the design of the storage cell.

Application Before the end of the century, most large-size memories were packaged as standalone ICs. With the advent of the system-on-a-chip and the integration of multiple functions on a single die, an ever larger fraction of memory is now integrated on the same die as the logic functionality. Memories of this type are called *embedded*. The colocation of these diverse functions has a severe impact on the memory design—not only on the overall memory architecture, but also on the its underlying technology and circuit techniques.

When massive amounts of storage are needed (multiples of Gigabytes and more), semiconductor memories tend to become too expensive. More cost-effective technologies such as magnetic and optical disk should be used. While these provide extensive storage capabilities at a low cost per bit, they tend to be either slow or provide limited access patterns. For instance, a magnetic tape generally allows for serial access only. For these reasons, such memories do not communicate directly with the computing processor, but are interfaced through a number of faster semiconductor memories. They are called *secondary* or *tertiary* memories and are beyond the scope of this textbook.

12.1.2 Memory Architectures and Building Blocks

When implementing an *N*-word memory where each word is *M* bits wide,[1] the most intuitive approach is to stack the subsequent memory words in a linear fashion, as shown in Figure 12-3a. One word at a time is selected for reading or writing with the aid of a *select* bit (S_0 to S_{N-1}), if we

---

[1] The length of a word varies between 1 and 128 bits. In commercial memory chips, the word length typically equals 1, 4, or 8 bits.

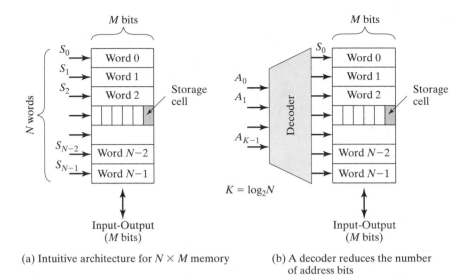

(a) Intuitive architecture for $N \times M$ memory   (b) A decoder reduces the number of address bits

Figure 12-3   Architectures for N-word memory (where each word is M bits).

assume that this module is a single-port memory. In other words, only one signal S_i can be high at any time. For simplicity, let us temporarily assume that each storage cell is a D flip-flop and that the select signal is used to activate (clock) the cell. While this approach is relatively simple and works well for very small memories, one runs into a number of problems when trying to use it for larger memories.

Assume that we would like to implement a memory that holds 1 million ($N = 10^6$) 8-bit ($M = 8$) words. The reader should be aware that 1 million is a simplification of the actual memory size, since memory dimensions always come in powers of two. In this particular case, the actual number of words equals $2^{20} = 1024 \times 1024 = 1,048,576$. For ease of use, it is common practice to denote such a memory as 1 Mword unit.

When implementing this structure using the strategy of Figure 12-3a, we quickly realize that 1 million select signals are needed—one for every word. Since these signals are normally provided from off-chip or from another part of the chip, this translates into insurmountable wiring and/or packaging problems. A *decoder* is inserted to reduce the number of select signals (Figure 12-3b). A memory word is selected by providing a binary encoded *address word* (A_0 to A_{K-1}). The decoder translates this address into $N = 2^K$ select lines, only one of which is active at a time. This approach reduces the number of external address lines from 1 million to 20 ($\log_2 2^{20}$) in our example, which virtually eliminates the wiring and packaging problems. The decoder is typically designed so that its dimensions are matched to the size of the storage cell and the connections between the two, in particular the S signals in Figure 12-3b, do not produce any area overhead. The value of this approach can be appreciated by interpreting Figure 12-3b as a physical floor plan of the memory module. By performing the pitch matching between decoder and memory core, the S wires can be very short, and no large routing channel is required.

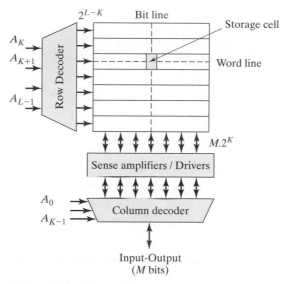

Figure 12-4 Array-structured memory organization.

While this resolves the select problem, it does not address the issue of the memory aspect ratio. Evaluation of the dimensions of the storage array of our token example shows that its height is approximately 128,000 times larger than its width ($2^{20}/2^3$), assuming the shape of the basic storage cell is approximately square which is almost always the case. Obviously, this results in a design that cannot be implemented. Besides the bizarre shape factor, the resulting design is also extremely slow. The vertical wires connecting the storage cells to the input/outputs become excessively long. Remember that the delay of an interconnect line increases at least linearly with its length.

To address this problem, memory arrays are organized so that the vertical and horizontal dimensions are of the same order of magnitude; thus, the aspect ratio approaches unity. Multiple words are stored in a single row and are selected simultaneously. To route the correct word to the input/output terminals, an extra piece of circuitry called the *column decoder* is needed. The concept is illustrated in Figure 12-4. The address word is partitioned into a *column address* (A_0 to A_{K-1}) and a *row address* (A_K to A_{L-1}). The row address enables one row of the memory for R/W, while the column address picks one particular word from the selected row.

Example 12.1 Memory Organization

An alternative choice would be to organize the memory core of our example as an array of 4000 by 2000 cells (to be more precise, 4096 × 2048), which approaches a square aspect ratio. Each of the 4000 rows stores 256 8-bit words. This results in a row address of 12 bits, while the column address measures 8 bits. It can be verified that the total address space still equals 20 bits.

Figure 12-4 introduces commonly used terminology. The horizontal select line that enables a single row of cells is called the *word line*, while the wire that connects the cells in a single column to the input/output circuitry is named the *bit line*.

The area of large memory modules is dominated by the size of the memory core. Thus, it is crucial to keep the size of the basic storage cell as small as possible. We could use one of the register cells introduced in Chapter 7 to implement a *R/W* memory. Such a cell easily requires more than 10 transistors per bit, and employing it in a large memory results in excessive area requirements. Semiconductor memory cells therefore reduce the cell area by trading off some desired properties of digital circuits, such as noise margin, logic swing, input/output isolation, fan-out, or speed. While a degradation of some of those properties is allowable within the confined domain of the memory core where noise levels can be tightly controlled, this is not acceptable when interfacing with the external or surrounding circuitry. The desired digital signal properties must be recovered with the aid of peripheral circuitry.

For example, it is common to reduce the voltage swing on the bit lines to a value substantially below the supply voltage. This reduces both the propagation delay and the power consumption. A careful control of the cross talk and other disturbances is possible within the memory array, ensuring that sufficient noise margin is obtained even for these small signal swings. Interfacing to the external world, on the other hand, requires an amplification of the internal swing to a full rail-to-rail amplitude. This is achieved by the *sense amplifiers* shown in Figure 12-4. The design of those peripheral circuits is discussed in Section 12.3. Relaxation of bounds on a number of the coveted digital properties makes it possible to reduce the transistor count of a single memory cell to between one and six transistors!

The architecture of Figure 12-4 works well for memories up to a range of 64 Kbits to 256 Kbits. Larger memories start to suffer from a serious speed degradation as the length, capacitance, and resistance of the word and bit lines become excessively large. Larger memories have consequently gone one step further and added one extra dimension to the address space, as illustrated in Figure 12-5.

The memory is partitioned into P smaller blocks. The composition of each of the individual blocks is identical to one of Figure 12-4. A word is selected on the basis of the row and column addresses that are broadcast to all the blocks. An extra address word called the *block address*, selects one of the P blocks to be read or written. This approach has a dual advantage.

1. The length of the local word and bit lines—that is, the length of the lines within the blocks—is kept within bounds, resulting in faster access times.
2. The block address can be used to activate only the addressed block. Nonactive blocks are put in power-saving mode with sense amplifiers and row and column decoders disabled. This results in a substantial power saving that is desirable, since power dissipation is a major concern in very large memories.

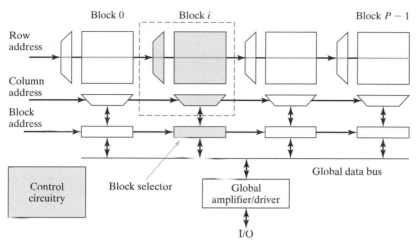

Figure 12-5   Hierarchical memory architecture. The block selector enables a single memory block at a time.

Example 12.2   Hierarchical Memory Architecture

As an example, a 4-Mbit SRAM can be designed [Hirose90] as a composition of 32 blocks, each of which contains 128 Kbits (Figure 12-6). Each block is structured as an

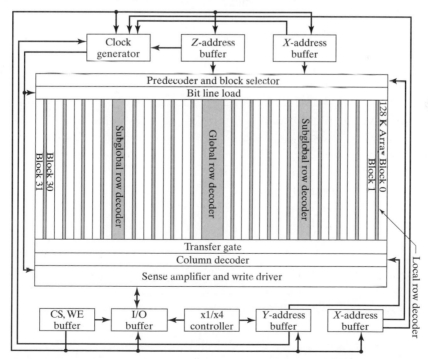

Figure 12-6   Block diagram of 4 Mbit memory (from [Hirose90]).

array with 1024 rows and 128 columns. The row address (X), column address (Y), and block address (Z) are 10, 7, and 5 bits wide, respectively.

Multiple variants of the proposed architecture are possible. Variations include the positioning of the sense amplifiers, the partitioning of the word and bit lines, and the styles of the decoders used. The underlying concept is the same—it is advantageous to partition a large memory into smaller subdivisions to combat the delay associated with extra long lines. The gains in performance and power consumption easily outweigh the overhead incurred by the partitioning.

The nature of *serial and contents-addressable memories* naturally leads to variations in the architecture and composition of the memory. Figure 12-7 shows an example of a 512-word CAM memory, which supports three modes of operation: read, write, and match. The read and write modes access and manipulate data in the CAM array in the same way as in an ordinary memory. The match mode is unique to associate memories. The *comparand* block is filled with the data pattern to match, and the *mask* word indicates which bits are significant. For example, to find all the words in the CAM array that have the pattern 0×123 in the most significant bits, we would fill the comparand with 0×12300000 and the mask with $0 \times \text{FFF00000}$. All 512 rows of the CAM array then simultaneously compare the 12 most significant bits of the comparand with the data contained in that row. Every row that matches the pattern is passed to the validity block. Since we do not care about rows that contain invalid data (which typically happens when the array is not full), only the valid rows that match are passed to the *priority encoder*. In the event that two or more rows match the pattern, the address of the row in the CAM array is used to break the tie. To do this, the priority encoder considers all 512 match lines from the CAM array, selects the one with the highest address, and encodes it in binary.[2] Since there are 512 rows in

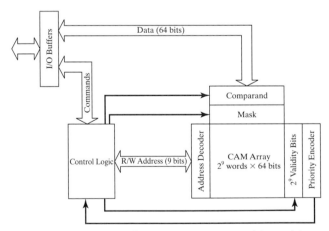

Figure 12-7   Architecture of 512-word contents-addressable memory.

---

[2]The highest address typically corresponds to the latest entry in the cache, and this the most desirable match.

the CAM array, 9 bits are required to indicate the highest row that matched. One additional 'match found' bit is provided, since it is possible that none of the rows matches the pattern.

Finally, a component of memory design that is often overlooked is the *input/output interface and control circuitry*. The nature of the I/O interface has an enormous impact on the global memory control and timing. This statement is illustrated by comparing the input–output behavior of typical DRAM and SRAM components with the associated timing structure.

Since the early days, DRAM designers have opted for a multiplexed addressing scheme. In this model, the lower and upper halves of the address words are presented sequentially on a single address bus. This approach reduces the number of package pins and has survived through the subsequent memory generations.DRAMs are generally produced in higher volumes. Lowering the pin count reduces cost and size at the expense of performance. The presence of a new address word is asserted by raising a number of strobe signals. (See Figure 12-8a.) Raising the *RAS* (row-access strobe) signal asserts that the *msb* part of the address is present on the address bus, and that the word-decoding process can be initiated. The *lsb* part of the address is applied next, and the *CAS* (column-access strobe) signal is asserted. To ensure correct memory operation, a careful timing of the *RAS–CAS* interval is necessary. In fact, the *RAS* and *CAS* signals act as clock inputs to the memory module, and are used to synchronize memory events, such as decoding, memory core access, and sensing.

The SRAM designers, on the other hand, have chosen a self-timed approach, as in Figure 12-8b. The complete address word is presented at once, and circuitry is provided to automatically detect any transitions on that bus. No external timing signals are needed. All internal timing events, such as the enabling of the decoders and sense amplifiers, are derived from the internally generated transition signal. This approach has the advantage that the cycle time of the SRAM is close or equal to its access time, while this is definitely not the case for the DRAM. The increased overall performance of compute systems that use DRAM as storage requires DRAM speeds to evolve at approximately the same pace. Several new approaches to improve the performance of the DRAM for read operations have been introduced. Examples are *Synchronous DRAM* (SDRAM) and *Rambus DRAM* (RDRAM). The main novelty in these new architectures, which are discussed later in the chapter, is not in the memory core, but in how they communicate with the outside world.

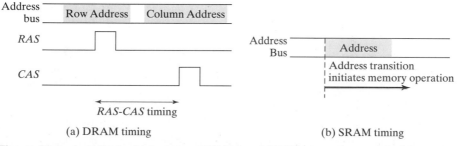

(a) DRAM timing   (b) SRAM timing

Figure 12-8 Input/output interface of DRAM and SRAM memories and their impact on memory control.

Designing the control and timing circuitry so that the memory is functional over a wide range of manufacturing tolerances and operating temperatures is a demanding task that requires extensive simulation and design optimization. It is an integral, but often overlooked, part of the memory-design process and has a major impact on both memory reliability and performance.

12.2 The Memory Core

This section concentrates on the design of the memory core and its composing cells for a variety of semiconductor memory types. While the most compelling issue in designing large memories is to keep the size of the cell as small as possible, this should be done so that other important design-quality measures such as speed and reliability are not fatally affected. In sequence, we discuss *read-only*, *nonvolatile*, and *read–write* memory cores. This section is concluded with a short discussion of the associative memory cell.

12.2.1 Read-Only Memories

While the idea of a memory that can only be read and never altered might seem odd at first, but a second glance reveals a large number of potential applications. Programs for processors with fixed applications such as washing machines, calculators, and game machines, once developed and debugged, need only reading. Fixing the contents at manufacturing time leads to small and fast implementations.

ROM Cells—An Overview

The fact that the contents of a ROM cell are permanently fixed considerably simplifies its design. The cell should be designed so that a 0 or 1 is presented to the bit line upon activation of its word line. Figure 12-9 shows several ways to accomplish this.

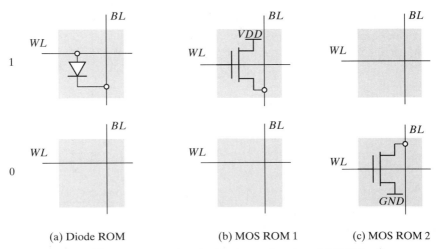

(a) Diode ROM   (b) MOS ROM 1   (c) MOS ROM 2

Figure 12-9   Different approaches for implementing 1 and 0 ROM cells.

Consider first the simplest cell, which is the diode-based ROM cell shown in Figure 12-9a. Assume that the bit line *BL* is resistively clamped to ground—that is, *BL* is pulled low through a resistor connected to ground lacking any other excitations or inputs. This is exactly what happens in the 0 cell. Since no physical connection between the word line *WL* and *BL* exists, the value on *BL* is low, independent of the value of *WL*. On the other hand, when a high voltage V_{WL} is applied to the word line of the 1 cell, the diode is enabled, and the word line is pulled up to $V_{WL} - V_{D(on)}$, resulting in a 1 on the bit line. In summary, the presence or absence of a diode between *WL* and *BL* differentiates between ROM cells storing a 1 or a 0, respectively.

The disadvantage of the diode cell is that it does not isolate the bit line from the word line. All current required to charge the bit line capacitance, which can be quite high for large memories, has to be provided through the word line and its drivers; therefore, this approach only works for small memories. A better approach is to use an active device in the cell, as proposed in Figure 12-9b. The diode is replaced by the gate-source connection of an NMOS transistor, whose drain is connected to the supply voltage. The operation is identical to that of the diode cell with one major difference: All output-driving current is provided by the MOS transistor in the cell. The word-line driver is only responsible for charging and discharging the word-line capacitance.

The improved isolation comes at the penalty of a more complex cell and a larger area. The latter is caused primarily by the extra supply contact. This contact must be provided in every cell, so that the supply rail must be distributed throughout the array. An example of a 4×4 array is shown in Figure 12-10. Notice how the overhead of the supply lines is reduced by sharing them between neighboring cells. This requires the *mirroring* of the odd cells around the horizontal axis, an approach that is extensively used in memory cores of all styles.

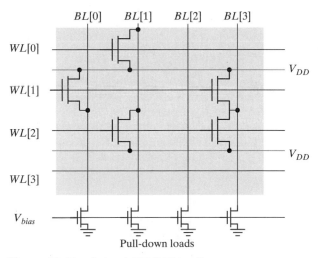

Figure 12-10 A 4×4 OR ROM cell array.

Problem 12.1   ROM Array

Determine the values of the data stored at addresses 0, 1, 2, and 3 in the ROM of Figure 12-10.

An alternative implementation is offered by the MOS cell of Figure 12-9c. To be operational, this cell requires the bit line to be resistively clamped to the supply voltage, or equivalently, the default value at the output must equal 1. The absence of a transistor between *WL* and *BL* thus means that a 1 is stored. The 0 cell is realized by providing an MOS device between bit line and ground. Applying a high voltage on the word line turns on the device, which in turn pulls down the bit line to GND. An example of a 4×4 MOS ROM array is shown in Figure 12-11. A PMOS load is used to pull up the bit lines in case none of the attached NMOS devices is enabled.

Problem 12.2   MOS NOR ROM Memory Array

Determine the values of the data stored at addresses 0, 1, 2, and 3 in the ROM of Figure 12-11.

Programming the ROM Memory

You may have noticed in the last example that the combination of a bit line, PMOS pull-up, and NMOS pull-downs constitutes nothing other than **a pseudo-NMOS NOR gate with the word lines as inputs**. An $N \times M$ ROM memory can be considered as a combination of M NOR gates with at most N inputs (for a fully populated column). It is therefore called a NOR ROM.[3] Under

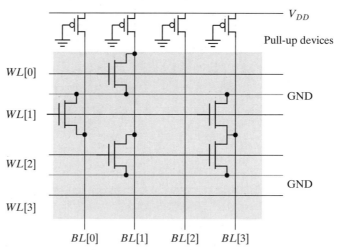

Figure 12-11   A 4×4 MOS NOR ROM.

---

[3]Similarly, the memory structure in Figure 12-10 implements an OR function, and hence is called an OR ROM.

normal operating conditions, only one of the word lines goes high, and, at most, one of the pull-down devices is turned on. This raises some interesting issues regarding the sizing of both the cell and pull-up transistors:

- To keep the cell size and the bit line capacitance small, the pull-down device should be kept as close as possible to minimum size.
- On the other hand, the resistance of the pull-up device must be larger than the pull-down resistance to ensure an adequate low level. In Chapter 6, we derived a factor of at least 4 for pseudo-NMOS gates. This large resistance has a detrimental effect on the low-to-high transitions on the bit lines. The bit line capacitance consists of the contributions of all connected devices and can be in the pF range for larger memories.

This is where memory and logic design differ. For the sake of density and performance, it is possible to relax some of the quality standards imposed on digital gates. In the NOR ROM, we can trade off noise margin for performance by letting the V_{OL} of the bit line stand at a higher voltage (e.g., 1 to 1.5 V for a 2.5-V supply). The pull-up device can now be widened, improving the low-to-high transition. The reduced noise margin is tolerable within the memory core, where the noise conditions and signal interferences can be carefully controlled. Going to the external world requires a restoration of the full voltage swing. This is accomplished by the peripheral devices—in this case, the sense amplifier. For instance, feeding the bit line into a complementary CMOS inverter with an appropriately adjusted switching threshold restores the full signal swing.

Figure 12-12 shows two possible layouts of the 4×4 NOR ROM array of Figure 12-11. The arrays are constructed by repeating the same cell in both the horizontal and vertical directions, mirroring the odd cells around the horizontal axis in order to share the *GND* wire. The difference between two artifacts lies in the way they are programmed. In the structure of Figure 12-12a, the memory is written (personalized) by selectively adding transistors when needed. This is accomplished with the aid of only the diffusion layer (the ACTIVE mask in the fabrication process). In the second approach (Figure 12-12b), the memory is programmed by the selective addition of metal-to-diffusion contacts. The presence of a metal contact to the bit line creates a 0 cell, while its absence indicates a 1 cell. Observe that only one mask layer, the CONTACT, is used to program the memory in this case.

A comparison of the ACTIVE and CONTACT implementations, based on identical design rules, reveals that the former results in an area savings of approximately 15%. On the other hand, the CONTACT programming strategy has the advantage that the contact layer is a later step in the manufacturing process. This delays the actual programming of the memory in the process cycle. Wafers can be prefabricated up to the CONTACT mask and stockpiled. The remaining fabrication steps can be executed quickly once a specific program is defined, reducing the turn-around time between order and delivery. In multilayer processes, programming is increasingly done in one of the VIA masks. Which programming approach is ultimately chosen depends on the dominant design metric—size/performance versus turnaround time.

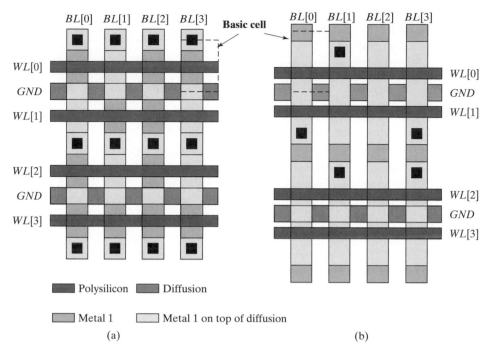

Figure 12-12   Possible layout of a 4×4 MOS NOR ROM. The bit lines are implemented in Metal 1 and are routed on top of the cell diffusion. *GND* lines are distributed horizontally in diffusion. The memory is programmed using (a) the ACTIVE layer, resulting in a cell size of $9.5 \lambda \times 7 \lambda$; (b) the CONTACT layer, yielding a cell of $11 \lambda \times 7 \lambda$.

Also, you may have observed that diffusion is used to route the *GND* signal. In general, this is an absolute "no-no." Yet, it is common practice in memories for the sake of increased density. A metal bypass with regularly spaced straps keeps the voltage drop over the ground wire within strict bounds.

It is important to note that the transistor occupies only a small ratio of the total cell size, which measures around 70 λ^2. It is actually possible to increase its size over the minimum dimensions without affecting the cell size. A transistor with a 4/2 aspect was chosen in the examples of Figure 12-12. A large part of the cell is devoted to the bit line contact and ground connection. One way to avoid this overhead is to adopt a different memory organization. Figure 12-13 shows a 4×4 ROM array based on a NAND configuration. All transistors constituting a column are connected in series. A basic property of a NAND gate is that all transistors in the pull-down chain must be on to produce a low value. The word lines must be operated in reverse-logic mode to make this memory function. All word lines are high by default with the exception of the selected row, which is set to 0. Transistors on nonselected rows are thus turned on. Now suppose that no transistor is present on the intersection between

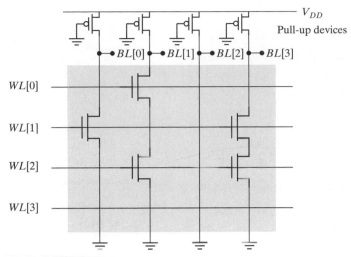

Figure 12-13 4×4 MOS NAND ROM.

the row and column of interest. Since all other transistors on the series chain are selected, the output is pulled low, and the stored value equals 0. On the other hand, a transistor present at the intersection is turned off when the associated word line is brought low. This results in a high output, which is equivalent to reading a 1.

Problem 12.3   MOS NAND ROM

Determine the values of the data stored at addresses 0, 1, 2, and 3 in the ROM of Figure 12-13.

The main advantage of the NAND structure is that the basic cell only consists of a transistor (or a lack of a transistor) and that no connection to any of the supply voltages is needed. This reduces the cell size substantially, as illustrated in the layouts of Figure 12-14. Again, different programming approaches can be employed. A first option is to use the METAL-1 layer to selectively short-circuit the transistor (a). The resulting cell reduces the cell size approximately 15% below the smallest NOR ROM cell. Even more impressive area reductions can be obtained if an extra implant step is available to program the memory. A threshold-lowering implant using n-type impurities turns the device into a depletion transistor, which is always on, regardless of the applied word-line voltage. It thus is equivalent to a shorted circuit. The resulting cell area of 30 λ^2 is more than two times smaller than the equivalent NOR ROM cell. This comes at a price, however. In the next section, we demonstrate that the NAND configuration results in a considerable loss in performance and is only useful for small memory arrays. Also, the extra implant step represents an additional process step and generally is not available to a designer.

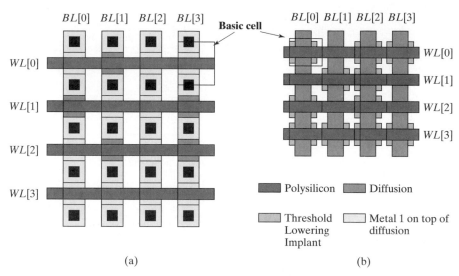

Figure 12-14 4 × 4 MOS NAND ROM (a) using metal-1 layer programming (cell size: 8 λ × 7 λ; (b) using threshold-lowering implants (cell size: 5 λ × 6 λ, assuming that the implant has to extend 1 λ in all directions from the gate).

Example 12.3   Voltage Swings in NOR and NAND ROMs

Assuming that the layouts of Figure 12-12 and Figure 12-14 are implemented using our standard 0.25–μm CMOS technology, determine the size of the PMOS pull-up device so that the worst case value of V_{OL} is never higher than 1.5 V (for a 2.5-V supply voltage). This translates into a bit line swing of 1 V. Determine the values for an 8 × 8 and a 512 × 512 array.

1. NOR ROM

Since, at most, one transistor can be on at a time, the value of V_{OL} is not a function of the array size nor the programming of the array. The low output voltage is computed by analyzing a pseudo-NMOS inverter with a single pull-down device of a (4/2) size. This circuit was analyzed in detail in Chapter 6 for a different operation region. It can be determined by inspection that both PMOS and NMOS are in velocity saturation for the aforementioned bias conditions:

$$\frac{(W/L)_p}{(W/L)_n} = \frac{k'_n[(V_{DD}-V_{Tn})V_{DSATn}-V_{DSATn}^2/2](1+\lambda_n V_{OL})}{-k'_p[(-V_{DD}-V_{Tp})V_{DSATp}-V_{DSATp}^2/2](1+\lambda_p(V_{OL}-V_{DD}))}$$

Solving for V_{DD} = 2.5 V and V_{OL} = 1.5 V leads to a PMOS/NMOS ratio of 2.62, or a required size for the PMOS device of $(W/L)_p$ = 5.24.

2. NAND ROM

Due to the series chaining, the value of V_{OL} is a function of both the size of the memory (number of rows) and the programming. The worst case occurs when all bits in a column are set to 1, which means N transistors are connected in series in the pull-down network. Making the simplifying assumption that the N transistors can be replaced by an N-times longer device, the values of $(W/L)_p$ can be derived for the (8×8) and (512×512) cases. Observe that the design of Figure 12-14b uses minimum-size devices of $(3/2)$:

$$(8 \times 8): (W/L)_p = 0.49$$

$$(512 \times 512): (W/L)_p = 0.0077$$

While the first case still produces acceptable results for the PMOS device, the second one would require an extremely long pull-up transistor, which is unacceptable. For this reason, NAND ROMs are rarely used for arrays with more than 8 or 16 rows.

ROM Transient Performance

The transient response of a memory array is defined as the time it takes from the time a word line switches until the point where the bit line has traversed a certain voltage swing ΔV. Since the bit line normally feeds into a sense amplifier, it is not necessary for it to traverse its complete swing. A voltage drop (or rise) of ΔV is sufficient to make the sense amplifier react. Typical values of ΔV range around 0.5 V.

One important difference between the analysis of the propagation delay of a logic gate and that of a memory array is that most of the delay is attributable to the interconnect parasitics. An accurate modeling of these parasitics is therefore of prime importance. This is illustrated in the example that follows, which extracts the parasitic resistance and capacitance of the word and bit lines of both the NOR and NAND ROM arrays introduced earlier. We cover this example quite extensively because the same approach also holds for other memory styles, such as SRAM and DRAM.

Example 12.4   Word- and Bit Line Parasitics

In this example, we first derive an equivalent model of the memory arrays of Figure 12-12b and Figure 12-14b, respectively. Taking into account all transistors in the array simultaneously quickly leads to intractable equations and models, especially for larger memories. Simplification and abstraction is the obvious approach to be followed. We consider only the case of the (512×512) array.

1. NOR ROM

Figure 12-15 shows a model that is appropriate for the analysis of the word- and bit line delay of the NOR ROM. The word line is best modeled as a distributed RC line, since it is implemented in polysilicon with a relatively high sheet resistance. The use of a silicided polysilicon is definitely advisable. The bit line, on the other hand, is implemented in

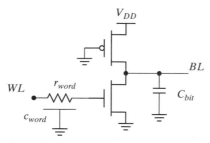

Figure 12-15   Equivalent transient model for the NOR ROM.

aluminum, and the resistance of the line only comes into play for very long lines. It is reasonable to assume for this example that a purely capacitive model is adequate and that all capacitive loads connected to the wire can be lumped into a single element.

Word-line parasitics (for the memory array of Figure 12-12b) are as follows:

Resistance/cell: $(7/2) \times 5\ \Omega/\text{sq} = 17.5\ \Omega$ (using the data of Table 4-5 on page 145)

Wire capacitance/cell: $(3\ \lambda \times 2\ \lambda)\ (0.125)^2\ 0.088 + 2 \times (3\ \lambda \times 0.125) \times 0.054 = 0.049$ fF
(from Table 4-2 on page 143)

Gate capacitance/cell: $(4\ \lambda \times 2\ \lambda)\ (0.125)^2\ 6 = 0.75$ fF.

The bit line parasitics are as follows:

Resistance/cell: $(11/4) \times 0.1\ \Omega/\text{sq} = 0.275\ \Omega$ (which is negligible)

Wire capacitance/cell: $(11\ \lambda \times 4\ \lambda)\ (0.125)^2\ 0.041 + 2 \times (11\ \lambda \times 0.125) \times 0.047 = 0.09$ fF
Drain capacitance/cell:

$((5\ \lambda \times 4\ \lambda)\ (0.125)^2 \times 2 \times 0.56 + 14\ \lambda \times 0.125 \times 0.28 \times 0.6 + 4\ \lambda \times 0.125 \times 0.31 = 0.8$ fF

The latter term deserves an explanation. The *drain capacitance* contributed by every cell connected to the bit line consists of the bottom junction, side wall, and overlap capacitances. It may be assumed that all cells, besides the one being switched, are in the off state, which explains why only the overlap part of the gate capacitance is taken into account. It is assumed further that the bit line swings between 1.5 V and 2.5 V. Under these conditions, the K_{eq} factor evaluates to 0.56 and 0.6 for area and side wall, respectively. For all other capacitance values, please refer to Table 3-5 on page 112.

2. NAND ROM

As in the approach taken for the NOR ROM, we can derive an equivalent model for the analysis of the delay of the NAND structure. While the word-line model is identical, modeling the bit line behavior is more complex due to the long chain of series-connected transistors. The worst case behavior occurs when the transistor at the bottom of the chain is switched and the column is completely populated with transistors. A model approximating the behavior for that case is shown in Figure 12-16. Each of the series transistors (which are normally in the on mode) is modeled as a resistance–capacitance combination. The entire chain can be modeled as a distributed *rc*-network for large memories.

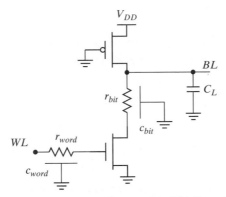

Figure 12-16 Equivalent model for word and bit line of NAND ROM.

Word-line parasitics (for the memory array of Figure 12-14) are as follows:

$$\text{Resistance/cell: } (6/2) \times 5 \ \Omega/\text{sq} = 15 \ \Omega$$

$$\text{Wire capacitance/cell: } (3 \ \lambda \times 2 \ \lambda) \ (0.125)^2 \ 0.088 + 2 \times (3 \ \lambda \times 0.125) \times 0.054 = 0.049 \ \text{fF}$$

$$\text{Gate Capacitance/cell: } (3 \ \lambda \times 2 \ \lambda) \ (0.125)^2 \ 6 = 0.56 \ \text{fF}$$

The bit line parasitics are as follows:

$$\text{Resistance/cell: } 13 \ / \ 1.5 = 8.7 \ \text{k}\Omega$$

$$\text{Wire capacitance/cell: Included in diffusion capacitance}$$

Finally, the source/drain capacitance/cell is given by

$$((3 \ \lambda \times 3 \ \lambda) \ (0.125)^2 \times 2 \times 0.56 + 2 \times 3 \ \lambda \times 0.125 \times 0.28 \times 0.6 + (3 \ \lambda \times 2 \ \lambda) \ (0.125)^2 \times 6$$
$$= 0.85 \ \text{fF}$$

The source/drain capacitance must include the gate-source and gate-drain capacitances, which means the complete gate capacitance. This is in contrast to the NOR case, in which only the drain-source overlap capacitance was included.

Determining the average transistor resistance is more complex. In fact, the resistance varies from device to device depending on the position in the chain, caused by differing gate-source voltages and body effects. In this first-order analysis, we simply use the equivalent resistance derived in Chapter 3. This approximation yielded reasonable results in the 4-input NAND-gate example of Chapter 6 (Example 6.4).

One might wonder why we bother with models at all, if simulations could readily produce more accurate results. One has to be aware that the memory modules can contain thousands to millions of transistors, making repeated simulations prohibitively slow. Models also help us to understand the behavior of the memory more readily. Computer simulations do not tell where the bottlenecks are located and how to address them.

Using the computed data and the equivalent model, an estimated value of the propagation delay of the memory core and its components can now be derived.

Example 12.5   Propagation delay of a 512 × 512 NOR ROM

1. The word-line delay of the distributed rc-line containing M cells can be approximated using the expressions derived in Chapter 4:

$$t_{word} = 0.38 \, (r_{word} \times c_{word}) \, M^2 = 0.38 \, (17.5 \, \Omega \times (0.049 + 0.75) \, \text{fF}) \, 512^2 = 1.4 \text{ ns}$$

2. For the bit line, its response time depends upon the transition direction. Assuming a (0.5/0.25) pull-down device and a (1.3125/0.25) pull-up transistor, as derived in Example 12.3, we can compute the propagation delay using the familiar techniques:

$$C_{bit} = 512 \times (0.8 + 0.09) \, \text{fF} = 0.46 \text{ pF}$$

$$t_{HL} = 0.69 \, (13 \text{ k}\Omega \, / \, 2 \, \| \, 31 \text{ k}\Omega \, / \, 5.25) \, 0.46 \text{ pF} = 0.98 \text{ ns}$$

The low-to-high response time can be computed using a similar approach:

$$t_{LH} = 0.69 \, (31 \text{ k}\Omega \, / \, 5.25) \, 0.46 \text{ pF} = 1.87 \text{ ns}$$

Inspection of the preceding results shows that the word-line delay dominates. The former is almost completely due to the large resistance of the polysilicon wire. Some of the techniques to reduce the delay of distributed rc-lines, as introduced in Chapter 9, come in handy here. Driving the address line from both sides and using metal bypass lines (often called *global word lines*) go a long way towards addressing the word-line delay problem (Figure 9-18). Yet, the most effective approach is to carefully partition the memory into sub-blocks of adequate size that balance word- and bit line delay. Partitioning also helps to reduce the energy consumption attributed to driving and switching the word lines.

If necessary, the bit line delay can be reduced as well. The approach most often used is to further reduce the voltage swing on the bit line and to let the sense amplifier restore the output signal to the full swing. Voltage swings around 0.5 V are quite common.

Example 12.6   Propagation Delay of NAND ROM

Using techniques similar to the ones used in Example 12.5, we determine the word-line and bit line delay of the 512 × 512 NAND ROM.

1. The word-line delay is quite similar to that of the NOR case:

$$t_{word} = 0.38 \, (r_{word} \times c_{word}) \, M^2 = 0.38 \, (15 \, \Omega \times (0.049 + 0.56) \, \text{fF}) \, 512^2 = 1.3 \text{ ns}$$

2. For the bit line delay, the worst case occurs when the complete column is populated with 0s except one, and the bottommost transistor is turned on. The distributed model offers a fair approximation of the delay (although this ignores the impact of the pull-up transistor):

$$t_{HL} = 0.38 \times 8.7 \text{ k}\Omega \times 0.85 \text{ fF} \times 511^2 = 0.73 \text{ μs}$$

The worst case for the low-to-high transition occurs when the bottommost transistor is turned off. Using the Elmore delay approach, we find that

$$t_{LH} = 0.69 \ (31 \text{ k}\Omega \ / \ 0.0077) \ (511 \times 0.85 \text{ fF}) = 1.2 \text{ μs}$$

These delays are clearly unacceptable in most cases. Partitioning the memory into smaller modules is the only plausible option.

Power Consumption and Precharged Memory Arrays

The proposed NAND and NOR structures inherit all the disadvantages of the pseudo-NMOS gate discussed in Chapter 6:

1. **Ratioed logic**. The V_{OL} is determined by the ratio of the pull-up and pull-down devices. This can result in unacceptable transistor ratios, as demonstrated earlier in the examples.
2. **Static power consumption**. A static current path exists between the supply rails when the output is low. This may cause severe power dissipation problems.

Example 12.7   Static Power Dissipation of NOR ROM

Consider the case of the (512×512) NOR ROM. It is reasonable to assume that, on the average, 50% of the outputs are low. The standby current for the design of Example 12.4 equals approximately 0.21 mA (for an output voltage of 1.5 V). This translates into a total static dissipation of $(512/2) \times 0.21 \text{ mA} \times 2.5 \text{ V} = 0.14$ W, which is consumed even when nothing happens. Obviously, this is far from desirable.

To address these two issues, one can fall back on the same practices used in designing digital gates. One approach would be to use fully complimentary NAND or NOR gates. The larger number of transistors and the connection to both supply rails makes this approach unattractive from an area perspective. A better approach is to use precharged logic, as shown in Figure 12-17. This approach eliminates the static dissipation as well as the ratioed logic requirements, while keeping the cell complexity the same. Since the logic structure of both the NAND and NOR ROM is simple and is only one level deep, it is possible to ensure that all pull-down paths are off during precharging. This allows us to eliminate the enabling NMOS transistor at the bottom of the pull-down network, keeping the cell simple.

The dynamic architecture enables independent control of the pull-up and pull-down timing. For instance, the PMOS precharge device can be made as large as necessary. Be aware that this transistor loads the clock driver, which might become increasingly hard to design.

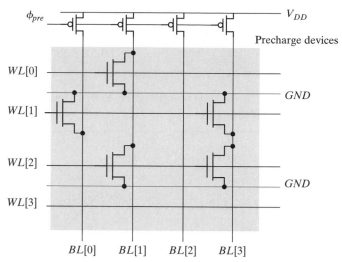

Figure 12-17   Precharged (4 × 4) MOS NOR ROM.

Problem 12.4   Precharged NAND ROM

While precharging works great for NOR ROMs, some severe problems emerge when it is applied to NAND ROMs. Explain why.

The excellent properties of the precharged approach have made it the memory structure of choice. Virtually all large memories currently designed, including NVRWM and RAMs, use dynamic precharging.

ROM Memories: A User Perspective

The reader should be aware that most of the static, dynamic, and power problems raised in the preceding sections are general in nature and apply to other memory architectures as well. Before addressing some of these structures, it is worth discussing the classification of ROM modules and programming approaches. The first class of ROM modules comprises the so-called *application-specific* ROMs, where the memory module is part of a larger custom design and programmed for that particular application only. Under these circumstances, the designer has all degrees of freedom and can use any mask layer (or combination thereof) to program the device.

A second class is formed by the *commodity* ROM chips, where a vendor mass-produces memory modules that are later customized according to customer specifications. Under these circumstances, it is essential that the number of process steps involved in programming be minimal and that they can be performed as a last phase of the manufacturing process. In this way, large amounts of unprogrammed dies can be preprocessed. This *mask-programmable* approach preferably uses the contact (or a metal) mask to personalize or program the memory, as was shown in some of the examples. The programming of a ROM module involves the manufacturer, which introduces an unwelcome delay in product development. It has consequently become increasingly unpopular. A variant of this approach is emerging in the system-on-a-chip arena, where the majority of the chip is preprocessed. Only a minor fraction of the die is mask programmed, preferably using one of the upper metal layers. This could be used, for instance, to program the microcontroller, embedded on the chip, for a variety of applications.

A more desirable approach is for the client to program the memory at his own facility. One technology that offers such capability is the PROM (Programmable ROM) structure that allows the customer to program the memory one time; hence, it is called a WRITE ONCE device. This is most often accomplished by introducing *fuses* (implemented in nichrome, polysilicon, or other conductors) in the memory cell. During the programming phase, some of these fuses are blown by applying a high current, which disables the connected transistor.

While PROMs have the advantage of being "customer programmable," the single write phase makes them unattractive. For instance, a single error in the programming process or application makes the device unusable. This explains the current preference for devices that can be programmed several times (albeit slowly). The next section explains how this can be achieved.

12.2.2 Nonvolatile Read–Write Memories

The architecture of the NVRW memories is virtually identical to the ROM structure. The memory core consists of an array of transistors placed on a word-line/bit line grid. The memory is programmed by selectively disabling or enabling some of those devices. In a ROM, this is accomplished by mask-level alterations. In an NVRW memory, a modified transistor that permits its threshold to be altered electrically is used instead. This modified threshold is retained indefinitely (or at least over a long lifetime) even when the supply voltage is turned off. To reprogram the memory, the programmed values must be *erased*, after which a new programming round can be started. The method of erasing is the main differentiating factor between the various classes of reprogrammable nonvolatile memories. The programming of the memory is typically an order of magnitude slower than the reading operation.

We start this section with a description of the floating-gate transistor, which is the device at the heart of the majority of the reprogrammable memories. The rest of the section is devoted to a number of alterations of the device, mainly with respect to the erase procedure. These modifications are at the source of the different NVRWM families. The section is concluded with a discussion of some emerging nonvolatile memories.

The Floating-Gate Transistor

Over the years, various attempts have been made to create a device with electrically alterable characteristics and enough reliability to support a multitude of write cycles. For example, the MNOS (metal nitride oxide semiconductor) transistor held promise, but has been unsuccessful until now. In this device, threshold-modifying electrons are trapped in a Si_3N_4 layer deposited on top of the gate SiO_2 [Chang77]. A more accepted solution is offered by the floating-gate transistor shown in Figure 12-18, which forms the core of virtually every NVRW memory built today. The structure is similar to a traditional MOS device, except that an extra polysilicon strip is

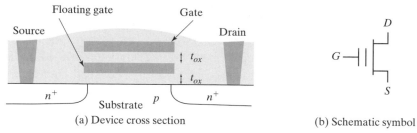

(a) Device cross section   (b) Schematic symbol

Figure 12-18   Floating-gate transistor (FAMOS).

inserted between the gate and channel. This strip is not connected to anything and is called a *floating gate*. The most obvious impact of inserting this extra gate is to double the gate oxide thickness t_{ox}, which results in a reduced device transconductance as well as an increased threshold voltage. Both these properties are not particularly desirable. From other points of view, this device acts as a normal transistor.

More important, this device has the interesting property that its threshold voltage is programmable. Applying a high voltage (above 10 V) between the source and gate-drain terminals creates a high electric field and causes avalanche injection to occur. Electrons acquire sufficient energy to become "hot" and traverse through the first oxide insulator, so that they get trapped on the floating gate. This phenomenon can occur with oxides as thick as 100 nm, which makes it relatively easy to fabricate the device. In reference to the programming mechanism, the floating-gate transistor is often called a *floating-gate avalanche-injection MOS* (or FAMOS) [Frohman74].

The trapping of electrons on the floating gate effectively drops the voltage on that gate. (See Figure 12-19a.) This process is self-limiting—the negative charge accumulated on the floating gate reduces the electrical field over the oxide so that ultimately it becomes incapable of accelerating any more hot electrons. Removing the voltage leaves the induced negative charge in place, and results in a negative voltage on the intermediate gate (Figure 12-19b). From a device point of view, this translates into an effective increase in threshold voltage. To turn on the device, a higher voltage is needed to overcome the effect of the induced negative charge (Figure 12-19c). Typically, the resulting threshold voltage is around 7 V; thus, a 5-V

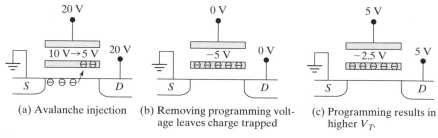

(a) Avalanche injection   (b) Removing programming volt-   (c) Programming results in
                               age leaves charge trapped   higher V_T.

Figure 12-19   Programming the floating-gate transistor.

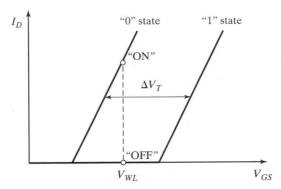

Figure 12-20 I–V curve shift caused by hot-electron programming. Applying a word-line voltage V_{WL} results in either current to flow ("0" state), or not ("1" state).

gate-to-source voltage is not sufficient to turn on the transistor, and the device is effectively disabled. The charge injected onto the floating gate effectively shifts the I–V curves of the transistor, as shown in Figure 12-20. Curve A is for the "0" state, while curve B stands for the transistor with the V_T shift, representing the "1" state. It can be established that the value of the V_T shift is expressed as

$$\Delta V_T = (-\Delta Q_{FG})/C_{FC} \tag{12.1}$$

with C_{FC} the capacitance between external gate contact and floating gate, and ΔQ_{FG} the charge injected onto the floating gate.

Since the floating gate is surrounded by SiO_2, which is an excellent insulator, the trapped charge can be stored for many years, even when the supply voltage is removed, creating a nonvolatile storage mechanism. One of the major concerns of the floating-gate approach is the need for high programming voltages. By tailoring the impurity profiles, technologists have been able to reduce the required voltage from the original 25 V to approximately 12.5 V in today's memories.

Virtually all nonvolatile memories are currently based on the floating-gate approach. Different classes can be identified, based on the erasure mechanism.

Erasable-Programmable Read-Only Memory (EPROM)

An EPROM is erased by shining ultraviolet light on the cells through a transparent window in the package. The UV radiation renders the oxide slightly conductive by the direct generation of electron-hole pairs in the material. The erasure process is slow and can take from seconds to several minutes, depending on the intensity of the UV source. Programming takes several (5–10) μs/word. Another problem with this approach is the *limited endurance*—the number of erase/program cycles is generally limited to a maximum of one thousand, mainly as a result of the UV erasing procedure. Reliability is also an issue. The device thresholds might vary with repeated programming cycles. Most EPROM memories therefore contain on-chip circuitry to control the value of the thresholds to within a specified range during programming. Finally, the injection always entails a large channel current, as high as 0.5 mA at a control gate voltage of 12.5 V. This causes high power dissipation during programming.

On the other hand, the EPROM cell is extremely simple and dense, making it possible to fabricate large memories at a low cost. EPROMs were therefore attractive in applications that do not require regular reprogramming. Due to the cost and reliability issues, EPROMs have fallen out of favor and have been replaced by Flash memories.

Electrically Erasable Programmable Read-Only Memory (EEPROM or E²PROM)

The major disadvantage of the EPROM approach is that the erasure procedure has to occur "off system." This means the memory must be removed from the board and placed in an EPROM programmer for programming. The EEPROM approach avoids this labor-intensive and annoying procedure by using another mechanism to inject or remove charges from a floating gate—namely, *tunneling*. A modified floating-gate device called the FLOTOX (floating-gate tunneling oxide) transistor is used as a programmable device that supports an electrical-erasure procedure [Johnson80]. A cross section of the FLOTOX structure is shown in Figure 12-21a. It resembles the FAMOS device, except that a portion of the dielectric separating the floating gate from the channel and drain is reduced in thickness to about 10 nm or less. When a voltage of approximately 10 V (equivalent to an electrical field of around 10^9 V/m) is applied over the thin insulator, electrons travel to and from the floating gate through a mechanism called *Fowler–Nordheim tunneling* [Snow67]. The I–V characteristic of the tunneling junction is plotted in Figure 12-21b.

The main advantage of this programming approach is that it is reversible; that is, erasing is simply achieved by reversing the voltage applied during the writing process. Injecting electrons onto the floating gate raises the threshold, while the reverse operation lowers the V_T. This bi-directionality, however, introduces a threshold-control problem: Removing too much charge from the floating gate results in a depletion device that cannot be turned off by the standard word-line signals. Notice that the resulting threshold voltage depends on the initial charge on the gate, as well as the applied programming voltages. It also is a strong function of the oxide thickness, which is subject to nonnegligible variations over the die. To remedy this problem, an extra transistor connected in series with the floating-gate transistor is added to the EEPROM cell to remedy this problem. This transistor acts as the access device during the read operation, while

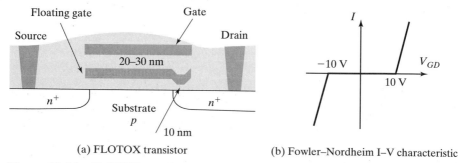

(a) FLOTOX transistor   (b) Fowler–Nordheim I–V characteristic

Figure 12-21   FLOTOX transistor, programmable by using Fowler–Nordheim tunneling.

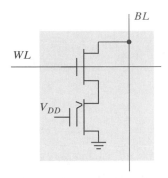

Figure 12-22 EEPROM cell as configured during a read operation. When programmed, the threshold of the FLOTOX device is higher than V_{DD}, effectively disabling it. If not, it acts as a closed switch.

the FLOTOX transistor performs the storage function. (See Figure 12-22.) This is in contrast to the EPROM cell, where the FAMOS transistor acts as both the programming and access device.

The EEPROM cell with its two transistors is larger than its EPROM counterpart. This area penalty is further aggravated by the fact that the FLOTOX device is intrinsically larger than the FAMOS transistor due to the extra area of the tunneling oxide. Additionally, the fabrication of the very thin oxide is a challenging and costly manufacturing step. EEPROM components thus pack less bits at a higher cost than EPROMs. On the positive side of the balance, EEPROMs offer a higher versatility. They also tend to last longer, as they can support up to 10^5 erase/write cycles.[4] Repeated programming causes a drift in the threshold voltages due to permanently trapped charges in the SiO_2. This finally leads to malfunction or the inability to reprogram the device.

Flash Electrically Erasable Programmable Read-Only Memory (Flash)

The concept of Flash EEPROMs was introduced in 1984 and has rapidly evolved into the most popular nonvolatile memory architecture. It combines the density of the EPROM with the versatility of the EEPROM structures, with cost and functionality ranging somewhere between the two.

Technically, the Flash EEPROM is a combination of the EPROM and EEPROM approaches. Most Flash EEPROM devices use the avalanche hot-electron-injection approach to program the devices. Erasure is performed using Fowler–Nordheim tunneling, as for EEPROM cells. The main difference is that erasure is performed in bulk for the complete chip, or for a subsection of the memory. While this represents a reduction in flexibility, it has the advantage that the extra access transistor of the EEPROM cell can be eliminated. Erasing the complete memory core at once makes it possible to carefully monitor of the device characteristics during erasure, guaranteeing that the unprogrammed transistor acts as an enhancement device. The monitoring control hardware on the memory chip regularly checks the value of the threshold during erasure,

---

[4]Thin oxides are not the only way to realize electron tunneling. Other approaches include the use of textured surfaces to locally enhance the surface field [Masuoka91].

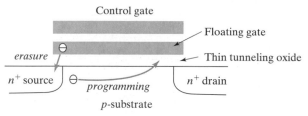

Figure 12-23   ETOX device as used in Flash EEPROM memories.

and adjusts the erasure time dynamically. This approach is only practical when erasing large chunks of memory at a time; hence the flash concept. The simpler cell structure results in a substantial reduction in cell size and an increased integration density.

For instance, Figure 12-23 shows the ETOX Flash cell introduced by Intel [Pashley89]. This is only one of the many existing alternatives. It resembles a FAMOS gate, except that a very thin tunneling gate oxide is utilized (10 nm). Different areas of the gate oxide are used for programming and erasure. Programming is performed by applying a high voltage (12 V) on gate and drain terminals for a grounded source, while erasure occurs with the gate grounded and the source at 12 V.

Figure 12-23 illustrates how this cell can be incorporated into a NOR ROM structure. A programming cycle starts with an *erase operation* (a): A 0 V gate voltage is applied, combined with a high voltage (12 V) at the source. Electrons, if any, at the floating gate are ejected to the source by tunneling. All cells are erased simultaneously. The different initial values of the cell threshold voltages, as well as variations in the oxide thickness, may cause variations in the threshold voltage at the end of the erase operation. This is remedied in two ways: (1) Before applying the erase pulse, all the cells in the array are programmed so that all the thresholds start at approximately the same value; (2) After that, an erase pulse of controlled width is applied. Subsequently, the whole array is read to ensure to check whether or not the cells have been erased. If not, another erase pulse is applied, followed by a read cycle. The algorithm is applied until all cells have threshold voltages that are below the required level. Typical erasing times are between 100 ms to 1 s. For the *write (programming) operation* (b), a high-voltage pulse is applied to the gate of the selected device. If a "1" is applied to the drain at that time, hot electrons are generated and injected onto the floating gate, raising the threshold (effectively turning it into an always-off device). If not, the floating gate remains in the previous state of no electrons, corresponding to a "0" state. To obtain the necessary threshold shift of 3 to 3.5 V, a pulse with typical values in the $1–10$-μs range must be applied. The *read operation* (c) proceeds as in any NOR ROM structure. To select a cell, its word line is raised to 5 V, causing a conditional discharge of the bit line.

The NOR architecture leads to fast random read access times. At the same time, erasure and programming times are slow due to the need for precise control of the thresholds. These properties make this style of Flash memory attractive for applications such as program-code storage. Other applications, such as video or audio file storage, do not need fast random access, but are better served by large storage density (reducing the memory cost), fast erasure and

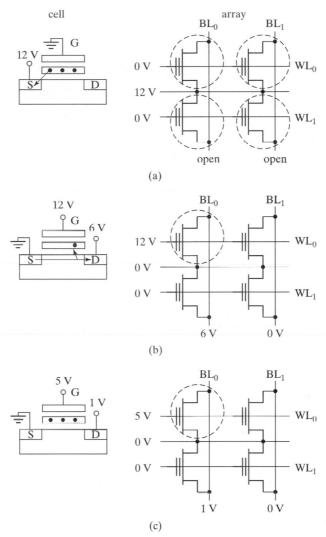

Figure 12-24 Basic operations in a NOR Flash memory [from Itoh01]. (a) Erase; (b) write; (c) read.

programming, and fast serial access. These requirements are more readily provided by the NAND ROM architecture described earlier. A range of Flash memory manufacturers have thus opted for this topology. The basic module consists of 8 to 16 floating-gate transistors connected in series, as shown in Figure 12-25. This chain is connected to the bit line and the source line (GND) with the aid of two select transistors. By eliminating all contacts between word lines, the resulting cell size is approximately 40% smaller than the NOR cell. The memory shown uses Fowler–Nordheim tunneling for both programming and erasure. During erasure, all cells in the

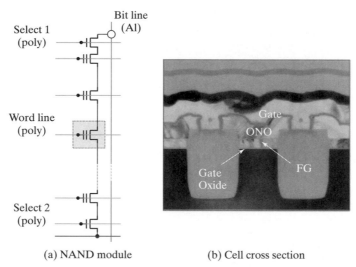

(a) NAND module   (b) Cell cross section

Figure 12-25   NAND-based Flash Module [Nakamura02]. The second gate uses a high-dielectric material (ONO or Oxide–Nitride–Oxide) to increase the C_{FG} capacitance, and hence the impact of charge injection on *VT*.

module are programmed to become depletion devices (this is, transistors with a negative threshold). This is accomplished by applying 20 V to the bit and source lines, and 0 V to the word lines. During programming, the selection transistors are set such that the bit line is connected and the source line isolated. To write a "1", the bit line is grounded and a high voltage (20 V) is applied to the word line of interest. Electrons tunnel into the floating gate, causing the threshold to increase. For a "0", the cell threshold is left unperturbed (by keeping the bit line high). The read operation proceeds just like in the NAND ROM, with both select transistors enabled. The use of tunneling as the primary mechanism reduces the current requirements, and allows for the parallel programming of many modules while keeping power consumption under control. The programming time per byte is as fast as 200–400 ns. By processing a complete module at a time, fast and reliable erasure can be obtained, leading to fast erasure of around 100 ms for a complete chip.

Many other architectures for flash memories have been defined, trading off read performance, erasure and programming speed, and density. All are based on the principles defined earlier. For a more detailed description, we refer the reader to the excellent overview in [Itoh01]. One issue deserves special attention. All nonvolatile memories need high voltages (> 10 V) for programming and erasure. Some of these supplies can be generated on chip using charge pumps, especially if the current requirements are low, such as is the case for Fowler–Nordheim tunneling. Hot-electron injection, on the other hand, draws a large current (> 0.5 mA), and an external supply is a necessity, adding to the cost and the complexity of the system. The search is definitely on for nonvolatile memories that do not need high voltages and external supplies.

New Trends in Nonvolatile Memories

All nonvolatile memories discussed so far are based on the floating-gate transistor concept. Over the years, a number of alternative approaches have been proposed. While most of those have languished and not made it to the commercial market, some novel cell structures are currently catching on, and may make a big impact in the foreseeable future. The two most prominent among those are the *FRAM* (or Ferroelectric RAM) and the *MRAM* (Magnetoresistive RAM). However, calling the floating-gate device at the end of its cycle is premature. Innovative approaches such as multilevel cells are creating opportunities for ever denser nonvolatile storage. We will briefly discuss the basic concepts of each of these exciting developments.

Multilevel Nonvolatile Memories   All memory cells considered so far are bistable, which means that they store either a 0 or a 1. The memory density could be increased substantially if more than two states could be stored in a single cell. The problem with this approach is that it requires a substantial improvement in the signal-to-noise ratio of the memory cell. By storing k bits in a memory cell, the available signal is reduced by $1/(2^k - 1)$. Fortunately, the flash cell has the advantage of providing internal gain. Also, the capability of V_T adjustment for each cell can help to overcome the signal-to-noise issue. Commercial memories, storing two bits per cell, have been demonstrated [Baur95]. Yet, with the continuing reduction of the supply voltages, it is highly unlikely that multilevel memories are destined for a lucrative future.

FRAM   Ferroelectric RAM has been "almost there" for quite some time, mainly based on optimistic expectations. After a lot of false starts, it seems like FRAM is finally close to fulfilling its promises. In contrast to the "programmable transistor" used in current NVRWMs, the FRAM is based on a "programmable capacitor." The dielectric material used in a ferroelectric capacitor belongs to a class of materials called *Perovskite crystals*. These crystals polarize when an electric field is applied across them. This polarization is maintained when the electrical field is removed. To depolarize the dielectric, an electrical field of the opposite direction has to be applied.

A memory cell built on this concept closely resembles the single-transistor DRAM circuit, to be discussed in a following section. During a write operation, the capacitor cell is polarized one way or the other. To read the contents, an electrical field is applied. Depending upon the state of the capacitor, a current is generated, which can be detected by a sense amplifier.

The advantage of the FRAM is a very high storage density, leading to cell sizes smaller than DRAMs, while offering nonvolatility at the same time. The number of read/write cycles for FRAMs is also orders of magnitude higher than EEPROMs and Flash memories. Its low power consumption makes it very attractive for applications such as smart cards and RF tags. The main challenge still is to generate large memories reliably using this technology.

MRAM   *Magnetoresistive random access memory* is a method of storing data bits using magnetic charges instead of the electrical charges used by DRAM. A metal is called *magnetoresistive* if it shows a slight change in electrical resistance when placed in a magnetic field. The concept is quite similar to the magnetic core memories used in the mainframe computers in the early days of electronics. The main differences are the scale (millions times smaller) and the materials used. Development of MRAM has followed two scientific schools: (1) spin electronics, the science behind giant magnetoresistive heads used in disk drives and (2) tunneling magnetic resistance, or TMR, which is expected to be the basis of future MRAM. Researchers at IBM demonstrated a 1 Kbit MRAM memory in 2000. Each cell consists of a magnetic tunnel junction and a FET, for a total area per cell of 3 μm^2 in a 0.25-μm CMOS process [Scheuerlein00]. Read and write times in the range of 10 ns were obtained. While it is obviously too early to predict the future of this particular device, it seems apparent that new materials and approaches are bound to change the field of nonvolatile storage as well as semiconductor memories in general.

Nonvolatile Read–Write Memories—Summary

Some numbers are useful to put the different nonvolatile technologies in perspective. Table 12-1 summarizes the current essential data for nonvolatile memories. The table confirms that the flexibility of the EEPROM structure comes at the expense of density and performance. EPROMs and Flash EEPROM devices are comparable in both density and speed. The versatility of the latter explains its explosive growth in a short time span.

A large number of design considerations raised in the section on read-only memories are valid for the NVRWMs as well. In addition, the (E)EPROM structures must cope with the extra complexity of the programming and erasure circuitry. Remember that all the proposed structures require the availability of high-voltage signals (12–20 V) on word and bit lines during the programming, while standard 3- or 5-V signals are used on the same wires during the read mode. The generation and distribution of those signals requires some interesting circuit design, which, unfortunately, is beyond the scope of this text.

Table 12-1 Comparison between nonvolatile memories ([Itoh01]). V_{DD} = 3.3 or 5 V; V_{PP} = 12 or 12.5 V.

	Cell— Nr. of Transistors	Cell Area (ratio wrt EPROM)	Mechanism		External Power Supply		Program/ Erase Cycles
			Erase	Write	Write	Read	
MASK ROM	1 T (NAND)	0.35–5	—	—	—	V_{DD}	0
EPROM	1 T	1	UV Exposure	Hot electrons	V_{PP}	V_{DD}	~100
EEPROM	2 T	3–5	FN Tunneling	FN Tunneling	V_{PP} (int)	V_{DD}	10^4–10^5
Flash Memory	1 T	1–2	FN Tunneling	Hot electrons	V_{PP}	V_{DD}	10^4–10^5
			FN Tunneling	FN Tunneling	V_{PP} (int)	V_{DD}	10^4–10^5

12.2.3 Read–Write Memories (RAM)

Providing a memory cell with roughly equal read and write performance requires a more complex cell structure. While the contents of the ROM and NVRWM memories are ingrained in the cell topology or programmed into the device characteristics, storage in RAM memories is based on either positive feedback or capacitive charge, similar to the ideas introduced in Chapter 6. These circuits would be perfectly suitable as R/W memory cells, but they tend to consume too much area. In this section, we introduce a number of simplifications that trade off area for either performance or electrical reliability. They are labeled as either SRAMs or DRAMs, depending on the storage concept used.

Static Random-Access Memory (SRAM)

The generic SRAM cell is introduced in Figure 12-26. It turns out to be quite similar to the static SR latch, shown in Figure 7-21. It requires six transistors per bit. Access to the cell is enabled by the word line, which replaces the clock and controls the two pass transistors M_5 and M_6, shared between the read and write operation. In contrast to the ROM cells, two bit lines transferring both the stored signal and its inverse are required. Although providing both polarities is not a necessity, doing so improves the noise margins during both read and write operations, as will become apparent in the subsequent analysis.

Problem 12.5   CMOS SRAM Cell

Does the SRAM cell presented in Figure 12-26 consume standby power? Explain. Draw an equivalent pseudo-NMOS implementation. How about the standby power in that case?

Operation of SRAM Cell   The SRAM cell should be sized as small as possible to achieve high memory densities. Reliable operation of the cell, however, imposes some sizing constraints.

To understand the operation of the memory cell, let us consider the read and write operations in sequence. While doing so, we also derive the transistor-sizing constraints.

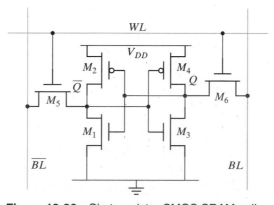

Figure 12-26   Six-transistor CMOS SRAM cell.

Example 12.8   CMOS SRAM—Read Operation

Assume that a 1 is stored at Q. We further assume that both bit lines are precharged to 2.5 V before the read operation is initiated. The read cycle is started by asserting the word line, enabling both pass transistors M_5 and M_6 after the initial word-line delay. During a correct read operation, the values stored in Q and $\overline{Q}$ are transferred to the bit lines by leaving BL at its precharge value and by discharging $\overline{BL}$ through M_1–M_5. A careful sizing of the transistors is necessary to avoid accidentally writing a 1 into the cell. This type of malfunction is frequently called a *read upset*.

This is illustrated in Figure 12-27. Consider the $\overline{BL}$ side of the cell. The bit line capacitance for larger memories is in the pF range. Consequently, the value of $\overline{BL}$ stays at the precharged value V_{DD} upon enabling of the read operation ($WL \rightarrow 1$). This series combination of two NMOS transistors pulls down the $\overline{BL}$ towards ground. For a small-sized cell, we would like to have these transistor sized as close to minimum as possible, which would result in a very slow discharge of the large bit line capacitance. As the difference between BL and $\overline{BL}$ builds up, the sense amplifier is activated to accelerate the reading process.

Initially, upon the rise of the WL, the intermediate node between these two NMOS transistors, $\overline{Q}$, is pulled up toward the precharge value of $\overline{BL}$. This voltage rise of $\overline{Q}$ must stay low enough not to cause a substantial current through the M_3–M_4 inverter, which in the worst case could flip the cell. It is necessary to keep the resistance of transistor M_5 larger than that of M_1 to prevent this from happening.

The boundary constraints on the device sizes can be derived by solving the current equation at the maximum allowed value of the voltage ripple ΔV. We ignore the body effect on transistor M_5 for simplicity and write

$$k_{n,M5}\left((V_{DD} - \Delta V - V_{Tn})V_{DSATn} - \frac{V_{DSATn}^2}{2}\right) = k_{n,M1}\left((V_{DD} - V_{Tn})\Delta V - \frac{\Delta V^2}{2}\right) \qquad (12.2)$$

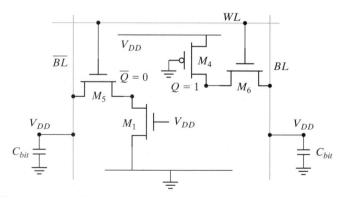

Figure 12-27   Simplified model of CMOS SRAM cell during read ($Q = 1$, $V_{precharge} = V_{DD}$).

which simplifies to

$$\Delta V = \frac{V_{DSATn} + CR(V_{DD} - V_{Tn}) - \sqrt{V_{DSATn}^2(1 + CR) + CR^2(V_{DD} - V_{Tn})^2}}{CR} \quad (12.3)$$

where CR is called the cell ratio and is defined as

$$CR = \frac{W_1/L_1}{W_5/L_5} \quad (12.4)$$

The value of voltage rise ΔV as a function of CR for our 0.25-μm technology is plotted in Figure 12-28. To keep the node voltage from rising above the transistor threshold (of about 0.4 V), the cell ratio must be greater than 1.2. For large memory arrays, it is desirable to keep the cell size minimal while maintaining read stability. If the transistor M_1 is minimum sized, the access pass transistor M_5 has to be made weaker by increasing its length. This is undesirable, because it adds to the load of the bit line. A preferred solution is to minimize the size of the pass transistor, and increase the width of the NMOS pull-down M_5 to meet the stability constraint. This slightly increases the minimum size of the cell. The designer must perform careful simulations to guarantee cell stability across all process corners [Preston01].

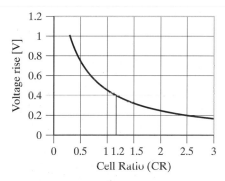

Figure 12-28 Voltage rise inside the cell upon read versus cell ratio (ratio of M_1/M_5). The voltage inside the cell does not rise above the threshold for $CR > 1.2$

The preceding analysis presents the worst case. The second bit line BL clamps Q to V_{DD}, which makes the inadvertent toggling of the cross-coupled inverter pair difficult. This demonstrates one of the major advantages of the dual bit line architecture.

Beyond adjusting the size of the cell transistors, the erroneous toggling can be prevented by precharging the bit lines to another value, such as $V_{DD}/2$. This effectively makes it impossible for Q to reach the switching threshold of the connecting inverter. Precharging to the midpoint of the voltage range has some performance benefits as well, since it limits the voltage swing on the bit lines.

Example 12.9   CMOS SRAM Write Operation

In this example, we derive the device constraints necessary to ensure a correct write operation. Assume that a 1 is stored in the cell (or $Q = 1$). A 0 is written in the cell by setting $\overline{BL}$ to 1 and BL to 0, which is identical to applying a reset pulse to an SR latch. This causes the flip-flop to change state if the devices are sized properly.

During the initiation of a write, the schematic of the SRAM cell can be simplified to the model of Figure 12-29. It is reasonable to assume that the gates of transistors M_1 and M_4 stay at V_{DD} and GND, respectively, as long as the switching has not commenced. While this condition is violated once the flip-flop starts toggling, the simplified model is more than accurate for hand-analysis purposes.

Note that $\overline{Q}$ side of the cell cannot be pulled high enough to ensure the writing of 1. The sizing constraint, imposed by the read stability, ensures that this voltage is kept below 0.4 V. Therefore, the new value of the cell has to be written through transistor M_6.

A reliable writing of the cell is ensured if we can pull node Q low enough—this is, below the threshold value of the transistor M_1.[5] The conditions for this to occur can be derived by writing out the *dc* current equations at the desired threshold point, as follows:

$$k_{n,\,M6}\left((V_{DD} - V_{Tn})V_Q - \frac{V_Q^2}{2}\right) = k_{p,\,M4}\left((V_{DD} - |V_{Tp}|)V_{DSATp} - \frac{V_{DSATp}^2}{2}\right) \qquad (12.5)$$

Solving for V_Q leads to

$$V_Q = V_{DD} - V_{Tn} - \sqrt{(V_{DD} - V_{Tn})^2 - 2\frac{\mu_p}{\mu_n}PR\left((V_{DD} - |V_{Tp}|)V_{DSATp} - \frac{V_{DSATp}^2}{2}\right)}, \quad (12.6)$$

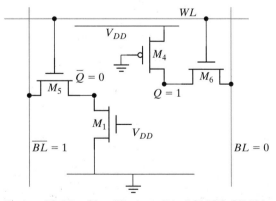

Figure 12-29   Simplified model of CMOS SRAM cell during write ($Q = 1$).

---

[5]In principle, it is sufficient to pull Q below the switching threshold of the inverter formed by M_1 and M_2 to ensure the initiation of the switching action. For noise margin purposes, it is safer to require that Q is pulled below the threshold of M_1.

where the *pull-up ratio of the cell*, *PR*, is defined as the size ratio between the PMOS pull-up and the NMOS pass transistor:

$$PR = \frac{W_4/L_4}{W_6/L_6}. \tag{12.7}$$

The dependence of V_Q on *PR* for a 0.25-μm process is plotted in Figure 12-30. The lower PR, the lower the value of V_Q. If we wish to pull the node below V_{Tn}, the pull-up ratio has to be below 1.8.

This constraint is met, by a large margin, when using a minimum-sized devices for both the PMOS pull-up M_4 and NMOS access transistor M_6. However, a designer must assure that the writeability constraint is met under all process corners. The worst case presents a process with strong PMOS devices, weak NMOS devices, and the memory operated at a higher than nominal supply voltage [Preston01].

Our initial assumption was that the transistors M_1 and M_2 do not participate in the writing process. This is not completely true in practice. As soon as one side of the cell starts switching, the other side eventually follows, engaging the positive feedback.

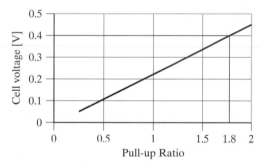

Figure 12-30   Voltage written into the cell versus pull-up ratio (size ratio between the PMOS pull-up and access transistor). *PR* should be less than 1.8.

Performance of SRAM Cell   When analyzing the transient behavior of the SRAM cell, one realizes that the read operation is the critical one. It requires the (dis)charging of the large bit linebit line capacitance through the stack of two small transistors of the selected cell. For instance, $C_{\overline{BL}}$ has to be discharged through the series combination of M_5 and M_1 in the example of Figure 12-27. The write time is dominated by the propagation delay of the cross-coupled inverter pair, as the drivers that force *BL* and $\overline{BL}$ to the desired values can be large. To accelerate the read time, SRAMs use sense amplifiers. As the difference in voltage between *BL* and $\overline{BL}$ builds up, the sense amplifier is activated, and it quickly discharges one of the bit lines.

Improved MOS SRAM Cells   The six-transistor SRAM cell, while simple and reliable, consumes a substantial area. Besides the devices, it requires the signal routing and connections to two bit lines, a word line, and both supply rails. Placing the two PMOS transistors in the *N*-well

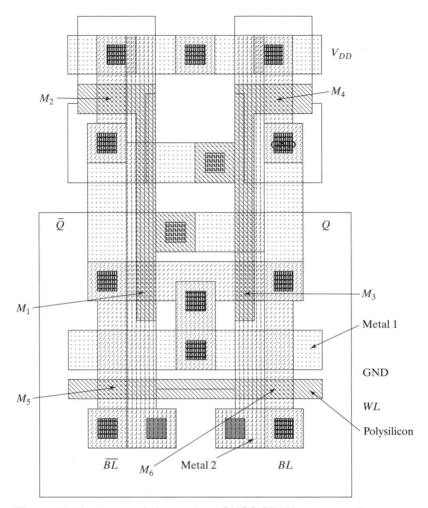

Figure 12-31   Layout of six-transistor CMOS SRAM memory cell.

significantly contributes to the area. Figure 12-31 shows a possible layout for such a cell. Its dimensions are dominated by the wiring and interlayer contacts (11.5 of them—the top and bottom ones only count for one-half, since they are shared with the neighboring cells), as well as the minimum spacing requirements for the well.

Designers of large memory arrays have therefore proposed other cell structures that are not only based on revised transistor topologies, but also on the presence of special devices and a more complex technology.

Consider the cell schematic of Figure 12-32, called the *resistive load* SRAM cell (also known as the four-transistor SRAM cell). The special feature of this cell is that the cross-coupled CMOS inverter pair is replaced by a pair of resistive-load NMOS inverters. The PMOS

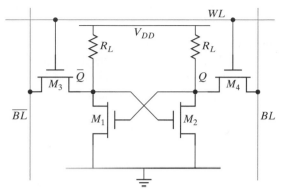

Figure 12-32 Resistive-load SRAM cell.

transistors are replaced by resistors, and the wiring is simplified. This reduces the SRAM cell size by approximately one-third, as illustrated in Table 12-2 for the example of a 1-Mbit SRAM.

From the SRAM cell writeability analysis, we found that there exists only a constraint on lower limit of the pull-up resistance. Since the bit lines are externally precharged, the cell is not involved in the pull-up process, and there is no penalty for having very large pull-ups, as opposed to conventional logic. Therefore, an advanced SRAM process technology includes special resistors that are made as large as possible to minimize static power consumption.

Keeping the static power dissipation per cell as low as possible is a prime design priority in SRAM cells. Consider a 1-Mbit SRAM memory operating at 2.5V and using a 10 kΩ resistor as the inverter load. With each cell sinking 0.25 mA in static current, a total standby dissipation of 250 W can be recorded! Therefore, the only obvious choice is to make the load resistance as large as possible. A very large, yet compact, resistor can be manufactured by using an undoped polysilicon, which has a sheet resistance of several TΩ/sq (Tera = 10^{12}!). The only additional constraint on the pull-up resistors is to maintain the state of the cell, that is to compensate for the leakage currents that typically range around 10^{-15} A/cell [Takada91]. The low leakage in recent

Table 12-2 Comparison of CMOS SRAM cells used in 1-Mbit memory (from [Takada91])

	Complementary CMOS	Resistive Load	TFT Cell
Number of transistors	6	4	4 (+2 TFT)
Cell size	58.2 μm^2 (0.7-μm rule)	40.8 μm^2 (0.7-μm rule)	41.1 μm^2 (0.8-μm rule)
Standby current (per cell)	10^{-15} A	10^{-12} A	10^{-13} A

technologies is achieved by using the devices with higher thresholds. The resistor current should be at least two orders of magnitude larger to accomplish that goal, or $I_{load} > 10^{-13}$ A. This puts an upper limit on the resistor value.

The realization that the pull-up devices are only needed for charge-loss compensation has resulted in a revised version of the six-transistor memory cell of Figure 12-26. Instead of using traditional, expensive PMOS devices, the pull-up transistors are realized as parasitic devices deposited on top of the cell structure using a thin-film technology. These PMOS *thin-film transistors* (TFTs) have inferior properties with respect to normal devices and are characterized by a current of approximately 10^{-8} A and 10^{-13} in the *ON* and *OFF* modes respectively for a 5 V gate-source voltage [Sasaki90, Ootani90]. The complimentary nature of the cell results in an increased cell reliability with less sensitivity to leakage and soft errors, yet at a lower standby current compared to the resistive load cell.

Note that a high-resistivity polysilicon and thin-film transistors are additional features that are not available in a standard logic process. Therefore, embedded SRAM cells, such as those used in microprocessor caches, stick to the conventional 6T cell of Figure 12-26.

Dynamic Random-Access Memory (DRAM)

While discussing the resistive-load SRAM cell, we noted that the only function of the load resistors is to replenish the charge lost by leakage. One option is to eliminate these loads completely and compensate for the charge loss by periodically rewriting the cell contents. This *refresh* operation, which consists of a read of the cell contents followed by a write operation, should occur often enough that the contents of the memory cells are never corrupted by the leakage. Typically, refresh should occur every 1 to 4 ms. For larger memories, the reduction in cell complexity more than compensates for the added system complexity imposed by the refresh requirement. These memories are called *dynamic*, since the underlying concept of these cells is based on charge storage on a capacitor.

Three-Transistor Dynamic Memory Cell   The first kind of dynamic cell is obtained by eliminating the load resistors in the schematic of Figure 12-33. The four-transistor cell can be further simplified by observing that the cell stores both the data value and its complement; hence, it contains redundancy. Eliminating one more device (e.g., M_1) removes this redundancy and results in the three-transistor (3T) cell of Figure 12-33 [Regitz70]. This cell formed the core of the first popular MOS semiconductor memories such as the first 1-Kbit memory from Intel [Hoff70]. While replaced by more area-efficient cells in the very large memories of today, it is still the cell of choice in many memories embedded in application-specific integrated circuits. This can be attributed to its relative simplicity in both design and operation.

The cell is written to by placing the appropriate data value on *BL1* and asserting the *write-word line* (*WWL*). The data is retained as charge on capacitance C_S once *WWL* is lowered. When reading the cell, the *read-word line* (*RWL*) is raised. The storage transistor M_2 is either on or off depending upon the stored value. The bit line *BL2* is either clamped to V_{DD} with the aid of a load device, for example, a grounded PMOS or saturated NMOS transistor, or is precharged to either

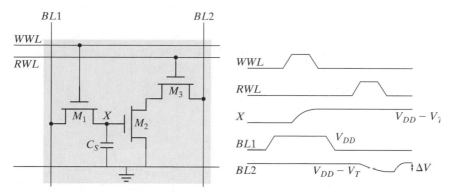

Figure 12-33 Three-transistor dynamic memory cell and the signal waveforms during read and write.

V_{DD} or $V_{DD} - V_T$. The former approach necessitates careful transistor sizing and causes static power consumption. Therefore, the precharged approach is generally preferable. The series connection of M_2 and M_3 pulls $BL2$ low when a 1 is stored. $BL2$ remains high in the opposite case. Notice that the cell is inverting; that is, the inverse value of the stored signal is sensed on the bit line. The most common approach to refreshing the cell is to read the stored data, put its inverse on $BL1$, and assert WWL in consecutive order.

The cell complexity is substantially reduced with respect to the static cell. This is illustrated by the example layout of Figure 12-34. The total area of the cell is 576 λ^2, compared to the 1092 λ^2 of the SRAM cell of Figure 12-31. These numbers do not take into account the potential area reduction obtained by sharing with neighboring cells. The area reduction is mainly due to the elimination of contacts and devices.

Further simplifications in the cell structure are possible at the expense of a more complex circuit operation. For instance, bit lines $BL1$ and $BL2$ can be merged into a single wire. The read and write cycles can proceed as before. The read–sense–write refresh cycle must be altered considerably, since the data value read from the cell is the complement of the stored value. This requires the bit line to be driven to both values in a single cycle. Another option is to merge the RWL and the WWL lines. Once again, this does not significantly change the cell operation. A read operation is automatically accompanied by a refresh of the cell contents. A careful control of the word-line voltage is necessary to prevent a writing of the cell before the actual value is read during refresh.

Finally, the following interesting properties of the 3T cell are worth mentioning:

1. In contrast to the SRAM cell, no constraints exist on the *device ratios*. This is a common property of dynamic circuits. The choice of device sizes is solely based on performance and reliability considerations. Observe that this statement is not valid when a static bit line load approach is employed.

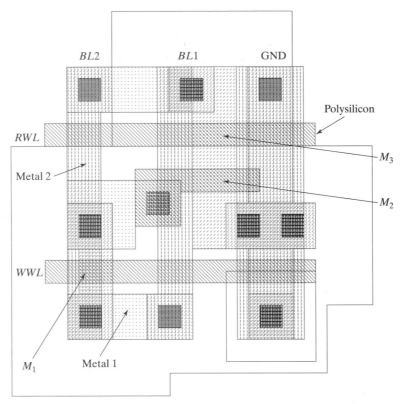

Figure 12-34   Example layout of three-transistor dynamic memory cell.

2. In contrast to other DRAM cells, reading the 3T cell contents is *nondestructive;* that is, the data value stored in the cell is not affected by a read.

3. No special process steps are needed. The storage capacitance is nothing more than the gate capacitance of the readout device. This is in contrast with the other DRAM cells, discussed next, and makes the 3T cell attractive for embedded memory applications.

4. The value stored on the storage node X when writing a 1 equals $V_{WWL} - V_{Tn}$. This threshold loss reduces the current flowing through M_2 during a read operation and increases the read access time. To prevent this, some designs *bootstrap* the word-line voltage, or in other words, raise V_{WWL} to a value higher than V_{DD}.

One-Transistor Dynamic Memory Cell   Another dramatic reduction in cell complexity can be obtained by a further sacrifice in some of the cell properties. The resulting structure, called the one-transistor DRAM cell (1T), is undoubtedly the most pervasive dynamic DRAM cell in commercial memory design.[6] A schematic is shown in Figure 12-35 [Dennard68]. Its basic opera-

---

[6]A DRAM cell containing only two transistors can also be conceived. It offers no substantial advantages over either the 3T or 1T cell and is, therefore, only rarely used.

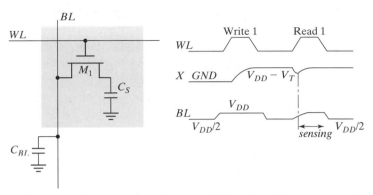

Figure 12-35 One-transistor dynamic RAM cell and the corresponding signal waveforms during read and write.

tional concepts are extremely simple. During a write cycle, the data value is placed on the bit line *BL,* and the word line *WL* is raised. Depending on the data value, the cell capacitance is either charged or discharged. Before a read operation is performed, the bit line is precharged to a voltage V_{PRE}. Upon asserting the word line, a charge redistribution takes place between the bit line and storage capacitance. This results in a voltage change on the bit line, the direction of which determines the value of the data stored. The magnitude of the swing is given by the expression

$$\Delta V = V_{BL} - V_{PRE} = (V_{BIT} - V_{PRE})\frac{C_S}{C_S + C_{BL}} \qquad (12.8)$$

where C_{BL} is the bit line capacitance, V_{BL} the potential of the bit line after the charge redistribution, and V_{BIT} the initial voltage over the cell capacitance C_S. As the cell capacitance is normally one or two orders of magnitude smaller than the bit line capacitance, this voltage change is very small, typically around 250 mV for state-of-the-art memories [Itoh90]. The ratio $C_S/(C_S + C_{BL})$ is called the *charge-transfer ratio* and ranges between 1% and 10%.

Amplification of ΔV to the full voltage swing is necessary if functionality is to be achieved. This observation marks a first major difference between the 1T and 3T, as well as other, DRAM cells.

1. A 1T DRAM requires the *presence of a sense amplifier* for each bit line to be functional. This is a result of the charge-redistribution-based readout. The read operation of all cells discussed previously relies on current sinking. A sense amplifier is only needed to speed up the readout, not for functionality considerations. It is also worth noticing that the DRAM memory cells are *single ended* in contrast to the SRAM cells, which present both the data value and its complement on the bit lines. This complicates the design of the sense amplifier, as will be discussed in the section on periphery.

Example 12.10   1T DRAM Readout

Assume a bit line capacitance of 1pF in correspondence to the numbers derived earlier in the chapter, and a bit line precharge voltage of 1.25 V. The voltage over the cell capacitance C_S (of 50 fF) equals 1.9 V and 0 V for a 1 and 0, respectively. This translates into a charge-transfer efficiency of 4.8% and the following voltage swings on the bit line during a read operation:

$$\Delta V(0) = -1.25 \text{ V} \times \frac{50 \text{ fF}}{50 \text{ fF} + 1 \text{ fF}} = -60 \text{ mV}$$

$$\Delta V(1) = 31 \text{ mV}$$

Other important differences are also worth enumerating:

2. The readout of the 1T DRAM cell is *destructive*. This means that the amount of charge stored in the cell is modified during the read operation. After a successful read operation, the original value must be restored. Read and refresh operations are therefore intrinsically intertwined in a 1T DRAM. Typically, the output of the sense amplifier is imposed onto the bit line during the readout. Keeping *WL* high ensures that the cell charge is restored during that period. This is illustrated in Figure 12-36, which plots a typical bit line voltage waveform during readout.

3. Unlike the 3T cell that relies on charge storage on a gate capacitance, the 1T cell requires the presence of an extra capacitance that must be explicitly included in the design. For reliability, the charge-transfer ratio is kept large, with the minimum value of the capacitance ranging around 30 fF. Fitting that large of a capacitance in as small an area as possible is one of the key challenges in DRAM designs. Some of the most popular ways to do so are briefly summarized in a following section.

4. Observe that when writing a 1 into the cell, a threshold voltage is lost, which reduces the available charge. This charge loss can be circumvented by bootstrapping the word lines to a value higher than V_{DD}. This is a common practice in state-of-the-art memory design.

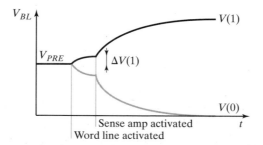

Figure 12-36   Bit line voltage waveform during read operation (for 1 and 0 data values).

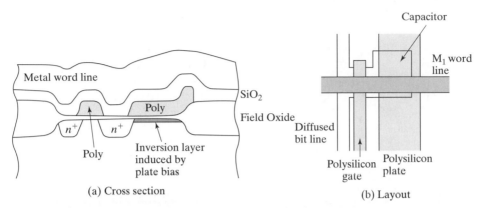

Figure 12-37 1T DRAM cell using a polysilicon-diffusion capacitance as storage node (from [Dillinger88]). The contact between word line and polysilicon gate is accomplished in the neighboring cell.

Figure 12-37 presents a first approach toward designing a 1T DRAM cell. The main advantage of this design is that it can be realized in a generic CMOS technology. The storage node in this cell is composed of the gate capacitance, sandwiched between a polysilicon plate and an inversion layer, induced into the substrate by applying a positive voltage bias on the polysilicon plate. When writing a 0 in the cell, the potential well of the storage node is filled with electrons, and the capacitor is charged. If a 1 is written in the cell (with a high voltage on the bit line) the electrons are removed from the induced inversion layer, and the surface area is depleted. The voltage over the capacitor is reduced. Observe that this represents the inverse of the scenario described before: The capacitor is charged for a 0 and discharged when storing a 1.

Implementing denser cells requires modifications in the manufacturing process. A first change is to add a second polysilicon layer, which serves as the second plate of the capacitor, with the first polysilicon layer forming the other plate. In the quest for ever-denser cells, DRAM technology as used in the 16-Mbit DRAMs and beyond, has focused on three-dimensional structures, where the storage capacitance is either implemented vertically in the substrate or on top of the access transistor [Lu89]. Cross sections of some of the most advanced cells are shown in Figure 12-38. The first cell shows the cross section of a *trench-capacitor* cell. In this structure, a vertical trench of up to 5 μm deep is etched into the substrate. The sidewalls and bottom of the trench are used for the capacitor electrode, which results in a large plate surface that occupies only a small die area. Figure 12-38b shows the cross section of a *stacked-capacitor cell* (STC), where the capacitance is superimposed on top of the access transistor and bit lines. Up to four polysilicon layers are employed to realize "fin-type" capacitors, reducing the effective cell area. Using these approaches, the cell area in a 64-Mbit DRAM ranges between 1.5 and 2.0 μm², yielding a storage capacitance between 20 and 30 fF in a 0.4 μm technology! Obviously, these size reductions are not free. The production and manufacturing of those esoteric devices to obtain a reasonable yield has become an increasingly difficult and expensive undertaking. More

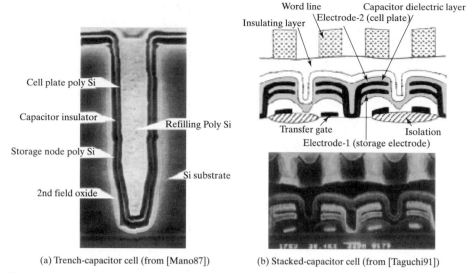

(a) Trench-capacitor cell (from [Mano87])   (b) Stacked-capacitor cell (from [Taguchi91])

Figure 12-38   Advanced 1T DRAM memory cells.

so than physics or implementation impediments, economics will determine whether there is life beyond 256-Mbit or 1-Gbit semiconductor memories.

12.2.4 Contents-Addressable or Associative Memory (CAM)

The concept of the content-addressable memory was described earlier in the chapter. A CAM is a special type of memory device that stores data, but also has the ability to compare all the stored data in parallel with incoming data in an efficient manner. Figure 12-39 shows a possible implementation of a CAM array. The cell combines a traditional 6T RAM storage cell ($M4$–$M9$) with additional circuitry to perform a 1-bit digital comparison ($M1$–$M3$). When the cell is to be written, complementary data is forced onto the bit lines, while the word line is enabled as in a standard SRAM cell. In the compare mode, the stored data (S and its complement $\overline{S}$) are compared to in the incoming data, which is provided on the complementary bit lines (Bit and $\overline{Bit}$). The *Match* line is tied to all the CAM cells in a given row, and is initially precharged to V_{DD}. If S and Bit match, the internal node *int* is discharged, and $M1$ is turned off, keeping the match line high. However, if the stored and incoming bit are different, *int* is charged to $V_{DD} - V_T$ causing the match line to discharge. For example, if $Bit = V_{DD}$ and $S = 0$, *int* charges up through $M3$. It is easily verified that the circuit performs nothing other than an XNOR function (or comparator).

It is important to note that the pull-down device in the comparator is connected to each of the CAM cells in a row in a wired-OR fashion. That is, even if only one of the bits in a given row mismatches, the match line is pulled low. For a memory with N rows, most rows (mismatches) will be pulled low in a given cycle. Clearly, not an enticing perspective from a power dissipation viewpoint. CAMs typically are not very power efficient! It is possible to

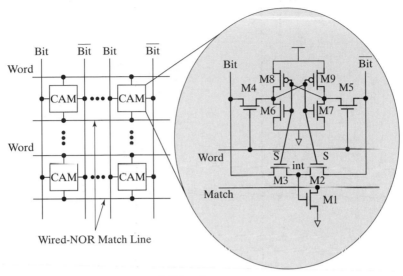

Figure 12-39 9-Transistor CAM cell.

re-arrange the logic such that only the match line switches, but this comes at a significant degradation in performance.

Example 12.11 Usage of Associate Memory in Cache Applications

The application of the CAM cell to a fully-associative *cache memory* is shown in Figure 12-40. A cache memory is a small, but fast memory that stores a small fraction of the overall memory required in a digital system. It works on the principle of spatial and temporal data locality. For example, if a data location is accessed at some point in time in a program, there is a high probability that it will be accessed again in the near future. The cache is placed between the processor and the large and dense main (DRAM) memory. It helps to defray the cost (in terms of time and power) of getting the data from the large memory. The cache memory array consists of two parts, the CAM array that stores addresses and a regular SRAM array that stores data. When the processor needs to write data, the address is written into the CAM and the data into the SRAM array. When a new address and data needs to be written and the cache is full, one entry must be displaced. This is done through a cache replacement policy. For instance, the least-accessed data word is a good candidate for displacement. In the read mode, the address of the data requested is presented to the CAM array, and parallel search takes places. A match indicates that the data is indeed available in the cache. The match signal acts as the word-line enable for the SRAM array which ultimately provides the data. In the case of a mismatch (this is, all match lines are low), the data must be obtained from the external slow system memory, and a cache miss has occurred.

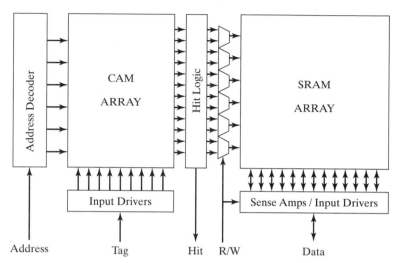

Figure 12-40 Application of CAM cell: high-performance on-chip cache memory.

Interested readers can find a lot more information about caches in the many textbooks covering computer architecture (for example, [Hennessy02]).

12.3 Memory Peripheral Circuitry*

Since the memory core trades performance and reliability for reduced area, memory design relies exceedingly on the peripheral circuitry to recover both speed and electrical integrity. While the design of the core is dominated by technological considerations and is largely beyond the scope of the circuit designer, it is in the design of the periphery where a good designer can make an important difference. In this section, we discuss the address decoders, I/O drivers/buffers, sense amplifiers, and memory timing and control.

12.3.1 The Address Decoders

Whenever a memory allows for random address-based access, address decoders must be present. The design of these decoders has a substantial impact on the speed and power consumption of the memory. In Section 12.1.2, we introduced two classes of decoders—the row encoders, whose task it is to enable one memory row out of 2^M, and the column and block decoders, which can be described as 2^K-input multiplexers, where M and K are the widths of the respective fields in the address word. While conceiving these decoders, it is important to keep the complete memory floorplan in perspective. These units are tightly coupled to the memory core, so that a geometry matching between the cell dimensions of decoders and the core is a must (*pitch matching*). Failing to do so would lead to a dramatic wiring overhead with its associated delay and power dissipation. Examples of pitch-matched decoders and memory arrays are shown in the case studies at the end of this chapter.

Row Decoders

A 1-out-of-2^M decoder is nothing less than a collection of 2^M complex, M-input, logic gates. Consider an 8-bit address decoder. Each of the outputs WL_i is a logic function of the 8 input address signals (A_0 to A_7). For example, the rows with addresses 0 and 127 are enabled by the following logic functions:

$$WL_0 = \overline{A}_0\overline{A}_1\overline{A}_2\overline{A}_3\overline{A}_4\overline{A}_5\overline{A}_6\overline{A}_7$$
$$WL_{127} = \overline{A}_0 A_1 A_2 A_3 A_4 A_5 A_6 A_7 \tag{12.9}$$

This function can be implemented in two stages, using a single 8-input NAND gate and an inverter. For a single-stage implementation, it can be transformed into a wide NOR using De Morgan's rules:

$$WL_0 = \overline{A_0 + A_1 + A_2 + A_3 + A_4 + A_5 + A_6 + A_7}$$
$$WL_{127} = \overline{A_0 + \overline{A}_1 + \overline{A}_2 + \overline{A}_3 + \overline{A}_4 + \overline{A}_5 + \overline{A}_6 + \overline{A}_7} \tag{12.10}$$

In essence, to implement this logic function, an 8-input NOR gate is needed per row. This poses several imposing challenges. First of all, the layout of the wide-NOR gate must fit within the word-line pitch. Secondly, the large fan-in of the gate has a negative impact on performance. The propagation delay of the decoder is a matter of prime importance, as it adds directly to both read- and write-access times. In addition, this NOR gate has to drive the large load presented by the word line, while not overloading the input addresses. Finally, the power dissipation of the decoder has to be kept in check. In the following paragraphs, we discuss static and dynamic implementation options.

Static Decoder Design Implementing a wide-NOR function in complementary CMOS is impractical. One possible solution is to go back to a pseudo-NMOS design style, which allows for an efficient implementation of wide NORs. Power-dissipation concerns make this approach not very attractive in today's overconstrained design world.

Fortunately, some of the principles introduced in Chapter 6 come to the rescue. Splitting a complex gate into two or more logic layers most often produces both a faster and a cheaper implementation. This decomposition concept makes it possible to build fast and area-efficient decoders in complementary CMOS, and is used effectively in most memories today. Segments of the address are decoded in a first logic layer called the *predecoder*. A second layer of logic gates then produces the final word-line signals.

Consider the case of an 8-input NAND decoder. The expression for WL_0 can be regrouped in the following way:

$$WL_0 = \overline{\overline{A}_0\overline{A}_1\overline{A}_2\overline{A}_3\overline{A}_4\overline{A}_5\overline{A}_6\overline{A}_7}$$
$$= \overline{(\overline{A_0 + A_1})(\overline{A_2 + A_3})(\overline{A_4 + A_5})(\overline{A_6 + A_7})} \tag{12.11}$$

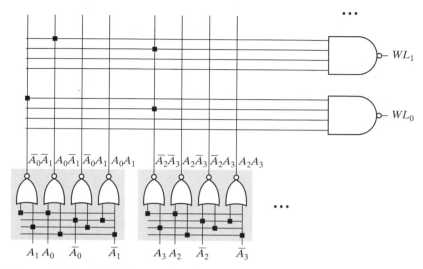

Figure 12-41   A NAND decoder using 2-input predecoders.

For this particular case, the address is partitioned into sections of 2 bits that are decoded in advance. The resulting signals are then combined using 4-input NAND gates to produce the fully decoded array of word-line signals. The resulting structure is pictured in Figure 12-41.

The use of a predecoder is advantageous in many ways:

- It reduces the number of transistors required. Assuming that the predecoder is implemented in complementary static CMOS, the number of active devices in the 8-input decoder equals $(256 \times 8) + (4 \times 4 \times 4) = 2,112$, which is 52% of a single stage decoder, which would require 4,096 transistors.
- As the number of inputs to the NAND gates is halved, the propagation delay is reduced by approximately a factor of 4. Remember the squared dependency between delay and fan-in.

Still, in this design, a 4-input NAND gate is driving the word line, which presents a large load. The best driver for large capacitances is an inverter—hence, the output of the NAND should be buffered. To bring the decoder design closer to optimal, the rules of the logical effort can be directly applied. It was concluded in Chapter 6 that, when driving very large loads, it is beneficial to have even more stages of logic. By performing more logic transformations similar to Eq. (12.11), the decoder can be broken into additional levels of logic, each of which consists of 2-input NANDs, 2-input NORs, or inverters.

Example 12.12   Decoder Design

Minimizing the delay of a memory decoder is a well-defined problem. The total capacitance of the word lines can be calculated, while the maximum capacitance that loads the

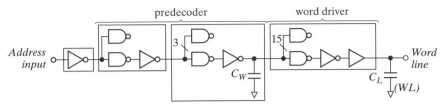

Figure 12-42   Logic path of an eight-bit decoder.

input address lines is constrained by the design. This specifies the effective fan-out of the decoder path, F. A number of logic structures can be derived that implement an 8-input AND function, one of which is shown in Figure 12-42. The chosen structure consists of inverters and 2-input NANDs. The input addresses are buffered first, and groups of 4 bits are predecoded. These predecoded signals are routed vertically, and connected to the final level of decoders. Since the decoders in the final level consist of just 2 input NANDs, they fit in the word line pitch.

The logical effort of this path equals $G = (4/3)^3 = 2.4$, and the intrinsic delay is $P = 12$. The total branching B equals 128 (= $2 \times 4 \times 16$) as indicated in Figure 12-42. By specifying the effective fan-out, F, a minimum delay and optimal gate sizing can be determined.

In this analysis, we neglected the wire capacitance C_W, between the predecoder and the final decoders, which is substantial in larger memories. This becomes clear when inspecting complete memory architectures, such as the one shown in Figure 12-6. By including this fixed capacitive load in the optimization, the problem becomes much more difficult, but solvable [Amrutur01].

A final optimization in performance is enabled by taking into account that the word lines are normally held low. Hence, we should only optimize the rising transition. The logical effort of all the gates in the path can be reduced by skewing the gate sizing to favor only one transition.

All large decoders are realized using at least a two-layer implementation. This predecoder–final decoder configuration has another advantage. Adding a select signal to each of the predecoders makes it possible to disable the decoder when the memory block in question is not selected. This results in important power savings.

Dynamic Decoders   Since only one transition determines the decoder speed, it is interesting to evaluate other circuit implementations. One option, dismissed earlier, is the use the pseudo-NMOS logic. Dynamic logic offers a better alternative. A first solution is presented in Figure 12-43, where the transistor diagram and the conceptual layout of a 2-to-4 decoder is depicted. Notice that this structure is geometrically identical to the NOR ROM array, differing only in the data patterns.

In a similar fashion, we can also implement the decoder as a NAND array, effectively realizing the inverse of the functions of Eq. (12.9). In this case, all the outputs of the array are high

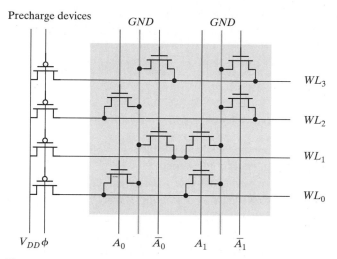

Figure 12-43   Dynamic 2-to-4 NOR decoder.

by default with the exception of the selected row, which is low. This "active low" signaling is in correspondence with the word-line requirements of the NAND ROM, as discussed in Section 12.2.1 Observe that the interface between decoder and memory often includes a buffer/driver that can be made inverting whenever needed. A 2-to-4 decoder in NAND configuration is shown in Figure 12-44.

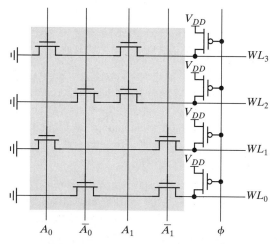

Figure 12-44   A 2-to-4 MOS dynamic NAND decoder. This implementation assumes that all address signals are low during precharge. An alternative approach is to provide evaluate transistors at the bottom of each transistor chain.

All the performance and density considerations raised in the discussion of NAND and NOR ROM arrays are valid. NOR decoders are substantially faster, but they consume more area than their NAND counterparts and dramatically more power. This is clear from the following observation: Only a single word line is being pulled down after the precharge in a NAND decoder, while only a single wire stays high in the NOR case. Similarly to the static decoder design, larger decoders are built using a multilayer approach.

Column and Block Decoders

Column decoders should match the bit line pitch of the memory array. The functionality of a column and block decoder is best described as a 2^K-input multiplexer, where K stands for the size of the address word. For read–write arrays, these multiplexers can be either separate or shared between read and write operations. During the read operation, they have to provide the discharge path from the precharged bit lines to the sense amplifier. When performing a write operation to a memory array, they have to be able to drive the bit line low to write a 0 in the memory cell.

Two implementations of this multiplexing function are in general use. Which one to choose depends upon area, performance, and architectural considerations.

One implementation is based on the CMOS pass-transistor multiplexer introduced in Chapter 6 (Figure 6-46). The control signals of the pass transistors are generated using a K-to-2^K predecoder, realized along the lines described in the previous section. The schematic of a 4-to-1 column decoder, using only NMOS transistors is shown in Figure 12-45. Complementary transmission gates must be used when these multiplexers are shared between the read and write operations to be able to provide a full swing in both directions. The main advantage of this approach is its speed. Only a single pass transistor is inserted in the signal path, which introduces only a minimal extra resistance. The column decoding is one of the last actions to be performed in the read sequence, so that the predecoding can be executed in parallel with other operations, such as the memory access and sensing, and can be performed as soon as the column address is

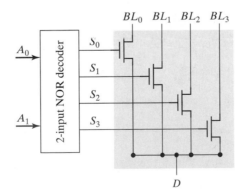

Figure 12-45 Four-input pass-transistor-based column decoder using a NOR predecoder.

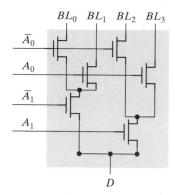

Figure 12-46 A 4-to-1 tree-based column decoder.

available. Consequently, its propagation delay does not add to the overall memory access time. Slower implementations such as NAND decoders might even be acceptable. The disadvantage of the structure is its large transistor count. $(K + 1)2^K + 2^K$ devices are needed for a 2^K-input decoder. For instance, a 1024-to-1 column decoder requires 12,288 transistors. One should also realize that the capacitance and thus the transient response at node D is proportional to the number of inputs of the multiplexer.

A more efficient implementation is offered by a *tree decoder* that uses a binary reduction scheme, as shown in Figure 12-46. Notice that no predecoder is required. The number of devices is drastically reduced, as is shown in the following equation (for a 2^K-input decoder):

$$N_{tree} = 2^K + 2^{K-1} + \dots + 4 + 2 = 2 \times (2^K - 1) \tag{12.12}$$

This means that a 1024-to-1 decoder requires only 2046 active devices, a reduction by a factor of 6! On the negative side, a chain of K series-connected pass transistors is inserted in the signal path. Because the delay increases quadratically with the number of sections, the tree approach becomes prohibitively slow for large decoders. This can be remedied by inserting intermediate buffers. A progressive sizing of the transistors is another option, with the transistor size increasing from bottom to top. A final option is to combine the pass-transistor and tree-based approaches. A fraction of the address word is predecoded (for instance, the *msb* side), while the remaining bits are tree decoded. This can reduce both the transistor count and the propagation delay.

Example 12.13   Column Decoders

Consider a 1024-to-1 decoder. Predecoding 5 bits results in the following transistor tally:

$$N_{dec} = N_{pre} + N_{pass} + N_{tree} = 6.$$
$$2^5 + 2^{10} + 2(2^5 - 1) = 1278!$$

The number of series-connected pass transistors is reduced to six.

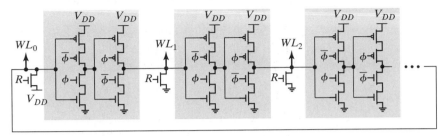

Figure 12-47 Decoder for circular shift register. The *R* signal resets the pointer to the first position.

Decoders for Non-Random-Access Memories

Memories that are not of the random-access class do not need a full-fledged decoder. In a serial-access memory, such as a video-line memory, the decoder degenerates into an *M*-bit shift register, with *M* the number of rows. Only one of the bits is high at a time and is called a *pointer*. The pointer moves to the next position every time an access is performed. An example of such a degenerated decoder implemented using a C^{2}MOS D-FF is shown in Figure 12-47. Similar approaches can be devised for other memory classes such as FIFOs.

12.3.2 Sense Amplifiers

Sense amplifiers play a major role in the functionality, performance and reliability of memory circuits. In particular, they perform the following functions:

- *Amplification*—In certain memory structures such as the 1T DRAM, amplification is required for proper functionality since the typical circuit swing is limited to 100 millivolts [Itoh01]. In other memories, it allows resolving data with small bit-line swings, enabling reduced power dissipation and delay.
- *Delay Reduction*—The amplifier compensates for the restricted fan-out driving capability of the memory cell by accelerating the bit line transition, or by detecting and amplifying small transitions on the bit line to large signal output swings.
- *Power reduction*—Reducing the signal swing on the bit lines can eliminate a substantial part of the power dissipation related to charging and discharging the bit lines.
- *Signal restoration*—Because the read and refresh functions are intrinsically linked in 1T DRAMs, it is necessary to drive the bit lines to the full signal range after sensing.

The topology of the sense amplifier is a strong function of the type of memory device, the voltage levels, and the overall memory architecture. Sense amplifiers are analog circuits by nature, and an in-depth analysis requires substantial analog expertise. Therefore, only a brief introduction to the design of such devices is given here. For an elaborate discussion on the design of amplifiers, the reader is referred to textbooks such as [Gray01][Sedra87].

Differential Voltage Sensing Amplifiers

A differential amplifier takes small-signal differential inputs (i.e., the bit-line voltages), and amplifies them to a large-signal single-ended output. It is generally known that a differential approach presents numerous advantages over its single-ended counterpart—one of the most important being the *common-mode rejection*. That is, such an amplifier rejects noise that is equally injected to both inputs. This is especially attractive in memories where the exact value of the bit line signal varies from die to die and even for different locations on a single die. In other words, the absolute value of a 1 or 0 signal is not exactly known and might vary over quite a large range. The picture is further complicated by the presence of multiple noise sources, such as switching spikes on the supply voltages and capacitive cross talk between word and bit lines. The impact of those noise signals can be substantial, especially when we realize that the amplitude of the signal to be sensed is generally small. The effectiveness of a differential amplifier is characterized by its ability to reject the common noise and amplify the true difference between the signals. The signals common to both inputs are suppressed at the output of the amplifier by a ratio called the *common-mode rejection ratio* (CMRR). Similarly, spikes on the power supply are suppressed by a ratio called the *power-supply rejection ratio* (PSRR). Differential sensing is therefore considered the technique of choice.

Unfortunately, the differential approach is only directly applicable to SRAM memories, since these are only the memory cells that offer a true differential output. Figure 12-48 shows the most basic differential sense amplifier. Amplification is accomplished with a single stage, based on the current mirroring concept. The input signals (*bit* and $\overline{bit}$) are heavily loaded and driven by the SRAM memory cell. The swing on those lines is small as the small memory cell drives a large capacitive load. The inputs are fed to the differential input devices (M_1 and M_2), and transistors M_3 and M_4 act as an active current mirror load. The amplifier is conditioned by the sense amplifier enable signal, *SE*. Initially, the inputs are precharged and equalized to a common value, while *SE* is low disabling the sensing circuit. Once the read operation is initiated, one of

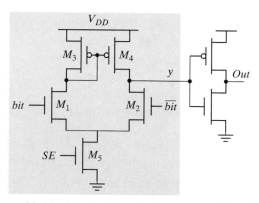

Figure 12-48   Basic differential sense amplifier circuit.

the bit lines drops. *SE* is enabled when a sufficient differential signal has been established, and the amplifier evaluates.

The gain of such the differential-to-single ended amplifier is given by

$$A_{sense} = -g_{m1}(r_{o2} \| r_{o4}) \tag{12.13}$$

where g_{m1} is the transconductance of the input transistors, and r_o the small-signal device resistance of the transistor. The r_o of the MOS transistor is very high in the saturation region of the MOSFET. The transconductance of the input devices can be increased by either widening the devices, or by increasing the bias current. The latter also reduces the output resistance of M_2, which limits the usefulness of this approach. A gain of around 100 can be achieved. However, the gain of sense amplifiers typically is set to around 10. The main goal of the sense amplifier is the rapid production of an output signal. Gain is hence secondary to response time. Multiple stages are required to achieve the desired full-swing signal.

Figure 12-49 shows a fully differential two-stage sensing approach along with the SRAM bit column structure. The bit lines are connected to the inputs x and $\bar{x}$ of the two-stage differential amplifier. A read cycle proceeds as follows.

1. In the first step, the bit lines are precharged to V_{DD} by pulling $\overline{PC}$ low. Simultaneously, the *EQ*-PMOS transistor is turned on, ensuring that the initial voltages on both bit lines are

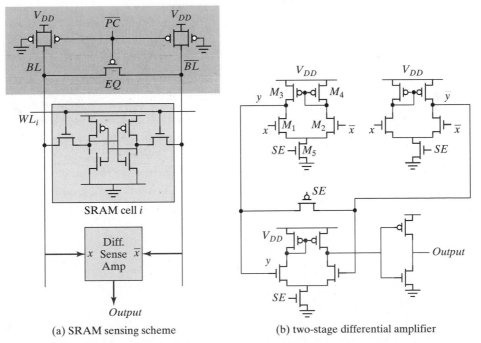

(a) SRAM sensing scheme   (b) two-stage differential amplifier

Figure 12-49   Differential sensing as applied to an SRAM memory column.

identical. This operation, called *equalization*, is necessary to prevent the sense amplifier from making erroneous excursions when turned on. In practice, every differential signal in the memory is equalized before performing a read. Equalization is critical when the bit lines are precharged through NMOS pull-ups, since the precharge value can differ due to the variations in the device threshold.

2. The read operation is started by disabling the precharge and equalization devices and enabling one of the word lines. One of the bit lines is pulled low by the selected memory cell. Notice that a grounded PMOS load, placed in parallel with the precharge transistor, limits the bit line swing and speeding up the next precharge cycle.

3. Once a sufficient signal is built up (typically around 0.5 V), the sense amplifier is turned on by raising *SE*. The differential input signal on the bit lines is amplified by the two stage amplifier and eventually a rail to rail output is produced at the output of the inverter.

The two-stage sense-amplifier circuit is shown in Figure 12-49b. While these circuits use PMOS loads and NMOS input devices, the dual configurations with PMOS input devices and NMOS loads are also regularly used, depending upon biasing conditions. Note that by pulsing the *SE* control signal to be active for short evaluation periods, static power in the amplifier can be reduced. It should also be noted that a single sense amplifier is shared between multiple columns by inserting the column decoder pass transistors between the memory cells and the amplifier (the input levels of the amplifier are reduced by a device threshold in this case). This results in area savings and power reduction.

The sense amplifier presented earlier decouples the inputs and outputs. That is, the bit-line swing is determined by the SRAM cells and the static PMOS load. A radically different sensing approach is offered by the circuit of Figure 12-50, where a CMOS cross-coupled inverter pair is used as a sense amplifier. A CMOS inverter exhibits a high gain when positioned in its transient region, as was established in Chapter 5. To act as a sense amplifier, the flip-flop is initialized in its metastable point by equalizing the bit lines. A voltage difference is built over the bit lines in the course of the read process. Once a large enough voltage gap is created, the sense amplifier is enabled by raising *SE*. Depending upon the input, the cross-coupled pair traverses to one of its stable operation points. The transition is swift as a result of the positive feedback.

While the flip-flop sense amplifier is simple and fast, it has the property that inputs and outputs are merged, so that a full rail-to-rail transition is enforced on the bit lines. This is exactly what is needed for a 1T DRAM, where a restoration of the signal levels on the bit lines is necessary for the refresh of the cell contents. The cross-coupled cell is, therefore, almost universally used in DRAM designs. How to turn a single-ended memory structure such as the DRAM cell into a differential one is discussed in the next section.

Single-Ended Sensing

While differential sensing is by far the preferred approach, memory cells used in ROMs, E(E)PROMs and DRAMs are inherently single ended. A first option to address this problem is to resort to single-ended amplification. Since the bit lines are typically precharged, an asymmetri-

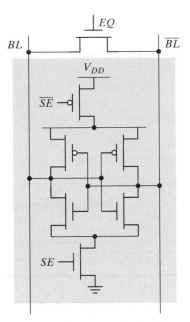

Figure 12-50 Cross-coupled CMOS inverter latch used as sense amplifier.

cally biased inverter, as introduced in Figure 9-35, is a good candidate for such. An interesting variant, called the *charge-redistribution amplifier*, is often used in small memory structures. (See Figure 12-51.) The basic idea is to exploit the imbalance between a large capacitance C_{large} and a much smaller component C_{small}. The two capacitors are isolated by the pass transistor M_1 [Heller75].

The initial voltages on nodes L and S (V_{L0} and V_{S0}) are precharged to $V_{ref} - V_{Tn}$ and V_{DD} by connecting node S to the supply voltage. Because of the voltage drop over M_1, V_L only

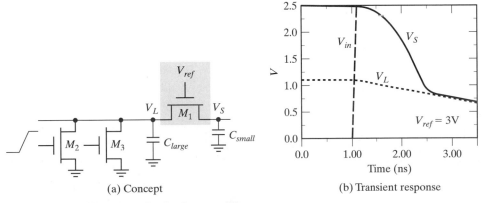

(a) Concept

(b) Transient response

Figure 12-51 Charge-redistribution amplifier.

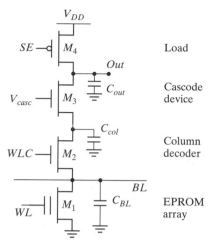

Figure 12-52   Charge-redistribution amplifier as used in EPROM memory.

precharges to $V_{ref} - V_{Tn}$. When one of the pull-down devices (e.g., M_2) turns on, node L with its large capacitance slowly discharges. As long as $V_L \geq V_{ref} - V_{Tn}$ transistor M_1 is off, and V_S remains constant. Once V_L drops below the trigger voltage $(V_{ref} - V_{Tn})$, M_1 turns on. A charge redistribution is initiated, and nodes L and S equalize. This can happen very fast due to the small capacitance on the latter node. A small voltage variation on node L translates into a large voltage drop on node S, as is illustrated in the simulated transient response of Figure 12-51b. The circuit thus acts as an amplifier. The resulting signal can be fed into an inverter with a switching threshold larger than $V_{ref} - V_{Tn}$ to produce a rail-to-rail swing.

The schematics of the charge-redistribution amplifier, as it is used in a memory, are presented in Figure 12-52. This structure has been very popular in EPROM memories. The disadvantage of the charge-redistribution amplifier is that it operates with a very small noise margin. A small variation of node L due to noise or leakage may cause an erroneous discharge of S. (See Figure 12-51.) Careful design and analysis is therefore necessary.

Single-to-Differential Conversion

Larger memories (>1 MBit) that are exceedingly prone to noise disturbances resort to translating the single-ended sensing problem into a differential one. The basic concept behind the single-to-differential conversion is demonstrated in Figure 12-53. A differential sense amplifier is con-

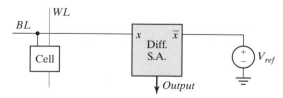

Figure 12-53   Single-to-differential conversion.

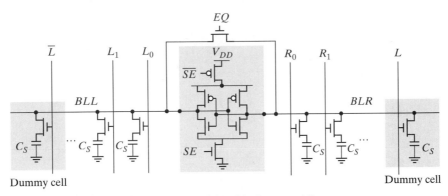

Figure 12-54 Open bit line architecture with dummy cells.

nected to a single-ended bit line on one side and a reference voltage, positioned between the 0 and 1 levels, at the other end. Depending on the value of *BL*, the amplifier toggles in one or the other direction. Creating a good reference source is not as easy as it sounds, since the voltage levels tend to vary from die to die or even over a single die. The reference source must therefore track those variations. A popular way of doing so is illustrated in Figure 12-54 for the case of a 1T DRAM. The memory array is divided into two halves, with the differential amplifier placed in the middle. On each side, a column of so-called *dummy cells* is added. These are 1T memory cells that are similar to the others, but whose sole purpose is to serve as reference. This approach is often called the *open bit line architecture*.

When the *EQ* signal is raised, both the bit lines *BLL* and *BLR* are precharged to $V_{DD}/2$. Enabling L and $\bar{L}$ at the same time ensures that the dummy cells are charged to $V_{DD}/2$. One of the word lines is enabled during the read cycle. Assume that a cell in the left half of the array is selected by raising WL_0, which causes a voltage change on *BLL*. The appropriate voltage reference is created by simultaneously selecting the dummy cell in the other memory half by raising L. Under the assumption that the left and right memory sides are perfectly matched, the resulting voltage on *BLR* resides between the 0 and 1 levels and causes the sense latch to toggle. Notice that maintaining perfect symmetry is important. Raising word lines WL_0 and L simultaneously turns the capacitive coupling between bit and word lines into a common-mode signal that is effectively eliminated by the sense amplifier. Observe that dividing the bit lines into two halves effectively reduces the bit line capacitance. This doubles the charge-transfer ratio and improves the signal-to-noise ratio.

Example 12.14 Sensing in 1T DRAM

The read-operation of a 1T DRAM implemented by using the open bit-line architecture, and the latch sense amplifier is simulated by using SPICE. The storage capacitance is set to 50 fF, while the bit line capacitance equals 0.5 pF. This is equivalent to a charge-transfer ratio of 9%. Assuming that the bit line is precharged to 1.25 V, charge redistribution results in a voltage drop of 110 mV for a 0 signal. Due to the threshold loss over the pass

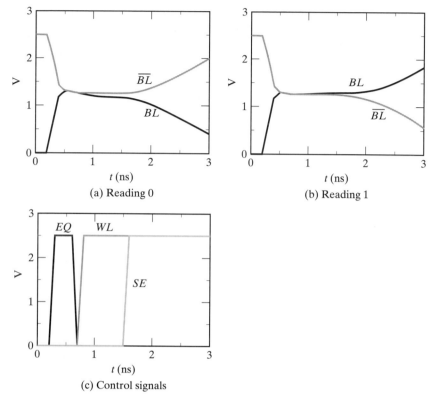

Figure 12-55 Simulation of the read process in an open bit line architecture with dummy cell. The dummy cell is connected to $\overline{BL}$ (C_S = 50 fF; C_{BL} = 0.5 pF).

transistor, the cell voltage associated with a 1 equals 1.9 V, which translates to a voltage increase of only 60 mV on the bit line after the enabling of the word line. These values are confirmed by the simulation, which shows values of 110 mV and 45 mV respectively, before the sense amplifier is turned on (Figure 12-55).

12.3.3 Voltage References

Most memories require some form of on-chip voltage generation. The operation of a sophisticated memory requires a number of voltage references and supply levels, including the following:

• *Boosted-Word-line Voltage*—In a conventional 1T DRAM cell, using an NMOS pass transistor, the maximum voltage level that can be written onto a cell equals $V_{DD} - V_T$, which negatively impacts the reliability of the memory. By raising the word-line voltage above V_{DD} (more specifically, to $V_{DD} + V_{Tn}$), a full-scale signal can be written. This "boosted-word-line" approach typically uses a charge pump to generate the elevated voltages.

- *Half-V_{DD}*: DRAM bit lines are precharged to $V_{DD}/2$, as discussed previously. This voltage must be generated on chip.
- *Reduced Internal Supply*: Most memory circuits operate at a lower power supply than the external supply. DRAM (and other memories) use internal voltage regulators to generate the required voltages, while being compliant with standard interface voltages.
- *Negative Substrate Bias*: An effective means to control the threshold voltages within a memory is to apply negative substrate biasing, augmented with a control loop. This approach has been used all recent generation of DRAM memories.

The design of voltage references falls under the category of analog circuit design. A short review of a couple of reference circuits is given next.

Voltage Down Converters

Voltage down converters are used to create low internal supplies, allowing the interface circuits to operate at higher voltages. Consequently, internal circuits can exploit aggressive voltage-scaling techniques using state-of-the-art scaled technologies, while remaining compatible with the outside world. The reduction of supply is in fact necessary to avoid breakdown in the deep-submicron devices. Level converters, discussed previously in Chapter 11, act as interfaces between the internal core and the external circuits. Regulators also serve to set a stable internal voltage while accepting a broad range on unregulated input voltages in battery-operated systems. This accommodates the fact that the battery voltage varies as a function of time.

Figure 12-56 shows the basic structure of a voltage down converter (also called a *linear regulator*). It is based on the operational amplifier that was described in the previous section. The circuit uses a large PMOS output driver transistor to drive the load of the memory circuit (an NMOS output device may also be used). The circuit uses negative feedback to set the output voltage V_{DL} to the reference voltage.

The converter must offer a voltage that is immune to variations in operating conditions. Slow variations, such as temperature changes, can be compensated by the feedback loop. Note that this is a feedback circuit, which can be unstable if improperly designed. In particular, the load current drawn by load varies wildly over time, and the converter must be designed to accommodate these wide variations.

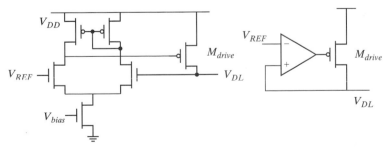

Figure 12-56   A voltage regulator and its equivalent representation.

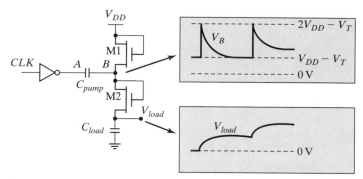

Figure 12-57   Simple charge pump and its signal waveforms.

Charge Pumps

Techniques such as word-line boosting and well biasing often need voltage sources that exceed the supply voltage, but do not draw much current. A charge pump is the ideal generator for this task. The concept is best explained with the simple circuit of Figure 12-57. Transistors M1 and M2 are connected in diode style. Assume initially that the clock *CLK* is high. During this phase, node A is at GND and node B at $V_{DD} - V_T$. The charge stored in the capacitor is given by

$$Q = C_{pump}(V_{DD} - V_T) \qquad (12.14)$$

During the second phase, *CLK* goes low, raising node A to V_{DD}. Node B rises in concert, effectively shutting off M1. When B is one threshold above V_{load}, M2 starts to conduct and charge is transferred to C_{load}. During consecutive clock cycles, the pump continues to deliver charge to V_{load} until the maximum voltage of $2(V_{DD} - V_T)$ is reached at the output. The amount of current that can be drawn from the generator is primarily determined by the capacitor's size and the clock frequency. The efficiency of generator, which measures how much current is wasted during every pump cycle, is between 30 and 50%. Charge pumps should hence only be used for generators that draw little current. The circuit presented here is quite simple and not that effective. A wide range of more complex charge pumps have been devised for larger voltage ranges and improved efficiency.

Voltage Reference

An accurate and stable voltage reference is an important component of the voltage down converter. The reference voltage is assumed to be relatively constant over power supply and temperature variations. Figure 12-58 shows an example of a V_T reference generator. The bottom devices (M_3 and M_4) act as a current mirror to force the same current through the drain of M_1 and the resistor R_1. By making the device M_1 large and keeping the current small enough, the source-to-gate voltage for M_1 can be made approximately equal to $|V_{Tp}|$, as can be derived from the equation

$$|V_{GS,M1}| = |V_{Tp}| + \sqrt{\left|\frac{2I_{M1}}{k_{p,M1}}\right|} \qquad (12.15)$$

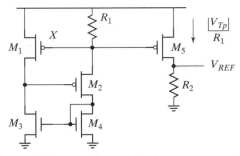

Figure 12-58 A simple V_T reference generator.

Also the currents going through the resistor and the drain current of M_1 both equal $|V_{Tp}|/R_1$. Note that M_2 acts as a biasing transistor. Since devices M_1 and M_5 experience the same gate-to-source voltage, the drain current of M_1 is mirrored to M_5. The reference voltage is thus given by

$$V_{REF} = |V_{Tp}| \cdot \frac{R_2}{R_1} \qquad (12.16)$$

Variations in threshold voltage can be compensated for by laser trimming of the resistors. V_{REF} also exhibits good temperature stability. By choosing the appropriate materials for the implementation of R_1 and R_2, the temperature dependence of V_{TP} can be cancelled by that of R_2/R_1. This circuit was used in a 16-Mb DRAM [Hidaka92], and displayed excellent stability: $\Delta V_{REF}/\Delta T = 0.15$ mV/°C, and $\Delta V_{REF}/\Delta V_{DD} = 10$ mV/V.

The preceding discussion is by no means complete. A more elaborate discussion of references is unfortunately beyond the scope of this text. The reader is referred to [Itoh01] and [Gray01] for a detailed treatment.

12.3.4 Drivers/Buffers

The length of word and bit lines increases with increasing memory sizes. Even though some of the associated performance degradation can be alleviated by partitioning the memory array, a large portion of the read and write access time can be attributed to the wire delays. A major part of the memory-periphery area is therefore allocated to the drivers, in particular the address buffers and the I/O drivers. The design of cascaded buffers was discussed at length in Chapter 5. All the issues raised there are valid here as well.

12.3.5 Timing and Control

The foregoing discussions present a picture of the memory module as a complex entity whose operation is governed by a well-defined sequence of actions such as address latching, word-line decoding, bit line precharging and equalization, sense-amplifier enabling, and output driving. To be operational, it is essential that this sequence be adhered to under all operational circumstances and over a range of device and technology parameters. A careful timing of the different events is necessary if maximum performance is to be achieved. Although the timing and control circuitry only occupies a minimal amount of area, its design is an integral and defining part of the memory design process. It requires careful optimization, and the execution of extensive and repetitive SPICE simulations over a range of operating conditions.

Over time, a number of different memory-timing approaches have emerged that can be largely classified as *clocked* and *self-timed*. A typical example of each class is discussed to illustrate the differences.

DRAM—A Clocked Approach

Since the early days, DRAM memories have opted for a multiplexed addressing scheme where the row address and column address are presented in sequence on the same address bus to save package pins. This approach has survived many memory generations and is still in use today, although it is getting increasingly awkward and in disparity with system-level requirements.

In the multiplexed addressing scheme, the user must provide two main control signals— *RAS* (row-address strobe) and *CAS* (column-address strobe)—that indicate the presence of the row and column addresses, respectively. (See Figure 12-8.) Another control signal (*W*) indicates if the intended operation is a read or a write. These signals can be interpreted as external clock signals, and are used to time the internal memory events. Similar to the synchronous clocking approach, the *RAS* and *CAS* signals must be sufficiently separated so that all the ensued operations have come to completion. Figure 12-59 shows a simplified timing diagram of a (1×4)-Mbit DRAM memory and some of the imposed timing constraints. The full specification involves more than 20 timing parameters, all of which have to be held within precise bounds to ensure proper operation. Obviously, generating the correct timing signals for these memories is a nontrivial task!

All the internal timing signals such as the *EQ*, *PC,* and *SE* signals are derivatives of the external control signals. To maximize performance, some of those signals have to fall within very precise intervals. For instance, the *SE* signal cannot be applied too early after word-line decoding is started, since this increases the sensitivity to noise. On the other hand, postponing it for too long slows down the memory. The separation of these events is a function of the word-line delay that might vary with temperature and process variations. It is a common practice to adjust the timing interval with operating parameters by including a model of the word line (*dummy word line*) in the timing-generation circuitry. This "delay module" tracks the delay of the actual word line accurately, making the memory more robust while preserving performance. It is thus fair to state that DRAMs combine a synchronous approach at the global level with self-timed techniques for the generation of some of the local signals.

Synchronous DRAM Memories

One of the main challenges of the DRAM memory is the large access time and the low data through-put. With the ever-increasing speed of microprocessors and their SRAM caches, the performance gap between processor core and DRAM main memory is getting larger. This gap is becoming one of the main challenges in high-performance system design. To address this problem, high-speed DRAM synchronous memories such as SDRAM (Synchronous DRAM)

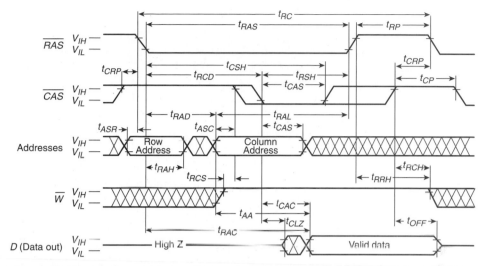

Figure 12-59 Read-cycle timing diagram for 4 M × 1 DRAM memory [Mot91]. The minimum length of the read cycle t_{RC} equals 110 ns. The spacing of the falling edges of *RAS* and *CAS* signals must range between 20 and 40 ns. A total of 24 timing constraints can be observed.

[Yoo95] and RDRAM (Rambus DRAM) [Crisp97] have been introduced. These memories present a major departure from the RAC/CAS timing model of traditional memories. While keeping the core of the memory relatively unchanged, synchronous DRAMs exploit the fact that core is highly parallel—in other words, for every given read and write cycle, a large number of bits can be read/written at the same time, albeit at a relatively low speed. Bringing these bits in or out of the memory chip can be accomplished with a high-speed synchronous interface, operating at a speed close to the processor clock frequency. This comes at the penalty of extra latches and buffers at the interface into the memory core, as well as high-speed circuitry to support the high-data rate I/O interface. The main contribution these memories offer is a redefinition of the interface into the memory. As long as data is written in large, sequential blocks, this approach has proven to be very effective.

Consider RDRAM as an example. Input/output data is transferred serially on a narrow bus, taking several clock cycles. The bus is operated at a very-high speed, however, and uses efficient packet protocols. Thus, a large amount of data can be transferred in a short period. Multiple memory chips can be connected to the bus, called the *Rambus channel*. Figure 12-60 shows a schematic diagram of the input/output circuitry of an RDRAM memory.

SRAM—A Self-Timed Approach

In contrast to DRAMs, SRAM memories have historically been designed from a globally static perspective. A memory operation is triggered by an event on the address bus or *R/W* signal (Figure 12-8b). No extra timing or control signals are needed. Such an approach seems to imply a fully static implementation for all circuitry such as decoders and sense amplifiers. A change in

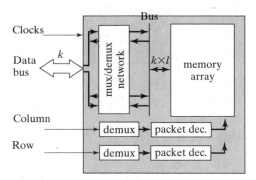

Figure 12-60   Schematic block diagram of RDRAM memory.

one of the input signals (data or address bus, *R/W*) then simply ripples through the subsequent layers of circuitry.

Our previous discussions made it apparent that such a fully static approach is not viable for larger memories for both area and power-dissipation considerations. SRAM memories, therefore, use a clever approach called *address-transition detection* (*ATD*) to automatically generate the internal signals such as *PC* and *SE* upon detection of a change in the external environment.

The *ATD* circuit plays an essential role in the architecture of SRAM and PROM modules. It acts as the source of most timing signals and is an integral part of the critical timing path. Speed is thus of utmost importance. A possible implementation of an *ATD* is shown in Figure 12-61. It consists of a number of transition-triggered one-shots (see Figure 7-50)—one per input bit—connected in a pseudo-NMOS NOR configuration. A transition on any of the input signals causes *ATD* to go low for a period t_d. The resulting pulse acts as the main timing reference for the rest of the memory, which results in a huge fan-out. Adequate buffering is thus recommended. A more elaborate view on SRAM timing can be found in the case studies at the end of this chapter.

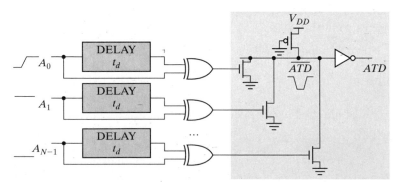

Figure 12-61   Address-transition detection circuitry. The delay lines are typically implemented as an inverter chain.

12.4 Memory Reliability and Yield*

Memories, both SRAM and DRAM, are operating under low signal-to-noise conditions. Maximizing the signals while minimizing the noise contributions is essential to achieve stable memory operation. Another problem plaguing memory design is the low yield due to structural and intermittent defects. This section discusses the nature of some of those problems, as well as potential solutions.

12.4.1 Signal-to-Noise Ratio

A tremendous effort is being made to produce memory cells that generate as large a signal as possible per unit area. Notwithstanding this effort, the produced signal quality decreases gradually with an increase in density, as is illustrated in Figure 12-62. For instance, the DRAM cell capacitance has degraded from approximately 70 fF for the 16 K memory to below 30 fF for the current generations. Simultaneously, the voltage levels have decreased for both power consumption and reliability purposes. Voltages at or below 1 V are the norm for newly designed memories, the main reason being the limited voltage stress that can be endured by the very thin oxides used in cells. Consequently, the signal charge stored on the capacitor has dropped with a factor of 5 from the 64 Kb to the 64 Mb generation. The fact that this drop is only that much, despite a cell area reduction of 100 over the same period of time, is solely due to various cell innovations.

At the same time, the increased integration density raises the noise level, due to the intersignal coupling. While the problem of the word-line-to-bit-line coupling capacitance was already an issue in the early 1980s, closer line spacing has brought the bit line–to–bit line coupling into the limelight. Contributing also to the problem are the higher speed requirements that result in an increasing switching noise for every new generation, and α-particle-

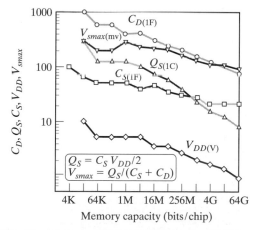

Figure 12-62 Trends in sensing parameters for DRAM memories (from [Itoh01]). C_D, V_{smax}, Q_s, and C_s stand for the bit line capacitance, sense signal, cell charge, and cell capacitance, respectively.

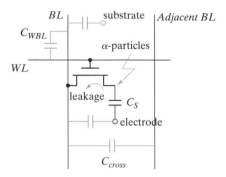

Figure 12-63   Noise sources in 1 T DRAM.

induced soft errors that can change the state of a high-impedance node at random and present an additional noise source. The latter problem was traditionally confined to dynamic memories, but now extends to static RAMs as well because the impedance of the SRAM storage nodes is rapidly increasing. Figure 12-63 enumerates the noise sources that impact the operation of the 1T DRAM cell, most of which are discussed in detail in the subsequent paragraphs.

Word-Line-to-Bit-line Coupling

Consider the open bit-line memory configuration of Figure 12-64. Word line WL_0 gets selected, which causes an amount of charge to be injected into the left bit line. We assume that the dummy cell on the right side, selected by WL_D, does not cause any charge redistribution, since it is precharged to $V_{DD}/2$. The presence of a coupling capacitance C_{WBL} between word line and bit line causes a charge redistribution to occur, whose amplitude approximately equals $\Delta_{WL} \times C_{WBL} / (C_{WBL} + C_{BL})$, where Δ_{WL} is the voltage swing on the word line and C_{BL} the bit line capacitance. If both sides of the memory array were completely symmetrical, the injected bit line noise (Δ_{BL}) would be identical on both sides and would appear as a common-mode signal to the sense amplifier. Unfortunately, this is not the case, because both coupling and bit line capacitance can vary substantially over the array.

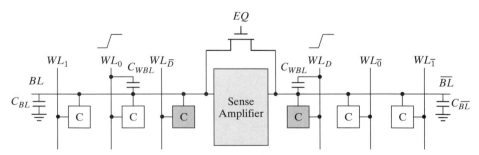

Figure 12-64   Word-line-to-bit-line coupling in open bit line architecture.

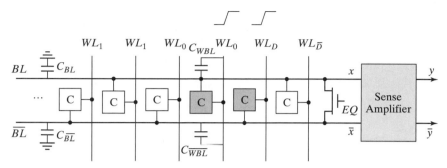

Figure 12-65 Folded bit line architecture for 1T DRAM.

This problem can be addressed by employing the *folded bit-line architecture* shown in Figure 12-65. Placing the sense amplifier at the end of the array and having *BL* and $\overline{BL}$ routed next to each other ensures a much closer matching between parasitic and bit line capacitances. The word-line signals (WL_0 and WL_D) cross both bit lines. The cross-coupling noise, hence, appears as a common-mode signal even if the voltage waveforms on WL_0 and WL_D differ considerably. While the folded bit line architecture has an obvious advantage in terms of noise suppression, it tends to increase the length of the bit line and consequently its capacitance. This overhead can be kept to a minimum by a clever interleaving of the cells.

The scenario just presented covers only a part of the total picture. A more complex case of word-line-to-bit-line coupling can be described as follows: During a read operation, a bit line *BL* undergoes a voltage transition ΔV. This voltage transition gets coupled to the nonselected word lines that are in precharged mode and thus represent high-impedance nodes. This noise signal, in turn, gets coupled back to other bit lines. The resulting cross talk depends on the data patterns stored in memory. Suppose that the majority of the bit lines are reading a 0. This causes a negative transition on the nonselected word lines, which in turn pulls down the bit lines. This is especially harmful for the (minority of) bit lines that are reading a 1 value. The fact that this noise signal is pattern dependent makes it particularly hard to detect. Testing a memory for functionality requires the application of a wide range of different data patterns. The folded bit line architecture helps to suppress this type of noise.

Bit-Line-to-Bit-Line Coupling

The impact of interwire cross coupling increases with reducing dimensions, as detailed in Chapter 9. This is especially true in the memory array, where the noise-sensitive bit lines run side by side for long distances. The art is to turn the noise signals into a common-mode signal for the sense amplifier. An ingenious way of doing so is represented by the *transposed (or twisted) bit-line architecture* illustrated in Figure 12-66. The first figure represents the straightforward implementation. Both *BL* and *BL* are coupled to the adjoining column (or row) by the capacitance C_{cross}. In the worst case, the signal swing observed at the sense amplifier can be reduced by

$$\Delta V_{cross} = 2\frac{C_{cross}}{C_{cross} + C_{BL}}V_{swing} \qquad (12.17)$$

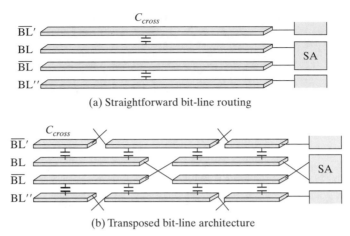

(a) Straightforward bit-line routing

(b) Transposed bit-line architecture

Figure 12-66   bit line-to-bit line coupling.

where V_{swing} is the signal swing on the bit lines. Up to one-fourth of the already weak signal can be lost due to this interference ([Itoh90]).

The *transposed bit line architecture* (Figure 12-66b) eliminates this source of disturbance by dividing the bit lines into segments that are connected in a cross-coupled fashion. This modification presents the interference signal introduced by a neighboring bit line in approximately the same way to both *BL* and $\overline{BL}$, turning it into a common-mode signal. An alternative approach is to use the capacitor-plate layer as an extra shielding between the data lines.

Leakage

Charge leakage is rapidly becoming one of the dominant noise sources in memories. This is no longer true only for dynamic memories, but also for static memories that increasingly feature high-impedance nodes. There are two leakage mechanisms that affect the storage node of a non-selected cell in a 1T DRAM: *pn*-junction leakage to the substrate, and subthreshold current to the bit line. This leakage causes a gradual degradation in the stored signal charge. The minimum bound on the signal-to-noise ratio sets an upper limit on the refresh time t_{REFmax}. Refresh operations consume power and represent a pure overhead from a systems point of view since they reduce the achievable memory bandwidth. Architectural modifications can help reduce the refresh overhead. A typical refresh procedure sequentially toggles each word line, refreshing all the cells connected to that line simultaneously. It is thus advantageous to keep the number of rows in the memory to a minimum. Dividing the memory into multiple blocks furthermore allows for the simultaneous refreshing of multiple rows. Even with these architectural improvements, leakage suppression is becoming more important with every new generation of memories.

Junction leakage can be improved in two ways: (1) by improving the quality of the fabrication process, for instance by changing the doping profile, and (2) by decreasing the ambient junction temperature as junction leakage is a strong function of the temperature. Low-power techniques reducing heat dissipation and the use of packages with low thermal resistance have proven to be quite effective. Yet, it is projected that the junction leakage should be smaller than 3.5 aA at 25C for a 64-Gb memory!

Keeping the subthreshold leakage at bay is an even larger challenge. The only reasonable solution is to increase the threshold of the access transistor, this when all the other voltages are being reduced! Maintaining sufficient drive current and performance under these conditions is hard. The only plausible solution is to scale down the refresh period more aggressively!

Critical Charge

Early memory designers were puzzled by the occurrence of *soft errors*, that is, nonrecurrent and nonpermanent errors, in DRAMs that could not be explained by either supply-voltage noise, leakage, or cross coupling. The impact of soft errors on system design is enormous. If no adequate protection against them is found, they can cause a computer system to crash in a non-reconstructible way making the task of the system debugger virtually hopeless. In a landmark paper, May and Woods identified *alpha particles* as the main culprits [May79].

Alpha particles are He^{2+} nuclei (two protons, two neutrons) emitted from radioactive elements during decay. Traces of such elements are unavoidably present in the device packaging materials. With an emitted energy of 8 to 9 MeV, alpha particles can travel up to 10 µm deep into silicon. While doing so, they interact strongly with the crystalline structure, generating roughly 2×10^6 electron-hole pairs in the substrate. The soft error problem occurs when the trajectory of one of these particles strikes the storage node of a memory cell. Consider the cell of Figure 12-67. When a 1 is stored in the cell, the potential well is empty. Electrons and holes generated by a striking particle diffuse through the substrate. Electrons that reach the edge of the depletion region before recombining are swept into the storage node by the electrical field. If enough electrons are collected, the stored value can change into a 0. The *collection efficiency* measures the percentage of electrons that make it into the cell.

Recently, it has been established that neutrons originating from cosmic rays are another important source of soft errors. Research has now indicated that the neutron-induced soft error rate (SER) in CMOS SRAMs is of the same order as the alpha-induced SER. Even more, the soft error rate of CMOS latch circuits seems to be dominated by neutrons [Tosaka97].

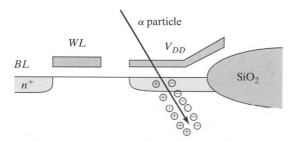

Figure 12-67   Alpha-particle induced soft errors.

The occurrence of soft errors can be reduced by keeping the cell charge larger than a *critical charge* Q_C. This puts a lower bound on the storage capacitance and cell voltage. As an example, a capacitance of 50 fF charge to a potential of 3.5 V holds 1.1×10^6 electrons. A single alpha particle striking such a cell with a collection efficiency of 55% can erase the complete charge. This is one of the main reasons why the cell capacitance of even the densest memories is kept higher than 30 fF. Another approach is to reduce the collection efficiency, which has been observed to be inversely proportional to the diagonal length of the depletion region of the cell. The value of the critical charge thus reduces with technology scaling! For instance, the critical charge equals 30 fC for a diagonal length of 2.5 μm which corresponds to a 64-Mbit memory. Chip coating and purification of materials are indispensable in reducing the number of alpha particles. For instance, a memory die can furthermore be covered with polymide to protect against alpha radiation.[7]

While the occurrence of soft errors can be kept to an absolute minimum by careful design, total elimination is hard to achieve. Free neutrons, originating from cosmic rays hitting the atmosphere, form another source of soft errors. The interaction of one of these neutrons interacting with a silicon nucleus—which is an unlikely, but possible, event—has a major probability of upsetting the stored data, as neutrons generate about 10 times as many charges as alpha particles.

The occasional occurrence of an error is thus hard to avoid, but it should not necessarily be fatal. System-level techniques such as error correction can help detect and correct most failures, and are becoming more and more indispensable in the memories of the future. This is briefly discussed in the next section.

12.4.2 Memory Yield

With increasing die size and integration density, a reduction in yield is to be expected, notwithstanding the improvements in the manufacturing process. Causes for malfunctioning of a part can be both material defects and process variations. Memory designers use two approaches to combat low yields and to reduce the cost of these complex components: redundancy and error correction. The latter technique has the advantage that it also addresses the occasional occurrence of a soft error.

Redundancy

Memories have the advantage of being extremely regular structures. Providing *redundant hardware* is easily accomplished. Defective bit lines in a memory array can be replaced by redundant ones, and the same holds for word lines. (See Figure 12-68.) When a defective column is detected during testing of the memory part, it is replaced by a spare one by programming the fuse bank connected to the column decoder. A typical way of doing so is to blow the fuses using a programming laser or pulsed current. Laser programming has a minimal impact on memory performance and occupies small chip area. It does require special equipment and increased

---

[7]It is worth mentioning that soft errors are also a problem in state-of-the-art SRAMs that are relying more heavily on capacitive storage on high-impedance nodes.

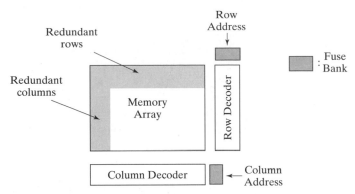

Figure 12-68   Redundancy in memory array increases the yield.

wafer handling time, however. The pulsed current approach can be executed by a standard tester, but bears larger overhead. A similar approach is followed for the defective word lines. Whenever a failing word line is addressed, the word redundancy system enables a redundant word line. In modern memories, as many as over 100 defective elements can be replaced by spare ones for an additional overhead of less than 5%. Even embedded SRAM memories, to be used in systems-on-a-chip, now come with redundancy.

Error Correction

Redundancy helps correct faults that affect a large section of the memory such as defective bit lines or word lines. It is ineffective when dealing with scattered point errors such as local errors caused by material defects. Achieving a reasonable fault coverage requires too much redundancy under these circumstances and results in a large area overhead. A better approach to address these faults is to use *error correction*. The idea behind this scheme is to use redundancy in the data representation so that an erroneous bit(s) can be detected and even corrected. Adding a parity bit to a data word, for instance, provides a way of detecting (but not correcting) an error. While a full discussion of error correction techniques would lead us astray, a simple example is used to illustrate the basic concept.

Example 12.15   Error Correction Using Hamming Codes

Consider a 4-bit number (B_i), encoded with 3 check bits (P_i),

$$P_1\, P_2\, B_3\, P_4\, B_5\, B_6\, B_7$$

with the P_i chosen such that

$$P_1 \oplus B_3 \oplus B_5 \oplus B_7 = 0$$
$$P_2 \oplus B_3 \oplus B_6 \oplus B_7 = 0$$
$$P_4 \oplus B_5 \oplus B_6 \oplus B_7 = 0$$

Suppose now that an error occurs in B_3. This causes the first two expressions to evaluate to 1 while the last one remains at 0. Binary encoded, this means that bit 011 or 3 is in error. This information is sufficient to correct the fault. In general, single error correction requires that

$$2^k \geq m + k + 1 \tag{12.18}$$

where m and k are the number of data and check bits, respectively. To perform single error correction on 64 data bits requires 7 check bits, resulting in a total word length of 71.

An important observation is that error correction not only combats technology-related faults, but is also an effective way of dealing with soft errors and time-variant faults. For instance, error correction is very effective in dealing with threshold variations in EEPROMs.

Error correction and redundancy address different angles of the memory yield problem. To cover all the bases, a combination of both is needed. This is convincingly illustrated in Figure 12-69, which plots the yield percentage for a 16-Mbit DRAM when the yield improvement techniques are used independently or combined [Kalter90].

One other issue is worth elaborating on in this discussion on memory reliability and yield. With the size and performance of memories ever increasing, testing a memory for defects takes ever more time on ever-more expensive testers. The situation is even more astute for embedded memories, as the memory ports may not be directly accessible through the I/O pins of the chip. A common approach to deal with this problem is to have the memory test itself. In this built-in self-test (BIST) approach, a small controller is added to the memory that generates input patterns, and verifies if the responses of the memory are correct. The BIST approach is very attractive: it eliminates the cost of the expensive tester, it can be performed at speed, and can be executed every time the chip is started up. More information on BIST can be found in Design Methodology Insert H at the end of this chapter.

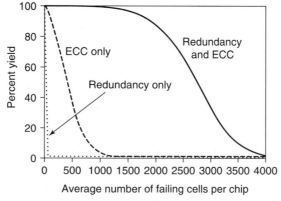

Figure 12-69   Yield curves for 16-Mbit DRAM using ECC and bit line redundancy [Kalter90].

12.5 Power Dissipation in Memories*

As in most of the arena of digital design, reduction of the power dissipation in memories is becoming of premier importance. Memory designers have been remarkably good in keeping dissipation in check, even while increasing the memory capacity with 6 orders of magnitude over the last 30 years. Yet, the challenges are mounting. Portable applications are lowering the bar on how much power memory may consume. At the same time, technology scaling with its reduction in supply and threshold voltages and its deterioration of the off-current of the transistor causes the standby power of the memory to rise. The introduction of innovative techniques is truly a necessity.

12.5.1 Sources of Power Dissipation in Memories

The power consumption in a memory chip can be attributed to three major sources: the memory cell array, the decoders (row, column, block), and the periphery. A unified active power equation for a modern CMOS memory of m columns and n rows is approximately given by [Itoh01] (see Figure 12-70). For a normal read cycle:

$$P = V_{DD}I_{DD}$$
$$I_{DD} = I_{array} + I_{decode} + I_{periphery} \tag{12.19}$$
$$= [mi_{act} + m(n-1)i_{hld}] + [(n+m)C_{DE}V_{int}f] + [C_{PT}V_{int}f + I_{DCP}]$$

The equation uses the following design parameters:

- i_{act}—the effective current of the selected (or active) cells
- i_{hld}—the data retention current of the inactive cells

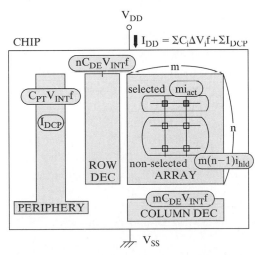

Figure 12-70   Sources of power dissipation in semiconductor memory [Itoh01].

- C_{DE}—the output node capacitance of each decoder
- C_{PT}—the total capacitance of the CMOS logic and periphery circuits
- V_{int}—the internal supply voltage
- I_{DCP}—the static (or quasi-static) current of the periphery. The major sources for this current are the sense amplifiers and the column circuitry. Other sources are the on-chip voltage generators.
- f—the operating frequency

As should be expected, the power dissipation is proportional to the size of the memory (n,m). Dividing the memory into subarrays, and keeping n and m small is essential to keep power within bounds. This approach is obviously only effective if the standby dissipation of inactive memory modules is substantially lower.

In general, the power dissipation of the memory is dominated by the array. The active power dissipation of the peripheral circuits is small compared with the other components. Its standby power can be high however, requiring that circuits such as sense amplifiers are turned off when not in action. The decoder charging current is also negligibly small in modern RAMs, especially if care is taken that only one out of the n or m nodes is charged at every cycle.

Reduction of power dissipation in memories is worth a chapter on its own. We limit ourselves to an enumeration of some techniques that are generally applicable, and refer the interested reader to [Itoh01] for more details and depth.

12.5.2 Partitioning of the Memory

A proper division of the memory in submodules goes a long way in confining active power dissipation to limited areas of the overall array. Memory units that are not in use should consume only the power necessary for data retention. Memory partitioning is accomplished by reducing m (the number of cells on a word line), and/or n (the number of cells on a bit line). By dividing the word line into several sub-word-lines that are enabled only when addressed, the overall switched capacitance per access is reduced. In some sense, this scheme is nothing more than a multistage hierarchical row decoder. This approach is quite popular in SRAM memories, as illustrated in the 4-Mbit SRAM memory discussed in the case studies of the next section.

In a similar way, partitioning of the bit line reduces the capacitance switched at every read/write operation. An approach that is often used in DRAM memories is the partially activated bit line. The bit line is partitioned in multiple sections (typically 2 or more). All these sections share a common sense amplifier, column decoder, and I/O module. This approach has helped to reduce CBL from more than 1 pF for the 16-KBit DRAM generation to approximately 200 fF for the 64-MBit DRAM.

12.5.3 Addressing the Active Power Dissipation

Similar to what we learned for logic circuits, reducing the voltage levels is one of the most effective techniques to reduce power dissipation in memories. In contrast to logic, however, voltage scaling might run out of steam earlier. Data retention and reliability issues make the scaling of the voltages to the deep sub-1-V level quite a challenge. In addition, careful control of the capacitance and switching activity, as well as minimization of the on time of the peripheral components is essential.

SRAM Active Power Reduction

To obtain a fast read, the voltage swing on the bit line is made as small as possible—typically between 0.1 V and 0.3 V. The resulting signal is transmitted to the sense amplifier for restoration. Since the signal is developed as a result of the ratio operation of the bit line load and the cell transistor, a current flows through the bit line as long as the word line is activated (Δt). Limiting Δt and the bit line swing helps to keep the active dissipation of SRAMs low. Signal-to-noise issues ultimately determine how small the bit line swing and the on time of the sense amplifier can be made. Self-timing can be a big help in determining exactly how long to keep the peripheral circuitry on.

The situation is worse for the write operation, since BL and $\overline{BL}$ have to make a full excursion. Reduction of the core voltage is the only remedy for this (in addition, of course, to the bit line partitioning approach described earlier). Ultimately, the reduction of the core voltage is limited by the mismatch between the paired MOS transistors in the SRAM cell. Even when designed to be completely identical, process and device variations cause the MOS transistors in the cell to be different. The ever-increasing V_T threshold mismatch is the most important culprit. It is caused by implant nonuniformities, channel length and width variations, and even random microscopic fluctuations of dopant atoms in very small devices. Even if process engineers manage to keep the threshold variation constant, its relative importance is rapidly increasing in light of decreasing supply and threshold voltages. The mismatch between the transistors causes the cell to become asymmetrical, biasing it toward either the 1 or 0 state. This makes the cell substantially less reliable in the presence of noise and during read operations. Stringent control of the MOSFET characteristics, either at process time or at run time using techniques such as body biasing, is essential in the low voltage operation mode. If this cannot be accomplished, extra redundancy and error correction is needed.

Reducing the supply voltage also impacts the memory access time. Lowering the device thresholds is only an option if the resulting increase in leakage current is dealt with effectively, especially in the nonactive cells (see *reduction of data-retention power*).

DRAM Active Power Reduction

The destructive readout process of a DRAM cell necessitates successive operations of readout, amplification and restoration for the selected cells. Consequently, the bit lines are charged and discharged over the full voltage swing (ΔV_{BL}) for every read operation. Care should thus be given to reduce the bit line dissipation charge $mC_{BL}\Delta V_{BL}$, since it dominates the active power. Reducing C_{BL} (the bit line capacitance) is advantageous from both a power and a signal-to-noise perspective (Eq. (12.8)), but is not simple given the trend toward larger memories. Reducing ΔV_{BL}, while very beneficial from a power perspective, negatively impacts the signal-to-noise ratio. This is apparent from Eq. (12.8), which shows that the charge stored in the cell—and hence the readout signal—is directly proportional to the voltage swing applied during the write or the refresh cycle. Voltage reduction thus has to be accompanied by either an increase in the size of the storage capacitor and/or a noise reduction. A number of techniques have proven to be quite effective:

- **Half-V_{DD} precharge**—Precharging the bit lines to $V_{DD}/2$ helps to reduce the active power dissipation in DRAM memories by a factor of almost 2. After amplification and restoration of the readout voltages on the bit lines, precharging is accomplished by a simple shorting of the two bit lines. Assuming that the two bit lines are balanced, this results in an exact voltage of $V_{DD}/2$.
- **Boosted word line**—Raising the value of the word line above V_{DD} during a write operation eliminates the threshold drop over the access transistor, yielding a substantial increase in stored charge.
- **Increased capacitor area or value**—Vertical capacitors such as those used in trench and stacked cells are very effective in increasing the capacitance value, albeit at a major cost in processing and manufacturing. In addition, keeping the "ground" plate of the storage capacitor at $V_{DD}/2$ reduces the maximum voltage over C_s, making it possible to use thinner oxides.
- **Increasing the cell size**—Ultimately, ultra-low-voltage DRAM memory operation might require a sacrifice in area efficiency, especially for memories that are embedded in a system-on-a-chip. At that juncture, however, it becomes worthwhile exploring whether cells that provide gain, such as the 3T DRAM, offer a better alternative.

12.5.4 Data-Retention Dissipation

Data Retention in SRAMs

In principle, an SRAM array should not have any static power dissipation (i_{hld}). Yet, the leakage current of the cell transistors is becoming a major source of retention current. While this current historically has been kept low, a considerable increase has been observed in recent embedded memories due to subthreshold leakage. This is illustrated in Figure 12-71, which plots the standby current of the same 8-kBit embedded memory, implemented in a 0.18-μm and a 0.13-μm CMOS process. An increase of the static current with a factor of almost 7 for the same supply voltage can be observed.

Techniques to reduce the retention current of SRAM memories are thus necessary, including the following:

- **Turning off unused memory blocks**. Memory functions such as caches do not fully use the available capacity for most of the time. Disconnecting unused blocks from the supply rails using high-threshold switches reduces their leakage to very low values. Obviously, the data stored in the memory is lost in this approach.

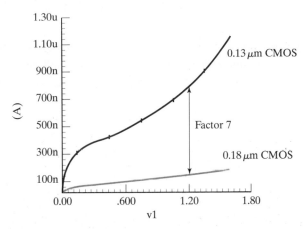

Figure 12-71   Leakage current in embedded eight-kbit SRAM memory as a function of voltage.

- **Increasing the thresholds by using body biasing**. Negative bias of the nonactive cells increases the thresholds of the devices and reduces the leakage current.
- **Inserting extra resistance in the leakage path**. When data retention is necessary, the insertion of a low-threshold switch in the leakage path provides a means to reduce leakage current while keeping the data intact (see Figure 12-72a). While the low-threshold device leaks on its own, which is sufficient to maintain the state in the memory. At the same time, the voltage drop over the switch introduces a "stacking effect" in the memory cells connected to it: A reduction of V_{GS} combined with a negative V_{BS} results in a substantial drop in leakage current.
- **Lowering the supply voltage.** Figure 12-71 indicates that the leakage current is a strong function of V_{DD}. An effective means to reduce leakage during standby is to lower the

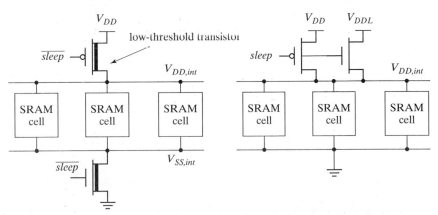

Figure 12-72   Standby leakage suppression techniques for SRAM memories: (a) inserting extra resistance; (b) lowering the supply voltage. (V_{DDL} can be as low as 100 mV.)

supply rails to a value that keeps the leakage within bounds, while ensuring data retention (see Figure 12-72b). The lower bound of the data retention voltage—the voltage that still maintains the stored value—is determined by device variations. A supply voltage as low as 100 mV (excluding noise margin) is sufficient to maintain retention in a standard 0.13-μm CMOS process. The combined effect of reduced supply voltage and leakage current is a powerful way of addressing standby power dissipation in SRAM memories.

DRAM Retention Power

Standby power in DRAMs is attributed to leakage, just as in SRAM memories. This is where the similarity ends, however. To combat leakage and loss of signal, DRAMs have to be refreshed continuously when in data-retention mode. The refresh operation is performed by reading the *m* cells connected to a word line and restoring them. This operation is repeated for each of the *n* word lines in sequence. The standby power is therefore proportional to the bit line dissipation charge, defined earlier, and the *refresh frequency*. The latter is a strong function of the leakage rate. Keeping the junction temperatures low is an effective means to keeping leakage within bounds.

Yet, the shrinking of the cell sizes and the charge stored in the cell, combined with a reduction in voltages, is forcing the refresh frequency upward, and causing the standby power of DRAMs to increase. This is reflected in Figure 12-73, which plots the evolution of the active and standby currents for different DRAM generations. The data retention current is overtaking the active current at the 1-Gbit DRAM generation and will be the dominant source of power dissipation unless some leakage suppression techniques are introduced.

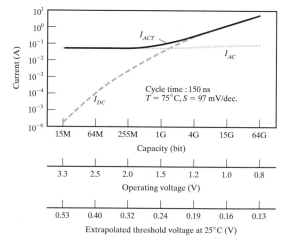

Figure 12-73   Estimated current distribution for consecutive DRAM generations (from [Itoh01]).

The secret to leakage minimization in DRAM memories is V_T control. This can be accomplished at design time (the fixed V_T approach), or dynamically (the variable V_T technique). One option to reduce leakage through the access transistor in the DRAM cell is to turn off the device hard by applying a negative voltage ($-\Delta V_{WL}$) to the word line of nonactive cells. Alternatively, we can raise the bit line voltage of unused cells by a certain amount ($+\Delta V_{BL}$). This approach, called the *boosted sense ground*, results in a negative gate-source voltage and a slight increase in the threshold voltage.

The variable V_T techniques raise the thresholds of the access transistors in inactive cells through well biasing. Control can be performed at different levels of granularity (chip, module, and individual transistor) with increasing levels of overhead. Careful attention should be paid to the generation of the well-voltages, as this can lead to instabilities.

12.5.5 Summary

SRAM memories have a distinctive advantage over DRAM from a power perspective. Yet, the increasing impact of leakage combined with reduced supply voltages might reduce this gap to a certain extent. When standby power is a dominant concern, nonvolatile memories offer a viable and attractive alternative.

12.6 Case Studies in Memory Design

While the previous sections have introduced the individual components of a semiconductor memory, it is worthwhile to study a number of examples to understand how it all comes together. We will use two case studies to achieve that goal—the programmable logic array and a 4-Mbit SRAM memory.

12.6.1 The Programmable Logic Array (PLA)

In our discussion on design implementation strategies in Chapter 8, we introduced the concept of the Programmable Logic Array (or PLA). The PLA became popular in the early 1980s as a regular approach for the implementation of random logic. (See "Historical Perspective: The Programmable Logic Array" on page 388.)

Now follows the reason why the issue of two-level logic implementation is raised in this chapter on memories. When discussing the ROM memory core, we realized that the ROM array is nothing more than a collection of large fan-in NOR (NAND) gates implemented in a very regular format. Similarly, the address decoder is also a collection of NOR (NAND) gates. The same approach can thus be used to implement any two-level logic function—hence the PLA. The only difference between a ROM and PLA is that the decoder (AND-plane) of the former enumerates all possible minterms (m inputs yield 2^m minterms), while the AND-plane of the latter only realizes a limited set of minterms, as dictated by the logic equations. Topologically, both structures are identical.

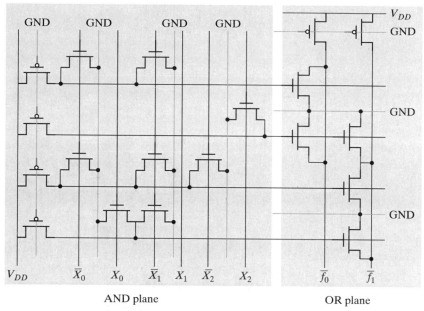

AND plane   OR plane

Figure 12-74   Pseudo-NMOS PLA.

A NOR–NOR implementation of a PLA using the pseudo-NMOS circuit style is shown in Figure 12-74. Although compact and fast, its power dissipation makes this realization unattractive for larger PLAs where a dynamic approach is better. As we know, a direct cascade of the dynamic planes is out of the question. Solutions could be to introduce an inverter between the planes in the DOMINO style, or implement the OR-plane with PMOS transistors and use pre-discharging in the np-CMOS fashion. The former solution causes pitch-matching problems, while the latter slows down the structure if minimum-size PMOS devices are used.

We will take this opportunity to examine the memory-timing issue more thoroughly. Figure 12-75 uses a more complex clocking scheme to resolve the plane-connection problem [Weste93]. Assume that the pull-up transistors of the AND plane and the OR plane are clocked by signals ϕ_{AND} and ϕ_{OR}, respectively. These signals are defined so that a clock cycle starts with the precharging of both planes. After a time interval long enough to ensure that precharging of the AND plane has ended, the AND plane starts to evaluate by raising ϕ_{AND}, while the OR plane remains disabled (see Figure 12-76a). Once the AND plane outputs are valid, it is safe to enable the OR plane by raising ϕ_{OR}.

The timing of these events is precarious, since it depends upon the PLA size and programming and technology variations. Although the clock signals can be provided externally, self-timing is the recommended approach to achieve maximum performance at the minimum risk. How to derive the appropriate timing signals from a single clock ϕ is shown in Figure 12-76b. The

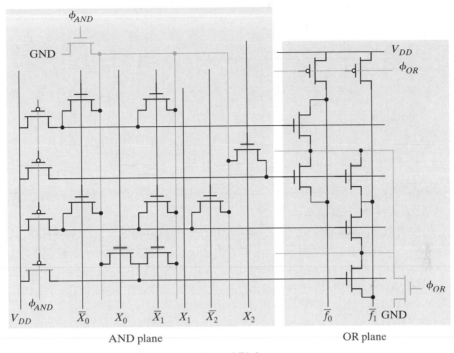

Figure 12-75 Dynamic implementation of PLA.

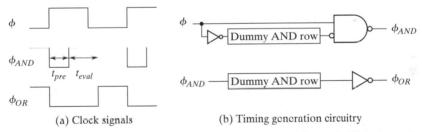

(a) Clock signals   (b) Timing generation circuitry

Figure 12-76 Generation of clock signals for self-timed dynamic PLA. t_{pre} and t_{eval} represent the worst case precharge and evaluation times of an AND row, respectively. These are obtained with the aid of dummy AND rows, which are fully populated; all transistors in the NOR network except one are turned off. This represents the worst case discharge scenario.

clock signal ϕ_{AND} is derived from ϕ using a monostable one-shot. The delay element consists of a dummy AND row with maximum loading—a fully populated row that provides the worst case estimate of the required precharge time. ϕ_{OR} is derived in a similar way by using a dummy AND row clocked by ϕ_{AND}. This provides a worst case estimate of the discharge time, including some extra safety margin. Although it requires additional logic, this approach ensures reliable operation of the array under a range of conditions.

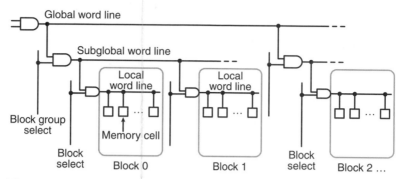

Figure 12-77   Hierarchical word-line selection scheme.

12.6.2   A 4-Mbit SRAM

In Figure 12-6, we have shown the block diagram of a 4-Mbit SRAM memory [Hirose90]. This memory has an access time of 20 ns for a single supply voltage of 3.3 V and is fabricated in a quadruple polysilicon, double metal 0.6-μm CMOS technology. The memory is organized as 32 blocks, each block counting 1024 rows and 128 columns. The row address (X), column address (Y), and block address (Z) are 10, 7, and 5 bits wide, respectively.

To provide fast and low-power row decoding, the memory uses an interesting approach called the *hierarchical word-decoding* scheme, shown in Figure 12-77. Instead of broadcasting the decoded X address to all blocks in polysilicon, it is distributed in metal and called the *global word line*. The local word line is confined to a single block and is only activated when that particular block is selected by using the block address. To further improve the performance and power consumption, an extra word-line hierarchy level called the *subglobal word line* is introduced. This approach results in a row-decoding delay of only 7 ns.

The bit line peripheral circuitry is presented in Figure 12-78. The memory cell is a four-transistor resistive-load cell occupying 19 μm². The load resistance equals 10 TΩ. A triggering of the *ATD* pulse causes the precharging and the equalizing of the bit lines of the selected block with the aid of the *BEQ* control signal. Notice that the bit line load is a composition of static and dynamic devices. Lowering *BEQ* starts the read process. One of the word lines is enabled, and the appropriate bit lines start to discharge. These are connected to the sense amplifiers after passing through the first layer of the column decoder (*CD*). Only 16 sense amplifiers are needed per block of 128 columns in this scheme. The amplifier consists of two stages (see Figure 12-78b). The first is a cross-coupled stage that provides a minimal gain, but also acts as a level shifter. This permits the second amplifier, which is of the current-mirror type, to operate at the maximum-gain point. A push–pull output stage drives the highly capacitive data lines that lead to the tristable input/output drivers. The approximative signal waveforms are displayed in Figure 12-78c. These demonstrate the subtlety of the timing and clocking strategy necessary to operate a large memory at high speed. To write into the memory, the appropriate value and its complement are imposed on the *I/O* and *I̅/̅O̅* lines, after which the appropriate row, column, and block addresses are enabled. It is worth

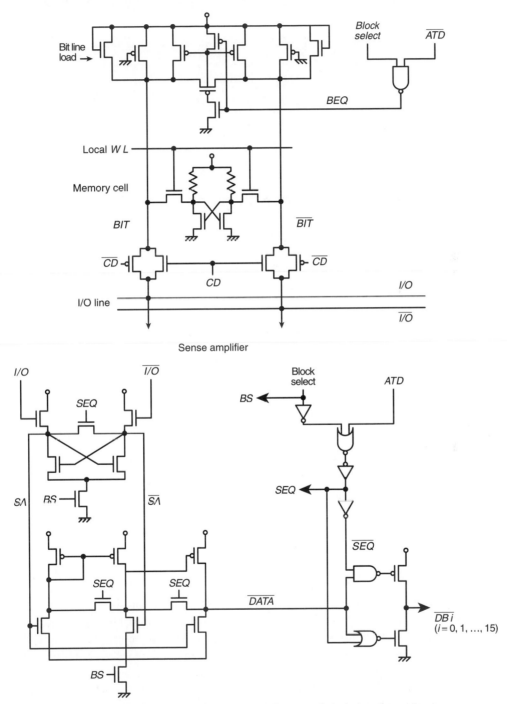

Figure 12-78   The bit line peripheral circuitry and the associated signal waveforms.

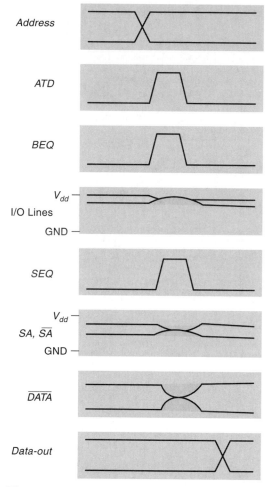

Figure 12-78 *(cont.)*

mentioning that the memory consumes only 70 mA when operated at 40 MHz. The standby current equals 1.5 μA.

12.6.3 A 1-Gbit NAND Flash Memory

The block diagram of a 1-Gb NAND Flash memory [Nakamura02] is shown in Figure 12-79. The memory is structured as two blocks of 500 Mb each. A NAND organization in the style of Figure 12-25 with 32 bits/block was chosen as the basic memory module, mainly for cell size considerations. Each bit line connects to 1024 of these modules. A block of 500 Mb combines 16,895 of these bit lines in parallel. The word lines are driven from both sides of the block. The page size of the memory, this is the number of bits that can be read/written in one cycle, equals 2 KByte. The large size of the page enables the high programming rate of 10 MByte/s. To further

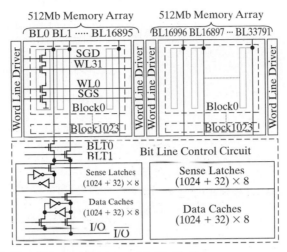

Figure 12-79 Chip architecture of 1-Gb Flash memory.

speed up the programming, an extra "cache" cell is provided per bit line, which allows for new data to be read into the memory, while the previous data is being written/verified.

Figure 12-80 illustrates the erasure/writing process of the Flash memory. During erasure, all bits in a single block are programmed to become a depletion device. During write, the threshold of selective devices is raised by using a programming/verify cycle. It takes at least four of these cycles for all of the thresholds to be above 0.8 V, which is the necessary requirement for the device to act as a correct switch (for the nominal word-line voltages). The distribution of the thresholds after the four cycles is plotted in Figure 12-80b.

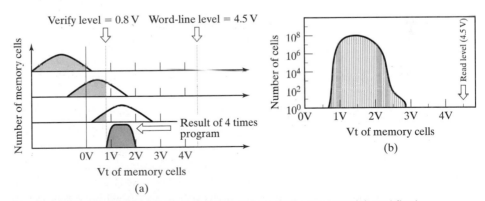

Figure 12-80 Evolution of cell threshold during writing process (a) and final threshold distribution (b) after four programming cycles.

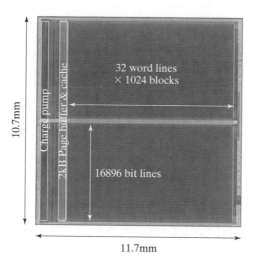

	Characteristics
Cell Size	$0.077\ \mu m^2$
Chip Size	$125.2\ mm^2$ in $0.13\ \mu m$ CMOS
Organization	$2112 \times 8b \times 64$ page $\times$ 1kb block
Power Supply	2.7-3.6 V
Cycle Time	50 ns
Read Time	$25\ \mu s$
Program Time	$200\ \mu s$
Erase Time	2 ms/block

Figure 12-81   Die microphotograph and chip characteristics of 1-Gb Flash memory [Nakamura02].

The chip microphotograph of the memory module and an enumeration of its characteristics are shown in Figure 12-81. The memory is implemented in a 0.13-µm CMOS triple well process with 1 poly, 1 polycide, 1 W, and 2 Al layers.

12.7 Perspective: Semiconductor Memory Trends and Evolutions

To conclude this chapter, some evolutionary trends are worth discussing. The first semiconductor memories introduced in the 1960s used bipolar technology and could store no more than 100 bits. A major breakthrough was achieved in the early 1970s when the first MOS DRAM memory was introduced with a capacity of 1 KBit ([Hoff70]). This was the beginning of a dramatic evolution. Currently, the 1-Gbit memory is here. This means that in 30 years, the amount of memory that can be integrated on a single die has increased by six orders of magnitude! It is fair to say that the number of memory cells that can be integrated on a single die quadruples approximately every three years (which generally corresponds to a memory generation). An overview of the trends in capacity for various memory styles is shown in Figure 12-82.

It is interesting to understand what gave rise to these spectacular improvements. Figure 12-83 plots the evolution of the cell sizes over the subsequent generations. Both DRAM and SRAM cells have been miniaturized at a pace of about one-fiftieth every 10 years (or a factor of somewhat higher than 3 every 3 years). In addition, die size has increased with a factor of 1.4 every generation. Combining the two factors yields the increase in density by a factor of around 4 between generations. One would be inclined to attribute the reduction in cell size solely to technology scaling, but this is only partially true. For instance, the L_{eff} of the transistors used in the 4-K SRAM memory equaled 3 µm and has decreased from generation to generation so that the 64-

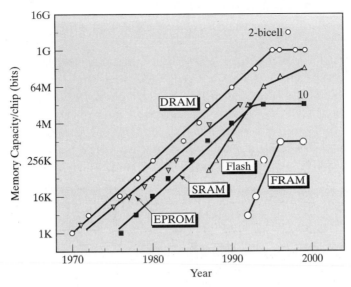

Figure 12-82 Trends in the memory capacity of VLSI memories [Itoh01].

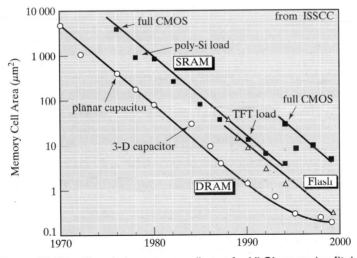

Figure 12-83 Trends in memory cell area for VLSI memories [Itoh01].

M SRAM uses transistors with an effective length of 0.25 μm. This suggests that scaling reduces the cell size by a factor of $1.5 \times 1.5 = 2.25$ between generations, which is not sufficient to account for the observed trends. Additional reductions are obtained by an ever-increasing sophistication in memory cell technology. Important events in this category include the introduction of the polysilicon gate, denser isolation using LOCOS and U-groove, polysilicon load resistors for SRAM cells, and three-dimensional DRAM cells using trench or stacked capacitors. The continuous introduction of such novelties is essential if the current trends in integration are to be prolonged.

This might not be trivial, as some major challenges and roadblocks have become apparent in recent years. This is already visible in Figure 12-82, which shows a distinctive saturation in the capacity of SRAM and DRAM memories at the end of the 1990s. In fact, Flash memory is about to pull even with DRAM as we speak.This saturation is mainly due to the extreme challenge and cost of further cell miniaturization. This is not the only cause, however, and other considerations should be taken into account. We briefly enumerate a number of issues that impact the further integration of semiconductor memories.

- High-density memories need very specialized technologies that are extremely differentiated from logic technologies. Their development is becoming more and more of a financial hazard.
- Maintaining correct operation of SRAM and DRAM memories in the presence of reduced supply and threshold voltages is very hard. Effects of leakage and soft errors are posing a challenge that has been unanswered so far. Important innovations in the memory cell ate necessary. Novel approaches such as nonvolatile FRAMs and MRAMs might offer an alternative.
- As stated in an earlier section, power consumption is becoming a limiting factor on how much memory can be integrated on a single die.
- But most importantly, a shift in business climate has emerged. The unprecedented hunger for more memory was mostly driven by the computer industry. With the shift of the center of mass of the semiconductor industry toward consumer, multimedia, and communications applications, different needs in terms of memory have arisen. For instance, mobile semiconductor video and audio players feed the need for nonvolatile memory. Also, the size and cost constraints of these applications make integrated system-on-a-chip solutions desirable. Most of the memory should be embedded on the die, together with the processor and logic components. This drives the needs for memory technologies that are compatible with standard logic processes.

12.8 Summary

In this chapter, we have discussed the design of semiconductor memories in extensive detail. While the art of designing memories is vastly different from the traditional logic world, it is obvious that many of the problems addressed by memory designers are similar to those encountered by logic designers. Moreover, with the advent of Systems-on-a-Chip, embedded memories are becoming a necessary component of virtually any complex design. Finally, due to the fact that memory design operates at the frontier of technological ability, such as the smallest device dimensions, many of the problems and solutions considered in contemporary memory design might very well surface in logic design at a later time. As an example, power dissipation is one of the major challenges facing high-density memory designers. Low-voltage operation around 1.0 V or below is one of the solutions being employed in state-of-the-art memories. Other approaches include self-timing and local power-down. Most of these techniques can be applied to the logic design world as well.

In conclusion, the memory-design problem can be summarized as follows:

- The *memory architecture* has a major impact on the ease of use of the memory, its reliability and yield, its performance, and, power consumption. Memories are organized as arrays of cells. An individual cell is addressed by a block, column, and row address.
- The *memory cell* should be designed so that a maximum signal is obtained in a minimum area. While the cell design is mostly dominated by technological considerations, a clever circuit design can help maximize the signal value and transient response. The cell topologies have not varied much over the last decades. Most of the improvement in density results from technology scaling and advanced manufacturing processes. We discussed cells for read-only, nonvolatile, and read–write memories.
- The *peripheral circuitry* is essential to operate the memory in a reliable way and with a reasonable performance, given the weak signaling characteristics of the cells. Decoders, sense amplifiers, and I/O buffers are an integral part of every memory design. The discussion presented in this chapter only touched the surface. Besides the mentioned functions, other circuitry in the periphery performs functions such as bootstrapping, voltage generation in EEPROMs, and voltage regulation ([Itoh01]). Although interesting and challenging, the discussion of these circuits would have lead us too far. Suffice it to say that the design of memory periphery is still the playground of the hard-core digital circuit designer.
- A memory must operate correctly over a variety of operating and manufacturing conditions. To increase the integration density, memory designers give up on signal-to-noise ratio. This makes the design vulnerable to a whole range of noise signals, which are normally less of an issue in logic design. Identifying the potential sources of malfunction and providing an appropriate model is the first requirement when addressing *memory reliability*. Circuit precautions to deal with potential malfunctions include redundancy and error correction.

12.9 To Probe Further

A vast amount of literature on memory design is presented in the leading circuit design journals. Most ground-breaking innovations in memory design are reported in either the *Journal of Solid State Circuits* (November issue in particular), the ISSCC conference proceedings, and the VLSI circuit symposium. Yet, until recently, not many textbooks were available on this topic. A number of interesting tutorial papers have been published over the years, some of which are included in the reference list that follows. Fortunately, a couple of great books on semiconductor memory have finally been published. [Itoh01] and [Keeth01] provide a wealth of background information for the memory aficionado.

References

[Amrutur01] B.S. Amrutur, M.A. Horowitz, "Fast Low-Power Decoders for RAMs," *IEEE Journal of Solid-State Circuits*, vol. 36, no. 10, pp. 1506–1515, October 2001.

[Baur95] M. Baur et al., "A Multilevel-Cell 32-Mb Flash Memory," *ISSCC Digest of Technical Papers*, pp. 13–33, February 1995.

[Chang77] J. Chang, "Theory of MNOS Memory Transistor," *IEEE Trans. Electron Devices*, ED-24, pp. 511–518, 1977.

[Crisp97] R. Crisp, "Direct Rambus Technology: The New Main Memory Standard," *IEEE Micro*, pp. 18–28, November/December 1997.

[Dennard68] R. Dennard, "Field-effect transistor memory," U.S. Patent 3 387 286, June 1968.

[Dillinger88] T. Dillinger, *VLSI Engineering*, Chapter 12, Prentice Hall, 1988.

[Frohman74] D. Frohman, "FAMOS—A New Semiconductor Charge Storage Device," *Solid State Electronics*, vol. 17, pp. 517–529, 1974.

[Gray01] P. Gray, P. Hurst, S. Lewis, and R. Meyer, *Analysis and Design of Analog Integrated Circuits,* 4th ed., John Wiley and Sons, 2001.

[Heller75] L. Heller et al., "High-Sensitivity Charge-Transfer Sense Amplifier," *Proceedings ISSCC Conf.*, pp. 112–113, 1975.

[Hennessy02] J. Hennessy, D. Patterson, and D. Goldberg, *Computer Architecture—A Quantitavie Approach,* 2d ed, Morgan Kaufman Publishers, 2002.

[Hidaka92] H. Hidaka et al., "A 34-ns 16-Mb DRAM with Controllable Voltage-Down Converter," *IEEE Journal of Solid State Circuits*, vol. 27, no. 7, pp. 1020–1027, July 1992.

[Hirose90] T. Hirose et al., "A 20-ns 4-Mb CMOS SRAM with Hierarchical Word Decoding Architecture," *IEEE Journal of Solid State Circuits*, vol. 25, no. 5, pp. 1068–1074, October 1990.

[Hoff70] E. Hoff, "Silicon-Gate Dynamic MOS Crams 1024 Bits on a Chip," *Electronics*, pp. 68–73, August 3, 1970.

[Itoh90] K. Itoh, "Trends in Megabit DRAM Circuit Design," *IEEE Journal of Solid State Circuits*, vol. 25, no. 3, pp. 778–798, June 1990.

[Itoh01] K. Itoh, *VLSI Memory Chip Design*, Springer-Verlag, 2001.

[Johnson80] W. Johnson et al., "A 16 Kb Electrically Erasable Nonvolatile Memory," *ISSCC Digest of Technical Papers*, pp. 152–153, February 1980.

[Kalter90] H. Kalter et al., "A 50-ns 16-Mb DRAM with a 10-ns Data Rate and On-Chip ECC," *IEEE Journal of Solid State Circuits*, vol. 25, no. 5, pp. 1118–1128, October 1990.

[Keeth01] B. Keeth and R. Baker, *DRAM Circuit Design—A Tutorial*, IEEE Press, 2001.

[Lu89] N. Lu, "Advanced Cell Structures for Dynamic RAMs," *IEEE Circuits and Devices Magazine*, pp. 27–36, January 1989.

[Mano87] T. Mano et al., "Circuit Technologies for 16-Mbit DRAMs," *ISSCC Digest of Technical Papers*, pp. 22–23, February 1987.

[Masuoka91] F. Masuoka et al., "Reviews and Prospects of Non-Volatile Semiconductor Memories," *IEICE Transactions*, vol. E74, no. 4, pp. 868–874, April 1991.

[May79] T. May and M. Woods, "Alpha-Particle-Induced Soft Errors in Dynamic Memories," *IEEE Transactions on Electron Devices*, ED-26, no. 1, pp 2–9, January 1979.

[Mot91] Motorola, "Memory Device Data," Specification Data Book, 1991.

[Nakamura02] H. Nakamura et al., "A 125-mm2 1-Gb NAND Flash Memory with 10-Mb/s Program Troughput," *ISSCC Digest of Technical Papers*, pp. 106-107, February 2002.

[Ootani90] T. Ootani et al., "A 4-Mb CMOS SRAM with PMOS Thin-Film-Transistor Load Cell," *IEEE Journal of Solid State Circuits*, vol. 25, no. 5, pp. 1082–1092, October 1990.

[Pashley89] R. Pashley and S. Lai, "Flash Memories: The Best of Two Worlds," *IEEE Spectrum*, pp. 30–33, December 1989.

[Preston01] R.P. Preston, "Register Files and Caches," in A. Chandrakasan, W.J. Bowhill and F. Fox (eds.), *Design of High-Performance Microprocessor Circuits*, Piscataway, NJ: IEEE Press, 2001.

[Regitz70] W. Regitz and J. Karp, "A Three-Transistor Cell, 1,024-bit 500 ns MOS RAM," *ISSCC Digest of Technical Papers*, pp 42–43, 1970.

[Sasaki90] K. Sasaki et al., "A 23-ns 4-Mb CMOS SRAM with 0.2-mA Standby Current," *IEEE Journal of Solid State Circuits*, vol. 25, no. 5, pp. 1075–1081, October 1990.

[Scheuerlein00] R. Scheuerlein et al., "A 10-ns Read-and-Write Non-Volatile Memory Array using a Magnetic Tunnel Junction and FET Switch in Each Cell," *ISSCC Digest of Technical Papers*, pp. 128–129, February 2000.

[Sedra87] A. Sedra and K. Smith, *Microelectronic Circuits*, 2d ed., Holt, Rinehart and Winston, 1987.

[Snow67] E. Snow, "Fowler-Nordheim Tunneling in SiO$_2$ Films," *Solid State Communications*, vol. 5, pp. 813–815, 1967.

[Taguchi91] M. Taguchi et al., "A 40-ns 64-Mb DRAM with 64-b Parallel Data Bus Architecture," *IEEE Journal of Solid State Circuits*, vol. 26, no. 11, pp. 1493–1497, November 1991.

[Takada91] M. Takada and T. Enomoto, "Reviews and Prospects of SRAM Technology," *IEICE Transactions*, vol. E74, no. 4, pp. 827–838, April 1991.

[Tosaka97] Y. Tosaka et al., "Cosmic Ray Neutron-Induced Soft Errors in Sub-Half Micron CMOS Circuits," *IEEE Electron Device Letters*, vol. 18, no. 3, pp. 99–101, March 1997.

[Weste93] N. Weste and K. Eshraghian, *Principles of CMOS VLSI Design*, 2d ed., Addison-Wesley, 1993.

[Yoo95] H. Yoo et al., "A 159-MHz 8-Banks 256-M Synchronous DRAM with Wave Pipelining Methods," *ISSCC Digest of Technical Papers*, pp. 374–375, 1995.

Exercises and Design Problem

The web site of the book (*http://bwrc.eecs.berkeley.edu/IcBook*) contains up-to-date and challenging exercises and design problems on memory.

Validation and Test
of Manufactured Circuits

> *Manufacturing test*
>
> *Design for testability*
>
> *Test pattern generation*

H.1 Introduction

While designers tend to spend numerous hours on the analysis, optimization, and layout of their circuits, one issue is often overlooked: When a component returns from the manufacturing plant, how does one know if it actually works? Does it meet the functionality and performance specifications? The customer expects a delivered component to perform as described in the specification sheets. Once a part is shipped or deployed in a system, it is expensive to discover that it does not work. The later a fault is detected, the higher the cost of correction. For instance, replacing a component in a sold television set means replacement of a complete board as well as the cost of labor. Shipping a nonworking or partially functional device should be avoided if at all possible.

A correct design does not guarantee that the manufactured component will be operational. A number of manufacturing defects can occur during fabrication, either due to faults in the base material (for instance, impurities in the silicon crystal), or as a result of variations in the process, such as misalignment. Other faults might be introduced during the stress tests that are performed after the manufacturing. These tests expose a part to cycles of temperature and mechanical stress to ensure its operation over a wide range of working conditions. Typical faults include short circuits between wires or layers and broken interconnections. This translates into network nodes that are either shorted to each other or to the supply rails, or that may be floating.

Making sure a delivered part is operating correctly under all possible input conditions is not as simple as it would seem at a first glance. When analyzing the circuit behavior during the design phase, the designer has unlimited access to all the nodes in the network. He is free to apply input patterns and observe the resulting response at any node he desires. This is not the case once the part is manufactured. The only access one has to the circuit is through the input-output pins. A complex component such as a microprocessor is composed of tens to hundreds of millions of transistors and contains an uncountable number of possible states. It is a very lengthy process—if it is possible at all—to bring such a component into a particular state and to observe the resulting circuit response through the limited bandwidth offered by the input-output pads. Hardware testing equipment tends to be very expensive and every second a part spends in the tester adds to its price.

It is therefore advisable to consider the testing early in the design process. Some small modifications in a circuit can help make it easier to validate the absence of faults. This approach to design has been dubbed *design for testability (DFT)*. While often despised by circuit designers who prefer to concentrate on the exciting aspects of design, such as transistor optimization, DFT is an integral and important part of the design process and should be considered as early as possible in the design flow. *"If you don't test it, it won't work! (Guaranteed)"* [Weste93]. A DFT strategy contains two components:

1. Provide the necessary *circuitry* so that the test procedure can be swift and comprehensive.
2. Provide the necessary *test patterns* (excitation vectors) to be employed during the test procedure. For reasons of cost, it is desirable that the test sequence be as short as possible while covering the majority of possible faults.

In the subsequent sections, we briefly cover some of the most important issues in each of these domains. Before doing so, a short description of a typical test procedure helps to put things in perspective.

H.2 Test Procedure

Manufacturing tests fall into a number of categories depending upon the intended goal:

- **The diagnostic test** is used during the debugging of a chip or board and tries to accomplish the following: Given a failing part, identify and locate the offending fault.
- **The functional test** (also called *go/no go* test) determines whether or not a manufactured component is functional. This problem is simpler than the diagnostic test since the only answer expected is yes or no. As this test must be executed on every manufactured die and has a direct impact on the cost, it should be as simple and swift as possible.
- **The parametric test** checks on a number of nondiscrete parameters, such as noise margins, propagation delays, and maximum clock frequencies, under a variety of working conditions, such as temperature and supply voltage. This requires a different set-up from

the functional tests that only deal with 0 and 1 signals. Parametric tests generally are sub-divided into static (dc) and dynamic (ac) tests.

A typical manufacturing test proceeds as follows. The predefined test patterns are loaded into the tester that provides excitations to the *device under test* (DUT) and collects the corresponding responses. The test patterns are defined in a *test program* that describes the waveforms to be applied, voltage levels, clock frequency, and expected response. A probe card, or DUT board, is needed to connect the outputs and inputs of the tester to the corresponding pins on the die or package.

A new part is automatically fed into the tester. The tester executes the test program, applies the sequence of input patterns to the DUT, and compares the obtained response with the expected one. If differences are observed, the part is marked as faulty (e.g., with an ink spot), and the probes are automatically moved to the next die on the wafer. During the scribing process that divides the wafer into the individual dies, spotted parts will be automatically discarded. For a packaged part, the tested component is removed from the test board and placed in a good or faulty bin, depending upon the outcome of the test. The whole procedure takes in the range of a few seconds per part, making it possible for a single tester to handle thousands of parts in an hour.

Figure H-1 shows a picture of a high-end tester for logic circuits. Automatic testers are very expensive pieces of equipment. The increasing performance requirements, imposed by the high-speed ICs of today, have aggravated the situation, causing the cost of the test equipment to skyrocket. Reducing the time that a die spends on the tester is the most effective way to reduce the test cost. Unfortunately, with the increasing complexity of ICs, an opposite trend can be observed. Design approaches to reduce the test burden are thus very desirable.

H.3 Design for Testability

H.3.1 Issues in Design for Testability

As mentioned, a high-speed tester that can adequately handle state-of-the-art components comes at an astronomical cost. Reducing the test time for a single component can help increase the

Figure H-1   Automatic tester (Courtesy Shlumberger).

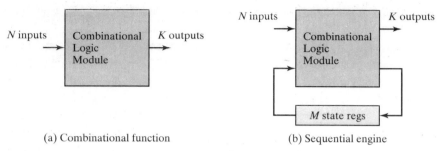

(a) Combinational function   (b) Sequential engine

Figure H-2   Combinational and sequential devices under test.

throughput of the tester, and has an important impact on the testing cost. By considering testing from the early phases of the design process, it is possible to simplify the whole validation process. In this section, we describe some approaches that achieve that goal. Before detailing these techniques, we should first understand some of the intricacies of the test problem.

Consider the combinational circuit block of Figure H-2a. The correctness of the circuit can be validated by exhaustively applying all possible input patterns and observing the responses. For an N-input circuit, this requires the application of 2^N patterns. For $N = 20$, more than 1 million patterns are needed. If the application and observation of a single pattern takes 1 μsec, the total test of the module requires 1 sec. The situation gets more dramatic when considering the sequential module of Figure H-2b. The output of the circuit depends not only upon the inputs applied, but also upon the value of the state. To exhaustively test this finite state machine (FSM) requires the application of 2^{N+M} input patterns, where M is the number of state registers [Williams83, Weste93]. For a state machine of moderate size (e.g., $M = 10$), this means that 1 billion patterns must be evaluated, which takes 16 minutes on our 1 μsec/pattern testing equipment. Modeling a modern microprocessor as a state machine translates into an equivalent model with over 50 state registers. Exhaustive testing of such an engine would require over a billion years!

Obviously, an alternative approach is required. A more feasible testing approach is based on the following premises.

- An exhaustive enumeration of all possible input patterns contains a substantial amount of *redundancy;* that is, a single fault in the circuit is covered by a number of input patterns. Detection of that fault requires only one of those patterns, while the other patterns are superfluous.
- A substantial reduction in the number of patterns can be obtained by relaxing the condition that all faults must be detected. For instance, detecting the last single percentage of possible faults might require an exorbitant number of extra patterns, and the cost of detecting them might be larger than the eventual replacement cost. Typical test procedures only attempt a 95–99% fault coverage.

By eliminating redundancy and providing a reduced fault coverage, it is possible to test most combinational logic blocks with a limited set of input vectors. This does not solve the sequential problem, however. To test a given fault in a state machine, it is not sufficient to apply the correct input excitation; the engine must be brought to the desired state first. This requires that a sequence of inputs be applied. Propagating the circuit response to one of the output pins might require another sequence of patterns. In other words, testing for a single fault in an FSM requires a sequence of vectors. Once again, this might make the process prohibitively expensive.

One way to address the problem is to turn the sequential network into a combinational one by breaking the feedback loop in the course of the test. This is one of the key concepts behind the *scan-test* methodology described later. Another approach is to let the circuit test itself. Such a test does not require external vectors and can proceed at a higher speed. The concept of *self-test* will be discussed in more detail later. When considering the testability of designs, two properties are of foremost importance:

1. **Controllability,** which measures the ease of bringing a circuit node to a given condition using only the input pins. A node is easily controllable if it can be brought to any condition with only a single input vector. A node (or circuit) with low controllability needs a long sequence of vectors to be brought to a desired state. It should be clear that a high degree of controllability is desirable in testable designs.

2. **Observability,** which measures the ease of observing the value of a node at the output pins. A node with a high observability can be monitored directly on the output pins. A node with a low observability needs a number of cycles before its state appears on the outputs. Given the complexity of a circuit and the limited number of output pins, a testable circuit should have a high observability. This is exactly the purpose of the test techniques discussed in the sections that follow.

Combinational circuits fall under the class of easily observable and controllable circuits, since any node can be controlled and observed in a single cycle.

Design-for-test approaches for the sequential modules can be classified in three categories: ad hoc test, scan-based test, and self-test.

H.3.2 Ad Hoc Testing

As suggested by the title, ad hoc testing combines a collection of tricks and techniques that can be used to increase the observability and controllability of a design and that are generally applied in an application-dependent fashion.

An example of such a technique is illustrated in Figure H-3a, which shows a simple processor with its data memory. Under normal configuration, the memory is only accessible through the processor. Writing and reading a data value into and out of a single memory position requires a number of clock cycles. The controllability and observability of the memory can be dramatically improved by adding multiplexers on the data and address busses (Figure H-3b).

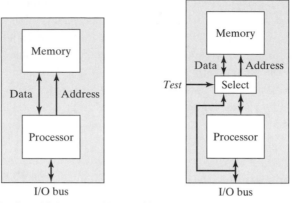

(a) Design with low testability   (b) Adding a selector improves testability

Figure H-3   Improving testability by inserting multiplexers.

During normal operation mode, these selectors direct the memory ports to the processor. During test, the data and address ports are connected directly to the I/O pins, and testing the memory can proceed more efficiently. The example illustrates some important design-for-testability concepts.

- It is often worthwhile to introduce *extra hardware* that has no functionality except improving the testability. Designers are often willing to incur a small penalty in area and performance if it makes the design substantially more observable or controllable.
- Design-for-testability often means that extra I/O pins must be provided besides the normal functional I/O pins. The *test* port in Figure H-3b is such an extra pin. To reduce the number of extra pads that would be required, one can multiplex test signals and functional signals on the same pads. For example, the I/O bus in Figure H-3b serves as a data bus during normal operation and provides and collects the test patterns during testing.

An extensive collection of ad hoc test approaches has been devised. Examples include the partitioning of large state machines, addition of extra test points, provision of reset states, and introduction of test busses. While very effective, the applicability of most of these techniques depends upon the application and architecture at hand. Their insertion into a given design requires expert knowledge and is difficult to automate. Structured and automatable approaches are more desirable.

H.3.3   Scan-Based Test

One way to avoid the sequential-test problem is to turn all registers into externally loadable and readable elements. This turns the circuit-under-test into a combinational entity. To control a node, an appropriate vector is constructed, loaded into the registers and propagated through the logic. The result of the excitation propagates to the registers and is latched, after which the con-

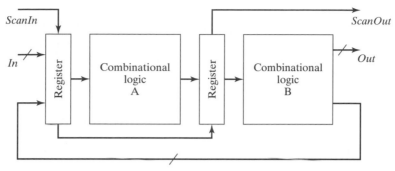

Figure H-4 Serial-scan test.

tents are transferred to the external world. Connecting all the registers in a design to a test bus regrettably introduces an unacceptable amount of overhead. A more elegant approach is offered by the serial-scan approach illustrated in Figure H-4.

The registers have been modified to support two operation modes. In the normal mode, they act as N-bit-wide clocked registers. During the test mode, the registers are chained together as a single serial shift register. A test procedure now proceeds as follows.

1. An excitation vector for logic module A (and/or B) is entered through pin *ScanIn* and shifted into the registers under control of a test clock.
2. The excitation is applied to the logic and propagates to the output of the logic module. The result is latched into the registers by issuing a single system-clock event.
3. The result is shifted out of the circuit through pin *ScanOut* and compared with the expected data. A new excitation vector can be entered simultaneously.

This approach incurs only a minimal overhead. The serial nature of the scan chain reduces the routing overhead. Traditional registers are easily modified to support the scan technique, as demonstrated in Figure H-5, which shows a 4-bit register extended with a scan chain. The only addition is an extra multiplexer at the input. When *Test* is low, the circuit is in normal operation

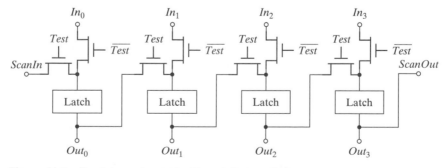

Figure H-5 Register extended with serial-scan chain.

mode. Setting *Test* high selects the *ScanIn* input and connects the registers into the scan chain. The output of the register *Out* connects to the fan-out logic, but also doubles as the *ScanOut* pin that connects to the *ScanIn* of the neighboring register. The overhead in both area and performance is small and can be limited to less than 5%.

Problem H.1   Scan-Register Design

Modify the static, two-phase master-slave register of Figure 7-10 to support serial scan.

Figure H-6 depicts the timing sequence that would be employed for the circuit in Figure H-4 under the assumption of a two-phase clocking approach. For a scan chain *N* registers deep, the *Test* signal is raised, and *N* clock pulses are issued, loading the registers. *Test* is lowered, and a single clock sequence is issued, latching the results from the combinational logic into the registers under normal circuit-operation conditions. Finally, *N* extra pulses (with *Test* = 1) transfer the obtained result to the output. Note again that the scan-out can overlap with the entering of the next vector.

Many variants of the serial-scan approach can be envisioned. A very popular one, which was actually the pioneering approach, was introduced by IBM and is called *level-sensitive scan design* (LSSD) [Eichelberger78]. The basic building block of the LSSD approach is the *shift-register latch* (SRL) shown in Figure H-7. It consists of two latches *L1* and *L2,* the latter being

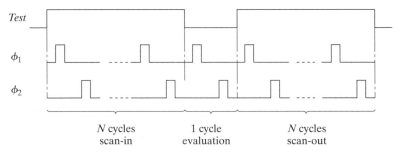

Figure H-6   Timing diagram of test-sequence. *N* represents the number of registers in the test chain.

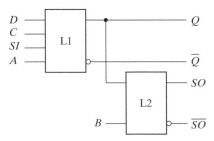

Figure H-7   Shift-register latch.

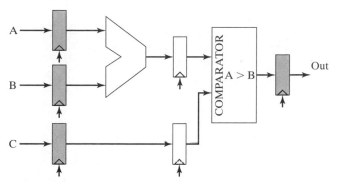

Figure H-8   Pipelined datapath using partial scan. Only the shaded registers are included in the chain.

present only for testing purposes. In normal circuit operation, signals D, Q (Q), and C serve as latch input, output, and clock. The test clocks A and B are low in this mode. In scan mode, SI and SO serve as scan input and scan output. Clock C is low, and clocks A and B act as nonoverlapping, two-phase test clocks.

The LSSD approach represents not only a test strategy, but also a complete clocking philosophy. By strictly adhering to the rules implied by this methodology, it is possible to automate to a large extent the test generation and the timing verification. This is why the use of LSSD was obligatory within IBM for a long time. The prime disadvantage of the approach is the complexity of the SRL latch.

It is not always necessary to make all the registers in the design scannable. Consider the pipelined datapath of Figure H-8. The pipeline registers in this design are only present for performance reasons and do not strictly add to the state of the circuit. It is, therefore, meaningful to make only the input and output registers scannable. During test generation, the adder and comparator can be considered together as a single combinational block. The only difference is that during the test execution, two cycles of the clocks are needed to propagate the effects of an excitation vector to the output register. This approach is called *partial scan* and is often employed when performance is of prime interest. The disadvantage is that deciding which registers to make scannable is not always obvious and may require interaction with the designer.

H.3.4   Boundary-Scan Design

Until recently, the test problem was most compelling at the integrated circuit level. Testing circuit boards was facilitated by the abundant availability of test points. The through-hole mounting approach made every pin of a package observable at the back side of the board. For test, it was sufficient to lower the board onto a set of test probes (called "bed-of-nails") and apply and observe the signals of interest. The picture changed with the introduction of advanced packaging techniques such as surface-mount or multichip modules (Chapter 2). Controllability and observability are not as readily available anymore, because the number of probe points is dramatically

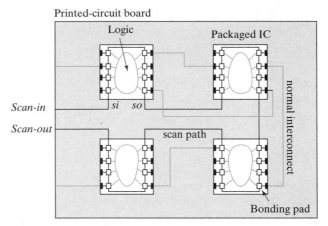

Figure H-9 The boundary-scan approach to board testing.

reduced. This problem can be addressed by extending the scan-based test approach to the component and board levels.

The resulting approach is called *boundary scan* and has been standardized to ensure compatibility between different vendors ([IEEE1149]). In essence, it connects the input-output pins of the components on a board into a serial scan chain, as shown in Figure H-9. During normal operation, the boundary-scan pads act as normal input-output devices. In test mode, vectors can be scanned in and out of the pads, providing controllability and observability at the boundary of the components (hence, the name). The test operation proceeds along similar lines as described in the previous paragraph. Various control modes allow for testing the individual components as well as the board interconnect. The overhead incurred includes slightly more complex input-output pads and an extra on-chip test controller (an FSM with 16 states). Boundary scan is now provided in most commodity components.

H.3.5 Built-in Self-Test (BIST)

An alternative and attractive approach to testability is having the circuit itself generate the test patterns instead of requiring the application of external patterns [Wang86]. Even more appealing is a technique where the circuit itself decides if the obtained results are correct. Depending upon the nature of the circuit, this might require the addition of extra circuitry for the generation and analysis of the patterns. Some of this hardware might already be available as part of the normal operation, and the size overhead of the self-test can be small.

The general format of a built-in self-test design is illustrated in Figure H-10 ([Kornegay92]). It contains a means for supplying test patterns to the device under test and a means of comparing the device's response to a known correct sequence.

There are many ways to generate stimuli. Most widely used are the *exhaustive* and the *random* approaches. In the exhaustive approach, the test length is 2^N, where N is the number of inputs to the circuit. The exhaustive nature of the test means that all detectable faults will be

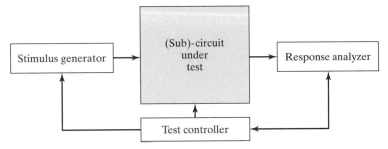

Figure H-10 General format of built-in self-test structure.

detected, given the space of the available input signals. An N-bit counter is a good example of an exhaustive pattern generator. For circuits with large values of N, the time to cycle through the complete input space might be prohibitive. An alternative approach is to use random testing that implies the application of a randomly chosen sub-set of 2^N possible input patterns. This subset should be selected so that a reasonable fault coverage is obtained. An example of a pseudorandom pattern generator is the *linear-feedback shift register* (or LFSR), which is shown in Figure H-11. It consists of a serial connection of 1-bit registers. Some of the outputs are XOR'd and fed back to the input of the shift register. An N-bit LFSR cycles through $2^N - 1$ states before repeating the sequence, which produces a seemingly random pattern. Initialization of the registers to a given seed value (different from 0 for our example circuit) determines what will be generated, subsequently.

The response analyzer could be implemented as a comparison between the generated response and the expected response stored in an on-chip memory, but this approach represents too much area overhead to be practical. A cheaper technique is to compress the responses before comparing them. Storing the compressed response of the correct circuit requires only a minimal amount of memory, especially when the compression ratio is high. The response analyzer then

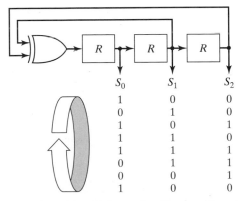

Figure H-11 Three-bit linear-feedback shift register and its generated sequence.

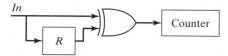

Figure H-12   Single bit-stream signature analysis.

consists of circuitry that dynamically compresses the output of the circuit under test and a comparator. The compressed output is often called the *signature* of the circuit, and the overall approach is dubbed *signature analysis.*

An example of a signature analyzer that compresses a single bit stream is shown in Figure H-12. Inspection reveals that this circuit simply counts the number of $0 \rightarrow 1$ and $1 \rightarrow 0$ transitions in the input stream. This compression does not guarantee that the received sequence is the correct one; that is, there are many different sequences with the same number of transitions. Since the chances of this happening are slim, it may be a risk worth taking if kept within bounds.

Another technique is illustrated in Figure H-13a. It represents a modification of the linear-feedback shift register and has the advantage that the same hardware can be used for both pattern generation and signature analysis. Each incoming data word is successively XOR'd with the contents of the LFSR. At the end of the test sequence, the LFSR contains the signature, or *syndrome*, of the data sequence, which can be compared with the syndrome of the correct circuit. The circuit not only implements a random-pattern generator and signature analyzer, but also can

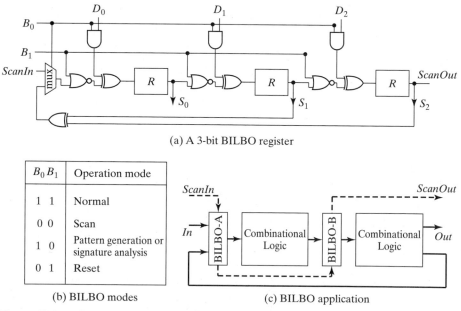

Figure H-13   Built-in logic block observation, or BILBO.

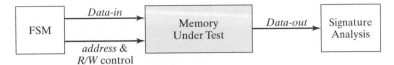

Figure H-14 Memory self-test.

be used as a normal register and scan register, depending on the values of the control signals B_0 and B_1 (Figure H-13b). This test approach, which combines all the different techniques, is known as *built-in logic block observation*, or *BILBO* [Koeneman79]. Figure H-13c illustrates the typical use of BILBO. Using the scan option, the seed is shifted into the BILBO register A while BILBO register B is initialized. Next, registers A and B are operated in the random pattern-generation and signature-analysis modes, respectively. At the end of the test sequence, the signature is read from B using the scan mode.

Finally, it is worth mentioning that self-test is extremely beneficial when testing regular structures such as memories. It is not easy to ensure that a memory, which is a sequential circuit, is fault free. The task is complicated by the fact that the data value read from or written into a cell can be influenced by the values stored in the neighboring cells because of cross coupling and other parasitic effects. Memory tests, therefore, include the reading and writing of a number of different patterns into and from the memory using alternating addressing sequences. Typical patterns can be all zeros or ones, or checkerboards of zeros and ones. Addressing schemes can include the writing of the complete memory, followed by a complete read-out or various alternating read-write sequences. With a minimal overhead compared with the size of a memory, this test approach can be built into the integrated circuit itself, as illustrated in Figure H-14. This approach significantly improves the testing time and minimizes the external control. Applying self-test is bound to become more important with the increasing complexity of integrated components and the growing popularity of embedded memories.

The advent of the systems-on-a-chip era does not make the test job any easier. A single IC may contain micro- and signal processors, multiple embedded memories, ASIC modules, FPGAs and on-chip busses and networks. Each of these modules has its own preferred way of being tested, and combining those into a coherent strategy is quite a challenge. Built-in self-test is really the only way out. A structured test-methodology for systems-on-a-chip, based on BIST is shown in Figure H-15. Each of the modules composing the system connects to the on-chip network through a "wrapper." This is a customized interface between the block and the network, supporting functions such as synchronization and communication. This wrapper can be extended to include a test support module. For instance, for an ASIC module that includes a scan chain, the test support module provides the interface to the scan chain and a buffer for the test patterns. This buffer can be directly written and read through the system bus. Similarly, a memory module can be equipped with a pattern generator and signature analysis. All of this would still not suffice if there were no general test orchestrator. Fortunately, most of these SOCs include a programmable processor which can be used at start-up time to direct the test and the verification of the other modules. Test patterns and signatures can be stored

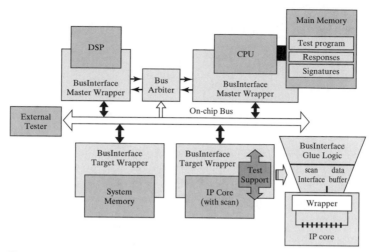

Figure H-15   System-on-a-chip test methodology.

in the main memory of the processor, and supplied to the module under test at test time. This approach has the advantage that it uses the resources that are already available on the die. Furthermore, the BIST self-test approach allows for the tests to be performed under actual clock speeds, decreasing the test time. The requirements for the external tester are minimized, reducing the tester cost. The development of a structured approach to self-test, as pictured in Figure H-15, is a topic of active research [Krstic01].   ■

H.4 Test-Pattern Generation

In the preceding sections, we have discussed how to modify a design so that test patterns can be effectively applied. What we have ignored so far is the complex task of determining what patterns should be applied so that a good fault coverage is obtained. This process was extremely problematic in the past, when the test engineer—a different person than the designer—had to construct the test vectors after the design was completed. This invariably required a substantial amount of wasteful reverse engineering that could have been avoided if testing had been considered early in the design flow. An increased sensitivity to design for testability and the emergence of automatic test-pattern generation (ATPG) has substantially changed this picture.

In this section, we delve somewhat deeper into the ATPG issue and present techniques to evaluate the quality of a test sequence. Before doing so, we must analyze the fault concept in more detail.

H.4.1   Fault Models

Manufacturing faults can be of a wide variety and manifest themselves as short circuits between signals, short circuits to the supply rails, and floating nodes. In order to evaluate the effectiveness of a test approach and the concept of a good or bad circuit, we must relate these faults to the circuit model, or, in other words, derive a *fault model*. The most popular approach is called the *stuck-at* model. Most testing tools consider only the short circuits to the supplies. These are

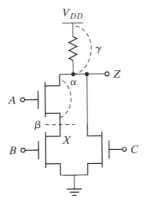

Figure H-16 Resistive-load gate, annotated with a number of stuck-at-open (β) and stuck-at-short (α, γ) faults.

called the *stuck-at-zero* (sa0) and *stuck-at-one* (sa1) faults for short circuits to *GND* and V_{DD}, respectively.

It can be argued that the sa0-sa1 model does not cover the complete range of faults that can occur in a state-of-the-art integrated circuit, and that *stuck-at-open* and *stuck-at-short* faults should also be introduced. However, adding these faults complicates the test pattern generation process. Moreover, a large number of these faults are covered by the sa0-sa1 model. To illustrate this observation, consider the resistive-load MOS gate of Figure H-16. All shorts to the supplies are modeled by the introduction of sa0 and sa1 faults at nodes *A, B, C, Z,* and *X*. The figure has been annotated with some stuck-at-open (β) and stuck-at-short faults (α, γ). It can be observed that these faults are already covered by the sa0 and sa1 faults on the various nodes. For example, fault α is covered by A_{sa1}, β is covered by A_{sa0} or B_{sa0}, while γ is equivalent to Z_{sa1}.

Even so, shorts and open-circuit faults can cause some interesting artifacts to occur in CMOS circuits that are not covered by the sa0-sa1 model and are worth mentioning. Consider the two-input NAND gate of Figure H-17, where a stuck-at-open fault α has occurred. The truth table of the faulty circuit is shown in the figure as well. For the combination (*A* = 1, *B* = 0), the output node is floating and retains its previous value, while the correct value should be a 1. Depending upon the previous excitation vector, this fault may or may not be detected. In fact, the circuit behaves as a sequential network. To detect this fault, two vectors must be applied in sequence. The first one forces the output to 0 (or *A* = 1 and *B* = 1), while the second applies the *A* = 1, *B* = 0 pattern. Also stuck-at-short faults are troublesome in CMOS circuits since they can cause dc currents to flow between the supply rails for certain input values, which produces undefined output voltages.

Even though the sa0-sa1 fault model is not perfect, its ease of use and relatively large coverage of the fault space have made it the de facto standard model. It is often supplemented with other techniques, such as functional test, *IDDQ test*—which measures the change in quiescent current of a CMOS circuit due to short circuits—and delay test. No single test model is completely foolproof, and a combination of several testing methods is often required [Bhavsar01].

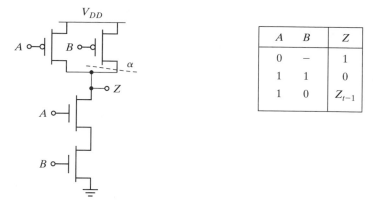

A	B	Z
0	–	1
1	1	0
1	0	Z_{t-1}

Figure H-17   Two-input complementary CMOS NAND gate and its truth table in the presence of a stuck-at-open fault.

H.4.2   Automatic Test-Pattern Generation (ATPG)

The task of the automatic test-pattern generation (ATPG) process is to determine a minimum set of excitation vectors that cover a sufficient portion of the fault set as defined by the adopted fault model. One possible approach is to start from a random set of test patterns. Fault simulation then determines how many of the potential faults are detected. With the obtained results as guidance, extra vectors can be added or removed iteratively. An alternative and potentially more attractive approach relies on the knowledge of the functionality of a Boolean network to derive a suitable test vector for a given fault. To illustrate the concept, consider the example of Figure H-18. The goal is to determine the input excitation that exposes an sa0 fault occurring at node U at the output of the network Z. The first requirement of such an excitation is that it should force the fault to occur (*controllability*, again). In this case, we look for a pattern that would set U to 1 under normal circumstances. The only option here is $A = 1$ and $B = 1$. Next, the faulty signal has to propagate to output node Z, so that it can be *observed*. This phase is called *path sensitizing*. For any change in node U to propagate, it is necessary for node X to be set to 1 and node E to 0. The (unique) test vector for U_{sa0} can now be assembled: $A = B = C = D = 1$, $E = 0$.

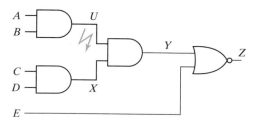

Figure H-18   Simple logic network, with sa0 fault at node U.

This example is extremely simple, and the derivation of a minimum test-vector set for more complex circuits is substantially more complex. A number of excellent approaches to address this problem have been developed. Landmark efforts in this domain are the D [Roth66] and PODEM algorithms [Goel81], which underlie many current ATPG tools. It suffices to say that ATPG is currently in the mainstream of design automation, and powerful tools are available from many vendors.

H.4.3 Fault Simulation

A fault simulator measures the quality of a test program. It determines the *fault coverage*, which is defined as the total number of faults detected by the test sequence divided by two times the number of nodes in the network—each node can give rise to an sa0 and sa1 fault. Naturally, the obtained coverage number is only as good as the fault model employed. In an sa0-sa1 model, some of the bridge and short faults are not covered and will not appear in the coverage statistics.

The most common approach to fault simulation is the parallel fault-simulation technique, in which the correct circuit is simulated concurrently with a number of faulty ones, each of which has a single fault injected. The results are compared, and a fault is labeled as detected for a given test vector set if the outputs diverge. This description is overly simplistic, and most simulators employ a number of techniques, such as selecting the faults with a higher chance of detection first, to expedite the simulation process. Hardware fault-simulation accelerators, based on parallel processing and providing a substantial speedup over pure software-based simulators, are available as well [Agrawal88, pp. 159–240].

H.5 To Probe Further

For an in-depth treatment of the Design-for-Testability topic, please refer to [Agrawal88] and [Abramovic91]. A great overview of the testing challenges and solutions for high-performance microprocessors can be found in [Bhavsar01].

References

[Abramovic91] M. Abramovic, M. Breuer, and A. Friedman, *Digital Systems Testing and Testable Design*, IEEE Press, Piscataway, NJ, 1991.

[Agrawal88] V. Agrawal and S. Seth, Eds. *Test Generation for VLSI Chips*, IEEE Computer Society Press, 1988.

[Bhavsar01] D. Bhavsar, "Testing of High-Performance Microprocessors," in *Design of High-Performance Microprocessor Circuits*, A. Chandrakasan et al., Ed., pp. 523–44, IEEE Press, 2001

[Eichelberger78] E. Eichelberger and T. Williams, "A Logic Design Structure for VLSI Testability," *Journal on Design Automation of Fault-Tolerant Computing,* vol. 2, pp. 165–78, May 1978.

[Goel81] P. Goel, "An Implicit Enumeration Algorithm to Generate Tests for Combinational Logic Circuits," *IEEE Trans. on Computers*, vol. C-30, no. 3, pp. 26–268, June 1981.

[IEEE1149] IEEE Standard 1149.1, "IEEE Standard Test Access Port and Boundary-Scan Architecture," IEEE Standards Board, New York.

[Koeneman79] B. Koeneman, J. Mucha, and O. Zwiehoff, "Built-in Logic-Block Observation Techniques," in *Digest 1979 Test Conference*, pp. 37–41, October 1979.

[Kornegay92] K. Kornegay, *Automated Testing in an Integrated System Design Environment,* Ph. D dissertation., Mem. No. UCB/ERL M92/104, Sept. 1992.

[Krstic01] A. Krstic, W. Lai, L. Chen, K. Cheng, and S. Dey, "Embedded Software-Based Self-Testing for SOC Design," *http://www.gigascale.org/pubs/139.html*, December 2001.

[Roth66] J. Roth, "Diagnosis of Automata Failures: A Calculus and a Method," *IBM Journal of Research and Development*, vol. 10, pp. 278–291, 1966.

[Wang86] L. Wang and E. McCluskey, "Complete Feedback Shift-Register Design for Built-in Self Test," *Proc. ICCAD 1986,* pp. 56–59, November, 1986.

[Weste93] N. Weste and K. Eshraghian, *Principles of CMOS VLSI Design—A Systems Perspective*, Addison-Wesley, 1993.

[Williams83] T. Williams and K. Parker, "Design for Testability—A Survey," *Proceedings IEEE*, vol. 71, pp. 98–112, Jan. 1983.

Problem Solutions

Problem 3.1—Bias the transistor into the desired operation mode by applying an appropriate gate voltage. Apply a current source I to the drain node. The current flowing into the drain capacitance I_C can be found by subtracting the transistor current I_D from I. The capacitance is easily found once we know V_{DS} and $I_C(V_{DS})$.

Figure P3.1 Simulating the drain capacitance of an NMOS.

Problem 3.2—Assuming a quadratic dependence of current leads to the following revised table entries:

	Full	General	Fixed
I_{SAT}	$1/S$	S/U^2	S
R_{ON}	1	U/S	$1/S$
t_p	$1/S$	U/S^2	$1/S^2$
P	$1/S^2$	S/U^3	S

Problem 4.1—See Figure P4.1.

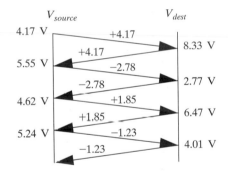

Figure P4.1 The parallel-termination cases proceed along a similar path.

Problem 5.1—For V_{DD} substantially larger than V_T, both PMOS and NMOS are saturated around the V_M operation point. The expression for V_M is obtained by equating the currents through the transistors. The channel-length modulation factor is ignored in this derivation.

$$\frac{k_n}{2}(V_M - V_{Tn})^2 = -\frac{k_p}{2}(V_M - V_{DD} - V_{Tp})^2$$

Problem 5.2—Assume that both PMOS and NMOS are saturated around V_M. Write down the current equation under that condition, differentiate, and solve for dV_{out}/dV_{in}. This yields

$$\frac{dV_{out}}{dV_{in}} = -\frac{k_n(V_{in} - V_{Tn})(1 + \lambda_n V_{out}) + k_p(V_{in} - V_{DD} - V_{Tp})(1 + \lambda_p V_{out} - \lambda_p V_{DD})}{(\lambda_n k_n(V_{in} - V_{Tn})^2 + \lambda_p k_p(V_{in} - V_{DD} - V_{Tp})^2)/2}$$

Ignoring some second-order terms and setting $V_{in} = V_M$ produces

$$g = -\frac{k_n(V_M - V_{Tn}) + k_p(V_M - V_{DD} - V_{Tp})}{I_D(V_M)(\lambda_n - \lambda_p)}$$

Once the gain is known, V_{IH} and V_{IL} are easily derived using Eq. (5.7).

739

Problem 5.3—Using the current equation of the MOS device in the subthreshold region, Eq. (3.39), ignoring the channel-length modulation factor, and assuming that PMOS and NMOS are of equal strength, we find the gain as $-(dI_D/dV_{GS})/(dI_D/dV_{DS})$. This leads directly to Eq. (5.12).

Problem 5.4—For the sake of brevity, we focus on t_{pHL} only. Under the current-source assumption, t_{pLH} is approximated as

$$t_{pLH} = \frac{C_L(V_{DD}/2)}{I_{av}}$$

with I_{av} the average discharge current. We compute I_{av} as the average between the points of interests, this is $V_{out} = V_{DD}$ and $V_{out} = V_{DD}/2$. With the NMOS transistor in velocity-saturation for most of the discharge cycle, this yields

$$I_{av} = \frac{I_{Dsat}(1 + \lambda V_{DD}) + I_{Dsat}(1 + \lambda(V_{DD}/2))}{2} = I_{Dsat}\left(1 + \lambda\frac{3V_{DD}}{4}\right).$$

$$\text{with } I_{Dsat} = k_n((V_{DD} - V_{Tn})V_{Dsat} - V_{Dsat}^2/2)$$

This produces the following expression for the propagation delay, which is quite close to Eq. (5.22).

$$t_{pHL} = \frac{C_L V_{DD}}{2I_{Dsat}\left(1 + \lambda\frac{3V_{DD}}{4}\right)}$$

Problem 5.5—Partial derivation of Eq. (5.32) with respect to the input capacitances of the three stages, and taking into account the extra fanout of 4 for stages 1 and 2, yields the following relationship:

$$\frac{4C_{g,2}}{C_{g,1}} = \frac{4C_{g,3}}{C_{g,2}} = \frac{C_L}{C_{g,3}}$$

It also holds that $C_L = 64\ C_{g,1}$. This set of equation is easily solved, and yields the following size factors: $s_1 = 1$; $s_2 = 2^{4/3} = 2.52$; $s_3 = 2^{8/3} = 6.35$. This leads to a minimum delay between input and output of $33.24t_{p0}$ (assuming that $\gamma = 1$).

Problem 5.6—The impact of input slope decreases with reduced supply voltage (for fixed thresholds). Decreasing the supply voltage reduces the time interval that both devices are on simultaneously, and thus the current that is lost.

Problem 6.1—Critical transistors are those on series connections. When N transistors are connectied in series, we multiply their width by N to obtain a composite device with an on-resistance equal to that of a single transistor. Transistors in parallel are not modified, as their worst-case resistance is equal to each of the individual devices. This technique is used recursively for every branch in the network. Device sizes are annotated on the schematic in Figure P6.1. From a delay perspective, this is equivalent to an inverter with an NMOS of 0.5μm/0.25μm and a PMOS of 1.5μm/0.25μm.

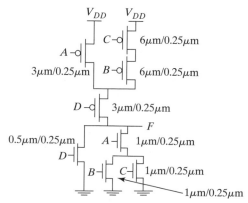

Figure P6.1 Sizing transistors in complex CMOS.

Problem 6.2—The branching efforts are $b_1 = 4$, $b_2 = 4$, $b_3 = 1$. Therefore $B = 16$. The path effective fan-out $F = 64$. The path logical effort $G = 1$. Therefore the total path effort is $H = GFB = 1024$. The gate effort that minimizes delay is $(H)^{1/N}$ which is equal to 10.079. $C_{in3} = 64 \ C_{g1} / 10.079 = 6.35 \ C_{g,1}$ and $C_{in2} = 4 * 6.35 \ C_{g,1} / 10.079 = 2.52 \ C_{g,1}$. This results in $s_1 = 1$; $s_2 = 2.52$; $s_3 = 6.35$.

Problem 6.3—Since the inputs to the N-input XOR gate are uncorrelated and uniformly distributed, the transition probability is given by:

$$\alpha_{0 \rightarrow 1} = \frac{N_0}{2^N} \cdot \frac{N_1}{2^N} = \frac{\dfrac{2^N}{2} \cdot \dfrac{2^N}{2}}{2^{2N}} = \frac{1}{4}$$

Problem 6.4—These transition probabilities are easily derived from the following observations. The chance that an AND gate is in the 1-state equals the probability that input A is high AND input B is high, or $P_{AND}(1) = P_A \cdot P_B$. From this, the transition probability can be computed. As $P_{0 \rightarrow 1} = (1 - P(1))P(1)$. Similarly, the probability that the output of the XOR gate is high equals the chance that A is high or B is high but not both of them at the same time. This is expressed as follows: $P_{XOR}(1) = P_A + P_B - 2P_AP_B$. Observe that the transition probabilities of the complementary gates (NAND, NOR, NXOR) are identical to their noninverting counterparts.

Probem 6.5—In contrast to complementary CMOS, the NOR structure is the preferred topology for pseudo-NMOS. The NOR structure completely avoids series transistors.

Problem 6.6—Putting 6 transistors in series increases the pulldown resistance by a factor of 6 relative to a single device with a size of 0.5μm/0.25μm. To pull node x to the switching threshold of the inveter ($V_{DD}/2$), the pulldown will also have an effective W/L of 1. Since the minimum device width is 0.375μm, the maximum W/L is 0.375μm/0.375μm (the length should be increased to give some margin).

Problem 6.7—The simulation of NMOS and PMOS resistance for the discharge case is plotted in Figure P6.7. The effective resistance is approximately 4K (averaged over the end points)

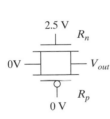

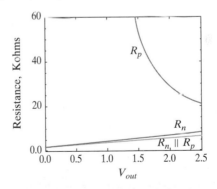

Figure P6.7 Discharge resistance of transmission gate.

Problem 6.8—The probability that the output discharges is the equal to the probability that all inputs are high. This also equals the probability of a 0 to 1 transition since the output is always precharged on the low clock phase.

$$\alpha_{0 \rightarrow 1} = P_A \cdot P_B \cdot P_C \cdot P_D = 0.012$$

Problem 6.9—Setting the two drain currents equal

$$I_S 10^{\frac{0}{S}} \left(1 - 10^{-\frac{nV_x}{S}} \right) = I_S 10^{\frac{-V_x}{S}} \left(1 - 10^{-n\frac{(V_{DD} - V_x)}{S}} \right)$$

Assuming that V_x is small, the second term in the brackets on the right can be ignored. This simplifies to:

$$\left(1 - 10^{-\frac{nV_x}{S}}\right) = 10^{\frac{-V_x}{S}} \quad \text{or as} \quad e^{-\frac{V_x}{nV_{th}}} = 1 - e^{-\frac{V_x}{V_{th}}}$$

Defining $\alpha = \dfrac{V_x}{V_{th}}$ and assuming α is small, we can use Taylor expansion around 0 and obtain:

$$1 - \frac{\alpha}{n} = 1 - (1 - \alpha) = \alpha$$

Reordering the elements in this equation: $1 - \alpha = 1 - \dfrac{n}{n+1} = \dfrac{1}{n+1}$

From where, using again Taylor expansion of the exponential on the left side and taking logarithms we arrive to: $V_x = V_{th} \ln(1 + n)$

Problem 7.1—Functionally, it is possible to remove the two inverters. The input inverter isolates the inputs from the internal nodes and provides a fixed input load to the flip-flop.

Problem 7.2—The non-overlap relationship is shown in Figure P7.2. Delays t_{d1} and t_{d2} can be added to increase the two non-overlap times.

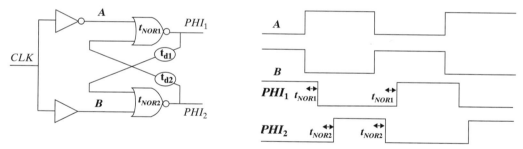

Figure P7.2 Generating non-overlapping clocks.

Problem 7.3—The diagrams in Figure P7.3 show the case where either the high threshold NMOS or PMOS in the latch is removed. In both cases, sneak leakage paths exists. Therefore, both high threshold NMOS and PMOS are needed to eliminate leakage.

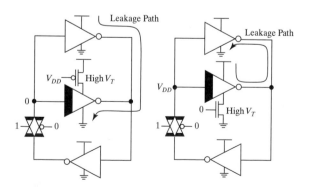

Figure P7.3 Dual-threshold registers for low-voltage operation.

Problem 7.4—The truth table for a NAND flip-flop is shown in Figure P7.4. The input 00 is a forbidden state.

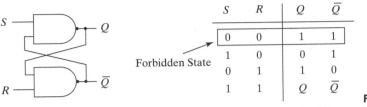

S	R	Q	$\overline{Q}$
0	0	1	1
1	0	0	1
0	1	1	0
1	1	Q	$\overline{Q}$

Forbidden State

(a) Schematic diagram (b) Characteristic table

Figure P7.4 NAND-based flip-flop.

Problem 7.5—For a fixed through-put, the clock rate can be reduced by a factor two for a dual-edge triggered system. To first order this translates to a factor of two lower power in the clock distribution network. Careful analysis shows that the physical load attached to the clock signal is double for a dual-edge triggered flip flop. As a result the clock distribution network would have to be made larger offsetting some power gains. The power used to drive the gate load itself remains unchanged going to a dual edge triggered solution since it has twice the physical capacitance switched at half the frequency.

Problem 7.6—A glitch register with enable is obtained by simply adding three device in series with the D input as shown in Figure P7.6.

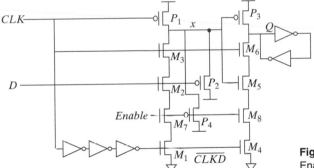

Figure P7.6 Glitch-register with Enable.

Problem 7.7—For a low-to-high transition, transistor M_2 only turns on when node X, the output voltage of the saturated-load inverter M_1—M_5 reaches the following value: $V_X = V_{in} - V_{Tn}$. We use this value of V_{in} as an approximation of V_{M-}. Ignoring the body effect and assuming both devices are in saturation, the following expression for V_{M-} can be derived:

$$k'_n\left(\frac{W}{L}\right)_1\left((V_{M-} - V_{Tn})V_{DSATn} - \frac{V_{DSATn}^2}{2}\right) = k'_n\left(\frac{W}{L}\right)_5\left((V_{DD}-(V_{M-} - V_{Tn}) - V_{Tn})V_{DSATn} - \frac{V_{DSATn}^2}{2}\right)$$

$$V_{M-} = \frac{V_{Tn} \cdot V_{DSATn} + \frac{V_{DSATn}^2}{2} + \left((V_{DD} \cdot V_{DSATn})-\frac{V_{DSATn}^2}{2}\right)(W_5/W_1)}{(1 + W_5/W_1) \cdot V_{DSATn}}$$

Similarly, we can derive V_{M+}.

$$V_{M+} = \frac{(V_{DD} + V_{Tp}) \cdot V_{DSATp} + \frac{V_{DSATp}^2}{2} + \left(-\frac{V_{DSATp}^2}{2}\right)(W_6/W_4)}{(1 + W_6/W_4) \cdot V_{DSATp}}$$

Problem 8.1—

$$\overline{f}_0 = \overline{\overline{x_0 x_1} \cdot x_2}$$

$$\overline{f}_1 = \overline{\overline{x_0 x_1 x_2} \cdot \bar{x}_2 \cdot \overline{\bar{x}_0 x_1}}$$

Problem 9.1—The correct approach is to scale down the last stage only (b). This is where the largest current change occurs, accounting for most of the *Ldi/dt* effect. The performance impact of scaling on only a single stage is small.

Problem 9.2—The energy consumption for a 0-1-0 transition equals $C_L V_{DD}(V_{DD} - V_{Tn} + V_{Tp})$. The energy reduction over the full-swing case hence equals $(V_{DD} - V_{Tn} + V_{Tp})/V_{DD}$. For $V_{DD} = 2.5$ V and device thresholds of 0.4 V, this amounts to a savings of 32%.

Problem 10.1—Assume that the FIFO is initially empty and that *En1*, *En2*, and *En3* are 0 as well as *Ack0*. We will input new data values without removing any old ones until the FIFO fills up. Notice that each data word automatically moves as deep in the FIFO as it can.

$$Req_i\uparrow \rightarrow En1\uparrow \rightarrow Done1\uparrow \rightarrow En2\uparrow \rightarrow Done2\uparrow \rightarrow En3\uparrow \rightarrow Req0\uparrow$$
$$\rightarrow Ack_i\uparrow \rightarrow Req_i\downarrow \rightarrow En1\downarrow \rightarrow Done1\downarrow \rightarrow En2\downarrow \rightarrow Done2\downarrow$$
$$\rightarrow Ack_i\downarrow \rightarrow Req_i\uparrow \rightarrow En1\uparrow \rightarrow Done1\uparrow$$
$$\rightarrow Ack_i\uparrow \rightarrow Req_i\downarrow$$

We can now see that the FIFO is full if the *Enable* Signals (or the *Ack*'s for the next stage) alternate between 0's and 1's. In the above case, we have $En1 = 1$, $En2 = 0$, $En3 = 1$, and $Ack0 = 0$, while $Req_i = 0$. On the other hand, the FIFO is empty if all enable signals are equal (either 0 or 1).

Problem 10.2—See Figure P10.2.

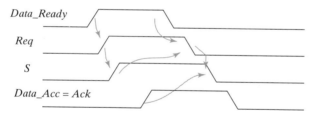

Figure P10.2. Four-phase protocol.

Problem 12.1—(0) 0100; (1) 1001; (2) 0101; (3) 0000

Problem 12.2—(0) 1011; (1) 0110; (2) 1010; (3) 1111

Problem 12.3—(0) 0100; (1) 1001; (2) 0101; (3) 0000

Problem 12.4—Charge sharing is a major concern in precharged NAND ROMs. It is possible to implement them, but the designer must be extremely careful.

Problem 12.5—No stand-by power is consumed. This is in contrast with the pseudo-NMOS cell, where there is always conduction in one side of the cell.

Index

A

Abrupt junction, 75, 83
Absolute coordinates, 69
Absolute jitter, 500
Abstraction, 8, 10–11
Acceptor impurities, 75
Acid etching, 39
Active regions (areas), 42, 48
Ad hoc testing, 724–25
Adders, 180, 561–86
  binary adders, 561–64, 571–86
    carry-bypass (carry-skip)
      adders, 571–73
    carry-lookahead adders, 578–86
    linear carry-select adders,
      574–75
    square-root carry-select adders,
      575–77
  carry-lookahead adders, 561–64,
    578–86, 620
    implementing in dynamic logic,
      581–83
    logarithmic lookahead adders,
      579–81, 584–86
    monolithic lookahead adders,
      578–79
  energy-delay tradeoffs in, 609
  full adders, 562, 564–70
    Manchester carry-chain adders,
      568–70
    mirror adder design, 566–67
    static adder circuit, 564–66
    transmission-gate-based adders,
      567–68
    truth table for, 562
  ripple-carry adders, 563–64
    expressed using the dot
      operator, 580
    inverting property, 564
    propagation delay of, 563–64
Address decoders, 672–79
  block decoders, 677–78

column decoders, 629, 630,
  677–78, 710
row decoders, 673–77
  dynamic decoders, 675–77
  static decoder design, 673–75
Address-transition detection (ATD),
  367, 692
Address word, 628
Advanced interconnect techniques,
  480–87
  current-mode transmission
    techniques, 486–87
  networks-on-a-chip, 487–88
  reduced-swing circuits, 480–85
    dynamic reduced-swing
      networks, 484–85
    static reduced-swing networks,
      481–83
Allocation function, architecture
  synthesis, 439
Alpha microprocessors, 511–14
  Alpha 21164 processor, 511–13
  Alpha 21064 processor, 511–13
  Alpha 21264 processor, 513–14
Alpha particles, 697
Altera MAX series, 416–18
Alternative register styles, 354–58
  pulse registers, 354–56
  sense-amplifier based registers,
    356–58
Amplifiers:
  charge-redistribution, 683–84
  flip-flop sense, 682
  sense, 481, 630, 679–86
Analytical Engine, 4
Annealing, 41
Antifuses, 413
Application-specific designs, 11
Application-specific ROMs, 646
Arbiters, 538–39
Architecture synthesis, 439–43

of wireless-communications
  processor, 441–43
Area capacitance, 139
Arithmetic building blocks:
  adders, 561–86
    binary adder, 561–64, 571–86
    carry lookahead adder, 561–64
    full adders, 562, 564–70
    ripple-carry adder, 563–64
  arithmetic-logic unit (ALU), design
    case study, 598–600
  comparators, 596–97
  counters, 596–97
  datapaths, in digital processor
    architectures, 560–61
  designing, 559–622
  dividers, 596–97
  goniometric operators, 596–97
  multiplier, 586–94
    array multiplier, 587, 589–91
    definitions, 586–87
    final addition, 593–94
    partial product accumulation,
      589–93
    partial-product generation,
      587–89
    partial products, 586
    shift-and-add algorithm, 586
  run-time power management,
    611–17
    dynamic supply voltage scaling
      (DVS), 611–14
    dynamic threshold scaling
      (DTS), 614–17
  shifter, 594–96
    barrel shifter, 595–96
    logarithmic shifter, 596
Arithmetic-logic unit (ALU), design
  case study, 598–600
Arithmetic operators, 379
ARM-7 embedded microprocessor,
  421

745

Array-based design, 399–420
    prediffused (mask-programmable)
        arrays, 399–404
    embedded gate-array approach,
        403
    gate-array approach, 399–401
    sea-of-gates approach, 399–401
    prewired arrays, 404–20
    programmable logic, 405–13
Array-based programmable logic,
    406–10
Array-based programmable wiring,
    413–14
Array multiplier, 587, 589–91
Arrays:
    prediffused, 399–404
    prewired, 404–20
Ashing, 40
Assignment function, architecture
        synthesis, 439
Associative memory, 626, 670–71
Astable circuits, 328, 368–70
Astable multivibrators, 372
Asynchronous design, 492
Asynchronous interconnect, 494
Asynchronous-synchronous
        interface, 534–35
Automatic pitch matching, 69–70
Automatic test-pattern generation
        (ATPG), 736–37
Available evaluation time, 288
Avalanche breakdown, 84

B

Backgate coupling, 295
Background memory, foreground
        memory vs., 328–30
Balance clock paths, 515
Ball grid arrays (BGAs), 56–58, 458
Barrel shifter, 595–96
Batch DRC, 70
Bed-of-nails, 729
Behavioral mode, VHDL, 314–15
Behavioral synthesis, 439–43
Berkeley FinFET dual-gated
        transistor, 127
Berkeley MIS tool, 437
Berkeley Short-Channel IGFET
        Model (BSIM), 118, 133
Berkeley SPICE3 simulator, 170
BiCMOS technology, 6
BILBO (built-in logic block
        observation), 733

Bin, 118
Binary adders, 571–86
    carry-bypass (carry-skip) adders,
        571–73
    carry-lookahead adders, 578–86
    linear carry-select adders, 574–75
    square-root carry-select adders,
        575–77
Bistability principle, 330–32
Bistable circuits, 328, 330–32, 371
Bit line, 630
Bit-sliced datapath, 561
Bits, 625
Black box view (model), 8
Bleeder transistors, 292, 299
Block address, 630
Block decoders, 677–78
Body-effect coefficient, 90
Bonding wires, 53–54
Boolean variables, 19–20
Boosted sense-ground, 707
Boosted word-line, 704
Boosted-word-line voltage, 686
Booth's recoding, 588–89
Bootstrapping, 666
Bottom-plate junction, 111
Boundary scan, 729–30
Branching effort, 255–56
Breakdown voltage, 84
Brent-Kung tree, 584–85
BSIM3-V3 SPICE model, 118
Built-in logic block observation
        (BILBO), 733
Built-in potential, 76
Built-in self-test (BIST), 730–34
Bulk CMOS process, 61
Bus, 458
Bypass, 465
Bypass signal, 555
Bytes, 625

C

C²MOS (clocked CMOS) register,
        346–50
Cache applications, use of associate
        memory in, 671–72
CAM, *See* Contents-addressable
        memories (CAM)
Capacitance, 138–44
    area, 139
    channel, 107–10
    depletion-region, 81–83
    diffusion, 110–11, 195–96

effective, 216
    extrinsic load, 199
    fan-in, 248–49
    fan-out, 248–49
    fringing, 140–41
    gate, 107
    interwire, 138–44
    intrinsic, 199, 249
    junction, 83, 110–11
    of metal wire, 144
    overlap, 107
    parallel-plate, 139
    parasitic, 137
Capacitive coupling, 25, 295
Capacitive cross talk, 446–49, 488
Capacitive device model, 111–12
Capacitive parasitics, 446–59
    capacitance and performance in,
        449–59
        design challenge, 458
        driver circuits, 458–59
        driving off-chip capacitances
            (case study), 453–57
    cross talk, 446–49
Carry-bypass (carry-skip) adders,
        571–73, 577, 620
Carry-lookahead adders, 578–86,
        620
    implementing in dynamic logic,
        581–83
    logarithmic lookahead adders,
        579–81, 584–86
    monolithic lookahead adders,
        578–79
Carry-save multiplier, 591–92
Carry-select adders, 574–75, 620
    linear carry-select, 574–75
    square-root, 575–77
CAS (column-access strobe), 633
Cascade of buffer stages, 488
Cascading dynamic gates, 295–303
    domino logic, 297–303
Cell-based design methodology,
        384–99
    compiled cells, 390–91
    intellectual property, 394–95
    macrocells (megacells), 392–96
        hard macro, 392–93
        soft macro, 393–95
    semicustom design flow, 396–99
    standard-cell approach, 385–90
Cell-based programmable logic,
        410–13

Cell libraries, 10
Channel capacitance, 107–10
Channel-length modulation, 93–94
Channel-stop implant, 43
Characteristic impedance, 162
Characteristic table, flip-flops, 341
Characterization:
  library, importance/challenge of, 427
  of logic cells, 428–31
  of registers, 431–33
Charge leakage, 290–92
Charge pumps, 544–45, 688
Charge-redistribution amplifiers, 683–84
Charge sharing, 292–95
Charge storage, 328
Charge-transfer ratio, 667
Chemical-mechanical planarization (CMP), 42
Chemical-mechanical polishing (CMP), 46
Chemical vapor deposition (CVD), 42
Chip designer, 625
Circuit designer, 625
Circuit simulation, 131–34
  device models, 133–34
  properties of, 132
  simulation modes, 132–33
    DC sweep, 133
    transient analysis, 133
Circuit synthesis, 435–36
  transistor schematics, 436
  transistor sizing, 436–37
CLK module, 362
Clock, *See also* Clock distribution;
  Clock-distribution techniques; Clock jitter; Clock skew; Clock signals
  activity factor, 332
  alignment, 13–14
  cycle time, 27
  domains, 492
  feed-through, 295
  frequency, 27
  gating, 601
  jitter, 492, 500–8, 515, 550
  misalignment, 13–14
  networks, 508–11
  nonoverlapping, 338
  overlap, 337
  perfect, 502

period, 327
skew, 27, 337, 492, 496–500, 550
slope of, 353
Clock-delayed domino, 533–34, 582
Clock distribution:
  and grid structure, 509
  optical, 548–49
  synchronous vs. asynchronous design, 549–50
  using Delay Locked Loops (DLLs), 546–48
Clock-distribution techniques, 508–15
Clock jitter, 492, 500, 550
  absolute jitter, 500
  combined impact of clock skew and, 501–2
  cycle-to-cycle jitter, 500
  data-dependent, 507–8
  design techniques for dealing with, 515
  sources of, 502–8
    capacitive coupling, 507
    clock-signal generation, 503
    environmental variations, 506–7
Clock signals, 284, 329
  non-ideal, 337–39
Clock skew, 27, 337, 492, 496–500, 550
  combined impact of clock jitter and, 501–2
  design techniques for dealing with, 515
  sources of, 502–8
    environmental variations, 506–7
    interconnect variations, 504–6
    manufacturing device variations, 503–4
Clock wires, shielding, 515
Clocking style, 372
Clustered voltage-scaling technique, 606
CMOS, 6, *See also* CMOS integrated circuits; CMOS inverter; CMOS SRAM; Dynamic CMOS design; Static CMOS design
  complementary, 237–63
CMOS integrated circuits, 36–46
  photolithography, 37–41
  recurring process steps, 41–42
  silicon wafer, 37

simplified CMOS process flow, 42–46
CMOS inverter, 179–233, *See also* Static CMOS design
  capacitances, computing, 194–99
  circuit diagram, 180
  diffusion capacitances, 195–96
  direct-path currents, dissipation due to, 220–23
  dynamic behavior, 193–213
  dynamic power consumption, 214–23
  evaluating the robustness of, 184–93
    device variations, 191
    supply voltage, scaling, 192–93
  fan-out capacitance, 196–97
  gate-drain capacitance, 194–95
  noise margins, 188–91
  performance of, 193–213
  propagation delay, 199–203
    choosing the number of stages in an inverter chain, 208–11
    delay in presence of interconnect wires, 212–13
    NMOS/PMOS ratio, 203–5
    rise/fall time of the input signal, 211–12
    sizing an inverter network, 207–11
    sizing inverters for performance, 205–7
  properties of, 181–82
  static power consumption, 223–25
  SPICE description of, 119–20
  switch model of dynamic behavior of, 181, 184
  switching threshold, 185–87
  wiring capacitance, 196
CMOS latchup, MOS transistors, 116–17
CMOS SRAM, 658–61
  read operation, 658–59
  write operation, 660–61
Collection efficiency, 697
Column address, 629
Column decoders, 629, 630, 677–78, 710
Combinational logic, sizing for minimum delay, 256
Combinational logic circuits, 236
Combinational networks, optimizing performance, 251–57

Commodity ROM chips, 646
Common-mode rejection, 680
Common-mode rejection ratio (CMRR), 680
Compactor, 69
Comparators, 596–97
Compiled cells, 390–91
Complementary CMOS, 237–63
  combinational networks, optimizing performance in, 251–57
  concept, 237–41
  dynamic or glitching transitions, 260–63
  gates:
    propagation delay of, 242–51
    static properties of, 241–42
  inter-signal correlations, 259–60
  logic function, 257–58
  power consumption, 257
  signal statistics, 258–59
Complementary pass-transistor logic (CPL), 272–73
Completion-signal generation, 522–26
  dual-rail coding, 522–24
  replica delay, 524–25
Complex gate layout techniques, 319–24
  Euler path approach, layout planning using, 320–23
  standard-cell layout technique, 320
  Weinberger approach, 319
Complex logic circuit:
  event-driven simulator, 311
  functional simulation, 313–15
  gate-level simulation, 312
  higher-level data models, using, 315–17
  logic simulation, 313
  representing digital data:
    as a continuous entity, 310
    as a discrete entity, 310–15
  simulating, 309–17
  switch-level simulation, 312
  timing simulators, 310
Complex logic gates, 439
Compound domino, 301
Computer-aided design (CAD), 10
Configurable hardware, 423
Configurable Logic Block (CLB), 412–13
Constant electrical field scaling, 124

"Constant resistance" scaling of wire properties, 173
Contact layers, 48
CONTACT-programming strategy, 637
Contact resistance, 146
Contact rules, 49
Contamination delay, 327–28
Contents-addressable memories (CAM), 626, 632, 670–72
Continuous waveforms, 132
Control module, digital processors, 379
Copper and low-k dielectrics, 61
Counters, 596–97
Coupling:
  backgate, 295
  capacitative, 295
  inductive, 549
  output-to-input, 295
  word-line-to-bit-line, 694–95
CPL, See Complementary pass-transistor logic (CPL)
Critical charge, 698
Critical input signal, 250
Critical length, 467
Critical path, 250
Critical-path replica, 524–25
Cross talk, 549
  capacitive, 446–49, 488
  design technique, 448–49
  driven lines, 448
  floating lines, 446–48
  interwire capacitance and, 447
  and performance, 449–51
  wire delay, 451–53
Current crowding, 146
Current-mode transmission techniques, 486–87
Current sensing, completion detection circuit using, 525–26
Current-source model, 104
Current-starved inverter, 369–70
Custom design, 67, 383–84, 423
Cut-off region, 129
Cycle time, 625
Cycle-to-cycle jitter, 500

D

Data-retention dissipation, 706–7
Datapaths:
  bit-sliced, 561
  defined, 560

in digital processor architectures, 379, 560–61
  power and speed trade-offs in, 600–618
DC transfer characteristic, 20–21
DCVSL, See Differential Cascade Voltage Switch Logic (DCVSL)
De Morgan's theorems, 239
Decoders, 628
  address, 672–79
  column, 677–78
  design, 674–75
  for non-random-access memories, 679
  predecoders, 673–74
  row, 673–77
Decoupling capacitances:
  impact of, 473
  on-chip, in Compaq's Alpha processor family, 473–74
Decoupling capacitors, on-chip, 515
Delay Locked Loops (DLLs), 550
  distributed clocking using, 546–48
Delay-locked loops, feedback circuits based on, 515
Delay of the wire, 428
Dense wire fabric, 452
Depletion region, 76
Depletion-region capacitance, 81–83
Deposition, 41–42
Design Compiler (Synopsys), 428, 438
Design for testability (DFT), 722
Design-Rule Checking (DRC), 50
Design rule set, 36
Design rules, 47–50, 64
  defined, 47
  interlayer constraints, 49
  intralayer constraints, 48
  layer representation, 48
  minimum line width, 47
  scalable, 47
  verifying the layout, 49–50
Design synthesis, 435–44
  architecture synthesis, 439–43
  circuit synthesis, 435–36
  logic synthesis, 437–39
Design verification, 553–57
  electrical verification, 554
  functional (formal) verification, 556–57
  timing verification, 554–56

Designing for manufacturability, 122
Destructive reading, cell contents, 668
Device capacitors, 129
Device engineering, 127
Device ratios, 665
Device under test (DUT), 723
Devices, 73–130
  diodes, 74–87
    depletion region, 75–77
  models, 74
  MOSFET transistors, 87–120
  process variations, 120–22
  technology scaling, 122–28
DIBL, *See* Drain-induced barrier lowering (DIBL)
Die yield, 17–18
Dielectric materials, relative permittivity of, 139
Dies, 16–18
Difference Engine, 4–5
Differential Cascade Voltage Switch Logic (DCVSL), 267–68
Differential logic, 267
Differential pass-transistor logic (DPL), 272
Differential registers, 515
Differential signaling techniques, 481
Diffusion, 76, 160
Diffusion capacitance, 110–11, 195–96
Diffusion equation, 156
Diffusion implantation, 41
Diffusion regions, 48
Digital circuits:
  arbiters, 538–39
  array-based design/implementation approaches, 384, 399–420
    prediffused (mask-programmable) arrays, 399–404
    prewired arrays, 404–20
  cell-based design methodology, 384–99
    compiled cells, 390–91
    intellectual property, 394–95
    macrocells (megacells), 392–96
    semicustom design flow, 396–99
    standard-cell approach, 385–90
  custom circuit design, 383–84
  custom design approaches, 382–83

phase-locked loop:
  basic concept, 540–46
  building blocks of, 542–46
  clock synthesis and synchronization using, 539–46
  defined, 539
  functional composition of, 541
  semicustom design approaches, 382–83
  structured array design approaches, 382–83
  synchronizers, 534–39
    asynchronous-synchronous interface, 534–35
    defined, 535
    design, 538
    and mean time to failure, 538
    synchronization failures, 535
    trajectories, 536
  synchronous design, 495–519
    clock-distribution techniques, 508–15
    latch-based clocking, 516–19
    sources of skew and jitter, 502–8
    synchronous timing basics, 495–502
  timing classifications, 492–95
    asynchronous interconnect, 494–95
    mesochronous interconnect, 493
    plesiochronous interconnect, 493–94
    synchronous interconnect, 492–93
  timing issues, 491–552
    future directions, 546–50
  validation and test of manufactured circuits, 721–38
    design for testability, 723–34
    diagnostic test, 722
    functional test, 722
    parametric test, 722–23
    test procedure, 722–23
Digital data manipulation, historical perspective, 4–6
Digital design:
  energy dissipation of first-order *RC* network, 31
  functionality and robustness, 18–27
    directivity, 25
    fan-in and fan-out, 25

    ideal digital gate, 25–26
    noise immunity, 24
    noise margins, 21
    regenerative property, 22–23
    voltage-transfer characteristic, 20–21, 26–27
  integrated circuit (IC), cost of, 16–18
    fixed cost, 16
    indirect costs, 16
    variable cost, 16–18
  latency-constrained design, 600
  performance, 27–30
    clock cycle time, 27
    clock frequency, 27
    propagation delay, 27–30
    ring oscillator circuit, 28–29
    rise and fall times, 28
  power and energy consumption, 30–31
    power-delay product (PDP), 30–31
  quality metrics of, 15–31
  through-put-constrained design, 600
Digital ICs, *See* Integrated circuits (ICs)
Digital integrated circuit design, issues in, 6–15
Digital processors, 378–80
  control module, 379
  datapaths, 379, 560–61
  input-output circuitry, 380
  interconnect network, 380
  memory module, 379
  system-on-chip, 380–81
Diodes, 74–87, 129
  analysis of diode network, 80
  depletion-region capacitance, 81–83
  dynamic (transient) behavior, 80–83
  exponential dependence, 78
  forward-bias mode, 77
  ideal diode equation, 77–79
  large-signal depletion-region capacitance, 83
  models for manual analysis, 79–80
  reverse-bias mode, 78
  saturation current, 79
  secondary effects, 84–85
  SPICE diode model, 85–87
  static behavior, 77–80

Direct current, 462–63
Directivity, 25
Distributed *rc* line, 156–59
Distributed *rc* model, 152
Distributed *RC*-network, 249
Dividers, 596–97
DLLs, *See* Delay Locked Loops
    (DLLs)
Domino logic, 297–303
    compound domino, 301
    concept, 297–99
    gates, optimization of, 300–303
    multiple output domino, 300
    noninverting property, 299–300
    np-CMOS logic, 301–3
Donor impurities, 75
Double-ended signaling techniques,
    481
Drain capacitance, 642–43
Drain-induced barrier lowering
    (DIBL), 114–15
DRAM, *See* Dynamic random-access
    memory (DRAM)
Drift, 76
    over time, 115
Dry etching, 42
Dual Damacene process, 61
Dual-edge registers, 349–50
Dual-in-line package (DIP), 55
Dual-rail coding, 522–24
Dual-rail domino, 299
Dual-well approach, 37
Dummy cells, 685
Dummy fills, 505, 515
Dummy word lines, 690
DUT board, 723
Dynamic behavior:
    diodes, 80–83
    MOSFET transistors, 106
Dynamic circuits, 100
Dynamic CMOS design, 284–303
    basic principles, 284–86
    cascading dynamic gates, 295–303
    domino logic, 297–303
    signal integrity issues, 290–95
        capacitive coupling, 295
        charge leakage, 290–92
        charge sharing, 292–95
        clock feed-through, 295
    speed/power dissipation of
        dynamic logic, 287–90
Dynamic decoders, 675–77
Dynamic hazards, 260, 289

Dynamic latches/registers, 344–50
    C^2MOS (clocked CMOS) register,
        346–50
    dual-edge registers, 349–50
    dynamic transmission-gate edge-
        triggered registers, 344–46
Dynamic logic, 284
    evaluation phase, 284, 285–86
    precharge phase, 284–85
    speed/power dissipation of,
        287–90
Dynamic memory, 372, 379, 626
    static memory vs., 328–29
Dynamic random-access memory
    (DRAM), 633, 664–70, 690
    active power reduction, 703–4
    data-retention dissipation, 706–7
    one-transistor dynamic memory
        cell, 666–70
    three-transistor dynamic memory
        cell, 664–66
Dynamic reduced-swing networks,
    484–85
Dynamic registers, 344–46, 372
Dynamic threshold scaling (DTS),
    614–17
Dynamic transmission-gate edge-
    triggered registers, 344–46
Dynamic voltage scaling (DVS),
    611–14

E

E^2PROM, *See* EEPROM (electrically
    erasable programmable read-
    only memory)
ECL (Emitter-Coupled Logic), 5
Edge-triggered registers, 329, 330,
    495
    master-slave, 333–39
    negative, 326–27, 329, 334
    positive, 326–27, 329, 334, 431
EEPROM (electrically erasable pro-
    grammable read-only mem-
    ory), 405, 626, 650–51, 717
Effective capacitance, 216
Effective channel width and length,
    92
Effective fan-out, 207, 252
ELDO, 131
Electrical effort, 252, 254
Electrical verification, 554
Electrical wire models, 150–69
    distributed *rc* line, 156–59

ideal-wire model, 151
lumped capacitance model,
    151–53
lumped *RC* model, 152–56
transmission line, 159–69
Electrically Programmable Logic
    Devices (EPLDs), 416–17
Electromigration, 460
Electrostatic discharge (ESD), 458
Electrostatic protection circuit,
    458–59
Elmore delay formula, 153, 156, 245,
    467
Embedded memories, 627
Emission coefficient, 85
Energy-delay product (E-D), 31,
    226–27
Energy efficiency, 381–82
Energy per operation, 225–26
ENIAC, 4
Epitaxial layer, 37
EPLDs (Electrically Programmable
    Logic Devices), 416–17
EPROM (erasable-programmable
    read-only memory), 626,
    649–50
Equalization, 682
Equipotential region, 151
Equivalent resistance model, 104–6
Erase operation, programming cycle,
    652
Error correction, 699–700
    using Hamming codes, 699–700
Etching, 42
ETOX Flash cell (Intel), 623
Evaluation phase, dynamic logic
    gate, 284, 285–86
Event-driven simulator, 311
Exhaustive approach, to generating
    stimuli, 730–31
Exponential dependence, diodes, 78
External design entities, 12
Extrinsic load capacitance, 199

F

Fairchild Corporation, 5, 6
Fairchild Micrologic family, 5
False paths, 555
FAMOS transistor, 648, 651
Fan-in and fan-out, 25
Fan-in capacitance, 248–49
Fan-out capacitance, 248–49
    CMOS inverter, 196–97

Fault models, 734–36
Fault simulation, 737
Feedback, positive, 267, 328
Field implant, 43
Field oxide, 43
Field-programmable gate array
  (FPGA), 404–6, 437, 452
  complexity/performance, 419–20
  nonvolatile FPGA, 405–6, 416
  volatile (RAM-based) FPGA, 406
  write-once (fuse-based) FPGA,
    405
FIFO (first-in first-out), 549, 626,
  679
Finite state machine (FSM), 326,
  379, 410, 557
Fixed cost, 16
Fixed-voltage scaling, 125 26
Flash EEPROM, 651
Flash memories, 405, 626, 651–55
Flip-chip mounting, 55, 58
Flip-flop sense amplifier, 682
Flip-flops, 330, 432, 625, *See also*
    Bistable circuits
  characteristic table, 341
  set-reset (SR), 341–42
  static SR, 341–44
Floating capacitors, 142
Floating gate, 648
Floating-gate avalanche-injection
  (MOS), 648
Floating-gate transistor, 647–49
FLOTUX (floating-gate tunneling
  oxide) transistor, 650–51
Folded bit-line architecture, 695
Foot switch, 581
Foreground memory, background
  memory vs., 328–30
Formal verification, 556–57
  equivalence proving, 557
  property proving, 557
Forward-bias mode, diodes, 77
Forward binary tree, 584
Four-phase signaling, 529
Four-transistor SRAM cell, 662–63
Fowler Nordheim tunneling, 650–51,
  653–54
FPGA, *See* Field-programmable gate
    array (FPGA)
FRAM (ferroelectric RAM), 655
Fringing capacitance, 140–41
FSM, *See* Finite state machine (FSM)
Full adders, 564–70, 592–93

Manchester carry-chain adder,
  568–70
  mirror adder design, 566–67
  static adder circuit, 564–66
  transmission-gate-based adder,
    567–68
  truth table for, 562
Full scaling, 124
Function generators, using
    multiplexers as, 410–11
Functional simulation, 313–15
Functional verification, 556–57
  equivalence proving, 557
  property proving, 557
Fuse-based FPGA, 405
Fuses, 647

G

Gain factor, 92
Gate-array approach, 399–401
Gate arrays, standard cells vs., 404
Gate capacitance, 107
Gate complexity, 18
Gate-drain capacitance, CMOS
  inverter, 194–95
Gate effort, 254
Gate isolation, 401–2
Gate-level simulation, 312
Gate oxide, 43
Gate threshold voltage, 20
Gates, 43
General scaling, 126–27
Geometry isolation, 401
Gigabytes (Gbyte), 625
Glitch concept, 372
Glitch propagation, 532
Glitch registers, 329
Glitches, 260, 355–56, 532
Glitching, 257, 289
  reducing through path balancing,
    611
Global clock grid, 514
Global clocks, 547–48
Global design entities, 12
Global interconnections, 171
Global reference, clocks, 508
Global word lines, 644, 707
Globally-asynchronous locally-syn-
    chronous approach, 550
Goniometric operators, 596–97
Grading coefficient, 83
Grid structure, and clock distribution,
  509

Ground bounce, 488–89
Guard rings, 117, 458

H

H-tree network, 509
Half-adder (HA), 592–93
Half-V_{DD}, 687
  precharge, 704
Hamming codes, error correction
  using, 699–700
Handshaking protocol, 521
  four-phase, 529–30
Hard macro, 392–93
Hard-timing edge, 583
Heat flow equation, 59–60
Heil, O., 6
Hierarchical-word decoding scheme,
  710
High-impedance state, 285
High-level synthesis, 439–43
Hold mode, latches, 329
Hold time, 327
  multiplexer-based master-slave
    registers, 335
  registers, 371, 431–33
Hold-time constraints, 347
Hot-carrier effects, MOS transistors,
  115–16
HSPICE, 131
Hybrid implementation platforms,
  examples of, 421–22

I

ICs, *See* Integrated circuits (ICs)
IDDQ test, 735–36
Ideal digital gate, 25–26
Ideal diode equation, 77–79
Ideal scaling, 171–73
Ideal-wire model, 151
IGFET, *See* MOSFET transistors
I^2L (Integrated Injection Logic), 6
Impedance:
  characteristic, 162
  low output, 181
Indirect costs, 16
Inductance, 137, 148–50
Inductive coupling, 549
Inductive parasitics, 469–79
  inductance and performance,
    475–79
  inductance and reliability, 469–75
  Ldi/dt voltage drop, 469–75
  transmission-line effects, 475–79

Input-output circuitry, digital
  processors, 380
Input resistance, 181
Input slope, 428
Integrated circuits (ICs), *See also*
  Digital circuits
  digital processors, 378–80
    control module, 379
    datapaths, 379, 560–61
    input/output circuitry, 380
    interconnect network, 380
    memory module, 379
    system-on-chip, 380–81
  fixed cost, 16
  implementation strategies,
    377–425
  indirect costs, 16
  invention of, 5
  layout, 67–72
    circuit extraction, 70–71
    design-rule checker (DRC), 70
    layout editor, 68–69
    MAGIC tool, 68
    MicroMagic tool, 68
    symbolic, 69
  packaging, 51–61, 64
    cost, 52
    electrical requirements, 52
    interconnect levels, 53–59
    mechanical and thermal
      properties, 52
    package materials, 52
    requirements, 52
    thermal considerations in,
      59–61
  packaging options, 36
  variable cost, 16–18
Integration density, 4
Intellectual property (IP) modules,
  394–95, 423
Interconnect:
  advanced techniques, 480–87
    current-mode transmission
      techniques, 486–87
    networks-on-a-chip, 487–88
    reduced-swing circuits, 480–85
  capacitive parasitics, 446–59
    capacitance and performance in
      CMOS, 449–59
    cross talk, 446–49
  coping with, 445–90
  inductance of, 488
  inductive parasitics, 469–79

inductance and performance,
    475–79
  inductance and reliability,
    469–75
  Ldi/dt voltage drop, 469–75
  transmission-line effects,
    475–79
  resistive parasitics, 460–69
    electromigration, 462–64
    ohmic voltage drop, 460–62
    RC delay, 464–69
    resistance and performance,
      464–69
    resistance and reliability,
      460–62
Interconnect architecture,
    optimizing, 468–69
Interconnect levels:
  level 1 (die-to-package substrate),
    53–55
  level 2 (package substrate to
    board), 55–57
  multi-chip modules—die-to-board,
    57–58
Interconnect network, digital
    processors, 380
Interconnect parasitics, 12
Interlayer constraints, 49
Interlock element, 538–39
Intermediate buffers, 466–68
Interwire capacitances, 138–44
  design data, 143–44
Intralayer constraints, 48
Intrinsic capacitance, 199, 249
Intrinsic delay, 206, 428
Intrinsic propagation delay, 246
Inverse binary tree, 584–85
Inverter metrics, impact of technol-
    ogy scaling on, 229–32
Inverter voltage-transfer
    characteristic, 20
Inverting property, ripple-carry
    adder, 564
Ion Implantation, 41, 44–46
IT DRAM readout, 668

J
Junction:
  abrupt, 75
  bottom-plate, 75
  linearly graded, 83
  step, 75
Junction capacitances, 83, 110–11

K
Kilby, Jack, 5
Kilobytes, 625
Kirchhoff's law, 477
Kogge-Stone tree, 581, 584–85, 611

L
Large-signal depletion-region
    capacitance, 83
Latch-based clocking, 516–19
Latch- vs. register-based pipelines,
    360
Latches, 330
  dynamic, 344–50
  hold mode, 329
  negative, 329
  positive, 329
  registers vs., 329–30
  shift-register latch (SRL), 728–29
  static, 330–44
  transparent mode, 329
Latency-constrained design, 600
Lattice diagram, 165
Layer representation, 48
Layout, 50
  ICs, 67–72
    circuit extraction, 70–71
    design-rule checker (DRC), 70
    layout editor, 68–69
    MAGIC tool, 68
    MicroMagic tool, 68
    symbolic, 69
    verifying, 49–50
LDD (Lightly Doped Drain), 61
Ldi/dt voltage drop, 469–75
  design techniques, 471–73
Leadless chip carrier, 56
Level converters, 604
Level restorer, 273–76
Level-sensitive circuits, 329, 371
Level-sensitive scan design (LSSD),
    728–29
Library characterization, impor-
    tance/challenge
    of, 427
LIFO (last-in first-out), 626
Lilienfeld, J., 6
Linear carry-select adder, 574–75
Linear dependence, 96
Linear-feedback shift register
    (LFSR), 731
Linear region, 92, 129

Linear regulators, 687
Linear scaling, 47
Linearly graded junction, 83
Load device, 264
Load-line plot, 182–83
Local clock grids, 515
Local clocks, 514
Local distribution, clocks, 508
Local interconnections, 171
Lock range, 541
Lock time, 541
Logarithmic lookahead adder,
   579–81, 584–86
   Brent-Kung tree, 584–85
   Kogge-Stone tree, 581, 584–85
   radix-4 adder, 585
Logarithmic shifter, 596
Logic:
   cell-based programmable logic,
      410–13
   complementary pass-transistor
      logic (CPL), 272–73
   differential logic, 267
   domino logic, 297–303
   dynamic logic, 284
   multilevel logic, 437
   multiple output domino logic, 300
   pass-transistor logic, 269–84
   post-charge logic, 532–33
   programmable logic, 405–13
   ratioed logic, 181, 263–69
   self-timed logic, 519–22, 531–34
   transmission gate logic, 277–78
Logic Array Block (LAB), 416
Logic cells, characterization of,
   428–31
Logic function, 264
Logic gates, 180
Logic operators, 379
Logic optimization systems, stages
   of, 437
Logic (signal) swing, 20
Logic simulation, 313
Logic style, how to choose, 303
Logic synthesis, 437–39
Logic-synthesis tool, 388
Logical effort, 252–55
Logical variables, 19–20
Long-channel devices, scaling of,
   125–26
Lookup tables (LUTs), 411
   delay models using, 430–31

Loop filter, 541
Lossless transmission line, 160–62
Lossy transmission line, 160, 167–69
Low energy/power design
   techniques, 217
Low output impedance, 181
Low-voltage static latches, 339–41
Lumped capacitance model, 151–53
Lumped RC-ladder network, 157
Lumped RC model, 152–56
Lumped RC network, 170
LUTs, See Lookup tables (LUTs)

M

Macrocells, 392–96, 408, 423
   programmed, 408–10
   programmed, example of, 409–10
MAGIC tool, 68
Major clock grid, 514
Manchester carry-chain adder,
   568–70
   carry bypass in, 572
   sizing of, 570
Manhattan-style routing, 465–66
Manual analysis model for 0.25 μm
   CMOS process, 102–3
Manufactured circuits:
   design for testability, 723–34
      ad hoc testing, 724–25
      boundary-scan design, 729–30
      built-in self-test (BIST), 730–33
      issues in, 723–24
      scan-based test, 726–29
   diagnostic test, 722
   functional test, 722
   go/no-go test, 722
   parametric test, 722–23
   system-on-chip, testing, 733–34
   test-pattern generation, 734–37
      automatic test-pattern genera-
         tion (ATPG), 736–37
      fault models, 734–36
      fault simulation, 737
   test procedure, 722–23
   validation/test of, 721–38
Manufacturing process, 35–65
   CMOS integrated circuits, 36–46
      photolithography, 37–41
      recurring process steps, 41–42
      silicon wafer, 37
      simplified CMOS process flow,
         42–46

Mask-programmable arrays,
   399–404
   gate-array approach, 399–401
   sea-of-gates approach, 399–401
Master-slave configuration, 329,
   333–34
Master-slave edge-triggered
   registers, 333–39
   non-ideal clock signals, 337–39
   timing properties, 335–37
MATCH flag, 626
Matched paths, 502–3
Mathworks Simulink environment,
   441–43
Maximum fan-out, 25
Mean time to failure, and
   synchronizers, 538
Megabytes (Mbyte), 625
Megacells, 392–96
Memory:
   architectures and building blocks,
      627–34, 717
      address word, 628
      bit line, 630
      block address, 630
      column address, 629
      column decoders, 629
      comparand, 632
      decoder, 628
      mask, 632
      row address, 629
      word line, 630
   associative, 626, 670–72
   classification, 617–18
      access pattern, 626–27
      application, 627
      function, 625–26
      input/output architecture, 627
      size, 625
      timing parameters, 625
   contents-addressable memories
      (CAM), 626, 670–72
   data-retention dissipation, 704–7
      in DRAMS, 706–7
      in SRAMS, 704–6
   drivers/buffers, 689
   hierarchical memory architecture,
      631–32
   memory core, 634–72
   memory design case studies,
      707–13
      4-Mbit SRAM, 710–12

Memory *(cont.)*
    1 Gbit NAND flash memory,
        712–13
    programmable logic array
        (PLA), 707–10
    memory peripheral circuitry,
        672–92
        address decoders, 672–79
        block decoders, 677–78
        column decoders, 677–78
        decoders for non-random-access
            memories, 679
    nonvolatile read-wire (NVRW)
        memories, 647–56
        EEPROM (electrically erasable
            programmable read-only
            memory), 650–51
        EPROM (erasable-programma-
            ble read-only memory),
            649–50
        Flash memories, 651–55
        floating-gate transistor, 647–49
        trends in, 655–56
    organization, 629
    power dissipation in, 701–7
        addressing the active power
            dissipation, 702–4
        DRAM active power reduction,
            703–4
        partitioning the memory, 702
        sources of, 701–2
        SRAM active power reduction,
            703
    random-access memory (RAM),
        657–70
        dynamic random-access
            memory (DRAM), 664–70
        static random-access memory
            (SRAM), 657–64
    read-only memories (ROM),
        634–47
        application-specific ROMs, 646
        commodity ROM chips, 646
        NAND ROM, 641–43
        NOR ROM, 640–42
        power consumption and pre-
            charged memory arrays,
            645–46
        programming, 636–46
        ROM cells, 634–36
        transient performance, 641–45
        user perspective, 646–47
    reliability, 693–98, 717

semiconductor memory trends and
    evolutions, 713–16
sense amplifiers, 679–86
    differential voltage sensing
        amplifiers, 680–82
    functions performed by, 679
    sensing in lT DRAM, 685–86
    single-ended sensing, 682–84
    single-to-differential conver-
        sion, 684–85
    signal-to-noise ratio, 693–98
        bit-line-to-bit-line coupling,
            695–96
        critical charge, 697–98
        leakage, 696–97
        word-line-to-bit-line coupling,
            694–95
    timing and control, 689–92
        dynamic random-access mem-
            ory (DRAM), 690
        static random-access memory
            (SRAM), 691–92
        synchronous DRAM memories,
            690–91
    voltage references, 686–89
    yield, 698–700
        error correction, 699–700
        redundancy, 698–99
Memory cells, 624, 717
Memory module, digital processors,
    379
Memory peripheral circuitry, 672–92
    address decoders, 672–79
    block decoders, 677–78
    column decoders, 677–78
    decoders for non-random-access
        memories, 679
MEMS (Micro Electro-Mechanical
    Systems), 63
Mesh interconnect structures, 380
Mesochronous interconnect, 493
Metal interconnect layers, 48
Metastable operation points, 331
MicroMagic tool, 68
Micron rules, 48
Microprocessors, 7–10, 180
    Alpha, 511–14
    ARM-7 embedded, 421
Microstrip line, 141, 162
Miller effect, 194
Minimum line width, 47
Minterms, 389

Mirror adder, 566–67
    circuit schematics, 567
Mismatched paths, 502–3
MNOS (Metal Nitride Oxide Semi-
    conductor) transistor, 647
Mobility, 92
Models, 11, 74
Modified Booth's recoding, 588–89
Module compiler, 392
Module libraries, design and imple-
    mentation of, 11
ModuleCompiler tool (Synopsys),
    394
Monolithic lookahead adders,
    578–79
Monostable circuits, 328
Monostable sequential circuits, 367,
    372
Monte Carlo analysis, 122
Moore, Gordon, 6–7
Moore's law, 7
MOS transistors, 114–17
    capacitances, 112
    circuit symbols for, 88
    CMOS latchup, 116–17
    hot-carrier effects, 115–16
    process variations, 120–22
    SPICE models for, 117–20
        BSIM3V3 SPICE model, 118
        transistor instantiation, 119–20
    threshold variations, 114–15
MOSFET transistors, 6, 87–120, 129,
    477
    capacitive device model, 111–12
    circuit symbols for, 88
    defined, 87
    dynamic behavior, 106
    MOS structure capacitances,
        107–12
        channel capacitance, 107–10
        junction capacitances, 110–11
    NMOS transistor, 87
    PMOS transistor, 87
    source-drain resistance, 113
    SPICE models for, 117–20
    under static conditions, 88–117
        channel-length modulation,
            93–94
        drain current vs. voltage charts,
            97–99
        models for manual analysis,
            101–14
        resistive operation, 91–93

saturation region, 93
subthreshold conduction, 99–101
threshold voltage, 88–91
velocity saturation, 94–97
MRAM (magnetoresistive random access memory), 655–56
Muller C-element, 526–30
implementations, 527–28
Multi level logic synthesis, 388
Multichip module technique (MCM), 58, 58–59
Multilevel logic, 437
Multilevel logic synthesis, 437
Multilevel nonvolatile memories, 655
Multiple output domino logic, 300
Multiple supply voltages, 603–7
module-level voltage selection, 604–5
multiple device thresholds, 607
multiple supplies inside a block, 605–10
reducing glitching through path balancing, 611
reducing switching activity, by resource allocation, 610
reducing switching capacitance:
through logic and architecture optimizations, 610
through transistor sizing, 608–9
Multiple threshold transistors, 276–77
Multiplexer-based latches, 332–33
Multiplexer-based master slave registers:
hold time, 335
propagation delay, 335
setup time, 335
timing properties of, 335–37
Multiplexers, using as function generators, 410–11
Multiplier, 180, 586–94, 620
array mulltiplier, 587
definitions, 586–87
partial product accumulation, 589–93
array multiplier, 589–91
carry-save multiplier, 591–92
tree multiplier, 592–93
partial-product generation, 587–89
partial products, 586
shift-and-add algorithm, 586

Multiport memories, 627
Multivibrator circuits, 328
Mutual-exclusion circuit, 538–39

N

n-p-n-p structures, 116–17
n-region, 76
NAND decoders, 675–77
NAND gates, 180
NAND ROM, 641–46
precharged, 645–46
propagation delay of, 644–45
voltage swings in, 641
word- and bit-line parasitics, 642–43
NanoSim tool set, 310
Narrow-channel effects, 115
National Semiconductors, 5
Negative edge-triggered registers, 326–27, 334
Negative latches, 329
Negative photoresist, 39
Negative substrate bias, 687
Netlists, 396
Networks-on-a-chip, 380, 487–88
New design issues, 12
NMOS transistor, 182, 321, 477
Noise immunity, 24
Noise margins, 242, 264
CMOS inverter, 188–91
inverter gates, 21
Noise signals, 19
Non-bistable sequential circuits, 364–70
astable circuits, 368–70
monostable sequential circuits, 367
Schmitt trigger, 364–67
Non-ideal clock signals, 337–39
Non-random-access memories, decoders for, 679
Nondestructive reading, cell contents, 666
Nonlinear dependence, 82
Nonoverlapping clocks, 338
Nonrecurring expense (NRE), 382
Nonregenerative circuits, 236
Nonvolatile FPGA, 405–6, 416
Nonvolatile memories, 626
Nonvolatile read-wire (NVRW) memories, 647–56

EEPROM (electrically erasable programmable read-only memory), 650–51
EPROM (erasable-programmable read-only memory), 649–50
Flash memories, 651–55
floating-gate transistor, 647–49
trends in, 655–56
FRAM (ferroelectric RAM), 655
MRAM (magnetoresistive random access memory), 655–56
multilevel nonvolatile memories, 655
Nonvolatile read-write (NVRWM) memories, 626
NOR decoders, 675–77
NOR gates, 180
NOR ROM, 640–46, 652–54
precharged, 645–46
propagation delay of, 644
static power dissipation of, 645
voltage swings in, 640–41
word- and bit-line parasitics, 641–42
NORA-CMOS, 361–63
np-CMOS logic, 301–3
Nyquist criterion, 534

O

On-chip busses, 380
On-chip decoupling capacitors, 515
On line DRC, 50, 70
One-shot circuits, 367
One-transistor dynamic memory cell, 666–70
Open bit-line architecture, 685
Optical clocks, 548–49
Optical mask correction (OPC), 41
Optical masks, 36, 64
Output-to-input coupling, 295
Overall effective fan-out, 208
Overglass, 46
Overlap capacitance, 107
Overlap period, 338
Oxidation layering, 39
Oxide isolation, 401

P

p-channel MOS, See PMOS transistor
p-region, 76

Packaging, 64
  cost, 52
  electrical requirements, 52
  interconnect levels, 53–59
  mechanical and thermal properties,
    52
  package materials, 52
  requirements, 52
  thermal considerations in, 59–61
Parallel-plate capacitor model, 139
Parallel termination, 476–77
Parallelism, minimizing power con-
    sumption using, 601–3
Parallelization, 600
Parasitic capacitance, 137
Partial product accumulation, 589–93
  array multiplier, 589–91
  carry-save multiplier, 591–92
  tree multiplier, 589–93
Partial-product generation, 587–89
Partial products, 586
Partial scan, 729
Partitioning drivers, 453
Pass-transistor logic, 269–84
  differential pass transistor logic,
    272–73
  pass-transistor basics, 269–72
  pass-transistor performance,
    280–84
  robust and efficient pass-transistor
    design, 273–80
  transmission gate logic, 280–84
Path balancing, reducing glitching
    through, 611
Path branching effort, 255
Path effective fan-out, 254
Path logical effort, 254
Path resistance, 153
Path sensitizing, 736
Pentium 4 microprocessor, 8–9
Perfect clock, 502
Performance, and cross talk, 449–51
Peripheral circuitry, 624, 630, 717
Perovskite crystals, 655
Phase detectors, 542–44
  phase frequency detector, 543–44
  XOR phase detector, 542–43
Phase frequency detector (PFD),
    543–44
Phase-locked loops, 550
  applications of, 540
  basic concept, 540–46
  building blocks of, 542–46

  charge pump, 544–45
  phase detectors, 542–44
  voltage controlled oscillator
    (VCO), 542–46
  clock synthesis and synchroniza-
    tion using, 539–46
  defined, 539
  functional composition of, 541
  transient behavior of, 545
Phases, periodic signals, 540
Photolithography, 37–41
  process steps, 39–40
Photoresist coating, 39
Photoresist development and resist,
    39
Photoresist removal, 40
Planarization, 42
Pin-grid-array (PGA) package,
    55–56
Pinched-off channel, 93
Pipelined datapath, 519–20, 530–31
Pipelining, 358–63, 372, 439, 468,
    600
  latch- vs. register-based pipelines,
    360
  NORA–CMOS, 361–63
  pipelined computations, example
    of, 359
Pitch matching, 672
  automatic, 69–70
Planarization, 42
PLAs, *See* Programmable logic
    arrays (PLAs)
Plasma etching, 42
Plastic leaded package (PLCC), 56
Plesiochronous interconnect, 493–94
PMOS, 6
PMOS transistor, 182, 321, 477
  thin-film, 664
pn-junction, built-in voltage of, 76
pn-junction diode, 75
Pointers, 679
Polycide, 146
Polygon pushing, 69
Polysilicon, 42, 43, 45, 48, 146
Polysilicon layers, 48
Positive edge-triggered registers,
    326–27, 334, 431
Positive feedback, 267, 328
Positive latches, 329
Post-charge logic, 532–33
Power consumption, 620
  analyzing using SPICE, 227–29

Power-delay product (PDP), 30–31,
    225–26
Power dissipation, 264
Power distribution network design,
    14–15
Power-down, 225
Power minimization techniques, 600
  design time techniques, 601–11
    reducing the supply voltage,
      601–3
    using multiple supply voltages,
      603–7
    using parallelism, 601–3
    using pipelining, 603
    reducing the power in standby
      (sleep) mode, 617–18
Power-supply rejection ratio (PSRR),
    680
Precharge phase, dynamic logic gate,
    284–85
Predecoders, 673–74
Prediffused (mask-programmable)
    arrays, 399–404
  gate-array approach, 399–401
  sea-of-gates approach, 399–401
Prewired arrays:
  programmable interconnect,
    413–16
    array-based programmable
      wiring, 413–14
    switch-box-based programma-
      ble wiring, 414–16
  programmable logic, 405–13
    array-based programmable
      logic, 406–10
    cell-based programmable logic,
      410–13
Probe card, 723
Process conductance, 120
Process technology trends, 61–64
  copper and low-k dielectrics, 61
  silicon-on-insulator (SOI) CMOS,
    61–62
  three-dimensional integrated cir-
    cuits, 63–64
Process transductance parameter, 92
Process variations, MOS transistors,
    120–22
Product terms, 389
Production cost per part, 382
Programmable array logic devices
    (PALs), 407–8

Programmable interconnect, 413–16
 array-based programmable
  wiring, 413–14
 switch-box-based programmable
  wiring, 414–16
Programmable Interconnect Array
 (PIA), 417
Programmable logic, 405–13
 array-based programmable logic,
  406–10
 cell-based programmable logic,
  410–13
Programmable logic arrays (PLAs),
 378, 388–89, 406–10, 437,
 707–10
Programmable logic devices (PLDs),
 407–8
PROM (Programmable ROM), 647
 architecture, 407
Propagation delay, 27–30, 249–50,
 264, 265, 327–28, 554
 CMOS inverter, 199–203
 complementary CMOS gates,
  242–51
 first-order RC network, 29–30
 intrinsic (unloaded), 246
 multiplexer-based master-slave
  registers, 335
 registers, 335, 371
Pseudo-static registers, 338
PSPICE, 131
Pull-down network (PDN), 237–41,
 284, 321
Pull-up network (PUN), 237–41, 321
Pulse concept, 372
Pulse registers, 354–56
Pulse signals, 354
Punch-through, 115

R

Race condition, 348
Radix-2 trees, 581
Radix-4 adder, 585
RAM-based FPGA, 406
Rambus channel, 691
Rambus DRAM (RDRAM), 633,
 691
Random-access memory (RAM),
 657–70
 dynamic random-access memory
  (DRAM), 664–70
 one-transistor dynamic memory
  cell, 666–70

three-transistor dynamic memory cell, 664–66
 static random-access memory
  (SRAM), 657–64
 improved MOS SRAM cells,
  661–64
 operation of SRAM cell,
  657–61
 performance of SRAM cell, 661
Random-access (RAM) memory, 626
Random approach, to generating
 stimuli, 730–31
Random errors, 502
RAS (row-access strobe), 633
Ratioed logic, 181, 263–69, 645
 building better loads, 267–69
 concept, 263–67
Ratioless gates, 181
Ratioless logic styles, 264
RC effect, 464
RC trees, 153
 matched, 509
RDRAM, *See* Rambus DRAM
 (RDRAM)
Read-access time, 625
Read-only memories (ROM), 625,
 634–47
 application-specific ROMs, 646
 commodity ROM chips, 646
 NAND ROM, 641–46
 NOR ROM, 640–46
 power consumption and precharged memory arrays,
  645–46
 programming, 636–46
 ROM cells, 634–36
 transient performance, 641–45
 user perspective, 646–47
Read operation, programming cycle,
 652
Read-word line, 664–65
Read-write memory (RWM), 625–26
Receivers, 481, 526
Reconfiguration, volatile (RAM-
 based) FPGAs, 406
Reduced internal supply, 687
Reduced rise/fall times, design of
 output driver with, 472
Reduced-swing circuits, 480–85
 dynamic reduced-swing networks,
  484–85
 static reduced-swing networks,
  481–83

Redundancy, 724
Redundant hardware, 698–99
Reference circuit, 688–89
Reflection coefficient, 162
Refresh frequency, 706
Regenerative circuits, 236
Regenerative property, 22–23
Registers, 625
 alternative styles, 354–58
  pulse registers, 354–56
  sense-amplifier based registers,
   356–58
 characterization of, 431–33
 differential, 515
 dual-edge, 349–50
 dynamic, 344–50
 edge-triggered, 329
 latches vs., 329–30
 master-slave edge-triggered,
  333–39
 negative edge-triggered, 326–27,
  329
 positive edge-triggered, 326–27,
  329, 334, 431
 pseudo-static, 338
 pulse, 354–56
 sense-amplifier based, 356–58
 static, 330–44
 true single-phase clocked register
  (TSPCR), 350–53
Relative coordinates, 69
Relative phase, periodic signals, 540
Relaxation-based approach, to circuit
 simulation, 310
Rent, E., 51
Rent's rule, 51
Repeaters, 466–68
Replica delay, 524–25
Resistance, 137, 144–48
 contact, 146
 input, 181
 interconnect, 144–48
 path, 153
 shared path, 154–55
 sheet, 113, 145
 source-drain, 113, 144–48
 static, 330–44
Resistive load SRAM cell, 662–63
Resistive parasitics:
 resistance and performance,
  464–69
  better interconnect materials,
   464–66

Resistive parasitics *(cont.)*
    better interconnect strategies,
       465–66
    interconnect architecture,
       optimizing, 468–69
    repeaters, 466–68
Resistive region, 92
Return-to-zero (RTZ), 529
Reverse-biased mode, 129
    diodes, 78, 290
Reverse conduction, 337
Ring oscillator, 28–29, 31
Ripple-carry adders, 563–64, 620
    expressed using the dot operator,
       580
    inverting property, 564
    propagation delay of, 563–64
Rise and fall times, 28
Routing channels, 320
Row address, 629
Row decoders, 673–77
    dynamic decoders, 675–77
    static decoder design, 673–75
Rule-based expert systems, 554
Run-time power management,
    611–17
    dynamic supply voltage scaling
       (DVS), 611–14
    dynamic threshold scaling (DTS),
       614–17

S

Saturation current, diodes, 79
Saturation region, 93, 129
Scalable design rules, 47
Scaling analysis, 123
Scaling, ideal, 171–73
Scan-test methodology, 725
Scheduling function, 439
Schmitt triggers, 364–67, 372
    CMOS implementation, 365–67
    defined, 364–65
Sea-of-gates approach, 399–401
Secondary memories, 627
Select layer, 49
Select regions, 48
Self-aligned process, 46
Self-duality, 567
Self-loading effect, 203
Self-test, 725
Self-timed adder circuit, 523–24
Self-timed circuit design, 519–34,
    550

completion-signal generation,
    522–26
    dual-rail coding, 522–24
    replica delay, 524–25
self-timed logic, 519–22, 531–34
    clock-delayed domino, 533–34
    glitch reduction, 532
    post-charge logic, 532–33
    practical examples of, 531–34
self-timed signaling, 526–31
Semiconductor memory trends/evo-
    lutions, 713–16
Semiconductor wire, inductance of,
    150
Semicustom design approaches,
    382–83
Semicustom design flow, 396–99
    design capture, 396
    extraction, 397
    floorplanning, 396
    logic synthesis tools, 396
    placement, 397
    post-layout simulation and
       verification, 397
    prelayout simulation and
       verification, 396
    routing, 397
    tape-out, 397
Sender module, 526
Sense-amplifier based registers,
    356–58
Sense amplifiers, 481, 630, 679–86
    differential voltage sensing
       amplifiers, 680–82
    functions performed by, 679
    sensing in 1T DRAM, 685–86
    single-ended sensing, 682–84
    single-to-differential conversion,
       684–85
Sequential logic circuits, 236
    alternative register styles, 354–58
       pulse registers, 354–56
       sense-amplifier based registers,
          356–58
    classification of memory elements,
       328–30
       foreground vs. background
          memory, 328–30
       latches vs. registers, 329–30
       static vs. dynamic memory,
          328–29
    defined, 326
    designing, 325–73

dynamic latches/registers, 344–50
    C^2MOS (clocked CMOS)
       register, 346–50
    dual-edge registers, 349–50
    dynamic transmission-gate
       edge-triggered registers,
       344–46
non-bistable sequential circuits,
    364–70
    astable circuits, 368–70
    monostable sequential circuits,
       367
    Schmitt trigger, 364–67
pipelining, 358–63
    latch- vs. register-based
       pipelines, 360
    NORA-CMOS, 361–63
    pipelined computations,
       example of, 359
static latches/registers, 330–44
    bistability principle, 330–32
    low-voltage static latches,
       339–41
    master-slave edge-triggered
       registers, 333–39
    multiplexer-based latches,
       332–33
    static SR flip-flops, 341–44
timing metrics for sequential
    circuits, 327–28
true single-phase clocked register
    (TSPCR), 350–53
Serial memories, 626
Set-reset flip-flop, 341–42
Setup time:
    multiplexer-based master–slave
       registers, 335
    registers, 327, 328, 371, 431–33
Shared path resistance, 154–55
Sheet resistance, 113, 145
Shichman-Hodges model, 117
Shift-and-add algorithm, 586
Shift register, 626
Shift-register latch (SRL), 728–29
Shifter, 594–96
    barrel shifter, 595–96
    logarithmic shifter, 596
    programmable, 594
Short-channel devices, 94
Side-wall junction, 111
Signal integrity issues, 290–95
    capacitive coupling, 295

charge leakage, 290–92
charge sharing, 292–95
clock feed-through, 295
Signal return path, 477
Signal swing, 20
Signal-to-noise ratio, 693–98
bit-line-to-bit-line coupling,
695–96
critical charge, 697–98
leakage, 696–97
word-line-to-bit-line coupling,
694–95
Signals, 132
Signature analysis, 732
Signature, circuits, 732
Silicidation, 113
Silicide, 61
Silicon, 58
Silicon-on-insulator (SOI) CMOS,
61–62
Silicon-on-silicon, 58
Silicon wafer, 16–17, 37
Simplified CMOS process flow,
42–46
Simulation approach, to design, 553
Simulink environment (Mathworks),
441–43
Single-ended sensing, 682–84
Single-ended signaling techniques,
481
Single-to-differential conversion,
684–85
Skin depth, 147–48
Skin effect, 147–48
Slack, 604
Slack borrowing, 516–19
Slope engineering, 212
Slope factor, 100
Slope sensitivity factor, 428
Soft error rate (SER), 697
Soft errors, 697
Soft macro, 393–95
Source-drain resistance, 113, 144–48
Space-charge region, 76
SPECTRE, 131
SPICE model(s):
diodes, 85–87
first-order parameters, 86
MOSFET transistors, 117–20
SPICE program, analyzing power
consumption using, 227–29
SPICE wire models, 170–71
distributed rc lines, 170

transmission line models, 170–71
Spin, rinse and dry (SRD), 39
Split-output, 352
Sputtering, 42
Square-root carry-select adder,
575–77
Squared dependency, 93
SR flip-flop, 341–42
SR flip-flops, 357
SRAM, *See* Static random-access
memory (SRAM):
Stacked-capacitor cell (STC), 669
Standard cells, 10
gate arrays vs., 404
Standby, 225
State encoding, 437
State minimization, 437
Static adder circuit, 564–66
Static behavior:
diodes, 77–80
of a gate, 19
Static CMOS design, 236–84
complementary CMOS, 237–63
pass-transistor logic, 269–84
ratioed logic, 263–69
Static CMOS inverter, *See* CMOS
inverter
Static decoder design, 673–75
Static latches/registers, 330–44
bistability principle, 330–32
low-voltage static latches, 339–41
master–slave edge-triggered registers, 333–39
multiplexer-based latches, 332–33
static SR flip-flops, 341–44
Static memory, 626
dynamic memory vs., 328–29
Static mismatches, 502–3
Static power consumption, 645
Static random-access memory
(SRAM), 549, 633, 657–64,
691–92
active power reduction, 703
data-retention dissipation, 704–6
improved MOS SRAM cells,
661–64
operation of SRAM cell, 657–61
performance of SRAM cell, 661
Static reduced-swing networks,
481–83
Static registers, 330–44, 372
Static SR flip-flop, 341–44
Steady-state parameters of a gate, 19

Step junction, 75
Stepper exposure, 39
Stick diagrams, 69, 321
Strong inversion, 89
Structural generators, 393
Structural mode, VHDL, 314
Stuck-at model, 734
Stuck-at-one (sa1) fault, 735
Stuck-at-open fault, 735
Stuck-at-short fault, 735
Stuck-at-zero (sa0) fault, 735
Sub-threshold leakage, 290
Subglobal word line, 710
Subnets, 240–41
Substrates, 48
contacts, 49
Subthreshold conduction, 99, 129
Sum-of-products format, 407
Sum selection, 582–83
Sum terms, 389
Supply network, 488
Supply voltages:
designing logic for reduction in,
303–6
scaling, 192–93
Surface-mount packaging, 56
Swing-restored pass transistor logic,
276
Switch-box-based programmable
wiring, 414–16
Switch-level simulation, 312
Switching activity, 216–17, 257–58
design techniques to reduce,
261–63
glitch reduction by balancing
signal paths, 263
input ordering, 262
logic restructuring, 261–62
time-multiplexing resources,
262–63
Switching threshold voltage, 20
Synchronizers, 534–39
asynchronous-synchronous
interface, 534–35
defined, 535
design, 538
and mean time to failure, 538
synchronization failures, 535
trajectories, 536
Synchronous design, 492, 495–519,
550
clock-distribution techniques,
508–15

Synchronous design *(cont.)*
  digital circuits, 495–519
    clock-distribution techniques, 508–15
    latch-based clocking, 516–19
    sources of skew and jitter, 502–8
    synchronous timing basics, 495–502
  latch-based clocking, 516–19
  self-timed circuit design, 519–34
    completion-signal generation, 522–26
    self-timed logic, 519–22, 531–34
    self-timed signaling, 526–31
  sources of skew and jitter, 502–8
  synchronous timing basics, 495–502
    clock jitter, 500
    clock skew, 496–500
Synchronous DRAM (SDRAM), 633, 690–91
Synchronous interconnect, 492–93
Synchronous sequential systems, 326
Syndrome, of a data sequence, 732
System C language, 314
System designer, 625
System Generator tool (Xilinx), 443
System-in-a-package (SIP) strategy, 59, 64
System-on-a-chip, 380–81, 395, 487
  testing, 733–34
Systematic errors, 502
Systems-on-a-die, 64

T

Tape Automated Bonding (TAB), 54–55, 58
Tape-out, 397
Targeted dissipation-source, 600
Technology file, 70
Technology-independent phase, logic optimization systems, 437
Technology-mapping phase, logic optimization systems, 437
Technology scaling, 122–28
  fixed-voltage scaling, 125–26
  full scaling, 124
  general scaling, 126–27
  impact on inverter metrics, 229–32
  scaling analysis, 123

verifying the model, 127–28
Terabytes (Tbyte), 625
Tertiary memories, 627
Test-pattern generation, 734–37
  automatic test-pattern generation (ATPG), 736–37
  fault models, 734–36
  fault simulation, 737
Test program, 723
Texas Instruments, 5
Thermal bounds on integration, 60
Thermal conduction, 60
Thermal voltage, 76
Thin-film transistors (TFTs), 664
THOR simulator, 313
Three-dimensional integrated circuits, 63–64
Three-transistor dynamic memory cell, 664–66
3-2 compressor, 592
3-input NAND gate cell, 428
Threshold variations, MOS transistors, 114–15
Threshold voltage, 90, 120
Through-hole mounting, 55–56
Through-put-constrained design, 600
Time-of-flight, 165
Time to market, 383
Time-varying mismatches, 502–3
Timing-closure, 398
Timing simulators, 310
Timing slack, 604
Timing verification, 554–56
  example of, 555–56
TL (Transistor-Transistor Logic), 5
Trajectories, synchronizers, 536
Transient behavior, diodes, 80–83
Transistor model for manual analysis, 103–4
Transistor rules, 49
Transistors, 4–5
  bleeder, 292, 299
  FAMOS, 648, 651
  floating-gate, 647–49
  gate, 43
  invention of, 4–5
  MOSFET, 117–20
  multiple threshold, 276–77
  NMOS, 182, 321, 477
  number of, in gate, 18
  PMOS, 182, 321, 477
    thin-film, 664

sizing, 453
thin-film transistors (TFTs), 664
vertical, 127
Transition state, 523
Transition width (TW), 21
Transmission delay (TD), 170
Transmission-gate-based adder, 567–68
Transmission gate logic, 277–78
Transmission gates, 277–84
  and pass-transistor performance, 280–84
  using to build complex gates, 279
  XOR, 280
Transmission line, 159–69
  effects, 168–69
  large source resistance, 164–65
  lossless, 160–62
  lossy, 160, 167–69
  matched source resistance, 166
  small source resistance, 165
  taps, 168
  termination of, 162–67
  transient response of, 164
Transmission-line effects, 475–79, 489
  shielding, 477–79
  termination, 475–77, 489
Transmission line model, 160
Transparent high/transparent low latches, 329
Transparent mode, latches, 329
Transposed (twisted) bit-line architecture, 695–96
Tree decoder, 678
Tree multiplier, 592–93
Trees, 508
Trench-capacitor cell, 669
Trench insulation, 43
Tri-state buffer, 458–59
Troubleshooting designs, 12
True single-phase clocked register (TSPCR), 350–53
  positive edge-triggered register in, 352
Tunneling, 650
Twisted bit-line architecture, 695–96
Two-level minimization tools, 437
Two-phase protocol, 526
Two-phase self-timed FIFO, 528
2.5D integration, 63–64

U

Undefined region, 20–21
Unidirectional gate, 25
Uniform-distributed *rc*-line model
    (URC), 170
Unit-delay model, 311
Unit time variable, 311
UNIVAC I, 4
Unloaded delay, 206, 249
Unloaded propagation delay, 246

V

Variable cost, 16–18
VCDL, *See* Voltage-controlled delay
    line (VCDL)
VCO, *See* Voltage-controlled
    oscillator (VCO)
Vector-merging adder, 591
Velocity saturation, 129
Velocity-saturation effect, 94
Verification:
    design, 553–57
    electrical, 554
    formal, 556–57
    functional, 556–57
    timing, 554–56
Verification approach, to design,
    553–54
Verifying the layout, 49–50
Verilog, 388, 395, 437
Vertical transistors, 127
VHDL (VHSIC Hardware Descrip-
    tion Language), 314, 388, 437
    behavioral mode, 314–15
    structural mode, 314
Via layers, 48
Via rules, 49
Video memories, 626
Virtex-II Pro (Xilinx), 421–22
Volatile memories, 626
Volatile (RAM-based) FPGA, 406
Voltage:
    boosted-wordline, 686

breakdown, 84
dynamic voltage scaling (DVS),
    611–14
gate threshold, 20
multiple supply, 603–7
switching threshold, 20
thermal, 76
threshold, 120
Voltage-controlled delay line
    (VCDL), 546
Voltage-controlled oscillator (VCO),
    369–71, 503, 540–46
Voltage domain, mapping logic levels
    to, 21
Voltage down-converters, 687
Voltage references, 686–89
    boosted-wordline voltage, 686
    charge pumps, 688
    half V_{DD}, 687
    negative substrate bias, 687
    reduced internal supply, 687
    reference circuit, 688–89
    voltage down-converters, 687
Voltage-transfer characteristic,
    20–21, 26–27
Voltage-transfer characteristic
    (VTC), 182

W

Wafer, 16–17, 37
Wallace tree multiplier, 593–94
Wave propagation equation, 160
Waves, 160
Weak-inversion conduction, 99
Well and substrate contacts, 49
Wells, 36, 48
    contacts, 49
Wet etching process, 42
Wire-bond attached chip capacitor
    (WACC), 474
Wire bonding, 53–54, 58
Wire models, 135–75, *See also*
    SPICE wire models

capacitance, 138–44
    electrical, 150–69, *See also* Elec-
        trical wire models
    future of, 171–74
    inductance, 148–50
    interconnect parameters, 138–50
    resistance, 144–48
    SPICE, 170–71
Wire pipelining, 468
Wire properties:
    "constant resistance" scaling of,
        173
    ideal scaling of, 172
Wiring capacitance, CMOS inverter,
    196
Word-line-to-bit-line coupling,
    694–95
Word lines, 630
    boosted, 707
    boosted-word-line voltage, 686
    dummy, 690
    global, 644, 707
    read, 664–65
    subglobal, 710
    write, 664–65
Wrappers, 733
Write-access time, 625
WRITE ONCE device, 647
Write-once (fuse-based) FPGA, 405
Write operation, 652
    programming cycle, 652
Write-word line, 664–65

X

Xilinx XC40xx series, 418–20
XOR gates, 180
XOR phase detector, 542–43

Z

Zener breakdown, 84
Zero-delay model, 311
Zero-threshold devices, 277